Amarjit S. Basra, PhD
Editor

Handbook of Seed Science and Technology

Pre-publication
REVIEWS,
COMMENTARIES,
EVALUATIONS . . .

"The scope and depth of the content of the *Handbook* are impressive. The four sections comprise chapters covering all the major directions of contemporary seed inquiry, and the linking of these by chapter arrangement and within chapters to older anatomical, morphological, and physiological literature is a most positive feature. The reviewing of ecological, gene bank, crop issues, and seed testing and seed health issues rounds out the scope to provide a reference which will be as valuable to researchers as it will to teachers at all levels of education.

The *Handbook* is also a valuable resource for students, teachers, and researchers in areas allied to seed biology and technology, particularly food technology, agricultural economics and bio- commerce, bio-engineering, epidemiology especially or crop diseases, toxicology and nutrition, nutrigenics, metabolomics, and all the other genetic technologies emergent in today's world. Seeds have always played a central role in the history of humankind and this book is a fine compendium, befitting in its scope and content."

David W. Fountain, PhD
Associate Professor,
Institute of Molecular BioSciences;
Director, Centre for Plant Reproduction and Seed Technology, Massey University, New Zealand

More pre-publication
REVIEWS, COMMENTARIES, EVALUATIONS . . .

"The *Handbook of Seed Science and Technology* is an up-to-date, comprehensive collection of information pertaining to seed development, dormancy, germination, ecology, and technology. Edited by Dr. Amarjit Basra, this handbook is an anthology of contributions from the most prominent seed scientists in the world today, whose research covers a wide range of species including *Arabidopsis,* a useful plant model, important agronomic crops such as cereals (rice and wheat), maize, grain legumes, cotton, and vegetable, tree, and weed seeds. It is a must-have for scientists who conduct seed research and for students who study seeds.

The book comprises twenty-six chapters organized into four sections. As a physiologist, I found the first two sections most interesting. These sections cover basic research related to seed development, dormancy, and germination, and offer a comprehensive review of the molecular genetic control, biochemical evidence, and functional significance of a wide range of topics including ovule development, genetically engineered seed nutritive value enhancement, and synthetic seed technology. Ecologists will find the most current basic and applied information on seed ecology. For people in the seed testing and regulating industry, the reviews of seed technology applications and detailed protocols are indispensable. The conservation of genetic resources for future generations, a major global concern, is discussed in detail at both genebank management and conservation methodology levels.

Each chapter provides a complete list of references and a discussion of future perspectives. The latter identify current knowledge gaps and propose future research directions needed to address these gaps. These perspectives are useful especially for research foundations, government, and the biotechnology industry that fund seed-related research.

I strongly recommend this handbook to all scientists who work with seeds—a majority of plant scientists—since seeds, together with soil and water, are one of the most important factors affecting plant growth and productivity. Dr. Amarjit Basra is both a productive scientist and a prolific author. With the publication of this comprehensive handbook, he has achieved a significant accomplishment for which he should be highly commended."

Tara VanToai, PhD
USDA-ARS-SDRU;
Visiting Scientist,
The Institute for Genomic Research,
Rockville, Maryland

"This book seems very useful and much needed. Because of the nature of seed developmental biology, seed ecology and technology are still of great interest. The structure and composition of included papers are of high quality, and the scientific value meets high standards.

The potential value for the target audience seems very promising, especially for those involved in agriculture education, research, and policymaking. This book is a good addition to the field of seed science."

Michal V. Marek, DSc
Professor, Institute of Systems Biology and Ecology, Academy of Sciences, Budejovice, Czech Republic

More pre-publication
REVIEWS, COMMENTARIES, EVALUATIONS . . .

"Throughout the world several thousand billion tons of seeds are produced yearly for food staples such as grains of cereals and legumes, for food industry such as oilseeds, and, last but not least, for sowing. A number of important properties have to be considered depending on their utilization in nutrition, industry, or plant production.

Handbook of Seed Science and Technology provides a state-of-the-art review of both the classical and the newly arisen fields of interest in seed science with integrated information on seed developmental biology and biotechnology, seed ecology, and seed technology. The authors present the latest research and advances in a wide range of subjects, including the biochemical processes involved in seed development, dormancy and germination, and their regulation on the level of genes; the fortification of the nutritive value of seeds by genetic engineering; seed quality, seed vigor, and their assessment; and technologies in germplasm conservation. The last section in each chapter addresses new perspectives on the field presented, followed by extensive, up-to-date bibliographies.

Exciting, inspiring, as science can be; useful to students, researchers, or professionals interested in seed science, and also a valuable item in the agricultural university library collection."

Krisztina R. Végh, PhD
Senior Researcher,
Research Institute for Soil Science and Agricultural Chemistry of Hungarian Academy of Sciences (RISSAC)

Handbook of Seed Science and Technology

Other Related Books

Crop Improvement and Mutation Breeding	*Sharma, A.K.*
Disease Problems in Vegetable Production	*Gupta, S.K.*
Diseases Management in Arid Land Crops	*Lodha, Satish*
Forage Legumes	*Joshi, U.N.*
Genetic Improvement of Field Crops	*Singh, C.B.*
Hand Book of Seed Industry (Prospects & its costing)	*Ramdeo, A.*
Legumes in Dry Areas	*Kumar, D.*
Nutritional and Physiological Disorders in Crop Plants	*Rajendran, C.*
Organic Seed: Indigenous Varieties and Technologies	*Vanangamudi, K.*
Principles and Practices of Seed Storage	*Justice, O.L.*
Seed Borne Diseases - Ecofriendly Management	*Arya, Arun*
Seed Economics: Commercial Considerations for Enterprise Management in Developing Countries	*Kugbei, S.*
Seed Identification Manual	*Martin, A.C.*
Seed Pathology (Vol. 1-2 set)	*Neergaard, P.*
Seed Production of Agricultural Crops	*Kelly, A.F.*
Seed Quality Enhancement : Principles and Practices	*Vanangamudi, K.*
Seed Technology (2nd Revised & Enlarged Ed.)	*Khare, D.*
Seed Technology in the Tropics	*McKay, D.B.*
Weed Seed Biology	*Vanangamudi, K.*

Handbook of Seed Science and Technology

Amarjit S. Basra, PhD
Editor

SCIENTIFIC PUBLISHERS (INDIA)

Published by

Scientific Publishers (India)
5 A, New Pali Road, P.O. Box 91
Jodhpur 342 001 (India)

E-mail: info@scientificpub.com
Website: www.scientificpub.com

Branch Office
Scientific Publishers (India)
4806/24, Ansari Road, Daryaganj
New Delhi - 110 002 (India)

Second Indian Reprint, 2014
First Indian Reprint, 2007

ISBN: 978-81-7233-573-1

Published under arrangement with Food Products Press®, an imprint of The Haworth Press, Inc., 10 Alice Street, Binghamton, NY 13904-1580 USA

Printed in India

CONTENTS

SECTION II: SEED DORMANCY AND GERMINATION

SECTION III: SEED ECOLOGY

SECTION IV: SEED TECHNOLOGY

ABOUT THE EDITOR

Dr. **Amarjit S. Basra** is an eminent plant physiologist at the Punjab Agricultural University in Ludhiana, India. He is currently a visiting scientist at the University of California at Davis. His outstanding work on seed quality and cotton fiber quality has been internationally recognized. Dr. Basra has more than 80 research publications to his credit, and is the lead editor of 11 books on topical subjects in plant/crop science. He is the Founding Editor in Chief of the *Journal of Crop Production* and the *Journal of New Seeds* (Food Products Press). He is a member of the American Society of Agronomy, the Crop Science Society of America, the American Society of Plant Biologists, the American Association for the Advancement of Sciences, the American Institute of Biological Sciences, the New York Academy of Sciences, the Australian Society of Plant Physiologists, the American Society for Horticultural Science, and the International Society of Horticultural Science. Dr. Basra is a decorated scientist who has received several coveted awards and honors and has made scientific visits to several countries fostering cooperation in agricultural research at the international level.

CONTRIBUTORS

L. Gabriela Abeledo, Departamento de Producción vegetal, Facultad de Agronomía, Universidad de Buenos Aires, Buenos Aires, Argentina.

Craig Atkins, School of Plant Biology (Botony), The University of Western Australia, Perth, Australia.

Sureshkumar Balasubramanian, Max-Planck Institute for Developmental Biology, Tuebingen, Germany.

V. A. Bapat, Plant Cell Culture Technology Section, Nuclear Agriculture and Biotechnology Division, Bhabha Atomic Research Centre, Trombay, Mumbai, India.

Jayne Bates, National Institute of Agricultural Botany (NIAB), Cambridge, United Kingdom.

Leonie Bentsink, Laboratory of Genetics, Wageningen University, Wageningen, the Netherlands.

Beverly J. Brown, Department of Biology, Nazareth College of Rochester, Rochester, New York.

Jorge Castro, Grupo de Ecología Terrestre, Departamento de Biología Animal y Ecología, Facultad de Ciencias, Universidad de Granada, Granada, Spain; Physiological Ecology Research Group, Botanical Institute, University of Copenhagen, Copenhagen, Denmark.

Prem S. Chourey, U.S. Department of Agriculture, Agricultural Research Service, and University of Florida, Gainesville, Florida.

Juan A. Delgado, Departamento de Ecología, Facultad de Biología, Universidad Complutense de Madrid, Madrid, Spain.

V. V. Deshpande, Plant Molecular Biology Unit, Division of Biochemical Sciences, National Chemical Laboratory, Pune, India.

S. D. Doijode, Principal Scientist (Horticulture) and Head, Division of Plant Genetic Resources, Indian Institute of Horticultural Research, Bangalore, India.

Sabry Elias, Department of Crop and Soil Science, Oregon State University, Corvallis, Oregon.

Neil Emery, Department of Biology, Trent University, Peterborough, Ontario, Canada.

E. P. Fuerst, Department of Crop and Soil Sciences, Washington State University, Pullman, Washington.

R. S. Gallagher, Department of Crop and Soil Sciences, Washington State University, Pullman, Washington.

T. R. Ganapathi, Plant Cell Culture Technology Section, Nuclear Agriculture and Biotechnology Division, Bhabha Atomic Research Centre, Trombay, Mumbai, India.

A. P. Giri, Plant Molecular Biology Unit, Division of Biochemical Sciences, National Chemical Laboratory, Pune, India.

Marc Goetz, CSIRO Plant Industry, Glen Osmond, South Australia.

José M. Gómez, Grupo de Ecología Terrestre, Departamento de Biología Animal y Ecología, Facultad de Ciencias, Universidad de Granada, Granada, Spain.

Fernanda G. González, Departamento de Producción vegetal, Facultad de Agronomía, Universidad de Buenos Aires, Buenos Aires, Argentina.

V. S. Gupta, Plant Molecular Biology Unit, Division of Biochemical Sciences, National Chemical Laboratory, Pune, India.

N. D. Hagan, CSIRO Plant Industry, Canberra, Australia.

D. Harris, Centre for Arid Zone Studies, University of Wales, Bangor, United Kingdom.

A. M. Harsulkar, Plant Molecular Biology Unit, Division of Biochemical Sciences, National Chemical Laboratory, Pune, India.

Tohru Hashimoto, Department of Life Sciences, Kobe Women's University, Kobe, Japan.

Eloit M. Herman, Plant Genetics Research Unit, U.S. Department of Agriculture, Agricultural Research Service, Donald Danforth Plant Science Center, St. Louis, Missouri.

T. J. V. Higgins, CSIRO Plant Industry, Canberra, Australia.

Henk W. M. Hilhorst, Laboratory of Plant Physiology, Wageningen University, Wageningen, the Netherlands.

José A. Hódar, Grupo de Ecología Terrestre, Departamento de Biología Animal y Ecología, Facultad de Ciencias, Universidad de Granada, Granada, Spain.

David Jaccoud, Departmento de Fitotecnia & Fitossanidade, Ponta Grossa State University, Ponto Grossa, Parana, Brazil.

Adriana G. Kantolic, Departamento de Producción vegetal, Facultad de Agronomía, Universidad de Buenos Aires, Buenos Aires, Argentina.

Maarten Koornneef, Laboratory of Genetics, Wageningen University, Wageningen, the Netherlands.

Gerhard Leubner-Metzger, Institut für Biologie II, Botanik, Albert-Ludwigs-Universität Freiburg, Freiburg, Germany.

Daniel J. Miralles, Departamento de Producción vegetal, Facultad de Agronomía, Universidad de Buenos Aires, Buenos Aires, Argentina.

A. Mottram, Centre for Arid Zone Studies, University of Wales, Bangor, United Kingdom.

A. J. Murdoch, Department of Agriculture, The University of Reading, Reading, United Kingdom.

Alison A. Powell, Department of Agriculture and Forestry, University of Aberdeen, Aberdeen, United Kingdom.

P. K. Ranjekar, Plant Molecular Biology Unit, Division of Biochemical Sciences, National Chemical Laboratory, Pune, India.

Thomas Roitsch, Institut für Pharmazeutische Biologie, Universität Würzburg, Würzburg, Germany.

M. N. Sainani, Plant Molecular Biology Unit, Division of Biochemical Sciences, National Chemical Laboratory, Pune, India.

Roxana Savin, Department de Producció vegetal i Ciència Forestal, Universitat de Lieída, Centre UdL-IRTA, Lleida, Spain.

Kay Schneitz, Plant Developmental Biology Unit, Technical University Munich, Freising, Germany.

Jose M. Serrano, Departamento de Ecología, Facultad de Biología, Universidad Complutense de Madrid, Madrid, Spain.

Chizuko Shichijo, Department of Biology, Faculty of Science, Kobe University, Kobe, Japan.

Gustavo A. Slafer, Departamento de Producción vegetal, Facultad de Agronomía, Universidad de Buenos Aires, Buenos Aires, Argentina.

P. Suprasanna, Plant Cell Culture Technology Section, Nuclear Agriculture and Biotechnology Division, Bhabha Atomic Research Centre, Trombay, Mumbai, India.

Osamu Tanaka, Department of Biology, Faculty of Science and Engineering, Konan University, Kobe, Japan.

David Tay, Director, Ornamental Plant Germplasm Center, Columbus, Ohio.

Emily Taylor, National Institute of Agricultural Botony (NIAB), Cambridge, United Kingdom.

Wei-Cai Yang, Institute of Genetics and Developmental Biology, Chinese Academy of Sciences, Beijing, China.

Gregory E. Welbaum, Department of Horticulture, Virginal Tech, Blacksburg, Virginia.

Elena M. Whitechurch, Departamento de Producción vegetal, Facultad de Agronomía, Universidad de Buenos Aires, Buenos Aires, Argentina.

Yong-Ling Ruan, CSIRO Plant Industry, Canberra, Australia.

SECTION I:
SEED DEVELOPMENTAL BIOLOGY AND BIOTECHNOLOGY

Chapter 1

Molecular Control of Ovule Development

Sureshkumar Balasubramanian
Kay Schneitz

INTRODUCTION

A flowering plant's life cycle contains two phases: a vegetative phase and a reproductive phase. The shoot apical meristem, which gives rise to all the aboveground parts of a plant, produces leaves at its flanks during the vegetative phase. Upon entry into the reproductive phase, it undergoes an irreversible phase change to become the inflorescence meristem that produces floral meristems which in turn form the flowers. External environmental as well as internal genetic cues control floral induction. More than 80 genetic loci have been implicated as playing a role in the induction of flowering. The genetics of floral induction is beyond the scope of this chapter, and the reader should look into the wealth of information available in several recent reviews (Levy and Dean, 1998; Simpson et al., 1999; Simpson and Dean, 2002).

A seed plant's flower contains two types of floral organs: the reproductive organs and the perianth organs (Esau, 1977). In a flower, stamens and the carpels represent the male and female reproductive structures, respectively. In dicots, sepals and petals are the usual perianth organs, and in monocots, they are represented by glumes, lemma, and palea. All the floral organs originate from the floral meristem. Molecular and genetic analysis using mainly *Arabidopsis thaliana* and *Antirrhinum majus* has given insight into the molecular mechanisms that control floral organ identity (Coen and Meyerowitz, 1991; Lohmann et al., 2001; Lohmann and Weigel, 2002;

We would like to thank David Chevalier for unpublished information and Markus Schmid for comments on the manuscript. S.B. was supported by a postdoctoral fellowship from Roche during the final stages of his stay in Schneitz lab at Zurich. The work in the Schneitz lab is supported by Swiss National Foundation, Kanton of Zurich at Zurich, and the Deutche Forschung Gemeinschaft at Munich.

Sessions et al., 1998; Weigel, 1995). The gynoecium, which bears the female reproductive organs, is made of carpels, and within the carpels ovules develop.

Ovules are the progenitors of the seeds and thus represent the major female reproductive organs. The molecular control of ovule development has picked up since the 1990s, and a network of genes that underlie various aspects of ovule development is being uncovered. In this chapter, we will summarize our present knowledge of the molecular control of ovule development, mainly focusing on the results obtained from the molecular and genetic analysis done in *Arabidopsis* and *Petunia.*

WHAT ARE OVULES?

Ovules are the female reproductive organs of seed plants (Esau, 1977). An ovule typically has three parts: the nucellus, the chalaza, and the funiculus. The nucellus at the distal end harbors the megaspore mother cell (mmc) that will give rise to the embryo sac. From the central region, referred to as chalaza, integuments originate and develop to envelop the growing embryo sac. The integuments eventually function as a seed coat. The number of integuments that are present can vary from species to species. Based on the number of the integuments, the ovules can be unitegmic (which have a single integument, e.g., *Petunia*) or bitegmic (which have two integuments, e.g., *Arabidopsis*) (Esau, 1977). At the proximal end, the funiculus connects the ovule to the placenta (Esau, 1977). Upon double fertilization, the ovule becomes the site of seed formation. Ovules have a characteristic shape as well. For example, *Arabidopsis* ovules show a characteristic curvature referred to as anatrophy. The anatrophy of the *Arabidopsis* ovules is a result of the differential initiation and growth of the outer integument along the adaxial-abaxial axis (adaxial: adjacent to the meristem; abaxial: away from the meristem). The outer integument initiates at the abaxial side and grows more on the abaxial side compared to the adaxial side. This leads to a curvature in the ovule.

EMERGENCE OF THE OVULE AS A MODEL TO STUDY ORGANOGENESIS

Since the 1990s, ovules have been recognized as a very good model to study organogenesis (Gasser et al., 1998; Grossniklaus and Schneitz, 1998; Schneitz, Balasubramanian, and Schiefthaler, 1998; Chevalier et al., 2001). Several factors make ovules a suitable system to study organ development.

First, ovules have a simple structure of a characteristic size and shape, providing a simple model to study organogenesis. Second, a defect in ovule development usually results in sterility, allowing easy identification of mutants in genetic screens. Third, all the ovules within a gynoecium follow a stereotypic developmental pattern. This allows for large numbers of ovules that can be studied, which is very useful in statistical analysis. Fourth, even mature ovules have only a few cell layers with a limited number of cell types, which allows for easy cytological inspection of ovules with organic clearing methods. Furthermore, being the progenitor of seeds, understanding the biology of ovule development has major agronomical implications.

Basic Aspects of Organogenesis

Organogenesis, the formation of an organ from a group of undifferentiated cells, is a precisely controlled process. This process includes several aspects: specification, initiation and outgrowth, pattern formation, and morphogenesis. These processes can theoretically be separated, though in reality they are interdependent and may occur simultaneously. During specification, a group of cells is set aside from a mass of undifferentiated cells to form an organ. This is followed by the initiation of organ formation and the outgrowth of the primordia. Pattern formation lays down the basic structural plan of the organ. Pattern formation refers to the process in which the cell activities are organized in a nonrandom manner to give rise to a definite pattern during development. Defined cell division and differentiation patterns during morphogenesis lead to a mature organ of the correct size and shape. Almost every organ formation would involve these basic aspects that occur repeatedly during development, though the mechanisms by which these processes occur might differ with respect to individual organs.

An Overview of Wild-Type Ovule Development

Ovule development in wild-type *Arabidopsis* is very well documented and is divided into distinct stages that can be distinguished by visible morphological markers (Schneitz et al., 1995). Ovules are formed within the carpel from the placental tissue as fingerlike protrusions around stage 1. Around stage 2-I, one hypodermal cell enlarges and differentiates into the mmc (see Figure 1.1, part B). After the specification of the mmc, the inner and outer integuments initiate from the epidermis of the chalaza, successively (see Figure 1.1, parts A and B). The outer integument shows asymmetry in its initiation and growth. It initiates and grows more on the abaxial side leading to the anatrophy of the *Arabidopsis* ovule. Simultaneously, the

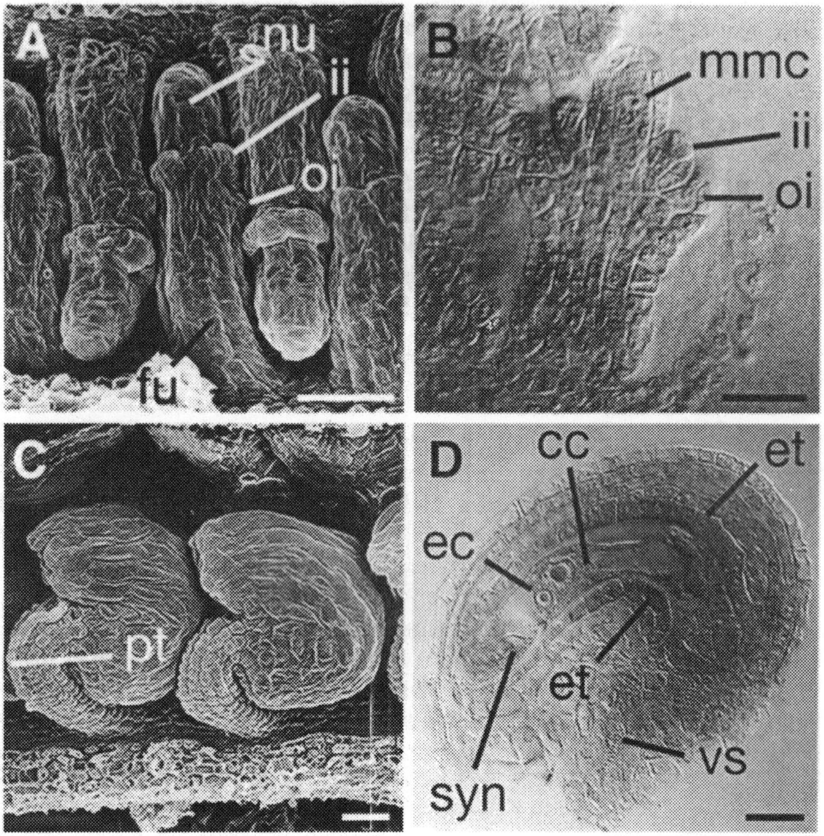

FIGURE 1.1. Ovule development in wild-type *(Ler) Arabidopsis.* A and C: Scanning electron micrographs (SEMs). B and D: Vertical optical sections through whole mount ovules. Stages: A and B: 2-III; C: 4-V; and D: Late 3-IV. nu-nucellus; ii-inner integument; oi-outer integument; mmc-megaspore mother cell; fu-funiculus; pt-pollen tube; cc-central cell; et-endothelium; ec-egg cell; vs-vascular strand; syn-synergids. The initiation of the outer integument at the abaxial side is visible (A). Note the presence of the megaspore mother cell (B) which develops into an embryo sac (D). Scale bars, 20 μM.

mmc undergoes meiosis and one of the meiotic products continues its development to form finally a seven-celled *polygonum* type embryo sac (see Figure 1.1, part D) (Maheswari, 1950). The growing integuments eventually envelop the embryo sac (see Figure 1.1, parts C and D).

OVULE SPECIFICATION

Although several genetic screens were done in different laboratories, no single mutant that fails to form an ovule could be isolated, suggesting that redundant factors are involved in the specification of an ovule. Tissue culture experiments in tobacco have revealed that the commitment to ovule identity follows a progressive pattern (Evans and Malmberg, 1989). Placental explants cultured in vitro showed that depending on the developmental age, the placentae can differentiate into either carpels or ovules. Although, under in vivo conditions, these placentae would have normally committed to form ovules, they seem to get to this commitment in a progressive way: first to a carpel and then to an ovule. At least two loci, *Mgr3* and *Mgr9*, affect this commitment to ovule identity in tobacco (Evans and Malmberg, 1989). These loci have not yet been characterized at a molecular level.

What is the molecular nature of genes that confer ovule identity? With the exception of *AP2*, all the ABC genes, which confer the identities to the floral organs, encode MADS domain transcription factors (Goto and Meyerowitz, 1994; Jack et al., 1992; Mandel, Gustafson-Brown, et al., 1992; Yanofsky et al., 1990). Not surprisingly, two MADS domain proteins, FLORAL BINDING PROTEIN 7 (FBP7) and FBP11 in *Petunia*, are expressed in the ovule primordium and play a role in the commitment to ovule identity (Angenent et al., 1995; Cheng et al., 2000; Colombo et al., 1995). Simultaneous cosuppression of *FBP7* and *FBP11* results in carpelloid structures in place of ovules, suggesting their involvement in the specification of the ovule. The ectopic expression of *FBP11* leads to ectopic ovule formation, corroborating its role in ovule specification (Colombo et al., 1995).

Several *MADS* box genes in *Arabidopsis* are expressed in ovules (Rounsley et al., 1995). Although the exact orthologues of *FBP7* and *FBP11* in *Arabidopsis* are not known, *AGL11* is a likely candidate based on sequence comparisons. Recently a loss-of-function allele of *agl11* has been isolated. Seeds formed in this mutant referred to as *seedstick (stk)* fail to detach from the placenta (Pinyopich et al., 2001). *STK* is closely related to two other *MADS* box genes, *SHATTERPROOF 1 (SHP1)* and *SHP2,* which show an expression pattern in the ovule similar to that of *STK*. A triple mutant of *stk shp1 shp 2* forms carpelloid tissue in place of ovules, showing the redundant involvement of *STK, SHP 1,* and *SHP 2* in the specification of an ovule (Pinyopich et al., 2001). *SHP* genes are negatively regulated by another *MADS* box gene, *FRUITFULL, (FUL),* in the carpel (Ferrandiz et al., 2000), and the carpelloid tissue that is formed in the *stk shp1 shp2* triple mutant shows an ectopic expression of *FUL* (Pinyopich et al., 2001).

Another *MADS* box gene that is expressed in ovules is *AGAMOUS (AG)* (Yanofsky et al., 1990). It has been difficult to study the role of *AG* in ovule development due to its early function that leads to a floral phenotype in *ag* mutants. The overexpression of the *AG* orthologue from *Brassica napus* causes homeotic conversion of sepals into carpels which bear carpelloid ovules (Mandel, Bowman, et al., 1992) in tobacco. These organs show resemblances to the *mgr* mutants described by Evans and Malmberg (1989). This suggests an inhibitory role of *AG* in ovule specification. Contrary to this, the A function gene *AP2* promotes ovule identity as several loss-of-function alleles of *ap2* show conversion of ovules into carpelloid structures (Modrusan, Reiser, Feldmann, et al., 1994). However, the *ap2 ag* double mutants form normal ovules (Bowman et al., 1991), although the percentage of normal ovules might be less in these double mutants compared to the single mutants (Western and Haughn, 1999). This suggests that neither *AG* nor *AP2* is absolutely required for the commitment to ovule fate.

INITIATION/OUTGROWTH OF THE OVULE PRIMORDIA

We refer to the emergence of the ovule primordia, as fingerlike protrusions, from the placental tissue as "outgrowth." Genetic analysis identified three loci that are involved in this process (Broadhvest et al., 2000; Schneitz, Baker, et al., 1998). In wild-type *Arabidopsis*, approximately 50 ovules develop within a gynoecium, which are spaced in a characteristic way to form interdigitating ovule primordia. *aintegumenta (ant)* mutants form fewer ovules that are distantly spaced compared to the wild type (Baker et al., 1997; Elliott et al., 1996; Klucher et al., 1996). *ANT* encodes a protein that belongs to the *AP2/EREBP* family of transcription factors (Okamuro et al., 1997; Riechmann and Meyerowitz, 1998). The ovules of *ant* mutants lack integuments and often show a reduction in the outgrowth of the ovule primordium. They do not form a viable embryo sac and, therefore, are sterile (Baker et al., 1997; Elliott et al., 1996; Klucher et al., 1996; Schneitz et al., 1997). *ANT* has pleiotropic roles in flower development and plays a role in controlling organ size (Krizek, 1999; Mizukami and Fischer, 2000). The expression pattern of *ANT* suggests that it is involved in cell proliferation and growth control.

In spite of the absence of *ANT*, few ovules still develop in *ant* mutants, suggesting the involvement of redundant factors. Plants that are mutant for *huellenloss (hll)* form fewer ovules that are smaller and more distantly spaced than in the wild type, similar to *ant* (Schneitz, Baker, et al., 1998; Schneitz et al., 1997). While the *ant* mutants show defects in many parts of the plant, the *hll* phenotype is restricted to ovules. The ovules of *hll* mutants lack integuments and often show cell death in the distal region of the ovule primordium. They appear small and short, suggesting a role for *HLL* in the outgrowth of the ovule primordium. Interestingly, the *ant hll* double mutants show a drastic reduction in the ovule primordium compared to that of either *ant* or *hll* single mutants. The analysis of these short outgrowths using molecular markers revealed that the morphological units that are present in an ovule can form independently of each other (Schneitz, Baker, et al., 1998). *HLL* encodes a mitochondrial ribosomal protein (Skinner et al., 2001), indicating the importance of energy requirements during organogenesis.

SHORT INTEGUMENT 2 (SIN2) is another locus that has been implicated in the outgrowth of the ovule primordium. Ovules of *sin2* mutants show a phenotype similar to that of *hll* mutants (Broadhvest et al., 2000). Integuments initiate normally in ovules of *sin2* mutants but are arrested in their growth shortly after initiation, implying a role for *SIN2* in the integument outgrowth. *sin2 hll* double mutants produce ovules that are arrested in their outgrowth and appear even shorter than the *ant hll* double mutants (Broadhvest et al., 2000). *sin2* mutants also form fewer ovules compared to the wild type, indicating a role for *SIN2* in the initiation of the ovule primordia. Taken together, these results indicate that *ANT, HLL,* and *SIN2* redundantly control the initiation and outgrowth of the ovule primordium through coordinating the energy requirements with cell proliferation.

PATTERN FORMATION

Pattern formation is the process by which the cell activities are organized in a nonrandom manner with respect to its position in a three-dimensional space. In a mature ovule, one can clearly distinguish two axes of polarity: the proximal-distal axis and the adaxial-abaxial axis (see Figure 1.2).

Proximal-Distal Pattern Formation

Along the proximal-distal axis (also referred to as chalazal-micropylar axis), three morphological units are present in an ovule (Esau, 1977; Schneitz et al., 1995): the nucellus at the distal end, chalaza in the central region, and the funiculus at the proximal end connecting the ovule to the placenta. This morphological pattern is typical of seed plants (Esau, 1977). It has been proposed that a corresponding initial prepattern is laid down which ultimately results in this structured arrangement along the proximal-distal axis (Schneitz et al., 1995).

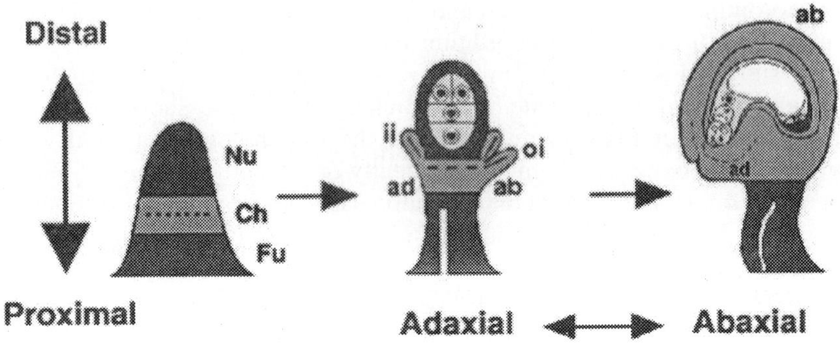

FIGURE 1.2. Schematic representation showing the different axes of polarity during ovule development in *Arabidopsis thaliana*. Note the onset of the outer integument at the abaxial side. Nu-nucellus; Ch-chalaza; Fu-funiculus; ii-inner integument; oi-outer integument; ad-adaxial; ab-abaxial. The dotted lines in the chalaza represent the possible subdivision of the chalaza that would give rise to the inner and the outer integuments.

Several lines of evidence support such a model of prepatterning. Several mutants show phenotypes restricted to distinct morphological units, which suggests the independence of each of these units. *bell (bel1)* mutants show a phenotype restricted to the chalaza and bear ovules that have bulbous outgrowths in place of integuments (Modrusan, Reiser, Fischer, et al., 1994; Robinson-Beers et al., 1992) (see Figure 1.3, part A). At very late stages in development, these outgrowths often show carpelloid features (Modrusan, Reiser, Feldmann, et al., 1994). *BEL1* was suggested to negatively regulate *AG* expression. Thus, the *bel1* phenotype could be due to an ectopic *AG* expression in the ovule (Ray et al., 1994). This view is not supported by the fact that *AG* and *BEL1* have overlapping expression domains in the ovule (Reiser et al., 1995). The molecular cloning of *BEL1* revealed that it encodes a homeodomain containing transcription factor (Reiser et al., 1995). Homeodomain proteins have been identified in animals as well as in plants and play vital roles in pattern formation (Wolpert et al., 1998). *BEL1* is expressed throughout the ovule primordia at the initial stages and gets restricted to the central region around stage 2-I (Reiser et al., 1995) (see Figure 1.3, part B) before the initiation of the integuments and thereby marks the central region at a molecular level. The expression pattern of *BEL1* and the biochemical nature of BEL1 protein make *BEL1* a candidate gene involved in pattern formation during ovule development. However, *bel1* mutants still show morphological distinctions along the proximal-distal axis.

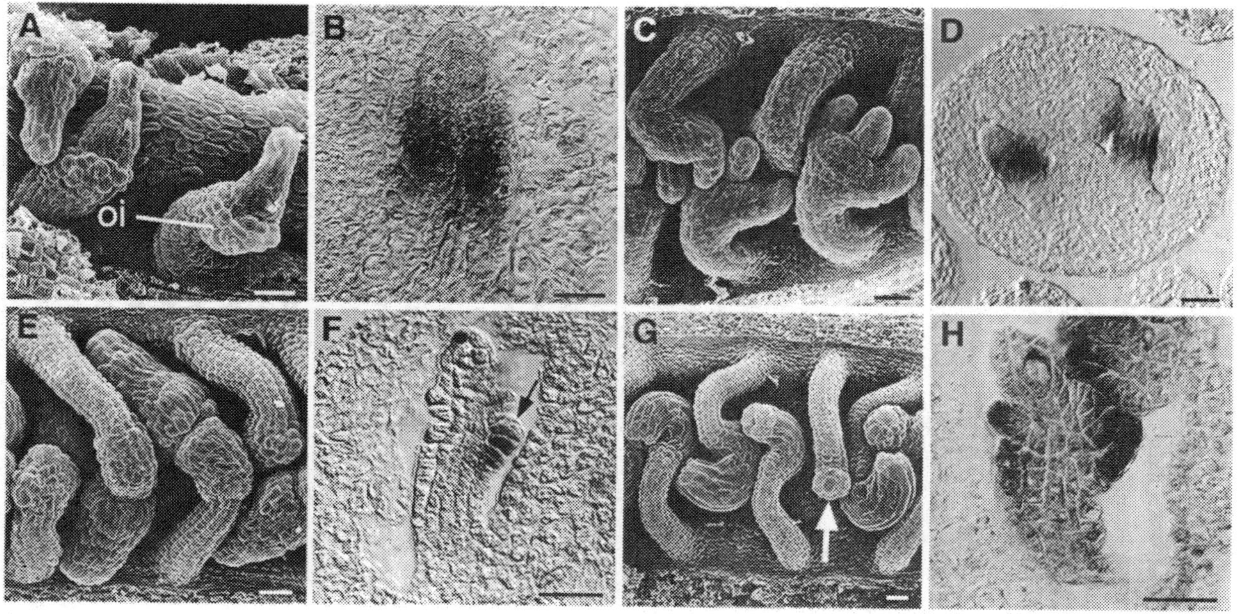

FIGURE 1.3. The ovule phenotypes of mutants and the expression patterns of the corresponding genes in the ovule. A, C, E, and G: SEMs of *ant, bel1, ino,* and *nzz,* respectively. B, D, F, and H: Expression patterns of *ANT, BEL1, INO,* and *NZZ,* respectively, in ovules of wild-type *Arabidopsis.* Stages: B and D: 2-I; F: 2-III; H: 2-IV; G: 3-IV; A, C, and E: 4-V. Note the absence of *INO* expression in the tip cell that undergoes initial division to give rise to the two cell layers of the outer integument (arrow F). Note a rare occassion of highly reduced integuments in *nzz* (arrow G). oi-outer integument. Scale bars = 20 μM.

The outgrowths appear only from the chalaza, and the chalazal restricted *BEL1* expression can be observed in *bel1-1460,* an *ems*-induced mutant allele. This suggests that *bel1* mutants still retain some chalazal identity.

The *ant* mutants show severe reduction in integument growth (see Figure 1.3, part C). *ANT* encodes a protein that belongs to the *AP2* family of transcription factors. The expression pattern of *ANT* in the ovule is very similar to that of *BEL1*. *ANT* is initially expressed throughout the ovule primordia. Around stage 2-I, *ANT* expression is excluded from the nucellus and the proximal region, and molecularly marks the central region (Elliott et al., 1996; Klucher et al., 1996) (see Figure 1.3, part D). Later in development, *ANT* is expressed in the developing integuments and the growing region (distal) of the funiculus.

The *inner no outer (ino)* mutants bear ovules that lack the outer integument (Baker et al., 1997; Schneitz et al., 1997) (see Figure 1.3, part E) and *INO* plays a vital role in the adaxial-abaxial pattern formation. The defects in the *ino* mutants are restricted to the proximal chalaza. From the flanks of the distal chalaza, the inner integument initiates and develops normally. Thus, the phenotype of *ino* mutants allows us to divide the central region into two subdomains, one that forms the inner integument and one that forms the outer integument (Schneitz, Balasubramanian, and Schiefthaler, 1998). Then what regulates the boundary between these two subdomains? This process seems to be controlled by *ABERRANT TESTA SHAPE (ATS),* since *ats* mutants show defects in the maintenance of the boundary between the inner and outer integuments (Léon-Kloosterziel et al., 1994). The ovules of *ats* mutants initiate both integuments normally as in wild type, but later in development, the inner cell layer of the outer integument and the outer cell layer of the inner integument fuse (K.S, unpublished). Consequently, at later stages, these integuments behave as a single integument (Léon-Kloosterziel et al., 1994). In addition to these mutants, several enhancer trap lines show region-restricted expression patterns (Grossniklaus and Schneitz, 1998), suggesting the existence of these pattern elements along this axis.

Recent studies have shown that *NOZZLE* (*NZZ,* also known as *SPL [SPOROCYTELESS]*), a gene encoding a novel protein, plays a vital role in patterning along the proximal-distal and the adaxial-abaxial axis (Balasubramanian and Schneitz, 2000, 2002; Schiefthaler et al., 1999; Yang et al., 1999). *nzz* mutants show defects throughout the ovule (see Figure 1.3, part G). In ovules of *nzz* mutants, the nucellus is reduced, the integuments show temporal alterations in their initiation and are often reduced, and the funiculus is enlarged. The expression pattern of *NZZ* matches with its phenotype (see Figure 1.3, part H). How does *NZZ* function during proximal-distal pattern formation? The best clue comes from the analysis of *nzz bel* double mutants (Balasubramanian and Schneitz, 2000). The *nzz bel* double mutants produce ovules that lack any integuments and have an extra-long funiculus. No sign of the presence of the central region can be seen in these double mutants, and the tissue that is formed in the central region resembles that of funiculus as seen by the epidermal cell morphology (see Figure 1.4, part A). Furthermore, *INO* expression, which is usually detected in only in few cells of the chalaza in wild type, cannot be detected in these double mutants, suggesting the absence of the chalazal identity of the tissue. Taken together, these results suggest that in the absence of *NZZ* and *BEL1* chalazal identity is lost and the tissue is replaced by funiculus, perhaps through a default pathway. Therefore, *NZZ* and *BEL1* function redundantly to specify the chalazal identity.

How does *NZZ* affect the nucellar development? A reduced nucellus is observed in the ovules of *nzz* mutants. This reduction in the nucellus is overcome in *ant* and *ino* mutant backgrounds, suggesting that *NZZ* functions antagonistically with *ANT and INO* in the nucellus. An ectopic distal expression can be observed for *ANT* as well as *BEL1*. Though a distal misexpression of *INO* is not observed in *nzz* mutants, an early onset of *INO* expression can be seen in *nzz* mutants. This early onset of *INO* expression interferes in the development of the nucellus. Furthermore, the expression of *INO* in *nzz* mutants appears to be distally shifted by a few cells. This seems to result in a distal shift of the central region, leading to a reduced nucellus and an enlarged funiculus. The role of *NZZ* in adaxial-abaxial patterning and in coordinating the patterning along these two axes is discussed in the next section.

Is the development of these morphological units dependent on each other? The analysis of *ant hll* double mutants provides a clue on this. The *ant hll* double mutants form shorter ovules (Schneitz, Baker, et al., 1998) in which the primordium outgrowth is arrested at different stages. The smallest primordia in *ant hll* double mutants show nucellar identity as seen by the presence of the mmc. When the primordia are bigger, in addition to the nucellus they form the chalaza as seen by the proximally restricted expression of *BEL1*. This suggests that the outgrowth of the primordium takes place in a sequential manner, with the nucellus being the first to be formed (Schneitz, Baker, et al., 1998). This analysis also suggests that the formation of nucellar pattern element is independent of chalaza or funiculus, since nucellus can be formed in their absence (Schneitz, Baker, et al., 1998). Likewise, the chalaza can also be formed in the absence of funiculus. However, whether the formation of the nucellus is an obligate requirement for the formation of the chalazal and funicular pattern elements remains to be tested.

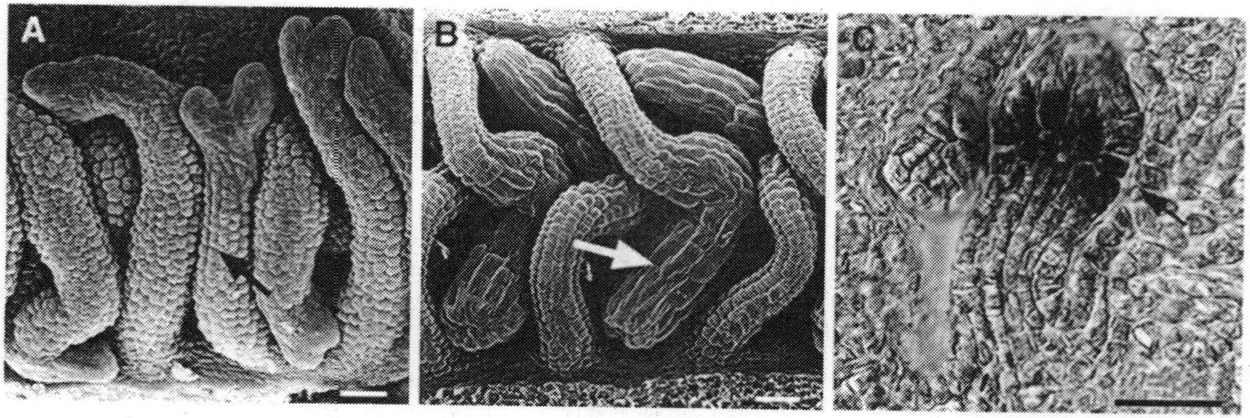

FIGURE 1.4. SEMs and expression analysis in double mutants. A and B: Ovule phenotype of the *nzz bel1* and the *nzz ats* double mutants. C: Expression pattern of *INO* in *nzz ats* double mutants. Stages: A and B, 4-V; C, 2-II. Note the presence of bulbous cells that bear a funicular-cell morphology in the central region of the *nzz bel1* double mutants (arrow A). Note the abaxial growth of the outer integument in *nzz ats* double mutants (arrow B) and the chalazal ectopic expression of *INO* in these mutants (arrow C).

Adaxial-Abaxial Pattern Formation

The mature ovule shows a polarity along the adaxial-abaxial axis. During the initial stages of development (until stage 2-III), the ovule does not exhibit any polarity along the adaxial-abaxial axis in its morphology (Schneitz et al., 1995). The polarity is visually established with the initiation of the outer integument at stage 2-III. The outer integument initiates at the abaxial side of the central region (Robinson-Beers et al., 1992). This is followed by an asymmetric growth of the outer integument along the adaxial-abaxial axis. The outer integument grows more on the abaxial side compared to the adaxial side and forces a curvature in the ovule that results in its anatropous shape. The nucellus and funiculus also show adaxial-abaxial polarity later in development (Schneitz et al., 1995). What is the molecular control of the formation of this axis?

A couple of mutants have been isolated that show alterations in the adaxial-abaxial axis. *ino* mutants fail to form the outer integument and do not show a typical anatropous ovule (Baker et al., 1997; Schneitz et al., 1997) (see Figure 1.3, part E). The cloning of *INO* revealed that it encodes a protein that belongs to the recently described YABBY family of putative transcription factors (Villanueva et al., 1999). The members of the YABBY family are involved in specifying the abaxial cell fate and are expressed in the abaxial side of the emerging lateral organs (Bowman, 2000; Siegfried et al., 1999). *INO* is expressed in a few cells at the abaxial, central epidermis that give rise to the outer integument (Villanueva et al., 1999) (see Figure 1.3, part F). The biochemical nature of *INO* and its abaxially restricted expression pattern suggest a role for *INO* in adaxial-abaxial pattern formation.

The *SUPERMAN (SUP)* locus has been implicated in the adaxial-abaxial pattern formation in the ovules, since the ovules of *sup* mutants show a symmetric growth of the outer integument resulting in a nonanatropous ovule (Gaiser et al., 1995; Schneitz et al., 1997). *sup* was initially isolated as a floral mutant that shows additional stamens at the expense of carpels, and *SUP* encodes a zinc-finger transcription factor (Sakai et al., 1995). During flower development, *SUP* plays a cadastral role at the boundary between the stamens and carpels (Bowman et al., 1992; Sakai et al., 1995). SUP is likely to play a similar cadastral role in ovules as well as restricting *INO* expression to the abaxial epidermis. *INO* expression can be detected throughout the central region of *sup* mutants (Balasubramanian and Schneitz, 2002; Villanueva et al., 1999).

Apart from *SUP* and *INO*, two more loci play a role in adaxial-abaxial pattern formation during ovule development. The ovules of *nzz ats* double mutants show a *sup*-like phenotype, suggesting the involvement of these two loci in the outgrowth of the outer integument. Similar to *sup*, *INO* expression can be detected throughout the central region in ovules of the *nzz ats* double mutants. Neither of the single mutants of *nzz* or *ats* show a chalazal misexpression of *INO,* highlighting the redundant functions of *NZZ* and *ATS* in restricting the *INO* expression to the abaxial epidermis. How does *SUP* relate to *nzz* and *ats?* Currently this remains an open question. The defects in *nzz ats* double mutants seem to be even more pronounced than in *sup* single mutants, since the alteration in *INO* expression can be seen at an earlier stage in these double mutants compared to *sup* single mutants. This suggests that *NZZ* and *ATS* may function earlier than *SUP* (Balasubramanian and Schneitz, 2002).

Proximal-Distal and Abaxial-Adaxial Pattern Formation Is Temporally Linked

Very little is known about how the pattern formation along various axes and growth are linked during plant development. Our present knowledge on ovule development throws some light on this aspect of plant development. The analysis of the *nzz* mutant phenotype shows that an early onset of the adaxial-abaxial axis interferes with the development along the proximal-distal axis. This situation in analogous with the situation in limb development in vertebrates. An early establishement of the anterior-posterior axis in the chick limb interferes with proximal-distal development. Thereofore, not only a spatial but also a proper temporal expression of genes is required for the proper growth and development of the organ. How is this achieved during ovule development at a molecular level? It appears that *NZZ* plays a vital role in linking patterning along two axes of symmetry during ovule development.

How does *NZZ* link proximal-distal and adaxial-abaxial pattern formation? A subtle but important aspect of *NZZ* function is the temporal negative regulation of *INO* expression. The onset of *INO* expression is the first sign of the adaxial-abaxial polarity establishment in the chalaza. In *nzz* mutants, *INO* is turned on earlier, and consequently, the outer integument initiates earlier than the inner integument at an incorrect time. Contrary to this, in *ant* mutants the onset of *INO* expression is delayed. Furthermore, *ANT* and *NZZ* function antagonistic to each other during ovule development. Taken together, these data suggest that the correct timing of the onset of *INO* expression requires coordinated function of *ANT* and *NZZ*. It may be recalled that *NZZ* functions together in *BEL1* in the specification of the chalaza. During this time, by negatively regulating *INO* expression in a temporal manner through its antagonistic interactions with *ANT*, *NZZ* ensures that the adaxial-abaxial axis is established at a correct time. Thus, it links the pattern formation along both these axes. The early onset of *INO* expression in *nzz* interferes with the development along the proximal-distal axis.

NZZ Couples Pattern Formation and Growth During Ovule Development

How is growth inter-linked with pattern formation? *ANT* plays a vital role in growth control during plant development. Plants that ectopically express *ANT* produce larger floral organs, showing that the level of *ANT* expression is crucial for the proper size of an organ (Krizek, 1999; Mizukami and Fischer, 2000). *NZZ* functions antagonistically with *ANT* during ovule development. An ectopic expression of *ANT* leads to a partial *nzz* phenocopy (Krizek, 1999). These findings suggest that *NZZ* plays a vital role in regulating the growth of the ovule through its interactions with *ANT*. Thus, *NZZ*—by playing a vital role in pattern formation and growth both in temporal and spatial manner—couples pattern formation and growth along the proximal-distal and the adaxial-abaxial axis during ovule development (Balasubramanian and Schneitz, 2002).

CELL-CELL COMMUNICATIONS DURING OVULE DEVELOPMENT

Although the morphological units can be formed independently of each other, cell-cell communications still are necessary for the growth and development of the ovule. *WUSCHEL (WUS)*, a gene encoding a homeodomain protein, is expressed in the nucellus; plants that are defective for *wus* show an ovule phenotype similar to *ANT*, suggesting a *WUS* function in integument development (Gross-Hardt et al., 2002). *WUS* expression, however, is restricted to the nucellus. Thus, there must be some communication between the nucellus and integuments through *WUS* that is necessary for the proper development of the integuments (Gross-Hardt et al., 2002).

In *nzz* mutants one can observe an early onset of *INO* expression, but spatially *INO* is still expressed in the chalazal region. Then how does it interfere with the development of the nucellus? Interestingly, in wild type *INO* is expressed only in the abaxial cell layer of the outer integument (Balasubramanian and Schneitz, 2000). Thus, the expression pattern and the *nzz* and *ino* mutant phenotypes suggest a non–cell-autonomous fuction of *INO*. Non–cell-autonomous mode of action has also been suggested for other *YABBY* genes.

Among the isolated ovule mutants, several mutants lack the integuments. These mutants never go on to make a viable embryo sac, suggesting that signals from the integuments are necessary for the proper growth and development of the embryo sac. This is true even in mutants that have a partially enveloped embryo sac. This suggests that proper embryo sac development also requires proper integument development. Embryo sac development is arrested at various stages in these mutants, suggesting the need for communication between integument development and embryo sac development (Baker et al., 1997; Schneitz et al., 1997).

MORPHOGENESIS

After an initial framework is set up by pattern formation, proper cell division and cell elongation occur that lead to an organ of a particular size and shape. This process is referred to as *morphogenesis*. In ovule development, morphogenesis essentially occurs due to the growth and development of the integuments. The integument morphogenesis can be divided into two phases: (1) initiation of integuments and (2) integument growth and development.

Initiation of Integuments

As mentioned earlier, several loci, including *ANT, BEL, WUS, INO, HLL, SIN2, ATS,* and *UNICORN* (Gasser et al., 1998; Schneitz et al., 1997), are involved in early aspects of integument development. The requirement of *ANT, HLL,* and *SIN2* at different stages of ovule development suggests that these factors are involved in a basic machinery that is repeatedly required throughout development. *BEL1* is also expressed in the developing integuments, suggesting its involvement in this process. *WUS* presents a unique case highlighting the necessity of signals from the nucellus for integument development. Thus, *WUS* seem to function in a non–cell-autonomous way (Gross-Hardt et al., 2002). Apart from specifying the adaxial-abaxial axis, *INO* is also required for the growth and development of the outer integument. During integument development, *INO* is expressed in a few cells at the growing end of the outer integument (Villanueva et al., 1999). It has been suggested that proper juxtaposition of adaxial-abaxial signals is necessary for the outgrowth of the lateral organs (Waites and Hudson, 1995). *INO* expression is observed in only the abaxial cell layer of the outer integument, and in *ino* mutants, due to the absence of *INO,* this juxtaposition of the adaxial-abaxial signals fails to take place, leading to an absence of integument outgrowth. *ats* single mutants show an alteration in seed shape due to the fusion of the two integuments (Léon-Kloosterziel et al., 1994). While all these genes have a positive effect on integument growth, *UCN* may have a negative effect. The ovules of *ucn* mutants form a bulge, which is suggested to be an ectopic integument (Schneitz et al., 1997). Therefore, *UCN* could be a negative regulator of integument outgrowth.

Integument Growth and Development

Several mutants have been isolated that show alterations in the growth and development of integuments, without affecting their initiation and outgrowth (Schneitz et al., 1997). These mutants often show pleiotropic defects throughout the whole plant, suggesting the involvement of the corresponding genes in a basic function during plant development. This group of mutants includes *strubbelig (sub), blasig (bag), mollig (mog), laelli (lal), sup, leunig (lug)* (Schneitz et al., 1997), *tso1* (Hauser et al., 2000; Liu et al., 1997; Song et al., 2000), *short integument 1 (sin1)* (Lang et al., 1994), and *tousled (tsl)* (Roe et al., 1993). Some of these mutants are characterized at the molecular level. *sub* mutants show protrusions arising from the outer integument at about stage 3-I. Apart from this, *sub* mutants often have ovules that are arrested in integument development (Schneitz et al., 1997). *SUB* has been cloned, and it encodes a putative leucine-rich repeat receptor kinase (Chevalier and Schneitz, unpublished). Apart from its function in ovules, *SUB* has a wider role in plant development. *SUB* has a role in controlling the meristem size and development since *sub* mutants show alterations in the size and shape of the SAM (Chevalier and Schneitz, unpublished). Another locus that encodes a kinase is *TSL* (Roe et al., 1993). *tsl* mutants have reduced or missing floral organs similar to *sub* mutants. *tsl* mutants also show a phenotype similar to that of *lal* (discussed next; Roe, Nemhauser, and Zambryski, 1997) with reduced outer and protruding inner integuments (Roe, Durfee, et al., 1997; Roe, Nemhauser, and Zambryski, 1997; Roe et al., 1993). The phenotype of *tsl* and *sub* and the biochemical nature of the corresponding gene products show the importance of cell signalling during morphogenesis and plant development.

LUG was initially reported to be a cadastral gene that regulates the expression of *AG* in the floral primordia (Liu and Meyerowitz, 1995). Isolation of new alleles of *lug* hinted at a function of *LUG* in ovule development (Schneitz et al., 1997). *lug* mutants lack the embryo sac and can often produce a protruding inner integument (Schneitz et al., 1997). *LUG* has been cloned, and it encodes a putative transcriptional corepressor (Conner and Liu, 2000). *TSO1* also encodes a possible corepressor (Hauser et al., 2000;

Song et al., 2000). *tso1* plants have ovules that show a phenotype similar to that of *lug*. The double mutants of *tso1 lug* show a synergistic effect, suggesting that these genes might function redundantly during ovule development. *TSO1* has been implicated in regulating cell division and directional cell expansion (Hauser et al., 1998; Liu et al., 1997).

lal and *sin1* show similar phenotypes in the ovule (Ray et al., 1996; Schneitz et al., 1997). They have reduced outer integument, and the nucellus, covered by the inner integument, appears protruded. *SIN1* has wider roles compared to *LAL* since the defects in *lal* mutants are restricted to the ovule compared to *sin1* mutants which show defects in other parts of the plant as well (Lang et al., 1994; Ray et al., 1994, 1996; Robinson-Beers et al., 1992). Preliminary analyses of *bag* and *mog* show that their gene products also regulate the late morphogenesis during ovule development (Schneitz et al., 1997).

PERSPECTIVES

It has been an exciting decade for people studying ovule development. Although much information was obtained during the 1990s, many new questions have also emerged. In spite of all the work, the molecular mechanisms have not been specified at the protein level. The genetic analysis has clearly laid out a foundation on which further experiments can be built. The biochemical function of NZZ protein remains elusive. The nuclear localization and the computational biochemical analysis suggests NZZ to be a putative orphan transcription factor (Schiefthaler et al., 1999; Yang et al., 1999). But the molecular basis of NZZ function is not clearly understood.

Genetic analysis had generated further interest in *ATS*, and the cloning of *ATS* is ongoing. Understanding the molecular nature of *ATS* shall provide additional insight into the regulation of adaxial-abaxial patterning and how the *NZZ/ATS/SUP* triad functions. Many loci that affect early ovule development encode transcription factors, but the target genes or the upstream activators for these loci have not been elucidated, although few potential candidates are being dissected through genetic analysis. Further molecular experiments are needed to address these questions. Several models have been proposed for ovule development, but none of them has been tested at a molecular level. While SUB and TSL kinases have been cloned, the ligand for SUB or the substrate for SUB and TSL have not yet been identified. On the whole one would expect to see a wealth of information coming out rather soon, and a bright time is surely ahead for understanding ovule development.

REFERENCES

Angenent, G. C., Franken, J., Busscher, M., van Dijken, A., van Went, J. L., Dons, H. J. M., and van Tunen, A. J. (1995). A novel class of *MADS* box genes is involved in ovule development in Petunia. *The Plant Cell* 7, 1569-1582.

Baker, S. C., Robinson-Beers, K., Villanueva, J. M., Gaiser, J. C., and Gasser, C. S. (1997). Interactions among genes regulating ovule development in *Arabidopsis thaliana*. *Genetics* 145, 1109-1124.

Balasubramanian, S. and Schneitz, K. (2000). *NOZZLE* regulates proximal-distal pattern formation, cell proliferation and early sporogenesis during ovule development in *Arabidopsis thaliana*. *Development* 127, 4227-4238.

Balasubramanian, S. and Schneitz, K. (2002). *NOZZLE* links proximal-distal and adaxial-abaxial pattern formation during ovule development in *Arabidopsis thaliana*. *Development* 129, 4291-4300.

Bowman, J. L. (2000). The *YABBY* gene family and abaxial cell fate. *Curr Opin Plant Biol* 3, 17-22.

Bowman, J. L., Drews, G. N., and Meyerowitz, E. M. (1991). Expression of the *Arabidopsis* floral homeotic gene *AGAMOUS* is restricted to specific cell types in flower development. *Plant Cell* 3, 749-758.

Bowman, J. L. and Meyerowitz, E. M. (1991). Genetic control of pattern formation during flower development in *Arabidopsis*. *Symp Soc Exp Biol* 45, 89-115.

Bowman, J. L., Sakai, H., Jack, T., Weigel, D., Mayer, U., and Meyerowitz, E. M. (1992). *SUPERMAN*, a regulator of floral homeotic genes in *Arabidopsis*. *Development* 114, 599-615.

Broadhvest, J., Baker, S. C., and Gasser, C. S. (2000). *SHORT INTEGUMENTS 2* promotes growth during *Arabidopsis* reproductive development. *Genetics* 155, 899-907.

Cheng, X.-F., Wittich, P. E., Kieft, H., Angenent, G., XuHan, X., and van Lammern, A. A. M. (2000). Temporal and spatial expressionof *MADS* box genes *FBP7* and *FBP11*, during initiation and early development of ovules in wild type and mutant *Petunia hybrida*. *Plant Biology* 2, 693-669.

Chevalier, D., Sieber, P., and Schneitz, K. (2001) . The genetic and molecular control of ovule development. In S. O'Neill and J. Roberts (eds.), *Annual plant reviews: Plant reproduction* (pp. 61-65). Sheffield, UK: Sheffield Academic Press.

Coen, E. S. and Meyerowitz, E. M. (1991). The war of whorls: Genetic interactions controlling flower development. *Nature* 353, 31-37.

Colombo, L., Franken, J., Koetje, E., van Went, J., Dons, H. J. M., Angenent, G. C., and van Tunen, A. J. (1995). The petunia *MADS* box gene *FBP11* determines ovule identity. *The Plant Cell* 7, 1859-1868.

Conner, J. and Liu, Z. (2000). *LEUNIG*, a putative transcriptional corepressor that regulates *AGAMOUS* expression during flower development. *Proc Natl Acad Sci USA* 97, 12902-12907.

Elliott, R. C., Betzner, A. S., Huttner, E., Oakes, M. P., Tucker, W. Q. J., Gerentes, D., Perez, P., and Smyth, D. R. (1996). *AINTEGUMENTA*, an *APETALA2*-like gene of *Arabidopsis* with pleiotropic roles in ovule development and floral organ growth. *The Plant Cell* 8, 155-168.

Esau, K. (1977). *Anatomy of seed plants.* New York: John Wiley & Sons.

Evans, P. T. and Malmberg, R. L. (1989). Alternative pathways of tobacco placental development: Time of commitment and analysis of a mutant. *Developmental Biology* 136, 273-283.

Ferrandiz, C., Gu, Q., Martienssen, R., and Yanofsky, M. F. (2000). Redundant regulation of meristem identity and plant architecture by *FRUITFULL, APETALA1* and *CAULIFLOWER*. *Development* 127, 725-734.

Gaiser, J. C., Robinson-Beers, K., and Gasser, C. S. (1995). The *Arabidopsis SUPERMAN* gene mediates asymmetric growth of the outer integument of ovules. *The Plant Cell* 7, 333-345.

Gasser, C. S., Broadhvest, J., and Hauser, B. A. (1998). Genetic analysis of ovule development. *Annual Review of Plant Physiology and Plant Molecular Biology* 49, 1-24.

Goto, K. and Meyerowitz, E. M. (1994). Function and regulation of the *Arabidopsis* floral homeotic gene *PISTILLATA*. *Genes and Development* 8, 1548-1560.

Gross-Hardt, R., Lenhard, M., and Laux, T. (2002). *WUSCHEL* signaling functions in interregional communication during *Arabidopsis* ovule development. *Genes Dev* 16, 1129-1138.

Grossniklaus, U. and Schneitz, K. (1998). The molecular and genetic basis of ovule and megagametophyte development. *Seminars in Cell and Developmental Biology* 9, 227-238.

Hauser, B. A., He, J. Q., Park, S. O., and Gasser, C. S. (2000). *TSO1* is a novel protein that modulates cytokinesis and cell expansion in *Arabidopsis*. *Development* 127, 2219-2226.

Hauser, B. A., Villanueva, J. M., and Gasser, C. S. (1998). *Arabidopsis TSO1* regulates directional processes in cells during floral organogenesis. *Genetics* 150, 411-423.

Jack, T., Brockman, L. L., and Meyerowitz, E. M. (1992). The homeotic gene *APETALA3* of *Arabidopsis thaliana* encodes a *MADS* box and is expressed in petals and stamens. *Cell* 68, 683-697.

Klucher, K. M., Chow, H., Reiser, L., and Fischer, R. L. (1996). The *AINTEGUMENTA* gene of *Arabidopsis* required for ovule and female gametophyte development is related to the floral homeotic gene *APETALA2*. *The Plant Cell* 8, 137-153.

Krizek, B. A. (1999). Ectopic expression of *Aintegumenta* gene results in increased growth of floral organs. *Dev. Genet* 25, 224-236.

Lang, J. D., Ray, S., and Ray, A. (1994). *sin1,* a mutation affecting female fertility in *Arabidopsis,* interacts with *mod1,* its recessive modifier. *Genetics* 137, 1101-1110.

Léon-Kloosterziel, K. M., Keijzer, C. J., and Koornneef, M. (1994). A seed shape mutant of *Arabidopsis* that is affected in integument development. *The Plant Cell* 6, 385-392.

Levy, Y. Y. and Dean, C. (1998). The transition to flowering. *The Plant Cell* 10, 1973-1990.

Liu, Z., Franks, R. G., and Klink, V. P. (2000). Regulation of gynoecium marginal tissue formation by *LEUNIG* and *AINTEGUMENTA*. *The Plant Cell* 12, 1879-1892.

Liu, Z. and Meyerowitz, E. M. (1995). *LEUNIG* regulates *AGAMOUS* expression in *Arabidopsis* flowers. *Development* 121, 975-991.

Liu, Z., Running, M., and Meyerowitz, E. M. (1997). *TSO1* functions in cell division during *Arabidopsis* flower development. *Development* 124, 665-672.

Lohmann, J. U., Hong, R. L., Hobe, M., Busch, M. A., Parcy, F., Simon, R., and Weigel, D. (2001). A molecular link between stem cell regulation and floral patterning in *Arabidopsis*. *Cell* 105, 793-803.

Lohmann, J. U. and Weigel, D. (2002). Building beauty: The genetic control of floral patterning. *Dev Cell* 2, 135-142.

Maheswari, P. (1950). *An introduction to the embryology of angiosperms.* New York: McGraw-Hill.

Mandel, M. A., Bowman, J. L., Kempin, S. A., Ma, H., Meyerowitz, E. M., and Yanofsky, M. F. (1992). Manipulation of flower structure in transgenic tobacco. *Cell* 71, 133-143.

Mandel, M. A., Gustafson-Brown, C., Savidge, B., and Yanofsky, M. F. (1992). Molecular characterization of the *Arabidopsis* floral homeotic gene *APETALA 1*. *Nature* 360, 273-277.

Mizukami, Y. and Fischer, R. L. (2000). Plant organ size control: AINTEGUMENTA regulates growth and cell numbers during organogenesis. *Proc Natl Acad Sci USA* 97, 942-947.

Modrusan, Z., Reiser, L., Feldmann, K. A., Fischer, R. L., and Haughn, G. W. (1994). Homeotic transformation of ovules into carpel-like structures in *Arabidopsis*. *The Plant Cell* 6, 333-349.

Modrusan, Z., Reiser, L., Fischer, R. L., Feldmann, K. A., and Haughn, G. W. (1994). *bel1* mutants are defective in ovule morphogenesis. In J. Bowman (ed.), *Arabidopsis: An atlas of morphology and development* (pp. 304-305). New York: Springer Verlag.

Okamuro, J. K., Caster, B., Villarroel, R., van Montagu, M., and Jofuku, K. D. (1997). The AP2 domain of *APETALA2* defines a large new family of DNA binding proteins in *Arabidopsis*. *Proceedings National Academy of Sciences, U.S.A.* 94, 7076-7081.

Pinyopich, A., Ditta, G. S., and Yanofsky, M. F. (2001). Roles of *SEEDSTICK MADS*-box gene during ovule and seed development. 12th Annual Arabidopsis Meeting, June 23-27, Madison, Wisconsin.

Ray, A., Lang, J. D., Golden, T., and Ray, S. (1996). *SHORT INTEGUMENT (SIN1),* a gene required for ovule development in *Arabidopsis*, also controls flowering time. *Development* 122, 2631-2638.

Ray, A., Robinson-Beers, K., Ray, S., Baker, S. C., Lang, J. D., Preuss, D., Milligan, S. B., and Gasser, C. S. (1994). *Arabidopsis* floral homeotic gene *BELL (BEL1)* controls ovule development through negative regulation of *AGAMOUS* gene *(AG)*. *Proc Natl Acad Sci USA* 91, 5761-5765.

Reiser, L., Modrusan, Z. L. M., Samach, A., Ohad, N., Haughn, G. W., and Fischer, R. L. (1995). The *BELL1* gene encodes a homeodomain protein involved in pattern formation in the *Arabidopsis* ovule primordium. *Cell* 83, 735-742.

Riechmann, J. L. and Meyerowitz, E. M. (1998). The AP2/EREBP familiy of plant transcription factors. *Biological Chemistry* 379, 633-646.

Robinson-Beers, K., Pruitt, R. E., and Gasser, C. S. (1992). Ovule development in wild-type *Arabidopsis* and two female-sterile mutants. *The Plant Cell* 4, 1237-1249.

Roe, J. L., Durfee, T., Zupan, J. R., Repetti, P., McLean, B. G., and Zambryski, P. C. (1997). TOUSLED is a nuclear serine/threonine protein kinase that requires a coiled-coil region for oligomerization and catalytic activity. *Journal of Biological Chemistry* 272, 5838-5845.

Roe, J. L., Nemhauser, J. L., and Zambryski, P. C. (1997). *TOUSLED* participates in apical tissue formation during gynoecium development in *Arabidopsis. The Plant Cell* 9, 335-353.

Roe, J. L., Rivin, C. J., Sessions, R. A., Feldmann, K. A., and Zambryski, P. C. (1993). The *Tousled* gene in *A. thaliana* encodes a protein kinase homolog that is required for leaf and flower development. *Cell* 75, 939-950.

Rounsley, S. D., Ditta, G. S., and Yanofsky, M. F. (1995). Diverse roles for *MADS* box genes in *Arabidopsis* development. *The Plant Cell* 7, 1259-1269.

Sakai, H., Medrano, L. J., and Meyerowitz, E. M. (1995). Role of *SUPERMAN* in maintaining *Arabidopsis* floral whorl boundaries. *Nature* 378, 199-203.

Schiefthaler, U., Balasubramanian, S., Sieber, P., Chevalier, D., Wisman, E., and Schneitz, K. (1999). Molecular analysis of *NOZZLE,* a gene involved in pattern formation and early sporogenesis during sex organ development in *Arabidopsis thaliana. Proc Natl Acad Sci USA* 96, 11664-11669.

Schneitz, K., Baker, S. C., Gasser, C. S., and Redweik, A. (1998). Pattern formation and growth during floral organogenesis: *HUELLENLOS* and *AINTEGUMENTA* are required for the formation of the proximal region of the ovule primordium in *Arabidopsis thaliana. Development* 125, 2555-2563.

Schneitz, K., Balasubramanian, S., and Schiefthaler, U. (1998). Organogenesis in plants: The molecular and genetic control of ovule development. *Trends in Plant Science* 3, 468-472.

Schneitz, K., Hülskamp, M., Kopczak, S. D., and Pruitt, R. E. (1997). Dissection of sexual organ ontogenesis: A genetic analysis of ovule development in *Arabidopsis thaliana. Development* 124, 1367-1376.

Schneitz, K., Hülskamp, M., and Pruitt, R. E. (1995). Wild-type ovule development in *Arabidopsis thaliana*: A light microscope study of cleared whole-mount tissue. *The Plant Journal* 7, 731-749.

Sessions, A., Yanofsky, M., and Weigel, D. (1998). Patterning the floral meristem. *Seminars in Cell and Developmental Biology* 9, 221-226.

Siegfried, K. R., Eshed, Y., Baum, S. F., Otsuga, D., Drews, G. N., and Bowman, J. L. (1999). Members of the YABBY gene family specify abaxial cell fate in *Arabidopsis. Development* 126, 4117-4128.

Simpson, G. G. and Dean, C. (2002). *Arabidopsis,* the Rosetta stone of flowering time? *Science* 296, 285-289.

Simpson, G. G., Gendall, A. R., and Dean, C. (1999). When to switch to flowering. *Annual Review of Cell and Developmental Biology* 15, 519-550.

Skinner, D. J., Baker, S. C., Meister, R. J., Broadhvest, J., Schneitz, K., and Gasser, C. S. (2001). The *Arabidopsis HUELLENLOS* gene, which is essential for normal ovule development, encodes a mitochondrial ribosomal protein. *The Plant Cell* 13, 2719-2730.

Song, J.-Y., Leung, T., Ehler, L. K., Wang, C., and Liu, Z. (2000). Regulation of meristem organisation and cell division by *TSO1,* an *Arabidopsis* gene with cysteine-rich repeats. *Development* 127, 2207-2217.

Villanueva, J. M., Broadhvest, J., Hauser, B. A., Meister, R. J., Schneitz, K., and Gasser, C. S. (1999). *INNER NO OUTER* regulates abaxial-adaxial patterning in *Arabidopsis* ovules. *Genes Dev* 13, 3160-3169.

Waites, R. and Hudson, A. (1995). *Phantastica:* A gene involved in dorsoventrality of leaves in *Antirrhinum majus. Development* 121, 2143-2154.

Weigel, D. (1995). The genetics of flower development: from floral induction to ovule morphogenesis. *Annual Review of Genetics* 29, 19-39.

Western, T. L. and Haughn, G. W. (1999). *BELL1* and *AGAMOUS* genes promote ovule identity in *Arabidopsis thaliana. Plant Journal* 18, 329-336.

Wolpert, L., Beddington, R., Brockes, J., Jessell, T., Lawrence, P., and Meyerowitz, E. (1998). *Principles of development.* Oxford, UK: Oxford University Press.

Yang, W.-C., Ye, D., Xu, J., and Sundaresan, V. (1999). The *SPOROCYTELESS* gene of *Arabidopsis* is required for initiation of sporogenesis and encodes a novel nuclear protein. *Genes Dev* 13, 2108-2117.

Yanofsky, M. F., Ma, H., Bowman, J. L., Drews, G. N., Feldmann, K. A., and Meyerowitz, E. M. (1990). The protein encoded by the *Arabidopsis* homeotic gene *agamous* resembles transcription factors. *Nature* 346, 35-39.

Chapter 2

Female Gametophyte Development

Wei-Cai Yang

INTRODUCTION

A fundamental phenomenon common to all higher plants is the alternation of a sporophytic diploid phase and a gametophytic haploid phase in their life cycle. This is achieved through meiosis that results in the reduction of chromosome sets from 2n to 1n. Instrumental to this process is the formation of sexual organs, namely, the ovule and anther. The development of ovules has been discussed in Chapter 1. The purpose of this chapter is to review our current knowledge on female gametophyte or embryo sac development.

The formation of the female gametophyte can be divided into two consecutive processes, megasporogenesis and megagametogenesis. Megasporogenesis refers to the period of development between the differentiation of archesporial cell and the completion of meiosis. Therefore, it consists of a premeiotic phase and meiosis. The premeiotic development is marked by the formation of the archesporial cell and subsequent megasporocyte or meiocyte that will go through the premeiotic interphase and meiosis. The completion of meiosis results in the formation of a linear or T-shaped tetrad consisting of four haploid megaspores. Megagametogenesis refers to the developmental period between the formation of the megaspore and embryo sac ready to be fertilized. Megagametogenesis is characterized by several rounds of mitoses and cell fate specification within the developing embryo sac, typically resulting in the formation of a seven-celled female gametophyte in most flowering plant species with four different cell types, namely, the egg, synergid, central cell, and antipodal (see Figure 2.1). In recent years, significant progress has been made to understand the genetic and molecular mechanisms during female gametophyte development. This chapter aims to provide an overview of the developmental processes with emphasis on the genetic and molecular aspects. For specialized reviews, readers are encouraged to consult several excellent publications (Kapil and Bhatnagar,

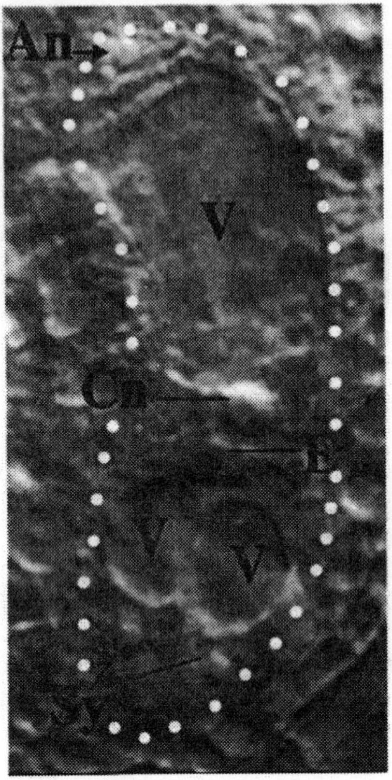

FIGURE 2.1. A Nomarski micrograph showing a seven-celled embryo sac with the micropylar end pointing downward in *Arabidopsis thaliana*. An: antipodal cells; Cn: central cell nucleus; E: egg cell; V: central vacuole; Sy: synergid.

1981; Willemse and van Went, 1984; Huang and Russell, 1992; Russell, 1993; Grossniklaus and Schneitz, 1998).

ORGANIZATION OF THE FEMALE GAMETOPHYTE

In the majority of higher plants, the female gametophyte is composed of an egg cell, two synergids, a central cell, and three antipodals (see Figure 2.1). However, the developmental paths leading to it vary. Differences exist in terms of cell wall formation during meiosis, the selection of functional megaspore, number of megaspores taking part, and mitotic cycles. According to whether there is cell wall formation during meiosis I and II, and the number of megaspore nuclei involved, the female gametophyte can be divided into three major groups, namely, monosporic, bisporic, and tetrasporic. They can be further subdivided into 11 types named after the genus where it was first described (see Figure 2.2).

Monosporic Embryo Sac

A characteristic of the monosporic embryo sac is that all the cell types of the female gametophyte are derived from the same megaspore; therefore, they are genetically identical. Cell walls are formed at the end of meiosis I and II, giving rise to four megaspores. The monosporic embryo sacs are further divided into *Polygonum* and *Oenothera* types based on the selection of the functional megaspore and number of postmeiotic, mitotic cycles. In the most prevalent *Polygonum* type, the embryo sac is formed from the chalazal megaspore of the tetrad that undergoes three successive rounds of nuclear divisions to give rise to an eight-nucleate embryo sac. The coenocytic embryo sac cellularizes and forms a seven-celled female gametophyte that consists of a three-celled egg apparatus, a central cell with two polar nuclei, and three antipodal cells (see Figure 2.2). On the contrary, in the *Oenothera* type, the embryo sac is derived from the micropylar megaspore of the tetrad and mitosis III does not occur. Therefore, this type of embryo sac is four- nucleate. The cellularized embryo sac is composed of only four cells, namely, the three-celled egg apparatus and a haploid central cell. Antipodal cells are missing (Figure 2.2). This type of female gametophyte occurs mainly in the Onagraceae family, with only one exception from *Schisandra chinensis*, where the embryo sac is from the chalazal functional megaspore instead of the micropylar megaspore. As the *Polygonum* type accounts for over 70 percent of the higher plant species investigated so far, including most food crops and the model plant *Arabidopsis*, this chapter will focus mainly on *Polygonum* embryo sac development in the following discussion.

Bisporic Embryo Sac

In plants belonging to this group, the first meiotic division is followed by cell wall formation, resulting in the formation of a dyad. However, there is no wall formation at the end of the second meiotic division in the functional dyad, giving rise to a two-nucleate megaspore. Subsequently, both nuclei undergo two mitotic divisions followed by cellularization to form the seven-celled embryo sac seen in *Allium* and *Endymion* species (Figure 2.2). These two types of female gametophytes differ only in the selection of functional dyad. In the *Allium* type, the ehalazal dyad is functional, whereas it is the micropylar dyad that is functional in the *Endymion* type. Another

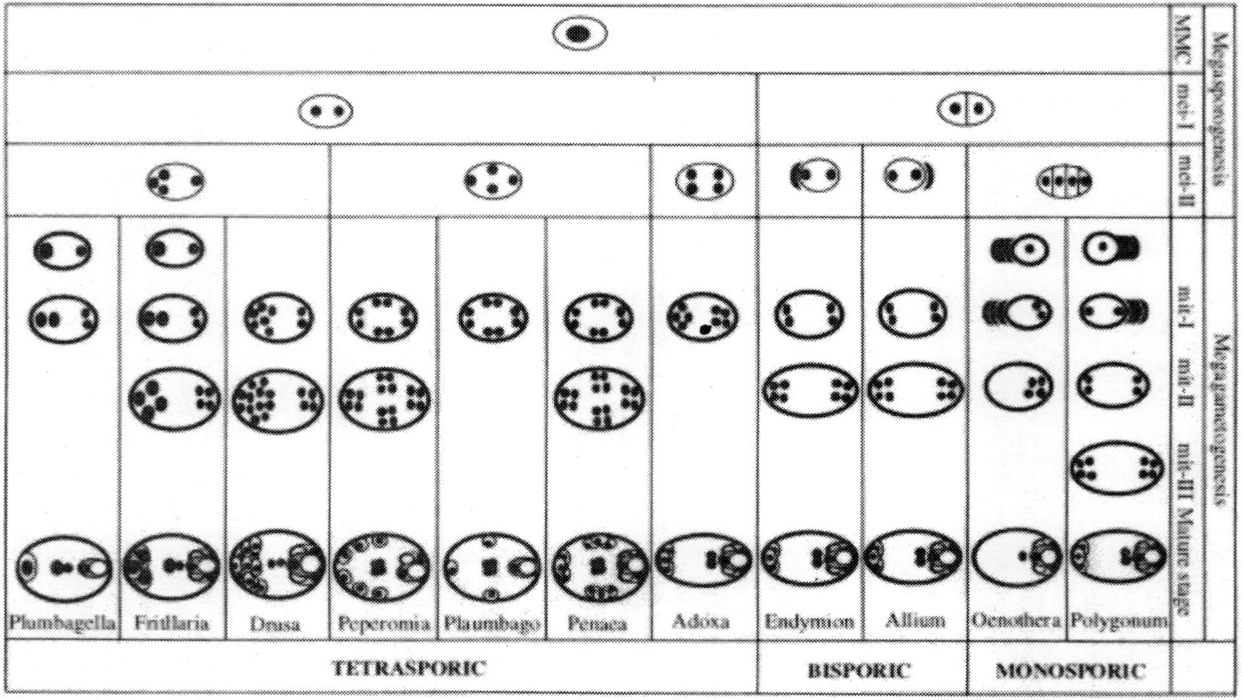

FIGURE 2.2. Diagrammatic representation of various types of embryo sacs. Mei-I: meiosis I; mei-II: meiosis II; mit-I: the first mitosis; mit-II: the second mitosis; mit-III: the third mitosis; MMC: meiotcyte. (*Source:* Adapted from Maheshwari, 1950.)

characteristic in this group is that cells in the mature embryo sac are genetically different since they come from two meiotic products.

Tetrasporic Embryo Sac

In plants bearing this type of embryo sac, there is no cell wall formation during meiosis, resulting in a four-nucleate coenomegaspore. All four nuclei participate in the formation of the female gametophyte; therefore, it is genetically more heterogeneous than the bisporic embryo sac. The nuclear behavior in the tetrasporic embryo sac is quite variable in terms of positioning, the number of postmeiotic nuclear divisions, and nuclear fusion during megagametogenesis (Figure 2.2). The four nuclei can be positioned polarly in a 2n + 2n pattern as seen in the *Adoxa* embryo sac, 1n + 1n + 1n + 1n in *Penaea, Plumbago,* and *Peperomia* embryo sacs, or a 1n + 3n pattern in *Drusa, Fritillara,* and *Plumbagella* embryo sacs (Figure 2.2). The nuclei in the coenomegaspore can undergo one mitotic division as seen in *Adoxa, Fritillara,* and *Plumbago* types, two nuclear divisions in *Penaea, Peperomia,* and *Drusa* types, or no postmeiotic nuclear division but directly forming a three-celled embryo sac as seen in *Plumbagella*. In *Fritillara* and *Plumbagella* types, the embryo sacs assume the 1n + 3n nuclear configuration, and the three nuclei at the chalazal end of the coenomegaspore fuse to form a triploid nucleus before the first mitotic division takes place, but the micropylar nucleus remains haploid. Consequently, the central cell contains a haploid nucleus and a triploid nucleus (Figure 2.2).

Apart from the embryo sacs discussed previously, there exist intermediate or specialized embryo sacs, although these pathways are exceptional. In *Chrysanthemum cineraiaefolium,* the four nuclei of the coenomegaspore adopt a 1n + 2n + 1n configuration with one nucleus at each pole and two in the center. The two central nuclei remain tightly associated and do not undergo further division, whereas nuclei at each pole continue mitotic cycles and form an embryo sac with a three-celled egg apparatus, a triploid central cell, and four haploid antipodals. In some cases, the two central nuclei can fuse and then undergo two more mitotic divisions. Before cellularization, the embryo sac contains 16 nuclei, which are positioned in a 4n + 1n + 4n configuration with 4 central diploid nuclei. As a result, the embryo sac contains a three-celled egg apparatus, a triploid central cell, and four haploid and three diploid antipodal cells (Maheshwari, 1950). The variation does occur within the same species or even in the same individual, though not frequently, and sometimes it is sensitive to growth conditions, such as temperature and climate changes.

ARCHESPORIAL CELL FORMATION

At the early stage of nucellus development, primordial cells undergo mitotic cycles, and no morphological differences could be seen among these meristematic cells (Figure 2.3, part A). Progressively, a single subepidermal cell (L2) at the tip of nucellar primordium starts to differentiate and becomes distinct from its neighboring cells as visualized by its hypodermal position, gradual increase in size, prominent nuclear staining, and being rich in cytoplasm (Figure 2.3, parts B and C). This cell is now designated the primary archesporial cell. In some species, more than one archesporial cell may be formed, often resulting in polyembryony. In the majority of angiosperms, the primary archesporial cell may undergo further periclinal mitotic division to give rise to an outer primary parietal cell and an inner

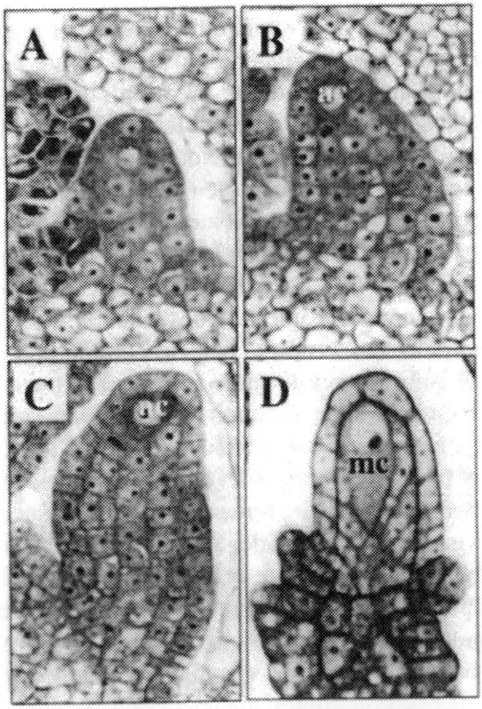

FIGURE 2.3. Micrographs showing transition stages from nucellar primordial cell to megasporocyte in *Arabidopsis thaliana*. A: A nucellar primodrium—Note a hypodermal cell (*) at the top of the nucellus; B: an archesporial cell (ac) becomes obvious; C: Archesporial cell increased in size; D: A megasporocyte (mc).

primary sporogenous cell. The primary parietal cell undergoes further periclinal and anticlinal divisions to form multiple-layered parietal tissue in crassinucellate ovules. The primary sporogenous cell normally does not undergo further division and differentiates directly into the megasporocyte. However, in some taxa, including Casuarinaceae, Rosaceae, and Liliaceae, more than one megasporocyte may be formed as a consequence of either differentiation of several archesporial cells or mitotic division of the primary sporogenous cell (Bouman, 1984). They are often associated with polyembryony and apomixis. In tenuinucellate ovules, such as *Arabidopsis*, the primary archesporial cell functions directly as a sporogenous cell and results in only one megasporocyte (Figure 2.3, parts C and D) (Misra, 1962; Webb and Gunning, 1990; Reiser and Fischer, 1993; Schneitz et al., 1995; Christensen et al., 1997; Bajon et al., 1999).

How is the number of achesporial cells and megasporocytes controlled in a given species? The fixed pattern of archesporium in most species suggests that it is genetically determined. Most likely, cell-to-cell interactions tell each other who will differentiate to form the archesporial cell, or, alternatively, it is just a random selection process. A mutation in the *MULTIPLE ARCHESPORIAL CELLS (MAC1)* gene in maize resulted in the differentiation of several hypodermal cells into archesporial cells instead of a single one in wild type (Sheridan et al., 1996). In the *mac1* mutant, the multiple archesporial cells undergo normal differentiation to form megasporocytes that are able to complete meiosis and form multiple embryo sacs in each ovule, indicating that *MAC1* plays a sporophytic role in determining archesporial cell fate. The recessive nature of the gene suggests that *MAC1* most likely regulates archesporial cell fate by controlling a signal that inhibits neighboring nucellar cells to become archesporial cells (Sheridan et al., 1996; Yang and Sundaresan, 2000). Interestingly, microsporocytes in anthers fail to complete meiosis which causes male sterility in the *macl* mutant (Sheridan et al., 1999), and this may be attributed to ontogenetie and/or developmental differences between the male and female sporogenesis (Sheridan et al., 1999). Nevertheless, the cloning of *MAC1* gene would provide clues about how the archesporial cell fate is controlled.

MEGASPOROGENESIS

As development proceeds, the primary sporogeneous cell increases its size and becomes the megasporocyte. Morphologically, the megasporocyte can be distinguished from sporogeneous cells by its size and its chromosomal and organelle behavior. Although little is known about the megasporocyte at the molecular level, cytoplasmic changes in the megasporocyte have been studied extensively and are thought to play a role in determining megaspore cell fate (Willemse, 1981; Huang and Russell, 1992). The cellular organization of the megasporocyte before meiosis is similar to that of meristematic cells. Polar distribution of organelles along the chalazal-micropylar axis becomes obvious at the initiation of meiosis. Generally, inheritable organelles such as plastids and mitochondria are concentrated at the functional pole of the megasporocyte, which is the chalazal pole in *Polygonum* type, whereas rough endoplasmic reticulum (rER) are mainly concentrated at the nonfunctional pole and often lie parallel to the micropylar wall and perpendicular to the chalazal-micropylar axis of the megasporocyte (Russell, 1979; Willemse and Bednara, 1979; Bajon et al., 1999). In contrast, Golgi and vesicles are either dispersed in the cytoplasm or grouped in the nonfunctional pole. The morphology of mitochondria and plastids also undergo dramatic changes in the megasporocyte during meiosis and become morphologically different from that of nucellar cells (Russell, 1979; Dickinson and Heslop-Harrison, 1977; Medina et al., 1981; Willemse and Franssen-Verheijen, 1978; Schulz and Jensen, 1981, 1986). Ribosome population undergoes dynamic changes and decreases significantly from leptotene to metaphase I (Medina et al., 1981; De Boer-de Jeu, 1978a; Dickinson and Andrews, 1977; Dickinson and Heslop-Harrison, 1977; Willemse and Bednara, 1979; Schulz and Jensen, 1986). The nucleus is often located in the middle or slightly toward the nonfunctional pole of the megasporocyte. A single nucleolus becomes prominent and stains heavily with Toluidine blue, indicating that active transcription occurs. The polarity establishment is also reflected in the cell wall, manifested by the asymmetrical distribution of callose deposition and plasmodesmata. In many cases, callose deposition starts in walls of the megasporocyte at the nonfunctional pole only and not in walls facing the functional pole (Willemse and Bednara, 1979). In some cases, it may be transient and incomplete in the form of a sieve at the functional pole (Rodkiewicz, 1975; Rodkiewicz and Bednara, 1976; Russell, 1979; Kapil and Bhatnagar, 1981). It is also observed in some species that the megasporocyte is surrounded in callose-deposited walls. In this case, plasmodesmata are exclusively present in the chalazal wall (at the functional pole) of the megasporocyte and may be cut off by callose impregnation during development (Russell, 1979; Willemse and Bednara, 1979). No plasmodesmata are observed in the micropylar wall (the nonfuctional pole) of the megasporocyte. It seems that callose deposition is quite dynamic and its role remains to be elucidated.

The pattern of callose impregnation and the polar distribution of cytoplasmic components and plasmodesmata indicate that polarity exists and may play a role during megasporocyte development and in the selection of functional megaspore by providing favorable structures and nutritional advantages to the functional megaspore over the other megaspores. No experimental data except structural observations are available to support this hypothesis. Alternatively, they are solely an expression of a regulated development of the female gametophyte in relation to ovular tissues (Willemse, 1981). It is unknown whether the polarity is maintained during meiosis I and II. In other words, the dynamics of organelles and cell wall modifications during meiosis and the biological significance of the polarity of the megasporocyte in terms of megaspore fate need to be addressed. This can be investigated now using gene knock-outs and organelle-specific GFP markers together with current confocal scanning laser microscopy (CSLM) technology.

During prophase I of meiosis, inheritable organelles, mitochondria, and plastids undergo dramatic changes in structure and dedifferentiation (Huang and Russell, 1992). Furthermore, the concentration of ribosomes is reduced and autophagic vacuoles are formed from the endoplasmic reticulum and preexisting vacuoles. These autophagic vacuoles often contain ribosomes and membraneous and organelle-like structures (Russell, 1979; Schulz and Jensen, 1981); they also have hydrolytic enzyme activities, demonstrated by the presence of acid phosphatase (Schulz and Jensen, 1981; Willemse and Bednara, 1979) and esterase activity (Willemse and Bednara, 1979). This suggests that the selected degradation of cytoplasmic components due to autophagic activities happened during prophase and may result in organelle selection and cytoplasmic regeneration in the megasporocyte. Nevertheless, the biological significance of these modifications remains to be resolved.

The transition from the archesporial cell to the megasporocyte is marked by an obvious increase in cell size and cytoplasmic changes as discussed previously. However, the genetic and molecular mechanisms controlling these changes are not understood. The gene *SPOROCYTELESS (SPL)/ NOZZLE (NZL)* that controls the transition has recently been characterized in *Arabidopsis* (Yang, Ye, et al., 1999; Schiefthaler et al., 1999). In the *spl/nzz* mutant, the archesporial cell is formed but fails to differentiate into the megasporocyte (Yang, Ye, et al., 1999), suggesting that the *SPL/NZZ* gene controls the transition of the archesporial cell to the megasporocyte. Because *SPL/NZZ* encodes for a nuclear protein, it is most likely that *SPL/NZZ* together with other transcription factors regulates a subset of genes essential for megasporocyte formation.

Interestingly, it was hypothesized that the sporocyte cell fate may be a default pathway during microsporogenesis in *Arabidopsis*, and a receptor protein kinase signaling pathway is most likely involved (Zhao et al., 2002). The mutation in the *EXCESS MICROSPOROCYTES1 (EMS1)* gene resulted in a change of tapetal cell fate into the microsporocyte fate. The

EMS1 gene encodes a leucine-rich repeat receptor protein kinase (LRR-RPK) homologous to ERECTA and CLAVATA1. The *EMS1* is expressed at a high level in the archesporial cell at an early stage and in both the sporogeneous cell and the parietal cell and tapetal cell at a later stage. This suggests that the *EMS1* is required for tapetal cell fate and is not essential for microsporocyte fate per se. However, it may be involved in the control of the number of microsporocytes during anther development. It would be extremely interesting to know whether such a mechanism is also functioning during megasporogenesis.

Once the fate of archesporial cells in the teuinucelate ovules or sporogenous cells in the crassinucellate ovules is committed, they undergo a period of rapid growth to reach a critical size at the leptotene stage (Willemse and Bednara, 1979) and enter a state of meiotic readiness. Now they may be designated as megasporocyte or meiocyte. In this premeiotic state (G2 phase) they are still capable of reverting to a mitotic division if preconditions for meiosis to occur are not satisfied. Earlier work in *Lilium* and *Trillium* showed that premeiotic megasporocytes could be reverted to mitosis as long as they do not enter the leptotene stage (Stern and Hotta, 1969; Ito and Takegami, 1982). This is further supported by recent genetic and cytological studies in maize and *Arabidopsis*. Several *ameiotic* loci were identified in maize (Golubovskaya, 1989; Golubovskaya et al., 1997) and *Arabidopsis* (Motamayor et al., 2000) that control the fate of megasporocyte and microsporocyte. Both megasporocyte and microsporocyte in the *am1-1* mutant are able to proceed into the prophase I of meiosis but are unable to enter meiosis as manifested by their chromosome behavior and microtubule organization. Instead, they undergo one or more successive mitotic divisions (Palmer, 1971; Golubovskaya et al., 1992; Staiger and Cande, 1992). The mitotic products of the mega- and microsporocyte degenerate eventually. This indicates that the *AM1-1* gene product most likely controls the entry of the mega- and microsporocyte into meiosis, or a departure from mitosis (Golubovskaya et al., 1997). In the *prophase I arrest (pra1)* mutant (allelic to *am1*) the megasporocyte is able to initiate meiosis and proceeds to the zygotene stage of the prophase I, then degenerates (Golubovskaya et al., 1993). While in the *switch1* mutant of *Arabidopsis*, megagametogenesis is delayed by the occurrence of additional mitotic divisions of the megasporocyte, resulting in embryo sacs with one, two, or several cells (Motamayor et al., 2000). Detailed studies showed that both chromatid cohesion at the leptotene stage and bivalent formation were affected, suggesting that SWI1 may be required for the establishment of sister chromatid cohesion, but not for its maintenance (Mercier et al., 2001). It becomes evident that chromatid cohesion is a critical point for meiotic commitment. An equational division, instead of an reductional division, occurs if chromatid cohesion can not proceed as seen in yeast mutants defective in chromatid cohesion, such as *REC8* (Watanabe and Nurse, 1999) and *SPO13/SLK19* (Zeng and Şaunders, 2000) mutants. Recently, the *SWI1* gene was cloned and encodes a novel, nuclear protein with no homology to cohesins in yeast and plants (Mercier et al., 2001; Bai et al., 1999; Bhatt et al., 1999). Similarly, the megasporocyte *afd1* mutant in maize can complete meiosis I, resulting in a nonreductional, mitotic division (Golubovskaya, 1989). A similar mutation, *dyad,* has also been identified in *Arabidopsis* in which meiosis is arrested after meiosis I (Siddiqi et al., 2000). While in the *el1* mutant, meiosis II is affected, resulting in the formation of nonreduced spores (Rhoades, 1956).

In yeast, the switch from mitotic cycle to meiosis is controlled by two genetic systems that respond to the mating type and nutritional conditions (Malone, 1990). It is not known whether similar mechanisms are also present in plants. Meiosis II and mitosis are very similar in terms of molecular machineries shared, while meiosis I is unique in that specific genes such as *Rec8* are evolved.

Mutations affecting chromosome pairing and segregation often cause abnormal megaspores, although meiosis can be completed. Disruption in *AtDMC1* (Couteau et al., 1999) and *SPO11* (Grelon et al., 2001) in *Arabidopsis* affects synapsis and bivalent formation at metaphase I and causes random segregation of univalent chromatids that consequently results in the formation of nonfunctional spores. However, meiosis does complete to form tetrads in anthers and even forms a few functional embryo sacs. The phenotypic difference in the *dmc1* mutations between plant and yeast indicates that some aspects of meiosis unique to plants exist (Couteau et al., 1999). Several other genes controlling chromosome behavior during male meiosis have also been identified. Mutations in *SYNAPSIS 1/ DETERMINATE INFERTILE 1 (SYN1/DIF1),* a homologue of the yeast meiosis-specific Rec8/Rad21 cohesin, are defective in chromosome condensation and pairing, cause abnormal segregation of sister chromatids throughout meiosis I and II, resulting in nondisjunction and nonfunctional spores (Bai et al., 1999; Bhatt et al., 1999). Several other genes involved in male meiosis have also been cloned, including a synaptonemal complex gene *MALE STERILITY 5* (Glover et al., 1998), *MEI1* (He and Mascarenhas, 1999), and a meiotic cyclin gene *SOLO DANCERS* (Azumi et al., 2002). Similar genes functioning during megasporogenesis have not been identified.

A mutation in *ASK1,* involved in targeting proteins for ubiquitin-mediated degradation, affects chromosome segregation at anaphase I in male meiosis and results in the formation of microspores that contain different numbers of chromosomes (Yang, Hu, et al., 1999). It is most likely that the

ASK1 gene plays a role in the degradation of proteins essential for chromosome association and dissociation. Although the defects in the *ask1* mutant are seen only in microsporogenesis, it is quite possible that its homologues may function similarly during megasporogenesis (Drouaud et al., 2000).

Apart from those defective in nuclear divisions, mutations affecting cytokinesis during meiosis are also identified. The *STUD (STD)* gene, for example, controls cell wall formation at the end of meiosis during microsporogenesis. Disruption of the *STD* gene function results in a coenocytic, tetranucleate microspore that gives rise to a large pollen with four vegetative nuclei and up to eight sperm nuclei (Hülskamp et al., 1997). Strikingly, postmeiotic mitosis is not affected irrespective of the defective cell wall formation, demonstrating that postmeiotic development is a separable process from cytokinesis during microgametogenesis. It would be interesting to know whether such genes also play a role during megagametogenesis. Several mutants defective mostly in cytokinesis during microsporogenesis have also been reported in maize (Beadle, 1932), alfalfa (Castetter, 1925; McCoy and Smith, 1983), soybean (Albertson and Palmer, 1979; Kennell and Horner, 1985b), *Arabidopsis* (Chen and McCormick, 1996; Spielman et al., 1997), and *Kniphofia* (Moffett, 1932). The characterization of these genes would further provide clues about possible mechanisms of meiotic and postmeiotic development during megasporogenesis and megagametogenesis.

In addition, other sporophytic tissues, such as integuments, are also required for megasporocyte function. Mutations affecting integument development, such as *aintegumenta* and *sap* also cause arrest of the megasporocyte and meiosis (Klucher et al., 1996; Elliot et al., 1996; Byzova et al., 1999). Therefore, sporophytic tissues may have a role in supporting the gametophyte development (Gasser et al., 1998; Yang and Sundaresan, 2000).

FORMATION OF THE FUNCTIONAL MEGASPORE

At the end of meiosis I, a dyad is formed in monosporic megasporogenesis. Cytokinesis often results in the formation of two unequal cells. The cell at the functional pole is larger and receives more cytoplasm of the megasporocyte. However, the one at the nonfunctional pole is smaller and receives less cytoplasm, although all types of organelles found in the megasporocyte are present (Schulz and Jensen, 1981). Both cells contain multiple concentric autophagic vacuoles with ribosomes and organelles (Schulz and Jensen, 1986). The transversal wall is irregular and has no plasmodesmata, and callow impregnation takes place soon after cytokinesis.

The completion of meiosis II of the megasporocyte usually results in the formation of a tetrad of four haploid megaspores. Meiosis II may be delayed in the dyad cell at the nonfunctional pole (Russell, 1979). The four newly formed megaspores showed no visible structural differences in the cytoplasm. Compared to the megasporocyte, megaspores contain elongated plastids and shorter rER (Bajon et al., 1999). Soon, callow impregnates the newly formed cell walls and separates megaspores from one another (Willemse and van Went, 1984). Often aberrant cell walls with no plasmodesmata are formed between nonfunctional megaspores, and callow impregnation persists throughout the degeneration process (Rodkiewicz, 1970; Russell, 1979; Schulz and Jensen, 1986; Webb and Gunning, 1990). The functional megaspore is linked with nucellar cells through numerous plasmodesmata (De Boer-de Jeu, 1978; Russell, 1979; Schulz and Jensen, 1981, 1986; Bajon et al., 1999), that may serve as a kind of super highway for nutrient flow from nucellar tissue to the megaspore.

The first sign of degeneration of nonfunctional megaspores is the aggregation of chromatin at the nuclear periphery and increase in ribosome concentration (Russell, 1979). One unique feature of the degenerating megaspores is the appearance of concentric laminae rER structure. The three degenerating megaspores are very electron dense in cytoplasm and contain hydrolytic activity revealed by peroxidase and esterase activities (Willemse and Bednara, 1979). High levels of acid phosphatase activity were also observed in the degenerating megaspore, although limited activity is present in the functional megaspore as well (Schulz and Jensen, 1986; Willemse and Bednara, 1979). Autophagic vacuoles appear, and eventually the plasma membrane is lost and hydrolytic activity prevails. The degraded products of the degenerating megaspores and nucellar tissues are presumably absorbed by the functional megaspore. Therefore, it may provide more space and nutrients for the expansion of the functional megaspore. In the meantime, the functional megaspore undergoes rapid restructuration and prepares for megagametogenesis. Plastids and mitochondria redifferentiate and multiply, and cytoplasm becomes rich in ribosomes and other organelles (De Boer-de Jeu, 1978; Russell, 1979; Schulz and Jensen, 1986). Starch and lipid accumulate (Willemse and Bednara, 1979) and microvillus-like membrane structure appears and may facilitate nutrient uptake from the digested substance of neighboring nucellar cells (Schulz and Jensen, 1986). The size of the functional megaspore increases dramatically before the first mitotic nuclear division takes place.

In the majority of higher flowering plants, only one megaspore survives and goes on to megagametogenesis, whereas the other three megaspores undergo cell death (Figure 2.2). Which of the four megaspores becomes functional is often fixed in a given species. It is the chalazal megaspore in *Polygonum* type and the micropylar megaspore in *Oenothera* type that becomes functional and goes through megagametogenesis. The asymmetrical division of the megasporocyte and their relative positions of the megaspore along the chalazal-micropylar axis presumably determine the fate of megaspores. Functional megaspores are generally connected through plasmodesmata to the nucellar tissues, and callow disappears early in cell walls of nonfunctional megaspores (Rodkiewicz, 1970; Willemse and Bednara, 1979).

A phenomenon common to most monosporic and bisporic megasporogenesis is the transient accumulation of callose, a β-1,3-polyglucan in cell walls. The pattern of callose deposition varies among species, and its role is not well understood. In most cases, callose deposition is often seen in walls specifically at the abortive or nonfunctional pole of the megasporocytes and less or not at all in walls at the functional pole. Plasmodesmata occur in walls that lack callose deposition (Russell, 1979; Bajon et al., 1999). The impregnation of callose in the entire wall of the megasporocyte begins at diakinesis and persists throughout meiosis I and II in some species (Rodkiewicz, 1970; Russell, 1979). At the dyad stage, callose deposition is obvious in the cross wall between the two daughter cells and the longitudinal walls at the chalazal end, much less at the micropylar end. At the early tetrad stage, callose is still present in cross walls between longitudinal walls of all megaspores. Less or no callose impregnation can be seen in walls at the chalazal megaspore. At the late tetrad stage, the megaspore at the chalazal end is free of callose staining, while there is still callose deposition in the walls of the three megaspores at the micropylar end. Callose deposition is gradually reduced and finally disappears on the lateral walls of the tetrad. Based on the pattern and timing of callose deposition, it has been postulated that the callose impregnated wall serves as a physical barrier to isolate the megasporocyte and newly formed haploid megaspore from the influence of neighboring nucellar cells, thereby assuming a different cell fate (Rodkiewicz, 1970). It is worthy to note that callose is absent in the wall in tetrasporic megasporogenesis where none of the meiotic products undergoes cell death and plasmodesmata are restricted to the wall at the functional pole (Rodkiewicz, 1978).

With the help of genetic and molecular approaches, it is now possible to address the role of callose deposition during megagametogenesis through the knockout of β-3-1,3-polyglucan synthase genes, for example. Alternatively, the degradation of callose by introducing β-1,3-glucanase would also address this. Premature degradation of callose in walls of microsporocytes causes male sterility in transgenic tobacco expressing a tapetum-specific

A9 promoter driving a β-3-1,3-glucanase gene. Dissolution of callose in cell walls does not affect meiotic divisions but results in the formation of microspores with aberrant cell walls that are sensitive to osmosis (Worrall et al., 1992). It is not known whether callose plays a similar role in megasporogenesis.

Meiosis results in the formation of four haploid megaspores that assume a linear or T-shaped tetrad configuration in most plant species. In monosporic embryo sacs, cytokinesis occurs at the end of both meiosis I and II. A dyad and tetrad are formed consequently, whereas cytokinesis happens at the end of meiosis I but not at meiosis II in bisporic development. Cytokinesis does not occur at all in tetrasporic female gametophyte development, resulting in the formation of a four-nucleated megaspore (Figure 2.2). The impaired pattern of nuclear division and cytokinesis is species specific, and the underlying mechanism is not yet known. Theoretically, it is believed that monosporic development is the most inherently stable form of megasporogenesis where the successful megaspore is determined spatially and there is no need to directly compete among megaspores to form the embryo. Therefore, it avoids genetic conflicts among megaspores and reduction in their vigor (Haig, 1990). The tetrasporic megasporogenesis, on the other hand, is the least stable form and provides the most variability in embryo sac organization. However, genetic conflict between the embryo and endosperm should be reduced because the fused central nucleus would be the product of all four of the megaspore nuclei.

MEGAGAMETOGENESIS

After rapid growth, the functional megaspore undergoes dramatic changes in cytoplasmic components, increases its volume dramatically, and elongates along the chalazal-micropylar axis to form an ovoid or teardrop cell (Russell, 1979; Christensen et al., 1997). The ribosome population increases and reaches levels comparable to that of the nucellar cells. Mitochondria and plastids become polymorphic in appearance and divide. Open whorls of rER also present in the functional megaspore, and rER is often associated with nuclear membrane, plasmalemma, and vesicle. Vacuolation takes place throughout the cytoplasm soon after the formation of the functional megaspore. Fusion of vacuoles containing hydrolytic enzymes is obvious throughout the cytoplasm, leading to the formation of a large central vacuole. Plasmodesmata are present only in walls at the chalazal end, not in the micropylar wall (Russell, 1979). Microtubules ramify randomly in the cytoplasm at an early developmental stage of the functional megaspore and emanate from the nucleus as it increases in size (Webb and Gunning, 1990;

Hu and Zhu, 2000). Finally, a mitotic spindle is formed along the chalazal-micropylar axis and nuclear division takes place to give rise to a two-nucleate megagametophyte (Webb and Gunning, 1990; Christensen et al., 1997).

Two-Nucleate Megagametophyte

In the two-nucleate megagametophyte, the concentration of ribosomes reaches high levels, with aboundant rER and Golgi apparatus. Concentric whorls of rER containing ribosomes are obvious and seem to reorganize into vesicles. Mitochondria and plastids multiply and are very abundant (Russell, 1979; Willemse and van Went, 1984). Shortly after the first mitosis, small vesicles coalesce to form a large central vacuole containing hydrolytic activities to separate the two nuclei physically to the chalazal and micropylar poles (Russell, 1979; Christensen et al., 1997). Toward the end of the two-nucleate stage, frequently an additional smaller vacuole forms at the chalazal pole. The central vacuole, in the meantime, expands further to occupy nearly one-third of the megagametophyte. Soon, the second mitotic division takes place to give rise to a four-nucleate megagametophyte.

Four-Nucleate Megagametophyte

The nucleus at the chalazal side divides orthogonally to the chalazal-micropylar axis. Interestingly, the two daughter nuclei realign along the chalazal-micropylar axis at a later stage in *Arabidopsis* (Christensen et al., 1997). However, the nucleus at the micropylar side of the two-nucleate megagametophyte always divides perpendicular to the axis. In the four-nucleate megagametophyte, the central vacuole and the chalazal vacuole continue to increase in size through fusions with smaller vesicles, starch-containing plastids, lipid bodies, ER, and mitochondria (Russell, 1979; De Boer-de Jeu, 1978a). The central vacuole occupies most of the megagametophyte space. At the same time, a new vacuole forms at the micropylar pole as well. Hydrolytic activity is obvious in the central vacuole (Russell, 1979). There is no clear difference in terms of the size of the four nuclei. The embryo sac expands as the micropylar nucellar cell degenerates. Plasmodesmata appear less frequently in cell walls of the four-nucleate embryo sac.

Eight-Nucleate Megagametophyte

One more round of nuclear division occurs to give rise to an eight-nucleate megagametophyte. The eight nuclei are separated by a large central vacuole to form a 4n + 4n configuration. Initially, the eight nuclei are similar in size. Soon after the final nuclear division, however, the two polar nuclei become larger than the other six nuclei. Often the micropylar nucleus at the chalazal pole travels a long distance through cytoplasm along the central vacuole toward the micropyle. Similarly, the chalazal nucleus at the micropylar pole migrates chalazally for a short distance. Consequently, the two polar nuclei meet at the micropylar half of the embryo sac, close to the future egg cell.

Dramatic changes in organelle morphology and abundancy occur during the three mitosis (Kapil and Bhatnagar, 1981). Among them, the dedifferentiation and redifferentiation of mitochondria and plastids and the completion of hydrolytic activity in the central vacuole have been proposed as indicators for the transition from vegetative to reproductive differentiation (Huang and Russell, 1992).

Cellularization of the Embryo Sac

At the end of the three free nuclear divisions, the embryo sac contains eight nuclei that adopt a 4n + 4n configuration. This pattern is physically brought about by the formation of the central vacuole at the end of the first nuclear division. The central vacuole therefore plays an important role in the separation of mitotic products of the first two daughter nuclei. The fusion of the polar nuclei often occurs after the cellularization takes place. The cellularization of the embryo sac results in the formation of a seven-celled embryo sac (Figure 2.2). The three cells at the chalazal end are relatively small and are named antipodal cells. The antipodal cells may undergo further mitotic division to form multicellular antipodals in some species. Cellularization at the micropylar part of the embryo sac results in the formation of two synergids and the egg cell. The egg and the two synergids are designated as the egg apparatus. The central cell contains the large central vacuole and a fused diploid nucleus and occupies the most space in the sac (see Figure 2.4). Now, cells at different positions within the embryo sac adopt different cell fates. The antipodals frequently degenerate before fertilization in most species. Fertilization of the central cell with sperm results in the formation of a triploid endosperm, while fertilization of the egg gives rise to the embryo. In summary, cells in the embryo sac adopt four fates, egg, synergid, central cell, and antipodal, although they share the same clonal origin.

The origin of gametophytic cells within the embryo sac can be traced back to the second nuclear division. The two synergids are most likely derived from the transversal division of one of the two nuclei at the micropylar

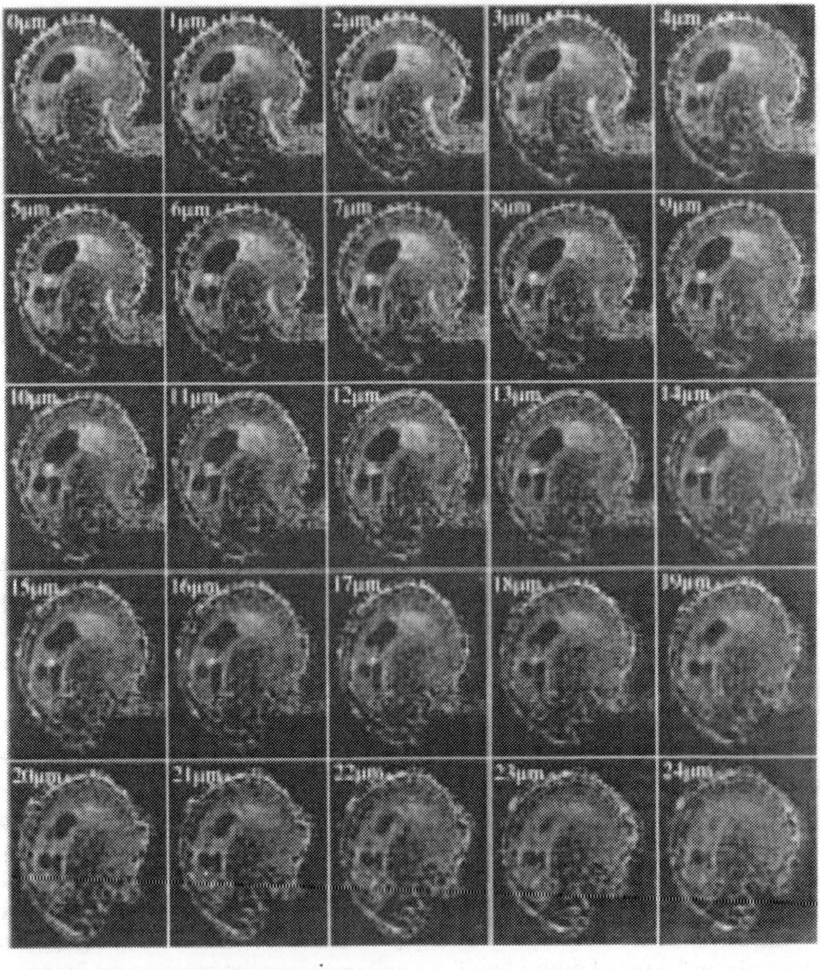

FIGURE 2.4. Serial confocal sections of an embryo sac from *Arabidopsis*. Ovules were prepared according to Christensen et al., 1997. 1 μm serial optical sections were made with a Zeiss LSM 510 microscope. The nucleus showing strong fluorescence is seen as a bright spot. The central vacuole is seen as a dark area.

end in the four-nucleate embryo sac. The first vertical cell plate is the origin of the future common wall between the two synergids, and the two synergid nuclei are sisters. The egg and one polar nucleus are from the other micropylar nucleus, where the horizontal cell plate of the second nuclear division in the four-nucleate embryo sac forms the wall that separates the egg and the central cell (Cass et al., 1986; Folsom and Cass, 1990; Russell, 1993). Different models have been proposed to explain the wall formation of the egg cell (Russell, 1993). The simplest situation may be that there are four cell plates separating the four micropylar nuclei; therefore, the formation of the egg apparatus involves four cell plates. However, no evidence exists to support the four-cell plate model (Russell, 1993). Bhandari and Chitralekha (1989) proposed a three-cell plate model in which a single curved cell plate separates the egg nucleus from the two synergid nuclei as is simultaneously seen in *Ranunculus.* Alternatively, the cell plate between the egg and the micropylar polar nucleus may bifurcate to form the walls that enclose the egg (Cass et al., 1985, 1986). However, the question remains unclear how phragmoplasts between sister and nonsister nuclei are formed. Evidence from barley megagametogenesis studies suggests that the phragmoplast between the sister nuclei are directly organized by microtubules from the interzone of the previous spindle body (Cass et al., 1986). On the other hand, it is not clear at all how the phragmoplast between nonsister nuclei could be formed. They may arise from "cytoplasmic domains," rich in microtubules emanating from the nucleus (Brown and Lemmon, 1991), as seen during microsporogenesis in certain monocots (Brown and Lemmon, 1991). In this case, cellularization occurs in the absence of the preprophase band and the cell plate appears to be determined by cytoplasmic domains (Brown and Lemmon, 1992). More sensitive and dynamic techniques, such as in vivo imaging of dynamic microtubule behavior, high-pressure freezing/freeze-substitution techniques, and TEM reconstruction, are essential to resolve these issues. Using these method, a unique type of "mini-phragmoplast" and a syncytial-type cell plate have been observed between both sister and nonsister nulcei during endosperm cellularization in *Arabidopsis* (Otegui and Staehelin, 2000a,b). It would be interesting to know whether such structures also occur during the cellularization of the female gametophyte. The cellularization at the chalazal pole typically occurs simultaneously with the formation of three antipodal cells and may involve three cell plates similar to that of the egg apparatus (Bhandari and Chitralekha, 1989).

The morphology of the nuclei in the free eight-nucleate embryo sac is readily distinct, indicating that the separation of vegetative and reproductive developmental programs takes place at the four-nucleate stage; cellularization may finally enforce the developmental program (Huang and Russell, 1992). Although it is not clear how such a program is established, studies in soybean embryo sac development suggest that the egg and synergid fates are related to the chalazal movement of one of the two micropylar nuclei (Folsom and Cass, 1990). This suggests that nuclear positioning rather than cell lineage plays an important role in the determination of gametophytic cell fate during the maturation of the embryo sac (Herr, 1972; Ehdaie and Russell, 1984). In this context, it has been demonstrated that positional information rather than cell lineage determines cell fate during embryo and root development (Scheres, 1996). Similarly, the orientation of the mitotic spindle and the cell plate may also play a role, as implicated in wheat embryo sac development (Cass et al., 1985; Russell, 1993). Alternatively, some sort of polarity and positional information, intrinsic or external to the embryo sac, are present in the embryo sac to determine the cell fate (Grossniklaus and Schneitz, 1998). This is supported by the identification of gametophytic cell-specific gene expression and their polar expression within the embryo sac (Grossniklaus and Schneitz, 1998).

Cellularization of the embryo sac and the mitotic cycle may be regulated independently, and gametophytic cell fates may depend on the completion of mitotic cycles. Nuclear division is disrupted in the *hadad (hdd)* mutant resulting in embryo sacs with two, four, or eight nuclei, and the nuclei at the two poles divide asynchronously. However, cellularization sometimes occurs prematurely before the mitotic cycles are completed, suggesting that the cellularization of the embryo sac can take place whether mitotic cycle is completed or not. This type of premature cellularization and asynchronous nuclear divisions at the two poles has also been observed in maize mutants (Vollbrecht and Hake, 1995). However, the prematurely cellularized cells fail to differentiate into a particular gametophytic cell type (Moore et al., 1997), implying that cell fate specification requires proper nuclear divisions and correct spatial information.

Seven-Celled Megagametophyte

The completion of the cellularization process results in formation of the egg apparatus consisting of the egg and two synergids located at the micropylar extreme of embryo sac. The cells of the egg apparatus are often arranged in a triangular fashion and share several vertical walls (Figure 2.4). The three cells are initially not differentiated, small and wedge-shaped, and indistinguishable from one another (Cass et al., 1986; Folsom and Cass, 1990; Bhandari and Chitralekha, 1989). The differentiation of the synergid first becomes obvious morphologically by the initiation of the filiform cell wall ingrowths, characteristic of transfer cells, at their micropylar wall (Cass et al., 1986; Huang and Russell, 1992). In contrast, the egg cell lacks such modification and is located more to the chalazal pole of the egg apparatus.

Egg Cell

The egg cell is similar to the synergids in shape but lacks the filiform cell wall modification. Walls of the synergid and the central cell surround the egg cell. Typically, the thickest wall of the egg is at its micropylar pole and gradually thins toward the chalazal pole (Schulz and Jensen, 1968; Kapil and Bhatnagar, 1981; Willemse and van Went, 1984; Sumner and van Caeseele, 1989; Huang and Russell, 1992). Often, some discrete homogenous, amorphous material stained by periodic acid and Schiff's reagent for insoluble carbohydrates are present (Schulz and Jensen, 1968a,b; Olson and Cass, 1981; Mansfield et al., 1991). Plasmodesmata are present in walls bordering the egg, synergid, and central cell at the micropylar portion, and sometimes pores may be present in the micropylar walls between the egg and synergid before fertilization (Schulz and Jensen, 1968a,b). Often the egg cell contains a hook region or apical pocket of the central cell extending micropylar to the attachment site between the egg and the embryo sac wall (Cass et al., 1986; Russell, 1979, 1992, 1993). The egg cell is attached to the embryo sac wall slightly above the two synergids; therefore, its chalazal portion extends beyond the synergids, though they are similar in height.

In most species, the ultrastructural features are distinct between the egg and the synergids. The egg cell is polarized with its nucleus and most of its cytoplasm located chalazally and a large central vacuole micropylarly (Figure 2.4). A large chalazal nucleus with numerous nuclear pores assumes a regular round shape and contains a large dense nucleolus. Compared to other cells, the cytoplasm is relatively quiescent with lower ribosome concentration and frequencies of Golgi and ER (Mansfield et al., 1991). Plastids are irregular in size and shape and are poorly developed. Large plastids with starch granules are observed located around the nucleus (Wilms, 1981; Schulz and Jensen, 1968a; Sumner and van Caeseele, 1989; Mansfield et al., 1991). In addition, during the maturation of the egg, mitochondria elongate and divide and become clustered around the nucleus at the micropylar portion of the egg. Compared to the synergid and central cell, the egg cell is relatively quiescent in cytoplasmic movements (Jensen, 1965a,b; Huang et al., 1992), although it has potential for highly synthetic activity. In a few species, such as cotton, the polarity of the egg is established after fertilization (Jensen, 1965a,b), with the regrouping of ER, ribosome, and accumulation of mitochondria and plastids around the nucleus. In *Zea mays* and *Papaver nudicaule,* the nucleus of the egg is located toward the micropylar pole but relocates chalazally upon fertilization (Olson and Cass, 1981; van Lammeren, 1981). It is generally believed that the polarity of the egg cell is important (or even a prerequisite) for subsequent embryo development

(Mansfield et al., 1991), although its role remains to be elucidated experimentally.

It has been hypothesized that the egg cell may serve other roles, such as a sink for carbohydrates prior to fertilization and providing nutrients during fertilization and early embryo development, since starch accumulates in plastids prior to fertilization (Sumner and van Caeseele, 1981; Mansfield et al., 1991). In *Plumbago*, where synergids are absent, the egg cell develops a filiform apparatus, a characteristic feature of the synergid, and the cytoplasmic composition is also similar to that of the synergid (Cass and Karas, 1974; Huang et al., 1990, 1993). Furthermore, the egg cell also contains microbodies (peroxisomes and glycoxysomes) not seen in the egg from other species. The egg cell therefore may have combined both gametic and synergid function together in *Plumbago* (Willemse and van Went, 1984).

Synergid

The synergid nucleus is slightly smaller than that of the egg cell and is positioned slightly to the micropylar pole (Mansfield et al., 1991) (Figure 2.4). A large central vacuole occupies nearly half of the cell volume and is associated with smaller vesicles present at their chalazal pole (Figure 2.4; Jensen, 1965a,b; Mansfield et al., 1991; Schulz and Jensen, 1968a,b). It was proposed that the central vacuole may serve as a storage site for calcium (Jensen, 1965b) or other chemotropic compounds necessary for pollen tube guidance (Mansfield et al., 1991).

The cytoplasm is located micropylarly and is rich in organelles with abundant rER and mitochondria which often accumulate around the filiform ingrowth. Smooth ER is not observed in the synergid cytoplasm. Plastids are scarce and poorly developed that often contain starch granules. Mitochondria plays an important role in the viability of the synergid cell (Christensen et al., 2002). The Golgi apparatus is abundant and well developed (Jensen, 1965a,b; Schulz and Jensen, 1968a,b; Mogensen, 1972; Wilms, 1981; Mansfield et al., 1991), indicative of active vesicle production and secretion. Overall, the abundance of organelles suggests that synergids are very active in metabolism. This is further demonstrated by histochemical studies showing that the synergids are rich in protein, RNA, and enzymatic activities of cytochrome oxidase, succinate dehydrogenase, acid phospatase, ATPase, phosphorylase, and lipase (Malik and Vermani, 1975; Mogensen, 1981).

The two synergids occupy the extreme of the micropyle and share the same vertical wall with plasmodesmal connections. One characteristic of the synergid is the formation of the filiform apparatus as mentioned earlier (Huang and Russell, 1992). The filiform apparatus is an extensive labyrinth of embryo sac wall extending deep into the synergid cytoplasm and can be viewed even under light microscope in wholemount preparations. Structurally, the filiform apparatus typically consists of a tightly stacked electron-dense, microfibril inner layer and a loosely attached peripheral layer of microfibrils in an electron-lucent matrix (Schulz and Jensen, 1968a,b; Mogensen and Suthar, 1979). The composition of the filiform apparatus differs among species and preparations. Generally, it contains cellulose, hemicellulose, polyglucans, proteins, and lipids. It may also contain callose as seen in *Gasteria* (Willemse and Franssen-Verheijen, 1988) and *Torenia* (Tiwari, 1982). The occurrence of callose, however, is believed to be an indication of zygotic abortion (Williams et al., 1982). The universal occurrence of the filiform apparatus and its restriction to synergids suggest that it may play a role in synergid functions such as chemotropic attraction of the pollen tube, pollen tube receptance, and nutrition of the egg apparatus. Because the wall ingrowth increases the surface area of the plasma membrane tremendously, it is believed to facilitate diffusion or uptake of nutrients in transfer cells (Gunning and Pate, 1969). It is quite possible that the role of the synergid is to absorb nutrients through the filiform apparatus from the mother tissue and transport to the egg cell and, later, the zygote (Newcomb and Steeves, 1971; Huang and Russell, 1992).

Plasmodesmata are present within the synergid cell walls in most species studied, although the number and position may vary among species (Huang and Russell, 1992; Russell, 1993). In soybean, plasmodesmata have been observed in walls bordering the synergid and the egg as well as the central cell (Folsom and Peterson, 1984; Dute et al., 1989). The plasmodesmatal connections between the synergid and the egg seem common to some species (Mogensen and Suthar, 1979), and rarely present in other species (Jensen, 1965a,b; Mogensen, 1972; Newcomb, 1973). The wide occurrence of plasmodesmata in megagametogenesis and its role in other cells (Ding et al., 1999) suggests it most likely plays a role in cell-to-cell communications, though this is yet to be proven.

The role of the synergid in attraction of pollen tubes has been demonstrated recently by laser ablation experiments in *Torenia fournieri* (Higashiyama et al., 2001). Laser ablation of the micropylar nucellar cells, the egg, and the central cell does not affect the ability of the embryo sac to attract and guide pollen tube growth, indicating that they are not essential for pollen tube guidance. However, ovules are absolutely unable to attract pollen tubes when both synergids are ablated. Ablation of one of the two synergids reduces the frequency of successful pollen tube attraction. This demonstrated elegantly that one synergid cell is sufficient to produce the attraction signal and that two synergids enhance it (Higashiyama et al., 2001). This is consistent with the variability of synergid degeneration observed in some species. In soybean, for example, some ovules have both synergids at fertilization, and others only have the persistent one (Kennell and Horner, 1985b). These data suggest that synergids generate a diffusable signal molecule that is sufficient to attract the pollen tube. Intriguingly, once the ovule is fertilized, it no longer attracts the pollen tube, although an intact synergid is still present, implying that fertilization may terminate the production of the attracting signal molecule in the synergid. This may be an effective way to block polyspermy.

Central Cell

The so-called central cell is located at the center of the embryo sac and is very large compared to other gametophytic cells. A large central vacuole restricts the cytoplasm of the cell to peripheral areas, except the perinuclear region adjacent to the egg cell. The two polar nuclei start to fuse shortly after cellularization and complete the fusion before fertilization or upon arrival of the sperm (Jensen, 1965a,b; Schulz and Jensen, 1973; Newcomb, 1973; Wilms, 1981; Mansfield et al., 1991), resulting in the formation a large nucleus with a large ringlike nucleolus. The nuclear envelope is creosote and contains numerous nuclear pores. The cell wall is continuous except at a region adjacent to the egg and synergid. The central cell shares a common wall with the antipodal cell, where plasmodesmata are present. The cytoplasm of the central cell accumulates starch-containing, elongated plastids that show variable structures and shapes (Mogensen and Suthar, 1979; Folsom and Peterson, 1984; Kennell and Horner, 1985b). The starch-containing plastids of the central cell are extremely abundant in *Arabidopsis* (Mansfield et al., 1991) but disappear after fertilization, suggesting that the central cell accummulates starch to provide nutrients for the initial endosperm development. Other organelles, such as mitochondria and Golgi apparatus, are abundant throughout the cytoplasm. The central cell also serves as a receptive cell for one of the male sperms and forms a triploid endosperm upon fertilization. In addition, the central cell plays a role to support the entire embryo sac with its turgor pressure, since all of the gametophytic cells in the embryo sac break down if the central cell is ablated (Higashiyama et al., 2001).

Antipodal Cells

The three small cells at the extreme chalazal pole of the embryo sac are named antipodal cells. These cells are configured in a triangular manner similar to the egg apparatus. The ephemeral antipodal cells contain all types of organelles similar to other gametophytic cells except that the central vacuole and Golgi apparatus are absent. One striking feature is that huge numbers of plasmodesmata are present in walls bordering the antipodals and central cell. Fewer plasmodesmata are present in walls bordering antipodal and the chalazal nucellar cells. These symplastic connections may provide a pathway for metabolite allocation from the maternal nucellar tissue to the megagametophyte (Newcomb, 1973; Willemse and Kapil, 1981; Mansfield et al., 1991), and possibly other signaling compounds as well. In many dicotyledons, the antipodal cell degenerates before fertilization (Schulz and Jenesen, 1973; Mansfield et al., 1991; Christensen et al., 1997), and the degraded products of antipodals may also serve as nutrients for the embryo sac. In other species, such as maize, the antipodal cell may divide and persist throughout the double fertilization process. The degeneration of the antipodal cell is manifested by the irregular appearance of its nucleus and nucleolus, and condensed heterochromatin (Willemse and Kapil, 1981). The early degeneration of antipodal cells before fertilization implies that they do not play a role in the function of the female gametophyte.

THE FEMALE GERM UNIT

The "female germ unit" (FGU) refers to the minimum complement of cells required to accomplish double fertilization in vivo (Dumas et al., 1984; Huang and Russell, 1992). A typical FGU in angiosperms consists of an egg, two synergids, and the central cell, which fulfills the role of pollen tube attraction, sperm discharge to the receptive cells, transportation of sperm, and double fertilization (Russell, 1992; Dumas and Mogenesen, 1993; Higashiyama et al., 2001). The intact FGU should fulfill the following functions as outlined by Russell (1993):

1. the micropylar structure formed by the inner and outer integuments that allows the entry of only one pollen tube;
2. the presence of a cytoskeleton that assists the transportation of sperm nuclei to the egg and central nucleus;
3. the presence of naked or poorly developed walls near the sites of the sperm transfer in the egg and the central cell;
4. the proximity of the egg and polar nuclei at the time of fusion, which should promote nuclear migration.

Recently it was demonstrated that synergids are solely responsible for the generation of diffusible signals to attract pollen tubes, and fertilization blocks their ability to attract the pollen tube. Therefore, polyspermy can be avoided. The egg and central cell function as receptive cells to receive the sperm. In addition, the central cell plays an essential role to provide sufficient turgor pressure for the embryo sac (Higashiyama et al., 2001). Other cells, such as the antipodal cell and the micropylar nucellar cell, seem to be dispensable in terms of pollen tube guidance and fertilization. Furthermore, the distance of the central nucleus relative to the egg is important for successful fertilization as shown in the soybean *partial-sterile-1(ps-1)* mutant (Pereira et al., 1997). The distance between the egg and the polar nuclei falls into three distinct classes in *ps-1* mutant ovules: the nearest (~86.5 μm), intermediate (~111 μm), and the farthest (~171.1 μm). Only ovules in the nearest class are fertilized. This indicates that the minimal requirement of the FGU has to be met to allow it to function normally.

GENETIC CONTROL OF MEGAGAMETOPHYTE DEVELOPMENT

Megagametophyte development involves many fundamental phenomena, such as nuclear migration and fusion, polarity, cell death, asymmetric division, and cell fate specification as discussed previously. We know very little about genes or gene products that regulate these cellular processes during megagametophyte development (Drews et al., 1998; Grossniklaus and Schneitz, 1998). This is partly because the female gametophyte, comprising a few cells, is embedded within the multicellular ovule in the floral organ, carpel, or ovary; therefore, experimental work is difficult. However, studies on the transmission of chromosomal deletions in maize (Patterson, 1978; Coe et al., 1988; Buckner and Reeves, 1994; Vollbrecht and Hake, 1995) and *Arabidopsis* (Vizir et al., 1994), and the high occurrence of gametophytic mutations (~1 percent) in T-DNA/transposon insertion lines (Drews et al., 1998; Moore et al., 1997) suggest that a large number of genes are required for female gametophyte development, since most chromosomal deletions are defective in embryo sac development. Analyses of embryo sac defective mutants in maize suggest that some of the embryo sac expressed genes are involved in controlling several basic cellular processes, such as nuclear migration, polarity, and the coordination of developmental events during female gametophyte development (Vollbrecht and Hake, 1995). However, isolation of these genes was extremely difficult in maize, and in fact, none of them has been characterized.

Recently, progress in *Arabidopsis* model plants, has given some insight into the genetic control of female gametophyte development (Drews et al., 1998; Grossniklaus and Schneitz, 1998; Yang and Sundaresan, 2000).

A genetic screen based on non-Mendelian segregation of gametophytic mutations and phenotypic analysis becomes possible thanks to the development of insertional mutagenesis and wholemount clearing and CLSM imaging techniques (Moore et al., 1997; Christensen et al., 1997) (Figure 2.4). Systematic screens for gametophytic mutants have been carried out in *Arabidopsis* by looking for distorted Mendelian segregations of antibiotic markers carried by either T-DNA (Feldmann et al., 1997; Howden et al., 1998) or transposon (Grossniklaus et al., 1995; Moore et al., 1997), and chromosome markers (Grini et al., 1999). A large number of gametophytic mutants were isolated and have been reviewed recently (Drews et al., 1998; Grossniklaus and Schneitz, 1998; Yang and Sundaresan, 2000; Christensen et al., 2002).

Genes Controlling Nuclear Divisions

To produce an eight-nucleate embryo sac, the functional megaspore has to undergo three nuclear divisions. Thus, it is expected that the basic cell cycle machinery is required, and any mutations in these cell cycle genes would likely disrupt nuclear divisions. Mutants defective in *CELL DIVISION CYCLE 16 (CDC16)*, a component of the anaphase-promoting complex, and *PROLIFERA (PRL)*, as a member of the *MCM2-3-5* family required for the initiation of the DNA replication, have been identified (Yang and Sundaresan, 2000). The embryo sac was arrested at the two-nucleate stage in the *cdc16* mutant and at the four-nucleate stage in the *prl* mutant, indicating that *CDC16* and *PRL* are required for the second and third nuclear division, respectively.

The majority of female gametophytic mutants reported so far are defective in the control of nuclear divisions. In five mutants from *Arabidopsis* and one from maize, embryo sacs are arrested at the one-nucleate stage (Drews et al., 1998; Yang and Sundaresan, 2000). These include the *female gametophyte2 (fem2)*, *fem3*, *gametophytic factor (Gf)*, *gfa4*, and *gfa5* from *Arabidopsis*, and *lethal ovule2 (lo2)* from maize. No detailed cytological phenotypic characterization is available for these mutants; therefore, it is not known what cellular processes are affected. Furthermore, the corresponding genes have not yet been isolated. Nevertheless, it clearly suggests that they are required very early in female gametophyte development.

In addition to *CDC16* and *PRL*, the *HDD* gene product is also required for the second nuclear division since embryo sacs are arrested at the two-nucleate stage in *hdd* mutant. This suggests that *CDC16* and *HDD* are essential for the second division. The knockout of *PRL* causes the embryo sac to arrest at the four-nucleate stage and is embryo lethal, indicating that *PRL* is an essential gene for all cell types. It is worthy to note that *hdd* and *lo2* are pleiotropic and causes embryo arrest at the one-, two-, and four nucleate stages. More interesting, as discussed earlier, is that some *hdd* mutant embryo sacs show asynchronous nuclear divisions at the two poles, indicating that *HDD* plays an additional role in coordinating nuclear division between micropylar and chalazal poles (Moore et al., 1997). Similarly, asynchronous nuclear divisions are also seen in the maize mutant *indetermined gametophyte (ig).* In addition, embryo sacs in the *ig* mutant undergo additional nuclear divisions resulting in the formation of excessive number of nuclei (Kermicle, 1971; Lin, 1978, 1981; Huang and Sheridian, 1996). This suggests that *IG* plays a role in maintaining a proper number of mitotic cycles during female gametophyte development. Furthermore, embryo sacs in *hdd, ig,* and *lo2* also have secondary defects in nuclear migration positioning and microtubule organization, indicating that *HDD, IG,* and *LO2* may control multiple steps to make sure that different developmental events are spatially and temporally coordinated during embryo sac formation.

Genes Controlling Other Cellular Events

Several megagametophytic mutations affecting steps other than nuclear divisions have also been isolated in *Arabidopsis*. They include *gfa2, gfa3,* and *gfa7*. In these mutants, polar nuclei migrate properly and lie side by side, but fail to fuse with one another (Drews et al., 1998). This indicates that *GFA2, GFA3,* and *GFA7* are most likely involved in polar nuclear fusion and their functions will provide information about molecular mechanisms of nuclear fusion.

The cellularization process seems to be defective in the *Arabidopsis* megagametophytic mutants *fem1* and *fem4* (Drews et al., 1998). In *fem1*, the female gametophyte lacks the central vacuole and disintegrates just before cellularization takes place, whereas nuclear division occurs normally. However, in *fem4*, cellularization seems to be affected as manifested by the abnormal morphology of the egg cell and synergids that exhibit abnormal vacuoles. These data imply that the vacuole or endomembranous system play an important role in female gametophyte development.

CONCLUSIONS AND PERSPECTIVES

The formation of the female gametophyte is a complex developmental program involving many basic cellular processes as discussed in this chapter. Our knowledge about female development is more on the cytological description of the event; little is known about genes and molecules controlling the process. Recent developments in molecular genetics and imaging technologies begin to shed light on female gametophyte development. Mutants defective in different steps of the developmental program are being identified and corresponding molecules are being characterized. The understanding of these processes will benefit not only our understanding of female gametophyte development but also plant developmental biology as a whole.

REFERENCES

Albertson, M.C. and R.G. Palmer (1979). A comparative light- and electron-microscopic study of microsporogenesis in male sterile *(ms$_1$)* and male fertile soybean (*Glycine max* (L.) Merr.). *Am. J. Bot.* 66:253-265.

Azumi, Y., D. Liu, D. Zhao, W. Li, G. Wang, Y. Hu, and H. Ma (2002). Homolog interaction during meiotic prophase I in *Arabidopsis* requires the *SOLO DANCERS* gene encoding a novel cyclin-like protein. *EMBO J.* 21:3081-3095.

Bai, X., B.N. Peirson, F. Dong, X. Cai, C.A. Makaroff, X.F. Bai, C. Xue, and F.G. Dong (1999). Isolation and characterization of *SYN1*, a *RAD21*-like gene essential for meiosis in *Arabidopsis*. *Plant Cell* 11:417-430.

Bajon, C., C. Horlow, J.C. Motamayer, A. Sauvanet, and D. Robert (1999). Megasporogenesis in *Arabidopsis thaliana* L.: An ultrastructural study. *Sex. Plant Reprod.* 12:99-119.

Beadle, G.W. (1932). A gene in *Zea mays* for failure of cytokinese during meiosis. *Cytologia* 3:142-155.

Bhandari, N.N. and P. Chitralekha (1989). Cellularization of the female gametophyte in *Ranunculus scleratus*. *Can. J. Bot.* 67:1325-1330.

Bhatt, A.M., C. Lister, T. Page, P. Fransz, K. Findlay, G.H. Jones, H.G. Dickinson, and C. Dean (1999). The *DIF1* gene of *Arabidopsis* is required for meiotic chromosome segregation and belongs to the *REC8/RAD21* cohesin gene family. *Plant J.* 19:463-472.

Bouman, F. (1984). The ovule. In B.M. Johri (ed.), *Embryology of angiosperm* (pp. 123-157). Berlin: Springer-Verlag.

Brown, R.C. and B.E. Lemmon (1991). Pollen development in orchids: 1. Cytoskeleton and the control of division plane in irregular patterns of cytokinesis. *Protoplasma* 163:1-18.

Brown, R.C. and B.E. Lemmon (1992). Cytoplasmic domain: A model for spatial control of cytokinesis in reproductive cells of plants. *EMSA Bull.* 22:48-53.

Buckner, B. and S.L. Reeves (1994). Viability of female gametophytes that possess deficiencies for the region of chromosome 6 containing the *Y1* gene. *Maydica* 39:247-254.

Byzova, M.V., J. Franken, M.G.M. Arts, J. De Almeida-Engler, G. Engler, C. Mariani, M.M. Van Lookeren Campagne, and G.C. Angenent (1999). *Arabidopsis STERILE APETALA*, a multifunctional gene regulating inflorescence, flower, and ovule development. *Genes Dev.* 13:1002-1014.

Cass, D.D. and I. Karas (1974). Ultrastructural organization of the egg of *Plumbago zeylanica*. *Protoplasma* 81:49-62.

Cass, D.D., D.J. Peteya, and B.L. Robertson (1985). Megagametophyte development in *Hordeum vulgare:* 1. Early megagametogenesis and the nature of cell wall formation. *Can. J. Bot.* 63:2164-2171.

Cass, D.D., D.J. Peteya, and B.L. Robertson (1986). Megagametophyte development in *Hordeum vulgare:* 2. Late stages of wall development and morphological aspects of megagametophyte cell differentiation. *Can. J. Bot.* 64:2327-2336.

Castetter, E.F. (1925). Studies on the comparative cytology of the annual and biennial varieties of *Melilotus alba*. *Am. J. Bot.* 12:270-286.

Chen, Y.C.S. and S. McCormick (1996). *sidecar pollen,* an *Arabidopsis thaliana* male gametophytic mutant with aberrant cell divisions during pollen development. *Development* 122:3243-3252.

Christensen, C.A., S.W. Gorsich, R.H. Brown, L.G. Jones, J.G. Brown, J.M. Shaw, and G.N. Drews (2002). Mitochondrial GFA2 is required for synergid cell death in *Arabidopsis*. *Plant Cell* 14:2215-2233.

Christensen, C.A., E.J. King, J.R. Jordan, and G.N. Drews (1997). Megagametogenesis in *Arabidopsis* wild type and the *Gf* mutant. *Sex. Plant Reprod.* 10: 49-64.

Coe, E.H., M.G. Neuffer, and D.A. Hoisington (1988). The genetics of corn. In G.F. Sprague and J.W. Dudley (eds.), *Corn and corn improvement,* Third edition (pp. 81-258). Madison, WI: American Society of Agronomy.

Couteau, F., F. Belzile, C. Horlow, O. Grandjean, D. Vezon, and M.-P. Doutriaux (1999). Random chromosome segregation without meiotic arrest in both male and female meiocytes of a *dmc1* mutant of *Arabidopsis*. *Plant Cell* 11:1623-1634.

De Boer-de Jeu, M.J. (1978). Ultrastructural aspects of megasporogenesis and initiation of megagametogenesis in *Lilium*. *Actual. Bot.* 125:175-181.

Dickinson, H.G. and L. Andrews (1977). The role of membrane-bound eytoplasmic inclusions during gametogenesis in *Lilium longiflorum* Thumb. *Planta* 134: 229-240.

Dickinson, H.G. and J. Heslop-Harrison (1977). Ribosomes and organelles during meiosis in angiosperms. *Philos. Trans. R. Soc. London B.* 227:327-342.

Ding, B., A. Itaya, and Y.-M. Woo (1999). Plasmadesmata and cell-to-cell communication in plants. *Int. Rev. Cytol.* 190:251-317.

Drews, G.N., D. Lee, and C.A. Christensen (1998). Genetic control of female gametophyte development and function. *Plant Cell* 10:1-15.

Drouaud, J., K. Marrocco, C. Ridel, G. Pelletier, and P. Guerche (2000). A *Brassica napus skp1*-like gene promoter drives GUS expression in *Arabidopsis thaliana* male and female gametophytes. *Sex. Plant Reprod.* 13:29-35.

Dumas, C., R.B. Knox, C.A. McConchie, and S.D. Russell (1984). Emerging physiological comcepts in fertilization. *What's New Plant Physiol.* 15:17-20.

Dumas, C. and H.L. Mogensen (1993). Gametes and fertilization: Maize as a model system for experimental embryogenesis in flowering plants. *Plant Cell* 5:1337-1348.

Dute, R.R., C.M. Peterson, and A.E. Rushing (1989). Ultrastructural changes of the egg apparatus associated with fertilization and proembryo development of soybean, *Glycine max* (Fabaceae). *Ann. Bot.* 64:123-135.

Ehdaie, M. and S.D. Russell (1984). Megagametophyte development in *Nandina domestica* and its taxonomic implications. *Phytomorphology* 34:221-225.

Elliott, R.C., A.S. Betzner, E. Huttner, M.P. Oakes, W.O.J. Tucker, D. Gerentes, P. Perez, and D.R. Smyth (1996). *AINTEGUMENTA*, an APETALA2-like gene of *Arabidopsis* with pleiotropic roles in ovule development and floral organ growth. *Plant Cell* 8:155-168.

Feldmann, K.A., D.A. Coury, and M.L. Christianson (1997). Exceptional segregation of a selectable marker (KanR) in *Arabidopsis* identifies genes important for gametophytic growth and development. *Genetics* 147:1411-1422.

Folsom, M.W. and D.D. Cass (1990). Embryo sac development in soybean: Cellularization and egg apparatus expansion. *Can. J. Bot.* 68:2135-2147.

Folsom, M.W. and C.M. Peterson (1984). Ultrastructural aspects of the mature embryo sac of soybean, *Glycine max* (L.) Merr. *Bot. Gaz.* 145:1-10.

Gasser, C.S., J. Broadhvest, and B.A. Hauser (1998). Genetic analysis of ovule development. *Annu. Rev. Plant Physiol. Plant Mol. Biol.* 49:1-24.

Glover, J., M. Grelon, S. Craig, A. Chaudhury, and L. Dennis (1998). Cloning and characterization of *MS5* from *Arabidopsis:* A gene critical in male meiosis. *Plant J.* 15:345-356.

Golubovskaya, I. (1989). Meiosis in maize: *mei* genes and conception of genetic control of meiosis. *Adv. Genet.* 26:149-192.

Golubovskaya, I., N.A. Avalkina, and W.F. Sheridan (1992). The effect of several meiotic mutations on female meiosis in maize. *Dev. Genet.* 13:411-424.

Golubovskaya, I., N.A. Avalkina, and W.F. Sheridan (1997). New insight into the role of the maize ameiotic 1 locus. *Genetics* 147:1339-1350.

Golubovskaya, I., Z.K. Grebennikova, N.A. Avalkina, and W.F. Sheridan (1993). The role of *ameiotic 1* gene in the initiation of meiosis and in subsequent meiotic events in maize. *Genetics* 135:1151-1166.

Grelon, M., D. Vezon, G. Gendrot, and G. Pelletier (2001). AtSP011-1 is necessary for efficient meiotic recombination in plants. *EMBO J.* 20:589-600.

Grini, P.E., A. Schnittger, H. Schwarz, I. Zimmermann, B. Schwab, G. Jürgens, and M. Hülskamp (1999). Isolation of ethyl methanesulfonate-induced gametophytic mutants in *Arabidopsis thaliana* by a segregation distortion assay using the multimarker chromosome l. *Genetics* 151:849-863.

Grossniklaus, U., J. Moore, and W. Gagliano (1995). Analysis of *Arabidopsis* ovule development and megagametogenesis using enhancer detection. In *Signaling in plant development* (p. 138), September 27-October 1. Cold Spring Harbor, NY: Cold Spring Harbor Laboratory Press.

Grossniklaus, U. and K. Schneitz (1998). The molecular and genetic basis of ovule and megagametophyte development. *Seminars Cell Dev. Biol.* 9:227-238.

Gunning, B.E.S. and J.S. Pate (1969). Transfer cells: Plant cells with wall ingrowths, specialized in relation to short distance transport of solutes—Their occurrence, structure, and development. *Protoplasma* 68:107-133.

Haig, D. (1990). New perspectives on the angiosperm female gametophyte development. *Bot. Rev.* 56:236-275.

He, C. and J.P. Mascarenhas (1999). *MEI1:* An *Arabidopsis* gene required for male meiosis: Isolation and characterization. *Sex. Plant Reprod.* 11:199-207.

Higashiyama, T., S. Yabe, N. Sasaki, Y. Nishmura, S. Miyagishima, H. Kuroiwa, and T. Kuroiwa (2001). Pollen tube attraction by the synergid cell. *Science* 293:1480-1483.

Howden, R., S.K. Park, J.M. Moore, J. Orme, and U. Grossniklaus (1998). Selection of T-DNA-tagged male and female gametophytic mutants by segregation distortion in *Arabidopsis. Genetics* 149:621-631.

Hu, S. and C. Zhu (2000). *Atlas of sexual reproduction in angiosperms.* Beijing: Chinese Academic Press.

Huang, B.Q., E.S. Pierson, S.D. Russell, A. Tiezzi, and M. Cresti (1992). Video microscopic observations of living, isolated embryo sacs of *Nicotiana* and their component cells. *Sex. Plant Reprod.* 5:156-162.

Huang, B.Q., E.S. Pierson, S.D. Russell, A. Tiezzi, and M. Cresti (1993). Cytoskeleton organization and modificationin the process of fertilization of *Plumbago zeylancia. Zygote* 1:143-154.

Huang, B.Q. and S.D. Russell (1992). Female germ unit: Organization, reconstruction and isolation. *Int. Rev. Cytol.* 140:233-293.

Huang, B.Q., S.D. Russell, G.W. Strout, and L.-J. Mao (1990). Organization of isolated embryo sacs and eggs of *Plumbago zeylancia* (Plumbaginaceae) before and after fertilization. *Am. J. Bot.* 77:1401-1410.

Huang, B.-Q. and W.F. Sheridian (1996). Embryo sac development in the maize *indeterminate gametophyte1* mutant: Abnormal nuclear behavior and defective microtubule organization. *Plant Cell* 8:1391-1407.

Hülskamp, M., N.S. Parekh, P. Grini, K. Schneitz, I. Zimmermann, S.J. Lolle, and R.E. Pruitt (1997). The *STUD* gene is required for male-specific cytokinesis after telophase II of meiosis in *Arabidopsis thaliana. Dev. Biol.* 187:114-124.

Ito, M. and M.H. Takegami (1982). Commitment of mitotic cells to meiosis during G2 phase of premeiosis. *Plant Cell Physiol.* 23:943-952.

Jensen, W.A. (1965a). The ultrastructure and composition of the egg and central cell of cotton. *Am. J. Bot.* 52:781-797.

Jensen, W.A. (1965b). The ultrastructure and histochemistry of the synergids of cotton. *Am. J. Bot.* 52:238-256.

Kapil, R.N. and A.K. Bhatnagar (1981). Ultrastructure and biology of female gametophyte in flowering plants. *Int. Rev. Cytol.* 70:291-341.

Kennell, J.C. and H.T. Horner (1985a). Influence of the soybean male-sterile gene *(ms_1)* on the development of the female gametophyte. *Can. J Genet Cytol.* 27:200-209.

Kennell, J.C. and H.T. Horner (1985b). Megasporogenesis and megagametogenesis in soybean, *Glycine max. Am. J. Bot.* 72:1553-1564.

Kermicle, J.L. (1971). Pleiotropic effects on seed development of the indeterminate gametophyte gene in maize. *Am. J. Bot.* 58:1-7.

Klucher, K.M., H. Chow, L. Reiser, and R.L. Fischer (1996). The *AINTEGUMENTA* gene of *Arabidopsis* required for ovule and female gametophyte development is related to the floral homeotic gene APETAL2. *Plant Cell* 8:137-153.

Lin, B.-Y. (1978). Structural modifications of the female gametophyte associated with the indeterminate gametophyte *(ig)* mutation in maize. *Can. J. Genet. Cytol.* 20:249-257.

Lin, B.-Y. (1981). Megagametogenetic alterations associated with the indeterminate gametophyte *(ig)* mutation in maize. *Rev. Bras. Biol.* 41:557-563.

Maheshwari, P. (1950). *An introduction to the embryology of angiosperms.* New York: McGraw-Hill Book Co., Inc.

Malik, C.P. and S. Vermani (1975). Physiology of sexual reproduction I. A histochemical study of the embryo sac development in *Zephyranthes rosea* and *Lagenaria vulgaris. Acta Histochem.* 53:244-280.

Malone, R.E. (1990). Dual regulation of meiosis. *Cell* 61:375-378.

Mansfield, S.G., L.G. Briarty, and S. Erni (1991). Early embryogenesis in *Arabidopsis thaliana:* I. The mature embryo sac. *Can. J. Bot.* 69:447-460.

McCoy, T.J. and L.Y. Smith (1983). Genetics, cytology and crossing behavior of an alfalfa *(Medicago sativa)* mutant resulting in failure of the postmeiotic cytokinesis. *Can. J. Genet. Cytol.* 25:390-397.

Medina, F.J., M.C. Risueno, and M.I. Rodriguez-Garcia (1981). Evolution of cytoplasmic organelles during female meiosis in *Pisum sativum* L. *Planta* 151: 215-225.

Mercier, R., D. Vezon, E. Bullier, J.C. Motamayor, A. Sellier, F. Lefevre, G. Pelletier, and C. Horlow (2001). SWITCH1 (SWI1): A novel protein required for the establishment of sister chromatid cohesion and for bivalent formation at meiosis. *Genes Dev.* 15:1859-1871.

Misra, R.C. (1962). Contribution to the embryology of *Arabidopsis thaliana* (Gay & Monn.). *Agra. Univ. J Res.* 11:191-199.

Moffett, A.A. (1932). Studies on the formation of multi-nuclear giant pollen grains in *Kniphofia. J. Genet.* 25:315-337.

Mogensen, H.L. (1972). Fine structure and composition of the egg apparatus before and after fertilization in *Quercus gambelii:* The functional ovule. *Am. J. Bot.* 59:931-941.

Mogensen, H.L. (1981). Ultracytochemical localization of adenosine triphosphatase in the ovules of *Saintpaulia ionantha* (Gesneriaceae) and its relation to synergid function and embryo sac nutrition. *Am. J. Bot.* 68:183-194.

Mogencsen, H.L. and H.K. Suthar (1979). Ultrastructure of the egg apparatus of *Nicotiana tabacum* (Solanaceae) before and after fertilization. *Bot. Gaz.* 140: 168-179.

Moore, J.M., J.-P. Vielle Calzada, W. Gagliano, and U. Grossniklaus (1997). Genetic characterization of *hadad,* a mutant disrupting female gametogenesis in *Arabidopsis thaliana. Cold Spring Harbor Sym. Quant. Biol.* 62:35-47.

Motamayor, J.C., D. Vezon, C. Bajon, A. Sauvanet, O. Grandjean, M. Marchand, N. Bechtold, G. Pelletier, and C. Horlow (2000). *Switch (swi1),* an *Arabidopsis thaliana* mutant affected in the female meiotic switch. *Sex. Plant Reprod.* 12:209-218.

Newcomb, W. (1973). The development of the embryo sac of sunflower *Helianthus annus* before fertilization. *Can. J. Bot.* 132:367-371.

Newcomb, W. and T.A. Steeves (1971). *Helianthus annus* embryogenesis: Embryo sac wall projections before and after fertilization. *Bot. Gaz.* 132:367-371.

Olson, A.R. and D.D. Cass (1981). Changes in megagametophyte structure in *Papaver nudicaule* following in vitro placental pollination. *Am. J. Bot.* 68:1338-1341.

Otegui, M. and L.A. Staehelin (2000a). Cytokinesis in flowering plants: More than one way to divide a cell. *Curr. Opin. Plant Biol.* 3:493-502.

Otegui, M. and L.A. Staehelin (2000b). Syncytial-type of cell plates: A novel kind of cell plate involved in endosperm cellularization of *Arabidopsis. Plant Cell* 12:933-947.

Palmer, R.G. (1971). Cytological studies of ameiotic and normal maize with respect to premeiotic pairing. *Chromosoma* 35:233-246.

Patterson, E.B. (1978). Properties and uses of duplicate deficient chromosome compliments in maize. In D.Walden (ed.), *Maize breeding and genetics* (pp. 693-710). New York: John Wiley and Sons.

Pereira, T.N., N.R. Lersten, and R.G. Palmer (1997). Genetic and cytological analysis of a partial-female-sterile mutant (PS-1) in soybean (*Glycine max;* Leguminosae). *Am. J. Bot.* 84:781-791.

Reiser, L. and R.L. Fischer (1993). The ovule and the embryo sac. *Plant Cell* 5: 1291-1301.

Rhoades, M.M. (1956). Genic control of chromosomal behaviour. *Maize Genet. Coop. News Lett.* 30:38-48.

Rodkiewicz, B. (1970). Callose in cell walls during megasporogenesis in angiosperms. *Planta* 90:39-47.

Rodkiewicz, B. (1975). Sieve-like distribution of callose in meiocyte chalazal wall in ovules in orchid *Epipactis. Bull. Acad. Pol. Sci. C Ser. Sci. Biol.* 23:707-711.

Rodkiewicz, B. and J. Bednara (1976). Cell wall ingrowths and callose distribution in megasporogenesis in some Orhidaceae. *Phytomorphology* 26:276-281.

Russell, S.D. (1979). Fine structure of megagametophyte development in *Zea mays. Can. J. Bot.* 57:1093-1110.

Russell, S.D. (1992). Double fertilization. *Int. Rev. Cytol.* 142:357-358.

Russell, S.D. (1993). The egg cell: Development and role in fertilization and early embryogenesis. *Plant Cell* 5:1349-1359.

Scheres, B. (1996). Embryo patterning genes and reinforcement cues determine cell fate in the *Arabidopsis thaliana* root. *Seminars Cell Dev. Biol.* 7:857-865.

Schneitz, K., M. Hulskamp, and R.E. Pruitt (1995). Wild-type ovule development in *Arabidopsis thaliana:* A light microscope study of cleared whole-mount tissue. *Plant J.* 7:731-749.

Schulz, P. and W.A. Jensen (1968a). *Capsella* embryogenesis: The egg, zygote and young embryo. *Am. J. Bot.* 55:807-819.

Schulz, P. and W.A. Jensen (1968b). *Capsella* embryogenesis: The synergids before and after fertilization. *Am. J. Bot.* 55:541-552.

Schulz, P. and W.A. Jensen (1973). *Capsella* embryogenesis: The central cell. *J. Cell Sci.* 12:741-763.

Schulz, P. and W.A. Jensen (1981). Pre-fertilization ovule development in *Capsella:* Ultrastructure and ultracytochemical localization of acid phosphatase in the meiocyte. *Protoplasma* 107:27-45.

Schulz, P. and W.A. Jensen (1986). Prefertilization ovule development in *Capsella:* The dyad, tetrad, developing megaspore and two-nucleate gametophyte. *Can. J. Bot.* 64:875-884.

Sheridan, W.F., N.A. Avalkina, I.I. Shamrov, T.B. Batygina, and I.N. Golubovskaya (1996). The *mac1* gene: Controlling the commitment to the meiotic pathway in maize. *Genetics* 142:1009-1020.

Sheridan, W.F., E.A. Golubeva, L.I. Abrhamova, and I.N. Golubovskaya (1999). The *mac1* mutation alters the developmental fate of the hypodermal cells and their cellular progeny in the mazie anther. *Genetics* 153:933-941.

Siddiqi, L, G. Ganesh, U. Grossniklaus, and V. Subbiah (2000). The dyad gene is required for progression through female meiosis in *Arabidopsis. Development* 127:197-207.

Spielman, M., D. Preuss, F.-L. Li, W.E. Browne, R.J. Scott, and H.G. Dickinson (1997). *TETRASPORE* is required for male meiotic cytokinesis in *Arabidopsis thaliana. Development* 124:2645-2657.

Staiger, C.J. and W.Z. Cande (1992). *Ameiotic,* a gene that controls meiotic chromosome and cytoskeleton behavior in maize. *Dev. Biol.* 154:226-230.

Stern, H. and Y. Hotta (1969). Biochemistry of meiosis. In A. Lima-de-Faria (ed.), *Handbook of molecular cytology* (pp. 520-539). Amsterdam: North-Holland.

Sumner, M.J., and L. van Caeseele (1989). The ultrastructure and cytochemistry of the egg apparatus of *Brassica campestris. Can. J. Bot.* 67:177-190.

Tiwari, S.C. (1982). Callose in the walls of mature embryo sac of *Torenia fournieri. Protoplasma* 110:1-4.

van Lammeren, A.A.M. (1981). Early events during embryogenesis in *Zea mays* L. *Acta. Soc. Bot. Pol.* 50:289-290.

Vizir, I.Y., M.L. Anderson, Z.A. Wilson, and B.J. Mulligan (1994). Isolation of deficiencies in the *Arabidopsis* genome by γ-radiation of pollen. *Genetics* 137: 1111-1119.

Vollbrecht, E. and S. Hake (1995). Deficiency analysis of female gametogenesis in maize. *Dev. Genet.* 16:44-63.

Watanabe, Y. and P. Nurse (1999). Cohesin *Rec8* is required for reductional chromosome segregation at meiosis. *Nature* 400:461-464.

Webb, M.C. and B.E.S. Gunning (1990). Embryo sac de elopment in *Arabidopsis thaliana: LI.* Megasporogenesis, including the micr ubular cytoskeleton. *Sex. Plant Reprod.* 3:244-256.

Willemse, M.T.M. (1981). Polarity during megasporogenesis and megagametogenesis. *Phytomorphology* 31:124-134.

Willemse, M.T.M. and J. Bednara (1979). Polarity during megasporogenesis in *Gasteria verrucosa. Phytomorphology* 29:156-165.

Willemse, M.T.M. and M.A.W. Franssen-Verheijen (1978). Cell organelle changes during megasporogenesis and megagametogenesis in *Gasteria verrucosa* (Mill.) Haw. *Acta. Bot.* 125:187-191.

Willemse, M.T.M. and R.N. Kapil (1981). Antipodals of *Gasteria verrucosa* (Liliaceae)—An ultrastructural study. *Acta Bot. Neerl.* 30:25-32.

Willemse, M.T.M. and J.L. van Went (1984). The female gametophyte. In B.M. Johri (ed.), *Embryology of angiosperms* (pp. 159-196). Berlin: Springer-Verlag.

Williams, E.G., R.B. Knox, and J.L. Rouse (1982). Pollination subsystems distinguished by pollen tubes after incompatible interspecific crosses in *Rhododendron* (Ericaceae). *J. Cell. Sci.* 53:255-277.

Wilms, H.J. (1981). Ultrastructure of the developing embryo sac of spinach. *Acta Bot. Neerl.* 30:75-99.

Worrall, D., D.L. Hird, R. Hodge, W. Paul, J. Draper, and R. Scott (1992). Premature dissolution of the microsporocyte callose wall causes male sterility in transgenic tobacco. *Plant Cell* 4:759-771.

Yang, M., Y. Hu, M. Lodhi, W.R. McCombie, and H. Ma (1999). The *Arabidopsis SKP1-LIKE* gene is essential for male meiosis and may control homologue separation. *Proc. Natl. Acad. Sci. USA* 96:11416-11421.

Yang, W.-C. and V. Sundaresan (2000). Genetics of gametophyte biogenesis in *Arabidopsis. Curr. Opin. Plant Biol.* 3:53-57.

Yang, W.-C., D. Ye, J. Xu, and V. Sundaresan (1999). The *SPOROCYTELESS* gene of *Arabidopsis* is required for initiation of sporogenesis and encodes a novel nuclear protein. *Genes Dev.* 13:2108-2117.

Zeng, X. and W.S. Saunders (2000). The *Saccharomyces cerevisiae* centromere protein slk19p is required for two successive divisions during meiosis. *Genetics* 155:577-587.

Zhao, D.-Z., G.-F. Wang, B. Speal, and H. Ma (2002). The *EXCESS MICROSPOROCYTES1* gene encodes a putative leucine-rich repeat receptor protein kinase that controls somatic and reproductive cell fates in the *Arabidopsis* anther. *Genes Dev.* 16:2021-2031.

Chapter 3

Cytokinins and Seed Development

Neil Emery
Craig Atkins

INTRODUCTION

Developing seeds have played a prominent part in cytokinin (CK) research from its very beginning; the first CK was discovered in immature corn kernels. Indeed, developing seeds have so far proven to be the richest source of this family of growth regulators, and there is good physiological evidence that they are important sites of CK synthesis and metabolism. Not surprisingly, then, developing seeds have been the source for a number of the newly discovered CK enzymes and their encoding genes. Seeds are nourished by assimilates of C, N, and other nutrients delivered in phloem, and to a lesser extent in xylem, attracting plant growth regulators, including CK, that are also present in these channels. Thus, a role for translocated CK as "signals" that are in some way linked to "sink strength" and consequently seed yield has been inferred. At the cellular level the long-standing hypothesis continues to be that CK improve seed set and development through their positive influence on cell division and cell cycle activity and that this results in increased ability to attract assimilates for growth. This thinking has come largely from cereal research which indicates CK accumulation is coincident with seed cell division profiles. It is less clear whether this correlation is also true for eudicots. Nevertheless, a number of attempts have been made to control and enhance seed yields through physiological or molecular manipulation of CK abundance.

The most recent authoritative review of hormonal regulation in seed development was made by Morris in 1997. The complexity of CK participation in the early development of seeds was emphasized and five questions were posed to focus further study in this area. Is there a causal relationship between CK content and endosperm cell division? Are CK synthesized locally in the grain shortly after pollination? If CK synthesis is localized to the grain, where in the grain does it occur? What is the biological role of CK oxidation in developing grains? Can CK content be manipulated to alter grain development?

This present review will update progress made toward answering these and other important questions. Topics include the synthesis, metabolism, and functional significance of CK, specifically in relation to fruit and seed set, seed development, and seed filling. The context for Morris's review (1997) was cereal grain physiology, and although this remains the best characterized developmental system for understanding CK dynamics, the less straightforward eudicot systems are proving to have novel and unexpected CK dynamics. They have also proved to be the source of new and powerful molecular genetic tools.

CYTOKININ BIOSYNTHESIS IN PLANTS

Synthesis and metabolism of CK in plants is complex and, in many respects, poorly understood (Auer, 2002) (Figures 3.1 and 3.2). Despite the apparently ubiquitous occurrence and central growth-regulating role for CK, until quite recently, there was no clearly defined pathway for their synthesis in plants. The enzyme isopentenyl transferase, which adds the isopentenyl side chain to adenine, had been characterized and its encoding gene (*ipt*) cloned from a number of bacteria but, until very recently, not from vascular plants. Because of the "apparent" absence of this activity, hypotheses were even developed around the idea that specific associative bacteria colonizing the phylloplane and rhizosphere provided the plant with all of its CK (Holland and Polacco, 1994). However, the *Arabidopsis* genome sequence database has revealed the presence of a small multigene family (*AtIPT1-AtIPT8*) that is structurally related to both bacterial adenylate isopentenyltransferase and tRNA isopentenyltransferase (Kakimoto, 2001; Takei et al., 2001). Although there appear to be differences between the plant and bacterial enzymes with respect to adenylate substrate specificity (Sakakibara and Takei, 2002), evidence for the adenyl-isopentenylation reaction in plants now seems unequivocal. Zubko et al. (2002) exploited T-DNA activation tagging to recover a mutant line in *Petunia hybrida* that yielded an orthologous gene (*Sho*) encoding an enzyme that enhances the levels of free CK. In wild-type petunia *Sho* gene expression is higher in roots than leaves, but, just as for the *AtIPT* genes, there are no data yet for expression in reproductive organs. Isopentenyl transferase activity has been detected in extracts from immature maize kernels (Blackwell and Horgan, 1994), and it seems likely that the recently cloned *ipt* genes or their orthologs will be found to express in developing seeds. Given large differences, both in CK concentration and the forms of CK that accumulate

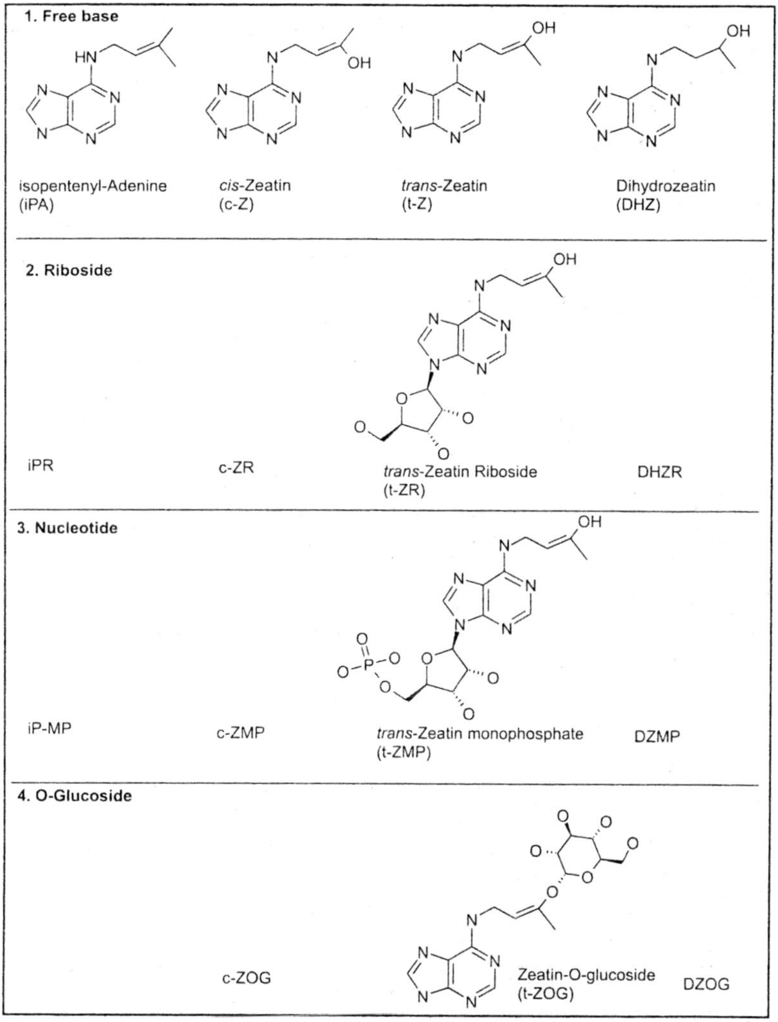

FIGURE 3.1. Common or well-studied forms of cytokinins (CK) found in plants and discussed in detail in this chapter. For the sake of clarity, four groups of CK are designated: free bases, ribosides, nucleotides, and O-glucosides. The general chemical structure is shown for each group once, and others are mentioned only in short form and can be inferred from their analogs in other groups that show structures in detail. Within each group, the presence of *cis*-hydroxylation, *trans*-hydroxylation, or side chain saturation (dihydro) is shown. Other forms of CK that exist, but are not shown, include O-acetyl-, O-xylosyl- and ring glucosylated conjugates.

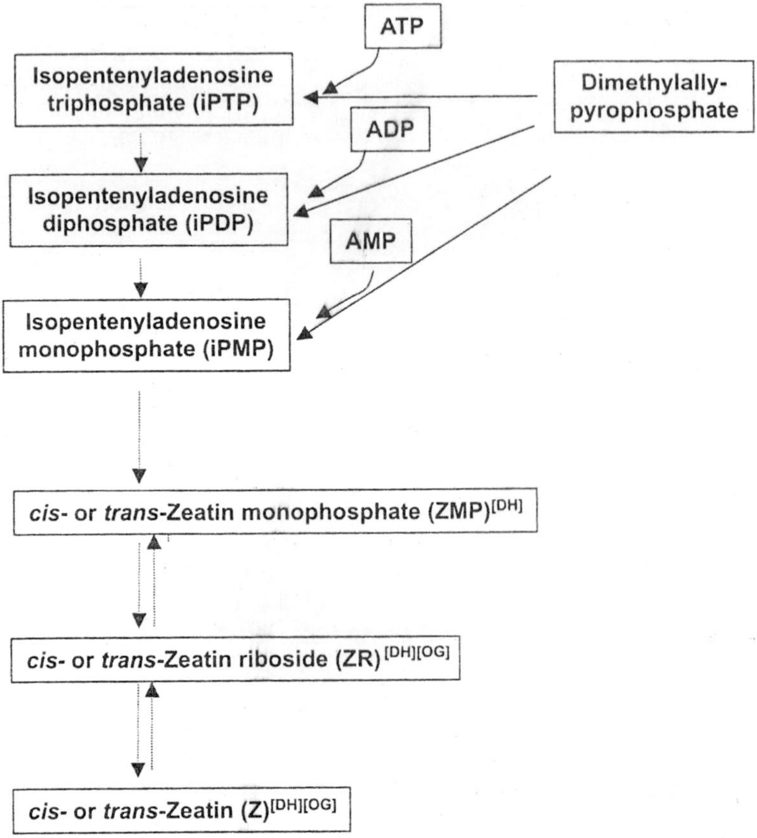

FIGURE 3.2. A general scheme of CK biosynthesis based on genetic or protein evidence (solid arrows) or inference and speculation (dashed arrows). [OG] indicates a CK which can be O-glucosylated; [DH] indicates a compound that can be side-chain saturated by a reductase (dihydronated); [OX] indicates a compound that can be degraded by loss of its side-chain by CK oxidase. Other reactions include O-acetylation, O-xylosylation, and ring glucosylation. These are not shown because they cannot be placed informatively into the scheme. (*Source:* Scheme adapted from Sakakibara and Takei, 2002.)

between different tissues as the seed develops (Emery et al., 2000), it will be interesting to find out where the initial steps of the synthetic pathway are most highly expressed.

Detailed pathways of CK formation and interconversion among forms may be found in recent papers and reviews (Chen, 1997; Kakimoto, 2001; Mok and Mok, 2001; Haberer and Kieber, 2002; Auer, 2002). For simplicity, the most commonly studied CK may be organized into four major groups comprising free bases, ribosides, ribotides (nucleotides), and O-glucosides (Figure 3.1). The outline in Figure 3.2 emphasizes what is known about CK pathways but omits much of the speculation that is necessary when trying to assemble more comprehensive pathways of synthesis and metabolism. Specifically, Figure 3.2 shows pathways for which there is adequate enzymological and molecular evidence. It indicates that CK formation can occur between dimethyl-allyl phosphate (as the source of the isopentenyl side chain) and three possible cosubstrates, AMP, ADP, or ATP (as the purine ring source) (Sakakibara and Takei, 2002). Regardless of the initial reaction, the first CK are thought to be nucleotide isopentenyl-types (Figure 3.1). However, Astot et al. (2000) have labeling evidence that in some cases the initial forms are in fact zeatin nucleotides. From this point onward, speculation abounds as to what groups are important in early reactions. Is there a nucleotide pathway or do phosphatases convert early CK nucleotides to ribose forms? At what point do conversions to the free base forms take place? There are also questions over reactions at the isopentenyl side chain. It is generally assumed that isopentenyl forms must be side-chain hydroxylated to become active. However, this hydroxylation can be stereospecific, resulting in either *cis*- or *trans*-CK isomers (Figure 3.1). To date, *trans*-isomers are assumed the major early pathway forms and *cis*-conversion is considered the result of later steps in metabolism (i.e., Haberer and Kieber, 2002), but no data support this contention. On the contrary, radioactive [^{3}H]-*trans*-ZR (zeatin riboside) feeding experiments indicate no formation of *cis*-ZR (Nandi and Palni, 1997), and there is evidence for independently regulated *cis*- and *trans*-CK pathways (Martin et al., 2001; Mader et al., 2003) or even *cis* to *trans* conversion (Bassil et al., 1993). Furthermore, the side chain can be saturated, giving rise to dihydro forms (Figure 3.1) in any of the CK groups, except for the isopentenyl-CK.

Interconversion among the free base, ribotide, and various riboside forms of the purines constitutes an important "housekeeping" pathway for purine (adenine) salvage, and there have been questions about the extent that the enzymes of this pathway contribute to CK interconversion (Chen, 1997; Mok and Mok, 2001; Auer, 2002). Although considerable evidence indicates that the salvage enzymes (adenine phosphoribosyl transferase, adenosine kinase, nucleosidase, etc.) are less reactive with CK as substrates compared to their adenine counterparts, they could nevertheless contribute to CK interconversion (reviewed by Auer, 2002). The only CK-specific enzyme with nucleosidase activity has been purified from *Mercurialis annua* (Yang et al., 1998), and interestingly this was stereospecific, hydrolyzing only *trans*-ZR. Given that the conversion of ribotides and ribosides to "active" free base forms may well constitute a central regulatory mechanism

Auer (2002) has questioned the likelihood that constitutive expression of the adenine salvage pathway could achieve the distinct tissue and developmentally specific accumulation that has been recorded (e.g., in lupin fruits, Emery et al., 2000). Despite our current state of knowledge, these questions are equally valid for synthetic pathways in seeds as well as in plants as a whole.

In addition to the group of isoprenoid CK described, it is becoming increasingly clear that plants also contain a complex group of CK that have an aromatic side chain attached to the N^6-position of adenine (Strnad, 1997; Tarkowska et al., 2003). Some evidence suggests that these CK are metabolized by separate biosynthetic and degradation pathways, that they have considerable biological activity in a number of CK bioassay systems, and that they have regulatory functions in plant development that are different from the isoprenoid types (Tarkowska et al., 2003). No studies have been conducted to date on the occurrence or possible function of the aromatic-CK in seeds, but it seems likely that as their significance becomes more obvious they too will be found to play important roles in seed development.

DISCOVERY AND CHARACTERIZATION OF CKs IN DEVELOPING SEEDS

Although the presence of CK in the liquid endosperm of the coconut seed was not appreciated at the time, coconut milk had been used routinely to stimulate the division and growth of cultured cells (Caplin and Steward, 1948). It was a seed source, immature kernals of *Zea mays,* that provided the first naturally occurring CK identified, zeatin (Letham, 1963). In hindsight, this is perhaps not surprising given the great abundance of CK in seed tissues (see Table 3.1). Davey and van Staden (1979) documented levels in developing white lupin *(Lupinus albus)* that appeared to be so "exceptionally high" that they were criticized as being artifacts of bioassay techniques (Zhang and Letham, 1990). However, a more recent study on white lupin seeds using GC-MS for quantification indicated that such high levels were not so unrealistic, concentrations in the endospermic fluid exceeding 0.6 $\mu g \cdot g\ FW^{-1}$ (Emery et al., 2000) (Figure 3.3). Although, to our knowledge, this is the greatest concentration of CK isolated from any plant tissue, developing seeds of other plants have similar impressively high levels of CK (Table 3.1). The variability among studies is not unique to lupin seeds, as similar discrepancies were observed for rice and corn (Table 3.1). Clearly, results depend on the stage of seed development, the tissues extracted, the analytical techniques used, cultivar differences, and whether the

TABLE 3.1. Examples of major CK found and their relative concentrations among developing seeds of various species.

Study	Species and tissue	Analytical method	Major CK[a]	Concentration (nmol·g FW^{-1})
Davey and van Staden, 1979	*L. albus* whole seed	Callus growth bioassay	Total CK	569[b]
Emery et al., 2000	*L. albus* endospermic fluid	GC-MS	*trans*-ZR	240
Morris, 1997	*Oryza sativa* grain	Immunoassay	iPA	135
Arnau et al., 1999	*Prunus persica* whole seed	GC-MS	DZR	60
Review of nine studies (Morris, 1997)	*Zea mays* kernals	Various	*trans*-ZR	0.11 to 34
Emery et al., 1998	*Cicer arietinum* endospermic fluid	GC/MS	*cis*-ZNT	12.4
Emery and Atkins, unpublished	*L. angustifolius* endospermic fluid	GC/MS	*trans*-ZR	7.38
Yang et al., 2000	*Oryza sativa* spikelets	ELISA	*trans*-ZR	1.07[b]
Zhang and Letham, 1990	*L. angustifolius* endospermic fluid	RIA	DZR	0.404
Yang et al., 2001	*Oryza sativa* dehulled grain	ELISA	*trans*-ZR	0.365[b]
Ma, Zhang, et al., 1998	*Nicotiana tabacum* whole seeds	RIA	*trans*-Z	0.27[b]
Quesnelle and Emery, unpublished	*L. polyphylus* endospermic whole seed	LC/MSMS	*cis*-ZR	0.149
Quesnelle and Emery, unpublished	*Pisum sativum* endospermic whole seed	LC/MSMS	*cis*-ZR	0.045
Takagi et al., 1985	*Oryza sativa* ears	GC-MS	*trans*-Z	0.044[b]
Banowetz et al., 1999	*Triticum aestivum* whole kernals	Immunoassay	Z plus DZ	0.025[b]

[a]Abbreviations: CK—cytokinin; iPA—isopentenyl adenosine; GC—gas chromatography; MS—mass spectrometry; LC—liquid chromatography; Z—zeatin; ZR—zeatin riboside; DZ—dihydrozeatin; DZR—dihydrozeatin riboside; NT—nucleotide; ZNT—zeatin nucleotide.
[b]Values recalculated to match present units.

major CK were appropriately detected. In the latter case, *cis*-CK and nucleotide forms are often overlooked but may represent predominant constituents in seeds (Table 3.1).

SOURCES OF CYTOKININ IN SEEDS

Since the review by Morris (1997) some progress has been made in establishing seeds as a site of significant CK synthesis. Piagessi et al. (1997) obtained evidence that CK is synthesized in *Sechium edule* seed by incubating endosperm with [U-^{14}C] adenine and [^{13}C] mevalonic acid. Incorporation of the labeled mevalonate into iP and ZR was detected by GC-MS. Also, isopentenyl transferase activity (isopentenyl group of dimethlylallyl

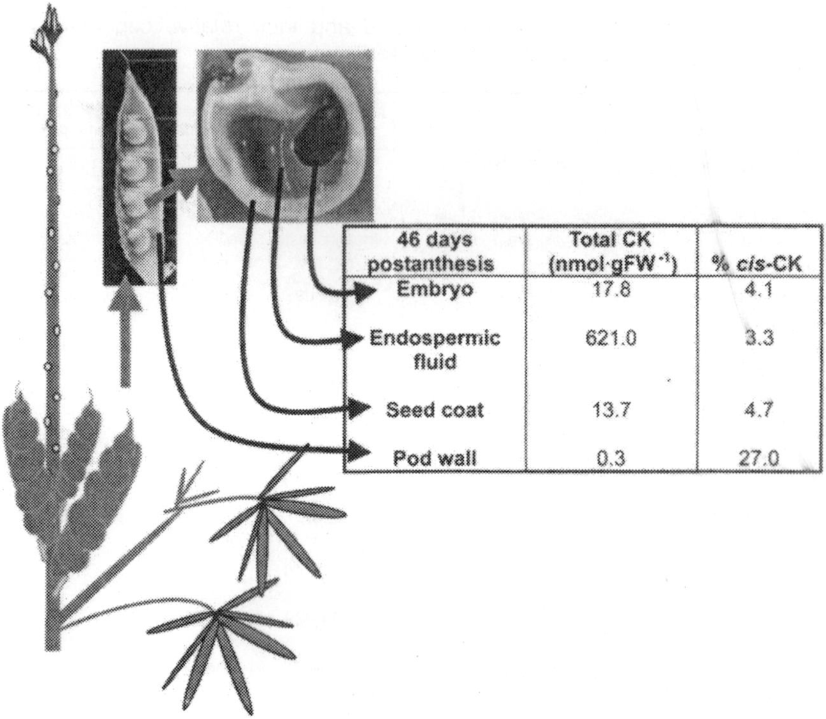

46 days postanthesis	Total CK ($nmol \cdot gFW^{-1}$)	% *cis*-CK
Embryo	17.8	4.1
Endospermic fluid	621.0	3.3
Seed coat	13.7	4.7
Pod wall	0.3	27.0

FIGURE 3.3. Accumulation of CK and percentage *cis*-CK in different seed tissues of a rapidly filling *Lupinus albus* fruit 46 days postanthesis.

diphosphate transferred to AMP to make iPMP) has been detected in extracts from immature maize kernels (Blackwell and Horgan, 1994), but the enzyme(s) responsible was never purified to homogeneity or characterized. These data do not indicate the extent to which CK is formed in seeds since, even if seeds express the synthetic pathway(s), significant quantities of CK may still be derived from translocation in the vasculature (Letham, 1994).

There is no doubt that roots are a site of CK synthesis and that they may be a major supplier of CK to the rest of the plant through xylem translocation (reviewed in Emery and Atkins, 2002). CK have been detected in both xylem and phloem exudates and, although their quantitative significance in phloem has been difficult to establish, they would nevertheless be delivered to sink organs by long-distance transport. In some instances, CK accumulation in situ far outweighs the amount that might be furnished by translocation from roots (Davey and Van Staden, 1979; Brenner and Cheikh, 1995; Morris, 1997; Emery et al., 2000). Calculations based on estimates for the water economy of developing *Lupinus albus* fruits predicted that the accumulated CK could not possibly be supplied by the CK content of xylem sap leaving the root system (Zhang and Letham, 1990). Our recent study took into account potential contributions of the combined fluxes of CK in xylem and phloem and showed that translocation accounted for less than 1 percent of CK accumulating in lupin fruit (Emery et al., 2000). It was concluded that, at this site of intense CK accumulation, in situ seed synthesis was the major contributor. Considering that diurnally reverse xylem flow to leaves has been documented in developing fruits of cowpea (*Vigna unguiculata*) (Pate et al., 1985; Peoples et al., 1985), this represents a possible mechanism for CK produced in seeds being eventually loaded onto phloem in vegetative organs. In this way the seed would be not only a major site of CK synthesis but also a source of these regulators for the plant as a whole, potentially exerting some regulatory "influence" of the developing seeds on events elsewhere.

DEVELOPING SEEDS: A RICH MINE OF CYTOKININ ENZYMES

Regardless of how the large quantities of CK accumulate in developing seeds, whether it is in situ synthesis or translocation from sites of synthesis in other organs, the seed appears to hold a great capacity for CK metabolism and interconversion between base, riboside, and ribotide forms. In fact, although even basic pathways of CK synthesis have proven difficult to establish, what is known about CK metabolic enzymes and the genes that encode them comes largely from work with developing seeds (Mok and Mok, 2001; Galuszka et al., 2000).

CK Conjugation Enzymes

CK conjugation on the side-chain hydroxyl group may help determine bioactive CK levels in the seed since it is thought that O-glycosyl-CKs are inactive, stable storage forms. Glycosylation may reversibly inactivate CK, assuming the O-glucosyl-CK do not have activity per se. The presence of a glycosyl moiety may also protect CK from degradation by CK oxidase (CKX). Progress to date on the conjugation enzymes has come exclusively from work on developing seeds. Detection of activity and subsequent isolation of enzymes for side-chain O-xylosylation and O-glucosylation was derived from studies of developing *Phaseolus lunatus* and *P. vulgaris* seeds in the mid to late 1980s. Eventually, this led to the cloning of two genes; the first, *ZOG1* (zeatin O-glucosyltransferase), was identified by screening expression of a cDNA library from developing *P. lunatus* seeds with monoclonal antibodies (Martin et al., 1999b). The second gene, *ZOX1* (zeatin O-xylosyltransferase), was cloned by inverse PCR with primers based on the *ZOG1* sequence (Martin et al., 1999a). These clones led to the surprising discovery of a third O-glucosyltransferase that is highly specific for *cis*-zeatin, *cis*ZOG1 (Martin et al., 2001). Subsequently *cisZOG1* was cloned using the *ZOG1* sequence to identify candidate genes from a maize EST (expressed sequence tag). The enzyme encoded by *cisZOG1* does not recognize *trans*-Z, DZ, or *cis*-ZR as substrates. Its substrate specificity indicates a high precision of CK metabolism and suggests that *cis* isomers of CK may deserve more attention than historically they have received.

CK Catabolism: CK Oxidase/Dehydrogenase

CK oxidase (CKX) activity was discovered in crude extracts of tobacco tissue cultures (Paces et al., 1971), and subsequently in a range of tissues from a number of species. However, developing seeds of maize and wheat have played a major role in studies of the enzyme (McGaw and Horgan, 1983; Burch and Horgan, 1989; Kaminek et al., 1994; Laloue and Fox, 1989), including its more recent molecular characterization. Until recently CKX was the only known plant enzyme capable of degrading naturally occurring CK (Galuszka et al., 2000). However, Galuszka et al. (2001) characterized a CK dehydrogenase from developing wheat and barley seeds. The protein contained a flavin (like CKX) and transferred electrons to a number of artificial acceptors as well as a ubiquinone precursor. In both cases the oxidation catalyzed causes irreversible loss of the CK molecule by cleaving the isopentenyl side chain from the purine ring. That CK oxidation is likely to be a key factor in determining CK balance in developing seeds is supported by its pattern of activity, which closely parallels CK accumulation in developing cereal grains following fertilization (Chiekh and Jones, 1994; Morris, 1997). Z and ZR are the main substrates for CKX, and they are also the principal CK that accumulate in developing grains of a number of cereal species (Morris, 1997). In eudicots, however, we have found that CK forms can be quite different. In chickpea, for example, CK-NT (nucleotide) and *cis*-ZR accumulate (Emery et al., 1998). The former are not substrates for CKX (Laloue and Fox, 1989) and the latter are metabolized with a higher Km and lower Kcat (Bilyeu et al., 2001). In lupin, the situation is similar to the cereals once the fruit is set and seeds begin to grow rapidly. Earlier, however, when the fate of seeds and fruits is not yet established, the main CK are *cis*-CKs and CK-NTs, which are not as susceptible to CKX catabolism (Emery et al., 2000). Although there is no established role for the aromatic CK in seeds, it is significant that they too are not readily oxidized (Tarkowska et al., 2003).

The greatest activities of CKX are in developing seeds (Galuszka et al., 2000), although levels vary among tissues and between species. For example, in lupin and soybean seed coats CK degradation is negligible (Zang and Letham, 1990; Singh et al., 1988), whereas that of bean seed coats is relatively high (Turner et al., 1985). In maize seeds there was much greater CKX localized in the pericarp and embryo compared to the endosperm (Jones et al., 1992).

Despite considerable attention to the role of CKX in the more than 30 years since its discovery, genetic characterization has just begun to unfold. To this end, developing seeds have played the leading role. Independently two laboratories cloned a *CKX* from maize kernels that encode the same protein. Morris et al. (1999) obtained 100 μg of relatively pure enzyme from 40 kg of kernels and designed degenerate oligonucleotide primers based on peptide sequences that were used for PCR (polymerase chain reaction) amplification and isolation of a *CKX* fragment. The fragment was, in turn, hybridized to a genomic library to isolate the full-length gene, *ckx1*. Houba-Herin et al. (1999) obtained peptides from only 6 μg of protein to design primers for PCR of mRNA purified from maize cobs. The cDNA was subsequently cloned and sequenced. Availability of this gene led to isolation of *Arabidopsis* orthologs (Bilyeu et al., 2001; Galuszka et al., 2000) that have already provided a significant breakthrough, confirming a number of roles for CK *in planta* that were previously suggested by physiological experiments (Werner et al., 2001). The *Arabidopsis* CK oxidase/ dehydrogenase gene family comprises seven members *(AtCKX1-AtCKX7)*. Four of these contain signal peptides that in two cases target a cell secretory pathway, the other two target mitochondria, and with just one transcript predicted to have a cytosolic location (Bilyeu et al., 2001; Werner et al., 2001).

Tobacco transformed to express one of four *AtCKX* genes exhibited dramatic phenotypes that reflected greatly reduced CK content (Werner et al., 2001). Leaf cell production was all but eliminated, while roots showed faster growth and hyper-branching. The plants were dwarfed with shorter internodes and had fewer, late-blooming flowers. The meristems were much reduced, but the size of meristematic cells was unaffected. No results have been reported for seed development, but with the seed-specific promoters that are available suitable *AtCKX* constructs could provide a major breakthrough in establishing whether CK accumulation has a causal role in increased mitotic indices of the endosperm or embryo.

CK Reductase

Relevant to the discussion of CKX, a zeatin reductase has been isolated, also from developing seeds *(Phaseolus vulgaris)* (Mok and Mok, 2001). The enzyme reduces the isopentenyl side chain and is highly specific for *trans*-Z (zeatin). The DZ (dihydrozeatin) product is considered an active CK that is resistant to CKX.

CK Isomerase

Once assumed to be artifacts or ignored because they were thought inactive, *cis*-CKs have been neglected in the majority of CK research, despite their predominance in many plant tissues (Morris, 1997; Emery et al., 1998). However, none of the bioassays used to determine activity relate to seed development. Their significance has also been confounded by the fact that these isomers are normal constituents of tRNA and that their occurrence might simply reflect RNA degradation during extraction (see Emery et al., 1998). Considering that *cis*-CK are known to predominate in developing seeds of chickpea (Emery et al., 1998) and rice (Takagi et al., 1985), and transiently during early embryogenesis in white lupin (Emery et al., 2000), more study into these aspects is needed. It has become apparent in developing buds of chickpea that *cis*- and *trans*-CK levels were independently regulated (Mader et al., 2003), which appears to match results of independent *cis*- and *trans*-CK regulation in roots (Emery and Atkins, 2002). Moreover, as discussed previously, an enzyme specific to *cis*-CK, *cis*ZOG1, has been isolated and the encoding gene cloned from developing maize kernals (Martin et al., 2001). Interestingly, the discovery of *cis*ZOG1 precedes any published reports for the occurrence of the substrate, *cis*-Z CK, in maize despite extensive studies of this species over the years (Auer, 1997).

Dobrev et al. (2002) have recently followed endogenous CK and activity of CKX during cell cycle progression in large-scale synchronized cultures of a tobacco cell line. They found two distinct transient peaks in CK, one at the beginning of the S phase and the second at the onset of mitosis. The first peak comprised transient increases in isopentenyl adenine and iPadenosine iP, the dihydro derivatives and *trans*-Z. Interestingly the premitotic peak included equal amounts of *cis*- and *trans*-Z along with Z-O-glucoside and its riboside. The authors speculate that the high levels of CKX are possibly responsible for the transient nature of these CK "peaks," but the pathways for the interconversion of forms and the source of the *cis* isomer were not indicated. These data, together with the emerging picture that *cis*-CK may be very common in developing seeds among many species, suggest that the ratio of *cis/trans* isomers may be an important determinant in the interaction between CK and regulation of cell division.

A question apart from biological activity of *cis*-CK as such is how much the tRNA-derived CKs could be utilized in the budget of active CKs after *cis-trans* conversion. Although there is evidence for enzymatic and non-enzymatic *cis-trans* isomerization (but not vice versa) (Bassil et al., 1993; Mok et al., 1992), the extent of conversion in vivo is unknown. In fact, evidence so far suggests there is very little in vivo isomerization of *cis-* to *trans*-CK (Wang et al., 1997; Suttle and Banowetz, 2000). Nevertheless, a partial purification of a *cis-trans*-isomerase from developing seeds of *Phaseolus vulgaris* has been reported (Bassil et al., 1993) but unfortunately has apparently not been followed up.

CK Transporters

Transport studies in *Arabidopsis* cells (Buerkle et al., 2003) have demonstrated that a common H^+-coupled purine transport system transports both adenine and CK. These purine permeases (AtPUP1 and AtPUP2) mediate energy-dependent high-affinity uptake of radiolabeled *trans*-Z as well as purine alkaloids (caffeine) and ribosylated purine derivatives. Promoter-reporter gene expression studies with *AtPUP2* in *Arabidopsis* have localized a site at the stigma surface of siliques, while *AtPUP1* was found in the phloem of leaves consistent with a long-distance CK transport role. Expression of a third purine permease *(AtPUP3)* was associated specifically with pollen. Although these studies have not been extended to developing or germinating seeds, the strong physiological evidence for translocation of CK to seeds early in their development suggests that PUP transporters might function in unloading from vascular channels.

BIOLOGICAL ACTIVITY OF CYTOKININ IN DEVELOPING SEEDS

Although it is becoming more clear that developing seeds are an extremely rich source of CK and that they express a number of CK metabolic enzymes, the exact functions that the diverse range of CK accumulating in seeds have in their development are still not clear. Recent advances in understanding how CK are perceived as extracellular signals and how their reception is linked to generic signal transduction pathways are beginning to reveal the diverse roles of CK in roots and leaves. Though yet to be studied in relation to seed development, there seems little doubt that a similar link between signaling and the diverse responses that result from a transduction pathway will be found in seed tissues.

Already good evidence is available that CK regulation of seed development may be quite complex, involving a number of response elements. For example, the *pasticcino (pas)* mutants of *Arabidopsis* are putative CK sensitivity mutants that have altered cell division leading to ectopic cell proliferation (Faure et al., 1998). The *pas* mutants also show altered embryo development, in which the cotyledon primordia do not form correctly after the heart stage, leading to a flat embryonic apex. At later stages, abnormally short and wide hypocotyl development prevents normal curvature of the embryo within the seed. Although it is still not certain if the *pas* mutant phenotype is uniquely, or even largely, due to impaired CK perception, these important mutants must be further characterized.

Relationship Between CK Content and Seed Sink Strength

Considerable attention has been given to a simple hypothesis that invokes a role for CK in mechanisms that determine the ability of a developing seed to attract assimilates and grow. It is from this basis that researchers have sought to manipulate assimilate "sink strength" as a means to increase crop yield. Most attempts have used application of exogenous CK and in general have not been successful. Greater knowledge of how to manipulate endogenous levels of CK and other plant growth regulators and the cellular, as well as organ, responses they mediate is needed before these outcomes can be realized.

How many CK increase the sink strength of a developing seed? Stimulation of cell division was the first CK response identified, and this is the basis for the most obvious hypothesis. Cell division increases seed organ sink size (Davies, 1977; Weber et al., 1997; Munier-Jolain and Ney, 1998) and in this way potentially affects whole-plant source-sink relations. The idea is simple; more cells provide an increased capacity for storage of assimilates. This relationship is striking in developing cereal grains. In barley, maize, rice, and wheat, transient increases in CK content occur shortly after pollination (reviewed in Morris, 1997), the peak in CK accumulation coinciding with the peak in endosperm mitotic index in each case. However, which of the peaks, CK or mitotic index, is the cause and which is the response? Correlative evidence, together with emerging knowledge about the response of cell cycle genes to CK, indicates that CK level is likely to strongly influence cell division rather than the reverse (Morris, 1997). There seems little doubt that genes associated with cell cycle activity are among the early responses to CK. *CycD3* encodes a cyclin protein involved in the G_1/S transition and, in *Arabidopsis;* its expression stimulates transcription within one hour (Riou-Khamlichi et al., 1999). Similarly, both auxin and CK enhance transcription of the *cdc2* gene, a histone kinase that mediates the G_2/M transition (Zhang et al., 1996). In large-scale cultures of a tobacco cell line, synchronized by blocking the cycle at the G_1/S transition and at G_2/M, Dobrev et al. (2002) found two distinct transient peaks in CK, one at the beginning of the S phase and the second at the onset of mitosis.

Other recent studies continue to support the cell-division hypothesis. For example, rice genotypes with higher rates of grain filling and percentage of filling grains had higher Z and ZR content (Yang et al., 2000). Also in rice, CK accumulation was not affected if water deficit stress was administered in a manner that allowed full rehydration and recovery of the filling grain (Yang et al., 2001). Similarly, Banowetz et al. (1999) showed that reduction in wheat kernel number and mass caused by high-temperature treatment coincides with a decline in CK. Interestingly, however, exogenous CK application mitigated the negative effects of high temperature on kernel number but not on kernel mass.

In eudicots the situation seems similar, but the documentation has been less clear and there have been few efforts to record changes in specific CK during seed development. In both chickpea (Emery et al., 1998) and lupin seeds (Davey and van Staden, 1979; Emery et al., 2000) CK levels were greatest during endospermic fluid stages of growth, presumably when rapid cell division was occurring in the embryo. However, more frequent sampling of tissues from developing *L. albus* seeds indicated peak CK content occurred after maximum endosperm volume was reached and the embryo had begun to expand (Emery et al., 2000). Thus, transient CK changes coincided more with the onset of seed filling than cell division, and it is possible that CK enhances sink strength in ways other than promotion of mitosis. Although such a conclusion seems reasonable on the basis of the total CK content of the developing *L. albus* seed, this research indicated a diversity of CK forms in the endosperm and surrounding tissues. Which of these is the effector and at what cellular site at any particular time is unknown.

CK enhancement of seed sink strength could result from a combination of cell division promotion, phloem unloading (Clifford et al., 1986), and the activities of enzymes involved in the transfer of assimilate to the developing embryo. In developing legume seeds, assimilates pass symplastically from the phloem to the inner epidermis of the seed coat where they are unloaded to the apoplast (and endosperm) surrounding the embryo (reviewed in Weber, Borisjuk, and Wobus, 1998; Weber, Heim et al., 1998; Herbers and Sonnewald, 1998) and from here are taken up by carrier-mediated transport at the cotyledon epidermis (McDonald et al., 1996). However, in *Vicia faba* when the embryo is still engaged in cell division, and before significant stored reserves accumulate, sucrose is first hydrolyzed to hexose by an extracellular invertase in the seed coat (Weber et al., 1996; Borisjuk et al., 1998). As a consequence, hexose levels of the order of 100 mM accumulate during early seed development (Weber et al., 1997) and a hexose-specific transporter (VfSTP1) is expressed in the outer epidermal cells of the embryo. Coordinated expression of invertase and hexose transporter genes has also been reported for *Arabidopsis* (Truernit et al., 1996) and cell cultures of *Chenopodium rubrum* (Ehneß and Roitsch, 1997). In *C. rubrum* expression of both genes was specifically induced by CK. Weber et al. (1996) suggest that hexoses are essential to maintain meristematic activity in the developing embryo and that a decline in invertase and the hexose transporter signals a switch from cell division to expansion and storage. Later when stored carbohydrate and protein reserves are being laid down sucrose is unloaded to the apoplast and is taken up as such by the embryo. In *V. faba*, the cells of the abaxial surface of the cotyledon differentiate into transfer cells and express a sucrose transporter (VfSUT1). Sugar hydrolysis in the embryo during the storage phase of development is mediated largely by sucrose synthase.

Thus, invertase-mediated unloading, supported by CK-stimulated expression of both invertase and the hexose transporter genes, may be interpreted as a mechanism to provide sink strength for continued phloem delivery of assimilates during the early phase of seed development. Because both CK and hexoses stimulate the cell cycle at the same point (between G1 and S phases) through enhanced expression of the D-type cyclins required to activate cell cycle kinases (Riou-Khamlichi et al., 1999), it is difficult to determine whether sugar signaling mediates CK signaling or the reverse. Rolland et al. (2002) recently described a molecular link between hexose and CK signaling through *ARR* (*Arabidopsis* response regulator; a transcription factor) gene expression using a glucose-insensitive mutant, *gin2*. It will be interesting to assess the CK phenotype of this and similar mutants.

CK Form and Function

If CK is responsible for determining seed size through its promotion of cell division, the variety of CK forms found among seeds (Table 3.1 and Emery et al., 2000) of different species raises some interesting questions. In white lupin, for example, as many as 18 different CK have been found in the endospermic fluid bathing the developing embryo (Emery et al., 2000) and it is reasonable to ask which of these is bioactive. The mass of CK literature indicates that the free bases and their ribosides, along with dihydro derivatives, are likely to be the most active. However, there have been no direct assays of cell division in seed tissues that confirm the dogma. Although no doubt some of the forms present during embryo development are not likely to stimulate cell division, other functional roles could be possible.

Pea *(Pisum sativum)* provides a particularly interesting example of the deficiencies in our knowledge about which CK forms that occur in seeds and their bioactivity in relation to seed development. As early as 1985, Wang and Sponsel pointed out an historical anomaly: no CK had been identified in seeds of any pea genotype, although seeds of other species were rich sources of these hormones. Furthermore, although early bioassays of crude pea-seed extracts showed positive results, follow-up studies revealed no biological activity (Wang and Sponsel, 1985). Since then, even more sophisticated and robust techniques such as immunoassay or mass spectrometry have apparently not succeeded in assaying the presence of any known CK in pea seeds. Instead, a plausible scenario has been developed that invokes gibberellins as the major controlling hormonal agents of pea seed development (Swain et al., 1995, 1997). Depending on the degree of deficiency, mutants low in seed gibberellins show stunted embryo growth and even seed abortion. It is envisaged that gibberellins may promote assimilate uptake by developing seeds or promote seed growth, thus increasing the use of more of the available assimilates. This model has similarities to those hypothesizing that CKs augment seed development through an increase in cell division and assimilate demand (Morris, 1997). Perhaps it is not surprising that attention has been diverted away from CK physiology. But do pea seeds really completely lack CK? Recent preliminary evidence using LC/MSMS reveals the presence of significant quantities of several CK in developing pea seeds (P. E. Quesnelle and R. J. N. Emery, unpublished results). However, like seeds of the closely related species chickpea, the CK profile of developing pea seeds is predominantly composed of *cis*-isomers. This may explain low or undetectable bioactivity of early bioassay attempts to quantify CK. *cis*-CK show greatly reduced activity in all widely used bioassays (Kaminek, 1982). When they chromatograph differently from *trans*-CK they might remain undetected if the proper retention times are not observed using techniques such as GC-MS or LC-MS. Conversely, Van Rhijn et al. (2001) have shown that rapid separation procedures, frequently used in high sample throughput LC/MSMS, have resulted in coelution of the isomers with consequent considerable overestimation of *trans*-isomers and a failure to identify *cis*-CK.

The significance of the *cis*-CK profile could mean several things. First, like chickpea, developing seeds of pea may have very low CK activity levels. No bioassay has yet been developed in which *cis*-CK have been shown to have high activity. Further evidence that *cis*-CK are inactive comes from characterization of the CRE1 CK receptor from *Arabidopsis* (Inoue et al., 2001). When incorporated into an osmosensing pathway in yeast, CRE1 must bind to *trans*-CK (or benzyl aminopurine) to activate the pathway and suppress lethality on glucose-containing media. *cis*-Z in this assay is ineffective. It may seem strange that chickpea and pea would be loaded with inactive *cis*-CK, especially if CK were required to stimulate cell division and augment the growth of the seed. However, lesser quantities of *trans*-CK or dihydro derivatives may be sufficient to establish sink strength at the seeds. For example, seeds of crop lupins *(L. albus, L. angustifolius)* are loaded with the extremely high concentrations of *trans*-CK, but these plants set and fill very few fruits with seeds—usually three to five pods on the lowermost region of a raceme that typically produces 30 to 90 self-fertilizing flowers. The strong assimilate sink due to the large-seeded fruits may preclude the set and filling of seeds further up the raceme. *Lupinus polyphylous*, by comparison, sets many more fruits. These contain relatively very small seeds and the endospermic CK profile is dominated by *cis*-CK (P. E. Quesnelle and R. J. N. Emery, unpublished results). Quite clearly our lack of knowledge about the bioactivity of these different CK forms in plants and especially in developing seeds precludes any useful explanation at this time.

Pod Set

Observed patterns of CK isomerization during pod set also support the hypothesis that *cis*-isomers are less active. One obvious change in the CK composition between anthesis and four days after anthesis (daa) was the significant transient increase in the proportion of *cis*-forms in ovaries in the two zones where flowers were destined to abort (Figure 3.4), while in those destined to set pods, the proportion as *cis*-forms did not change. If it can be assumed that the "active" forms in lupin ovary tissue are among the free bases and ribosides then the ratio of *cis*:*trans* forms in this group changes progressively from more than 17:1 at anthesis to less than 1:1 by 10 daa at positions 1 to 3, while at positions 4 to 6 there is a progressive decline from 16:1 to 3:1, and at positions 7 to 10 the ratio remains high throughout the 10-day period, ranging from 7:1 to more than 19:1.

When active CK (benzyl aminopurine or *trans*-Z) is applied exogenously to all pedicels of *L. angustifolius* flowers, 100 percent pod set can result on the raceme (Atkins and Pigeaire, 1993). However, the resulting seeds are significantly smaller and total seed yield does not surpass that of the larger-seeded controls (Ma, Longnecker, and Atkins, 1998; Palta and Ludwig, 1997). In soybean a similar response to exogenous CK was observed (Dybing et al., 1986). In that case it was concluded that ovary

Four days postanthesis

Raceme position	Probability of pod set	Total CK (nmol·gFW^{-1})	% *cis*-CK
1 to 3	≈100 percent	2.08	38
4 to 6	≈70 percent	1.59	71
7 to 10	0 percent	0.94	85

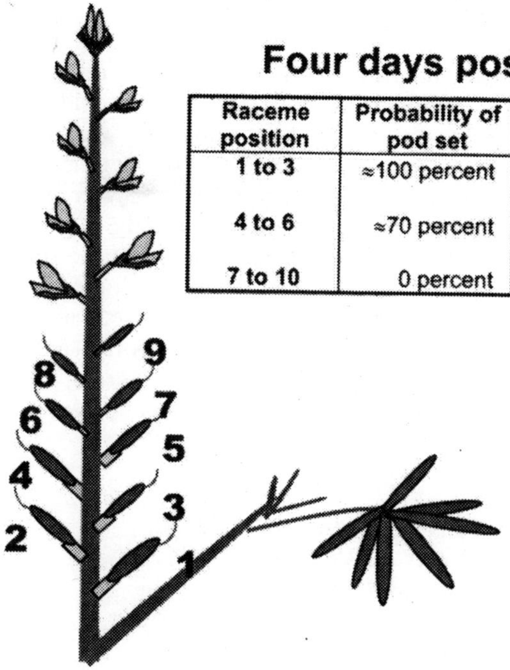

FIGURE 3.4. Concentrations of total CK and percentage *cis*-CK among different pod positions with varying probabilities of successful fruit set along a *Lupinus albus* main stem raceme.

abortion is initiated by events that slow growth at or within 1 daa. Thus, changes in abundance of *cis*- and *trans*-isomers observed for lupin could be a consequence of earlier events rather than a cause of flower abortion. Nevertheless, *cis*-isomers may well act as competitive inhibitors of processes that involve the binding of or reaction with *trans*-isomers (Kuraishi et al., 1991), and in this way exert a negative effect on embryogenesis. The partially purified Z *cis-trans* isomerase from developing bean (*Phaseolus vulgaris* L.) seeds (Bassil et al. 1993) leads to the tempting speculation that regulation of the expression or activity of an equivalent enzyme in lupin might be one of those "early events."

Seed CK and Pericarp Development

As discussed previously, CK budgets indicate developing seeds may actually be a significant source of CK for other organs of a plant (Emery et al.,

2000). This may include a positive effect on pericarp development. The pericarp is not considered a significant site of CK biosynthesis (Letham, 1994), yet CK are a requirement for fruit development in several species (Lewis et al., 1996; Arnau et al., 1999; Yu et al., 2001). For example, some naturally parthenocarpic fruits have a specific requirement for CK, which cannot be replaced by gibberellins or auxin (Yu et al., 2001). Synthetic CK can also be used to artificially induce parthenocarpic fruit growth in kiwi *(Actinidia deliciosa)*, replacing the positive influence normally exerted by fertilized ovules (Lewis et al., 1996). Whether the seed represents a source of this positive CK signal is not entirely clear, but it may be linked to the production of Z, which is reduced in smaller fruits with a low seed number. In peach the developing seed and pericarp have distinct CK profiles and dynamics that change critically with the growth of each (Arnau et al., 1999), but there is no solid reason, at this point, to rule out translocation from the former to the latter.

CAN CK CONTENT BE MANIPULATED TO ALTER GRAIN DEVELOPMENT?

Following the hypothesis discussed previously that CK boosts seed growth through cell division in the developing seed, a number of attempts have been made to exploit this idea and boost yields both in the field and the laboratory. Morris (1997) reviewed 75 cases of CK application to cereals but discounts all but a half dozen as not being carefully controlled. Although overall the results have been mixed or confusing, it appears that exogenous application of CK, particularly long-lasting artificial CK such as BAP, have potential to increase grain set by overcoming some key developmental or nutritional block (reviewed in Morris, 1997). However, it is far from clear what the nature of this block might be. In several instances the CK application had to be made at a very specific tissue location or developmental stage to obtain any sort of beneficial result. For example, stem infusion of BAP into maize at pollination may increase grain set by up to 30 percent, yet infusions only three days after pollination brought about no change (Dietrich et al., 1995) and a similar situation has been described for narrow-leafed lupin (Atkins and Pigeaire, 1993).

One of the main problems limiting the use of exogenous CK application as a study tool or even as a potential agrochemical is timing and the ability to get the CK to the target tissue. Even if widespread use of CK applications was economically feasible in a cropping situation, it is highly unlikely that application could be sufficiently precise to achieve the desired result. The best-studied example is that of narrow-leafed lupin.

Precise application of CK to the pedicels or the basal sepal tissues of flowers at or around anthesis can improve pod set to 93 percent (Atkins and Pigeaire, 1993). However, a blanket foliar spray causes ovule abortion and fruit abscission to the point of entirely de-podding the plant (unpublished data). In fact, even if the number of fruit sinks is dramatically increased by meticulous CK (benzyl aminopurine) application, seed yield is not necessarily improved (Ma, Longnecker, and Atkins, 1998; Palta and Ludwig, 1998). The number of seeds per pod and individual seed weights tend to decline, even when supplied with extra N fertilizer (Ma, Longnecker, and Atkins, 1998). Moreover, although carbon fertilization (ambient [CO_2] = 700 vpm) increased the occurrence of large seeds and total seed yield (but not harvest index), this effect was independent of pod set manipulation by CK application (Palta and Ludwig, 2000). With all the complexity over timing and method of application, it is not surprising that the only attempt at field application of benzyl amonipurine to lupin did not result in a beneficial effect on grain yield (Seymour, 1996). Likewise, even though xylem-fed CK could increase soybean seed yield by 79 percent in a greenhouse, no significant differences were detected in a parallel field trial (Nagel et al., 2001).

Transgenic Approaches to Seed Hormone Manipulation

Exploiting mutants and genetic transformation might reduce problems of studying CK effects on seed development. Genes encoding the enzyme responsible for early CK formation in bacteria, isopentenyl transferase (IPT), or *trans*-zeatin synthase (TZS) have been available for some time. Transgenic plants with these genes under the control of constitutive promoters have yielded little insight into seed development, probably because, like exogenous applications, target tissues are missed or effects swamped out by constitutive expression and greatly elevated CK levels elsewhere in the plant. Thus, several attempts have already been made to target seed development by using seed-specific promoters to drive expression. One such chimeric gene combined the seed-specific pea *vicilin* promoter with the *ipt* coding sequence and was incorporated into tobacco (Ma, Zhang, et al., 1998). The expression of *ipt* in at least one of the transformants appeared to be seed specific, and total CK as determined by radio-immunoassay (one replicate) was approximately double that of controls, but, unfortunately, the phenotypes were poorly characterized. Protein content was said to increase in seeds from a transformed *vicilin-ipt* plant compared to one that was transformed with the "vector only," but there appeared to be no significant difference from an untransformed control. Moreover, seed weights were not determined in the latter, so no conclusion regarding seed yield could be reached.

A second study involved transformed oilseed rape and tobacco with a construct comprised of an *Arabidopsis* seed-specific promoter, *AT2S1*, with *ipt* (Roeckel et al., 1997). Once again, it appeared that expression of *ipt* was limited to seeds (although leaves were the only other tissue tested) and CK content of both species was more than double that of untransformed controls (although data for other tissues was not presented). Total seed yields per plant showed a promising increase in both species. However, the yield increase was due to increased inflorescence branching resulting in significantly more total capsules and siliques in *AT2S1-ipt N. tabaccum* and *B. napus* plants, respectively. In fact, individual seed weight (both species) and number of seeds per silique (*B. napus*) both declined. Reduced root growth was also observed. These are traits normally associated with *ipt* transgenics that have either nonspecific or leaky promoters that increase CK content throughout the plant (Schmulling et al., 1989; Smigocki, 1991; Medford et al., 1989; Smigocki et al., 1993; Harding and Smigocki, 1994; Thomas et al., 1995). Indeed, a follow-up study of transgenic *B. napus* incorporating a leaky heat-shock promoter of *Zea mays, hsp70*, with *tzs* showed that an increase of CK in vegetative tissues also led to increased seed yield (Roeckel et al., 1998). Transgenic plants had reduced root systems, increased branching, and delayed flowering. Moreover, the total seed yield increase resulted from significant increases in the number of seeds per silique and individual seed weights.

Ma et al. (2002) transformed tobacco with *ipt* driven by the seed-specific pea vicilin promoter and found increased CK levels (by ELISA) in seeds together with a significant change in their development. The transgenic embryo showed increased plerome cell division and as a consequence enlarged diameter and small, though significant, increases in seed fresh and dry weight. Germination frequency was not altered, but the *ipt* transgenics showed greater rates of seedling growth up to 35 days after sowing; after this time there were no differences. Whether the *ipt* gene was expressed exclusively in seeds and that only seed CK levels were elevated was not determined. However, this appears to be the first study in which overexpression of the *ipt* gene has resulted in a clear phenotype consistent with a regulatory role for CK in seed development.

MOLECULAR BASIS OF A SIGNALING ROLE FOR CYTOKININS

Recent advances have begun to provide a molecular basis for the signaling role of CK in plants (Hutchison and Kieber, 2002; Schmulling, 2002). There is now compelling genetic and biochemical evidence for transmembrane CK receptor proteins localized to the plasmalemma, and two CK binding proteins have recently been cloned from oat and wheat seeds (Kaminek et al., 2003). The functional significance of these proteins is that they have been linked to a two-component signaling pathway to provide a cellular transduction mechanism for the extracellular CK signal. Sequence analysis of the *Arabidopsis* genome indicates at least 16 genes encoding putative His (histidine) protein kinases and related proteins (Hwang et al., 2002). These are clustered into three distinct families: ethylene receptors, phytochrome receptors, and the AHK family that includes at least one CK receptor (CRE1/AHK4/WOL1) and probably others. These are members of the *Arabidopsis* his-kinase family that typically contain an input or signal-binding domain and a transmitter domain that links the signal to a cytosolic phosphorylation cascade (Schmulling, 2002). Binding assays in yeast revealed that AHK4 proteins directly bind "active" CK such as isopentenyladenine, *trans*-zeatin, and benzyl aminopurine with K_m values of 4.5 nM (Yamada et al., 2001). The AHK4 protein is expressed predominantly in roots, and Hwang et al. (2002) suggest that CK perception in shoot organs involves other members of the AHK family. Indeed, *CKI1*, encoding a similar though different type of His protein kinase but believed to be also implicated in CK signaling, is expressed in the ovule and endosperm but not in the embryo, based on expression of a *CKI1:GUS* reporter construct in transgenic plants. A group of His phosphotransfer proteins (AHP1-5) serve to relay the signal to downstream nuclear responses. AHPs are transiently translocated to the nucleus where they activate response regulators (type-B ARRs). In response to CK these ARRs are believed to enhance transcription of target genes. Schmulling (2002) has suggested that the multiple elements of this signaling cascade may provide some of the diversity necessary to account for the many effects that CK have on different developmental processes in plants. However, beyond the phosphorelay signaling pathway the links to CK function remain obscure. Among genes that have been described as CK responsive is *CycD3*, encoding cyclin D3 that is believed to have a role in regulating cell cycle at the G_1-S transition, a function consistent with the effects of CK on cell division and the activity of meristems. Smalle et al. (2002) have found that the *Arabidopsis* 26S proteasome, responsible for degradation of many short-lived cell regulator proteins, also functions in controlling the stability of factors involved in CK response, including CYCD3.

FUTURE PROSPECTS

Despite this new and unfolding story that is linking CK signaling to cellular pathways of signal transduction and possibly to cell cycle regulation, a number of serious questions, particularly as they apply to seed development, remain. There is no doubt that CK levels in seeds are dynamic and that the component tissues are not simply sites for CK accumulation. To address this question adequately a transgenic seed-specific approach is required. The availability of newly discovered genes of CK metabolism as well as those associated with cellular responses makes such studies feasible for the first time. However, the diversity of CK forms found from the more inclusive analyses that are beginning to appear raises the prospect that studies of gene expression need to readdress the question of which of the endogenous forms of CK are bioactive. This applies especially to the bioactivity of *cis* and *trans* forms as well as to the aromatic CK. Because developing seeds appear to be intense sites of CK synthesis and metabolism, they offer the ideal system in which to unravel this mystery.

REFERENCES

Arnau, J.A., F.R. Tadeo, J. Guerri, and E. Primo-Millo (1999). Cytokinins in peach: Endogenous levels during early fruit development. *Plant Physiology and Biochemistry* 37: 741-750.

Astot, C., K. Dolczal, A. Nordström, O. Wang, T. Kunkel, T. Moritz, N.H. Chua, and G. Sandberg (2000). An alternative cytokinin biosynthesis pathway. *Proceedings of National Academy of Sciences USA* 97: 14778-14783.

Atkins, C. A. and A. Pigeaire (1993). Application of cytokinins to flowers to increase pod set in *Lupinus angustifolius*. *Australian Journal of Agricultural Research* 44: 1799-1819.

Auer, C.A. (1997). Cytokinin conjugation: Recent advances and patterns in plant evolution. *Journal of Plant Growth Regulation* 23: 17-32.

Auer, C.A. (2002). Discoveries and dilemmas concerning cytokinin metabolism. *Journal of Plant Growth Regulation* 21: 24-31.

Banowetz, G.M., K. Ammar, and D.D. Chen (1999). Postanthesis temperatures influence cytokinin accumulation in wheat kernal weight. *Plant Cell and Environment* 22: 309-316.

Bassil, N.V., D.W.S. Mok, and M.C. Mok (1993). Partial purification of a *cis-trans*-isomerase of zeatin from immature seed of *Phaseolus vulgaris* L. *Plant Physiology* 102: 867-872.

Bilyeu, K.D., J.L. Cole, J.G. Laskey, W.R. Riekhof, T.J. Esparza, M.D. Kramer, and R.O. Morris (2001). Molecular and biochemical characterization of a cytokinin oxidase from maize. *Plant Physiology* 125: 378-386.

Blackwell, J.R. and R. Horgan (1994). Cytokinin biosynthesis by extracts of *Zea mays*. *Phytochemistry* 35: 339-342.

Borisjuk, L., S. Walenta, H. Weber, W. Mueller-Klieser, and U. Wobus (1998). High resolution histographical mapping of glucose concentrations in developing cotyledons of *Vicia faba* in relation to mitotic activity and storage processes: Glucose as a possible developmental trigger. *The Plant Journal* 15: 583-591.

Brenner, M.L. and N. Cheikh (1995). The role of hormones in photosynthate partitioning and seed filling. In P.J. Davies (ed.), *Plant hormones: Physiology biochemistry and molecular biology,* Second edition (pp. 649-670). Dordrecht, Netherlands: Kluwer Academic Press.

Buerkle, L., A. Cedzich, C. Doepke, H. Stransky, S. Okumoto, B. Gillissen, C. Huehn, and W.B. Frommer (2003). Transport of cytokinins mediated by purine transporters of the PUP family expressed in phloem, hydathodes, and pollen of *Arabidopsis*. *The Plant Journal* 34: 13-26.

Burch, L.R. and R. Horgan (1989). The purification of cytokinin oxidase from *Zea mays* kernals. *Phytochemistry* 28: 1313-1319.

Caplin, S.M. and F.C. Steward (1948). Effect of coconut milk on the growth of the explants from carrot root. *Science* 108: 655-657.

Chen, C.-M. (1997). Cytokinin biosynthesis and interconversion. *Physiologia Plantarum* 101: 665-673.

Chiekh, N.C. and R.J. Jones (1994). Disruption of kernal growth and development by heat stress: Role of cytokinins/ABA balance. *Plant Physiology* 106: 45-51.

Clifford, P.E., C.E. Offler, and J.W. Patrick (1986). Growth regulators have rapid effects on photosynthate unloading from seed coats of *Phaseolus vulgaris* L. *Plant Physiology* 80: 635-637.

Davey, J.E. and J. van Staden (1979). Cytokinin activity in *Lupinus albus:* IV. Distribution in seeds. *Plant Physiology* 63: 873-877.

Davies, D.R. (1977). DNA contents and cell number in relation to seed size in the genus *Vicia. Heredity* 39:153-163.

Dietrich, J.T., M. Kaminek, D.G. Blevins, T.M. Reinbott, and R.O. Morris (1995). Changes in cytokinins and cytokinin oxidase activity in developing maize kernals and the effects of exogenous cytokinin on kernal development. *Plant Physiology and Biochemistry* 33: 327-336.

Dobrev, P., V. Motyka, A. Gaudinova, J. Malbeck, A. Travnickova, M. Kaminek, and R. Vankova (2002). Transient accumulation of *cis*- and *trans*-zeatin type cytokinins and its relation to cytokinin oxidase activity during cell cycle of synchronized tobacco BY-2 cells. *Plant Physiology and Biochemistry* 40: 333-337.

Dybing, C.D., H. Ghiasi, and C. Paech (1986). Biochemical characterization of soybean ovary growth from anthesis to abscission of aborting ovaries. *Plant Physiology* 81: 1069-1074.

Ehneß, R. and T. Roitsch (1997). Co-ordinated induction of mRNAs for extracellular invertase and a glucose transporter in *Chenopodium rubrum* by cytokinins. *The Plant Journal* 11: 539-548.

Emery, R.J.N. and C.A. Atkins (2002). Roots and cytokinins. In Y. Waisel, A. Eshel, and U. Kafkafi (eds.), *Plant roots: The hidden half,* Third edition (pp. 417-433). New York: Marcel Decker.

Emery, R.J.N., L. Leport, J.E. Barton, N.C. Turner, and C.A. Atkins (1998). *cis*-isomers of cytokinins predominate *Cicer arietinum* throughout their development. *Plant Physiology* 117: 1515-1523.

Emery, R.J.N., Q. Ma, and C.A. Atkins (2000). The forms and sources of cytokinins in developing *Lupinus albus* seeds and fruits. *Plant Physiology* 123: 1-12.

Faure, J.D., P. Vittorioso, V. Santoni, V. Fraisier, E. Prinsen, I. Barlier, H. Van Onckelen, M. Caboche, and C. Bellini (1998). The PASTICCINO genes of *Arabidopsis thaliana* are involved in the control of cell division and differentiation. *Development* 125: 909-918.

Galuszka, P., I. Frebort, M. Sebela, and P. Pec (2000). Degradation of cytokinins by cytokinin oxidases in plants. *Plant Growth Regulation* 32: 315-327.

Galuszka, P., I. Frebort, M. Sebela, P. Sauer, S. Jacobsen, and P. Pec (2001). Cytokinin oxidase or dehydrogenase? Mechanism of cytokinin degradation in cereals. *European Journal of Biochemistry* 268: 450-461.

Haberer, G. and J.J. Kieber (2002). Cytokinins: New insights into a classic phytohormone. *Plant Physiology* 128: 354-362.

Harding, S.A. and A.C. Smigocki (1994). Cytokinins modulate stress response genes in isopentenyl transferase-transformed *Nicotiana plumbaginifolia* plants. *Physiologia Plantarum* 90: 327-333.

Herbers, K. and U. Sonnewald (1998). Molecular determinants of sink strength. *Current Opinion in Plant Biology* 1: 207-216.

Holland, M.A. and J.C. Polacco (1994). PPFMs and other covert contaminants: Is there more to plant physiology than just plant? *Annual Review of Plant Physiology and Plant Molecular Biology* 45: 197-209.

Houba-Herin, N., C. Pethe, J. d'Alayer, and M. Laloue (1999). Cytokinin oxidase from *Zea mays*: Purification, cDNA cloning and expression in moss protoplasts. *The Plant Journal* 17: 615-626.

Hutchison, C.E. and J.J. Kieber (2002). Cytokinin signaling in *Arabidopsis. The Plant Cell* 14: S47-S59.

Hwang, I., H.-C. Chen, and J. Sheen (2002). Two-component signal transduction pathways in *Arabidopsis. Plant Physiology* 129: 500-515.

Inoue, T., M. Higuchi, Y. Hashimoto, M. Seki, T. Kobayashi, S. Kato, S. Tabata, K. Shinozaki, and T. Kakimoto (2001). Identification of CRE1 as a ctyokinin receptor from *Arabidopsis. Nature* 409: 1060-1063.

Jones, R.J., B.M. Schreiber, K. McNeill, and M.L. Brenner (1992). Cytokinin levels and oxidase activity during maize kernal development In M. Kaminek, D.W.S. Mok, and E. Zazimova (eds.), *Physiology and biochemistry of cytokinins in plants* (pp. 235-239). The Hague, the Netherlands: SBP Academic Publishing.

Kakimoto, T. (2001). Identification of plant cytokinin biosynthetic enzymes as dimethylallyl diphosphate:ATP/ADP isopentenyltranferases. *Plant and Cell Physiology* 42: 677-685.

Kaminek, M. (1982). Mechanisms preventing the interference of tRNA cytokinins in hormonal regulation. In P.F. Wareing (ed.), *Plant growth substances* (pp. 215-224). New York: Academic Press.

Kaminek, M., M. Trckova, J.E. Fox, and A. Gaudinova (2003). Comparison of cytokinin-binding proteins from wheat and oat grains. *Physiologia Plantarum* 117: 453-458.

Kaminek, M., M. Trckova, V. Motyka, and A. Gaudinova (1994). Role of cytokinins in control of wheat grain development and utilization of nutrients. *Biologia Plantarum* 36: 135.

Kuraishi, S., K. Tasaki, N. Sakuri, and K. Sadatoku (1991). Changes in levels of cytokinins in etiolated squash seedlings after illumination. *Plant and Cell Physiology* 32: 585-591.

Laloue, M. and J.E. Fox (1989). Cytokinin oxidase from wheat: Partial purification and general properties. *Plant Physiology* 90: 899-906.

Letham, D.S. (1963). Zeatin, a factor inducing cell division from *Zea mays*. *Life Sciences* 8: 569-573.

Letham, D.S. (1994). Cytokinins as phytohormones—Sites of biosynthesis, translocation and function of translocated cytokinin. In D.W.S. Mok and M.C. Mok (eds.), *Cytokinins chemistry, activity and function* (pp. 57-80). Boca Raton, FL: CRC Press.

Lewis, D.H., G.K. Burge, M.E. Hopping, and P.E. Jameson (1996). Cytokinins and fruit development in the kiwifruit *(Actinidia deliciosa)*: II. Effects of reduced pollination and CPPU application. *Physiologia Plantarum* 98: 187-195.

Ma, Q.F., N. Longnecker, and C. Atkins (1998). Exogenous cytokinin and nitrogen do not increase grain yield in narrow-leafed lupins. *Crop Science* 38: 717-721.

Ma, Q.-H., Z.-B. Lin, and D.-Z. Fu (2002). Increased seed cytokinin levels in transgenic tobacco influence embryo and seedling development. *Functional Plant Biology* 29: 1107-1113.

Ma, Q.-H., R. Zhang, C.H. Hocart, D.S. Letham, and T.J. Higgins (1998). Seed-specific expression of isopentenyl transferase gene *(ipt)* in transgenic tobacco. *Australian Journal of Plant Physiology* 25: 53-59.

Mader, J.C., C.G.N. Turnbull, and R.J.N. Emery (2003). Spatial and temporal changes in multiple hormone groups during lateral bud release shortly following apex decapitation of chickpea (*Cicer arietinum* L.) seedlings. *Physiologia Plantarum* 119: 295-308.

Martin, R.C., M.C. Mok, J.E. Habben, and D.W.S. Mok (2001). A maize cytokinin gene encoding an O-glucosyltransferase specific to *cis*-zeatin. *Proceedings of the National Academy of Science (USA)* 98: 5922-5926.

Martin, R.C., M.C. Mok, and D.W.S. Mok (1999a). A gene encoding the cytokinin enzyme zeatin O-xylosyltransferase of *Phaseolus vulgaris*. *Plant Physiology* 120: 553-557.

Martin, R.C., M.C. Mok, and D.W.S. Mok (1999b). Isolation of a cytokinin gene, ZOG1, encoding zeatin O-glucosyltransferase from *Phaseolus lunatus*. *Proceedings of the National Academy of Science (USA)* 96: 284-289.

McDonald, R., S. Fieuw, and J.W. Patrick (1996). Sugar uptake by the dermal transfer cells of developing cotyledons of *Vicia faba* L.: Experimental systems and general transport properties. *Planta* 198: 54-63.

McGaw, B.A. and R. Horgan (1983). Cytokinin oxidase from *Zea mays* kernals and *Vinca rosea* crown gall tissue. *Planta* 159: 30-37.

Medford, J.I., R. Horgan, Z. El-Sawi, and H.J. Klee (1989). Alterations of endogenous cytokinins in transgenic plants using a chimeric isopentenyl transferase gene. *The Plant Cell* 1: 403-413.

Mok, D.W.S. and M.C. Mok (2001). Cytokinin metabolism and action. *Annual Review of Plant Physiology and Plant Molecular Biology* 52: 89-118.

Mok, M.C., R.C. Martin, D.W.S. Mok, and G. Shaw (1992). Cytokinin activity, metabolism and function in *Phaseolus*. In M. Kaminek, D.W.S. Mok, and E. Zazimova (eds.), *Physiology and biochemistry of cytokinins in plants* (pp. 41-46). The Hague, the Netherlands: SBP Academic Publishing.

Morris, R.O. (1997). Hormonal regulation of seed development. In B.A. Larkins and I.K. Vasil (eds.), *Cellular and molecular biology of plant seed development* (pp. 117-149). Dordrecht, Netherlands: Kluwer Academic.

Morris, R.O., K.D. Bilyeu, J.G. Laskey, and N.N. Cheikh (1999). Isolation of a gene encoding a glycosylated cytokinin oxidase from maize. *Biochemical and Biophysical Research Communications* 255: 328-333.

Munier-Jolain, N.G. and B. Ney (1998). Seed growth rate in grain legumes: II. Seed growth rate depends on cotyledon cell number. *Journal of Experimental Botany* 49: 1963-1969.

Nagel, L., R. Brewster, W.E. Riedell, and R.N. Reese (2001). Cytokinin regulation of flower and pod set in soybeans (*Glycine max* (L.) Merr). *Annals of Botany* 88: 27-31.

Nandi, S.K. and L.M.S. Palni (1997). Metbolism of zeatin riboside in a hormone autonomous genetic tumour line of tobacco. *Plant Growth Regulation* 23: 159-166.

Paces, V., E. Werstiuk, and R.H. Hall (1971). Conversion of N6-isopentenyladenosine to adenosine by enzyme activity in tobacco tissue. *Plant Physiology* 48: 775-778.

Palta, J.A. and C. Ludwig (1997). Pod set and seed yield as affected by cytokinin application and terminal drought in narrow-leafed lupin. *Australian Journal of Agricultural Research* 48: 81-90.

Palta, J.A. and C. Ludwig (1998). Yield response of narrow-leafed lupin plants to variations in pod number. *Australian Journal of Agricultural Research* 49: 63-68.

Palta, J.A. and C. Ludwig (2000). Elevated CO_2 during pod filling increased seed yield but not harvest index in indeterminate narrow-leafed lupin. *Australian Journal of Agricultural Research* 51: 279-286.

Pate, J.S., M.B. Peoples, A.J.E. van Bel, J. Kuo, and C.A. Atkins (1985). Diurnal water balance of the cowpea fruit. *Plant Physiology* 77: 148-156.

Peoples, M.B., J.S. Pate, C.A. Atkins, and D.R. Murray (1985). Economy of water, carbon and nitrogen in the developing cowpea fruit. *Plant Physiology* 77: 142-147.

Piagessi, A., P. Picciarelli, N. Ceccarelli, and R. Lorenzi (1997). Cytokinin biosynthesis in endosperm of *Sechium edule. Plant Science* 129: 131-140.

Riou-Khamlichi, C., R. Huntley, A. Jacqmard, and J.A.H. Murray (1999). Cytokinin activation of *Arabidopsis* cell division through a D-type cyclin. *Science* 283: 1541-1544.

Roeckel, P., T. Oancia, and J.R. Drevet (1997). Effects of seed-specific expression of a cytokinin biosynthetic gene on canola and tobacco phenotypes. *Transgenic Research* 6: 133-141.

Roeckel, P., T. Oancia, and J.R. Drevet (1998). Phenotypic alterations and component analysis of seed yield in transgenic *Brassica napus* plants expressing the *tzs* gene. *Physiologia Plantarum* 102: 243-249.

Rolland, F., B. Moore, and J. Sheen (2002). Sugar sensing and signaling in plants. *The Plant Cell* 14: S185-S205.

Sakakibara, H. and K. Takei (2002). Identification of cytokinin biosynthesis genes in *Arabidopsis*: A breakthrough for understanding the metabolic pathway and the regulation in higher plants. *Journal of Plant Growth Regulation* 21: 17-23.

Schmulling, T. (2002). New insights into the functions of cytokinins in plant development. *Journal of Plant Growth Regulation* 21: 40-49.

Schmulling, T., S. Beinsberger, J. De Greef, J. Schell, H. Van Onckelen, and A. Spena (1989). Construction of a heat inducible chimaeric gene to increase the cytokinin content in transgenic plant tissue *FEBS Letters* 249: 401-460.

Seymour, M. (1996). Response of narrow leafed lupin (*Lupinus angustifolius* L.) to foliar application of growth regulators in Western Australia. *Australian Journal of Agricultural Research* 36: 473-478.

Singh, S., D.W.S. Letham, P.E. Jameson, R. Zhang, C.W. Parker, J. Badenoch-Jones, and L.D. Nooden (1988). Cytokinin biochemistry in relation to leaf senescence: IV. Cytokinin metabolism in soybean explants. *Plant Physiology* 88: 788-794.

Smalle, J., J. Kurepa,. P.Z. Yang, E. Babiychuk, S. Kushnir, A. Durski, and R.D. Vierstra (2002). Cytokinin growth responses in *Arabidopsis* involve the 26S proteasome subunit RPN12. *The Plant Cell* 14: 17-32.

Smigocki, A.C. (1991). Cytokinin content and tissue distribution in plants transformed by a reconstructed isopentenyl transferase gene. *Plant Molecular Biology* 16: 105-115.

Smigocki, A., J.W.J. Neal, I. McCanna, and L. Douglass (1993). Cytokinin-mediated insect resistance in *Nicotiana* plants transformed with the *ipt* gene. *Plant Molecular Biology* 23: 325-335.

Strnad, M. (1997). The aromatic cytokinins. *Physiologia Plantarum* 101: 674-688.

Suttle, J.C. and G.M. Banowetz (2000). Changes in *cis*-zeatin and *cis*-zeatin riboside levels and biological activity during potato tuber dormancy. *Physiologia Plantarum* 109: 68-74.

Swain, S.M., J.B. Reid, and Y. Kamiya (1997). Gibberellins are required for embryo growth and seed development in pea. *The Plant Journal* 12: 1329-1338.

Swain, S.M., J.J. Ross, J.B. Reid, and Y. Kamiya (1995). Gibberellins and pea seed development. Expression of the *lh,i ls* and le^{5839} mutations. *Planta* 195: 426-433.

Takagi, M., T. Yokota, N. Murofushi, Y. Ota, and N. Takahashi (1985). Fluctuation of endogenous cytokinin contents in rice during its life cycle—Quantification of cytokinins by selected ion monitoring using deuterium-labeled internal standards. *Agricultural and Biological Chemistry* 49: 3271-3277.

Takei, K., H. Sakakibara, and T. Sugiyama (2001). Identification of genes encoding adenylate isopentenyltransferase, a cytokinin biosynthesis enzyme, in *Arabidopsis thaliana. Journal of Biological Chemistry* 276: 26405-26410.

Tarkowska, D., K. Dolezal, P. Tarkowski, C. Astot, J. Holub, K. Fuksova, T. Schmulling, G. Sandberg, and M. Strnad (2003). Identification of new aromatic cytokinins in *Arabidopsis thaliana* and *Populus* × *canadensis* leaves by LC-(+)ESI-MS and capillary liquid chromatography/frit-fast atom bombardment mass spectrometry. *Physiologia Plantarum* 117: 579-590.

Thomas, J.C., A.C. Smigocki, and H.J. Bohnert (1995). Light-induced expression of ipt from *Agrobacterium tumefaciens* results in cytokinin accumulation and osmotic stress symptoms in transgenic tobacco. *Plant Molecular Biology* 27: 225-235.

Truernit, E., J. Schmidt, P. Epple, J. Illig, and N. Sauer (1996). The sink-specific and stress-regulated *Arabidopsis* STP4 gene: Enhanced expression of a gene encoding a monosaccharide transporter by wounding elicitors and pathogen challenge. *The Plant Cell* 8: 2169-2182.

Turner, J.E., M.C. Mok, and D.W.S. Mok (1985). Zeatin metabolism in fruits of *Phaseolus:* Comparison between embryos, seed coat, and pod tissues. *Plant Physiology* 79: 321-322.

Van Rhijn, J.A., H.H. Heskamp, E. Davelaar, W. Jordi, M.S. Leloux, and U.A.Th. Brinkman (2001). Quantitative determination of glycosylated and aglycon isoprenoid cytokinins at sub-picomolar levels by microcolumn liquid chromatography combined with electrospray tandem mass spectrometry. *Journal of Chromatography A* 929: 31-42.

Wang, J., D.S. Letham, E. Cornish, K. Wie, C.H. Hocart, M. Michael, and K.R. Stevenson (1997). Studies of cytokinin action and metabolism using tobacco plants expressing either the *ipt* or the GUS gene controlled by a chalcone synthase promoter: II. *ipt* and GUS gene expression, cytokinin levels and metabolism. *Australian Journal of Plant Physiology* 24: 673-683.

Wang, T.L. and V.M. Sponsel (1985). Pea-fruit development—a role for plant hormones. In P.D. Hebblewaithe, M.C. Heath, and T.L.K. Dawkins (eds.), *The pea crop: A basis for improvement* (pp. 339-348). London, England: Butterworths.

Weber, H., L. Borisjuk, and U. Wobus (1996). Controlling seed development and seed size in *Vicia faba*: A role for seed coat-associated invertases and carbohydrate. *The Plant Journal* 10: 823-834.

Weber, H., L. Borisjuk, and U. Wobus (1997). Sugar import and metabolism during seed development. *Trends in Plant Science* 2: 169-174.

Weber, H., L. Borisjuk, and U. Wobus (1998). Controlling development and seed size in *Vicia faba*: A role for seed coat-associated invertases and carbohydrate status. *The Plant Journal* 10: 823-834.

Weber, H., U. Heim, S. Golombek, L. Borisjuk, and U. Wobus (1998). Assimilate uptake and the regulation of seed development. *Seed Science Research* 8: 331-345.

Werner, T., V. Motyka, M. Strnad, and T. Schmulling (2001). Regulation of plant growth by cytokinin. *Proceedings of the National Academy of Sciences of the United States of America* 98: 10487-10492.

Yamada, H., T. Suzuki, K. Terada, K. Takei, K. Ishikawa, K. Miwa, T. Yamashino, and T. Mizuno (2001). The *Arabidopsis* AHK4 histidine kinase is a cytokinin-binding receptor that transduces cytokinin signals across the membrane. *Plant and Cell Physiology* 42: 1017-1023.

Yang, J., S. Peng, R.M. Visperas, A.L. Sanico, Q. Zhu, and S. Gu (2000). Grain filling pattern and cytokinin content in the grains and roots of rice plants. *Plant Growth Regulation* 30: 261-270.

Yang, J., J. Zhang, Z. Wang, Q. Zhu, and W. Wang (2001). Hormonal changes in the grains of rice subjected to water stress during grain filling. *Plant Physiology* 127: 315-323.

Yang, Z., J.E. Aidi, T. Ait-Ali, C. Augur, G. Teller, F. Schoentgen, R. Durand, and B. Durand (1998). Sex-specific marker and trans-zeatin ribosidase in female annual Mercury. *Plant Science* 139: 93-103.

Yu, J.Q., Y. Li, Y.R. Qian, and Z.J. Zhu (2001). Cell division and cell enlargement in fruit of *Lagenaria leucantha* as influenced by pollination and plant growth substances. *Plant Growth Regulation* 33: 117-122.

Zhang, K., D.S. Letham, and P.C. John (1996). Cytokinin controls the cell cycle at mitosis by stimulating the tyrosine dephosphorylation and activation of $p34^{cdc2}$ like H1 histone kinase. *Planta* 200: 2-12.

Zhang, R. and D.S. Letham (1990). Cytokinin translocation and metabolism in lupin species: III. Translocation of xylem cytokinin into the seeds of lateral shoots of *Lupinus angustifolius*. *Plant Science* 70: 65-71.

Zubko, E., C.J. Adams, I. Machaekova, J. Malbeck, C. Scollan, and P. Meyer (2002). Activation tagging identifies a gene from *Petunia hybrida* responsible for the production of active cytokinins in plants. *The Plant Journal* 29: 797-808.

Chapter 4

Grain Number Determination in Major Grain Crops

Gustavo A. Slafer
Fernanda G. González
Adriana G. Kantolic
Elena M. Whitechurch
L. Gabriela Abeledo
Daniel J. Miralles
Roxana Savin

INTRODUCTION

Yield is frequently the most important attribute for designing strategies for both breeding and management of such major grain crops as cereals, pulses, and oilseed crops. Despite the large visible differences among grain crops in both morphological and physiological attributes, in most cases yield is strongly related to the number of grains per unit land area (Slafer, 1994; Egli, 1998; Abeledo et al., 2003). In other words, it seems universal that to manipulate yield the number of grains per m^2 must be manipulated, either genetically through breeding or environmentally through management. This general assertion applies even when the final average weight of the individual grains has been reported to be negatively correlated to the number of grains per m^2 (e.g., Calderini et al., 1999). It is implicit here that this negative relationship is not entirely due to competition among growing grains for assimilates, a fact clearly documented, at least for wheat (Slafer and Savin, 1994; Miralles and Slafer, 1995; Kruk et al., 1997). In this crop the capacity of the canopy to provide assimilates to the growing grains is more than adequate to allow the grains to fill completely in most situations (Savin and Slafer, 1991; Richards, 1996). Although different crops may differ in the degree in which grain growth is limited by the availability of resources, cases of mutually exclusive competition among grains are rather exceptional (Borrás, 2002), which is expected from the strong relationship almost unequivocally found for any major grain crop between yield and number of grains per m^2.

Thus, in order to understand yield (a prerequisite to functionally manipulate it) the determination of the number of grains per unit land area has to be understood. In this chapter, we discuss the two main approaches to understand grain number per m^2 in major crops. First, the traditional and most frequently used approach of dividing grain number per m^2 into simpler numerical components is described and its weaknesses discussed, illustrating the analysis with two contrasting crops: a temperate monocot cereal grass (wheat) and a subtropical dicot oilseed legume (soybean); however, the analysis discussed is valid for any grain crop. Then, an alternative and slightly more complex approach of understanding number of grains per m^2 as the consequence of growth and partitioning during critical periods, which avoids the inconveniences of the traditional numerical components approach, is discussed. For this approach, we illustrate the analysis with wheat in some detail but then provide evidence that the main criteria supporting the analysis are applicable to many other grain crops.

GRAIN NUMBER DETERMINATION

Numerical Components

It is traditional in physiological studies to analyze the number of grains per m^2 in terms of its numerical components, probably due to its implicit simplicity. This approach may slightly differ between crops depending on their reproductive structure (whether the crop has branches-tillers or is uniculm, whether the inflorescences are determined by a terminal structure or not, and so on). The principle is to divide the number of grains per m^2 into simpler components, and consequently the number of grains per unit land area is the result of multiplying these components.

For instance, in the case of wheat, number of grains may be considered to be the product of plants per unit land area, spikes per plant, spikelets per spike, and grains per spikelet (Figure 4.1, left panel). In the case of soybean it may be plants per unit land area, nodes per plant, pods per node, and grains per pod (Figure 4.1, right panel). Clearly, the components into which the number of grains per m^2 may be divided can be organized in a slightly different way from that illustrated in Figure 4.1, but the principles, advantages, and weaknesses would not vary from what is herein discussed.

Although the decomposition of grain number per unit land area into its numerical components is doubtlessly correct from a logic point of view (the product of multiplying each simple component by the others unequivocally results in the number of grains per m^2), it has a major problem that makes it unsound from a physiological point of view, at least for predicting the effect of manipulating a simple component on the final number of grains per m^2

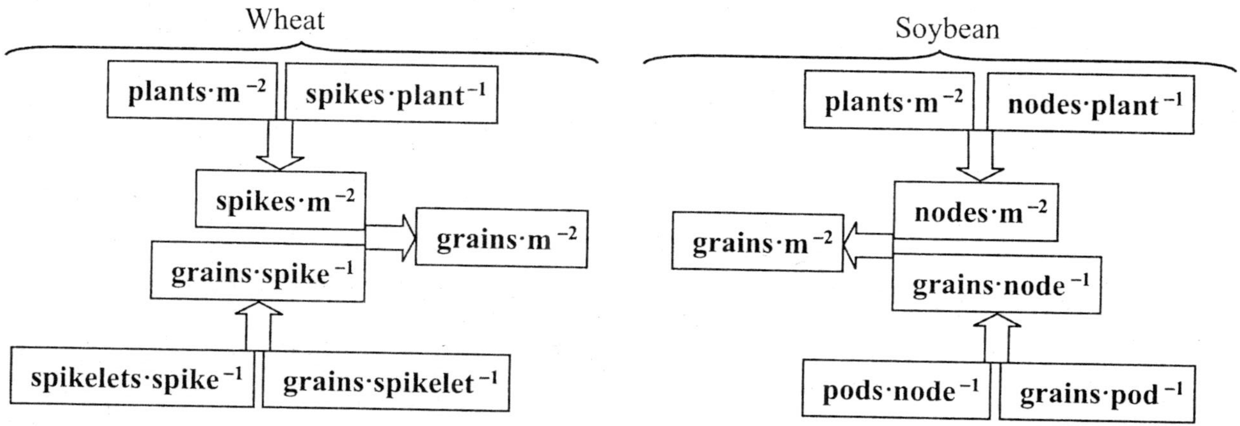

FIGURE 4.1. Scheme of the numerical components of a grain crop, illustrated for the case of wheat (left panel) and soybean (right panel).

(e.g., Fischer, 1984; Slafer et al., 1996). The inconvenience is that these yield components are, almost invariably, negatively related to one another (e.g., Slafer, 2003). For instance the number of spikes or nodes per plant is negatively related to the number of plants per m^2, and the number of spikelets per spike or pods per node is negatively related to the average number of grains per spikelet or pod. Similarly, at the next level of complexity, the number of spikes per m^2 is negatively related to the average number of grains per spike in wheat as well as the number of nodes per m^2 is negatively related to the average number of grains per node in soybean. Consequently, although the mathematical logic of the analysis is beyond any questioning, its usefulness in predictive analysis of the hypothetical effect of manipulating one component on the final number of grains per m^2 is, in the best of cases, rather speculative.

The consistently negative relationships among the components of number of grains per m^2 may be due, at least in part, to the fact that their determination through crop growth and development is to a great extent overlapped (Figure 4.2), and then might be attributed to feedback processes determining a sort of compensation between them. A simplified schematic diagram of developmental progress on an arbitrary time scale (the timing of these stages strongly depends on the genotype × temperature × photoperiod interactions; Hadley et al., 1984; Slafer and Rawson, 1994; García del Moral et al., 2002) is presented in Figure 4.2, where the overlapping between the processes determining the level of each component of the final number of grains per m^2 is evidenced.

Therefore, despite its simplicity and logic, the decomposition of the number of grains per m^2 into numerical components is of little value for any predictive use, and an alternative approach has to be used.

Crop Growth and Partitioning During a Critical Phase

The main weakness of the numerical component approach to understand grain number per m^2 dividing it into simpler attributes is based on the unpredictability of the outcome of modifying (through breeding or management) any of the simplest components on the final number of grains per unit land area. Any alternative, necessarily less simple, must solve this inconvenience, identifying attributes that while being actually simpler than the number of grains per m^2, straightforwardly determine it.

It is clear that the number of grains per m^2 that can be found at the end of the growing season is the final output of the establishment of a potential

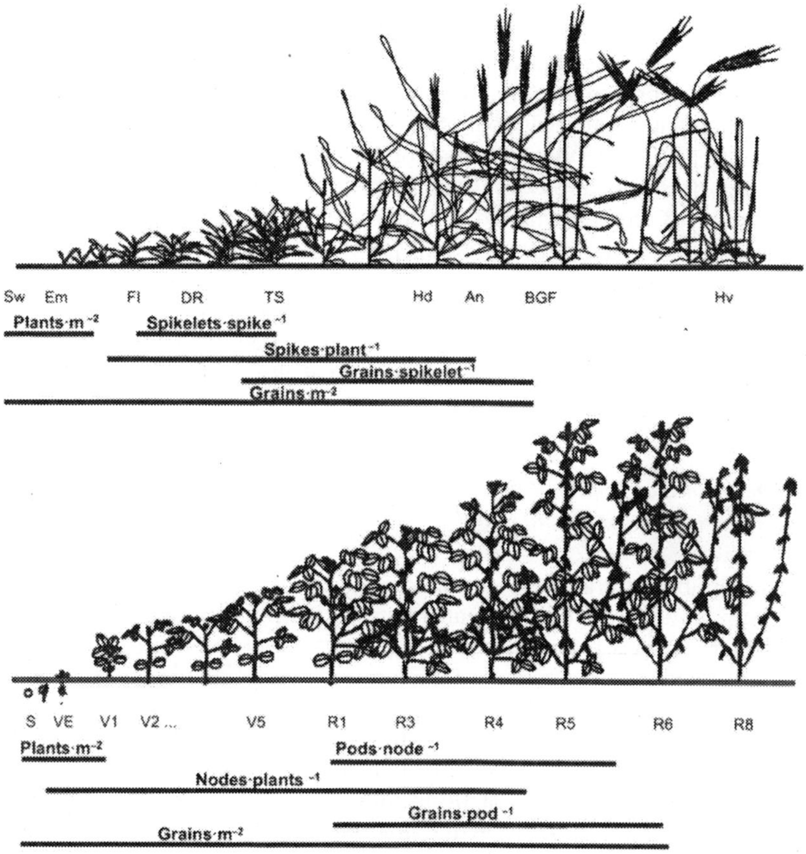

FIGURE 4.2. Schematic diagram of wheat (top panel) and soybean (bottom panel) growth and development with the approximate timing when different components of grain number per m^2 are produced. Some stages of development are indicated: sowing (Sw), emergence (Em), initiation of the first double ridge (DR), terminal spikelet initiation (TS), heading (Hd), anthesis (An), beginning of the grain filling period (BGF), physiological maturity (PM), and harvest (Hv) for wheat. In soybean, stages correspond to those described by Fehr and Caviness (1977). (Source: Adapted for wheat from Slafer and Rawson, 1994, and Miralles and Slafer, 1999, and for soybean from Kantolic et al., 2001.)

number of structures, were all these structures to survive, being several times greater than the maximum number of grains that are finally set and grow during grain filling. The most successful approach to understand the physiological bases of grain number per m^2 determination has been

1. to analyze the dynamics of generation and degeneration of structures that may finally render grains, if growth and development of each structure proceeded without any limitation, and
2. to find out whether number of grains per m^2 does exhibit any differential sensitivity to the generation or degeneration processes, and then to identify a critical process, which may in fact be related to a particular timing of crop growth and development: the critical period for the determination of number of grains per m^2.

For the sake of simplicity, and taking advantage of the fact that it is one of the most widely studied crop species in this type of yield analysis, we will first discuss these points in detail for wheat and then will show how the main conclusions are extrapolated to other major crops.

Dynamics of Development of Structures Finally Determining Grain Number in Wheat

The number of grains per unit land area is being formed throughout the whole preanthesis period in wheat (Figure 4.2, top panel) as a consequence of a rather complex process through which the structures bearing grains are first generated and then a rather large proportion of the initiated structures degenerate, so that only a fraction of the total number of structures produced by the crop actually end up setting grains. The issue is illustrated in Figure 4.3, where the processes of tiller initiation and tiller death are described together with those of floret initiation and floret death and compared with stem and spike growth dynamics.

Cereals develop the capacity to produce a tiller in each phytomer. The process of emergence and growth of tillers, termed tillering, starts when the first tiller-bud is mature to grow, approximately when the third leaf is expanded (Miralles and Slafer, 1999). From then on, the dynamics of tiller emergence follows closely that of leaf emergence (Masle, 1985; Porter, 1985; Miralles and Slafer, 1999), in absence of restrictions of assimilates availability. As tiller phytromers also have tiller buds, tillers of higher order may appear, i.e., tillers of tillers. Thus, the overall pattern of tillering seems exponential (Masle, 1985), while resources for tiller growth are not severely restricted, which is commonly the case in early stages of development (Figure 4.3). Wheat crops normally experience a limited offer of resources after the initial part of the growing season, and therefore actual tillering becomes increasingly slower than the potential, until there are not enough resources to maintain growth of all tillers and some die, in the reverse order they had appeared (contributing to the synchrony and convergence of development in a

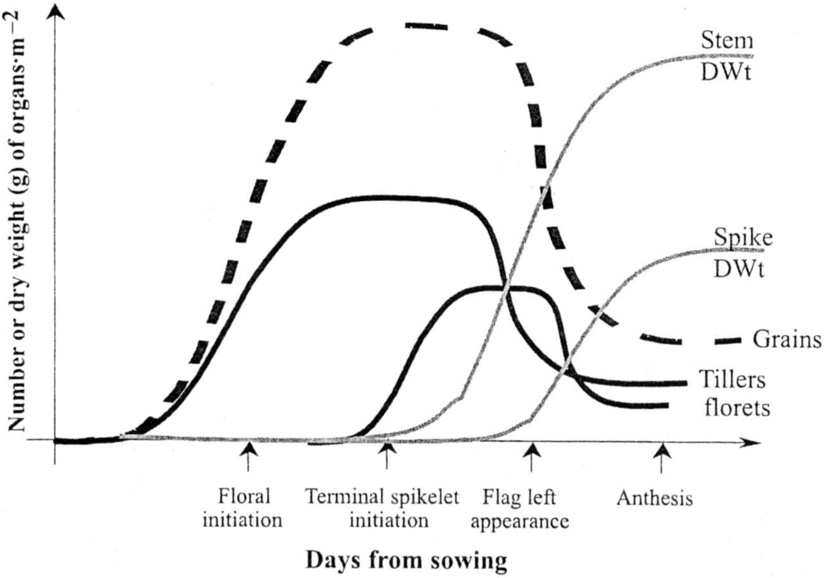

FIGURE 4.3. Diagram of the dynamics of tillering and tiller death, floret initiation and survival, and the product of both (plus the number of spikelets per spike). The figure is drawn from a large body of evidences in the literature that can be seen with detail in the reviews and descriptive works by several authors. (*Sources:* Aitken, 1975; Evans et al., 1975; Waddington et al., 1983; Fischer, 1984, 1985; Gardner et al., 1985; Kirby, 1988; Hay and Kirby, 1991; Sibony and Pinthus, 1988; Slafer et al., 1996; Miralles and Slafer, 1999.)

crop; Hay and Kirby, 1991). In most cases, the onset of tiller mortality coincides with the beginning of stem elongation. This is a consequence of an abrupt change in dry matter partitioning due to the increase in demand for resources of the elongating internodes. After the youngest and smallest tillers die, tiller mortality stops and the number of spikes per unit land area is finally established (Figure 4.3).

Some time during the tillering period, the apex switches from producing leaf primordia to the reproductive stage, initiating spikelet primordia (the timing of this switch varies largely with environment and genotype, and may occur at the beginning, in the middle or at the end of tillering; Miralles and Slafer, 1999). From then to the initiation of the terminal spikelet, all spikelet primordia are developed, at a rate substantially higher than that of leaf initiation (e.g., Delécolle et al., 1989; Kirby, 1990; González et al., 2002). During the spikelet initiation phase (floral initiation—terminal spikelet initiation), all spikelets are developed producing the potential number of structures able to bear florets. But, as each spikelet has an undetermined habit (i.e., there is not a physical limit to the production of floret primordia per spikelet), the process of spikelet initiation is not critical in determining the number of fertile florets per spike. Before the initiation of the terminal spikelet, floret initiation begins in the spikelets of the middle third of the spike (those first initiated; Gardner et al., 1985), a process that then continues toward the top and the bottom spikelets. Within each spikelet florets are intiated first in the basal positions, adjacent to the rachis, and then they progress toward the distal positions (e.g., Sibony and Pinthus, 1988). Floret initiation continues approximately until the appearance of the flag leaf (Kirby, 1988; Miralles and Slafer, 1999), when the maximum number of florets per spikelet (and then per spike) is achieved: a value ranging from about 6 to 12 depending on the spikelet position and growing environment (Sibony and Pinthus, 1988; Youssefian et al., 1992). Then, during the spike growth period, sometimes at the beginning of this phase, many of the initiated florets do not progress rapidly enough to reach the stage of fertile floret (Waddington et al., 1983) at anthesis. Only few (0 to 4 florets per spikelet) of the many floret primordial initiated slightly earlier become fertile florets at anthesis. The period of floret mortality does partly overlap with and continue after that of tiller mortality.

As a consequence of the dynamics of tillers (and then spikes) per m^2 and florets (and then grains) per spike, the number of potential grains (organs that if progressing without restriction may yield a grain) rises from the sowing density, under agronomical conditions, to a maximum that may be as high as 100,000 to 200,000 "potential grains" per m^2 by the end of tillering-beginning of stem elongation, and then decreases mostly during stem elongation from that maximum to a final number of grains that, within the best field conditions may be about 25,000 to 30,000$\cdot m^{-2}$, and in average may be 12,000 to 20,000$\cdot m^{-2}$. The question for agronomists and breeders is how to manipulate the final number of grains per m^2 with simple traits putatively determining it. A key point in this question is whether the final number of grains is more sensitive to changes during the generation or degeneration of structures finally determining it.

Sensitivity of Grain Number to Different Growth Conditions Throughout Development in Wheat

The period of mortality of structures which finally determine number of grains per m^2 matches quite clearly with the period of rapid growth of stems and spikes (Figure 4.3; see Kirby, 1988 for a detailed description of growth of each internode and spike in relation to floret death). This fact led to the hypothesis that the greater the growth of the spikes, the more abundant the availability of assimilates, which may prevent the mortality of these structures and then yield a higher final number of grains per m^2 (Fischer, 1984).

Experiments in which the assimilate supply was made deficient by means of stresses applied in different moments during the developmental cycle of the crop were used here to illustrate the point mentioned. The relationship between solar radiation and the number of grains per m^2 was studied by Fischer (1985). In that study wheat plants were subjected to artificial shading treatments in different stages. Grain number was closely related to the incident solar radiation during the 30 days preceding anthesis, which coincides with the stem and spike growth period. The radiation deficiency therefore reduced assimilates, which in turn reduced the spike growth and thus the number of potential grains. This period of stem and spike growth, previous to anthesis, is frequently termed the *critical* period. During this period the spike has been reported to compete for assimilates with the other actively growing organs, as the last appeared leaves and elongating internodes (Fischer, 1985; Kirby, 1988; Savin and Slafer, 1991). Moreover, a high correlation has been found experimentally between the spike dry weight at anthesis and the number of grains per m^2 (Fischer, 1985; Thorne and Wood, 1987; Savin and Slafer, 1991; Slafer and Andrade, 1993).

The fact that during the critical period spikes and stems are competing strongly for assimilates enhances the effect of any deficiency or stress on the growth of the spike and thus on the number of grains per m^2. Experiments using different N supply levels during the critical stem and spike growth period have shown that the radiation use efficiency increased with higher N availability during this period (Fischer, 1993; Dreccer et al., 2000), indicating that N fertilization should be applied so as to maximize the cumulative light absorption during the critical period (Dreccer et al., 2000) in order to increase the number of grains per m^2.

The importance of spike growth during preanthesis in grain number determination (and thus in grain yield) can be illustrated by studying the changes in physiological attributes that occurred in wheat breeding for higher yield potential. Many studies around the world were conducted to identify those physiological attributes, and in almost all of them it seems that genetic improvement in wheat yield was achieved through an increased harvest index (proportion of total biomass partitioned to reproductive organs) without modifying the total aboveground biomass produced (see review of Slafer and Andrade, 1991; Calderini et al., 1995). This increased capacity of partitioning assimilates to reproductive organs was evident at anthesis. Modern cultivars have heavier spikes (and lighter stems associated with a reduced height, see Calderini et al., 1995) and thus, higher spike to stem ratio at anthesis than old cultivars (Siddique et al., 1989; Slafer et al., 1990). The overall result of the heavier spikes at anthesis was an increased number of fertile florets and grains per spike that led to higher number of grains per m^2 (Siddique et al., 1989; Slafer and Andrade, 1989; Slafer et al., 1990; Slafer and Miralles, 1993). Another example of the importance of preanthesis partitioning of assimilates to the spikes for wheat yield was the introduction of the dwarf genes (mainly *Rht1* and *Rht2*). These genes reduced the stem height (and thus the ability of the stem to compete for assimilates) and consequently, the partitioning of assimilates to the spike prior to anthesis increased, yielding higher number of fertile florets per spikelet and grains per spike (Siddique et al., 1989; Youssefian et al., 1992; Miralles and Slafer, 1995; Miralles et al., 1998). The increased number of fertile florets at anthesis has been found to be more associated with the survival of floret primordia than with the total number of florets initiated per spike (Siddique et al., 1989; Miralles et al., 1998).

As stated earlier, the stem and spike growth period previous to anthesis has been proven to be of paramount importance in determining the number of grains per m^2. An alternative to increase assimilate acquisition by spikes may be to alter the relative duration of the vegetative and reproductive phases, without modifying the total time to anthesis, which is already optimal in most production areas. Extending the duration of the stem and spike growth phase would increase spike weight at anthesis and thereby the number of grains produced (Slafer et al., 1996). Evidence supporting this idea, was obtained in studies where duration of stem an spike growth phase was modified through artificial manipulation of the photoperiodic conditions exclusively during that phase (Miralles et al., 2000; Slafer et al., 2001; González et al., 2003b). In a phytotron experiment, in which wheat and barley plants were grown in different constant and interchanged photoperiods so as to attain different durations of the spike growth phase, an increased duration of the late reproductive phase from terminal spikelet to anthesis resulted in heavier spikes and more fertile florets per spike (Miralles et al., 2000). In a field experiment, in which wheat and barley plants were sown in different sowing dates, the longer the duration of the stem and spike growth phase, the more fertile florets per spikelet (Miralles et al., 2001). In another field experiment with wheat, in which photoperiod was extended artificially with portable lighting structures, the shorter photoperiod treatments caused longer durations of the stem and spike growth phase, which in turn increased the spike weight at anthesis and the number of fertile florets and grains per spike (González et al., 2003b; Whitechurch et al., unpublished). The increased number of fertile florets due to the heavier spikes at anthesis was associated with a higher survival of floret primordia without changes in the total number of floret primordia differentiated (Miralles et al., 2000; González et al., 2003a).

However, in order to achieve modifications in the duration of phases without changing the time to anthesis, developmental sensitivity of these durations must be independent of each other, at least in part. Halloran and Pennell (1982) suggested the possibility of manipulating the duration of certain developmental phases, independently of others. Furthermore, evidence of likely genetic variation in the duration of the stem and spike growth phase has been reported for both wheat (Kirby et al., 1999) and barley (Kernich et al., 1995). In an extensive screening of Argentine wheat and barley cultivars, Whitechurch et al. (unpublished) found variability in the duration of the spike growth phase within groups of cultivars with a similar duration of the complete cycle to anthesis (see Slafer et al., 2001; Slafer, 2003).

All the examples provided here seem to reinforce the hypothesis presented at the beginning of the section. Wheat yield is highly dependent on the number of florets that survive and are fertile at anthesis, and this survival is mainly determined by the amount of assimilates partitioned to the spikes prior to anthesis. However, this is not the only variable that may modify the number of grains per spike and per m^2 and thus wheat yield. The number of grains produced per unit of spike weight (grains per spike weight) has been shown to be also associated with grain number per m^2 and to vary between cultivars (Abbate et al., 1998). Among other reasons, part of the differences may be related to a different partitioning of assimilates within the spike between the spikelets and the rachis (e.g., Slafer and Andrade, 1993).

EXTRAPOLATIONS TO OTHER MAJOR CROPS

The analysis of the determination of grain number described for wheat is also valid for other grain crops. In the next paragraphs we summarized the physiological basis of grain determination corresponding to maize, rice, soybean, and sunflower.

Maize

As in other cereals, maize grain yield is more closely related to grain number per m^2 than to average grain weight. The number of grains per m^2 is, from a physiological point of view, strongly determined during a critical period of grain set. The mechanisms associated with the determination of grain number are related to morphogenic processes and the physiological conditions of the crop during this critical period, which depends on the duration of the period, the plant growth rate, the partitioning of dry matter to the spikes, and the number of grains fixed per unit spike biomass (e.g., see review by Otegui and Andrade, 2000, and several papers quoted therein).

The definition of the critical period for the determination of grain number was established by experiments inducing water stress (Hall et al., 1982; Grant et al., 1989; Schussler and Westgate, 1991) or shading (Reed et al., 1988; Andrade, Uhart, and Frugone, 1993) as treatments applied during different phenological stages of the crop. At variance with what occurs in wheat, the critical period for grain set in maize, although coinciding with the spike growth period, is also extended toward the lag phase of the grain filling period (Cirilo and Andrade, 1994), determining that the number of grain is affected by environmental conditions well after silking stage, i.e., up to 20 days after silking (Andrade et al., 1996). For instance, evidence has been reported under contrasting environments that spike growth occurs during a period from –227 °C day to +100 °C day around silking (Otegui and Melón, 1997; Otegui and Bonhomme, 1998). Therefore, while the bottleneck for grains setting in wheat is the number of floret primordia that reach the stage of fertile florets at anthesis, in maize fertilization of fertile florets and grain abortion are rather important processes able to impose a restriction to the final number of grains per m^2. Environmental factors (i.e., temperature) may alter the duration of the ear growth period and thereby modify the amount of solar radiation intercepted by the crop during the critical period, altering consequently the strength of the source for growth associated with grain set. In the definition of the number of grains per m^2 in maize two processes must be harmonized: fertile florets must be fertilized and then the grains initially formed have to proceed with growth and development avoiding abortion.

One characteristic of maize is its dichogamous reproductive system, in which receptivity of the stigmas and pollen shedding occur at different times; grains are set in the axillary female inflorescence fertilized with pollen from the apical inflorescence (Ritchie et al., 1993). The most common case in maize hybrids is protandry, with pollen shedding commencing before the stigmas are receptive (though in some cases both events might be synchronous or even in some exceptional combinations of genotypes and environments the opposite pattern—protogyny—may be found). Thus, the dynamics of male and female flowering (anthesis and silking, respectively) may play a decisive role in determining the likelihood of a fertile floret to be fertilized. Delay in the synchrony between anthesis and silking and variations in the rate of silk appearance determine differences in the number of fertile spikelets (Hall et al., 1982; Jacobs and Pearson, 1991; Bolaños and Edmeades, 1993, 1996; Otegui and Melón, 1997; Kiniry and Otegui, 2000). A lack of synchronization in the development of the spikelets along the ear and among ears affects the number of grains that are set per unit of crop growth due to the different time of ovary pollination among structures (Cárcova et al., 2000). However, in most conditions protandry hardly imposes the main restriction to the number of grains per unit land area a maize hybrid produces, the main restriction frequently being imposed through the abortion of fertilized florets (Westgate and Boyer, 1986; Kiniry and Otegui, 2000).

The number of grains established at maturity is sharply related to plant growth rate during this critical period (Tollenaar and Daynard, 1978; Kiniry and Ritchie, 1985; Andrade, Uhart, and Frugone, 1993), which is affected by genotype (Tollenaar et al., 1992) and environmental and management conditions such as temperature and radiation (Muchow et al., 1990; Andrade, Uhart, and Cirilo, 1993; Lafitte and Edmeades, 1997), plant density (Echarte et al., 2000), and water and nitrogen availability (Grant et al., 1989; Uhart and Andrade, 1995, Andrade et al., 2000). The relationship between the number of grains established and the plant growth rate during the critical period follows a curvilinear function, which is characterized by (1) a threshold of plant growth rate below which the grains are not established, (2) the grain set per unit of plant growth rate, and (3) the existence of a superior threshold for establishing a second ear in prolific genotypes (Tollenaar et al., 1992; Andrade et al., 1999; and see also Figure 11.1 in Maddonni et al., 2000).

Partitioning of dry matter between vegetative and reproductive biomass is strongly controlled by the amount of available resources and consequently influenced by management factors such as plant density (Echarte et al., 2000). Genetic improvement in maize yield has been associated with a higher number of grains due to an increase in the number of grains established per unit of plant growth rate (Echarte et al., 2000; Maddonni et al., 2000).

Rice

As was previously described for wheat and maize, increases in rice yield are associated with increases in grain number per m^2 (Peng et al., 1999). The number of spikelets per m^2 is the main determinant of grain yield but, as in wheat, increasing this component is not easy because a strong compensation exists between number of panicle per plant and number of spikelets per panicle (Ying et al., 1998).

Panicle differentiation takes place about 25 days before heading, while the late differentiation stage occurs around 10 days before heading. The maximum number of spikelets per panicle is observed at the late differentiation stage (Sheehy et al., 2001). At anthesis, only a fraction of these differentiated spikelets complete their development and have the capacity of being pollinated, but then only a portion of these pollinated spikelets become a grain at maturity.

Rice differentiates an important number of spikelets primordia per panicle, but only about 40 to 50 percent of them survive. A modern rice genotype may produce about 210 spikelets primordia per panicle, decreasing at maturity to about 140 filled spikelets (Sheehy et al., 2001). Spikelet mortality varies with the hierarchy of the tiller. Thus, although in an early tiller about 65 percent of the initiated spikelet primordia may survive, in a tiller appearing later in the crop cycle only about 50 percent of initiated spikelet primordia may survive. Similar to what occurs in wheat, grain setting efficiency in rice is higher than floret primordia survival, reaching values of about 70 percent of grain setting (Sheehy et al., 2001). Although there are important differences in tillering capacity among cultivars (i.e., indica cultivars have more tillering capacity than japonica cultivars; Ying et al., 1998), the proportion of potential grains that abort is very similar (Sheehy et al., 2001).

Difference in sink size has been strongly dependent upon the capacity of the crop to generate biomass in the period prior to anthesis, in particular during the late reproductive phase of the crop (Horie et al., 1977). The length of the crop cycle and the crop growth rate (Ying et al., 1998) prior to anthesis have been associated with higher yielding capacity. Therefore, potential yield of rice depends on achieving an optimum length of the preflowering period (Yin, Kropff, Aggarwal, et al., 1997), which is controlled genetically by the response of each cultivar to temperature and photoperiod (Yin, Kropff, and Goudriaan, 1997; Yin and Kropff, 1998).

As in wheat, introgression of semidwarf genes in the rice breeding programs contributed significantly to increase potential (McKenzie et al., 1994). In the same way as seen in wheat, dwarfing genes contributed to produce more favorable assimilate acquisition to the panicle, increasing not only harvestable yield by reducing plant height, but also yield potential.

In a way similar to maize, along the panicle of rice a difference in the development among the spikelets and the synchronization of the pollination (Mohapatra and Sahu, 1991; Mohapatra et al., 1993) due to hormonal control along the panicle (Patel and Mohapatra, 1992; Mohapatra et al., 2000) is important in the determination of final number of grains, but without implying an intrinsic restraint on the supply of carbohydrates (Mohapatra and Sahu, 1991; Mohapatra et al., 1993).

Soybean

Seed number is determined in soybean during a very long period, beginning before emergence and ending late in the reproductive period (see Figure 4.2). Although the number of reproductive nodes may be limiting under many circumstances (Board et al., 1999), the potential number of sites where flowers can be developed is extremely high. At each node, soybean plants have at least three axillary buds: the middle one can develop into a primary branch or a main raceme, while the others may (or may not) develop into subracemes or subbranches on reproductive nodes (Borthwick and Parker, 1938; Carlson, 1973; Thomas and Raper, 1977; Gai et al., 1984). Shoot apices and axillary meristems also contribute to flower production, conferring a high plasticity to this crop (Thomas, 1994).

Genotypic response to environmental factors, mainly photoperiod and temperature, regulate the time of flower initiation and the subsequent reproductive development. The extension of the flowering stimulus throughout the entire plant, transforming all axillary and terminal apices into floral primordia, is quite dependent upon the exposure to inductive cycles of photoperiod and temperature (Thomas and Raper, 1977). Management practices such as planting date or cultivar selection have strong influences in regulating timing of reproductive events and plant size and, consequently, also regulate seed number production.

Although the number of seeds primarily depends upon the number of floral primordia that are initiated, a large proportion of the ovaries abort before developing into immature pods (Wiebold et al., 1981). High rates (about 50 to 70 percent) of flower and young pod abortion can occur even in high-yielding environments (Jiang and Egli, 1993) and pod set can be decreased further by stress. Abortion caused by preanthesis water deficits is not attributed to impairment of pollen as in the case of maize, and is probably associated with failures in the structure and/or function of ovules (Kokubun et al., 2001). Most reproductive abortion occurs at early stages of embryo development after fertilization (Peterson et al., 1992; Westgate and Peterson, 1993). Abscission of large pods is generally a consequence of abortion of the seeds developing inside them. Seed abortion does not occur after seeds have reached a certain critical size and their water content has begun to decrease and started to fill at a high rate (Munier-Jolain et al., 1993; Egli and Bruening, 2002a).

Under most circumstances, the availability of flowers does not limit pod or seed production and seed number is ultimately dependent upon the success of these flowers to set pods (Bruening and Egli, 1999, 2000). The causes of flower and pod abortion are not completely understood, but allocation of assimilates to reproductive organs and their utilization by developing pods seem to be the main features that regulate the fate of young reproductive organs (Egli and Bruening, 2002a).

At a raceme level, flowers are initiated sequentially. Progressive development of flowers begins first at the base of the raceme and proceeds acropetaly (Peterson et al., 1992). Even in high-yielding environments, pod set of the proximal positions of a raceme may be high, whereas distal flowers usually fail to produce pods (Kokubun et al., 2001; Peterson et al., 1992). These observations have led to hypotheses suggesting that young distal flowers cannot compete successfully with older proximal flowers for limited supplies of assimilates. However, it is difficult to explain how assimilates could limit the growth of the flowers and small pods that typically have extremely low assimilate requirements (for a general discussion about this topic see Egli and Bruening, 2002a). Moreover, pod and seed set per node fail to respond to very high assimilate availability, suggesting that other processes, in addition to assimilate supply, may be involved in causing a pod or a seed to set (Egli and Bruening 2002a,b).

Understanding the environmental and functional regulation of the fate of a single flower or the mechanisms regulating pod set at a single node level highlights the importance of assimilation availability in regulating seed number but appears insufficient at a plant or crop level. Soybeans have a complex morphological and reproductive organization, and this complexity increases with developmental time. For example, upper nodes of the main stem or distal ones in the branches can be flowering while lower nodes (or proximal ones in the branches) are producing pods with growing seeds. Moreover, at the same single node there may be large fruits together with recently opened flowers. This intra- and inter-nodal complexity cannot be ignored when trying to understand pod set in order to manipulate seed number.

The dynamics of flowering and pod development seem to play an important role in pod set as source/sink balance of assimilates changes during reproductive development (Bruening and Egli, 1999, 2000; Egli and Bruening 2002a). The first flowers and pods to develop may not initially provide a sink large enough to use all the assimilates (Bruening and Egli, 2000), but the assimilates available to flowers and pods developing later may be limited when seeds in the early ones enter the rapid growth phase and consume large amounts of assimilate. The rapid growth of the older structures can trigger abortion of flowers and small pods, and hormones are presumably involved in signals contributing to their failure to produce mature pods (Nagel et al., 2001).

The continuous change of source/sink balance during ontogeny highlights the importance of approaching the understanding of pod and seed set using the "critical period" concept. Abortive structures are present in plants from preflowering stages until the moment seeds of the last appearing pods begin to accumulate dry matter at a high rate. Moreover, increased abortion at one stage of development may be compensated by increased set at another stage. But the possibility of compensation decreases as the crop approaches maturity. As the availability of assimilates is the major cause of pod set and abortion, experiments that impose crop growth limitations by shading or defoliation have led to definition of the critical period for pod and seed set in soybean (Egli and Zhen-wen, 1991; Board and Tan, 1995; Jiang and Egli, 1995). This period starts at flowering and extends through pod set (Jiang and Egli, 1995). Board and Tan (1995) have suggested that this critical period ends 10 to 12 days after the beginning of seed growth, but changes in seed number have also been found to occur in response to shading at the beginning of the full seed stage (Egli, 1997). If the canopy photosynthetic rate during this period is increased by carbon or light enrichment, more seeds are produced (Hardman and Brun, 1971; Schou et al., 1978). If the crop is exposed to optimal environmental conditions during the critical period producing a high photosynthesis and crop growth rate, source limitations are attenuated and seed number is increased (Egli, 1998). Conversely, growth limitations such as those imposed by water stress after flowering reduce pod and seed number (Andriani et al., 1991; Kantolic et al., 1994, 1995), and reductions in individual plant growth rate during the critical period imposed by high populations also result in lower number of seeds produced per plant (Vega et al., 2001).

Seed number is also positively associated with the length of the flowering and pod or seed set period in soybean (Egli and Bruening, 2000; Kantolic and Slafer, 2001). In indeterminate soybeans, exposure to long photoperiods during the critical period promoted node production and increased their fertility, leading to increases in pod and seed number (Kantolic and Slafer, 2001). The mechanisms underlying the relationships between seed number and duration of the critical period are not completely understood, but it is clear that a longer critical period is associated with more intercepted radiation and a high crop growth rate during the period when seed number is being determined.

Duration of the critical period can be modified by manipulating plant responses to the environmental factors controlling development, mainly temperature and photoperiod. Plant responses to photoperiod during postflowering stages are under genetic control (Summerfield et al., 1998; Ellis et al., 2000), and genetic variability in sensitivity to photoperiod during reproductive period has been found to be partially independent of preflowering response (Kantolic and Slafer, 2001). Thus, it is possible to speculate that breeding may further increase seed number by improving the balance between pre- and postflowering phases manipulating photoperiod sensitivity. In this way, lengthening the critical period at the expense of the preflowering period will result in the exposure of the final reproductive stages to high radiation and adequate temperature conditions, optimizing both crop growth rate and duration of the critical period.

Sunflower

Seed number in sunflower is determined during a broad period that includes floret differentiation, floret growth up to anthesis, fertilization, and embryo abortion (Connor and Hall, 1997). The analysis of yield components in sunflower is related to the understanding of seed number determination, as only one capitulum or head is developed by every plant and the number of seeds per plant is negatively correlated to plants per unit land area over a broad range of plant populations (Villalobos et al., 1994). Therefore, efforts in trying to manage seed number determination should include an understanding of seed set at the capitulum and crop levels.

Depending upon genotypic response to temperature and photoperiod, some time after emergence the terminal apex progresses into the reproductive stages (Marc and Palmer, 1976). The upper limit to seed number per plant is set by the number of florets per head that are differentiated during the early floral stages, as primordia of sterile ray florets appear in the periphery of the floral disc until floret primordia cover the entire disc. The number of primordia on the floral disc depends on the number of floret rows (parastichies) initiated, their length, and the degree of amalgamation of rows that occurs toward the center of the disc (Connor and Hall, 1997). The number of florets per head is quite variable (700 to 3,000) and is associated with the duration (the integral across time) of the generative area, an inner disc on the receptacle limited by the front of floret initiation (Palmer and Steer, 1985). Factors affecting duration of the generative area have, consequently, a direct effect over floret differentiation that may limit potential number of seeds from early stages of development.

The integral of generative area is sensitive to assimilation deficits. For example, shading sunflower crops during these early stages reduced the integral of generative area by reducing its maximal size, without affecting the duration of the period of floret generation (Cantagallo and Hall, 2002). These reductions in the generative area were reflected in significant reductions in the total number of florets (Cantagallo and Hall, 2002). Low availability of nitrogen also limits the number of florets by reducing the size of generative area and floret differentiation rate (Hocking and Steer, 1989). The integral of the generative area and potential number of florets were reduced by temperature increments between 14 and 25°C in association with a reduction of the duration of floret differentiation period (Chimenti and Hall, 2001). Although low temperatures (14 to 20°C) during floret differentiation increase the potential number of flowers, maximal number of seeds were attained when crops were exposed to intermediate (20 to 25°C) temperatures. This results from a residual impact of low temperatures during differentiation on the functionality of the flowers during anthesis and fertilization (Chimenti and Hall, 2001).

Once floret primordia have differentiated, they continue an orderly process of development and growth that culminates in anthesis (Connor and Hall, 1997), and this development is sensitive to reductions in assimilates. Shading sunflower crops from the end of floret differentiation onward delayed floret development, particularly in the center of the head (Cantagallo and Hall, 2002). The effects of stresses such as shading, water, or nitrogen deficits are most clearly seen in smaller fertility ratios in the central portion of the capitulum. The physiological basis of the so-called empty center syndrome remains uncertain, but dominance effects, presumably mediated by hormonal signals, of early-established seeds or fruits over later-established ones may be operating (Connor and Hall, 1997).

Brief exposures to drought or shade after anthesis reduce the number of seeds (Hall et al., 1985; Chimenti and Hall, 1992). This reduction can be mediated by loses in flower and pollen viability and embryo setting (Cantagallo and Hall, 2002). The final period for seed number adjustment in sunflower appears to include the early stages of embryo growth (Chimenti and Hall, 1992; Cantagallo and Hall, 2002).

As in the other crops, exposure to brief stresses during the sunflower cycle has been an effective technique for identifying the periods of sensitivity of seed number to environmental changes, the critical period for seed number determination. Chimenti and Hall (1992) found that there is a broad window (from 20 days before to 20 after first anthesis) in which the number of full fruits can be reduced by shading. Cantagallo and Hall (2002) expanded this window including early stages of primordia differentiation. Seed number correlates well with a photothermal quotient [mean radiation/(mean temperature – base temperature)] calculated from 30 days before to 20 days after first anthesis (Cantagallo et al., 1997). An interval from 180°Cd (with a base temperature of 4°C) before and 230°Cd after first anthesis, is used for predicting grain number per plant as a function of plant biomass increase rates, in a crop simulation model that reasonably simulates performance of sunflower over a wide range of conditions (Villalobos et al., 1996).

As other crop species, sunflower produces more florets per capitulum during initiation than are likely to set seeds, and that loss of florets and fertilized ovules is part of a normal adjustment to environmental conditions during the crop cycle (Connor and Hall, 1997). The maximum number established at floret differentiation is unlikely to present a barrier to seed number in most commercial production situations and great emphasis should be put on minimizing their mortality in pre- and postflowering stages. However, and mainly in high-yielding environments where abortion can be minimized, morphogenic constraints to potential seed number should be recognized.

CONCLUDING REMARKS

In this chapter two main approaches were discussed to understand grain number determinations per unit land area in major grain crops: (1) numerical components and (2) crop growth and partitioning during the critical phase. Although numerical component analysis is probably the most common approach due to its simplicity, it is unsound from a physiological point of view for predicting the effects of manipulating a simple component on the final number of grains. The main inconvenience is that subcomponents are generally negatively related.

As in the major crops included in this chapter, the final number of grains per unit area is largely a consequence of the survival of a potential number of structures (usually far greater than the maximum number of grains that may be attained even under stress-free conditions); the alternative approach is based on the crop attributes that determine the dynamics of generation and degeneration of structures that finally determine grains.

Despite the important differences in reproductive structures among major crops described in this chapter, the physiology of grain number is framed in a similar scheme in which it is possible to identify a "critical period" around flowering time during which this main component—grains/seeds per unit land area—is largely determined. Environmental conditions altering the duration (e.g., temperature and photoperiod) and/or crop growth rate (e.g., radiation, water, and nitrogen availability) during that "critical period" will modify the number of grains/seeds that will be set per m^2. Using this alternative approach taking into account the crop growth rate and the assimilate partitioning during the critical period, the number of grains per unit area, using wheat as example, may be redefined as a product of the duration of the spike growth period ($Length_{SE}$), the crop growth rate during that critical period (CGR_{SE}), the proportion of assimilate partitioned between reproductive and vegetative organs (spike:stem ratio), and the fertility factor of the reproductive organs represented in wheat as the number of grains set per spike dry weight at maturity (grains/SpDWt):

$$GN = Length_{SE} \times CGR_{SE} \times \text{Spike to stem ratio} \times \text{Grains/SpDWt}$$

This simple model for the wheat crop may be extrapolated to other major crops, identifying in each particular crop to be analyzed the critical period, defining its duration and the CGR of that period. Because each term in the equation appears to be predominantly independent of the other terms, this alternative model seems to be useful for predictive use and is logical from a physiological viewpoint.

REFERENCES

Abbate, P.E., F.H. Andrade, L. Lázaro, H.G. Berardocco, V.H. Inza, and F. Martuano (1998). Grain yield increase in recent Argentine wheat cultivars. *Crop Science* 38: 1203-1209.

Abeledo, L.G., D.F. Calderini, and G.A. Slafer (2003). Genetic improvement of barley yield potential and its physiological determinants in Argentina (1944-1998). *Euphytica* 130: 325-334.

Aitken, Y. (1975). *Flowering time, climate and genotype*. Melbourne: Melbourne University Press.

Andrade, F.H., A.G. Cirilo, and L. Echarte (2000). Factors affecting kernel number in maize. In M.E. Otegui and G.A. Slafer (eds.), *Physiological bases for maize improvement* (pp. 59-74). Binghamton, NY: Food Product Press.

Andrade F.H., A.G. Cirilo, S.A. Uhart, and M.E. Otegui (1996). *Ecofisiología del cultivo de maíz*. Balcarce, Argentina: La Barrosa, Cerbas and Dekalb Press.

Andrade F.H., S.A. Uhart, and A.G. Cirilo (1993). Temperature affects radiation use efficiency in maize. *Field Crops Research* 32: 17-25.

Andrade F.H., S.A. Uhart, and M.I. Frugone (1993). Intercepted radiation at flowering and kernel number in maize: Shade versus plant density effects. *Crop Science* 33: 482-485.

Andrade F.H., C.R. Vega, S.A. Uhart, A.G. Cirilo, M. Cantarero, and O. Valentinuz (1999). Kernel number determination in maize. *Crop Science* 39: 453-459.

Andriani, J.M., F.H. Andrade, E.E. Suero, and J.L. Dardanelli (1991). Water deficits during reproductive growth of soybeans: I. Their effects on dry matter accumulation, seed yield and its components. *Agronomie* 11: 737-746.

Board, J.E., M.S. Kang, and B.G. Harville (1999). Path analyses of the yield formation process for late-planted soybean. *Agronomy Journal* 91: 128-135.

Board, J.E. and Q. Tan (1995). Assimilatory capacity effects on soybean yield components and pod number. *Crop Science* 35: 846-851.

Bolaños, J. and G.O. Edmeades (1993). Eight cycles of selection for drought tolerance in lowland tropical maize: II. Responses in reproductive behavior. *Field Crops Research* 31: 253-268.

Bolaños J. and G.O. Edmeades (1996). The importance of the anthesis-silking interval in breeding for drought tolerance in tropical maize. *Field Crops Research* 48: 65-80.

Borrás, L. (2002). Diferencias entre especies en la respuesta a variaciones en la relación fuente/destino en postfloración. Actas de la XI Reunión Latinoamericana de Fisiología Vegetal, Punta del Este, Uruguay.

Borthwick, H.A and M.W. Parker (1938). Influence of photoperiod upon the differentiation of meristems and the blossoming of Biloxi soy beans. *Botanical Gazette* 99: 825-839.

Bruening, W.P. and D.B. Egli (1999). Relationship between photosynthesis and seed number at phloem isolated nodes in soybean. *Crop Science* 39:1769-1775.

Bruening, W.P. and D.B. Egli (2000). Leaf starch accumulation and seed set at phloem-isolated nodes in soybean. *Field Crops Research* 68: 113-120.

Calderini, D.F., M.F. Dreccer, and G.A. Slafer (1995). Genetic improvement wheat yield and associated traits: A re-examination of previous results and the latest trends. *Plant Breeding* 114: 108-112.

Calderini, D.F., M.P. Reynolds, and G.A. Slafer (1999). Genetic gains in wheat yield and main physiological changes associated with them during the 20th century. In E.H. Satorre and G.A. Slafer (eds.), *Wheat: Ecology and physiology of yield determination* (pp. 351-377). Binghamton, NY: Food Product Press.

Cantagallo, J.E., C.A Chimenti, and A.J. Hall (1997). Number of seeds per unit area in sunflower correlates well with a phototermal quotient. *Crop Science* 37: 1780-1786.

Cantagallo, J.E. and A.J. Hall (2002). Seed number in sunflower as affected by light stress during the floret differentiation interval. *Field Crops Research* 74: 173-181.

Cárcova J., M. Uribelarrea, L. Borrás, M.E. Otegui, and M.E. Westgate (2000). Synchronous pollination within and between ears improves kernel set in maize. *Crop Science* 40: 1056-1061.

Carlson, J.B. (1973). Morphology. In B.E. Caldwell (ed.), *Soybeans: Improvement, production and uses,* First edition (pp. 17-95). Agronomy Monograph 16. Madison, WI: ASA, CSSA, SSSA.

Chimenti, C.A. and A.J. Hall (1992). Sensibilidad del número de frutos por capítulo de girasol (*Helianthus annuus* L.) a cambios en el nivel de radiación durante la ontogenia del cultivo. In Proceedings XIX Reunión Agentina de Fisiología Vegetal, Huerta Grande, Córdoba, Argentina, pp. 27-28.

Chimenti, C.A. and A.J Hall (2001). Grain number responses to temperature during floret differentiation in sunflower. *Field Crops Research* 72: 177-184.

Cirilo A.G. and F.H. Andrade (1994). Sowing date and maize productivity: II. Kernel number determination. *Crop Science* 34: 1044-1046.

Connor, D.J. and A.J. Hall (1997). Sunflower physiology. In *Sunflower technology and production* (pp. 113-181). Agronomy Monograph 35. Madison, WI: ASA CSSA, SSSA.

Delécolle, R., R.K.M. Hay, M. Guerif, P. Pluchard, and C. Varlet-Grancher (1989) A method of describing the progress of apical development in wheat based on the time course of organogenesis. *Field Crops Research* 21:147-160.

Dreccer, M.F., A.H.C.M Schapendonk, G.A. Slafer, and R. Rabbinge (2000). Comparative response of wheat and oilseed rape to nitrogen supply: Absorption and utilitarian efficiency of radiation and nitrogen during the reproductive stages determining yield. *Plant and Soil* 220: 189-205.

Echarte L., S. Luque, F.H. Andrade, V.O. Sadras, A. Cirilo, M.E. Otegui, and C.R.C. Vega (2000). Response of maize kernel number to plant density in Argentinean hybrids released between 1965 and 1993. *Field Crops Research* 68: 1-8.

Egli, D.B. (1997). Cultivar maturity and response of soybean to shade stress during seed filling. *Field Crops Research* 52: 1-8.

Egli, D.B. (1998). *Seed biology and the yield of grain crops.* Wallingford, UK: CAB International.

Egli, D.B. and W.P. Bruening (2000). Potential of early-maturing soybean cultivars in late plantings. *Agronomy Journal* 92: 532-537.

Egli, D.B. and W.P. Bruening. (2002a). Flowering and fruit set dynamics at phloem-isolated- nodes in soybean. *Field Crops Research* 79: 9-19.

Egli, D.B. and W.P. Bruening. (2002b). Synchronous flowering and fruit set at phloem-isoleted nodes in soybean. *Crop Science* 42: 1535-1540.

Egli, D.B. and Y. Zhen-wen (1991). Crop growth rate and seeds per unit area in soybeans. *Crop Science* 31: 439-442.

Ellis, R.H., H. Asumadu, A. Qi, and R.J. Summerfield (2000). Effects of photoperiod and maturity genes on plant growth, partitioning, radiation use efficiency, and yield in soyabean [*Glycine max* (L.) Merrill] 'Clark'. *Annals of Botany* 85: 335-343.

Evans, L.T., I.F. Wardlaw, and R.A. Fischer (1975). Wheat. In L.T. Evans (ed.), *Crop physiology* (pp. 101-149). Cambridge, UK: Cambridge University Press.

Fehr, W. R. and C.E. Caviness (1977). *Stages of soybean development.* Special Report 80. Ames: Iowa State University.

Fischer, R.A. (1984). Wheat. In W.H. Smith and S.J. Banta (eds.), *Symposium on potential productivity of field crops under different environments* (pp. 129-153). Los Baños: IRRI.

Fischer, R.A. (1985). Number of kernels in wheat crops and the influence of solar radiation and temperature. *Journal of Agricultural Science* 100: 447-461.

Fischer, R. A. (1993). Irrigated spring wheat and timing and amount of nitrogen fertilizer: II. Physiology of grain and yield response. *Field Crops Research* 33: 57-80.

Gai, J.G., R.G. Palmer, and W.R. Fehr (1984). Bloom and pod set in determinate and indeterminate soybeans grown in China. *Agronomy Journal* 76: 979-984.

García del Moral, L., D.J. Miralles, and G.A. Slafer (2002). Phasic and foliar development. In G.A. Slafer, J.L. Molina-Cano, R. Savin, J.L. Araus, and I. Romagosa (eds.), *Barley science: Recent advances from molecular biology to agronomy of yield and quality* (pp. 243-268). Binghamton, NY: Food Product Press.

Gardner, J.S., W.M. Hess, and E.J. Trione (1985). Development of a young wheat spike: A SEM study of Chinese Spring wheat. *American Journal of Botany* 72: 548-559.

González, F.G., G.A. Slafer, and D.J. Miralles (2002). Vernalization and photoperiod responses in wheat pre-flowering reproductive phases. *Field Crops Research* 74: 183-195.

González, F.G., G.A. Slafer, and D.J. Miralles (2003a). Floret development and spike growth as affected by photoperiod during stem elongation in wheat. *Field Crops Research* 81: 29-38.

González, F.G., G.A. Slafer, and D.J. Miralles (2003b). Grain and floret number in response to photoperiod during stem elongation in fully and slightly vernalized wheats. *Field Crops Research* 81: 17-27.

Grant, R.F., B.S. Jackson, J.R. Kiniry, and G.F. Arkin (1989). Water deficit timing effects on yield components in maize. *Agronomy Journal* 81: 61-65.

Hadley, P., E.H. Roberts, R.J. Summerfield, and F.R. Minchin (1984). Effects of temperature and photoperiod on flowering in soya bean [*Glycine max* (L.) Merrill]: A quantitative model. *Annals of Botany* 53: 669-681.

Hall, A.J., C.A. Chimenti, F. Villela, and G. Frier (1985). Timing of water stress on yield components in sunflower. In Proceedings 11th International Sunflower Conference, Mar del Plata, Argentina, March 10-13, pp. 131-136.

Hall, A.J., F. Vilella, N. Trapani, and C.A. Chimenti (1982). The effects of water stress and genotype on the dynamics of pollen-shedding and silking in maize. *Field Crops Research* 5: 349-363.

Halloran, G.M. and A.L Pennell (1982). Duration and rate of development phases in wheat in two environments. *Annals of Botany* 49: 115-121.

Hardman, L.L. and W.A. Brun (1971). Effects of atmospheric carbon dioxide enrichment at different development stages on growth and yield components of soybeans. *Crop Science* 11: 886-888.

Hay, R.K.M. and E.J.M. Kirby (1991). Convergence and synchrony—A review of the coordination of development in wheat. *Australian Journal of Agricultural Research* 42: 661-700.

Hocking, P.J. and B.T. Steer (1989). Effects on seed size, cotyledon removal and nitrogen stress on growth and on yield components of oilseed sunflower. *Field Crops Research* 22: 59-75.

Horie T., M. Ohnishi, J.F. Angus, L.G. Lewin, T. Tsukaguchi, and T. Matano (1997). Physiological characteristics of high yielding rice inferred from cross location experiments. *Field Crops Research* 52: 55-57.

Jacobs B.C. and C.J. Pearson (1991). Potential yield of maize determined by rates of growth and development of ears. *Field Crops Research* 27: 281-298.

Jiang, H. and D.B. Egli (1993). Shade induced changes in flower and pod number and flower and fruit abscission in soybean. *Agronomy Journal* 85: 221-225.

Jiang, H. and D.B. Egli (1995). Soybean seed number and crop growth rate during flowering. *Agronomy Journal* 87: 264-267.

Kantolic, A.G., P.I. Giménez, and M. Charaf (1994). Respuesta de dos isolíneas de soja, determinada e indeterminada, al estrés hídrico durante floración. *Oleaginosos* 8: 6-10.

Kantolic, A.G, P.I. Giménez, C. Gutiérrez Hachard, and J. Saráchaga (1995). Tolerancia a la sequía durante el período reproductivo: Comparación del comportamiento de dos isolíneas de soja con diferente tipo de crecimiento. In Proceedings II Reunión Nacional de Oleaginosas—Congreso Nacional de soja. Pergamino, Argentina, October 24-27, pp. 143-150.

Kantolic, A.G., D.J. Miralles, R. Savin, D.F. Calderini, and G.A. Slafer (2001). Guía de manejo de Soja. Suplemento especial *Revista SuperCampo* 84: 67-97.

Kantolic, A.G. and G.A. Slafer (2001). Photoperiod sensitivity after flowering and seed number determination in indeterminate soybean cultivars. *Field Crops Research* 72: 109-118

Kernich, G.C., G.A. Slafer, and G.M. Halloran (1995) Barley development as affected by rate of change of photoperiod. *Journal of Agricultural Science* 124: 379-388.

Kiniry, J.R. and M.E. Otegui (2000). Processes affecting maize grain yield potential in temperate conditions. In M.E. Otegui and G.A. Slafer (eds.), *Physiological bases for maize improvement* (pp. 31-46). Binghamton, NY: Food Product Press.

Kiniry, J.R. and J.T. Ritchie (1985). Shade-sensitive interval of kernel number in maize. *Agronomy Journal* 77: 711-715.

Kirby, E.J.M. (1988). Analysis of leaf, stem and ear growth in wheat from terminal spikelet stage to anthesis. *Field Crops Research* 18: 127-140.

Kirby, E.J.M. (1990). Co-ordination of leaf emergence and leaf and spikelet primordium initiation in wheat. *Field Crops Research* 25: 253-264.

Kirby, E.J.M., J.H. Spink, D.L. Frost, R. Sylvester-Bradley, R.K. Scott, M.J. Foulkes, R.W. Clare, and E.J. Evans (1999). A study of wheat development in the field: Analysis by phases. *European Journal of Agronomy* 11: 63-82.

Kokubun, M., S. Shimada, and M. Takahashi (2001). Flower abortion caused by preanthesis water deficit is not attributed to impairment of pollen in soybean. *Crop Science* 41: 1517-1521

Kruk, B., D.F. Calderini, and G.A. Slafer (1997). Source-sink ratios in modern and old wheat cultivars. *Journal of Agricultural Science* 128: 273-281.

Lafitte H.R. and G.O. Edmeades (1997). Temperature effects on radiation use and biomass partitioning in diverse tropical maize cultivars. *Field Crops Research* 49: 231-247.

Maddonni, G.A., J. Cárcova, M.E. Otegui, and G.A. Slafer (2000). Recent research on maize grain yield in Argentina. In M.E. Otegui and G.A. Slafer (eds.), *Physiological bases for maize improvement* (pp. 191-204). Binghamton, NY: Food Product Press.

Marc, J. and J.H. Palmer (1976). Relationship between water potential and leaf and inflorescence initiation in *Helianthus annuus*. *Physiologia Plantarum* 36: 101-104.

Masle, J. (1985). Competition among tillers in winter wheat: Consequences for growth and development of the crop. In W. Day and R.K. Atkin (eds.), *Wheat growth and modelling* (pp. 33-54). New York: Plenum Press.

McKenzie K.S., C.W. Johnson, S.T. Tseng, J.J. Oster, and D.M. Brandon (1994). Breeding improved rice cultivars for temperate regions: a case study. *Australian Journal of Experimental Agriculture* 34: 897-905.

Miralles, D.J., B.C. Ferro, and G.A. Slafer (2001) Developmental responses to sowing date in wheat, barley and rapeseed. *Field Crops Research* 71: 211-223.

Miralles, D.J., S.D. Katz, A. Colloca, and G.A. Slafer (1998). Floret development in near isogenic wheat lines differing in plant height. *Field Crops Research* 59: 21-30.

Miralles, D.J., R.A. Richards, and G.A. Slafer (2000) Duration of the stem elongation period influences the number of fertile florets in wheat and barley. *Australian Journal of Plant Physiology* 27: 931-940.

Miralles, D.J. and G.A. Slafer (1995). Yield, biomass and yield components in dwarf, semidwarf and tall isogenic lines of spring wheat under recommended and late sowings dates. *Plant Breeding* 114: 392-396.

Miralles, D.J. and G.A. Slafer (1999). Wheat development. In E.H. Satorre and G.A. Slafer (eds.), *Wheat: Ecology and physiology of yield determination* (pp. 13-43). Binghamton, NY: Food Product Press.

Mohapatra P.K., P.K. Naik, and R. Patel (2000). Ethylene inhibitors improve dry matter partitioning and development of late flowering spikelets on rice panicles. *Australian Journal of Plant Physiology* 27: 311-323.

Mohapatra P.K., R. Patel, and S.K. Sahu (1993). Time of flowering affects grain quality and spikelet partitioning within the rice panicle. *Australian Journal of Plant Physiology* 20: 231-241.

Mohapatra P.K. and S.K. Sahu (1991). Heterogeneity of primary branch development and spikelet survival in rice panicle in relation to assimilates of primary branches. *Journal of Experimental Botany* 42: 871-879.

Muchow R.C., T.R. Sinclair, and J.M. Bennet (1990). Temperature and solar radiation effects on potential maize yields across locations. *Agronomy Journal* 82: 338-343.

Munier-Jolain, N., B. Ney, and C. Duthion (1993). Sequential development of flowers and seeds on the mainstem of an indeterminate soybean. *Crop Science* 33: 768-771.

Nagel, L., R. Brewster, W.E. Riedell, and R.N. Reese (2001). Cytokinin regulation of flower and pod set in soybeans (*Glycine max* (L.) Merr.). *Annals of Botany* 88: 27-31.

Otegui, M.E. and F.H. Andrade (2000). New relationships between light interception, ear growth and kernel set in maize. In S. Mickelson (ed.), *Physiology and modeling kernel set in maize* (pp. 89-102). Special publication number 29. Madison, WI: Crop Science Society of America.

Otegui M.E. and R. Bonhomme (1998). Grain yield components in maize: I. Ear growth and kernel set. *Field Crops Research* 56: 247-256.

Otegui M.E., and S. Melon (1997). Kernel set and flower synchrony within the ear of maize: I. Sowing date effects. *Crop Science* 37: 441-447.

Palmer, J.H. and B.T. Steer (1985). The generative area as the site of floret initiation in the sunflower capitulum and its integration to predict floret number. *Field Crops Research* 11: 1-12.

Patel R. and P.K. Mohapatra (1992). Regulation of spikelet development in rice by hormones. *Journal of Experimental Botany* 43: 257-262.

Peng S., K.G. Cassman, S.S. Virmani, J. Sheehy, and G.S. Khush (1999). Yield potential trends of tropical rice since the release of IR8 and the challenge of increasing rice yield potential. *Crop Science* 39:1552-1559.

Peterson, C.M., C.O'H. Mosjidis, R.R. Dute, and M.E. Westgate (1992). A flower and pod staging system for soybean. *Annals of Botany* 69: 59-67.

Porter, J.R. (1985). Approaches to modelling canopy development in wheat. In W. Day and R.K. Atkin (eds.), *Wheat growth and modelling* (pp. 69-81). New York: Plenum Press.

Reed A.J., G.W. Singletary, J.R. Schussler, D.R. Williamson, and A.L. Christy (1988). Shading effects on dry matter and nitrogen partitioning, kernel number, and yield of maize. *Crop Science* 28: 819-825.

Richards, R.A. (1996). Increasing yield potential in wheat—source and sink limitations. In M.P. Reynolds, S. Rajaram, and A. McNab (eds.), *Increasing yield potential in wheat: Breaking the barriers* (pp. 134-149). Mexico DF: CIMMYT.

Ritchie, S.W., J.J. Hanway, and G.O. Benson (1993). *How a corn plant develops.* Iowa State University, Special Report No. 48.

Savin, R. and G.A. Slafer (1991). Shading effects on the yield of an Argentinian wheat cultivar. *Journal of Agricultural Science* 116: 1-7.

Schou, J.B., D.L Jeffers, and J.G. Streeter (1978). Effects of reflectors, black boards, or shades applied at different stages of plant development on yield of soybeans. *Crop Science* 18: 29-34.

Schussler, J.R. and M.E. Westgate (1991). Maize kernel set at low water potential: I. Sensitivity to reduced assimilates during early kernel growth. *Crop Science* 31: 1189-1195.

Sheehy, J.E., M.J.A. Dionora, and P.L. Mitchell (2001). Spikelet number, sink size and potential yield in rice. *Field Crops Research* 71: 77-85.

Sibony, M. and M.J. Pinthus (1988). Floret initiation and development in spring wheat (*Triticum aestivum* L). *Annals of Botany* 61: 473-479.

Siddique, K.H.M., E.J.M. Kirby, and M.W. Perry (1989). Ear-stem ration in old and modern wheat varieties, relationship with improvement in number of grains per ear and yield. *Field Crops Research* 21: 59-78.

Slafer, G.A. (1994). *Genetic improvement of field crops.* New York: Marcel Dekker, Inc.

Slafer, G.A. (2003). Genetic basis of yield as viewed from a crop physiologist's perspective. *Annals of Applied Biology* 142: 117-128.

Slafer, G.A., L.G. Abeledo, D.J. Miralles, F.G. González, and E.M. Whitechurch (2001). Photoperiod sensitivity during stem elongation as an avenue to raise potential yield in wheat. *Euphytica* 19: 191-197.

Slafer, G.A. and F.H. Andrade (1989). Genetic improvement in bread wheat *(Triticum aestivum)* yield in Argentina. *Field Crops Research* 21: 289-296.

Slafer, G.A. and F.H. Andrade (1991). Changes in physiological attributes of the dry matter economy of bread wheat *(Triticum aestivum)* through genetic improvement of grain yield potential at different regions of the world. *Euphytica* 58: 37-49.

Slafer, G.A. and F.H. Andrade (1993). Physiological attributes related to the generation of grain yield in bread wheat cultivars released at different eras. *Field Crops Research* 31: 351-367.

Slafer, G.A., F.H. Andrade, and E.H. Satorre (1990). Genetic improvement effects on pre-anthesis physiological attributes related to wheat grain yield. *Field Crops Research* 23: 255-263.

Slafer, G.A., D.F. Calderini, and D.J. Miralles (1996). Yield components and compensation in wheat: Opportunities for further increasing yield potential. In M.P. Reynolds, S. Rajaram, and A. McNab (eds.), *Increasing yield potential in wheat: Breaking the barriers* (pp. 101-133). Mexico DF: CIMMYT.

Slafer, G.A. and D.J. Miralles (1993). Fruiting efficiency in three bread wheat *(Triticum aestivum)* cultivars released at different eras: Number of grains per spike and grain weight. *Journal of Agronomy and Crop Science* 170: 251-260.

Slafer, G.A. and H.M. Rawson (1994). Sensitivity of wheat phasic development to major environmental factors: A re-examination of some assumptions made by physiologists and modellers. *Australian Journal of Plant Physiology* 21: 393-426.

Slafer, G.A. and R. Savin (1994). Sink-source relationships and grain mass at different positions within the spike in wheat. *Field Crops Research* 37: 39-49.

Summerfield, R.J., H. Asumadu, R.H. Ellis, and A. Qi (1998). Characterization of the photoperiodic response of post-flowering development in maturity isolines of soyabean [*Glycine max* (L.) Merrill] 'Clark.' *Annals of Botany* 82: 765-771.

Thomas, J.F. (1994). Morphological and developmental plasticity in legumes. In I.K. Ferguson and S. Tucker (eds.), *Advances in legume systematics* 6 (pp. 1-10). Kew, UK: Structural Botany Royal Botanic Gardens.

Thomas, J.F. and C.D. Raper Jr. (1977). Morphological response of soybean as governed by photoperiod, temperature and age at treatment. *Botanical Gazette* 138: 321-328.

Thorne, G.N. and D.W. Wood (1987). Effects of radiation and temperature on tiller survival, grain number and grain yield in winter wheat. *Annals of Botany* 59: 413-426.

Tollenaar, M. and T.B. Daynard (1978). Relationship between assimilate source and reproductive sink in maize grown in a short season environment. *Agronomy Journal* 70: 219-223.

Tollenaar, M., L.M. Dywer, and D.W. Stewart (1992). Ear and kernel formation in maize hybrids representing three decades of grain yield improvement in Ontario. *Crop Science* 32: 432-438.

Uhart, S.A. and F.H. Andrade (1995). Nitrogen deficiency in maize: I. Effects on crop growth, development, dry matter partitioning, and kernel set. *Crop Science* 35: 1376-1383.

Vega , C.R.C., F.H. Andrade, V.O. Sadras, S.A. Uhart, and O.R. Valentinuz (2001). Seed number as a function of growth: A comparative study in soybean, sunflower, and maize. *Crop Science* 41: 748-754.

Villalobos, F.J., A.J. Hall, J.T. Ritchie, and F. Orgaz (1996). OILCROP-SUN: A development, growth, and yield model of the sunflower crop. *Agronomy Journal* 88: 403-415.

Villalobos, F.J., V.O. Sadras, A. Soriano, and E. Ferreres (1994). Planting density effects on dry matter partitioning and productivity of sunflower hybrids. *Field Crops Research* 36: 1-11.

Waddington, S.R., P.M. Cartwright, and P.C. Wall (1983). A quantitative scale of spike initial and pistil development in barley and wheat. *Annals of Botany* 51: 119-130.

Westgate, M.E. and J.S. Boyer (1986). Reproduction at low silk and pollen water potentials in maize. *Crop Science* 26: 951-956.

Westgate, M.E. and C.M. Peterson (1993). Flower and pod development in water deficient soybeans (*Glycine max* L. Merr.). *Journal of Experimental Botany* 44: 109-117.

Wiebold, W. J., D.A. Ashley, and H.R. Boerma (1981). Reproductive abscission levels and patterns for eleven determinate soybean cultivars. *Agronomy Journal* 73: 43-46.

Yin, X. and M.J. Kropff (1998). The effect of photoperiod on interval between panicle initiation and flowering in rice. *Field Crops Research* 57: 301-307.

Yin, X., M.J. Kropff, P.K. Aggarwal, S. Peng, and T. Horie (1997). Optimal preflowering phenology of irrigated rice for high yield potential in three Asian environments: A simulation study. *Field Crops Research* 51: 19-27.

Yin, X., M.J. Kropff, and J. Goudriaan (1997). Changes in temperature sensitivity of development from sowing to flowering in rice. *Crop Science* 37: 1787-1794.

Ying, J., S. Peng, Q. He, H. Yang, C. Yang, R.M. Visperas, and K.G. Cassman (1998). Comparison of high-yield rice in tropical and subtropical environments: I. Determinants of grain yield and dry matter yields. *Field Crops Research* 57: 71-84.

Youssefian, S., E.J.M. Kirby, and M.D. Gale (1992). Pleiotropic effects of the G.A. insensitive Rht dwarfing gene in wheat: 2. Effects on leaf, stem, ear and floret growth. *Field Crops Research* 28: 191-210.

Chapter 5

Carbon Partitioning in Developing Seed

Yong-Ling Ruan
Prem S. Chourey

INTRODUCTION

Seeds are the principal sources of food and fiber for human society. For example, the seeds of wheat, rice, maize, and legume provide virtually all the starch and a substantial portion of protein required by humankind. Canola and cotton seeds are important sources of edible oils. The latter also produce cellulose-enriched mature fibers from their seed coat epidermis, making cotton the most important textile crop in the world.

Despite the diversity of seed storage products, i.e., starch, protein, oil, and cellulose, the synthesis of these biopolymers all utilizes sucrose, imported into seeds from photosynthetic leaves, as the initial and primary carbon source. Thus, understanding the delivery, metabolism, and utilization of sucrose or its derivates in the seed are of utmost importance for designing strategies to improve seed productivity through either conventional breeding or genetic engineering. Apart from their fundamental significance in agriculture, the developing seeds are also a well-defined system for studying several important aspects of plant science, for example, the role of sugar in seed development and the interaction between the maternal and filial tissues of the seed.

In this chapter, we will review our understanding on the control of carbon partitioning in seeds by sequentially addressing the questions of seed

---

We gratefully acknowledge the financial support to our research from Bayer-CropScience and Australian Cotton Research and Development Corporation (to Yong-Ling Ruan) and from the U.S. Department of Agriculture (to Prem S. Chourey). This was, in part, a cooperative investigation of the U.S. Department of Agriculture, Agricultural Research Service, and the Institute of Food and Agricultural Science, University of Florida. We thank Drs. R.T. Furbank, C. Jenkins, and S.M. Xu of CSIRO Plant Industry for their critical reading of the manuscript.

anatomy and the cellular pathway of sucrose transport, the spatial and temporal patterns of sucrose utilization, and the key genes/proteins controlling carbon partitioning in seed. Focus will be placed on two groups of model species: the monocots of maize, wheat, and rice, and dicots of legumes and cotton. The seeds of these species have attracted extensive studies on carbon partitioning over the past two decades. Where possible, comparison will also be made with other seed systems.

SEED ANATOMY AND CELLULAR PATHWAY OF SUGAR TRANSPORT

Seed anatomy provides a structural basis to evaluate the cellular pathway of postphloem transport of photoassimilate (principally sucrose) and potential control points for carbon partitioning within the seed. Developing seeds share some common anatomical characteristics. First, they all comprise maternal and filial tissues. The vein, composed of phloem and xylem, terminates in the maternal tissue, pedicel, in the seed of monocot and in the seed coat of dicot species (see Figure 5.1). Second, the filial tissue is symplastically isolated from the maternal tissue, as there are no plasmodesmatal connections between the two (Thorne, 1985). This necessitates membrane efflux of sugars and other nutrients from maternal tissue and subsequent membrane influx by the filial tissue (Patrick and Offler, 1995). The bulk of the developed filial tissues are endospermic in origin for cereal-wheat, maize, and rice, and embryonic in nature for all of the dicots, including legume and cotton seed (see Figure 5.1).

Monocot Seed

Maize

The maternally derived pedicel of maize consists of pedicel parenchyma, placento-chalazal pad, and nucellar cells (Felker and Shannon, 1980). The phloem, composed of a sieve element and companion cell (se-cc) complex and phloem parenchyma, terminates in the pedicel parenchyma at the base of the maize seed (see Figure 5.1, part A). Adjacent to the maternal nucellar cells are the modified aleurone/basal endosperm transfer cells (BETC) of the filial tissues, which are separated from the maternal tissue by the apoplast (Davis et al., 1990). These transfer cells develop cell wall ingrowths (Davis et al., 1990). However, it remains unknown whether they are engaged in active sucrose uptake (Patrick and Offler, 1995). Similar to the seeds of other cereal species, the embryo occupies only a small area of the

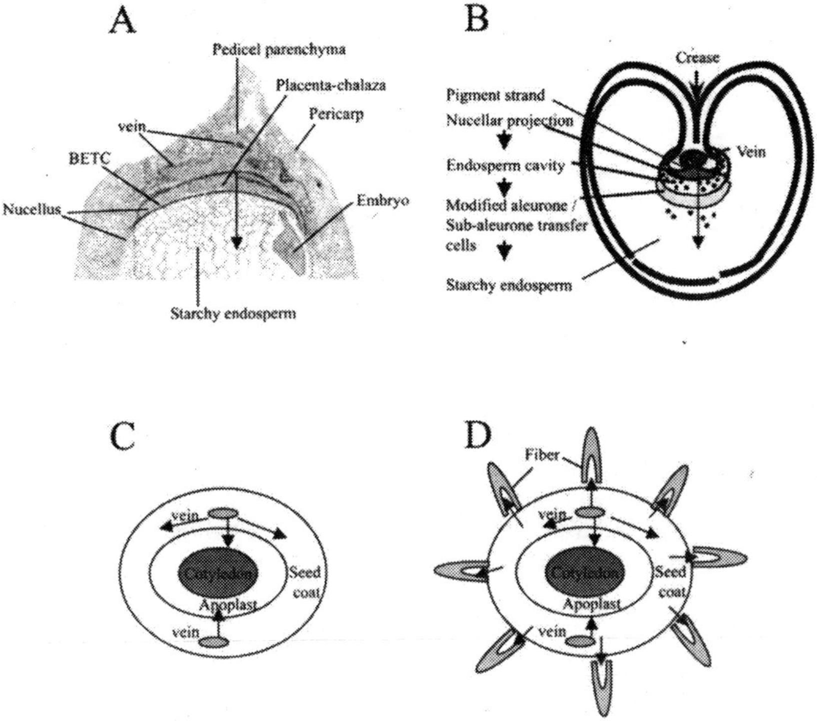

FIGURE 5.1. The anatomy of developing seeds. A: A longitudinal section of maize seed; B: A diagram of longitudinal-section view of wheat grain; C: A diagram of cross-section view of legume broad bean seed; D: A diagram of cross-section view of cotton seed; Arrow indicates the main direction of sucrose flow in the seed; BETC: Basal endosperm transfer cells.

filial tissue, which primarily comprises endosperm cells for starch synthesis (see Figure 5.1, part A).

Physiological studies indicate symplastic continuity from phloem to the placento-chalazal/nucellar cells of the pedicel and throughout the filial tissue (Felker and Shannon, 1980; Davis et al., 1990). Sucrose is unloaded from the se-cc complex of the pedicel and moves symplastically across the maternal tissue to the seed apoplast (Felker, 1992). Upon reaching the aqueous cell wall space of the modified aleurone/endosperm transfer cell at the basal endosperm, the sucrose is hydrolyzed into glucose and fructose by the cell wall-bound invertase located in this region (Miller and Chourey, 1992; Cheng et al., 1996). The resultant hexoses are then taken up by these cells and are likely resynthesized into sucrose by sucrose phosphate synthase for transport to the upper part of the endosperm. Finally, sucrose is degraded by sucrose synthase for synthesis of starch in the maize endosperm (Chen and Chourey, 1989).

Wheat and Rice

The maternal tissue of the wheat grain essentially consists of vein, pigment strand, and nucellar cells (see Figure 5.1, part B). The cell walls of the pigment strand become lignified at the onset of the grain filling stage, thus preventing solute transport through the apoplast of this region (Zee and O'Brien, 1970). The nucellus projects into an apoplastic endosperm cavity, which develops between the maternal and filial tissue (see Figure 5.1, part B). Numerous functional plasmodesmata interconnect the se-cc complex to the nucellar projection across the entire maternal tissue (Wang, Offler, and Patrick, 1994). The outer layer of the filial tissue facing the endosperm cavity develops modified aleurone and subaleurone layers. Both the nucelllar cells of the maternal tissue and the subaleurone layer of the filial tissue develop wall ingrowths, characteristic of transfer cells (Wang, Offler, Patrick, and Ugalde, 1994).

During the cell division/prefilling stage of wheat grain, the cell wall of the pigment strand is not lignified (Zee and O'Brien, 1970). Therefore, an apoplastic pathway of postphloem sugar transport is possible in the maternal seed tissues. This, together with the plasmodesmata connection throughout the pericarp (Wang, Offler, and Patrick, 1994), suggests that sucrose is likely transferred via a combination of apoplastic and symplastic pathway. The cellular pathway in the filial tissue at this stage remains to be explored.

During the filling stage, systematic structural and physiological studies, coupled with analysis of symplastic or apoplastic fluorescent dye movements, have provided strong evidence to support the following model of sucrose transport into the developing wheat grain. Sucrose moves via the symplastic pathway from the se-cc complex of the phloem to the nucellar projection transfer cells of the maternal tissue (Wang, Offler, and Patrick, 1994; Wang, Offler, Patrick, and Ugalde, 1994). Sucrose is then transferred across the amplified plasma membrane of the transfer cells into the endosperm cavity, possibly facilitated by a sucrose transporter (SUT) located on these membranes (Wang, Offler, and Patrick, 1994; Bagnall et al., 2000). Upon delivery to the endosperm cavity, sucrose is actively taken up by a SUT on the plasma membrane of the modified aleurone and subaleurone layer transfer cells (Bagnall et al., 2000). Sucrose is further transported symplastically to the remaining endosperm cells for the synthesis of starch (Wang, Offler, Patrick, and Ugalde, 1994).

For the rice seed, the anatomy resembles that of wheat in many respects. For example, the importing vein extends along the length of the grain in both rice and wheat (Oparka and Gates, 1981). However, some structural differences exist in grains between rice and other cereals. Rice seeds lack the endosperm cavity present in the wheat and barley. In addition, the aleurone transfer cells, which may facilitate sucrose uptake into endosperm, are present in wheat, barley, and maize but absent in rice.

The cellular pathway of sugar movement has not been thoroughly studied in the rice seed, but it could be similar to that in wheat (see Patrick and Offler, 1995). At the early cell division phase of rice grain development, the cell wall of pigment strand is not suberized, which would allow apoplastic transport of assimilate in the maternal tissue (Oparka and Gates, 1981). The expression of a cell wall invertase gene in the vascular region of the rice gain suggests its role in hydrolysis of apoplastically unloaded sucrose (Hirose et al., 2002). As the grain develops to the filling stage, the cell wall of the pigment strand is suberized, becoming an apoplastic barrier for solute transport (Oparka and Gates, 1982). This, together with the observed symplastic continuity between se-cc complexes and the nucellular cells (Oparka and Gates, 1981), confines the movement of assimilates in the symplast in the maternal tissues. Consistent with this view is the absence of cell wall invertase expression in the pedicel at the filling stage of the rice grain (Hirose et al., 2002). A sucrose transporter gene *(OsSUT1)*, expressed in the aleurone layer of the endosperm, may be responsible for the uptake of sucrose into the rice endosperm for starch biosynthesis (Furbank et al., 2001; Hirose et al., 2002).

Dicot Seed

Grain Legumes

The seed coat of the grain legumes contains, in sequence from the outermost of the tissue, epidermis, hypodermis, chlorenchyma, and ground parenchyma (Patrick et al., 1995; Weber et al., 1996). The inner side of the ground parenchyma develops into thin-walled parenchyma and thin-walled parenchyma transfer cells for *Vicia faba* and *Pisum sativum* and branch parenchyma for *Phaseolus vulgaris* (Patrick et al., 1995; Wang et al., 1995; Tegeder et al., 1999). The vein, embedded in the ground parenchyma, is either restricted in distribution and forms a limited number of circumferential bands on the seed coat (e.g., *Pisum,* Hardham, 1976; *Vicia,* Offler et al., 1989; also see Figure 5.1 part C), or is reticulated throughout the coat (e.g., *Phaseolus,* Patrick et al., 1995).

During the seed development of dicotyledonous species, including grain legumes, the endosperm is crushed and absorbed by the expanding cotyledons. The cotyledons make up the bulk of the embryonic tissue and are symplastically isolated from the seed coat by the apoplast at the interface between the maternal and filial tissues (Figure 5.1, part C). For *Vicia* and *Pisum,* the abaxial epidermal cells of the cotyledons, which abut the seed coat, differentiate to form transfer cells with extensive wall ingrowths on their outer periclinal walls immediately before the phase of fresh weight gain (Bonnemain et al., 1991; Tegeder et al., 1999). Except for their epidermis, the cotyledons are entirely comprised of storage parenchyma cells.

Sucrose unloading and postphloem transport in the seed coat, and further movement within the cotyledons, follow a symplastic pathway in all the legume species studied thus far, including *Vicia, Phaseolus,* and *Pisum* (Tegeder et al., 1999, and references therein). The principal cellular sites for sucrose efflux from the seed coat are thin-walled parenchyma transfer cells for *Vicia faba* and *Pisum sativum* and branch parenchyma for *Phaseolus vulgaris* (Wang et al., 1995; Tegeder et al., 1999). Sucrose influx to the cotyledons occurs across the plasma membrane of the epidermal transfer cells for *Vicia faba* and *Pisum sativum* and undifferentiated epidermal/subepidermal cells of *Phaseolus vulgaris* (Patrick and Offler, 1995; Tegeder et al., 1999).

Cotton

The developing cotton seed is unique because of the specialized hairlike fibers derived from the seed coat epidermis (Figure 5.1, part D). To our knowledge, this is the only seed type in which the major economically valuable tissue is maternal in origin. The anatomy of the remaining cotton seed is similar to other dicotyledonous seeds. The vein is distributed throughout the outer seed coat. The innermost layer of the inner seed coat develops into transfer cells with irregular wall ingrowth facing the seed apoplast, which encircles the filial tissues (Ryser et al., 1988; Ruan et al., 1997). The cellular pathway of sugar delivery in the coat and cotyledons of cotton seed has been studied to some extent (Ruan et al., 1997; Ruan, unpublished data) and is similar to that in the legume seed (see previous subsection). However, the pathway for sucrose transport to the fibers is characterized with some distinct features and is summarized as follows.

Each cotton fiber is a single cell, which initiates at or just before anthesis from the ovule epidermis. After initiation the single-celled fibers elongate rapidly to about 2.5 to 3.0 cm for about 16 d before they switch to intensive secondary cell wall cellulose synthesis. By maturity at about 60 days after anthesis (DAA), more than 94 percent of fiber dry weight is pure cellulose a ß-1,4 polymer of glucose (Basra and Malik, 1984). The rapid elongation and intensive cellulose synthesis of the fiber cells require an efficient supply of sufficient amounts of sugars and other solutes (Ruan et al., 1997, 2001).

During the initiation stage, plasmolysis studies coupled with confocal imaging of the membrane-impermeant fluorescent solute, carboxyfluorescein (CF), indicated a symplastic pathway of sucrose entry to the fiber cells (Ruan et al., 2000). This conclusion is supported by the earlier findings of the presence of plasmodesmatal structures at the fiber base interconnecting the underlying seed coat (Ryser, 1992) and the abundance of the cytoplasmic sucrose synthase and absence of cell wall-bound invertase proteins in the fiber initials (Ruan and Chourey, 1998). Remarkably, the initially "opened" plasmodesmata closed at about 10 DAA and reopened at 16 DAA (Ruan et al., 2001). Coincident with the transient closure of the plasmodesmata, the sucrose transporter genes were expressed maximally in the base of the fiber cells at 10 DAA to facilitate uptake of sucrose across the plasma membrane of fiber cells for further elongation (Ruan et al., 2001). Finally, the reopening of plasmodesmata and the diminished expression of the transporter genes of the fibers changed the cellular pathway of sucrose import back to the symplastic route.

This was the first demonstration that the cellular route of sugar transport can be developmentally reversible at a single-cell level. The changes in plasmodesmatal gating with coordinated expression of transporter genes could play a key role in controlling cotton fiber elongation by building up or releasing the turgor pressure within the cell (Ruan et al., 2001).

In summary, the symplastic pathway prevails for sugar transport within the maternal and filial tissues of seed during filling/storage phase. Sugar efflux to, and influx from, the seed apoplast occurs at the interface between the maternal and filial tissues. The apoplastic pathway may also operate in the seed maternal tissue of wheat (Wang, Offler, Patrick, and Ugalde, 1994) and bean (Weber et al., 1995) at cell division phase and in elongating fiber cells of the cotton seed (Ruan et al., 2001).

TEMPORAL AND SPATIAL PATTERNS OF CARBON PARTITIONING

Temporal Process of Carbon Partitioning

Seed development proceeds in a series of temporal steps, starting from cell division, followed by expansion or storage and maturation. Carbon partitioning plays pivotal roles in these developmental processes.

After fertilization, the young embryo undergoes a cell division phase, characterized by high mitotic activities. In the seed of *Vicia faba*, hexoses, derived from sucrose in the seed coat, are the major sugars nurturing the young embryo (Weber et al., 1995). Correlative evidence indicates that hexoses not only provide a carbon source for basic cell metabolism, but could also trigger and enhance cell division in embryo (Weber, Borisjuk, Heim, and Wobus, 1997). For example, high-resolution imaging of sugar concentration in *Vicia* cotyledons reveals that high glucose concentration is present in the mitotic region, whereas differentiated starch-accumulating regions contain particularly low concentrations of glucose (Borisjuk et al., 1998). Similarly, the hexose released metabolically by a cell wall-bound invertase appears to play an important role in the cell division of maize endosperm (Cheng and Chourey, 1999). Indeed, the endosperm cell number from a cell wall-invertase-deficient mutant, *mn1*, is only about half of that in the wild type (Vilhar et al., 2002). In addition, feeding hexoses to in vitro-cultured cotyledons of *Vicia faba* represses the expression of leguminB, a major storage protein (Weber et al., 1995). This indicates that hexoses may also play a role in maintaining cell division status by suppressing the expression of the storage protein genes during the early seed development.

As the *Vicia* seed develops, the embryo switches from cell division to the phase of cell expansion and storage, possibly mediated by a high sucrose to hexose ratio (Weber, Borisjuk, Heim, and Wobus, 1997). At this stage, sucrose is unloaded from the seed coat and moves toward the embryos symplastically without prior hydrolysis (Patrick and Offler, 1995; Weber et al., 1995). This, coupled with the decrease of seed coat cell wall invertase levels, results in high ratio of sucrose to hexose concentration in the embryo (Weber et al., 1995). High sucrose concentration correlates with high transcript levels of sucrose synthase and ADP-glucose pyrophosphorylase, two key enzymes involved in starch synthesis (Borisjuk et al., 2002). This indicates a possible signaling function of sucrose to induce starch biosynthesis at the gene expression level. Indeed, a high concentration of sucrose has been localized to the expanding and starch-accumulating region of the *Vicia faba* cotyledons (Borisjuk et al., 2002). At the cell expansion and storage stage, the imported sucrose is utilized by the seed endosperm to synthesize predominantly starch in cereals (e.g., Chourey et al., 1998) and by the cotyledons of grain legumes to synthesize starch and storage proteins (Weber, Borisjuk, Heim, and Wobus, 1997). For the cotyledons of canola (King et al., 1997) and cotton seed (Ruan et al., 1997), lipids and proteins are the main storage products synthesized from the imported carbon.

In the seed coat of most dicotyledonous seeds, starch is transiently synthesized during the prestorage phase. The starch is degraded into hexoses for remobilization into embryos at the storage phase in the seed of canola

(King et al., 1997), cotton (Ruan et al., 1997), and grain legume (Weber, Borisjuk, Heim, and Wobus, 1997). Upon maturation, the seed maternal tissue is desiccated and becomes a protective layer in all seeds.

Developmentally programmed carbon partitioning in seeds at the single-cell level may be best seen from the cotton fiber system. These single-celled hairs, located on the outer epidermis of the cotton seed coat, utilize sucrose for their initiation and rapid elongation from 0 to 16 DAA (Ruan et al., 2000, 2001). At the end of elongation, however, the imported sucrose is used by fiber cells to predominantly synthesize massive amounts of cellulose along the plasma membrane/cell wall of the fibers (Ruan et al., 1997; Haigler et al., 2001). This provides a unique example of the diverse destinations of sucrose utilization during the development of cells in seed.

Spatial Allocation of Carbon

Carbon partitioning in developing seeds is also under tight spatial control. Depending on species, the phloem-unloaded sucrose can move either unidirectionally for synthesis of one major storage product or multidirectionally for diverse destinations of carbon utilization. The following two examples are provided to highlight this issue.

In the maize seed, it has been estimated that more than 70 percent of imported sucrose is delivered to the developing endosperm from the phloem termini in the pedicel (Hanft and Jones, 1986; Schmalstig and Hitz, 1987). Once it reaches the basal endosperm, the sucrose is hydrolyzed by cell wall-bound invertase located in the basal endosperm transfer cells (Cheng et al., 1996). The resultant hexoses could play a dual role in the maize endosperm development. They may serve as signaling molecules for triggering and maintaining cell devision (Cheng and Chourey, 1999). The majority of the hexoses, however, are believed to be resynthesized into sucrose by sucrose phosphate synthase at the basal region of the endosperm. The sucrose is then utilized for synthesis of starch in the endosperm via the coordinated action of sucrose synthase, ADP-glucose pyrophosphorylase, and starch synthase. This polar flow of carbon from sucrose in the maternal tissue to starch in the endosperm also prevails in seeds of other cereals, such as wheat and rice.

In developing cotton seed, however, sucrose is utilized for synthesis of diverse products in spatially separated seed tissues (see Figure 5.2) (Ruan et al., 1997). After unloading from the phloem in the outer seed coat, sucrose is mobilized in two opposite directions: outward to the fiber cells on

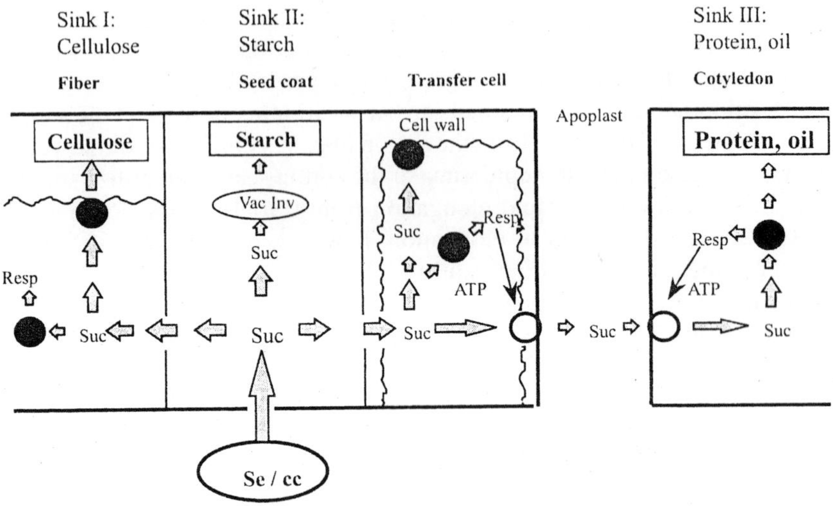

FIGURE 5.2. Diverse patterns of carbon partitioning in developing cotton seed. The differential expression of Sus protein in fiber cells and transfer cells of the seed coat plays a key role in mobilizing sucrose into fibers for cellulose biosynthesis and into transfer cells for possible energy coupled sucrose efflux into the apoplast. Sucrose is then taken up from the apoplast by cotyledonary cells and degraded by Sus for protein and oil biosynthesis. The remainder of unloaded sucrose moves into seed coat cells and is degraded by vacuolar invertase for starch biosynthesis in this tissue. Inv: Invertase; Resp: Respiration; Se/cc: Sievel element/companion cell complex; Suc: Sucrose; Vac: Vacuole; ●: Sucrose synthase (Sus); O: Putative sucrose transporter. The arrow indicates the main direction of carbon flow. (See Ruan et al., 1997, for more details.)

the seed epidermis and inward to the inner seed coat and embryo (See Figure 5.1, part D). At the onset of the secondary cell wall synthesis stage, about 80 percent of the carbon imported to fibers is directed to cellulose (Haigler et al., 2001). Sucrose moving toward the innermost seed coat is unloaded from the transfer cells to the embryos for synthesis of lipids and proteins. A small portion of sucrose is retained in the seed coat for temporary starch production (Ruan et al., 1997).

In this model (Figure 5.2), carbon partitioning within the seed may be under strong competition between the maternal tissue fibers for synthesizing cellulose and embryonic tissues for lipids and proteins. It remains to be tested whether reducing sucrose flow and utilization in embryos can enhance cellulose synthesis in fibers or the converse.

KEY GENES CONTROLLING CARBON PARTITIONING IN SEEDS

Invertase

After import to the seed, the sucrose must be degraded into hexoses, i.e., glucose and fructose, or their derivates prior to its use in metabolism and biosynthesis. Two enzymes are responsible for the cleavage of sucrose: invertase and sucrose synthase. The role in seed development of each enzyme in turn is evaluated in this and the following sections.

Depending on the cellular location, invertase (β-fructofuranosidases, EC 3.2.1.26; Sucrose + $H_2O \rightarrow$ Glucose + Fructose) can be divided into three classes: extracellular (cell wall), vacuolar, and cytosolic forms, which all hydrolyze sucrose into glucose and fructose. The optimum pH is about 4.5 to 5.5 for the cell wall-bound and vacuolar invertase and about 6.8 to 8.0 for the invertase in the cytoplasm (Lee and Sturm, 1996).

In developing seeds, the most detailed analyses on invertases are reported in maize. Tsai et al. (1970) reported the first correlative evidence between the high level of invertase activity and the cell division phase of developing endosperm in a seminal study on a detailed profile of various enzyme activities in this important sink tissue. Molecular, genetic and biochemical studies showed that the loss on *Mn1*-encoded cell wall invertase, *INCW2*, is the causal basis of the miniature *(mn1)* seed phenotype (Miller and Chourey, 1992; Cheng et al., 1996). Additional evidence that *INCW2* is critical in normal seed development of maize is that it is spatially and temporally the first enzyme that metabolizes the incoming sucrose in developing endosperm. It should be noted that the *mn1* mutant seed is highly endosperm specific as there is no detectable change in the embryo. Indeed, there is also no invertase activity in developing embryo. Recent analyses on the *mn-1* seed mutant have shown that the loss of seed weight is due to reduced mitotic activity and cell size in the developing endosperm (Vilhar et al., 2002). Further, gene-dose dependent enzyme activity analyses (Cheng et al., 1996) show a rate-limiting role of *INCW2* in the control of seed size and seed weight, especially in the low range, 2 to 10 percent of the wild type, of activity levels. In vitro kernel culture of *Mn1* (wild type) and invertase-deficient *mn1* seed mutant on sucrose and hexose media provided new insights on both the physiological role of *INCW2* in seed development and the mode of sucrose transport in developing seeds of maize (Cheng and Chourey, 1999). Remarkably, hexose-based medium was similar to the sucrose-based medium in promoting the normal development of *Mn1* kernels. However, the *mn1* kernels remained mutant in phenotype regardless of whether sucrose or hexose was in the medium. A lack of the wild-type seed phenotype of the *mn1* seed mutant in hexose media suggests that a metabolic release of hexose catalyzed by *INCW2* rather than an exogenous source is critical in normal seed development (Cheng and Chourey, 1999). These data also raise a possibility that sucrose turnover in the basal endosperm cells may regulate the net balance of carbohydrate flux entering the endosperm for cell division and storage.

There are several reports of soluble invertase in developing seeds of maize (see Carlson and Chourey, 1999, and references therein). The most recent detailed biochemical and molecular analyses (Carlson and Chourey, 1999), however, indicate that there is actually no detectable soluble invertase transcript and protein in developing maize endosperm. Overall, the collective evidence indicates that soluble invertase activity is actually due to a contamination of cell wall invertase protein during the extraction. Soluble invertase is, however, present in the early stages of developing caryopsis in maternal cells, i.e., nucellus and pedicel, and is shown to be critical under water-stressed conditions (Andersen et al., 2002). The contribution of the neutral and alkaline invertase to sucrose breakdown is thought to be negligible, since its activity is much lower than the acid invertase or sucrose synthase (Ruan et al., 1997, 2003; Weber, Borisjuk, Heim, and Wobus, 1997).

In the developing seed of *Vicia faba,* a gene encoding cell wall invertase *(VfCWINV1)* is specifically expressed in the chalazal vein and the thin wall parenchyma of the seed coat during prestorage phase (Weber et al., 1995). The thin wall parenchyma is the principal cellular site of sugar efflux to the seed apoplast bathing the encircled embryo (Wang et al., 1995). The resultant high concentration of hexose is believed to promote cell division and thus control seed size of the embryo through enhancing mitotic activities (Weber et al., 1995). Further evidence supporting this conclusion comes from comparative studies between two genotypes of *Vicia faba* differing in seed size (Weber et al., 1996). In comparison to that of the small seed genotype, the mRNA and activity of the cell wall invertase gene in the seed coat of a large seed genotype are expressed in more cell layers of the thin wall parenchyma over longer time period. This gene expression pattern correlates to the higher number of cells in the seed coat and embryo parenchyma of the larger seed genotype (Weber et al., 1996). Both in vivo measurement and in vitro feeding experiment show that high hexose conditions are associated with an extended mitotic activity of the larger cotyledons (Weber, Borisjuk, Heim, and Wobus, 1997). As the seed develops further, expression of the cell wall invertase declines, leading to a decreased level of hexoses in the *Vicia* cotyledons (Weber et al., 1996). This appears to signify the transition from the cell division to the cell expansion/storage phase of the seed (Weber, Borisjuk, Heim, and Wobus, 1997). Taken together, these results suggest that during the prestorage phase of *Vicia* seed development, sink strength could be provided by the seed coat-associated cell wall bound invertase, which controls the carbon status of the apoplastic environment and, consequently, the developmental phase of the embryos.

Recently, Hirose et al. (2002) cloned a cDNA encoding cell wall invertase *(OsCIN1)* from developing grain of rice. They showed that the transcript of *OsCIN1* is detectable only in s the very early stage of the rice grain development, at 1 to 4 d after anthesis, and is primarily localized in the vascular parenchyma of the dorsal vein and surrounding cells of the maternal tissue (Hirose et al., 2002). The temporal and spatial expression patterns of *OsCIN1* are similar to that of *VfCWINV1* in the seed coat of *Vicia faba* (Weber et al., 1995) but are different from that of *INCW2* in maize seed. In maize, *INCW2* is mainly expressed in the basal part of endosperm (Cheng et al., 1996). These findings suggest that *OsCIN1* may play an important role in cleavage of phloem-unloaded sucrose in the extracellular space of the maternal tissue of the rice seed (Hirose et al., 2002).

Sucrose Synthase

The enzyme sucrose synthase (EC 2.4.1.13; Sucrose + UDP ↔ Fructose + UDP-glucose) plays a major role in the degradation of sucrose for various metabolic and biosynthetic processes in developing seeds. Although the reaction is readily reversible, extensive evidence has shown that, *in planta,* this enzyme predominantly catalyzes the cleavage reaction (e.g., Chourey and Nelson, 1979; Geigenberger and Stitt, 1993). This is due to the rapid removal of the cleavage products fructose and UDP-glucose in plant cells. The former is effectively phosphorylated by high activity of fructokinase, while the latter is used both as a precursor or substrate to synthesize starch, cellulose or callose, and other storage products (e.g., Chourey et al., 1998; Haigler et al., 2001).

Sucrose synthase may play a major role in starch biosynthesis in seeds of many species. In maize, which has three sucrose synthase genes, *Sh1, Sus1,* and *Sus2* (Carlson et al., 2002), the loss of *Sh1*-encoded SuSy activity reduces about 95 percent of the total endosperm enzyme activity (Chourey and Nelson, 1976; Chourey, 1981). Such a drastic reduction in SuSy activity leads to early cell degeneration in developing endosperm, which ultimately leads to the shrunken *(sh1)* seed phenotype. The reduced levels of starch in the *sh1* mutant (about 22 percent less as compared to the normal *Sh1* endosperm) is believed to be a secondary effect due to greatly diminished cellular space for starch biosynthesis (Chourey et al., 1998). In the cotyledons of *Vicia faba,* sucrose concentration increases during the transition phase from cell division to the storage phase (Weber et al., 1995). This is likely due to the decreases of cell wall bound invertase in the seed coat and the increases of sucrose phosphate synthase in the embryo at this stage (Weber et al., 1996). High sucrose level is reported to induce greater steady state levels sucrose synthase transcripts (Koch, 1996). This seems to be the case in the seed of *Vicia faba,* where the rise of sucrose correlates with the increased expression of sucrose synthase and ADP-glucose pyrophosphorylase, leading to the biosynthesis of starch at the storage phase (Weber et al., 1996). A correlation between sucrose synthase activity and starch biosynthesis has also been observed in the developing seed of pea (Déjardin et al., 1997).

The grain legume seeds are also able to synthesize a wide range of amino acids from imported carbon (Weber, Borisjuk, Heim, and Wobus, 1997). Several studies have shown that the processes of starch and protein biosynthesis are integrated, probably mediated via the sucrose synthase pathway. For example, feeding hexose to in vitro cultured cotyledons of *Vicia faba* decreases the activity of sucrose synthase, which reduces the starch to sucrose ratio and also lowers the levels of free amino acid and transcript of the storage protein legumin B (Weber et al., 1995). The role of sucrose synthase in oil biosynthesis has been indicated by its high expression in the oil-accumulating embryos of canola (King et al., 1997) and cotton (Ruan et al., 1997).

It is important to note that sucrose synthase is not always involved in storage product synthesis in developing seeds, particularly in those nonstorage cell types. For example, the *Sus1*-encoded sucrose synthase protein, SS2, is expressed specifically in the aleurone layer and basal endosperm transfer cells of developing endosperm of maize (Chen and Chourey, 1989). None of these cell types accumulate starch. In cotton ovule or young seed, sucrose synthase is abundantly but specifically expressed in the initiating fibers on the seed coat epidermis (Ruan and Chourey, 1998). However, the young fibers show no detectable storage activities for either starch or proteins (Basra and Malik, 1984). Apparently, sucrose synthase could play a role in metabolism or sugar signaling, rather than in storage, in these cell types of developing seed.

Sucrose synthase traditionally has been considered to be a soluble protein located in the cytoplasm. However, recent studies have discovered that a substantial amount of the protein is tightly bound to the plasma membrane of cotton fiber (Amor et al., 1995), maize endosperm cells (Carlson and Chourey, 1996), and many other tissues (see Haigler et al., 2001, and references therein). The mechanism(s) of the membrane association is currently unknown but might involve phosphorylation or dephosphorylation of the serine residue(s) at the N-terminal of the protein and interaction with cytoskeleton (Winter and Huber, 2000; Haigler et al., 2001). The role of the membrane association of sucrose synthase is postulated to channel carbon directly from sucrose into cellulose via UDP-glucose, the immediate substrate for cellulose biosynthesis (Amor et al., 1995). Coupling between the plasma membrane associated sucrose synthase and glucan synthase has the advantage of

1. promoting synthesis of cellulose from sucrose with no additional energy input,
2. avoiding competition for use of UDP-glucose by other pathways, and
3. allowing immediate recycling of UDP, a compound that inhibits the reaction catalyzed by cellulose synthase (see Haigler et al., 2001 for a recent review).

Comparative genetic studies suggest that the membrane association of SS1, encoded by the *Sh1* gene, plays a major role in providing substrate for cellulose biosynthesis in developing endosperm of maize seed (Carlson and Chourey, 1996; Chourey et al., 1998). Indeed, 14 percent of the total SuSy activity is associated with plasma membrane during cell expansion, whereas the same fraction constitutes only 3 percent during the starch biosynthesis phase in developing seed (Carlson and Chourey, 1996). A similar role has been implicated for the fiber cells undergoing secondary cell wall cellulose synthesis (Amor et al., 1995). The definitive role of the plasma membrane association of sucrose synthase in cellulose synthesis is, however, yet to be demonstrated.

Significant insights into the role of sucrose synthase in seed development have been recently revealed from analysis of cotton plants transformed with suppression constructs of a sucrose synthase gene (Ruan et al., 2003). In the wild type, this gene is specifically expressed in the initiating and elongating fiber cells of the ovule/seed epidermis at 0 to 3 DAA, but also in the endosperm and embryo at 10 DAA onward (Ruan et al., 1997; Ruan and Chourey, 1998). Reduction of sucrose synthase activity by 70 percent or more by using an antisense or cosuppression approach results in a fiberless phenotype (Ruan et al., 2003). Indeed, there is a linear relationship between the level of the gene suppression and the degree of the inhibition of early fiber elongation among the transgenic lines. These results demonstrate, for the first time, the truly rate-limiting role of sucrose synthase in plant development at the single-cell (fiber) level (Ruan et al., 2003). By conducting further molecular, genetic, and biochemical analyses, Ruan et al. (2003) found that suppression of sucrose synthase only in the maternal seed tissue represses fiber development without affecting the embryo development and seed size. Additional suppression in the endosperm and embryo inhibits their own development, which blocks the formation of the adjacent seed coat transfer cells and arrests seed development entirely (Figure 5.3). These results show that the development of the seed maternal tissue is under metabolic control of the embryo and the seed coat transfer cell could be the cellular site to perceive regulatory signals from the embryo. This study provides novel information on the role of sucrose synthase in controlling plant cell and seed development and offers opportunities to modify fiber and seed development through genetic engineering of sucrose synthase gene expression (Ruan et al., 2002, 2003).

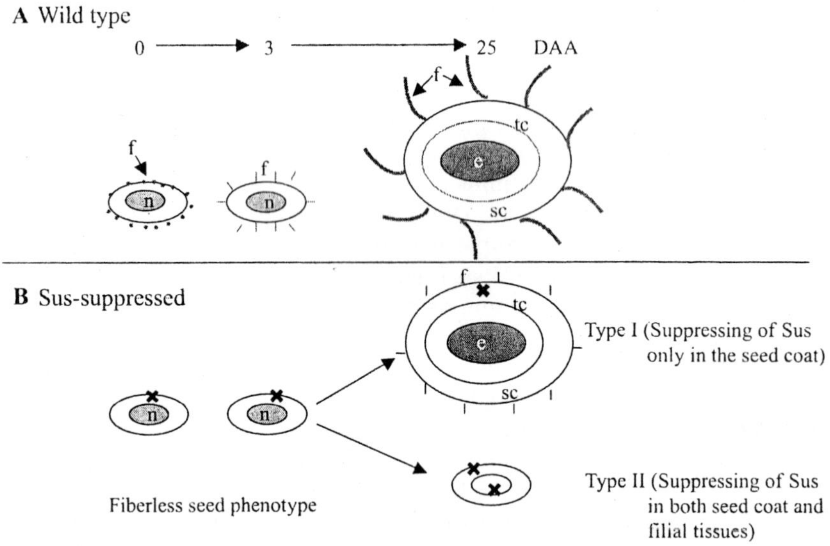

FIGURE 5.3. New insights on the crucial role of sucrose synthase (Sus) in cotton fiber cell and seed development. (A) In the wild type, Sus is expressed specifically in the fiber cells (f) from 0 to 3 days after anthesis (DAA). As the seed develops further, Sus is expressed both in the seed coat fiber and transfer cells (tc) and in the endosperm and embryo (e). (B) In the Sus-suppressed lines, reduction of Sus activity by 70 percent or more results in a fiberless seed phenotype at 0 to 3 DAA. As the seed develops to 25 DAA, a portion of the transgenic seeds had Sus suppression only in maternal tissue due to transgene segregation or inactivation in the filial tissues (Type I). These seeds are identical to the wild seed except for much reduced fiber growth. The remaining seeds, however, showed Sus suppression in both maternal and filial tissues (Type II). These seeds are shrunken and inviable. e: Endosperm and embryo; f: Fiber; n: Nucellus; sc: Seed coat; tc: Transfer cell. (See Ruan et al., 2003, for more details.)

Adenosine 5'-Diphosphoglucose (ADPG) Pyrophosphorylase (AGPase)

AGPase catalyzes a reversible synthesis of ADPG and PPi from ATP and glucose-1-phosphate. AGPase provides the substrate, ADPG, which constitutes the first step that commits a glucose moiety of sucrose to starch biosynthesis. There is substantial biochemical, genetic, and molecular evidence that AGPase plays a crucial role in regulating starch biosynthesis in plant cells, including developing seeds. Tsai and Nelson (1966) and Dickinson and Preiss (1969) provided the initial data to show that severe starch deficiency of the *shrunken-2 (sh2)* endosperm mutant is due to the loss of endosperm-specific AGPase activity in maize. The native AGPase enzyme is a heterotetramer in several diverse plant species, including *Arabidopsis*, barley, maize, spinach, and wheat (reviewed by Preiss, 1988). In maize, the enzyme is composed of two subunits, large and small, encoded by the *Sh2* and *brittle-2 (Bt2)* loci, respectively (Preiss et al., 1990). Both genes have been cloned and well analyzed in terms of gene structure and function relationships (Bae et al., 1990; Bhave et al., 1990; Giroux and Hannah, 1994). In regard to its intracellular localization, especially in developing endosperm, the results thus far indicate that the enzyme might be present in both the amyloplast and cytosol of the cell (Miller and Chourey, 1995; Denyer et al., 1996; Beckles et al., 2001). Several recent molecular studies clearly demonstrate the critical role of AGPase in starch biosynthesis in developing seeds. Specifically, Giroux et al. (1996) reported a mutant with two amino acid substitutions in the *Sh2*-encoded subunit of AGPase that led to increased starch content, and ultimately, 11 to 18 percent increase in seed weight in maize. Another mutant, *Sh2r6hs*, is altered in the allosteric domain of the *Sh2*-encoded protein such that it shows reduced sensitivity to phosphate, a long-known inhibitor of AGPase activity. Most remarkably, transgenic wheat expressing the maize *Sh2r6hs* gene also led to on average 38 percent more seed weight per plant due to enhanced sink strength (Smidansky et al., 2002). In maize, AGPase is believed to be one of the most heat-labile enzymes of the starch biosynthetic pathway. A heat-stable variant of the *Sh2* subunit led to enhanced interactions with the *Bt2* subunit (Greene and Hannah, 1998). Such a heterotetramer was heat stable, leading to increased AGPase activity and ultimately higher crop yields (Greene and Hannah, 1998). Overall, these results provide a direct link between AGPase activity and sink strength in developing seeds of maize.

Sucrose-Phosphate Synthase

The biosynthesis of sucrose is catalyzed by the sequential reactions of sucrose-phosphate synthase (UDP-glucose + Fructose-6-P $\leftrightarrow$ Sucrose-6'-P

+ UDP) and sucrose-6'- phosphate phosphatase (Sucrose-6'-P + $H_2O \rightarrow$ Sucrose + Pi). These two enzymes play a pivotal role in the sucrose synthesis of photosynthetic leaves. The activity of sucrose phosphate-synthase is highly regulated through reversible phosphorylation on multiple serine residues of the proteins (Winter and Huber, 2000, and literature therein). Sucrose-phosphate synthase is also expressed in many nonphotosynthetic tissues, including developing seeds. Its possible role in carbon partitioning within seed is discussed as follows.

During seed development, a resynthesis of sucrose from imported hexoses may occur in the embryo or endosperm via sucrose-phosphate synthase. For example, the activity of this enzyme, probably utilizing hexose derived from the activity of cell wall-bound invertase, increases sharply at the end of the cell division phase of *Vicia faba* cotyledons (Weber et al., 1995). The significance of a highly active sucrose-phosphate synthase at this stage may be related to the synthesis of sucrose, thereby decreasing the ratio of hexose to sucrose to a level that induces the storage activity (Weber, Borisjuk, Heim, and Wobus, 1997). Similarly, this enzyme has been localized to the basal region in the developing endosperms of maize where cell wall-bound invertase is expressed (Cheng and Chourey, 1999). It is speculated that sucrose is resynthesized at this region for further cleavage by cell wall bound invertase, as metabolic release of hexose through the invertase appears to be critical for the normal development of the maize kernel (Chen and Chourey, 1999). Sucrose is also metabolically more stable than hexoses. Thus, an alternative possibility is that the resynthesized sucrose may be suitable for being transported to the upper part of the maize endosperm for starch biosynthesis via the sucrose synthase pathway. Pertinently, resynthesis of sucrose from hexoses produced by apoplastic invertase has also been described for the storage parenchyma cells of sugar cane (Hatch et al., 1963).

Sucrose-phosphate synthase may also be coordinately expressed with sucrose synthase in some seeds. In the fiber cells of developing cotton seed, for example, the activities of both enzymes increase at the onset of secondary wall deposition (Haigler et al., 2001). In vitro experiments show that much of the glucose supplied to the cultured ovules is subsequently converted back to sucrose within the attached fibers, indicating cotton fibers do synthesize sucrose (Carpita and Delmer, 1981). Given that about 50 percent of sucrose synthase proteins are associated with plasma membrane (Amor et al., 1995), the concurrent sucrose degradation (via the membrane-bound sucrose synthase) and resynthesis (via sucrose-phosphate synthase) could be a manifestation of enhancing cellulose synthesis through recycling the fructose released (Haigler et al., 2001). Efficient removal of fructose by sucrose-phosphate synthase is advantageous for maintaining the activity of sucrose synthase, as fructose is an end-product inhibitor of sucrose synthase (Doehlert, 1987). A linkage between the expression of sucrose-phosphate synthase and sucrose synthase also exists in some fruits. Nguyen-Quoc et al. (1999) showed that over-expression of sucrose-phosphate synthase in young tomato fruit increases sucrose synthase activity, leading to enhanced sucrose import and cleavage and fruit yield. Unfortunately, in this report, no data were provided regarding the possible impact on the seed development within the fruit.

In conclusion, sucrose-phosphate synthase is expressed concurrently with invertase or sucrose synthase in developing seeds of many species. Together, these enzymes may perform a cycle of sucrose cleavage and resynthesis, also known as "futile cycle" (Geigenberger and Stitt, 1993). It is important to note, however, that direct evidence on the role of sucrose-phosphate synthase in seed development is lacking thus far. Genetic studies on this enzyme are impeded by the absence of mutants defective in sucrose-phosphate synthase expression. Further molecular studies are urgently needed to determine the role of this enzyme in carbon partitioning in developing seed.

Sugar Transporter

The seed filial tissue is symplastically isolated from the maternal tissue. This provides a structural basis for two mandatory transport steps of assimilates across plasma membranes in seed: efflux to the seed apoplast from the maternal tissue and subsequent uptake by the embryo or endosperm from the apoplast (see Figures 5.1 and 5.2). A large body of evidence points to the central role of sugar transporters in these membrane transport processes during seed development.

First, sugar transporter genes are abundantly expressed at the interface of the maternal and filial tissues of developing seed in a wide range of species. In *Vicia faba,* for example, physiological experiments established that sucrose efflux from the seed coat is facilitated by a putative sucrose-H^+ antiporter with the energy provided by a H^+-ATPase located on the plasma membrane of the thin wall parenchyma transfer cells of the seed coat (Fieuw and Patrick, 1993). Once loaded to the seed apoplast, sucrose is taken up into the filial cells by a sucrose-H^+ symporter (SUT), predominantly located on the plasma membrane of transfer cells at the epidermis of the developing cotyledons (McDonald et al., 1996). To understand the role of sugar transporter in seed development at the molecular level, Weber, Borisjuk, Heim, Sauer, and Wobus (1997) cloned two cDNAs from developing seed of *Vicia faba,* encoding one sucrose and one hexose transporter, designated *VfSUT1* and *VfSTP1*, respectively. The proteins encoded by both genes posses the typical 12 membrane-spanning domains and are characterized as sugar-H^+ symporters. The transcripts of *VfSUT1* and *VfSTP1* are detected in the thin wall parenchyma of the seed coat, indicating their roles in either retrieval or efflux of sucrose and hexoses (Weber, Borisjuk, Heim, Sauer, and Wobus, 1997). Both transporter genes are also expressed in the cotyledonary epidermis. However, the location and timing are different. The *VfSTP1* mRNA accumulates during the midcotyledon stage in epidermal cells covering the mitotically active parenchyma, whereas the *VfSUT1* transcript is specific to the outer epidermal cells showing transfer cell morphology and covering the storage parenchyma (Weber, Borisjuk, Heim, Sauer, and Wobus, 1997). The expression of *VfSUT1* in the epidermis of the cotyledons increases the sucrose concentration and starch accumulation underneath this cell layer (Borisjuk et al., 2002). These results support the view that a hexose carrier plays an important role in transport of hexoses for stimulating cell division, while expression of a sucrose carrier is required for delivery of sucrose for storage activity during the seed development of *Vicia faba*.

In cereals, gene expression of sucrose-H^+ symporter and plasma membrane H^+ -ATPase are detected in the nucellar projection transfer cells of the maternal tissues and also in the modified aleurone/subaleurone transfer cells of the wheat grain endosperm (Bagnall et al., 2000).

Interestingly, the function of sucrose-H^+ symporteris is detected only in the filial tissues, not in the maternal tissues (Bagnall et al., 2000). This raises the possibility that the sucrose carrier located in the nucellar projection transfer cells might function as a sucrose-H^+ antiporter. Genes encoding such an antiporter, however, have not been cloned thus far. The results show that sucrose uptake from the endosperm cavity into filial tissues is mediated by a SUT localized to the plasma membranes of the modified aleurone/subaleurone transfer cells (Bagnall et al., 2000). In rice, a cDNA encoding a sucrose-H^+ symporter *(OsSUT1)* has been cloned from developing grain (Hirose et al., 1997). Both transcript and protein of *OsSUT1* are localized to the aleurone cells of the endosperm and are also detected in the maternal tissues, particularly the nucellus, vascular tissue, and nucellar projection (Furbank et al., 2001). Although the role of the *OsSUT1* in the maternal tissue is less clear, it appears certain that the transporter plays an important role in uptake of sucrose into the developing endosperm of rice (Furbank et al., 2001).

Hexose transporters are frequently found to be coexpressed with cell wall invertases in developing seeds. For example, genes encoding two hexose transporters (*HvSTP1* and *HvSTP2*) and two cell wall-bound invertases (*HvCWINV1* and *HvCWINV2*) have been cloned from barley

(Weschke et al., 2003). *HvCWINV2* and *HvSTP2* are expressed in the pericarp at the very early stage of seed development, indicating their roles in the development of the maternal tissue (Weschke et al., 2003). As the seed develops to the onset of starch filling, *HvCWINV1* and *HvSTP1* are strongly expressed in the outermost area of the nucellar projection and in the endosperm transfer cells. This coexpression pattern suggests that hexoses liberated by the invertase within the endosperm cavity are taken up by the hexose transporters for supplying mitotically active endosperm cells with hexoses (Weschke et al., 2003). Coexpression of invertase and hexose transporters has also been reported in developing seeds of faba bean (Weber et al., 1995; Weber, Borisjuk, Heim, Sauer, and Wobus, 1997) and may also occur early in development of rice grain (Hirose et al., 2002). Such cooperation between the two genes could have dual roles in seed development: first, enhancing phloem unloading of sucrose into seed by rapid hydrolysis and removal of sucrose, and second, promotion of seed cell division by elevating hexose levels.

Expression of sugar transporters occurs not only at the symplastically disconnected interface between mater and filial tiisues, but also in some cell types where the existing plasmodesmata decrease their permeability for solute transport. In developing cotton seed, a sucrose transporter *(GhSUT1)* is highly expressed at the base of the fiber cells interconnecting the underlying seed coat, coincident with the transient closure of plasmodesmata (Ruan et al., 2001). This follows a rapid elongation of the fiber cells. It is proposed that this elevated expression of *GhSUT1* and the closure of plasmodesmata generates a higher turgor, which "inflates" the single-celled cotton fiber up to about 3.0 cm long (Ruan and Patrick, 1995; Ruan et al., 2001).

Although ample data are available regarding the expression patterns of sugar transporters in seed, only a few studies have been reported on the direct role of these gene in seed development using reverse or forward genetic approaches. To determine the role of *OsSUT1* in rice seed development, Scofield et al. (2002) transformed rice with an *OsSUT1* antisense construct, driven by the maize ubiquitin-1 promoter. They found that suppression of *OsSUT1* expression impaired grain filling and reduced final grain weight, but without affecting leaf photosynthesis. This provides direct evidence for the requirement of a sucrose transporter for grain filling in a cereal species. In developing seed of pea, the signal intensity of a sucrose-H^+ symporter *(PsSUT1)* protein is most pronounced in the thin wall parenchyma cells of seed coat and epidermal transfer cells of cotyledons (Tegeder et al., 1999). To determine the role of *PsSUT1* in cotyledon development, Rosche et al. (2002) transformed pea with an overexpression construct containing a potato sucrose transporter cDNA, driven by a cotyledon-specific vicilin promoter. The transgene increases transport activity of cotyledon storage parenchyma tissues by approximately twofold, leading to higher rates of biomass gain by excised cotyledons cultured on a sucrose medium (Rosche et al., 2002). This provides an example of the potential capacity for *SUT* gene expression in promoting the growth of seed filial tissue.

CONCLUSIONS AND FUTURE PERSPECTIVES

There is no doubt that carbon partitioning plays a central role in seed development. Sucrose is the predominant form of carbon imported to developing seed. Once unloaded from the phloem in the maternal tissues of the seed, sucrose or its derivatives can move to the rest of the seed through either symplastic or apoplastic pathways or a combination of both, depending on the developmental stages of and cellular locations within the seed. Sucrose has to be degraded into hexoses by invertase or sucrose synthase for the use of metabolism and biosynthetic processes. Within the seed, sucrose may be resynthesized from hexoses by sucrose phosphate-synthase at some stages. Sucrose and hexose carriers could play important roles in the movement of sugars within the seed, particularly at the interface between the maternal and filial tissues. Apart from their roles as substrates for metabolism and biosynthesis, sucrose and hexoses may also serve as signaling molecules to regulate the development of seeds.

Despite the significant accomplishments made over the past two decades or so, many important questions remain to be answered regarding the control of carbon partitioning in developing seed. What is the molecular nature of the sugar-H^+ antiporter? How important is the resynthesis of sucrose in seed development? Do increases in endogenous hexose and sucrose concentrations result in enhanced mitotic and storage activity of seed cells, respectively? What is the regulatory mechanism(s) for transfer cell development in seeds? What is the precise role of endosperm in providing carbon for young embryo during development of dicot seeds?

It is clear that answers to these questions may only come from studies using multidisciplinary approaches. Gain- or loss-of-function studies coupled with insightful analysis of the developmental and cell-specific seed phenotypes are now emerging to provide novel information on carbon partitioning in developing seeds at the cellular, biochemical, and molecular levels (e.g., Cheng et al., 1996; Rosche et al., 2002; Ruan et al., 2003).

REFERENCES

Amor, Y., C.H. Haigler, S. Johnson, M. Wainscott, and D.P. Delmer (1995). A membrane-associated form of Sus and its potential role in synthesis of cellulose and callose in plants. *Proceeding of National Academy of Science USA* 92: 9353-9357.

Andersen, M.N., F.A. Asch, Y. Wu, C.R. Jensen, H. Naesred, V.O. Mogensen, and K.E. Koch (2002). Soluble invertase expression is an early target of stress during the critical, abortion sensitive phase of young ovary development in maize. *Plant Physiology* 130: 591-604.

Bae J.M., M. Girou, and L.C. Hannah (1990). Cloning and characterization of the *brittle-2* gene of maize. *Maydica* 35: 317-322.

Bagnall, N., X.-D. Wang, G.N. Scofield, R.T. Furbank, C.E. Offler, and J.W. Patrick (2000). Sucrose transport-related genes are expressed in both maternal and filial tissues of developing wheat grains. *Australia Journal of Plant Physiology* 27: 1009-1020.

Basra, A. and C.P. Malik (1984). Development of the cotton fiber. *International Review of Cytology* 89: 65-113.

Beckles, D.M., A.M. Smith, and T. ap Rees (2001). A cytosolic ADP-glucose pyrophosphorylase is a feature of Graminaceous endosperms, but not of other starch-storing organs. *Plant Physiology* 125: 818-827.

Bhave M.R., S. Lawrence, C. Barton, and L.C. Hannah (1990). Identification and molecular characterization of the shrunken-2 cDNA clones of maize. *The Plant Cell* 2: 581-588.

Bonnemain, J.-L., S. Bourquin, S. Renault, C. Offler, and D.G. Fisher (1991). Transfer cells: Structure and physiology. In J. Bonnemain, S.S. Delrot, J. Dainty, and W.J. Lucas (eds), *Recent advances in phloem transport and assimilate compartmentation* (pp. 74-83). Nantes, France: Quest Editions, Presses Academiques.

Borisjuk, L., S. Walenta, H. Rolletschek, W. Mueller-Klieser, U. Wobus, and H. Weber (2002). Spatial analysis of plant metabolism: Sucrose imaging within *Vicia faba* cotyledons reveals specific developmental patterns. *The Plant Journal* 29: 521-530.

Borisjuk, L., S. Walenta, H. Weber, W. Mueller-Klieser, and U. Wobus (1998). High-resolution histographical mapping of glucose concentrations in developing cotyledons of *Vicia faba* in relation to mitotic activity and storage process: Glucose as a possible developmental trigger. *The Plant Journal* 15: 583-587.

Carlson, S.J. and P.S. Chourey (1996). Evidence for plasma membrane-associated forms of sucrose synthase in maize. *Molecular General Genetics* 252: 303-310.

Carlson, S.J. and P.S. Chourey (1999). A re-evaluation of the relative roles of two invertases, *INCW2* and *IVR1*, in developing maize kernels and other tissues. *Plant Physiology* 121: 1025-1035.

Carlson, S.J., P.S. Chourey, T. Helentjaris, and R. Datta (2002). Gene expression studies on developing kernels of maize sucrose synthase (SuSy) mutants show evidence for a third SuSy gene. *Plant Molecular Biolology* 49: 15-29.

Carpita, N.C. and D.P. Delmer (1981). Concentration and metabolic turnover of UDP-glucose in developing cotton fibers. *Journal of Biological Chemistry* 256: 308-315.

Chen, Y.-C. and P.S. Chourey (1989). Spatial and temporal expression of the two sucrose synthase genes in maize: Immunohistological evidence. *Theoretical and Applied Genetics* 78: 553-559.

Cheng, W.-H. and P.S. Chourey (1999). Genetic evidence that invertase-mediated release of hexoses is critical for appropriate carbon partitioning and normal seed development in maize. *Theoretical and Applied Genetics* 98: 485-495.

Cheng, W.-H., E.W. Taliercio, and P.S. Chourey (1996). The *Miniature* seed locus of maize encodes a cell-wall invertase required for normal development of endosperm and maternal cells in the pedicel. *The Plant Cell*: 971-983.

Chourey, P.S. (1981) Genetic control of sucrose synthetase in maize endosperm. *Molecular General Genetics* 184: 372-376.

Chourey, P.S. and O. Nelson (1976). The enzymatic deficiency conditioned by the *shrunken 1* mutations in maize. *Biochemical Genetics* 14: 1041-1055.

Chourey, P.S. and O. Nelson (1979). Interallelic complementation at the *sh* locus in maize at the enzyme level. *Genetics* 91: 317-325.

Chourey, P.S., E.W. Taliercio, S.J. Carlson, and Y.-L. Ruan (1998). Genetic evidence that the two isozymes of sucrose synthase present in developing maize endosperm are critical, one for cell wall integrity and the other for starch biosynthesis. *Molecular General Genetics* 259: 88-96.

Davis, R.W., J.D. Smith, and B.G. Cobb (1990). A light and electron microscope investigation of the transfer cell region of maize caryopses. *Canadian Journal of Botany* 68: 471-479.

Déjardin, A., C. Rochat, S. Wuillém, and J.-P. Boutin (1997). Contribution of sucrose synthase, ADP-glucose pyrophosphorylase and starch synthase to starch synthesis in developing pea seeds. *Plant Cell and Environment* 20: 1421-1430.

Denyer, K., F. Dunlap, T. Thorbjornsen, P. Keeling, and A.M. Smith (1996). The major form of ADP-glucose pyrophosphorylase in maize endosperm is extra-plastidial. *Plant Physiology* 112: 779-785.

Dickinson, D.B. and J. Preiss (1969). Presence of ADP-glucose pyrophosphorylase in *shrunken-2* and *brittle-2* mutants of maize endosperm. *Plant Physiology* 44:1058-1062.

Doehlert, D.C. (1987). Substrate inhibition of maize endosperm sucrose synthase by fructose and its interaction with glucose inhibition. *Plant Science* 52: 153-157.

Felker, F.C. (1992). Participation of cob tissue in the uptake of medium component by maize kernels cultured in vitro. *Journal of Plant Physiology* 139: 647-652.

Felker, F.C. and J.C. Shannon (1980). Movement of 14C-labelled assimilates into kernels of *Zea mays* L.: An anatomical examination and microautoradiographic study of assimilate transfer. *Plant Physiology* 65: 864-870.

Fieuw, S. and J.W. Patrick (1993). Mechanism of photosynthate efflux from *Vicia faba* L. seed coats: I. Tissue studies. *Journal of Experimental Botany* 44: 65-74.

Furbank, R.T., G.N. Scofield, T. Hirose, X.-D. Wang, J.W. Patrick, and C.E. Offler (2001). Cellular localization and function of a sucrose transporter OsSUT1 in developing rice grains. *Australian Journal of Plant Physiology* 28: 1187-1196.

Geigenberger, P. and M. Stitt (1993). Sucrose synthase catalyses a readily reversible reaction in vivo in developing potato tubers and other plant tissues. *Planta* 189: 329-393.

Giroux, M.J. and L.C. Hannah (1994). ADPglucose pyrophosphorylase in *sh2* and *bt2* mutants of maize. *Molecular General Genetics* 243: 400-408.

Giroux, M.J., J. Shaw, G. Barry, B.G. Cobb, T. Greene, T. Okita, and L.C. Hannah (1996). A single gene mutation that increases maize seed weight. *Proceeding of National Academy of Sciences USA* 93: 5824-5829.

Greene, T.W. and L.C. Hannah (1998). Enhanced stability of maize endosperm ADP-glucose pyrophosphorylaseis gained through mutants that alter subunit interactions. *Proceeding of National Academy of Sciences USA* 95: 13342-13347.

Haigler, C.H., M. Ivanova-Datcheva, P.S. Hogan, V.V. Salnikov, S. Hwang, K. Martin, and D.P. Delmer (2001). Carbon partitioning to cellulose synthesis. *Plant Molecular Biology* 47: 29-51.

Hanft, J.M. and R.J. Jones (1986). Kernel abortion in maize: II. Distribution of ^{14}C among kernel carbohydrates. *Plant Physiology* 81: 511-515.

Hardham, A.R. (1976). Structural aspects of the pathways of nutrient flow to the developing embryo and cotyledons of *Pisum sativum* L. *Australian Journal of Botany* 24: 711-721.

Hatch, M.D., J.M. Sacher, and K.T. Glaszious (1963). Sugar accumulation cycle in sugar cane: I. Studies in enzymes of the cycle. *Plant Physiology* 38:338-343.

Hirose, T., N. Imaizumi, G.N. Scofield, and R.T. Furbank, and R. Ohsugi (1997). CDNA cloning and tissue-specific expression of a sucrose transporter from rice. *Plant and Cell Physiology* 38: 1389-1396.

Hirose, T., M. Takano, and T. Terao (2002). Cell wall invertase in developing rice caryopsis: Molecular cloning of *OsCIN1* and analysis of its expression in relation to its role in grain filling. *Plant and Cell Physiology* 43: 452-459.

King, S.P., J.E. Lung, and R.T. Furbank (1997). Catbohydrate content and enzyme metabolism in developing canola siliques. *Plant Physiology* 114:153-160.

Koch, K.E. (1996). Carbohydrate-modulated gene expression in plants. *Annual Review of Plant Physiology and Plant Molecular Biology* 47: 509-540.

Lee, H.-S. and A. Sturm (1996). Purification and characterization of neutral and alkaline invertase from carrot. *Plant Physiology* 112: 1513-1522.

McDonald, R., S. Fieuw, and J.W. Patrick (1996). Sugar uptake by the dermal transfer cells of developing cotyledons of *Vicia faba* L.: Experimental systems and general transport properties. *Planta* 198: 54-63.

Miller, M.E. and P.S. Chourey (1992). The maize invertase-deficient *miniature-1* seed mutation is associated with aberrant pedicel and endosperm development. *The Plant Cell* 4: 297-305.

Miller, M.E. and P.S. Chourey (1995). Intracellular immunolocalization of adenosine 5'-diphosphoglucose pyrophosphorylase in developing endosperm cells of maize (*Zea mays* L.). *Planta* 197: 522-527.

Nguyen-Quoc, B., H. N'Tchobo, C.H. Foyer, and S. Yelle (1999). Overexpression of sucrose phosphate synthase increases sucrose unloading in transformed tomato fruit. *Journal of Experimental Botany* 50: 785-791.

Offler, C.E., S.M. Nerlich, and J.W. Patrick (1989). Pathway of photosynthate transfer in the developing seed of *Vicia faba* L.: Transfer in relation to seed anatomy. *Journal of Experimental Botany* 40: 769-780.

Oparka, K.J. and P. Gates (1981). Transport of assimilates in the developing caryopsis of rice (*Oryza sativa* L.): Ultrastructure of the pericarp vascular bundle and its connection with the aleurone layer. *Planta* 151: 561-573.

Oparka, K.J. and P. Gates (1982). Ultrastructure of the developing pigment strand of rice (*Oryza sativa* L.) in relation to its role in solute transport. *Protoplasma* 113: 33-43.

Patrick, J.P. and C.E. Offler (1995). Post-sieve element transport of sucrose in developing seeds. *Australian Journal of Plant Physiology* 22: 681-702.

Patrick, J.P., C.E. Offler, and X.-D. Wang (1995). Cellular pathway of photosynthate transport in coats of developing seed of *Vicia faba* L. and *Phaseolus vulgaris* L.: 1. Extent of transport through the seed coat symplast. *Journal of Experimental Botany* 46: 35-47.

Preiss, J. (1988). Biosynthesis of starch and its regulation. In J. Preiss (ed.), *The biochemistry of plants: A comprehensive treatise,* Volume 14 (pp. 182-249). New York: Academic Press.

Preiss, J., S. Donner, P.S. Summers, M. Morell, C.R. Barton, L.Yang, and M. Neider (1990). Molecular characterization of the *Brittle 1* effect on maize endosperm AGPase subunits. *Plant Physiology* 92: 881-885.

Rosche, E., D. Blackmore, M. Tegeder, T. Richardson, H. Schroeder, T.J. Higgins, W.F. Frommer, C.E. Offler, and J.W. Patrick (2002). Seed-specific overexpression of a potato sucrose transporter increases sucrose uptake and growth rates of developing pea cotyledons. *The Plant Journal* 30: 165-175.

Ruan, Y.-L. and P.S. Chourey (1998). A fiberless seed mutation in cotton is associated with lack of fiber cell initiation in ovule epidermis and alterations in sucrose synthase expression and carbon partitioning in developing seeds. *Plant Physiology* 118: 399-406.

Ruan, Y.-L., P.S. Chourey, P.D. Delmer, and L. Perez-Grau (1997). The differential expression of sucrose synthase in relation to diverse patterns of carbon partitioning in developing cotton seed. *Plant Physiology* 115: 375-385.

Ruan, Y.L., D.J. Llewellyn, and R.T. Furbank (2000). Pathway and control of sucrose import into initiating cotton fibers. *Australian Journal of Plant Physiology* 27: 795-800.

Ruan, Y.L., D.J. Llewellyn, and R.T. Furbank (2001). The control of single-celled cotton fiber elongation by developmentally reversible gating of plasmodesmata and coordinated expression of sucrose and K^+ transporters and expansin. *The Plant Cell* 13: 47-63.

Ruan, Y.L., D.J. Llewellyn, and R.T. Furbank (2002). Modification of sucrose synthase gene expression in plant tissue and uses therefor. Patent number: WO-0245485.

Ruan, Y.L., D.J. Llewellyn, and R.T. Furbank (2003). Suppression of sucrose synthase gene expression represses cotton fiber cell initiation, elongation and seed development. *The Plant Cell* 15: 952-964.

Ruan, Y.-L. and J.W. Patrick (1995). The cellular pathway of postphloem sugar transport in developing tomato fruit. *Planta* 196: 434-444.

Ryser, U. (1992). Ultrastructure of the epidermis of developing cotton (Gossypium) seeds: Suberin, pits, plasmodesmata, and their implication for assimilate transport into cotton fibers. *American Journal of Botany* 79: 14-22.

Ryser, U., M. Schorderet, U. Jauch, and H. Meier (1988). Ultrastructure of the "fringe-layer," the innermost epidermis of cotton seed coats. *Protoplasma* 147: 81-90.

Schmalstig, J.G. and W.D. Hitz (1987). Transport and metabolism of a sucrose analog (1'-fluorosucrose) into *Zea mays* L. endosperm without invertase hydrolysis. *Plant Physiology* 85: 902-905.

Scofield, G.N., T. Hirose, J.A. Gaudron, N.M. Upadhyaya, R. Ohsug, and R.T. Furbank (2002). Antisense suppression of the rice sucrose transporter gene, *OsSUT1*, leads to impaired grain filling and germination but does not affect photosynthesis. *Functional Plant Biology* 29: 815-826.

Smidansky, E.D., M. Clancy, F.D. Meyer, S.P. Lanning, N.K. Blake, L.E. Talbert, and M.J. Giroux (2002). Enhanced ADPglucose pyrophosphorylase activity in wheat endosperm increases seed yield. *Proceeding of National Academy of Sciences USA* 99: 1724-1729.

Tegeder, M., X.-D. Wang, W.F. Frommer, C.E. Offler, and J.W. Patrick (1999). Sucrose transport into developing seeds of *Pisum sativum* L. *The Plant Journal* 18: 151-161.

Thorne, J.H. (1985). Phloem unloading of C and N assimilates in developing seeds. *Annual Review of Plant Physiology* 36: 317-343.

Tsai, C.Y. and O.E. Nelson (1966). Starch-deficient maize mutant lacking adenosine diphosphate glucose pyrophosphorylase activity. *Science* 151: 341-343.

Tsai, C.Y., F. Salamini, and O.E. Nelson (1970). Enzymes of carbohydrate metabolism in the developing endosperm of maize. *Plant Physiology* 46: 299-306.

Vilhar, B., A. Kladnik, A. Blejec, P.S. Chourey, and M. Dermastia (2002). Cytometrical evidence that the loss of seed weight in the *miniature1* seed mutant of maize is associated with reduced mitotic activity in the developing endosperm. *Plant Physiology* 129: 23-30.

Wang, H.L., C.E. Offler, and J.W. Patrick (1994). Nucellar projection transfer cells in the developing wheat grain. *Protoplasma* 182: 39-52.

Wang, H.L., C.E. Offler, J.W. Patrick, and T.D. Ugalde (1994). The cellular pathway of photosynthate transfer in the developing wheat grain: I. Delineation of a potential transfer pathway using fluorescent dyes. *Plant, Cell and Environment* 17: 257-266.

Wang, X.-D., G. Harrington, J.W. Patrick, C.E. Offler, and S. Fieuw (1995). Cellular pathway of photosynthate transport in coats of developing seed of *Vicia faba* L. and *Phaseolus vulgaris* L.: II. Principal cellular site(s) of efflux. *Journal of Experimental Botany* 46: 49-63.

Weber, H., L. Borisjuk, U. Heim, P. Buchner, and U. Wobus (1995). Seed coat-associated invertase of fava bean control both unloading and storage functions: Cloning of cDNAs and cell type-specific expression. *The Plant Cell* 7: 1835-1846.

Weber, H., L. Borisjuk, U. Heim, and U. Wobus (1996). Controlling seed development and seed size in *Vicia faba:* A role for seed coat-associated invertase and carbohydrate state. *The Plant Journal* 10: 823-834.

Weber, H., L. Borisjuk, U. Heim, and U. Wobus (1997). Sugar import and metabolism during seed development. *Trends in Plant Science* 2: 169-174.

Weber, H., L. Borisjuk, U. Heim, N. Sauer, and U. Wobus (1997). A role for sugar transporters during seed development: Molecular characterization of hexose and sucrose carriers in fava bean seeds. *The Plant Cell* 9: 895-908.

Weschke, W., R. Panitz, S. Gubatz, W. Wang, R. Radchuk, H. Weber, and U. Wobus (2003). The role of invertases and hexose transporters in controlling sugar ratios: I. Maternal and filial tissues of barley caryopses during early development. *The Plant Journal* 33: 395-411.

Winter, H. and S.C. Huber (2000). Regulation of sucrose metabolism in higher plants: Localization and regulation of activities of key enzymes. *Critical Reviews in Plant Sciences* 19: 31-67.

Zee, S.Y. and T.P. O'Brien (1970). Studies on the ontogeny of the pigment strand in the caryopsis of wheat. *Australian Journal of Biological Sciences* 23: 1153-1171.

Chapter 6

Metabolic Engineering of Carbohydrate Supply in Plant Reproductive Development

Marc Goetz
Thomas Roitsch

THE IMPACT OF PLANT REPRODUCTION EVENTS ON AGRICULTURE

Harvest yields depend on pollination and fertilization events, which are strongly influenced by various environmental and endogenous factors. Fruit and seed can, with few exceptions, be formed only after pollination and fertilization have successfully taken place. Thus, much effort is spent to optimize plant sexual reproduction for successful farming and increase of crop yield.

One important way to increase harvest yields in agriculture is the use of hybrid plants. These hybrids are distinguished by being derived from genetically different parental lines and not emerging from self-pollination that is predominant in most agricultural species. Their special importance is based on the so-called heterosis effect. These hybrids show enhanced resistance against diseases, increased prevalence in different environmental conditions, and higher yields than the parental lines (Brewbaker, 1964; Mayo, 1980; Feistritzer and Kelly, 1987). However, the molecular basis of this heterosis effect is still unknown. To generate hybrids, self-pollination of flowers, which occurs in many species, has to be prevented to allow cross-pollination. Male sterile plants are of special agricultural interest in this context, since they offer an easy and inexpensive opportunity to prevent self-pollination and create hybrid plants. In recent years different strategies

---

The authors thank W. Tanner, University of Regensburg, Germany, for his continued interest and support. The fruitful collaboration with Anne Guivarc'h and Dominique Chriquie, Universite Pierre & Marie Curie, Paris, France, and U. Kahmann, University of Bielefeld, Germany, is also gratefully acknowledged.

have been employed to create male sterile mutants. Most studies include modifications within the tapetum, a specialized anther tissue, that lead to male sterility.

In addition to hybrid seed production, male sterile lines facilitate breeding programs and are considered to be an efficient safety measure to prevent outcrossing of transgenes of genetically engineered plants.

The present review will focus on the generation of male sterility by a novel approach, the interference with the carbohydrate supply to the developing male gametophyte.

THE ROLE OF THE TAPETUM IN MALE GAMETOPHYTE DEVELOPMENT

Plant reproductive tissues are formed within the flowers in a complex sequence of developmental events. Those are precisely timed to ensure that male and female gametophyte development are coordinated so that mature pollen can find receptive stigmata and ovules for pollination and fertilization. Male gametophyte development takes place within the anthers. The anthers consist of reproductive and nonreproductive tissues that are responsible for production and release of the pollen grains. One of the tissues having critical importance in the processes leading to pollen formation and development is the tapetum (Pacini et al., 1985). The tapetum surrounds the pollen sacs during early phases of development, degenerates at later stages of pollen development, and is degraded in mature anthers. Three main responsibilities of the tapetum are known that contribute to the development of the pollen.

1. It secretes nutrients into the loculus to nourish the symplastically isolated microspores (Clément and Audran, 1995).
2. It is important in releasing the young haploid microspores from the callose wall surrounding the meiotic tetrades by secreting an alpha-1,3 glucanase (Steiglitz, 1977; Bucciaglia and Smith, 1994).
3. It produces components used for the biosynthesis of the outer pollen wall. This happens during degeneration of the tapetum. The degeneration products, tryphine and pollenkit, are deposited on the maturing pollen grains (Mascarenhas, 1990; Pacini and Franchi, 1991). These substances are mostly lipoid and serve to protect the pollen grains from dehydration or to attract insects and stick to them to be carried to other flowers for pollination.

A number of other functions of the tapetum are not clearly proven, but it is evident that the tapetum has a critical role in the correct formation and development of pollen.

STRATEGIES TO GENERATE MALE STERILITY IN PLANTS

Pollen development is a complex developmental process that is sensitive to mutations. Many cytoplasmic and nuclear mutations leading to male sterility have been shown to interfere with tapetal development and function (Kaul, 1988; Chaudhury, 1993), supporting a critical function of this tissue for the production of functional pollen grains. Experiments to introduce male sterility in plants therefore aim to interfere with the specific functions of the tapetum outlined previously to prevent correct pollen formation and development.

One strategy used is the expression of certain degrading enzymes to disturb normal tapetum function. Mariani et al. (1990) expressed a ribonuclease from *Bacillus amyloliquefaciens* (barnase) in the tapetum of plants. As a result, the tapetum cannot develop properly and no pollen can be formed in these plants. In other experiments the premature degradation of the callose wall surrounding the tetrads by expression of an alpha-1,3 glucanase in the tapetum led to abortion of pollen development (Worral et al., 1992; Tsuchiya et al., 1995).

Other approaches disturbed pollen development by interfering with biosynthesis pathways of important substances needed for correct pollen formation such as flavonols (van der Meer et al., 1992; Napoli et al., 1999) or jasmonate (Stintzi and Browse, 2000), by expressing bacterial genes affecting normal pollen development (Schmülling et al., 1988), or by impairing mitochondrial function in the tapetum (Hernould et al., 1998). Recently it was shown that interference with the carbohydrate supply of pollen during development by reducing extracellular invertase activity in the tapetum leads to a block in pollen formation, causing male sterility (Goetz et al., 2001).

IMPORTANCE OF CARBOHYDRATES IN PLANT GROWTH AND DEVELOPMENT

Carbohydrate partitioning between the autotrophic source tissues and a variable number of sink tissues competing for a common pool of carbohydrates is a highly dynamic process that accompanies all stages of growth and development of higher plants. The supply of the transport sugar sucrose is a limiting step for the growth of sink tissues (Farrar, 1996), and sucrose-metabolizing enzymes are important determinants of sink capacity by generating a sucrose gradient to support unloading of sucrose from the phloem. For these reasons, the enzymes responsible for the first metabolic reaction of sucrose are likely critical links between photosynthate production in source leaves and growth capacity of sink organs (Farrar, 1996).

Sucrose is transported through the phloem along a concentration gradient from the place of synthesis to the receiving sink tissues and is driven by differences in osmotic potentials (Ho, 1988). The unloading of the sucrose from the phloem can either be symplastically via plasmodesmata or apoplastically via sucrose transporters. In symplastically isolated tissues such as embryos, meristems, stomata, or developing pollen, an apoplastic unloading mechanism is indispensable. After unloading into the apoplast via sucrose transporters the sucrose is cleaved into the hexose monomers glucose and fructose by extracellular, cell-wall-bound invertases, which in turn are taken up into the sink cells via high affinity hexose transporters (Roitsch and Tanner, 1996). The relevance of this functional coupling between extracellular invertases and hexose transporters is substantiated by their coordinated regulation after cytokinin induction (Ehneß and Roitsch, 1997). At the same time, extracellular invertases fulfill another important function, by building up and maintaining the sucrose concentration gradient between source and sink tissues that drives the carbohydrate distribution (Eschrich, 1980). The importance of extracellular cleavage of sucrose for assimilate partitioning, source-sink regulation and developmental processes has been proven in a number of studies in recent years, including the overexpression of a yeast invertase in the apoplast of transgenic tobacco plants (von Schaewen et al., 1990), the analysis of an invertase-deficient maize mutant (Miller and Chourey, 1992), the examination of developmental regulations during seed development (Weber et al., 1995), the study of the effect of antisense suppression of extracellular invertase in transgenic carrots (Tang et al., 1999), the regulation by various endogenous and exogenous stimuli (Roitsch et al., 2000), and the antisense repression of extracellular invertase during pollen development in tobacco (Goetz et al., 2001).

Extracellular invertases have been shown to be a major point of regulation by a great number of different stimuli, including phytohormones, sugars, elicitors, wounding, or pathogens (Sturm and Chrispeels, 1990; Ehneß and Roitsch, 1997; Tymowska-Lalanne and Kreis, 1998; Roitsch, 1999; Goetz et al., 2000; Roitsch et al., 2000). Moreover, the sugars involved are not only carbon and energy sources, but act also as signal molecules driving important developmental processes (Koch, 1996; Smeekens, 1998; Roitsch, 1999; Wobus and Weber, 1999).

It has been shown that carbohydrates play a critical role in anther and pollen development. They serve as nutrients for growth as well as signals that can influence developmental processes both in vivo and in vitro (Clément and Audran, 1995, 1999; Clément et al., 1996). Anthers possess the greatest sink strength within the plant (Wang and Breen, 1986), and it has been shown that the carbohydrate supply during growth is mainly in the form of sucrose, the main transport form of carbohydrates in plants.

CARBOHYDRATE SUPPLY AND MALE STERILITY

It has become clear that carbohydrate supply, the type of sugars available, and the functional coupling of invertases and sugar transporters are important in reproductive development in anthers. The main sugars found in anthers are sucrose, maltose, glucose, and fructose. Sucrose is cleaved to glucose and fructose in anthers by invertases (Miller and Ranwala, 1994; Dorion et al., 1996), leading to identical amounts of both monosaccharides inside the anthers throughout pollen development (Clément et al., 1996; Kawaguchi et al., 1996). During pollen maturation, a transient starch accumulation occurs and developing pollen seem to be autonomous and rely solely on stored carbohydrates.

Experiments with in vitro pollen cultures have also shown that the nature and amount of carbohydrates in pollen and the point at which the respective sugars are available to the pollen are critical for pollen development (Touraev and Herber le-Bors, 1999; Clément and Audran, 1999). The carbohydrate supply and its composition are fundamental for complete and correct pollen development both in vivo and in vitro. The importance of sucrose cleavage in anthers via invertase for providing carbohydrates to the male gametophyte was suggested for lily (Clément et al., 1996; Ranwala and Miller, 1998), tomato (Godt and Roitsch, 1997), and potato (Maddision et al., 1999). This was further supported by the identification of anther-specific expressed extracellular invertases in *Vicia faba* (Weber et al., 1996), *Lilium longiforum* (Singh and Knox, 1984; Miller and Ranwala, 1994), maize (Xu et al., 1996, Kim, Mahé, Brangeon, and Prioul, 2000), potato (Maddison et al., 1999), tobacco (Goetz et al., 2001), and tomato (Godt and Roitsch, 1997). The importance of supplying assimilates via an apoplasmic pathway involving cleavage of sucrose by invertase is further supported by the identification of anther-specific hexose transporters. In *Arabidopsis* a monosaccharide transporter was characterized that is expressed in developing pollen after the onset of symplasmic isolation of the microspore (Truernit et al., 1999). Likewise, monosaccharide transporters specifically expressed in anthers have been demonstrated in *Petunia* (Ylstra et al., 1998) and tobacco (Goetz and Roitsch, unpublished observations).

In accordance with the critical function of carbohydrates for pollen development and anther function, it was reported that various male sterile plants show a disturbed carbohydrate metabolism (Kaul, 1988; Hidalgo et al., 1999; Dorion et al., 1996, Datta et al., 2001; Biasi et al., 2001). A common effect of the different mutations characterized so far seems to be the perturbation of starch accumulation. In wheat and rice a water deficit during meiosis led to an arrest in pollen development that correlated with alterations in carbohydrate metabolism and a drastic decrease in invertase activity (Sheoran and Saini, 1996; Dorion et al., 1996; Saini, 1997). Likewise, a strong correlation between male sterility and invertase function was demonstrated for tobacco (Goetz et al., 2001). Combining the data available on the importance of carbohydrates for pollen formation and development and the distinctive involvement of invertases in carbohydrate partitioning leads to the conclusion that extracellular cleavage of sucrose by invertases is critically involved in these processes.

Inducing Male Sterility by Metabolic Engineering: Perturbation of the Carbohydrate Supply and Sugar Status During Pollen Development

The existence of specific extracellular invertase isoforms in anthers and the importance of carbohydrates for pollen formation and development offer a promising starting point to influence male fertility by metabolic engineering of the carbohydrate supply to the anthers and pollen. In tobacco an extracellular invertase was identified that shows a very specific spatial and temporal expression pattern. This invertase isoenzyme, Nin88, is expressed in the tapetum early in pollen development, until the tapetum is degraded, and found to be expressed in the pollen grain itself from the tetrad stage on, as shown by RNA in situ- and immuno-localization studies (Goetz et al., 2001).

Cloning of the Nin88 promoter allowed the cell-type-specific reduction of the extracellular invertase activity in these tissues by using an invertase-antisense construct expressed in transgenic tobacco plants under control of this promoter. The Nin88-antisense plants were identical to untransformed plants with respect to growth rate, height, morphology of vegetative and floral organs and tissues, time of flowering, and flower coloration pattern. However, a substantial portion of the antisense plants failed to produce seed capsules or showed a reduced seed number in the capsules. This inability or reduced ability to produce seeds correspond to the germination efficiency of pollen isolated from mature anthers in an in vitro germination assay. The plants that were unable to produce seeds showed a germination efficiency of less than 2 percent despite the fact that 90 percent of the pollen were viable.

Analysis of these plants showed a highly tissue-specific and significant reduction of extracellular invertase activity during pollen development. The reduction was most significant at the stage of microspore mitosis I, which shows a maximum of invertase activity in wild-type pollen development. This is supported by the finding that developmentally arrested pollen from the antisense plants are characterized by one nucleus, whereas mature pollen from wild-type plants possess two nuclei, as shown in Figure 6.1. Thus, the resulting carbohydrate deficiency during a critical stage of pollen development leads to a development block at the unicellular microspore stage. Further analysis revealed that the inability of pollen from NIN88-antisense plants to germinate is inversely correlate with the ability to accumulate starch.

Pollen from the Nin88-antisense plants is also characterized by a different morphology. Light microscopic inspection revealed that pollen from antisense plants represent a heterogeneous population with respect to pollen grains of different size and shape. The proportion of wild-type-like pollen of the various antisense lines corresponds to the germination efficiencies. The differences in morphology between wild-type pollen and pollen from antisense lines, in particular, are evident in analysis with electron microscopes, as shown in Figure 6.2. Analysis with the scanning electron

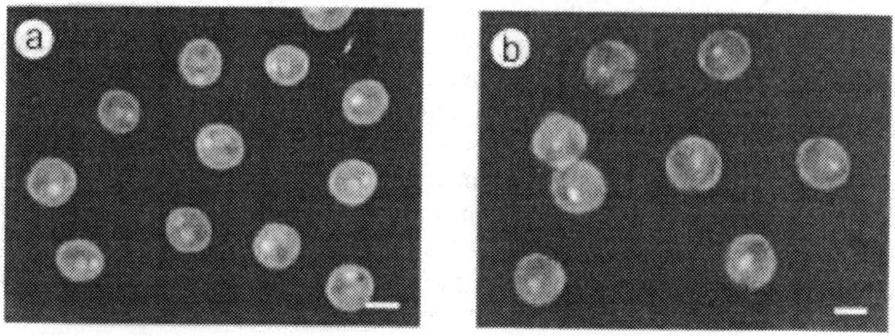

FIGURE 6.1. Antisense repression of extracellular invertase Nin88 results in a developmental arrest of pollen at the unicellular microspore stage. The nuclei of wild-type pollen (a) and pollen from antisense line NT23-81 (b) were visualized by staining the DNA with the fluorescent dye 4',-6-diamidinino-2-phenylindole (DAPI). Pollen of wild-type plants (a) are characterized by two nuclei, whereas pollen from line NT23-81, with a pollen germination efficiency of 0.6 percent, are characterized by one nucleus.

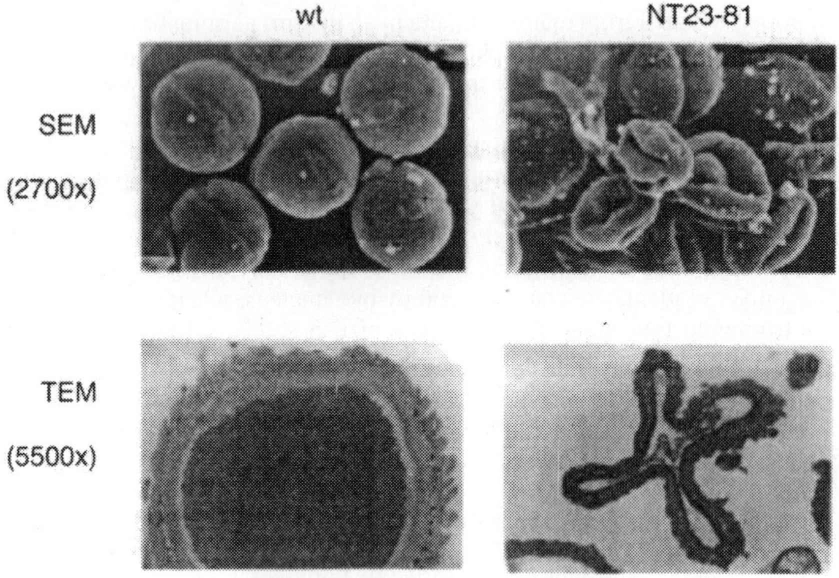

FIGURE 6.2. Effect of antisense repression of extracellular invertase Nin88 on pollen morphology. SEM and TEM electron microscopic pictures of pollen from wild type (wt) and antisense line NT23-81 with pollen germination efficiencies of 0.6 percent.

microscope revealed a distorted and invaginated morphology, indicating a reduced amount of intracellular material or turgor. This is further supported by cross-sections analyzed under the transmission electron microscope that demonstrate a highly reduced amount of intracellular material which apparently results in compression of the outer layer.

Using in vitro maturation assays, it could be demonstrated that this development can partially be overcome by supplying the pollen with external sugars as shown in Table 6.1. In vitro maturation assays of the transgenic pollen in the presence of either 250 mM glucose or sucrose led to cell division and further development with the typical starch accumulation. These results demonstrate that the developmentally arrested pollen are viable and in principle still possess the potential for further development. However, no complete maturation and subsequent germination could be achieved. The proportion of developing pollen from the invertase-antisense plants was about 60 percent lower when compared with wild-type pollen. Also, no significant difference was evident between the application of glucose and sucrose, despite the reduction in the sucrose-cleaving enzyme activity, showing that pollen are able to utilize different forms of carbohydrates.

TABLE 6.1. Effect of the sugar composition on in vitro pollen maturation.

Type of sugar used for in vitro pollen maturation	**Proportion of fully developed pollen**	
	Wild-type	**Line NT23-81**
500 mM Suc	53.4% (± 12.0%)	27.2% (± 3.4%)
250 mM Suc	57.9% (± 3.1%)	24.3% (± 0.6%)
100 mM Suc	18.8% (± 7.3%)	n.d.
50 mM Suc	8.3% (± 3.4%)	n.d.
500 mM Glc	44.9% (± 7.1%)	15.5% (± 0.7%)
250 mM Glc + 250 mM Suc	46.1% (± 5.2%)	15.3% (± 2.9%)
250 mM Glc + 250 mM Frc	37.0% (± 0.7%)	10.2% (± 0.7%)
In planta pollen development	81.4% (± 9.7%)	9.3% (± 3.3%)

n.d.: No data.

Accordingly, both sucrose and hexose transporters have been found in anthers and are assumed to be involved in anther/pollen development (Ylstra et al., 1998, Lemoine et al., 1999). These results indicate that the developmental block is not solely due to a perturbation of the carbohydrate supply but that the function of extracellular invertase is also critical for metabolic signalling required for pollen development. This assumption is backed up by other findings showing special signalling functions for hexoses and sucrose (Roitsch, 1999; Wobus and Weber, 1999). Sugars are known to be not only nutrients but also signal molecules which, like hormones and in combination with hormones and other stimuli, can regulate many aspects of the plant's life from gene expression to long-distance nutrient allocation. Sugar status, in particular the hexose/sucrose ratio, is important for cell division and development (Weber et al., 1996; Roitsch, 1999). Weber et al. (1996) showed that altered extracellular invertase activity in *Vicia faba* seed development and the changes in hexose/sucrose ratios influence seed size and development. A high hexose/sucrose ratio favors cell division activity, whereas a low hexose/sucrose ratio induces storage activities. The reduction of extracellular invertase activity in the pollen of the invertase-antisense plants therefore determines a low hexose/sucrose ratio interfering with the normal regulatory functions and thereby affecting cell division activity and normal development. The assumption that antisense repression of extracellular invertase also affects sugar sensing and signalling is further supported by the finding that antisense repression of the SnRK1 kinase also results in a developmental arrest of pollen development (Zhang et al., 2001). The SnRK1 has been proposed to be activated by high sucrose and/or low glucose levels, and the homologue of this kinase, SNF1, was shown to be a global regulator of carbon metabolism in yeast. Sugar regulation also was shown to be involved in the induction of extracellular invertase (Roitsch et al., 1995), resulting in a positive feed forward regulation to increase the sugar supply (Roitsch et al., 2000).

The data discussed suggest that tissue-specific antisense repression of extracellular invertase in the tapetum has a dual effect:

1. limiting or abolishing the carbohydrate supply and thus starving the developing pollen for the carbon source, which resulted in the inability to accumulate starch and other intracellular storage material, and
2. perturbation of the sugar status, thus affecting sugar sensing and signalling required as metabolic stimulus for regulating pollen development.

Metabolic Engineering of Carbohydrate Supply to the Anthers Is Specific and Does Not Affect Vegetative Plant Growth and Development

Unlike the situation in carrot plants, in which the antisense expression of an extracellular invertase under control of the 35S-promoter led to the interference of the regeneration of the transformed plants and resulted in highly pleiotropic effects (Tang et al., 1999), the invertase-antisense expression using the anther-specific promoter showed no effect on any other aspect of growth or development in the transgenic plants, except for failure to produce functional pollen (Goetz et al., 2001). Even anther development was not impaired and the sole effect was the developmental arrest of viable pollen. This makes the perturbation of the carbohydrate supply during reproductive development a subtle although highly efficient biotechnological method to engineer male sterility for practical applications in plants. Two lines of evidence suggest a broad applicability of the system. First, extracellular invertases are found and known to be central regulators in all plant species, and specific isoforms of extracellular invertases that are highly expressed in anthers and pollen have been found in many species. Second, it has been shown that the Nin88 promoter can function correctly in other plant species also. A *Nin88*-promoter-GUS fusion construct was also used for transformation of tomato plants to determine whether the Nin88 5' regulatory region also could function and be used in different plants. Histochemical detection of GUS activity in the transgenic plants demonstrated that the expression of the reporter gene driven by the *Nin88*-promoter was restricted to anthers and pollen and could not be detected in any other tissue (Goetz et al., 2001). This demonstrates that the 5' upstream sequences of the

Nin88 gene of tobacco are sufficient in other species to confer the same specific expression pattern as in tobacco and can be used to introduce the system into other agriculturally important species.

Applications for Inducing Male Sterility by Metabolic Engineering

The ability of inducing male sterility by metabolic engineering and the possibility to use this system in agricultural important species provide a new strategy to generate hybrid plants. Transferring this system to crop plants such as rape or maize should enable researchers to produce hybrids without mechanical emasculation of the plants. Although various systems have been already developed, various problems are still associated with the practical applications. The advantage would be that the perturbation by reducing the invertase activity is highly efficient while having minimal secondary effects.

Another application of interfering with pollen development through perturbing the carbohydrate supply be would be breeding programs where male sterile plants greatly facilitate the generation of improved lines for agriculture.

A further important possible application for this male sterility system is the use as biological safety measure for the increasing number of genetically modified organisms used in field trials and agricultural production. Male sterility excludes out-crossing of transgenes to wild-type species or to the same species growing in neighboring fields. Although the risk of out-crossing to wild-type species exists for only a few agronomically important species due to the lack of cross-pollinating plants in the natural ecosystems, the availability and use of such biological safety will help to cope with the increasing public awareness and fears of the potential spread of transgenes. However, in contrast to most crop species, the current efforts to genetically modify domestic woody plant species or medicinal herbs will require developing efficient methods to avoid possible outcrossing of transgenes to natural populations and forest ecosystems. Genetically engineered trees and pharmaceutically relevant species will be available in a few years for commercial applications, and their use will be preferentially in plantations. Specific lines of poplar that are characterized by dwarfism and early flowering should be valuable to evaluate the use of the tissue-specific promoters isolated from tobacco and tomato to perturb the carbohydrate supply by antisense repression of extracellular invertase to generate male sterile woody plant species.

FUTURE PERSPECTIVES

In plants in which the fruits are not the harvested products (e.g., lettuce, carrot), male sterile plants generated by perturbation of the carbohydrate supply can be crossed with any pollinator line to produce hybrid seeds. By contrast, in other plants in which the fruits are harvested (e.g., rice, corn, tomato), a restorer line is needed to restore full male fertility. For male sterile plants that were engineered by tissue-specific repression of an extracellular invertase, the restorer line could be plants that express a distantly related invertase, such as from bacteria or yeast that are not subject to the antisense repression through a plant invertase. Alternatively, one could use a repressor system. For this a prokaryotic regulatory element would have to be introduced in addition to the *Nin88* promoter in front of the antisense gene. The restorer plant would have to express a prokaryotic repressor protein under control of the *Nin88* promoter. After crossing, the repressor protein can bind to the regulatory element in the F1-hybrids, thereby suppressing the expression of the antisense gene and restoring normal extracellular invertase activity. A third possibility could be changing the way of supplying carbohydrate. The use of sucrose transporters could bypass the requirement of cleavage by extracellular invertase.

A further challenge will be the development of methods to efficiently and economically propagate the male-sterile female parent line, a problem that has been often overlooked in the development of male sterility systems (Perez-Prat and van Lookeren Campagne, 2002).

Although it has been shown that the NIN88 promoter is functional in other solanaceous species, the possible use for other plant species, in particular monocotyledonous plants, remains to be determined.

An alternative to antisense repression of extracellular invertase could be the used of an invertase inhibitor characterized in different species (Krausgrill et al., 1996). This protein inhibits the activity of invertase by physical interaction and complex formation. The advantage of the use of the invertase inhibitor to reduce the invertase activity during pollen development is that the interaction is less sequence specific compared to antisense repression.

Up to now no functional analysis of extracellular invertases have been performed mainly in the male gametophyte. Different expression patterns for extracellular invertases in ovaries (Kim, Mahé, Guy, et al., 2000) and gynoeceum (Godt and Roitsch, 1997) have been reported. Specific antisense approaches could help to clarify the function of extracellular invertases in those floral organs, especially with respect to the early development of the female gametophyte.

The potential broad applicability of engineering the carbohydrate supply to the male gametophyte in many plant species and the different and important possibilities to use the system indicate that metabolic engineering of the carbohydrate supply in male gametophyte development might have a high potential to exert a great impact on agriculture in the future.

REFERENCES

Biasi, R., G. Falasca, A. Speranza, A. De-Stadis, V. Scoccianti, M. Franceschetti, N. Bagni, and M.M. Alatmura (2001). Biochemical and ultrastructural features related to male sterilitiy in the dioecious species *Actinidia deliciosa*. *Plant Physiology and Biochemistry, Paris* 39:395-406.

Brewbaker, J.L. (1964). *Agricultural genetics*. Englewood Cliffs, NJ: Prentice Hall.

Bucciaglia, P.A. and A.G. Smith (1994). Cloning and characterization of Tag 1, a tobacco anther alpha-1,3 glucanase expressed during tetrad dissolution. *Plant Molecular Biology* 24:903-914.

Chaudhury, A.M. (1993). Nuclear genes controlling male fertility. *The Plant Cell* 5:1277-1283.

Clément, C. and J.C. Audran (1995). Anther wall layers control pollen sugar nutrition in *Lilium*. *Protoplasma* 187:172-181.

Clément, C. and J.C. Audran (1999). Anther carbohydrates during in vivo and in vitro pollen development. In S. Clément, E. Pacini, and J.C. Audran, (eds.), *Anther and pollen: From biology to biotechnology* (pp. 69-90). Heidelberg, Germany: Springer-Verlag.

Clément, C., M. Burrus, and J.C. Audran (1996). Floral organ growth and carbohydrate content during pollen development in *Lilium*. *American Journal of Botany* 83:459-469.

Datta, R., P.S. Chourey, D.R. Pring, and H.V. Tang (2001). Gene-expression analysis of sucrose-starch metabolism during pollen maturation in cytoplasmic male-sterile and fertile lines of sorghum. *Sexual Plant Reproduction* 14:127-134.

Dorion, S., S. Lalonde, and H.S. Saini (1996). Induction of male sterility in wheat by meiotic-stage water deficit is preceded by a decline in invertase activity and changes in carbohydrate metabolism in anthers. *Plant Physiology* 111:137-145.

Ehneß, R. and T. Roitsch (1997). Co-ordinated induction of mRNAs for extracellular invertase and a glucose transporter in *Chenopodium rubrum* by cytokinins. *The Plant Journal* 11:539-548.

Eschrich, W. (1980). Free space invertase, its possible role in phloem unloading. *Berichte der Deutschen Botanischen Gesellschaft* 93:363-378.

Farrar J. (1996). Regulation of shoot-root ratio is mediated by sucrose. *Plant and Soil* 185:13-19.

Feistritzer, W.R. and A.F. Kelly (1987). *Hybrid seed production of selected cereal oil and vegetable crops*. Rome: Food and Agriculture Organization of the United Nations.

Godt, D.E. and T. Roitsch (1997). Regulation and tissue-specific distribution of mRNAs for three extracellular invertase isoenzymes of tomato suggests an important function in establishing and maintaining sink metabolism. *Plant Physiology* 115:273-282.

Goetz, M., D.E. Godt, A. Guivarc'h, U. Kahmann, D. Chriqui, and T. Roitsch (2001). Induction of male sterility in plants by metabolic engineering of the carbohydrate supply. *Proceedings of the National Academy of Science (USA)* 98:6522-6527.

Goetz, M., D.E. Godt, and T. Roitsch (2000). Tissue-specific induction of the mRNA for an extracellular invertase isoenzyme of tomato by brassinosteroids suggests a role of steroid hormones in assimilate partitioning. *The Plant Journal* 22:515-522.

Hernould, M., S. Suharsono, E. Zabaleta, J.P. Carde, S. Litvak, A. Araya, and A. Mouras (1998). Impairment of tapetum and mitochondria in engineered male-sterile tobacco plants. *Plant Molecular Biology* 36:499-508.

Hidalgo, P.J., M. Hesse, and J.L. Ubera (1999). Microsporogenesis in male sterile *Rosmarinus officinalis* L. (Lamiaceae), an ultrastructural study. *Grana* 28:343-355.

Ho, L.C. (1988). Metabolism and compartmentation of imported sugars in sink organs in relation to sink strength. *Annual Reviews of Plant Physiology and Plant Molecular Biology* 39:355-378.

Kaul, M.L.H. (1988). *Male sterility in higher plants*. Monographs on theoretical and applied genetics, Volume 10. Berlin: Springer-Verlag.

Kawaguchi, K., N. Shibuya, and T. Ishii (1996). A novel tetrasaccharide, with a structure similar to the terminal sequence of an arabinogalactan-protein, accumulates in rice anthers in a stage specific manner. *The Plant Journal* 9:777-785.

Kim, J.Y., A. Mahé, J. Brangeon, and J.L. Prioul (2000). A maize vacuolar invertase, IVR2, is induced by water stress: Organ/tissue specifity and diurnal modulation of expression. *Plant Physiology* 124:71-84.

Kim J.Y, A. Mahé, S. Guy, J. Brangeon, O. Roche, P.S. Chourey, and J-L. Prioul (2000). Characterization of two members of the maize gene family, Incw3 and Incw4, encoding cell-wall invertases. *Gene* 245:89-102.

Koch, K.E. (1996). Carbohydrate-modulated gene expression in plants. *Annual Reviews of Plant Physiology and Plant Molecular Biology* 47:509-540.

Krausgrill S., A. Sander, S. Greiner, M. Weil, T. Rausch (1996). Regulation of cell-wall invertase by a proteinaceous inhibitor. *Journal of Experimental Botany* 47: 1193-1198.

Lemoine, R., L. Bürkle, L. Barker, S. Sakr, C. Kühn, M. Regnacq, C. Gaillard, S. Delrot, and W.B. Frommer (1999). Identification of a pollen-specific sucrose transporter-like protein NtSUT3 from tobacco. *FEBS Letters* 454:325-330.

Maddison, A.L., P.E. Hedley, R.C. Meyer, N. Aziz, D. Davidson, and G.C. Machray (1999). Expression of tandem invertase genes associated with sexual and vegetative growth cycles in potato. *Plant Molecular Biology* 41:741-751.

Mariani, C., M. DeBeuckeleer, J. Truettner, J. Leemans, and R. B. Goldberg (1990). Induction of male sterility in plants by a chimaeric ribonuclease gene. *Nature* 347:737-741.

Mascarenhas, J.P. (1990). Gene activity during pollen development. *Annual Reviews of Plant Physiology and Plant Molecular Biology* 41:317-338.

Mayo, H. (1980). *The theory of plant breeding*. Oxford, UK: Clarendon.

Miller, E.M. and P.S. Chourey (1992). The maize invertase-deficient miniature-1 seed mutant is associated with aberrant pedicel and endosperm development. *The Plant Cell* 4:297-305.

Miller, W.B. and A.P. Ranwala (1994). Characterization and localization of three soluble invertase forms from *Lilium longiflorum* flower buds. *Physiologia Plantarum* 92:247-253.

Napoli, C.A., D. Fahy, H.Y. Wang, and L.P. Taylor (1999). White anther: A petunia mutant that abolishes pollen flavonol accumulation, induces male sterility, and is complemented by a chalcone synthase transgene. *Plant Physiology* 120:615-622.

Pacini, E. and G.G. Franchi (1991). Diversification and evolution of the tapetum. In S. Blackmore and S.H. Barnes (eds.), *Microspores: Evolution and ontogeny* (pp. 213-237). San Diego: Academic Press.

Pacini, E., G.G. Franchi, and M. Hesse (1985). The tapetum: Its form, function, and possible phylogeny in embryophyta. *Plant Systematics and Evolution* 149:155-185.

Perez-Prat, E. and M. M. van Lookeren Campagne (2002). Hybrid seed production and the challenge of propagating male-sterile plants. *Trends in Plant Science* 7:199-203.

Ranwala, A.P. and W.B. Miller (1998). Sucrose-cleaving enzymes and carbohydrate pools in *Lilium longiflorum* floral organs. *Physiologia Plantarum* 103: 541-550.

Roitsch, T. (1999). Source-sink regulation by sugar and stress. *Current Opinion in Plant Biology* 2:198-206.

Roitsch T., M. Bittner, and D. Godt (1995) Induction of apoplastic invertase of *Chenopodium rubrum* by D-glucose and a glucose analogue and tissue-specific expression suggest a role in sink-source regulation. *Plant Physiology* 108:285-294.

Roitsch, T., R. Ehneß, M. Goetz, B. Hause, M. Hofmann, and A.K. Sinha (2000). Regulation and function of extracellular invertase from higher plants in relation to assimilate partitioning, stress responses and sugar signalling. *Australian Journal of Plant Physiology* 27:815-825.

Roitsch, T. and W. Tanner (1996). Cell wall invertase: Bridging the gap. *Botanica Acta* 109:90-93.

Saini, H.S. (1997). Effects of water stress on male gametophyte development in plants. *Sexual Plant Reproduction* 10:67-73.

Schmülling, T., J. Schell, and A. Spena (1988). Single genes from *Agrobacterium rhizogenes* influence plant development. *The EMBO Journal* 7:2621-2629.

Sheoran, I.S. and H.S. Saini (1996). Drought-induced male sterility in rice: Changes in carbohydrate levels and enzyme activities associated with the inhibition of starch accumulation in pollen. *Sexual Plant Reproduction* 9:161-169.

Singh, M.B. and R.B. Knox (1984). Invertases of *Lilium* pollen. *Plant Physiology* 74:510-515.

Smeekens, S. (1998). Sugar regulation of gene expression in plants. *Current Opinion in Plant Biology* 1:230-234.

Steiglitz, H. (1977). Role of alpha-1,3-glucanase in postmeiotic microspore release. *Developmental Biology* 57:87-97.

Stintzi, A. and J. Browse (2000). The *Arabidopsis* male-sterile mutant, opr3, lacks the 12-oxophytodienoic acid reductase required for jasmonate synthesis. *Proceedings of the National Academy of Science (USA)* 97:10625-10630.

Sturm, A. and M.J. Chrispeels (1990). cDNA cloning of carrot extracellular ß-fructosidase and its expression in response to wounding and bacterial infection. *The Plant Cell* 2:1107-1119.

Tang, G.Q., M. Lüscher, and A. Sturm (1999). Antisense repression of vacuolar and cell wall invertase in transgenic carrot alters early plant development and sucrose partitioning. *The Plant Cell* 11:177-189.

Touraev, A. and E. Heberle-Bors (1999). Microspore embryogenesis and in vitro pollen maturation in tobacco. In R.D. Hall (ed.), *Methods in molecular biology 111* (pp. 281-291). Totowa, NJ: Humana Press.

Truernit, E., R. Stadler, K. Baier, and N. Sauer (1999). A male gametophyt-specific monosaccharide transporter in *Arabidopsis*. *The Plant Journal* 17:191-201.

Tsuchiya, T., K. Toriyama, M. Yoshikawa, S. Ejiri, and K. Hinata (1995). Tapetum-specific expression of the gene for an endo-alpha-1,3-glucanase causes male sterility in transgenic tobacco. *Plant, Cell and Physiology* 36:487-494.

Tymowska-Lalanne, Z. and M. Kreis (1998). The plant invertases: Physiology, biochemistry and molecular biology. *Advances in Botanical Research* 28:71-117.

van der Meer, I.M., M.E. Stam, A.J. van Tunen, J.N.M. Mol, and A.R. Stuitje (1992). Antisense inhibition of flavonoid biosynthesis in petunia anthers results in male sterility. *The Plant Cell* 4:253-262.

von Schaewen, A., M. Stitt, R. Schmidt, U. Sonnewald, and L. Willmitzer (1990). Expression of yeast-derived invertase in the cell wall of tobacco and *Arabidopsis* leads to accumulation of carbohydrate and inhibition of photosynthesis and strongly influences growth and phenotype of transgenic tobacco plants. *The EMBO Journal* 9:3033-3043.

Wang, Y.T. and P.J. Breen (1986). Partitioning of ^{14}C assimilate in Easter lily as affected by growth stage and flower removal. *Scientific Horticulture* 29:273-281.

Weber, H., L. Borisjuk, U. Heim, P. Buchner, and U. Wobus (1995). Seed coat-associated invertases of faba bean control both unloading and storage functions: Cloning of cDNAs and cell type-specific expression. *The Plant Cell* 7:1835-1846.

Weber, H., L. Borisjuk, and U. Wobus (1996). Controlling seed development and seed size in *Vicia faba:* A role for seed coat-associated invertases and carbohydrate state. *The Plant Journal* 10:823-834.

Wobus, U. and H. Weber (1999). Sugars as signal molecules in plant seed development. *Biological Chemistry* 380:937-944.

Worral, D., D.L. Hird, R. Hodge, W. Paul, J. Draper, and R. Scott (1992). Premature dissolution of the microsporocyte callose wall causes male sterility in transgenic tobacco. *The Plant Cell* 4:759-771.

Xu, J., W.T. Avigne, D.R. McCarty, and K.E. Koch (1996). A similar dichotomy of sugar modulation and developmental expression affects both paths of sucrose metabolism: Evidence from maize invertase gene family. *The Plant Cell* 8:1209-1220.

Ylstra, B., D. Garrido, J. Busscher, and A.J. van Tunen (1998). Hexose transport in growing petunia pollen tubes and characterization of a pollen-specific, putative monosaccharide transporter. *Plant Physiology* 118:297-304.

Zhang, Y., P.R. Shewry, H. Jones, P. Barcelo, P.A. Lazzeri, and N. Halford (2001). Expression of antisense SnRK1 protein kinase sequence causes abnormal pollen development and male sterility in transgenic barley. *The Plant Journal* 28:431-441.

Chapter 7

Enhancing the Nutritive Value of Seeds by Genetic Engineering

N. D. Hagan
T. J. V. Higgins

To date, the majority of genetically engineered crops have been developed with production benefits, such as insect resistance or herbicide tolerance, in mind. In the future, genetically engineered crops have the potential to provide consumer benefits such as improved nutritional or quality traits. We review research using genetic engineering to improve the nutritional quality of seeds. Examples include altering seed amino acid composition in legumes and cereals to increase the levels of limiting essential amino acids and improvements in the fatty acid, vitamin, and mineral contents of seeds.

ENHANCING THE NUTRITIVE VALUE OF SEED PROTEIN

Cereal and legume grains are major sources of digestible protein for humans and form the basis of livestock feeds. However, individually these crops cannot supply the full complement of essential amino acids. The animal growth requirements for methionine and lysine are approximately 1.6 and 7 g per 100 g of total protein, respectively. Animals require an additional 1.9 g of cysteine per 100 g protein, but this can be met by either of the sulfur amino acids, as animals can convert methionine to cysteine. Most cereal grains contain sufficient methionine to meet the minimum growth requirement but fall well short of the requirement for lysine. Conversely, legume seeds contain enough lysine but insufficient methionine (see Table 7.1). Consequently, a diet based on a single cereal or legume results in amino acid deficiencies. These deficiencies can be partly overcome by mixing seed meals from legumes and cereals, but this alone is not sufficient for maximum animal growth. As a result, livestock feeds are usually supplemented with methionine and lysine, which adds to the cost of production considerably.

TABLE 7.1. Protein, lysine, and methionine contents of common cereal and legume seeds.

Seed crop	Lysine content ($g \cdot 100\ g^{-1}$ protein)[a]	Methionine content ($g \cdot 100\ g^{-1}$ protein)[a]
Cereals		
Wheat[b] *(Triticum aestivum)*	3.1	1.5
Rice[b] *(Oryza sativa)*	4.0	1.8
Maize[b] *(Zea mays)*	2.9	1.9
Legumes		
Pea[b] *(Pisum sativum)*	7.3	1.2
Broad bean[b] *(Vicia faba)*	7.4	1.0
Lupin[c] *(Lupinus angustifolius)*	4.4	0.6
Soybean[b] *(Glycine max)*	6.9	1.5

[a]The animal growth requirement for lysine and sulfur amino acids is approximately 7 and 1.6 $g \cdot 100g^{-1}$ protein, respectively.
[b]Data quoted in Salisbury and Ross, 1985.
[c]Data from Molvig et al., 1997.

Humans require a minimum quantity of the essential amino acids for normal physical and mental development. Cereal and legume seeds are a major source of essential amino acids, supplied as protein in the diet. These grains provide over 65 percent of the protein in developing countries and over 30 percent of the protein in developed countries (Friedman, 1996). However, eating a vegetarian diet that relies exclusively on either cereal or legume grains can result in a lysine or methionine deficiency and can affect normal growth and development.

Plant breeding has attempted to improve the balance of essential amino acids in seed proteins. Recently, genetic engineering has been used to introduce genes for proteins with specific amino acid compositions into cereals, legumes, and oilseeds and has shown potential to address the nutritional deficiencies in these crops (Altenbach et al., 1992; Pickardt et al., 1995; Molvig et al., 1997; Lai and Messing, 2002). Genetic engineering has also shown potential to produce feed with improved digestibility for ruminant animals by introducing rumen-stable proteins into seeds.

Modification of Endogenous Genes to Increase Essential Amino Acids

The amino acid composition of the most abundant proteins accumulated in the seed determines the value of that seed protein in animal diets. Among legumes, the major storage proteins are the 7S (vicilin) and 11S (legumin) globulins, which account for 70 to 80 percent of the seed protein. As both types of globulins have few sulfur-containing amino acids, they are primarily responsible for the nutritional deficiencies of legume grains. In most cereals, the prolamins account for the largest proportion of protein in the endosperm. These proteins generally contain low levels of lysine, threonine, and isoleucine, resulting in a deficiency of these amino acids in cereals.

Targeted approaches using genetic engineering are now being employed to increase essential amino acids in cereal and legume grains. One approach has been to alter the genes for endogenous storage proteins to contain higher levels of limiting essential amino acids. This approach was used to increase the nutritional value of a maize α-zein protein. Wallace et al. (1988) engineered tryptophan and lysine residues into the 19 kDa α-zein by site-directed mutagenesis. When injected with zein mRNA, *Xenopus* oocytes expressed and assembled the modified zein protein. The engineered zein gene was subsequently introduced into transgenic tobacco under the control of a β-phaseolin promoter, but the protein was not stable in the tobacco embryos (Ohtani et al., 1991).

The 2S albumins are generally highly conserved but contain a hypervariable region that has been the subject of protein modification. De Clercq et al. (1990) modified the hypervariable region of the *Arabidopsis* 2S albumin gene by replacing it with sequences for methionine-rich novel peptides to produce modified proteins containing 7 to 14 methionine residues. These proteins were stable in transgenic *Arabidopsis* seeds and were shown to accumulate to 1 to 2 percent of the total high salt-extractable seed protein.

The vicilin-like and legumin-like globulins of faba bean *(Vicia faba)* have been modified in an attempt to create methionine-rich storage proteins. A gene for a methionine free legumin subunit was altered by frameshift and site-directed mutagenesis to introduce four methionine residues in the new protein (Saalbach et al., 1988). The modified legumin gene was introduced into tobacco, but it was shown that the amino acid changes interfered with correct folding and the protein was rapidly degraded. When eight methionine codons were introduced into the *Vicia faba* vicilin gene, the variant protein was stable in transgenic tobacco seed but was produced at levels too low to change seed protein amino acid composition (Saalbach et al., 1995).

Storage proteins from bean *(Phaseolus vulgaris)* have also been targeted for modification. β-phaseolin, the 7S globulin from *Phaseolus vulgaris*, was modified by the addition of a methionine-rich region from the maize 15 kDa zein storage protein (Hoffman et al., 1988), which increased the number of methionine residues from three to nine in the new protein. The maize- derived peptide was predicted to form an α-helical structure and was inserted into an α-helical region of β-phaseolin. The level of mRNA transcript was shown to be high, but only a small amount of protein accumulated. It was concluded that despite the conserved α-helical insertion, the high methionine β-phaseolin was unstable in the developing seed. Phytohemagglutinin, an abundant lectin from *Phaseolus vulgaris* seeds, contains no cysteine or methionine residues and has also been targeted for modification. Four methionine residues were introduced by site-directed mutagenesis on the basis of the location of methionine residues in homologous lectin proteins. The engineered gene was introduced into tobacco, where it was found that the protein underwent correct modification and accumulated in the vacuoles of the cotyledons (Kjemtrup et al., 1994). However, the level of expression was too low to have any effect on overall amino acid composition.

An alternative approach has focused on increasing the expression of endogenous methionine-rich proteins that are normally expressed at low levels in the seed. Anthony et al. (1997) produced transgenic maize expressing *Dzs10,* a gene that encodes a sulfur-rich zein. They observed that methionine levels in some kernels could be increased by up to 30 percent, although this was not always correlated with the overexpressed zein. The same zein was found to be elevated in maize variants with increased seed methionine. The variable level of this protein observed in different maize varieties was subsequently found to be a result of posttranscriptional regulation of the *Dzs10* gene transcript (Cruz-Alvarez et al., 1991). When *Dzs10* was introduced back into maize with *cis*-acting regulatory sites removed, Lai and Messing (2002) described an increase in the level of this zein and a 15 percent increase in seed methionine content when expressed as a proportion of total protein.

Expressing Genes for Proteins from Other Plants to Increase Essential Amino Acids

A promising genetic engineering approach for addressing amino acid deficiencies has been to express genes for proteins that are rich in the required amino acid and that have been isolated from other plants. The introduced proteins are likely to be as stable in the new plant as they are in their plant of origin, particularly if they are deposited in the same organ and tissue type.

Although lysine-rich proteins have been identified, they are not normally found in high levels in seeds. An example is a chymotrypsin inhibitor from barley (CI-2) that contains 11.5 percent lysine (Williamson et al., 1987). Hyproyl barley mutants display a sevenfold increase in chymotrypsin-inhibitor activity, suggesting that overexpression of this particular biologically active protein may not hinder plant development and germination. Recent efforts have focused on engineering additional lysine residues into CI-2, based on the tertiary structure of the native protein (Roesler and Rao, 1999). A number of seed storage proteins have been identified that are rich in methionine. The 2S albumins from many plants are also rich in cysteine, containing eight conserved residues that contribute to inter- and intra-chain disulfide bonding. The most extensively studied of these have been the Brazil nut *(Bertholletia excelsa)* 2S albumin (BNA) containing 19 percent methionine and 8 percent cysteine residues (Ampe et al. 1986; Sun et al., 1987) and the sunflower *(Helianthus annuus)* seed albumin protein (SSA) containing 16 percent methionine and 8 percent cysteine residues (Lilley et al., 1989; Kortt et al., 1991).

Chimeric genes encoding the BNA protein have been introduced into a variety of plants, resulting in seed-specific expression. An early study expressed BNA in tobacco seeds under the control of the β-phaseolin promoter (Altenbach et al., 1989). The BNA protein accumulated to a level of 8 percent of total seed protein, resulting in a 30 percent increase in methionine. More recently, chimeric BNA genes have been introduced into large seed legumes such as the narbon bean (Pickardt et al., 1995; Müntz et al., 1997). Transgenic narbon bean lines that expressed BNA under control of the faba bean legumin promoter showed accumulation of up to 5 percent of total seed protein. The highest BNA expressing line showed a doubling of total methionine, putting it in the range required for optimal animal growth. In an approach that combined the use of BNA with protein modification, variants containing up to seven additional methionine residues were produced (Tu et al., 1998). The altered genes were expressed in potato tubers but at such low levels as to have no effect on overall tuber amino acid composition.

The sunflower seed albumin (SSA) gene encodes another well-studied sulfur-rich protein and has been introduced into a variety of legumes, including subterranean clover, pea, chickpea, and narrow leaf lupin (Khan et al., 1996; Tabe et al., 1997; Molvig et al., 1997). The SSA gene was placed under the control of the pea vicilin promoter and introduced into narrow leaf lupin *(Lupinus angustifolius)*, a legume commonly fed to stock because of its high level of protein and fiber. SSA accounted for 5 percent of total seed protein in the transgenic lupin and resulted in a doubling of seed methionine and an overall increase in sulfur-containing amino acids of 20 percent (Molvig et al., 1997). There was no significant change in total seed sulfur, with increases in sulfur amino acids coming largely at the expense of inorganic sulfur (sulfate) pools in the grain. In chicken feeding trials it was shown that when the high methionine lupins were used at 25 percent inclusion in the diet it was possible to reduce the level of added methionine by 22 percent without reducing animal growth (Ravindran et al., 2002).

High-methionine lupins have also been evaluated as an improved feedstock for ruminant animals. In ruminants, protein absorption depends not only on total intake but also on the relative proportion that escapes rumen fermentation and passes unaltered into the small intestine. As much as 40 percent of the fermented plant protein remains as ammonia and is lost to the animal (MacRae and Ulyatt, 1974; Beever and Siddons, 1986). However, individual proteins have different susceptibilities to degradation in the rumen. SSA is particularly rumen stable, as demonstrated in a study in which over 90 percent SSA remained undegraded after an eight-hour in vitro incubation in rumen fluid (McNabb, 1990). In feeding trials with sheep, the transgenic lupin seeds expressing SSA produced a 7 percent increase in live weight gain and an 8 percent increase in wool growth, compared to nontransgenic seeds (White et al., 2001). This improvement was likely to be due to the stability of SSA in the rumen as well as the relatively high-methionine content of the seeds. Transgenic lupins expressing SSA are currently undergoing further development with a view to commercial release.

Creating Synthetic Storage Proteins to Increase Essential Amino Acids

Synthetic genes encoding proteins with high levels of lysine, methionine, tryptophan, threonine, and isoleucine have been formed by random ligation of oligonucleotides containing a high proportion of codons for essential amino acids. Low-level accumulation of these proteins was demonstrated in potato tubers (Yang et al., 1989) but failed to significantly increase the levels of essential amino acids. This may have been due to inefficient gene expression or protein instability. It is not known whether the proteins produced would be likely to fold into stable structures.

To increase lysine and methionine content in seeds, Keeler et al. (1997) designed a gene for a synthetic protein containing 31 percent lysine and 20 percent methionine. The novel protein was predicted to have an α-helical coiled coil structure in order to promote the formation of stable multimers and overcome stabilization problems encountered in previous research. The gene for this protein was introduced into tobacco seeds under the control of the soybean β-conglycinin promoter, resulting in a 30 percent increase in seed methionine and a 47 percent increase in seed lysine. If expressed at a similar level in grains of cereals or legumes, such a protein could significantly increase their nutritive value.

ENHANCING THE FATTY ACID CONTENT OF SEEDS

Plant fatty acids* have important uses in cooking and food manufacturing and are a major component of human diets. The metabolic engineering of plant fatty acid profiles has recently become possible due to the cloning and characterization of genes that encode the major enzymes involved in fatty acid biosynthesis. These include Δ9- and Δ12-desaturases, enzymes that determine the relative proportions of saturated, monounsaturated, and polyunsaturated fatty acids. The overexpression or directed silencing of biosynthetic genes have been used to modify the fatty acid profile of seeds. In addition, research is underway to increase the content of essential fatty acids in commercially grown crops through the use of genes derived from non-oilseed species.

Modifying Fatty Acid Composition

There is a growing move away from animal fats, which contain a high level of cholesterol and saturated fatty acids, to vegetable oils with their high level of unsaturated fatty acids. Evidence indicates that myristic acid (C14:0; see previous footnote) and palmitic acid (C16:0), found at high levels in animal fats, raise low-density lipoprotein (LDL) cholesterol levels in the blood, increasing the risk of cardiovascular disease (reviewed in Katan et al., 1995). Vegetable oil from many of the major oilseed crops contains high levels of unsaturated fatty acids such as oleic acid (C18:1 Δ^{12}), linoleic acid (C18:2 $\Delta^{9,12}$), and α-linolenic acid (omega-3 linolenic acid; C18:3 $\Delta^{9,12,15}$) (see Table 7.2). These fatty acids have been shown to lower serum LDL cholesterol, thus reducing the risk of cardiovascular disease. The C18 saturated fatty acid, stearic acid (18:0), is present in some plant oils but does not significantly affect LDL cholesterol levels.

Unfortunately, polyunsaturated fatty acids such as linoleic acid and linolenic acid are not stable in frying and cooking applications. Currently, chemical hydrogenation is used increase the C18:1 and C18:0 fatty acid content, thereby improving the stability of the oil at the high temperatures used in frying. However, this process can result in the production of *trans*-fatty acids, which, similar to myristic acid and palmitic acid, have the

---

* In this chapter, fatty acids are designated C*x*:*y*, where *x* represents the number of carbon atoms in the fatty acid and *y* represents the number of double bonds it contains. Numbers following a Δ symbol indicate the position of the double bonds relative to the carboxyl group. Some unsaturated fatty acids are also classified according to the omega nomenclature, which describes the position of the closest double bond to the terminal methyl carbon. For example, an omega-3 fatty acid contains a double bond three carbons away from the methyl end.

TABLE 7.2. Fatty acid composition of the major oilseed crops.

	Fatty acid (% w/w)				
	Palmitic acid (16:0)	Stearic acid (18:0)	Oleic acid (18:1)	Linoleic acid (18:2)	Linolenic acid (18:3)
Soybean[a] *(Glycine max)*	11	4	22	53	7.5
Oil Palm[a] *(Elaeis guineensis)*	45	4.5	38	11	0.2
Canola[b] *(Brassica napus)*	4.5	1.5	61	22	10
Sunflower[a] *(Helianthus annuus)*	5.5	5	20	69	0.1
Cotton[b] *(Gossypium hirsutum)*	25.6	2.3	13.2	58.5	0.1

[a]Data from Murphy, 1999.
[b]Data from Liu et al., 2002.

undesirable property of raising LDL cholesterol (reviewed in Katan et al., 1995). For many food applications, stable vegetable oils with a reduced amount of *trans*-fatty acids are desirable. To this end, genetic engineering approaches have recently been used to increase levels of oleic acid and stearic acid in a variety of oilseed crops (reviewed in Thelen and Ohlrogge, 2002).

Seed-specific expression of an antisense Δ12-desaturase gene in transgenic canola resulted in reduced levels of the enzyme and a dramatic increase in stearic acid at the expense of oleic acid (Knutzon et al., 1992) (see Figure 7.1). However, this change was accompanied by a 45 percent decrease in total seed fatty acid content (Knutzon et al., 1992). In soybeans, suppression of the Δ12-desaturase gene (Kinney, 1996) resulted in oil containing 86 percent oleic acid and less than 1 percent linoleic acid. This oil is now in commercial production.

Recently, improved methods of gene silencing have been developed based on the finding that double-stranded RNA transcripts direct the degradation of homologous single-stranded RNAs (Smith et al., 2000). In particular, the production of self-complementary hairpin RNA transcripts has been shown to efficiently silence both transgenes and endogenous genes (Smith et al., 2000; Wesley et al., 2001). This technique has now been used to modify the fatty acid composition of cottonseed oil (Table 7.2) by down-regulating the expression of fatty acid desaturase genes in cotton seeds. Hairpin RNA-mediated silencing of the cotton Δ12-desaturase gene raised the oleic acid content of the transgenic cottonseed oil to 78 percent from the

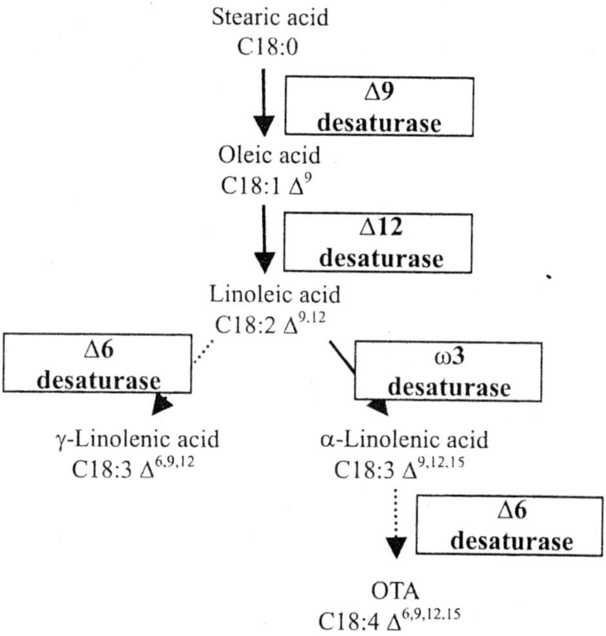

FIGURE 7.1. Biosynthetic pathway of selected C18 fatty acids. Solid arrows represent reactions that are common in higher plants, while open arrows represent reactions that occur in only a few plants. Stearic acid can be desaturated to form oleic acid in the plant plastid by a Δ9 desaturase. Oleic acid can then be transported to the endoplasmic reticulum where further desaturation can occur, first through the action of a Δ12 desaturase to form linoleic acid then through the action of a ω3 desaturase (Δ15 desaturase) to form α-linolenic acid. Δ6 desaturation (open arrows) occurs in only a small number of plant species such as evening primrose *(Oenothera biennis)* and borage *(Borago officinalis)*. OTA is octadecatetraenoic acid.

normal level of 13 percent, while silencing of the cotton Δ9-desaturase gene raised stearic acid levels from 2 percent to as high as 40 percent (Liu et al., 2002). Unexpectedly, palmitic acid was decreased to approximately 15 percent in both high-oleic and high-stearic lines (Liu et al., 2002).

Increasing the Level of Essential Polyunsaturated Fatty Acids

Several polyunsaturated fatty acids including linoleic acid and α-linolenic acid are recognized as essential fatty acids due to their role in preventing nutrition-related diseases (reviewed in López and García, 2000). It is hypothesized that these fatty acids serve as precursors for very long chain polyunsaturated fatty acids which subsequently form a variety of biologically active compounds known as eicosanoids (prostaglandins, leukotrienes, and thromboxanes) that influence a wide variety of human metabolic processes. Consumption of very long chain polyunsaturated fatty acids has also been associated with lowering total serum cholesterol as well as reducing the risk of coronary heart disease (reviewed in Hu et al., 2001).

In animals, members of the omega-6 class of polyunsaturated fatty acids are produced from linoleic acid. The first step in this process is the desaturation of linoleic acid to γ-linolenic acid (omega-6 linolenic acid; C18 $\Delta^{6,9,12}$) catalyzed by a Δ6-desaturase (Figure 7.1). However, decreased activity of this enzyme, as observed in aging and some disorders, may lead to a deficiency of γ-linolenic acid. As γ-linolenic acid is produced in only a small number of plant species, there is some interest in using genetic engineering to produce this fatty acid in the seeds of transgenic oilseed crops. To date, however, research has been confined to model systems. γ-linolenic acid accumulation in transgenic plants was first reported by Reddy and Thomas (1996), who expressed a cyanobacterial Δ6-desaturase gene in tobacco leaves. However, γ-linolenic acid accumulated to only 1.2 percent of C18 fatty acids in the leaf. Octadecatetraenoic acid (OTA; C18:4 $\Delta^{6,9,12,15}$), a highly unsaturated, industrially important fatty acid, was also found in the transgenic tobacco plants and constituted up to 2.9 percent of C18 fatty acids (Reddy and Thomas, 1996). Transgenic tobacco plants expressing a Δ6-desaturase gene from borage *(Borago officinalis)* resulted in the accumulation of γ-linolenic acid to 13.2 percent and OTA to 9.6 percent of the total leaf fatty acids (Sayanova et al., 1997).

Oilfish such as sardines accumulate relatively large quantities of long chain omega-3 polyunsaturated fatty acids such as eicosapentenoic acid (EPA, C20:5) and docosahexenoic acid (DHA, C22:6). The production of these fatty acids in plants now appears possible with the cloning if genes for most of the enzymes involved in their biosynthesis (reviewed in López and García, 2000). The cotransformation of yeast cells with a Δ5-desaturase gene and an "elongase" gene from the fungus *Mortierella alpina* resulted in the production of EPA from OTA (Parker-Barnes et al., 2000). These enzymes were also able to catalyze the production of arachidonic acid (C20:4) from γ-linolenic acid gene (Parker-Barnes et al., 2000), suggesting that these enzymes could be used in the bioproduction omega-6 as well as omega-3 polyunsaturated fatty acids.

ENHANCING THE VITAMIN AND MINERAL CONTENT OF SEEDS

Vitamin and mineral deficiency has been identified as a major cause of many health problems in developing countries. Although most vitamins and minerals can be supplied in a plant-based diet, not all plant foods contain adequate levels of these nutrients to meet human requirements. For example, milled rice is a good source of carbohydrate and protein but is deficient in many micronutrients. It may be possible to produce many different vitamins in seeds such as rice once the biosynthetic pathways have been elucidated and the relevant biosynthetic enzymes have been cloned. Good examples of this can be seen with the cloning of genes encoding key enzymes involved in provitamin A and vitamin E synthesis, which led to the production of transgenic plants with higher levels of these vitamins. The nutritional value of complexed minerals is reduced in many seeds by a high level of phytic acid, a potent inhibitor of mineral absorption. Therefore, increasing the level of seed minerals such as iron may need to coincide with changes in other factors that influence mineral absorption to improve seed nutrition.

Carotenoids: Vitamin A

Carotenoids comprise a group of lipid-soluble pigments with a diverse range of biological functions. Humans cannot synthesize these compounds de novo but require β-carotene (provitamin A, a C40 carotenoid) in the diet in order to produce vitamin A. Also known as retinol, vitamin A plays an essential role in normal human development and is required for normal vision (reviewed in Bendich, 1993). Carotenoids such as vitamin A have also been associated with a wide range of other health benefits because of their antioxidant activity. The recommended dietary allowance (RDA) of β-carotene is 6 mg β-carotene per day (National Research Council, 1989). Insufficient dietary β-carotene leads to severe clinical symptoms, including the eye disease xerophthalmia and an increased susceptibility to potentially fatal afflictions such as diarrhea and respiratory diseases (reviewed in Mayne, 1996).

In order to increase seed carotenoids, canola was transformed with a plastid-targeted bacterial *(Erwinia uredovora)* phytoene synthase, the enzyme that catalyzes the first step required to produce β-carotene from geranyl geranyl diphosphate (Shewmaker et al., 1999) (Figure 7.2). The enzyme was expressed only in the seeds of the transgenic plants and was targeted to the plastid. The resulting seeds were visibly orange and contained

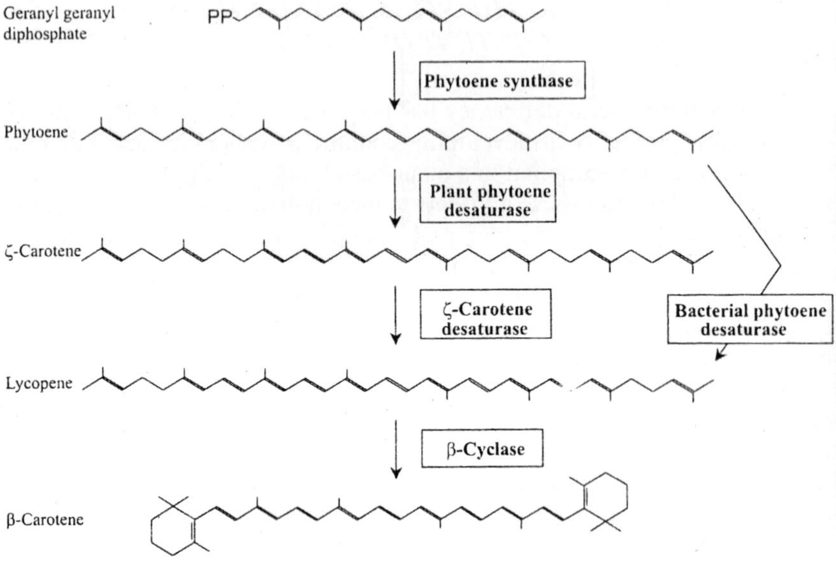

FIGURE 7.2. Summary of β-carotene synthesis. Carotenoids are synthesized by all photosynthetic organisms as well as many nonphotosynthetic bacteria and fungi. In β-carotene synthesis, two molecules of geranyl geranyl diphosphate are condensed followed by desaturation steps and the formation of β-ionone rings. Phytoene desaturase from the bacteria *Erwinia uredovora* can substitute for plant phytoene desaturatse and ζ-carotene desaturase to produce lycopene from phytoene.

50-fold more carotenoids than the untransformed control. Levels of phytoene were increased in the trànsgenic seeds, but a dramatic increase in α- and β-carotenes were also measured. In one transgenic line, β-carotene content increased from 3 to 949 $\mu g \cdot gFW^{-1}$ (Shewmaker et al., 1999). The same bacterial phytoene synthase was recently expressed in tomato fruit resulting in a two- to fourfold increase in total carotenoids (Fraser et al., 2002). Römer et al. (2000) transformed tomato with a bacterial phytoene desaturase gene, an enzyme capable of converting phytoene into lycopene (Figure 7.2). Expression of the transgene in the tomato fruit did not raise lycopene or total carotenoid levels but did raise β-carotene levels threefold (up to 50 $\mu g \cdot gFW^{-1}$). The researchers found that some endogenous carotenogenic enzymes, including ζ-carotene desaturase and β-cyclase, were up-regulated in the transgenic tomatos, possibly explaining why β-carotene rather than lycopene was increased (Römer et al., 2000).

Rice is the staple food in many areas where vitamin A deficiency is widespread but is devoid of β-carotene (Ye et al., 2000). Therefore, there has been considerable interest in fortifying rice with provitamin A through genetic engineering. Examination of the immature rice endosperm using labeled substrates revealed the presence of geranyl geranyl diphosphate, the general precursor necessary for C40 carotenoid biosynthesis (Burkhardt et al., 1997). To investigate the feasibility of producing carotenoids into endosperm tissue, rice was transformed with daffodil *(Narcissus pseudonarcissus)* phytoene synthase (Figure 7.2). Transgenic rice seeds accumulated 0.74 $\mu g \cdot gDW^{-1}$ phytoene, but no other carotenoids were detected (Burkhardt et al., 1997). This case differs from the transgenic canola seeds expressing a bacterial phytoene synthase, which showed large increases in α- and β-carotene (Shewmaker et al., 1999). To complete the β-carotene biosynthetic pathway in rice, Ye et al. (2000) transformed rice with a bacterial phytoene desaturase gene, a daffodil phytoene synthase gene, and a β-cyclase gene also from daffodil. Phytoene synthase and phytoene desaturase genes were sufficient for β-carotene production, implying that β-cyclase is constitutively expressed in rice endosperm or induced as a result of the transformation. Carotenoid measurements showed that β-carotene accumulated to a maximum level of 1.6 $\mu g \cdot gDW^{-1}$ in the transgenic rice grains ("golden rice"). This level is relatively low in comparison to the RDA of 6 mg β-carotene per day but is proof that provitamin A accumulation in rice is possible.

Tocopherols: Vitamin E

Tocopherols are a class of lipid-soluble molecules collectively known as vitamin E. The best-characterized function of tocopherols is to act as lipid antioxidants, protecting against oxidation of polyunsaturated fatty acids (Kamal-Eldin and Appelqvist, 1996). All four major tocopherols (α, β, γ, and δ) are absorbed during digestion but α-tocopherol is preferentially retained and distributed throughout the body (Traber and Sies, 1996). The vitamin E activity of α-tocopherol is also higher than the other family members, being ten times that of its precursor, γ-tocopherol. The maximum adult RDA of vitamin E is 10 mg α-tocopherol or equivalent per day (National Research Council, 1989). Vitamin E deficiency is associated with an increased risk of cardiovascular disease, cancer, and a variety of degenerative disorders (reviewed in Traber and Sies, 1996).

Genes for enzymes involved in the tocopherol biosynthetic pathway (Figure 7.3) have now been cloned (reviewed in Grusak and DellaPenna, 1999). Overexpression of these genes opens the possibility of increasing the

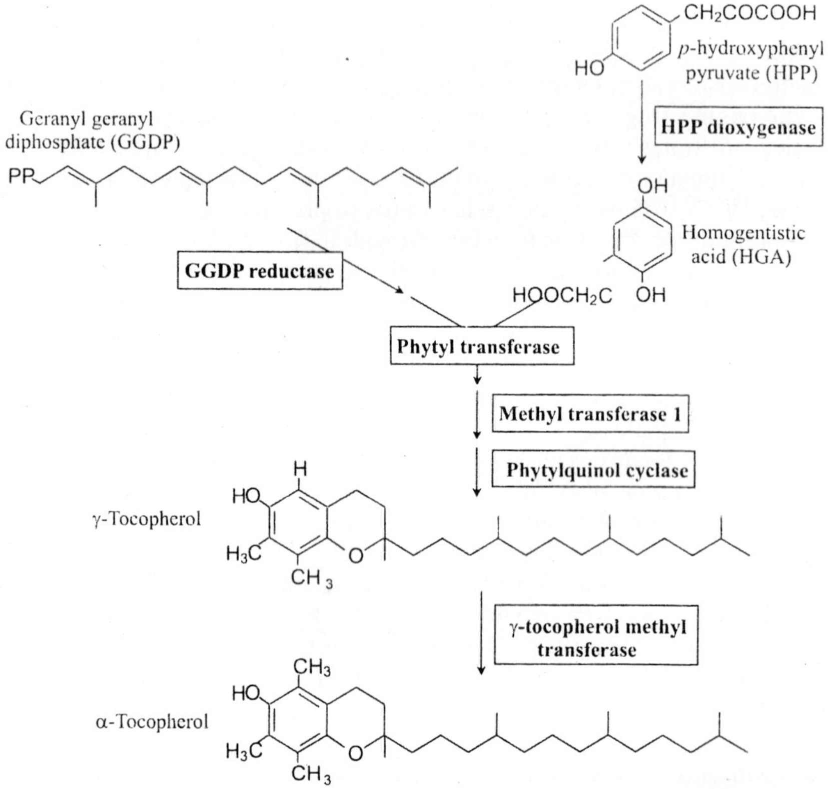

FIGURE 7.3. Summary of α-tocopherol synthesis. The tocopherol biosynthetic pathway is present in all photosynthetic organisms. Homogentisic acid (HGA) is produced from *p*-hydroxyphenyl pyruvate by the cytosolic enzyme HPP dioxygenase. All other catalytic steps are likely to be localized to the plastid:

flux through the pathway and increasing the total tocopherol concentration in specific plant tissues. Phytyl transferase, the enzyme that catalyzes the addition of a phytyl sidechain (produced from geranyl geranyl diphosphate reduction) to homogentisic acid (HGA), is a promising target in this regard as it represents the branch point of tocopherol and plastoquinone pathways (Figure 7.3). Research to date has focused on altering tocopherol composition. Despite a high concentration of tocopherols in most seeds, α-tocopherol is present as only a minor component. For example, *Arabidopsis* seeds contain approximately 350 μg total tocopherols per gFW, of which 95 percent is present as γ-tocopherol, 4 percent as δ-tocopherol, and 1 percent as α-tocopherol (Grusak and DellaPenna, 1999). Shintani and DellaPenna (1998) isolated and characterized an *Arabidopsis* gene for γ-tocopherol methyl transferase (γ-TMT), an enzyme which catalyzes the production of α-tocopherol from γ-tocopherol. The γ-TMT gene was overexpressed in transgenic *Arabidopsis* under the control of a seed-specific promoter. No change in total tocopherols was observed, but the level of α-tocopherol was increased 80-fold and constituted 85 to 95 percent of the tocopherol pool in the transgenic seed (Shintani and DellaPenna, 1998). This change in tocopherol composition represents a ninefold increase in vitamin E activity due to the higher vitamin E activity of α-tocopherol compared to γ-tocopherol.

Iron

Iron is necessary in human diets as it is an integral constituent of hemoglobin and myoglobin and is a component of many enzymes involved in redox processes. Iron deficiency is the most prevalent micronutrient deficiency worldwide and is particularly common among women in developing countries. Iron deficiency during pregnancy increases the risk of premature birth and increases perinatal mortality, while in infants lack of sufficient iron can impair mental and motor development. Iron deficiency has also been associated with decreased immune function and tiredness (reviewed in Martínez-Navarrete et al., 2002). Iron content is limited in many major crops, including rice. In addition, the bioavailability of iron in cereal and legume grains is generally low due to high levels of phytic acid, a potent inhibitor of iron absorption (Hurrell et al., 1992).

Increasing the store of bioavailable iron in crops by plant breeding or genetic engineering is a promising strategy to decrease in the incidence of iron deficiency. One successful approach has been to express the iron storage protein ferritin in transgenic plants. Ferritin, an iron storage protein in animals, plants, and bacteria, forms a large complex that stores ferrous iron in a central cavity (Laulhere et al., 1990). The use of ferritin to increase iron accumulation in transgenic plants was first demonstrated in tobacco leaves. Goto et al. (1998) expressed soybean ferritin in tobacco leaves and demonstrated iron increases of up to 30 percent. In each transformant, the increased iron content was correlated with increases in ferritin content (Goto et al., 1998). In subsequent research, rice was transformed with the same soybean ferritin gene, expressed in the endosperm under the control of a glutelin promoter (Goto et al., 1999). The ferritin accumulated in the endosperm and the iron content of T1 seeds was as much as threefold greater than that of their untransformed counterparts (Goto et al., 1999). In addition to acting as an iron store in the transgenic plants, the introduced ferritin protein may have increased the iron content by feedback activation of the plant iron transport systems.

The bioavailability of iron from plant sources may be improved through an increase in enhancers such as cysteine-rich peptides (Taylor et al., 1986) and the removal of phytic acid by the addition of phytase (Hurrell et al., 1992). In an attempt to increase both the level of iron and its absorption by the human body, Lucca et al. (2001) transformed rice with genes encoding a ferritin gene from *Phaseolus vulgaris*, an acid-stable phytase from *Aspergillus fumigatus,* and an endogenous cysteine-rich metallothionein-like protein. Rice transformed with the *Phaseolus* ferritin gene showed a doubling of seed iron (up to 22 $\mu g \cdot gFW^{-1}$), while plants transformed with the fungal phytase showed threefold lower seed phytic acid levels after incubation with simulated small-intestine conditions (Lucca et al., 2001). These three traits are now being combined. Any benefits to iron absorption mediated by the phytase and metallothionein proteins can then be assessed.

CURRENT LIMITATIONS

Major advances in our understanding of plant metabolism in recent years have led to the production of the first transgenic plants with nutritionally enhanced seeds. However, a number of problems remain that limit the benefit of some of these seeds. In some cases, the level of production of the protein or metabolite of interest is too low to effect significant nutritional improvement. This may require higher levels of gene expression or manipulation of the production and distribution of precursor metabolites. In other cases, nontarget metabolites are altered in the transgenic seed, potentially affecting desirable traits.

The allergenic properties of some storage proteins rich in essential amino acids have limited their use in transgenic plants. A number of 2S albumins have been identified as allergens, including the sulfur-rich Brazil nut albumin, a protein that has been introduced into a variety of species. Allergenicity was maintained in the seeds of transgenic soybean expressing BNA (Nordlee et al., 1996). Despite this limitation, predictive methods are now available to assess the allergenicity of storage proteins before their introduction into new host plants. Biotechnology approaches are now being used to eliminate the major allergenic proteins from some foods. Tada et al. (1996) have reported the reduction of allergenic proteins in rice using antisense gene constructs. However, the presence of multiple allergenic proteins in some species may make the production of allergen-free foodstuffs quite difficult.

One significant limitation to the improvement of seed nutritional quality has been low levels of accumulation of the nutritionally important protein or metabolite. More efficient seed-specific promoters have been identified in recent years which have helped increase the levels of transgenic proteins to useful levels. However, a strong promoter or efficient expression system does not guarantee a high level of accumulation of the protein or metabolite of interest, as precursor molecules may be limiting. In order to increase the amount of limiting amino acids, one approach has been to deregulate the biosynthetic pathway involved. Recently, feedback-insensitive metabolic enzymes have been introduced into transgenic plants to increase flux through the lysine biosynthetic pathway (Falco et al., 1995).

Trait improvement may also be limited due to changes in plant metabolism that occur in response to the transgenic protein or its product. For example, the overexpression of phytoene synthase in canola seeds resulted in an increase in carotenoids as expected but coincided with a decrease in other plastid-localized isoprenoids, such as tocopherols and chlorophyll (Shewmaker et al., 1999). This suggests that there is a limit to the amount of isoprenoid precursors available to the plastid. In another example, transgenic rice expressing SSA at 7 percent of total protein shows no increase in seed cysteine and methionine due to the down-regulation of endogenous sulfur-rich storage proteins (Hagan et al., 2003). This contrasts with the SSA-expressing lupins, which showed a 20 percent increase in seed sulfur amino acids (Molvig et al., 1997). These differences may be due to a smaller pool of inorganic sulfur available in the rice grains. In cases of low sulfur availability in the seed, engineering of sulfur uptake and distribution in the plant may be required in order to increase total sulfur amino acids.

CONCLUSIONS

The development of high-methionine seeds, high-oleic and stearic oils, and vitamin, and mineral-producing seeds are proof that genetic engineering can bring about a significant improvement in seed nutritional quality. In many cases, such as golden rice, this improvement could not have been achieved through plant breeding. However, we do not always understand how transgenes impact on plant metabolism. In order to produce nutritionally important compounds at useful levels and minimize the impact on other metabolic pathways, a greater understanding of metabolite transport, flux, and compartmentalization is required. Despite these limitations, the use of genetic engineering to modify seed composition has the potential to produce healthier foods and combat a wide range of nutrient deficiencies.

REFERENCES

Altenbach, S.B., Chiung-Chi, K., Staraci, L.C., Pearson, K.W., Wainwright, C., Georgescu, A., and Townsend, J. (1992). Accumulation of a Brazil nut albumin in seeds of transgenic canola results in enhanced levels of seed protein methionine. *Plant Molecular Biology* 18:235-245.

Altenbach, S.B., Pearson, K.W., Meeker, G., Staraci, L.C., and Sun, S.S.M. (1989). Enhancement of the methionine content of seed proteins by the expression of a chimeric gene encoding a methionine-rich protein in transgenic plants. *Plant Molecular Biology* 13:513-522.

Ampe, C., Van Damme, J., de Castro, L., Sampaio, M.J., Van Montagu, M., and Vandekerckhove, J. (1986). The amino acid sequence of the 2S sulfur-rich proteins from seeds of Brazil nut (*Bertholletia excelsa* H.B.K.). *European Journal of Biochemistry* 159:597-604.

Anthony, A., Brown, W., Buhr, D., Ronhovde, G., Genovesi, D., Lane, T., Yingling, R., Aves, K., Rosato, M., and Anderson, P. (1997). Transgenic maize with elevated 10 kD zein and methionine. In Cram, W.J., De Kok, L.J., Stulen, I., Brunold, C., and Rennenberg, H. (eds), *Sulphur metabolism in higher plants* (pp. 295-297). Leiden: Backhuys Publishers.

Beever, D.E. and Siddons, R.C. (1986). Digestion and metabolism in the grazing ruminant. In Milligan, L.P., Grovum, W.L., and Dobson, A. (eds), *Control of digestion and metabolism in ruminants* (pp. 479-496). Englewood Cliffs, NJ: Prentice Hall.

Bendich, A. (1993). Biological functions of dietary carotenoids. *Annals of the New York Academy of Sciences* 691:61-67.

Burkhardt, P.K., Beyer, P., Wunn, J., Kloti, A., Armstrong, G.A., Schledz, M., von Lintig, J., and Potrykus I. (1997) Transgenic rice *(Oryza sativa)* endosperm expressing daffodil *(Narcissus pseudonarcissus)* phytoene synthase accumulates phytoene, a key intermediate of provitamin A biosynthesis. *The Plant Journal* 11:1071-1078.

Cruz-Alvarez, M., Kirihara, J.A., and Messing, J. (1991). Post-transcriptional regulation of methionine content in maize kernels. *Molecular and General Genetics* 225:331-339.

De Clercq, A., Vandewiele, M., Van Damme, J., Guerche, P., Van Montagu, M., Vandekerckhove, J., and Krebbers, E. (1990). Stable accumulation of modified 2S albumin seed storage proteins with higher methionine contents in transgenic plants. *Plant Physiology* 94:970-979.

Falco, S.C., Guida, T., Locke, M., Mauvais, J., Sanders, C., Ward, R.T., and Webber, P. (1995) Transgenic canola and soybean seeds with increased lysine. *Biotechnology* 13:577-582.

Fraser, P.D., Römer, S., Shipton, C.A., Mills, P.B., Kiano, J.W., Misawa, N., Drake, R.G., Schuch, W., and Bramley, P.M. (2002). Evaluation of transgenic tomato plants expressing an additional phytoene synthase in a fruit-specific manner. *Proceedings of the National Academy of Sciences of the USA* 99:1092-1097.

Friedman, M. (1996). Nutritional value of proteins from different food sources: A review. *Journal of Agriculture and Food Chemistry* 44:6-29.

Goto, F., Yoshihara, T., and Saiki, H. (1998). Iron accumulation in tobacco plants expressing soyabean ferritin gene. *Transgenic Research* 7:173-180.

Goto, F., Yoshihara, T., Shigemoto, N., Toki, S., and Takaiwa, F. (1999). Iron fortification of rice seed by the soybean ferritin gene. *Nature Biotechnology* 17:282-286.

Grusak, M.A. and DellaPenna, D. (1999). Improving the nutrient composition of plants to enhance human nutrition and health. *Annual Review of Plant Physiology and Plant Molecular Biology* 50:133-161.

Hagan, N.D., Upadhyaya, N., Tabe, L.M., and Higgins, T.J.V. (2003). The redistribution of protein sulfur in transgenic rice expressing a gene for a foreign, sulfur-rich protein. *The Plant Journal* 34:1-11.

Hoffman, L.M., Donaldson, D.D., and Herman, E.M. (1988). A modified storage protein is synthesised, processed and degraded in the seeds of transgenic plants. *Plant Molecular Biology* 11:717-729.

Hu, F.B., Manson, J.E., and Willett, W.C. (2001). Types of dietary fat and risk of coronary heart disease: A critical review. *Journal of the American College of Nutrition* 20:5-19.

Hurrell, R.F., Juillerat, M.A., Reddy, M.B., Lynch, S.R., Dassenko, S.A., and Cook, J.D. (1992). Soy protein, phytate, and iron absorption in humans. *The American Journal of Clinical Nutrition* 56:573-578.

Kamal-Eldin, A. and Appelqvist, L.A. (1996). The chemistry and antioxidant properties of tocopherols and tocotrienols. *Lipids* 31:671-701.

Katan, M.B., Zock, P.L., and Mensink, R.P. (1995). Dietary oils, serum lipoproteins, and coronary heart disease. *The American Journal of Clinical Nutrition* 61: 1368S-1373S.

Keeler, S.J., Maloney, C.L., Webber, P.Y., Patterson, C., Hirata, L.T., Falco, S.C., and Rice, J.A. (1997). Expression of *de novo* high-lysine α-helical coiled-coil proteins may significantly increase the accumulated levels of lysine in mature seeds of transgenic tobacco plants. *Plant Molecular Biology*. 34:15-29.

Khan, M.R., Ceriotti, A., Tabe, L., Aryan, A., McNabb, W., Moore, A., Craig, S., Spencer, D., and Higgins, T.J.V. (1996). Accumulation of a sulphur-rich seed albumin from sunflower in the leaves of transgenic subterranean clover (*Trifolium subterraneum* L.). *Transgenic Research* 5:179-185.

Kinney, A.J. (1996). Designer oils for better nutrition. *Nature Biotechnology* 14: 946.

Kjemtrup, S., Herman, E.M., and Chrispeels, M.J. (1994). Correct post-translational modification and stable vacuolar accumulation of phytohemagglutinin engineered to contain multiple methionine residues. *European Journal of Biochemistry* 226:385-391.

Kortt, A.A., Caldwell, J.B., Lilley, G.G., and Higgins, T.J.V. (1991). Amino acid and cDNA sequences of a methionine-rice 2S albumin protein from sunflower seed (*Helianthus annus* L.). *European Journal of Biochemistry* 195:329-334.

Knutzon, D.S., Thompson, G.A., Radke, S.E., Johnson, W.B., Knauf, V.C., and Kridl, J.C. (1992). Modification of *Brassica* seed oil by antisense expression of a stearoyl-acyl carrier protein desaturase gene. *Proceedings of the National Academy of Sciences of the USA* 89:2624-2628.

Lai, J. and Messing, J. (2002). Increasing maize seed methionine by mRNA stability. *The Plant Journal* 30:395-402.

Laulhere, J.P., Laboure, A.M., and Briat, J.F. (1990). Photoreduction and incorporation of iron into ferritins. *The Biochemical Journal* 269:79-84.

Lilley, G.G., Caldwell, J.B., Kortt, A.A., Higgins, T.J., and Spencer, D. (1989). Isolation and primary structure for a novel methionine-rich protein from sunflower seeds (*Helianthus annus* L.). In Applewhite, T.H. (ed.), *Proceedings of the World Congress on Vegetable Protein Utilization in Human Foods and Animal Feedstuffs* (pp. 487-502). Champaign, IL: American Oil Chemists Society.

Liu, Q., Singh, S.P., and Green, A.G. (2002). High-stearic and high-oleic cottonseed oils produced by hairpin RNA-mediated post-transcriptional gene silencing. *Plant Physiology* 129:1732-1743.

López, A.D. and García, M.F. (2000). Plants as "chemical factories" for the production of polyunsaturated fatty acids. *Biotechnology Advances* 18:481-497.

Lucca, P., Hurrell, R., and Potrykus, I. (2001). Genetic engineering approaches to improve the bioavailability and the level of iron in rice grains. *Theoretical and Applied Genetics* 102:392-397.

MacRae, J.C. and Ulyatt, M.J. (1974). Quantitative digestion of fresh herbage by sheep: II. The sites of digestion of some nitrogenous constituents [*Lolium,* white clover]. *The Journal of Agricultural Science* 82:309-319.

Martínez-Navarrete, N., Camacho, M.M., Martínez-Lahuerta, J., Martínez-Monzó, J., and Fito, P. (2002). Iron deficiency and iron fortified foods—A review. *Food Research International* 35:225-231.

Mayne, S.T. (1996). Beta-carotene, carotenoids, and disease prevention in humans. *The Federation of American Societies for Experimental Biology Journal* 10:690-701.

McNabb, W.C. (1990). Digestion and metabolism of sulfur containing amino acids in sheep fed fresh forage diets. Doctoral thesis, Massey University, New Zealand.

Molvig, L., Tabe, L., Eggum, B.O., Moore, A., Craig, S., Spencer, D., and Higgins, T.J.V. (1997). Enhanced methionine levels and increased nutritive value of seeds of transgenic lupins (*Lupinus angustifolius* L.) expressing a sunflower seed albumin gene. *Proceedings of the National Academy of Sciences of the USA* 94:8393-8398.

Müntz, K., Christov, V., Jung, R., Saalbach, G., Saalbach, I., Waddell, D., Pickardt, T., and Schieder, O. (1997). Genetic engineering of high methionine proteins in grain legumes. In Cram, W.J., De Kok, L.J., Stulen, I., Brunold, C., and Rennenberg, H. (eds), *Sulfur metabolism in higher plants* (pp. 71-86). Leiden, Germany: Backhuys Publishers.

Murphy, D.J. (1999). Manipulation of plant oil composition for the production of valuable chemicals. In Shahidi, F., Kolodziejczyk, P., Whitaker, J.R., Munguia, A.L., and Fuller, G. (eds), *Chemicals via higher plant bioengineering* (pp. 21-35). New York: Kluwer Academic/Plenum Publishers.

National Research Council, Food and Nutrition Board (1989). *Recommended dietary allowances,* Tenth edition. Washington, DC: National Academy.

Nordlee, J.A., Taylor, S.L., Townsend, J.A., Thomas, L.A., and Bush, R.K. (1996). Identification of a Brazil nut allergen in transgenic soybeans. *The New England Journal of Medicine* 334:688-692.

Ohtani, T., Galili, G., Wallace, J.C., Thompson, G.A., and Larkins, B.A. (1991). Normal and lysine-containing zeins are unstable in transgenic tobacco seeds. *Plant Molecular Biology* 16:117-128.

Parker-Barnes, J.M., Das, T., Bobik, E., Leonard, A.E., Thurmond, J.M., Chaung, L.T., Huang, Y.S., and Mukerji, P. (2000). Identification and characterization of an enzyme involved in the elongation of n-6 and n-3 polyunsaturated fatty acids. *Proceedings of the National Academy of Sciences of the USA* 97:8284-8289.

Pickardt, T., Saalbach, I., Waddell, D., Meixner, M., Müntz, K., and Schieder, O. (1995). Seed specific expression of the 2S albumin gene from Brazil nut *(Bertholletia excelsa)* in transgenic *Vicia narbonensis. Molecular Breeding* 1:295-301.

Ravindran, V., Tabe, L.M., Molvig, L., Higgins, T.J.V., and Bryden, W.L. (2002). Nutritional elevation of transgenic high-methionine lupins (*Lupinus angustifolius* L.) with broiler chickens. *Journal of the Science of Food and Agriculture* 82:280-285.

Reddy, A.S. and Thomas, T.L. (1996). Expression of a cyanobacterial delta 6-desaturase gene results in gamma-linolenic acid production in transgenic plants. *Nature Biotechnology* 14:639-642.

Roesler, K.R. and Rao, A.G. (1999). Conformation and stability of barley chymotrypsin inhibitor-2 (CI-2) mutants containing multiple lysine substitutions. *Protein Engineering* 12:967-973.

Römer, S., Fraser, P.D., Kiano, J.W., Shipton, C.A., Misawa, N., Schuch, W., and Bramley, P.M. (2000). Elevation of the provitamin A content of transgenic tomato plants. *Nature Biotechnology* 18:666-669.

Saalbach, G., Christov, V., Jung, R., Saalbach, I., Manteuffel, R., Kunze, G., Brambarov, K., and Müntz, K. (1995). Stable expression of vicilin from *Vicia faba* with eight additional single methionine residues but failure of accumulation of legumin with an attached peptide in tobacco seeds. *Molecular Breeding* 1:245-258.

Saalbach, G., Jung, R., Saalbach, I., and Müntz, K. (1988) Construction of storage protein genes with increased number of methionine codons and their use in transformation experiments. *Biochemie und Physiologie der Pflanzen* 183:211-218.

Salisbury, F.B. and Ross, C.W. (1985) *Plant physiology,* Third edition. Belmont, CA: Wadsworth Publishing Company.

Sayanova, O., Smith, M.A., Lapinskas, P., Stobart, A.K., Dobson, G., Christie, W.W., Shewry, P.R., and Napier JA. (1997). Expression of a borage desaturase cDNA containing an N-terminal cytochrome b5 domain results in the accumulation of high levels of delta6-desaturated fatty acids in transgenic tobacco. *Proceedings of the National Academy of Sciences of the USA* 94:4211-4216.

Shewmaker, C.K., Sheehy, J.A., Daley, M., Colburn, S., and Ke, D.Y. (1999). Seed-specific overexpression of phytoene synthase: Increase in carotenoids and other metabolic effects. *The Plant Journal* 20:401-412.

Shintani, D. and DellaPenna, D. (1998). Elevating the vitamin E content of plants through metabolic engineering. *Science* 282:2098-2100.

Smith, N.A., Singh, S.P., Wang, M.B., Stoutjesdijk, P.A., Green, A.G., and Waterhouse, P.M. (2000). Total silencing by intron-spliced hairpin RNAs. *Nature* 407:319-320.

Sun, S.S., Altenbach, S.B., and Leung, F.W. (1987). Properties, biosynthesis and processing of a sulfur-rich protein in Brazil nut (*Bertholletia excelsa* H.B.K.). *European Journal of Biochemistry* 162:477-483.

Tabe, L., Molvig, L., Khan, R., Schroeder, H., Gollasch, S., Wardley-Richardson, T., Moore, A., Craig, S., Spencer, D., Eggum, B., and Higgins, T.J.V. (1997). Modifying the sulphur amino acid content of protein in transgenic legumes. In Cram, W.J., De Kok, L.J., Stulen, I., Brunold, C., and Rennenberg, H. (eds), *Sulphur metabolism in higher plants* (pp. 87-93). Leiden, Germany: Backhuys Publishers.

Tada, Y., Nakase, M., Adachi, T., Nakamura, R., Shimada, H., Takahashi, M., Fujimura, T., and Matsuda, T. (1996). Reduction of 14-16 kDa allergenic proteins in transgenic rice plants by antisense gene. *Federation of European Biochemical Societies Letters* 391:341-345.

Taylor, P.G., Martinez-Torres, C., Romano, E.L., and Layrisse, M. (1986) The effect of cysteine-containing peptides released during meat digestion on iron absorption in humans. *The American Journal of Clinical Nutrition* 43:68-71.

Thelen, J.J. and Ohlrogge, J.B. (2002). Metabolic engineering of fatty acid biosynthesis in plants. *Metabolic Engineering* 4:12-21.

Traber, M.G. and Sies, H. (1996). Vitamin E in humans: Demand and delivery. *Annual Review of Nutrition* 16:321-347.

Tu, H.M., Godfrey, L.W., and Sun, S.S.M. (1998). Expression of the Brazil nut methionine-rich protein and mutants with increased methionine in transgenic potato. *Plant Molecular Biology* 37:829-838.

Wallace, J.C., Galili, G., Kawata, E.E., Cuellar, R.E., Shotwell, M.A., and Larkins, B.A. (1988). Aggregation of lysine-containing zeins into protein bodies in *Xenopus* oocytes. *Science* 240:662-664.

Wesley, S.V., Helliwell, C.A., Smith, N.A., Wang, M.B., Rouse, D.T., Liu, Q., Gooding, P.S., Singh, S.P., Abbott, D., Stoutjesdijk, P.A., Robinson, S.P., Gleave, A.P., Green, A.G., and Waterhouse, P.M. (2001). Construct design for efficient, effective and high-throughput gene silencing in plants. *The Plant Journal* 27:581-590.

White, C.L., Tabe, L.M., Dove, H., Hamblin, J., Young, P., Phillips, N., Taylor, R., Gulati, S., Ashes, J., and Higgins, T.J.V. (2001). Increased efficiency of wool growth and live weight gain in Merino sheep fed transgenic lupin seed containing sunflower albumin. *Journal of the Science of Food and Agriculture* 81:147-154.

Williamson, M.S., Forde, J., Buxton, B., and Kreis, M. (1987). Nucleotide sequence of barley chymotrypsin inhibitor-2 (CI-2) and its expression in normal and high-lysine barley. *European Journal of Biochemistry* 165:99-106.

Yang, M.S., Espinoza, N.O., Nagpala, P.G., Dodds, J.H., White, F.F., Schnorr, K.L., and Jaynes, J.M. (1989). Expression of a synthetic gene for improved protein quality in transformed potato plants. *Plant Science* 64:99-111.

Ye, X., Al-Babili, S., Kloti, A., Zhang, J., Lucca, P., Beyer, P., and Potrykus, I. (2000). Engineering the provitamin A (β-carotene) biosynthetic pathway into carotenoid-free rice endosperm. *Science* 287:303-305.

Chapter 8

The Process of Accumulation of Seed Proteins and the Prospects for Using Biotechnology to Improve Crops

Eliot M. Herman

INTRODUCTION

Seeds, as represented by the major crops of corn, wheat, soybeans as well as other legumes and cereals throughout the world, are the central agricultural commodities. The domestication of these and other a few other crops has been hypothesized to be one of the most significant developments in the origins of civilization throughout the world. Human selection of plants has optimized seeds to accumulate large quantities of proteins and for high productivity of the seed plants that bear the seeds. Modern biology has resulted in a broad understanding of the molecular and cell biology of seed constituents. New tools and perspective are available to produce directed changes in seeds using biotechnology. A large literature on this subject is available, with many excellent reviews of the process of seed development, storage protein families, and how storage proteins are assembled into storage organelles. The cutting edge of seed biology involves using this base of information to conduct further experiments to understand how seeds regulate composition. This knowledge is now being used in conjunction with biotechnology to alter seeds to improve composition, allergenicity, and the production of industrial or pharmaceutical products.

SYNTHESIS AND ACCUMULATION OF SEED PROTEINS

Storage cells of seeds consist of two distinct functional types. The first type are those cells that survive into germination, including both the developing embryo, some endosperm or aleurone cells of dicotyledonous seeds, and the aleurone cells of cereals. Within these cells, the stored reserve substances are mobilized by de novo synthesis of hydrolases during germination, with the hydrolysis products transported out of the cell by active processes. The second type of storage cells accumulate reserve substances and then undergo programmed cell death at the end of seed maturation. An example of this are the starchy endosperm cells of cereal grains. The stored reserve substances are contained within dead disrupted cells that are mobilized by hydrolases secreted from aleurone and embryo cells that persist into germination, with the resulting hydrolysis products absorbed into the growing embryo. The specialized seed storage cells may contain one or both types of protein storage organelles, ER-derived protein bodies (PB) or Golgi-derived protein storage vacuoles (PSV). As detailed in the subsequent sections, the composition and ontogeny of these two distinct protein storage organelles are different, although in some plants the two paths of protein storage may merge, forming a single composite compartment.

THE PROTEIN STORAGE VACUOLE

Much of the research on the ontogeny of PSVs has been conducted on dicotyledonous embryonic cells. PSVs are transiently differentiated vacuoles that originate from the subdivision of the central vacuole during mid-maturation of seeds, producing small spherical organelles filled with protein deposits (Craig et al., 1979, 1980). The population of PSVs is maintained through seed maturation and is still present when the seed germinates. As storage protein mobilization proceeds and the PSV once again acquires an "empty" appearance, the individual PSVs merge and the vacuole resumes the morphology of a central organelle. The formation and disassembly of PSVs occurs coordinately with the respective deposition and mobilization of storage proteins. The process that controls and mediates the subdivision and reformation of the vacuole into and from PSVs remains unknown. Although the subdivision and reformation of the vacuole is correlated with the presence of storage proteins, it does not appear that the presence or absence of protein deposits induce this change in vacuole morphology. Protein storage vacuoles are also found in vegetative cells, including leaves, stems, bark, and tubers; in most of these cells, although the vacuole may be filled with protein deposits, the vacuole does not undergo the large-scale subdivision characteristic of seeds. The subdivision of the vacuole into numerous small vacuoles is likely an adaptation to preserve structural integrity during the dehydration and rehydration during the seed's life cycle that vegetative cells do not usually encounter. Therefore, it is likely the processes that control vacuole subdivision occur in parallel rather than being induced by storage protein deposition. Light and electron microscopy investigations have indicated that the central vacuole is subdivided coordinately with the deposition of storage proteins producing numerous protein-filled vacuoles (Craig et al., 1979, 1980). The process of forming a protein-filled vacuole appears to occur through several different mechanisms. For instance, in pea seeds electron microscopy and immunocytochemistry with antistorage protein antibodies have shown that storage proteins are sequestered and packaged into protein aggregates on the inner surface of the tonoplast that is budded off of the vacuole as a mature protein-filled PSV (Hoh et al., 1995). The various seed storage proteins within the PSV matrix appear to stratify, resulting in different domains enriched in each protein (Hinz et al., 1997; Hara-Nishimura et al., 1985). In other legumes as well as other dicotyledonous plants, the subdivision of the vacuole into incipient PSVs appears to precede the deposition of most of the storage proteins. This results in numerous small partially filled PSVs that are gradually filled with storage proteins—resulting in mature, protein-filled PSVs. Because this mechanism for PSV formation appears to occur over a protracted period of time, this allows for additional complexity and subdomains to be established within the protein matrix of the PSV. These subdomains can include not only amorphous deposits of storage proteins but also a protein crystalloid that is assembled during the course of PSV maturation. PSVs also often contain deposits of phytin (inositol polyphosphate) and other inclusions derived from autophagy of the cytosolic constituents (see Herman, 1994, for review).

The interpretation that PSVs arise directly from subdivision of the embryonic vegetative vacuole is not universally held. A competing model is that PSVs arise from a newly created population of vacuoles that is assembled alongside a preexisting population of vegetative vacuoles. Robinson et al. (1995) have proposed that PSVs might constitute an entirely new vacuole that is synthesized de novo and replaces the preexisting central vacuole (Robinson, Bäumer, et al., 1998; Robinson, Hinz, and Hostein, 1998; Robinson, Galili, et al., 1998; Robinson et al., 2000, for review). Among the evidence for this are data from micrographs that compellingly illustrate the presence of two distinctly different and coexisting vacuoles in storage parenchyma cells of maturing pea seeds (Hoh et al., 1995). One of the vacuole types observed appears to be incipient PSVs containing dense aggregates of storage proteins that appear to be in the process of releasing mature protein-filled PSVs by budding. The other vacuole type observed retains the "empty" appearance of the preexisting embryonic central vacuole (Robinson et al., 1995). The isolated membranes of the PSV and its vegetative vacuole precursor possess different intrinsic tonoplast polypeptides and density in equilibrium sucrose gradients (Hoh et al., 1995). This has been interpreted to indicate that the vegetative vacuole and PSV are entirely different structures having independent origins, with the PSV replacing the vegetative vacuole as seed maturation progresses. Support for this hypothesis has been obtained by microscopy using antibodies against α- and γ-TIP, seed-specific, and vegetative aquaporin isoforms (Paris et al., 1996). The observations showed that the cells contain two distinct vacuoles, which differ in the polypeptide composition of both tonoplast and matrix proteins. How these two distinct models, either the vegetative vacuole transformation into the PSV, or the PSV being formed independently and replacing the vegetative vacuole, can be reconciled still requires additional research. With the completion of the *Arabidopsis* genome and the availability of mutant reverse genetics, it may prove possible to select for seeds that provide key information about how the PSV formed and how it originates—whether entirely from a preexisting vacuole or with de novo organelle formation. It is possible that more than one general pattern of PSV ontogeny exists and the process used to form the PSV might be species variable or even developmentally or environmentally regulated. It is possible that processes that appear to be quite different in the various seeds are the consequence of experimental material and/or the conditions of plant growth.

Protein Storage Vacuole Contents

The primary and most prominent storage proteins sequestered in PSVs are the 7S and 11S seed storage globulins, which are members of large gene families (see Shewry et al., 1998, for review). The storage globulins are encoded for by multigene families (Casey, 1979; Müntz, 1996; Casey and Domoney, 1999). Storage proteins are classified into two groups by their sedimentation coefficient and/or by solubility in solutions of differing salt and pH. The 11S globulins of the legumin type (e.g., pea legumin and soybean glycinin) and the 7S globulins of the vicilin type (e.g., pea vicilin and bean phaseolin) are the two largest families. These proteins are found in diverse species of plants and appear to have an ancient origin, as indicated by their presence in conifers and *Ginkgo biloba* seeds (Häger et al., 1995). In PSVs that possess crystalloids that separate the two major proteins into distinct intra-PSV domains, the 7S proteins are in the peripheral matrix, while the 11S proteins are the primary constituent of the crystalloid (Hara-Nishimura et al., 1985; Jiang et al., 2001). In other species, the 7S and 11S proteins are uniformly codistributed throughout the PSV matrix. Why some 7S and 11S proteins form distinct domains while others are distributed together in an amorphous matrix remains unknown and may simply reflect subtle differences in the ability or inability of the proteins to interact with each other.

Storage proteins are synthesized on ER-bound ribosomes and are cotranslationally translocated into the ER lumen where ER resident chaperones and processing enzymes fold and, if required, form disulfide bonds and add an aspargine-linked glycan. Within the ER the precursor polypeptides of the 11S globulins form trimers with a sedimentation coefficient of about 9S and a molecular mass of 180 kDa. After import into the protein storage vacuole, the precursor polypeptides become proteolytically processed at a conserved arginine residue by highly conserved cysteine proteases (Müntz, 1996). Each 60 kDa monomer is cleaved into two chains with molecular masses of 40 (α-chain) and 20 (β-chain) kDa. These two chains remain linked via an intramolecular disulfide bridge which was formed in the ER. This disulfide bridge is not important for targeting, but processing and hexamer formation are impaired after deletion of these disulfide bridges (Jung et al., 1997). After processing the two trimers form a 11S hexamer with a molecular mass of 360 kDa. The 7S proteins are not usually posttranslationally processed but are often cotranslationally glycosylated, which may be further modified in the Golgi prior to deposition in the vacuole. The 7S globulins also form trimer oligomers in the ER and remain trimers after deposition in the protein storage vacuole.

The PSV contains many other proteins sequestered with the storage proteins, and some of these may be in sufficient concentration to constitute auxiliary storage proteins. The 2S albumin is a major auxiliary storage protein in diverse plants ranging from tree nuts such as Brazil nut, *Brassica,* pumpkin, castor bean, and sunflower. Another extensively characterized family of auxiliary storage proteins are the "classic" seed lectins, which in some legumes can account for 10 percent or more of the total protein (see Etzler, 1985, for review). Members of the lectin gene family include proteins with various functions including the α-amylase inhibitors that defend against insect feeding on mature dry seeds (see Chrispeels and Raikhel, 1991a, for review). Other PSV-sequestered defense proteins include Kunitz-type trypsin inhibitors (Horisberger and Volanthen, 1983) and P34 (Kalinski et al., 1992), a distantly related member of the papain superfamily that binds an elicitor derived from *Pseudomonas* (Cheng et al., 1998). Storing defense proteins in dry seeds may anticipate the potential of insects feeding when the seed is unable to respond with an inducible reaction. The other ancillary proteins stored in the PSVs are diverse acid hydrolases. Many of these hydrolases are accumulated in parallel with the storage proteins; these include glycosidases, phosphatases, phospholipase D, and nucleases (Nishimura and Beevers, 1978; Mettler and Beevers, 1979; Chappell et al., 1980; Herman and Chrispeels, 1980; Van der Wilden et al., 1980).

Protein Storage Vacuoles Are a Post-Golgi Compartment

The 7S and 11S storage proteins pass through the Golgi apparatus, where they become segregated from other proteins of the secretory pathway and are sorted into Golgi-derived transport vesicles. The transport of proteins from the ER to the Golgi apparatus appears to be mediated by the tubular domain of the ER network (Harris, 1979) with the Golgi apparatus closely aligned along the tubular ER. This suggests that a specific subdomain of the ER may be responsible for storage protein exit to the Golgi apparatus. Freeze-fracture EM yielded similar observations (Herman et al., 1984) showing the *cis*-Golgi aligned with the tubular ER. The first indication that storage proteins are deposited in the vacuole by the Golgi apparatus was based on the interpretation of conventional electron micrographs showing proteinaceous electron-dense vesicles that were postulated to carry storage proteins exiting the *trans*-face of the Golgi. This proposal was confirmed in the early 1980s, when several investigators used immunogold EM to show that PSV proteins are present in the ER, Golgi apparatus, and secretory vesicles forming PSVs (see Herman and Shannon, 1984; Herman et al., 1984; Craig and Goodchild, 1984; Greenwood and Chrispeels, 1985; Kim et al., 1988, for examples; reviewed in Robinson and Hinz, 1997).

The first biochemical evidence for storage protein trafficking through the Golgi apparatus was based on pulse-chase experiments using radioactive fucose and sucrose gradient subcellular fractionation (Chrispeels, Higgins, Craig, and Spencer, 1982; Chrispeels, Higgins, and Spencer, 1982; Chrispeels, 1983). Fucosylation is mediated by the Golgi, and the transfer of fucose to a PSV resident protein provides direct evidence for trafficking through the Golgi. The Golgi apparatus processes the ER-derived high mannose glycan side chains of vacuolar glycoproteins (Chrispeels, 1983; Faye et al., 1986; see Stahaelin and Moore, 1995, for review) that are sequentially trimmed to remove mannosyl residues followed by attachment of additional xylosyl, fucosyl, and/or galactosyl residues. Glycan processing is most often assayed by testing for the acquisition of endoglyosidase H resistance due to fucosyl attachment to N-acetylglucosamine #1 of the side chain. Subcellular fractionation showed that the UDP-sugar transfer occurs in Golgi apparatus fractions (Sturm et al., 1987). The position and accessibility of the glycan side chain is a significant factor in the subsequent glycan processing in the Golgi apparatus (Faye et al., 1986). Immunogold labeling of the plant Golgi apparatus with antibodies specific for various glycosyl residues showed that the *cis, medial,* and *trans* domains are distinct (see Moore et al., 1991) to mediate the ordered steps of glycan processing before protein packaging and targeting. Xylosyl residues on a storage protein of soybean aleurone cells was visualized by immunogold EM, and it appeared that attachment of the xylosyl residues occurs in the *medial* to *trans* domain of the ER (Yaklich and Herman, 1995).

Vacuolar proteins possess sequence-specific vacuolar sorting sequences, some of which are located either at the N-terminus or the C-terminus, while others are located internally (Neuhaus and Rogers, 1998; Matsuoka and Neuhaus, 1999; Vitale and Raikhel, 1999; Frigerio et al., 2001). The targeting signal of vacuolar cysteine proteases and some other vacuolar proteins consists of a larger, charged amino acid (often N) at the first position, a nonacidic amino acid (P) at the second position, a large hydrophobic amino acid (I, L) at the third position, the fourth position being not strictly conserved, and an amino acid with a large hydrophobic side chain (L, P) at the fifth position (Matsuoka and Nakamura, 1999; Matsuoka and Neuhaus, 1999). The N-terminal NPIR motif from barley cysteine protease aleurain binds to a putative vacuolar sorting receptor called BP80/AtELP that is associated with the *trans*-Golgi (Kirsch et al., 1994, 1996; Ahmed et al., 1997; Paris et al., 1996).

The various targeting signals identified in the storage proteins do not exhibit obvious sequence homology (Matsuoka and Neuhaus, 1999) to one another or to those of other PSV proteins. The receptor that recognizes other targeting sequences remains to be determined, and whether other isoforms of BP80/AtELP possess broader specificity will require further research. Complicating investigations on the targeting sequences of storage proteins and some ancillary storage proteins such as lectins is that there may be two or more distinct targeting sequences (Frigerio et al., 1998; Saalbach et al., 1991; von Schaewen and Chrispeels, 1993). This makes the design of deletion mutants to test targeting difficult if the sequences are possibly redundant or synergistic.

The PSV targeting of the 7S globulins is not dependent on glycosylation, shown by blocking glycosylation either by inhibitors or by mutating the *N*-glycosylation sites. Tunicamycin, an inhibitor that blocks core *N*-glycosylation, does not impede the storage protein progression from the ER and subsequent targeting to the PSV (Bollini et al., 1985). The phaseolin from *Phaseolus lunatus,* the lima bean, is specifically retained in the ER after tunicamycin treatment due to misfolding and aggregation (Sparvoli et al., 2000). Site-directed mutatgensis to eliminate one or both of the glycosylation sites of the seed lectin PHA expressed in transgenic tobacco seeds did not impede the correct targeting of glycan minus protein to the PSV (Sturm et al., 1988).

Autophagy into Protein Storage Vacuoles

Complicating and possibly confusing the observations on the ontogeny of PSVs are the numerous inclusions that originate by the secondary process of autophagy. That PSVs are autophagic organelles was first observed in germinating seeds, as the coyledon cells progress through storage protein mobilization and undergo programmed cell death (Van der Wilden et al., 1980; Herman et al., 1981). Because PSVs originate from a central vacuole during seed maturation and return to a central vacuole after protein mobilization, the PSV has a specialized and differentiated tonoplast that contains high levels of the aquaporin α-TIP that functions to conduct water out of the PSV in late maturation and into the PSV during germination. The α-TIP is found only in PSV and not in the central vacuole precursor or the reformed central vacuole resulting from protein mobilization. Melroy and Herman (1991) showed that the α-TIP is incorporated into the PSV tonoplast in parallel with storage protein deposition presumably carried by the same vesicles delivering PSV matrix. The removal of α-TIP from the PSV tonoplast during germination is mediated by autophagy during storage protein mobilization and as the individual PSVs merge to reform the vacuole. During the process of merging the PSVs there is a large net reduction in tonoplast area. Autophagy and degradation of the excess PSV tonoplast results in the removal of the preexisting PSV tonoplast and its replacement with new membrane containing different tonoplast proteins.

The most prominent PSV inclusion in many seeds is the crystalloid. The crystalloid has been assumed to be a self-assembly of 11S or 2S storage proteins delivered by the same Golgi-mediated trafficking mechanisms as the other PSV constituents. Jiang et al. (2001) have presented an alternative model for cystalloid formation. It has been assumed that the crystalloid self-assembles during seed maturation in by self-aggregation from the other storage proteins. Rogers and colleagues (Jiang et al., 2001) discovered that the crystalloid contains embedded tonoplast proteins that are the same as the tonoplast proteins observed on small membrane vesicles sequestered in the PSV matrix. They have developed a model for cystalloid formation that the cystalloid proteins are deposited into the PSV by autophagy of small vesicles containing the protein. The presence of tonoplast proteins distinct from the PSV tonoplast in the crystalloid would appear to support this model. What is still unclear if this model is correct is how the precursors of the crystalloid protein (either 11S or 2S) become accessible to the vacuolar processing enzyme necessary to form the mature storage protein.

PROTEIN BODIES

Seeds contain a second class of protein storage compartments termed protein bodies (PB) that are distinct from the PSVs in terms of their origin, contents, and biology. The ER-derived PBs of the maize endosperm are among the best-characterized ER storage compartments (for a review, see Herman and Larkins, 1999). The presence of polysomes on the limiting membranes of PBs and the continuity between the limiting membrane and the ER cisternae provide the best evidence that these storage compartments are part of, and derived from, the ER. Zeins belong to the class of seed proteins known as prolamins (proline and glutamine-rich polypeptides) and in the endosperm of some other cereals. All prolamins are stored in ER-derived PBs. In wheat endosperm, prolamins are also assembled in ER-derived protein bodies, but the PBs subsequently are sequestered in vacuoles of the cell by autophagy (Levanony et al., 1992). Once sequestered in the the vacuole, the limiting ER-derived membrane is degraded and the prolamin core aggregates with other prolamin cores, producing large storage protein aggregates.

EM observations show that PBs are formed at the distal end of ER cisternae (Larkins and Hurkman, 1978). Because the prolamin proteins aggregate shortly after synthesis at the site of PB formation, it is likely to be adjacent to the site of prolamin synthesis. The site of PB formation and prolamin synthesis appears to be regulated in part by the distribution of mRNAs encoding the prolamin polypeptides that are spatially segregated to discrete domains of the ER (see Okita and Rogers, 1996).

The formation of cereal seed ER-derived storage compartments can be mimicked in both seed and leaves of transgenic plants by the synthesis of prolamin proteins. The expression of zein genes in alfalfa leaves (Bagga et al., 1995, 1997) or tobacco seeds (Coleman et al., 1996) induces zein to accumulate in ER-derived PBs. At least two different zein genes have to be coexpressed to obtain substantial accumulation of zein. It appears that the formation of stable zein complexes requires the interaction of different zein polypeptides. The formation of these ER-derived PBs is not necessarily the end-KDpoint, and in tobacco seeds the zein-containing PBs apparently enter the PSVs by autophagy.

An alternate way to generate ER-derived protein bodies is to induce ER retention of a soluble transport-competent protein. An early example of this was the expression of the seed storage protein vicilin engineered to possess a carboxy-terminal KDEL motif (Wandelt et al., 1992). The C-terminal motifs KDEL or HDEL act as ER retention/retrieval signals. Vicilin polypeptides with the KDEL sequence should have a greatly enhanced residence time in the ER-lumen. Vicilin, similar to other legume seed storage proteins expressed in vegetative cells, is unstable, probably as a consequence of its sensitivity to vacuolar proteases after transport to the vacuole. The KDEL tail retards progression to the vacuole, and vicilin-KDEL greatly increased the level of vicilin accumulation and the stability of the vicilin. The leaves of the transgenic plants contained electron-dense aggregates in the ER cisternae and new electron-dense organelles in the cytoplasm. Immunocytochemistry showed these aggregates and organelles to contain vicilin. In pulse-chase experiments vicilin-KDEL was stable for 48hours and degraded only after a prolonged chase. Whether this breakdown was the consequence of ER-protein quality control or transfer of the protein to the vacuole was not determined. Similar experiments were done with the sunflower albumin, a sulfur-rich protein, which was expressed with a KDEL tail in the leaves of subterranean clover (Khan et al., 1996). Without KDEL, the protein was undetectable, and with KDEL it continued to accumulate in the leaves and reached 1.3 percent of extractable protein. Similar experiments in transgenic seeds using phaseolin-HDEL (Pueyo et al., 1995) and phytohemagglutinin-KDEL (Herman et al., 1990) did not result in the aggregation of the proteins. With both proteins the retention sequences retarded the exit from the ER and greatly increased the protein concentration in the ER lumen. However, neither modified protein formed aggregates that budded from the ER, and both proteins progressed to the vacuole by trafficking through the Golgi to the PSV.

ER-Derived Storage Protein Transport Vesicles and Protein Bodies As a Means of Delivering Storage Proteins to the PSV

An alternate pathway for deposition of storage proteins operates in some seeds, for instance in developing pumpkin cotyledons, by assembling electron-dense transport vesicles mediated by the ER rather than by the Golgi. Hara-Nishimura et al. (1998) described ER-derived dense vesicles that transport the storage protein 2S albumin to the PSVs. Albumin is synthesized as a precursor that is proteolytically processed after it arrives in the PSVs. The vesicles contain only precursor and are therefore called precursor accumulation (PAC) vesicles. These vesicles possess a central core of pro-2S albumin and are surrounded by an ER-derived membrane with bound ribosomes. The pro-2S albumin is processed to the mature form only after deposition in the PSV and as the consequence of the action of vacuole-specific enzyme(s). There may be further complexities in the biogenesis of PAC vesicles because the vesicles appear to contain proteins with complex glycans, indicating contact with Golgi enzymes even though the vesicle has an obvious ER origin. Further evidence of Golgi interaction with the PAC vesicles is seen in the presence of vacuolar sorting receptors in the PAC vesicle membrane (Shimada et al., 1997). Vacuolar sorting receptors are localized in the *trans*-Golgi. Whether this indicates a Golgi contribution to the PAC vesicles after ER-mediated assembly or whether Golgi-derived material is transported in a retrograde manner into the ER and then incorporated into the PAC vesicles still needs to be resolved. Vegetative expression under the control of the 35S promoter of the entire 2S albumin coding sequence fused to phosphinothricin acetyltransferase (PAT) or the N-terminal half of albumin fused to PAT in *Arabidopsis* plants replicates the intrinsic PAC body formation in pumpkin seeds (Hayashi et al., 1999). Aggregates of the pro-2S albumin fusion protein surrounded by an ER-derived membrane are present in the cytoplasm and may indicate that aggregation of the 2S protein is an intrinsic property of this protein.

The biology of PSV ontogeny, protein bodies, and storage protein deposition is further complicated in wheat. During early and intermediate stages of wheat grain development, the prolamin storage proteins are first assembled into ER-derived protein bodies similar to those in other cereals and maize. However the PBs, instead of remaining in the cytoplasm as discrete organelles, are deposited by autophagy into small, vegetative-like vacuoles. These vacuoles then fuse to form large central vacuoles containing large protein inclusions that appear to be aggregates of the initially sequestered protein bodies. The wheat endosperm cells die at the end of maturation and the protein aggregate forms a large protein complex in the cellular remnants that are mobilized by hydrolases secreted from the aleurone and embryo.

Altering storage protein composition of seeds can also induce the formation of ER-derived compartments. Cosuppression of the 7S conglycinin of soybean by a transgene that includes the conglycinin promoter eliminates almost all conglycinin mRNA, resulting in transgenic soybean seeds that accumulate the precursor of glycinin, proglycinin. Although glycinin is encoded by six expressed genes, only one of these genes contributes to the proglycinin that accumulates to compensate for the lost conglycinin. Ultrastructural observations show the presence of small (about 0.5 mm diameter), dense protein bodies that contain glycinin and are surrounded by an ER-derived membrane with bound ribosomes appearing identical to PBs and PAC vesicles in morphology. The membrane of the dense bodies is not labeled with antibodies to the PSV tonoplast protein α-TIP or complex glycans, showing that the glycinin protein bodies are ER-derived (Kinney et al., 2001). These protein bodies are not present in wild-type soybeans and they resemble the PAC vesicles of pumpkin seeds. Normally, glycinin and conglycinin are synthesized together and both are present in the ER. Why the absence of conglycinin causes glycinin-protein bodies to form from one of the six isoforms is not clear. The protein bodies accumulate in the cytoplasm of the storage parenchyma cells and remain as discrete organelles after germination of the seeds.

BIOTECHNOLOGY TO STUDY AND CHANGE SEED PROTEINS

Synthesis, Targeting, and Stability of Protein Products of Introduced Storage Protein Genes

Several genes encoding legume globulins and lectins have been expressed in transgenic plants. The stability of the storage proteins inside the transgenic seed PSVs indicates proteins must be assembled correctly, achieving a specific structure that is resistant to degradation by PSV-resident proteases. The collective experience of these experiments show that seed storage proteins expressed as heterologous products in the seeds of another plant are most often stable and accumulate at high levels in the PSVs of the transgenic plant. However, this general rule appears to apply only to the wild-type protein, as expressing modified storage proteins has resulted in a mixed result of some being accumulated as stable products while others have proven to be posttranslationally unstable and degraded by at least two independent processes. Hoffman et al. (1988) modified phaseolin by inserting a short amino acid sequence containing several methionine residues (HiMet phaseolin) in an attempt to increase the methionine content of the protein. Although it was expressed and synthesized at the same levels as the unmodified wild-type control, almost no protein was accumulated. Pueyo et al. (1995) showed that the vacuole is the intracellular site where the HiMet phaseolin is degraded. This result suggests that HiMet modification of the phaseolin caused exposure of some protein domains to proteolytic degradation by PSV-resident proteases. Interestingly, supplementing the HiMet phaseolin with a C-terminal ER-retention sequence K/HDEL resulted in the accumulation of HiMet protein in the ER and Golgi. This shows that HiMet is stable until it progresses to the PSV.

The stability of prolamins as transgene products was further investigated by expressing maize prolamins in seeds of transgenic tobacco and petunia plants. Maize zeins accumulate inside the ER, forming PBs (Larkins and Hurkman, 1978; Lending et al., 1988; Lending and Larkins, 1989). The α-zeins are mostly composed of small peptide repeats rich in glutamine and proline, while the other zeins have entirely different structures and are rich in methionine (β-zeins) and cysteine (γ- and δ-zeins). Although all zeins accumulate in the ER of maize endosperm, a 15 kDa β-zein, expressed in transgenic tobacco under control of the phaseolin promoter, accumulated in PSVs and not in the ER (Hoffman et al., 1987). In a subsequent study, the same 15 kDa β-zein was expressed under the 35S promoter, and it accumulated in both cytoplasmic protein bodies that were derived by fission from the ER, as well as in intravacuolar inclusions that likely originated by autophagy of the cytosolic protein bodies (Bagga et al., 1995). The 35S promoter differs from the phaseolin promoter by being expressed mainly at early stages of seed development. Thus, the intracellular site of accumulation of the 15 kDa β-zein may differ, depending on whether the protein is expressed at the early stages of seed maturation, i.e., in vegetative-type cells (35S promoter), or in embryonic cells formed at the later stages of seed maturation (phaseolin promoter). In other studies (Williamson et al., 1988; Ohtani et al., 1991), a maize α-zein expressed with a phaseolin promoter was unstable in developing seeds of transgenic petunia and tobacco plants. The problem of zein posttranslational instability in transgenic plants was partially solved by the discovery that that zein protein bodies contain more than one isoform of zeins that coaggregate during protein body formation, resulting in a complex structure of a matrix of some zein isoforms containing embedded locules of other zein isoforms. Attempts to replicate the coexpression on two zeins in transgenic plants by crossing plants expressing different zeins conferred considerable additional stability to zein expression in transgenic seeds (Coleman et al., 1996). However, although additional stability resulted from the cross—producing transgenic seeds that formed protein bodies containing two zeins—little zein remained by the time the seeds were mature. Storage proteins routing to PSVs apparently pass a "quality control" point inside the ER. To study this, Vitale and associates (Pedrazzini et al., 1997) characterized the fate of a wild-type and a monomeric deletion mutant of bean phaseolin in protoplasts and leaves of transgenic tobacco plants. Both the wild-type and malfolded phaseolin were unstable in leaf cells. However, while the turnover of the wild-type protein was inhibited by treatment with Brefeldin A, that of the malfolded protein was not (Vitale et al., 1995). As Brefeldin A is known to block the transport of secretory proteins from the ER to the Golgi, these results suggest that the wild-type and mutant phaseolin follow different intracellular routes for their disposal. The wild-type protein is apparently transported via the Golgi to vacuoles where it is degraded by vacuolar proteases, but the situation with the malfolded protein is not clear. Vitale et al. (1995) suggested that the Brefeldin A-insensitive turnover of the deletion mutant protein implies that it is degraded within the ER. However, because autophagy may not be inhibited by Brefeldin A, it is impossible yet to eliminate the possibility that the mutant phaseolin is transported from the ER to vacuoles by autophagy and that it is also degraded in this organelle. As this study was conducted with leaf cells, it would be intriguing to express the mutant phaseolin protein using a phaseolin promoter and test the stability of the malfolded protein and whether it is disposed within the ER or the PSVs of the seed embryonic cells. These results have now been extended by the expression of the monomeric phaseolin that includes a KDEL carboxy-terminal sequence (Frigerio et al., 2001). The results of these experiments show that the potential retention in the ER does not impede posttranslational degradation, indicating that the monomeric phaseolin deletion mutant is unstable by an entirely different process than is used to degrade HiMet. Whether this process would also occur in seeds as well as in the transient expression used in this study remains to be determined. The collective results of posttranslational instability observations indicate that a primary route is degradation in the vacuole or PSV after sequesteration by either Golgi-mediated or autophagy processes. An alternate route of posttranslational instability associated with the ER also occurs.

IDENTIFYING AND ALTERING MOLECULAR-BASED TRAITS IN SEEDS

Among the earliest experiments of plant molecular biology was the introduction of genes encoding seed storage proteins in transgenic seeds. Because seed protein mRNAs are highly abundant—often exceeding 50 percent of the mRNA population—the cDNAs encoding seed storage were among the first cloned genes and with its regulated development recognized as good models to develop the methods of molecular biology in plants. Complete cDNA and genome sequences for storage proteins were rapidly obtained in several labs and quickly exploited as models for regulation and control of gene expression, protein processing, and trafficking.

Seed proteins are utilized in a wide variety of ways as both primary foods for people and animals and extensively in processed foods for both human consumption and animal feed. The properties of seed proteins and impact of these properties on nutritional, health, and processing properties of seed proteins have all been addressed by molecular approaches, and with this explosion of information the opportunity to use biotechnology to alter and improve seed proteins is now feasible. Although a comprehensive review of all the food characteristics of seed proteins that have a molecular basis is beyond the current discussion, the initial attempts to identify and modify seed protein characteristics by biotechnology is pertinent. Properties of seed proteins that have been studied include composition, physical properties, and allergenicity. Emerging in the new generation of biotechnology products are those that have altered composition which changes nutritional value, allergenicity, and utilization characteristics. In parallel with altering seed proteins, the pathways and regulation that seeds use to produce and accumulate storage proteins can be exploited to redirect protein synthesis to industrial, medical, and other proteins where cheap production in field crops can prove useful.

Composition

Long before the advent of molecular biology, the overall amino acid composition of seed proteins was already well established. With the complete sequence of storage protein families, the contribution of each gene product to the total composition was elucidated. That cereals are generally deficient in the essential amino acid lysine, and legumes are generally deficient in essential sulfur amino acids, made both targets for directed alteration. Because transformation technology advanced more rapidly for dicotyledoneous plants, the initial attempts to modify amino acid composition was investigated with model plants, particularly tobacco. Experiments to increase essential methionine amino acid content of seeds was one of the early objectives of biotechnology. Legumes such as soybean are methionine deficient, limiting the use of 50 to 70 percent of diet. Two distinct approaches have been tested: the expression of a high methionine protein or site-directed mutagenesis of a homologous gene to increase sulfur amino acid content (see Müntz et al., 1998, for review).

Intrinsic high methonine proteins of plants, including the prolamins and the 2S albumins of many seeds, have been expressed in transgenic plants. The use of the 2S albumin of Brazil nut and its expression in transgenic plants is one of the best known and most widely cited example of the potential of adverse consequence of biotechnology. The Brazil nut 2S albumin is highly methionine rich and at moderate levels of accumulation could convert a somewhat methionine-deficient seed into one that possesses sufficient methionine to meet dietary requirements (Altenbach et al., 1992; Saalbach et al., 1994). However, what was not recognized when these experiments were conceived was that the Brazil nut 2S albumin is a potent human allergen that poses a significant risk for sensitive people (Nordlee et al., 1996). So although the expression of the Brazil nut protein solves the methionine deficiency problem, it is a solution that would not make the enhanced seeds available. Other 2S albumin proteins have been used to produce transgenic plants with the aim of increasing methionine content. For instance, the sunflower 2S albumin has been expressed in plants, but like the Brazil nut 2S protein, the sunflower protein is potentially a high-risk allergen for people (Khan et al., 1996). The aim of the experiments that placed the 2S albumin in transgenic plants was to enhance forage crops that would have no risk of accidental incorporation into the human food supply (Molvig et al., 1997; Tabe et al., 1997; Tabe and Droux, 2002). In such applications there would be little risk to people and might therefore be an acceptable means to enhance animal feed amino acid composition

The zeins have relatively little risk of introducing allergenicity if expressed in transgenic plants, and a high methionine content is an attractive possibility for gene transfer. Initial experiments to express various zeins in model transgenic seeds met with mixed results. Some of the zein isoforms were synthesized at high levels but were rapidly degraded, indicating posttranslational instability (Ohtani et al., 1991). Other zein isoforms, 15 kDa zein, were accumulated in transgenic seeds at the level of 1 to 2 percent of the total protein (Hoffman et al., 1987). As outlined previously, the autophagy of the zein PBs into the PSV exposed the sequestered organelles to PSV hydrolases, resulting in its degradation (Coleman et al., 1996). This approach is still promising, and other proteins may prove more stable in transgenic seeds and still not confer a high allergenic risk.

The alternative approach for increasing methionine content of proteins is to use directed substitution methods to add methionine residues or methionine-containing segments into low methionine proteins. One of the first attempts to use this strategy involved the insertion of an oligonucleotide segment encoding a high methionine alpha helix into a similar domain of phaseolin, producing a protein called HiMet (Hoffman et al., 1988). Although synthesized at a moderately high level, the protein was rapidly degraded by posttranslational processes. Further research showed that the protein was synthesized and correctly targeted for progression through the endomembrane system; it was only after the protein was delivered to the vacuole that the protein was degraded (Peuyo et al., 1995). Wild-type phaseolin is accumulated in transgenic seeds so the modification appeared to alter the proteins' structure, making it susceptible to the action of vacuolar hydrolases. This suggests that methionine alterations could be effective if done in a way that does not severely disrupt the structure of the altered protein. This approach has been tested by substituting methionine residues in the lectin phytohemagglutinin and expressing the wild type and higher methionine versions in transgenic seeds (Kjemtrump et al., 1994). The variants with enhanced methionine content proved as stable in the transgenic seeds as the wild type, indicating that with proper design and caution, enhancing methionine content would be permissible.

Using biotechnology to increase methionine content of forage crops has been investigated and shows promise to increase essential sulfur amino acid content by expressing foreign proteins (Saalbach et al., 1994; Khan et al., 1996; Molvig et al., 1997; Tabe et al., 1997; Tabe and Droux, 2002).

Seed Protein Allergies

Food allergies are a widespread and growing problem. A large and expanding literature identifies the various seeds and constituents that cause problems for people and animals, both pets and livestock. Allergies to the major seed crops are well-known, with wheat and soybean among the most prevalent offending protein sources. Other well-known seed allergies include peanut and tree nuts that may evoke dangerous and potentially lethal responses (see Breiteneder and Ebner, 2000). Food allergies have become a serious issue in the application of biotechnology-based approaches to alter plants (Helm, 2002; Lehrer and Reese, 1997; Taylor and Hefle, 2001). To assess allergenic risks in biotechnology-produced plant varieties, it is essential to understand the intrinsic allergenic risk of unaltered plants (Helm and Burks, 2000; Helm, 2002). Recent research has yielded detailed information on the allergenicity of major and minor food crops and their protein constituents for sensitive humans as well as farm and companion animals. Awareness has grown of how accidental exposure to allergenic food places sensitive people at risk. Recalls of processed foods have occurred because of unlabeled components in processed foods that may contain a possible allergen—often peanuts in baked goods. Concerns about allergenicity of genetically modified plants has also induced a very public recall of processed foods that may have contained a corn that expressed Starlink, a Bt unapproved for human consumption.

Three critical aspects characterize and mitigate a food allergen. Research on seed allergens involves the identification of allergens, elucidating the offending portions of the molecule that evoke the allergenic response and describing efforts to mitigate the allergenicity. This knowledge serves to assist in avoidance, treatment of sensitive people, plant breeding, and use of biotechnology to eliminate the allergen. Seed proteins are among the most frequently cited sources of human and animal allergens, in large part due to the prominence of seeds and seed-derived products in diets. A large number of crops have been either documented or suspected of inducing an allergenic response. The most prominent of these are the tree nuts such as walnuts (Teuber et al., 1998) and Brazil nuts (Bartolome et al., 1997), and peanuts (Barnett et al., 1983), which can induce allergenic responses that have a higher probability of a severe reaction or death. Peanuts in particular are suspected of causing 100 to 200 deaths per year. Other seeds may also cause severe or fatal reactions but the frequency of these reactions is significantly lower. Of the major crops, soybean (see Beardslee et al., 2000; Helm et al., 1998; Helm, Cockrell, Connaughton, Sampson, Bannon, Beilisnon, Livingstone, et al., 2000; Helm, Cockrell, Connaughton, Sampson, Bannon,

Beilinson, Neilsen, and Burks, 2000; Helm, Cockrell, West, et al., 2000, for examples), maize (Pastorello et al., 2000; Pastorello, Varin, et al., 2001; Pastorello, Pompei, et al., 2001; Chiung et al., 2000), and wheat (Weiss et al., 1997) are well-known sources of food allergies. Other examples of significant seed allergen sources include sesame (Pastorello et al., 2001), buckwheat (Yoshimasu et al., 2000), sunflower (Kelly and Hefle, 2000; Kelly et al., 2000; Khan et al., 1996), poppy (Vocks et al., 1993; Jensen-Jarolim et al., 1999), coconut (Teuber and Peterson, 1999), castor bean (Thorpe et al., 1988), rapeseed (Monsalve et al., 1997), pinenut (Garcia-Menaya et al., 2000), and pistachio (Parra et al., 1993). Among food allergies there are two broad categories of consumption and occupational allergies. Consumption allergies are primarily a consumer problem most often induced by eating, but other forms of stimulation including skin and respiratory system may also occur. Occupational allergies are induced by working in the food industry where exposure is most often to seed fragments and flour dust.

In order to critically identify a food allergy, specific immune responses are required in the form of IgE cross-reactivity with specific proteins derived from the food that produce the allergic reaction. Immunoblots are most frequently used to characterize proteins inducing allergenic responses, although fractionation by SDS/PAGE does limit detectable cross-reactivity to primary sequence epitopes as a result of denaturation destruction of conformational epitopes. ELISA (enzyme linked immunoabsorbant assay) is often used to detect IgE cross-reactivity with the added advantage that denaturation is not necessary, thereby preserving conformational epitopes. With a positive IgE response, the overt reaction by sensitive people and animals can be placed into the context of a food allergy. Sensitivity is manifested as overt physiological reactions that include gastric distress and skin reactions to far more several severe and life-threatening responses.

Among seed proteins inducing consumption allergies are the major storage proteins of the 7S (vicilin), 11S (legumin), and 2S albumin families, lectins, protease inhibitors, and other lytic enzymes. Among these there are wide variations of recorded instances of allergenicity within the gene families, with some of these proteins inducing severe and sometimes fatal responses, while with others there is little or no allergenic stimulation. For example, the 2S albumin family has examples such as Brazil nut and sunflower that have been shown to be among potent allergens, while other seed 2S proteins to not induce similarly strong reactions. The IgE epitopes have been mapped for several seed allergens, for example, soybean (Beardslee et al., 2000; Helm et al., 1998; Helm, Cockrell, Connaughton, Sampson, Bannon, Beilinson, Nielson, and Burks, 2000). The proteins inducing allergenic responses may differ in consumption and occupational allergies. Soybeans are a case in point, with consumption sensitivity resulting from proteins contained in the PSVs. The IgE cross-reactive proteins include both 7S and 11S storage protein subunits and isoforms, P34, and protease inhibitors. In contrast, the occupational allergies against soybean seed proteins are largely induced by low-molecular-weight seed coat proteins that are distinct from the storage vacuole allergenic proteins. The occupational exposure of seed coat allergens are most often as aeroallergens from dust and flour from food processing rather than from internal consumption.

Attempts thus far to use biotechnology to moderate or eliminate allergens from food have been limited. Determination of the IgE binding sites on allergens creates the opportunity to use site-directed mutagenesis to remove the allergenic epitopes. However, removing the naturally occurring allergenic gene and substituting the altered nonallergenic gene remains a significant problem. With highly allergenic proteins such as tree nut or peanut, providing assurance that the naturally occurring allergenic gene is completely suppressed is required before sensitive individuals could consider modifying avoidance as the primary health strategy. Modifying an allergen may be necessary if the protein that is allergenic is essential to food characteristics of the plant. In other instances the allergenic protein may not be essential for either the food or agronomic characteristics of the seed. Gene silencing or suppression can be used to eliminate the allergen accumulation. This approach has found limited application to date with one rice allergen (Tada et al., 1996) and an immunodominant soybean allergen (Herman et al., 2003) suppressed by this approach.

Inducing Immunological Responses and Immunological Reagents

The reciprocal of minimizing plant antigens is to facilitate additional directed immunological potential in seeds. The potential for expressing immunogenic proteins derived from pathogens in transgenic plants to immunize people and animals inexpensively and efficiently has received much attention (Walmsley and Arntzen, 2000, for review). The expression of immunizing proteins in seeds is more likely to prove useful in impeding the spread of farm animal diseases rather than diseases of people. Because most seeds used in products for people are cooked, this would denature potentially immunogenic sites. Most investigators have often focused on using fruit, such as bananas, that are eaten raw. Because animal feed often uses uncooked seeds or seed components, the opportunity for delivering protective immunogens with seeds is much greater.

Antibodies have been produced in plants, including seeds. Plantabodies, as these proteins are often termed, have several potential applications. Antibodies are the key part of a wide variety of immunologically based test kits and one of the costliest components to produce. Plant-produced antibodies should be a cost-effective means to produce antibodies cheaply and efficiently. Recovering antibodies from seed lysates would prove easy by adapting available purification schemes (Fiedler and Conrad, 1995; Fischer et al., 2000; Stoger et al., 2000). The production of antibodies in plants would remove a number of concerns that are always present in the use of animals or cell lines, including animal use and welfare as well as the potential of viruses in cell lines in in vitro bioreactor antibody production. Plant-produced antibodies would not carry any risk of being contaminated with pathogenic or carcinogenic viruses and could provide additional quality assurance when antibodies are used for therapeutic purposes. One of the primary questions that will emerge with adoption of this approach is the segregation of immunogenic varieties from the normal stream of seed commerce.

Physical Properties of Seed Proteins

The physical properties of seed proteins are central to the use of seeds in food preparation. For instance, the reaction of proteins to solution conditions such as salt solubility, melting, or coaggulation when heated, packing, and the ability to combine with other ingredients are only a few of the characteristics that play critical roles in how cooks make use of seed proteins. Among seed proteins, wheat stands apart from other seed proteins in its use in bread, pasta, and baking that requires unique physiochemical characteristics. The stretching of dough requires the desirable property of elasticity that allows wheat to be used in products such as bread and noodles. The high-molecular-weight glutenin subunit (HMW-GS) of wheat is the component that introduces elasticity into the dough used to make bread and bread products. These proteins are large polymers linked by disulfide bridges and noncovalent bonds producing one the largest examples of protein assembly known. Experiments have been conducted with transgenic plants by altering the polymerization pattern of the glutenin to alter the elasticity of dough (Barro et al., 1997). Shimoni et al. (1997) have shown decreased elasticity by introducing a gene encoding a glutenin that maintained a fraction of the glutenin as monomers. The reciprocal experiment has been performed by Popineau et al. (2001) who introduced a transgene that encoded a protein that would introduce increased glutenin crosslinking. They found the introduction of the higher cross-linking into glutenin and resulting dough caused an abnormal property of very high strength. Taken together, these two studies show the potential for alteration of dough-making properties by introducing a gene altering the polymerization pattern. These experiments may lead to using the polymerization density to design wheat that will perform with a predetermined elasticity required to optimally produce a specific product. Further, these experiments show that it is feasible to extend the range of polymerization and increase or decrease elasticity well beyond what might be selected by plant breeders to produce novel products.

The approach is not restricted to wheat. Alteration of any seed proteins' physiochemical properties, such as melting, solubility, aggregation, and coaggregation, might be mediated by expressing other proteins by removing one or more endogenous proteins through gene silencing. For instance, the processing properties of soybean proteins could be altered by silencing or selecting mutants that suppress one or more of the seed storage proteins. Soybean protein is widely used in processed foods, and the particular characteristics of how the soybean proteins might be used in conjunction with other constituents may be altered by changing seed protein composition.

Seeds As Protein Factories: The Emerging Future of Cheap Production of Valuable Commodities

One great potential of biotechnology and seed biology is to use seeds as protein factories to produce novel proteins in plants cheaply and efficiently that would otherwise be obtained from sources where the proteins are either in low abundance or expensive to produce. Technical issues need to be resolved to optimize the correct synthesis and assembly of transgene products as well as efficient accumulation that will vary from protein to protein to be synthesized. Important issues of production and commerce of seeds containing novel products will require creating new product streams to sequester the transgenic seeds, which will certainly require new forms of regulation and oversight.

Seed storage proteins are stored in either ER-derived PBs or Golgi-derived PSVs. These are the only two compartments within seed cells that possess the volume and cellular origin to make them obvious candidates to store large quantities of proteins. Many, if not most, proteins that are good candidates for production in seeds require the assistance of ER-localized molecular chaperones and processing enzymes to cotranslationally remove signal sequences, correctly fold the protein, produce disulfide bridges, and cotranslationally glycosylate the protein. Synthesis on the ER in order to have these processing events available means that transgene products will progress into a post-ER compartment with either the PBs or PSVs as the best candidate sites. Proteins can also default from the cell after ER synthesis and traffic through the Golgi, however the extracellular apoplast is a very small compartment and incorporating the transgene product into the ordered cell matrix may prove difficult. The technical difficulty of inducing the accumulation of novel products is to direct the product to a compartment where the correctly formed protein will be stably accumulated. Both the PBs and PSVs present distinct difficulties that must be overcome to accumulate a protein product. Because PBs are apparently the product of the self-aggregation of proteins in the ER lumen, leading self-assembly and budding of a PBs from the distal end of the ER to direct proteins to be aggregated and stored in PBs may prove difficult. Further destabilizing proteins to induce aggregation has been described, an indication that a protein is malformed and inactive. Therefore, the consequence of optimizing for storage in PBs may work against optimizing for protein activity. Yet storage in PBs may be the best candidate for accumulating novel proteins in seeds. The alternative of sequestering novel proteins in PSVs requires solving several different problems that will almost certainly be distinct for each new protein. Storage proteins progress from the ER synthesis site to the vacuole after trafficking through the Golgi where specific vacuole targeting sequences on the protein are recognized by a targeting receptor. Without these necessary and sufficient targeting sequences, proteins will default to extracellular space. Because most novel proteins will not be intrinsic vacuolar proteins, it will be necessary to add a polypeptide targeting sequence. Experiments have confirmed that attaching vacuolar targeting sequences to proteins will induce correct vacuolar targeting in seeds and vegetative cells. Still, for each new transgene product it will be necessary to confirm the alteration of the protein to incorporate vacuolar targeting sequences that do not alter the protein's structure, adversely affecting its activity. A potentially more significant impediment to storing novel proteins in PSVs is that the cell's lytic compartment contains array of acid hydrolases that are capable of degrading proteins not designed to be resistant to the enzymes.

One of the serious issues that needs to be resolved is the segregation of specific transgenics from the general stream of seed commerce. The recent experience of the Starlink corn, which was not intended to enter the human food supply, shows the consequences of unintended contamination. Starlink corn is from plants expressing a Bt form approved only for use in field corn to be used for animal feed. Once Starlink Bt was detected in processed foods intended for human consumption through a failure of segregation, it prompted fears of the protein's potential for allergenicity, and claims of adverse reactions from a few individuals provoked a costly and extensive recall. Fortunately, in retrospect, the Starlink did not seem to present significant risk, with later investigations indicating few if any possibly sensitive people exhibiting an IgE cross-reactivity with the Starlink Bt. However, this experience has shown the importance of developing appropriate strategies for segregation of some specialty transgenic seeds. At some time in the future it will be desirable to produce pharmaceuticals or other proteins such as commercially useful enzymes in plants that may be unsuitable for human consumption. Under these circumstances segregation is a critical issue that needs to be resolved in a way that provides reassurance that the food supply is not contaminated by the production of nonfood substances or other restricted substances such as vaccine immunogens.

REFERENCES

Ahmed, S.U., Bar-Peled, M., and Raikhel, N.V. (1997). Cloning and subcellular location of an *Arabidopsis* receptor-like protein that shares common features with protein-sorting receptors of eukaryotic cells. *Plant Physiology* 114: 325-336.

Altenbach, S.B., Kuo, C.C., Staraci, L.C., Pearson, K.W., Wainwright, C., Georgescu, A., and Townsend, J. (1992). Accumulation of a Brazil nut albumin in seeds of transgenic canola results in enhanced levels of seed protein methionine. *Plant Molecular Biology* 18: 235-245.

Bagga, S., Adams, H., Kemp, J.D., and Sengupta-Gopalan, C. (1995). Accumulation of the 15-kD zein in novel protein bodies in transgenic tobacco. *Plant Physiology* 107: 13-23.

Bagga, S., Adams, H.P., Rodriguez, F.D., Kemp, J.D., and Sengupta-Gopalan, C. (1997). Coexpression of the maize δ-zein and β-zein genes results in stable accumulation of δ-zein in endoplasmic reticulum-derived protein bodies formed by β-zein. *Plant Cell* 9: 1683-1696.

Bartolome, B., Mendez, J.D., Armentia, A., Vallverdu, A., and Palacios, R. (1997). Allergens from Brazil nut: Immunochemical characterization. *Allergology Immunopathology* (Madrid) 25: 135-144.

Barnett, D., Baldo, B.A., and Howden, M.E. (1983). Multiplicity of allergens in peanuts. *Journal Allergy Clinical Immunology* 72: 61-68.

Barro, F., Rooke, L., Bekes, F., Gras, P., Tatham, A.S., Fido, R., Lazzeri, P.A., Shewry, P.R., and Barcelo, P. (1997). Transformation of wheat with high molecular weight subunit genes results in improved functional properties. *Nature Biotechnology* 15: 1295-1299.

Beardslee, T.A., Zeece, M.G., Sarath, G., and Markwell, J.P. (2000). Soybean glycinin G1 acidic chain shares IgE epitopes with peanut allergen Ara h 3. *International Archives Allergy Immunology* 123: 299-307.

Bollini, R., Ceriotti, A., Daminati, M.G., and Vitale, A. (1985). Glycosylation is not needed for intracellular transport of phytohemagglutinin in developing *Phaseolus vulgaris* cotyledons and for maintenance of its biological activity. *Physiological Plantarium* 65: 15-22.

Breiteneder, H. and Ebner, C. (2000). Molecular and biochemical classification of plant-derived food allergens. *Journal Allergy Clinical Immunology* 106: 27-36.

Casey, R. (1979). Genetic variability in the structure of the α-subunits of legumin from *Pisum*—A two dimensional electrophoresis study. *Heredity* 43: 265-272.

Casey, R. and Domoney, C. (1999). Pea globulins. In P.R. Shewry and R. Casey (eds.), *Seed proteins* (pp. 171-208). Dordrecht, the Netherlands: Kluwer Academic Publishers.

Chappell, J., Van der Wilden, W., and Chrispeels, M.J. (1980). The biosynthesis of ribonuclease and its accumulation in protein bodies in the cotyledons of mung beans. *Developmental Biology* 76: 115-125.

Cheng, J.I., Boyd, C., Slaymaker, D., Okinaka, Y., Herman, E.M., and Keen, N.T. (1998). Purification and characterization of a 34 kDa syringolide binding protein from soybean. *Proceedings National Academy Sciences USA* 95: 3306-3311.

Chiung, Y.M., Lin, B.L., Yeh, C.H., and Lin, C.Y. (2000). Heat shock protein (hsp 70)-related epitopes are common allergenic determinants for barley and corn antigens. *Electrophoresis* 21: 297-300.

Chrispeels, M.J. (1983). The Golgi apparatus mediates the transport of phytohemagglutinin to the protein bodies in bean cotyledons. *Planta* 158: 140-151.

Chrispeels, M.J., Higgins, T.J.V., Craig, S., and Spencer, D. (1982). Role of the endoplasmic reticulum in the synthesis of reserve proteins and the kinetics of their transport to protein bodies in devleping pea cotyledons. *Journal Cell Biology* 93: 5-14.

Chrispeels, M.J., Higgins, J.V., and Spencer, D. (1982). Assembly of storage protein oligomers in the endoplasmic reticulum and processing of the polypeptides in the protein bodies of developing pea cotyledons. *Journal Cell Biology* 9: 306-313.

Chrispeels, M.J. and Raikhel, N.V. (1991a). Lectins, lectin genes, and their role in plant defense. *Plant Cell* 3: 1-9.

Chrispeels, M.J. and Raikhel, N.V. (1991b). Sorting of proteins in the secretory system. *Annual Review Plant Physiology Plant Molecular Biology* 42: 21-53.

Coleman, C.E., Herman, E.M., Takasaki, K., and Larkins, B.A. (1996). γ-Zein sequesters α-zein and stabilizes its accumulation in transgenic tobacco endosperm. *Plant Cell* 8: 2335-2345.

Craig, S. and Goodchild, D.J. (1984). Periodic-acid treatment of sections permits on-grid localization of pea seed vicilin in ER and Golgi. *Protoplasma* 122: 35-44.

Craig, S., Goodchild, D.J., and Hardham, A.R. (1979). Structural aspects of protein accumulation in developing pea cotyledons: I. Qualitative and quantitative changes in parenchyma cell vacuoles. *Australian Journal Plant Physiology* 6: 81-98.

Craig, S., Goodchild, D.J., and Miller, C. (1980). Structural aspects of protein accumulation in developing pea cotyledons: II. Three-dimensional reconstitution of vacuoles and protein bodies from serial sections. *Australian Journal Plant Physiology* 7: 329-337.

Etzler, M.E. (1985). Plant lectins: Molecular and biological aspects. *Annual Review Plant Physiology Plant Molecular Biology* 36: 209-234.

Faye, L., Strum, A., Bollini, R., Vitale, A., and Chrispeels, M.J. (1986). The position of the oligosaccharide side-chains on phytohemagglutinin and their accesibility to glycosidases determines their subsequent processing in the Golgi. *European Journal Biochemistry* 158: 741-745.

Fiedler, U. and Conrad, U. (1995). High-level production and long-term storage of engineered antibodies in transgenic tobacco seeds. *Biotechnology* 13: 1090-1093.

Fischer, R., Hoffmann, K., Schillberg, S., and Emans N. (2000). Antibody production by molecular farming in plants. *Journal Biology Regulatory Homeostasis Agents* 14: 83-92.

Frigerio, L., de Virgilio, M., Prada, A., Faoro, F., and Vitale, A. (1998). Sorting of phaseolin to the vacuole is saturable and requires a short C-terminal peptide. *Plant Cell* 10: 1031-1042.

Frigerio, L., Jolliffe, N.A., Di Cola, A., Felipe, D.H., Paris, N., Neuhaus, J.-M., Lord, J.M., Ceriotti, A., and Roberts, L.M. (2001). The internal propeptide of the ricin precursor carries a sequence-specific determinant for vacuolar sorting. *Plant Physiology* 126: 167-175.

Garcia-Menaya, J.M., Gonzalo-Garijo, M.A., Moneo, I., Fernandez, B., Garcia-Gonzalez, F., and Moreno, F. (2000). A 17-kDa allergen detected in pine nuts. *Allergy* 55: 291-293.

Greenwood, J.S. and Chrispeels, M.J. (1985). Imunocytochemical localization of phaseolin and phytohemagglutinin in the endoplasmic reticulum and Golgi complex of developing bean cotyledons. *Planta* 164: 295-302.

Häger, K.P., Braun, H., Czihal, A., Müller, B., and Bäumlein, H. (1995). Evolution of seed storage protein genes: Legumin genes of *Ginkgo biloba*. *Journal Molecular Evolution* 41: 455-466.

Hara-Nishimura, I., Nishimura, M., and Akazawa, T. (1985). Biosynthesis and intracellular transport of 11S globulin in developing pumpkin cotyledons. *Plant Physiology* 77: 747-752.

Hara-Nishimura, I., Shimada, T., Hatano, K., Takeuchi, Y., and Nishimura, M. (1998). Transport of storage proteins to protein storage vacuoles is mediated by large precursor-accumulating vesicles. *Plant Cell* 10: 825-836.

Harris, N. (1979). Endoplasmic reticulum in developing seeds of *Vicia faba:* A high voltage electron microscope study. *Planta* 146: 63-69.

Hayashi, M., Toriyama, K., Kondo, M., Hara-Nishimura, I., and Nishimura, M. (1999). Accumulation of a fusion protein containing 2S albumin induces novel vesicles in vegetative cells of *Arabidopsis*. *Plant Cell Physiology* 40: 263-272.

Helm, R.M. (2002). Biotechnology and food allergy. *Current Allergy and Asthma Reports* 2: 55-62.

Helm, R.M. and Burks, A.W. (2000). Mechanisms of food allergy. *Current Opinion Immunology* 12: 647-653.

Helm, R.M., Cockrell, G., Connaughton, C., Sampson, H.A., Bannon, G.A., Beilinson, V., Livingstone, D., Nielsen, N.C., and Burks, A.W. (2000). A soybean G2 glycinin allergen: 1. Identification and characterization. *International Archives Allergy Immunology* 123: 205-212.

Helm, R.M., Cockrell, G., Connaughton, C., Sampson, H.A., Bannon, G.A., Beilinson, V., Nielsen, N.C., and Burks, A.W. (2000). A soybean G2 glycinin allergen: 2. Epitope mapping and three-dimensional modeling. *International Archives Allergy Immunology* 123: 213-219.

Helm, R.M., Cockrell, G., Herman, E., Burks, A.W., Sampson, H. A., and Bannon, G.A. (1998). Cellular and molecular characterization of a major soybean allergen. *International Archives Allergy Immunology* 117: 29-37.

Helm, R.M., Cockrell, G., West, C.M., Herman, E.M., Sampson, H.A., Bannon, G.A., and Burks, A.W. (2000). Mutational analysis of the IgE-binding epitopes of P34/Gly m1. *Journal Allergy Clinical Immunology* 105: 378-384.

Herman, E.M. (1994). Multiple origins of intravacuolar protein accumulation of plant cells. *Advances in Structural Research* 3: 243-283.

Herman, E.M., Baumgartner, B., and Chrispeels, M.J. (1981). Uptake and apparent digestion of cytoplasmic organelles by protein bodies (protein storage vacuoles) in mungbean cotyledons. *European Journal of Cell Biology* 24: 226-235.

Herman, E.M. and Chrispeels, M.J. (1980). Phospholipase D and phosphatidic acid phosphatase: Acid hydrolases involved in phospholipid catabolism in mung bean cotyledons. *Plant Physiol*ogy 66: 1001-1007.

Herman, E.M., Helm, R., Jung, R., and Kinney, A.J. (2003). Targeted gene silencing removes an immunodominant allergen from soybean seeds. *Plant Physiology* 132: 36-43.

Herman, E.M. and Larkins, B.A. (1999). Protein storage bodies. *Plant Cell* 11: 601-613.

Herman, E.M., Platt-Aloeia, K.A., Thomson, W.W., and Shannon, L.M. (1984). Freeze fracture and filipin cytochemical observatons of developing soybean cotyledon protein bodies and Golgi apparatus. *European Journal Cell Biology* 35: 1-7.

Herman, E.M. and L.M. Shannon (1984). Immunocytochemical localization of concanavalin A in developing jack bean cotyledons. *Planta* 161: 97-104.

Herman, E.M., B. Tague, L.M. Hoffman, S.E. Kjemtrup, and M.J. Chrispeels (1990). Retention of phytohemagglutinin with carboxyterminal tetrapeptide KDEL in the nuclear envelope and endoplasmic reticulum. *Planta* 182: 305-312.

Hinz, G., Menze, A., Hohl, I., and Vaux, D. (1997). Isolation of prolegumin from developing pea seeds: Its binding to endomembranes and assembly into prolegumin hexamers in the protein storage vacuole. *Journal Experimental Botany* 48: 139-149.

Hoffman, L.M., Donaldson, D.D., Bookland, R., Rashka, K., and Herman, E.M. (1987). Synthesis and protein body deposition of maize 15-kd zein in transgenic tobacco seeds. *EMBO Journal* 6: 3213-3221.

Hoffman, L.M., Donaldson, D.D., and Herman, E.M. (1988). A modified storage protein is synthesized, processed, and degraded in seeds of transgenic plants. *Plant Molecular Biology* 11: 717-729.

Hoh, B., Hinz, G., Jeong, B.-K., and Robinson, D.G. (1995). Protein storage vacuoles form de novo during pea cotyledon development. *Journal Cell Science* 108: 299-310.

Horisberger, M. and Volanthen, M.T. (1983). Ultrastructural localization of Kunitz trypsin inhibitor on thin sections of *Glycine max* (soybean) cv Maple Arrow by the gold method. *Histochemistry* 77: 313-321.

Jensen-Jarolim. E., Gerstmayer, G., Kraft, D., Scheiner, O., Ebner, H., and Ebner, C. (1999). Serological characterization of allergens in poppy seeds. *Clinical Experimental Allergy* 29: 1075-1079.

Jiang, L.W., Phillips, T.E., Hamm, C.A., Drozdowicz, Y.M., Rae, P.A., Maeshima, M., Rogers. S.W., and Rogers, J.C. (2001). The protein storage vacuole: A unique compound organelle. *Journal Cell Biology* 155: 991-1002.

Jung, R., Nam, Y.W., Saalbach, I., Müntz, K., and Nielsen, N.C. (1997). Role of the sulfhydryl redox state and disulfide bonds in processing and assembly of 11S seed globulins. *Plant Cell* 9: 2037-2050.

Kalinski, A.J., Melroy, D.L., Dwivedi, R.S., and Herman, E.M. (1992). A soybean vacuolar protein (P34) related to thiol proteases which is synthesized as a glycoprotein precursor during seed maturation. *Journal Biology Chemistry* 267: 12068-12076.

Kelly, J.D. and Hefle, S.L. (2000). 2S methionine-rich protein (SSA) from sunflower seed is an IgE-binding protein. *Allergy* 55: 556-560.

Kelly, J.D., Hlywka, J.J., and Hefle, S.L. (2000). Identification of sunflower seed IgE-binding proteins. *International Archives Allergy Immunology* 121: 19-24.

Khan, M.R.I., Ceriotti, A., Table, L., Aryan, A., McNabb, W., Moore, A., Craig, S., Spencer, D., and Higgins, T.J.V. (1996). Accumulation of a sulphur-rich seed albumin from sunflower in the leaves of transgenic subterranean clover (*Trifolium subterraneum* L.). *Transgenic Research* 5: 179-185.

Kim, W.T., Franceschi, V.R., Krishnan, H.B., and Okita, T.W. (1988). Formation of wheat protein bodies: Involvement of the Golgi apparatus in gliadin transport. *Planta* 176: 173-182.

Kinney, A.J., Jung, R., and Herman, E.M. (2001). Cosuppression of the α subunits of β-conglycinin in transgenic seeds induces the formation of endoplasmatic reticulum-derived protein bodies. *Plant Cell* 13: 1165-1178.

Kirsch, T., Paris, N., Butler, J.M., Beevers, L., and Rogers, J.C. (1994). Purification and initial characterization of a potential plant vacuolar targeting receptor. *Proceedings National Academy Sciences USA* 91: 3403-3407.

Kirsch, T., Saalbach, G., Raikhel, N.V., and Beevers, L. (1996). Interaction of a potential vacuolar targeting receptor with amino- and carboxyl-terminal targeting determinants. *Plant Physiology* 111: 469-474.

Kjemtrup, S., Herman, E.M., and Chrispeels, M.J. (1994). Correct post-translational modification and stable vacuolar accumulation of phytohemagglutinin engineered to contain multiple methionine residues. *European Journal Biochemistry* 226: 385-391.

Larkins, B.A. and Hurkman, W.J. (1978). Synthesis and deposition of zein in protein bodies of maize endosperm. *Plant Physiology* 62: 256-263.

Lehrer, S.B. and Reese, G. (1997). Recombinant proteins in newly developed foods: Identification of allergenic activity. *International Archives Allergy Immunology* 113: 122-124.

Lending, C.R., Kriz, A.L., Larkins, B.A., and Bracker, C.E. (1988). Structure of maize protein bodies and imunocytochemical localization of zeins. *Protoplasma* 143: 51-62.

Lending, C.R. and Larkins, B.A. (1989). Changes in the zein composition of protein bodies during maize endosperm development. *Plant Cell* 1: 1011-1023.

Levanony, H., Rubin, R., Altshuler, Y., and Galili, G. (1992). Evidence of a novel route of wheat storage proteins to vacuoles. *Journal Cell Biology* 119: 1117-1128.

Matsuoka, K. and Nakamura, K. (1999). Large alkyl side-chains of isoleucine and leucine in the NPIR region constitute the core of the vacuolar sorting determinant of sporamin precursor. *Plant Molecular Biology* 41: 825-835.

Matsuoka, K. and Neuhaus, J.-M. (1999). *Cis*-elements of protein transport to the plant vacuoles. *Journal Experimental Botany* 50: 165-174.

Melroy, D.L. and Herman, E.M. (1991). TIP, an integral membrane protein of the soybean seed protein storage vacuole, undergoes developmentally regulated membrane insertion and removal. *Planta* 184: 113-122.

Mettler, I.J. and Beevers, H. (1979). Isolation and characterization of the protein body membrane of caster beans. *Plant Physiology* 64: 506-511.

Molvig, L., Tabe, L.M., Eggum, B.O., Moore, A.E., Craig, S., Spencer, D., and Higgins, T.J. (1997). Enhanced methionine levels and increased nutritive value of seeds of transgenic lupins (*Lupinus angustifolius* L.) expressing a sunflower seed albumin gene. *Proceedings National Academy Sciences USA* 94: 8393-8398.

Monsalve, R.I., Gonzalez, de la Pena, M.A., Lopez-Otin, C., Fiandor, A., Fernandez, C., Villalba, M., and Rodriguez, R. (1997). Detection, isolation and complete amino acid sequence of an aeroallergenic protein from rapeseed flour. *Clinical Experimental Allergy* 27: 833-841.

Moore, P.J., Swords, K.M.M., Lynch, M.A., and Staehelin, L.A. (1991). Spatial organization of the assembly pathways of glycoproteins and complex polysaccharides in the Golgi apparatus of plants. *Journal Cell Biology* 112: 589-602.

Müntz, K. (1996). Proteases and proteolytic cleavage of storage proteins in developing and germinating dicotyledenous seeds. *Journal Experimental Botany* 47: 605-622.

Müntz, K., Christov, V., Saalbach, G., Saalbach, I., Waddell, D., Pickardt, T., Schieder, O., and Wustenhagen, T. (1998). Genetic engineering for high methionine grain legumes. *Nahrung* 42: 125-127.

Neuhaus, J.M. and Rogers, J.C. (1998). Sorting of proteins to vacuoles in plants. *Plant Molecular Biology* 38: 127-144.

Nishimura, M. and Beevers, H. (1978). Hydrolases in vacuoles from castor bean endosperm. *Plant Physiology* 62: 44-48.

Nordlee, J.A., Taylor, S.L., Townsend, J.A., Thomas, L.A., and Bush, R.K. (1996). Identification of a Brazil-nut allergen in transgenic soybeans. *New England Journal Medicine* 334: 688-692.

Okita, T.W. and Rogers, J.C. (1996). Compartmentation of proteins in the endomembrane system of plant cells. *Annual Reviews of Plant Physiology and Plant Molecular Biology* 47: 327-349.

Ohtani, T., Galili, G., Wallace, J.C., Thompson, G.A., and Larkins, B.A. (1991). Normal and lysine-containing zeins are unstable in transgenic tobacco seeds. *Plant Molecular Biology* 16: 117-128.

Paris, N., Stanley, C.M., Jones, R.L., and Rogers, J.C. (1996). Plant cells contain two functionally distinct vacuolar compartments. *Cell* 85: 563-572.

Parra, F.M., Cuevas, M., Lezaun, A., Alonso, M.D., Beristain, A.M., and Losada, E. (1993). Pistachio nut hypersensitivity: Identification of pistachio nut allergens. *Clinical Experimental Allergy* 23: 996-1001.

Pastorello, E.A., Farioli, L., Pravettoni, V., Ispano, M., Scibola, E., Trambaioli, C., Giuffrida, M.G., Ansaloni, R., Godovac-Zimmermann, J., Conti, A., et al. (2000). The maize major allergen, which is responsible for food-induced allergic reactions, is a lipid transfer protein. *Journal Allergy Clinical Immunology* 106: 744-751.

Pastorello, E.A., Pompei, C., Pravettoni, V., Brenna, O., Farioli, L., Trambaioli, C., and Conti, A. (2001). Lipid transfer proteins and 2S albumins as allergens. *Allergy* 67: 45-47.

Pastorello, E.A., Varin, E., Farioli, L., Pravettoni, V., Ortolani, C., Trambaioli, C., Fortunato, D., Giuffrida, M.G., Rivolta, F., Robino, A., et al. (2001). The major allergen of sesame seeds *(Sesamum indicum)* is a 2S albumin. *Journal Chromatography B Biomedical Science Applications* 756: 85-93.

Pedrazzini, E., Giovinazzo, G., Bielli, A., de Virgilio, M., Frigerio, L., Pesca, M., Faoro, F., Bollini, R., Ceriotti, A., and Vitale, A. (1997). Protein quality control along the route to the plant vacuole. *Plant Cell* 9: 1869-1880.

Popineau, Y., Deshayes, G., Lefebvre, J., Fido, R., Tatham, A.S., and Shewry, P.R. (2001). Prolamin aggregation, gluten viscoelasticity, and mixing properties of transgenic wheat lines expressing 1Ax and 1Dx high molecular weight glutenin subunit transgenes. *Journal Agricultural Food Chemistry* 49: 395-401.

Pueyo, J.J., Chrispeels, M.J., and Herman, E.M. (1995). Degradation of transport-competent destabilized phaseolin with a signal for retention in the endoplasmic reticulum occurs in the vacuole. *Planta* 196: 586-596.

Robinson, D.G., Bäumer, M., Hinz, G., and Hohl, I. (1998). Vesicle transfer of storage proteins to the vacuole: The role of the Golgi apparatus and multivesicular bodies. *Journal Plant Physiology* 152: 659-667.

Robinson, D.G., Galili, G., Herman, E., and Hillmer, S. (1998). Topical aspects of vacuolar protein transport: Autophagy and prevacuolar compartments. *Journal Experimental Botany* 49: 1263-1270.

Robinson, D.G. and Hinz, G. (1997). Vacuole biogenesis and protein transport to the plant vacuole: A comparison with the yeast vacuole and the mammalian lysosome. *Protoplasma* 197: 1-25.

Robinson, D.G., Hinz, G., and Holstein, S.E.H. (1998). The molecular biology of transport vesicles. *Plant Molecular Biology* 38: 49-76.

Robinson, D.G., Hoh, B., Hinz, G., and Jeong, B.-K. (1995). One vacuole or two vacuoles: Do protein storage vacuoles arise de novo during pea cotyledon development? *Journal Plant Physiology* 145: 654-664.

Robinson, D.G., Rogers, J.C., and Hinz, G. (2000). Post-Golgi, prevacuolar compartments. In D.G. Robinson and J.C. Rogers (eds.), *Annual plant reviews,*

Volume 5: *Vacuolar compartments* (pp. 270-298). Sheffield, England: Sheffield Academic Press.

Saalbach, G., Jung, R., Kunze, G., Saalbach, I., Adler, K., and Müntz, K. (1991). Different legumin protein domains act as vacuolar targeting signals. *Plant Cell* 3: 695-708.

Saalbach, I., Pickardt, T., Machemehl, F., Saalbach, G., Schieder, O., and Muntz, K. (1994). A chimeric gene encoding the methionine-rich 2S albumin of the Brazil nut (*Bertholletia excelsa* H.B.K.) is stably expressed and inherited in transgenic grain legumes. *Molecular General Genetics* 242: 226-236.

Shewry, P.R., Napier, J.A., and Tatham, A.S. (1995). Seed storage proteins: Structures and biosynthesis. *Plant Cell* 7: 945-956.

Shimada, T., Kuroyanagi, M., Nishimura, M., and Hara-Nishimura, I. (1997). A pumpkin 72-kDa membrane protein of precursor-accumulating vesicles has characteristics of a vacuolar sorting receptor. *Plant Cell Physiology* 38: 1414-1420.

Shimoni, Y., Blechl, A.E., Anderson, O.D., and Galili, G. (1997). A recombinant protein of two high molecular weight glutenins alters gluten polymer formation in transgenic wheat. *Journal Biological Chemistry* 272: 15488-15495.

Sparvoli, F., Faoro, F., Gloria Damiati, M., Ceriotti, A., and Bollini, R. (2000). Misfolding and aggregation of vacuolar glycoproteins in plant cells. *Plant Journal* 24: 825-836.

Staehelin, L.A. and Moore, I. (1995). The plant Golgi apparatus: Structure, functional organization and trafficking mechanisms. *Annual Review Plant Physiology Plant Molecular Biology* 46: 261-288.

Stoger, E., Vaquero, C., Torres, E., Sack, M., Nicholson, L., Drossard, J., Williams, S., Keen, D., Perrin, Y., Christou, P., and Fischer, R. (2000). Cereal crops as viable production and storage systems for pharmaceutical scFv antibodies. *Plant Molecular Biology* 42: 583-590.

Sturm, A., Johnson, K.D., Szumilo, T., Elbein, A.D., and Chrispeels, M.J. (1987). Subcellular localization of glycosidases and glycosyltransferases involved in the processing of N-linked oligosaccharides. *Plant Physiology* 85: 741-745.

Sturm, A., Voelker, T.A., Herman, E.H., and Chrispeels, M.J. (1988). Correct glycosylation, Golgi-processing, and targeting to protein bodies of the vacuolar protein phytohemagglutinin in transgenic tobacco. *Planta* 175: 170-183.

Tabe, L.M. and Droux, M. (2002). Limits to sulfur accumulation in transgenic lupin seeds expressing a foreign sulfur-rich protein. *Plant Physiology* 128: 1137-1148.

Tabe, L.M., Wardley-Richardson, T., Ceriotti, A., Aryan, A., McNabb, W., Moore, A., and Higgins, T.J. (1997). A biotechnological approach to improving the nutritive value of alfalfa. *Journal Animal Science* 73: 2752-2759.

Tada, Y., Nakase, M., Adachi, T., Nakamura, R., Shimada, H., Takahashi, M., Fujimura, T., and Matsuda, T. (1996). Reduction of 14-16 kDa allergenic proteins in transgenic rice plants by antisense gene. *FEBS Letters* 391: 341-345.

Taylor, S.L. and Hefle, S.L. (2001). Will genetically modified food be allergenic? *Journal Allergy Clinical Immunology* 107: 765-771.

Teuber, S.S., Dandekar, A.M., Peterson, W.R., and Sellers, C.L. (1998). Cloning and sequencing of a gene encoding a 2S albumin seed storage protein precursor from English walnut *(Juglans regia)*, a major food allergen. *Journal Allergy Clinical Immunology* 101: 807-814.

Teuber, S.S. and Peterson, W.R. (1999). Systemic allergic reaction to coconut *(Cocos nucifera)* in 2 subjects with hypersensitivity to tree nut and demonstration of cross-reactivity to legumin-like seed storage proteins: New coconut and walnut food allergens. *Journal Allergy Clinical Immunology* 103: 1180-1185.

Van der Wilden, W., Herman, E.M., and Chrispeels, M.J. (1980). Protein bodies of mung bean cotyledons as autophagic organelles. *Proceedings National Academy Sciences USA* 77: 428-432.

Vitale, A., Bielli, A., and Ceriotti, A. (1995). The binding protein associates with monomeric phaseolin. *Plant Physiology* 107: 1411-1418.

Vitale, A. and Raikhel, N.V. (1999). What do proteins need to reach different vacuoles? *Trends Plant Sciences* 4: 149-155.

Vocks, E., Borga, A., Szliska, C., Seifert, H.U., Seifert, B., Burow, G., and Borelli, S. (1993). Common allergenic structures in hazelnut, rye grain, sesame seeds, kiwi, and poppy seeds. *Allergy* 48: 168-172.

von Schaewen, A. and Chrispeels M.J. (1993). Identification of vacuolar sorting information in phytohemagglutinin, an unprocessed vacuolar protein. *Journal Experimental Botany* 44: 339-342.

Walmsley, A.M. and Arntzen, C.J. (2000). Plants for delivery of edible vaccines. *Current Opinion Biotechnology* 11: 126-129.

Wandelt, C.I., Khan, M.R.I., Craig, S., Schroeder, H.E., Spencer, D., and Higgins, T.J.V. (1992). Vicilin with carboxy-terminal KDEL is retained in the endoplasmic reticulum and accumulates to high levels in the leaves of transgenic plants. *Plant Journal* 2: 181-192.

Weiss, W., Huber, G., Engel, K.H., Pethran, A., Dunn, M.J., Gooley, A.A., and Gorg, A. (1997). Identification and characterization of wheat grain albumin/globulin allergens. *Electrophoresis* 18: 826-833.

Williamson, J.O., Galili, G., Shaw, B.A., Larkins, B.A., and Gelvin, S.B. (1988). The synthesis of a 19 kD zein protein in transgenic petunia plants. *Plant Physiology* 88: 1002-1007.

Yaklich, B. and Herman, E.M. (1995). Protein storage vacuoles of soybean aleurone cells accumulate a unique glycoprotein as well as proteins thought to be embryo specific. *Plant Science* 107: 57-67.

Yoshimasu, M.A., Zhang, J.W., Hayakawa, S., and Mine, Y. (2000). Electrophoretic and immunochemical characterization of allergenic proteins in buckwheat. *International Archives Allergy Immunology* 123: 130-136.

Chapter 9

Synthetic Seed Technology

P. Suprasanna
T. R. Ganapathi
V. A. Bapat

INTRODUCTION

The discovery of totipotency of plant cells in the early part of the twentieth century and the utilization of this concept in further years marked the beginning of a new era of plant biotechnology (Gamborg 2002). In the intermediary decades, several plant-tissue-culture-related methods have explored the potential of totipotent plant cells. Synthetic seed (also referred to as syn seed or artificial seed), emanating from the encapsulation of somatic embryos, is an emerging field with a vast potential for plant propagation, delivery, and storage.

Seeds in most of the crop plants are vehicles for propagation and storage mainly because of the convenience in production and handling. In certain other crops, plants are propagated either vegetatively or by genetically nonuniform seeds. Application of in vitro techniques has offered great potential for propagation, storage, and genetic manipulation of crop plants (Rao et al., 1996, 1997, 1998; Rao and Suprasanna, 1999). Plant micropropagation offers an efficient method for propagating mass amounts of disease-free, genetically uniform plants in vitro (Honda et al., 2001). Somatic embryogenesis, adventitious shoot production, and axillary shoot production are the principal modes of this propagation system. Clonal propagation using plant explants in vitro ensures faster rates of plant multiplication. Somatic embryogenesis is the production of bipolar embryos that possess the required capability to develop into complete plants. Although micropropagation has been extensively used in several

---

Thanks are due to Dr. G. Padmaja, Department of Plant Sciences, University of Hyderabad, Hyderabad, India; Dr. Nadina Neives, Bio Plantas, Cuba; and Drs. A. Standardi and U. Piccioni, Italy, for their cooperation in sharing their publications and photographs.

horticultural crops, somatic embryogenesis has ushered in a new era in the clonal propagation of plants because of the higher production volumes compared to other methods.

The concept of artificial seed (syn seed) originated from a proposal by Murashige (1977) during a symposium that somatic embryos could be encapsulated, handled, and used like a natural seed for transport, storage, and sowing. Later, other researchers made efforts toward using somatic embryos for encapsulation. It was not until 1986 when Gray discussed, in a first symposium on artificial seeds, the issues of embryogeny, maturation, and desiccation of somatic embryos, bioreactor production, encapsulation, and germination. During the late 1980s, Bapat et al. (1987) explored the potential of axillary buds of mulberry for encapsulation and preparation of synthetic seeds. Subsequently, several researchers have conducted studies on using in vitro–derived propagules (axillary buds, shoot apices, buds, bulbs, or any form of meristems) that are used for in vitro propagation to prepare synthetic seeds. Hence, the current description of synthetic seed extends to an artificially encapsulated somatic embryo, shoot apex or any other meristematic tissue which can develop into a plant under in vitro or in vivo conditions. This includes microbulbs, microtubers, rhizomes, corms, microcuttings, shoot apices, axillary buds, meristemoids, cell aggregates, embryogenic clumps, and primordia which serve as source material for developing synthetic seeds. Based on the information on the available systems that have so far been employed, these can be categorized into embryogenic or bipolar and nonembryogenic or unipolar explants derived from in vitro cultures or in vivo mature plants or trees (Standardi and Piccioni, 1998). Mamiya and Sakamoto (2001) suggested the term *encapsulatable units* (EU) for the propagules used for encapsulation.

Syn seeds have several potentials such as those considered for micropropagation: a high-volume propagation system, ease in handling, and reduced need for space. The small-sized syn seeds will ensure handling, storage, transport, and planting of large numbers in a more economical way compared to other existing methods (Redenbaugh et al., 1988; Redenbaugh, 1990). The plant population resulting from the syn seed will be uniform, and the direct delivery of encapsulated material will save many subcultures generally practiced to obtain plantlets. In addition, the encapsulated material can be packed with beneficial adjuvants such as pesticides, nutrients, fertilizers, nitrogen-fixing bacteria, or even microscopic parasite-destroying worms so as to enable proper delivery, better growth, and resistance to microbial invasion.

One of the prerequisites that needs to be addressed for synthetic seed research is the production of a large number of propagules (somatic embryos or any other meristematic propagules) that can develop or regrow synchronously, withstand the stress of encapsulation, and retain the ability to develop into plantlets. It is necessary for the syn seeds to be competitive with seeds for germination and plant development. Different stages in the development of synthetic seed technology are illustrated in Figure 9.1. The three requirements necessary are (1) good in vitro culture systems with high efficiency of production of encapsulatable units, (2) encapsulation and coating methods for better viability and handling of the syn seeds, and (3) ex vitro and/or soil conversion of the syn seeds. Conceptually, synthetic seed should be analogous to normal seed (Table 9.1). Hence, drying to less moisture content needs to be thoroughly examined for good germination and prolonged storage. During the process of somatic embryo germination, there are two processes: *germination,* referred to as the process culminating in the emergence of radicle, and *conversion,* defined as the process of emergence of radicle and shoot (Lai et al., 1995).

Many plant systems are known to produce a large number of somatic embryos in tissue cultures, and such asexually produced embryos proceed through developmental stages similar to zygotic embryos obtained after fertilization (Thorpe and Stasolla, 2001). Somatic embryos possess root and shoot meristems connected by a common vascular system. Somatic embryo quality is very critical and significant for their conversion into plants. For achieving good quality embryos, it is necessary to have a synchronously developing cell culture with embryos at a particular developmental stage. In plant systems such as carrot, celery, and alfalfa, somatic embryogenesis has been well characterized, and this has led to the development of successful synthetic seed preparations using encapsulated somatic embryos. In addition, somatic embryo encapsulation has been demonstrated in several plant species, including brassica, cotton, lettuce, rice, Norway spruce, banana, mango, papaya, sandalwood, eggplant, and horseradish.

ENCAPSULATION METHODS

Coating Systems

One of the prime considerations of synthetic seed is coating or encapsulation of the embryo or other propagule. The primary function of such coating is to allow germination and conversion of the propagule and ensure protection during handling, transportation, storage, and planting. In addition, nutrients, growth regulators, and growth adjuvants are added to constitute an artificial endosperm, analogous to the seed endosperm. The necessity of an artificial endosperm or a nutrient-supplying system during the early stages of the conversion of an encapsulatable unit, especially for

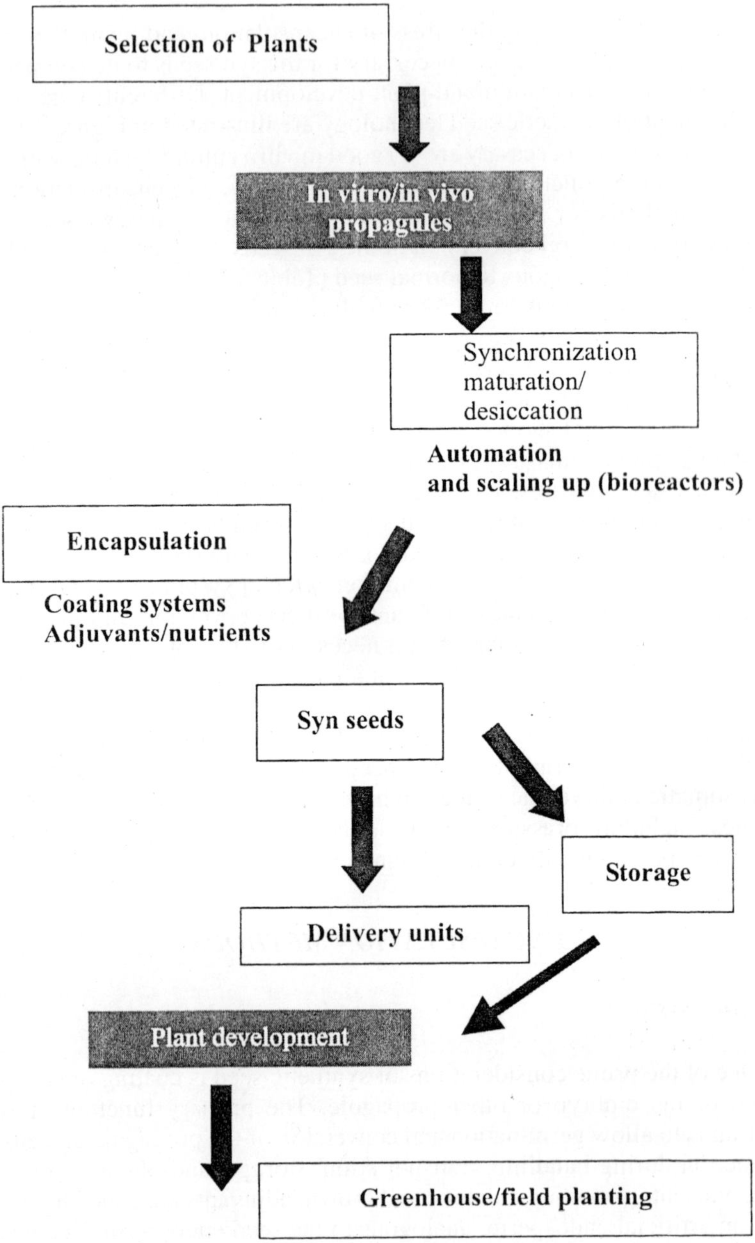

FIGURE 9.1. Different steps in the production of synthetic seeds.

TABLE 9.1. A comparison of seeds and syn seeds.

Seed	Syn seed
Somatic embryo develops from a somatic cell without fertilization and it has both shoot and root poles. Morphologically and functionally it is similar to seed embryo.	In vitro propagules, viz., axillary buds, shoot apices, cormlets, bulbs, nodes, protocorms, are produced in tissue cultures for plant propagation.
Seed consists of an embryo and endosperm encased with a seed coat.	Synthetic seed consists of either a somatic embryo or a vegetative propagule, encased in a gel matrix.
Embryo consists of one or two cotyledons attached to central axis. The upper part, the plumule, develops into a shoot and the lower, the radicle, grows into a root.	Uni- or bipolar structures
Endosperm is the food reservoir in the seed, which the embryo utilize during germination.	Artificical endosperm is the nutrient mixture provided for the development of the somatic embryo/ propagule.

albuminous species, has been suggested (Redenbaugh et al., 1984, 1987; Sanada et al., 1993).

Synthetic seeds can be prepared with or without a coating. Based on this, four types of synthetic seeds have been proposed:

1. Uncoated desiccated somatic embryos (dry artificial seeds)
2. Coated, desiccated embryos
3. Encapsulated, hydrated embryos
4. Hydrated embryos in a fluid-drilling gel (Redenbaugh et al., 1991)

Fluid drilling was initially used to coat citrus embryos with several compounds to develop a suitable synthetic seed coat; among them, polyoxyethylene was found to be suitable in carrot and celery synthetic seed preparations (Kittoo and Janick, 1980, 1985).

Several natural or synthetic polymers have been used for coating, and among them, natural polysaccharides such as alginate and carrageenan are mostly used because of their simple gellation property. Synthetic polymers often undergo gellation with an accompanying exothermic reaction and/or formation of free radicals, which can be toxic to the encapsulatable units. Subsequently, several gels such as agar, alginate, polyco2133, carboxy methyl cellulose, carrageenan, gelrite, guargum, sodium pectate, tragacanth gum, etc., have been tested for synthetic seed preparation (Redenbaugh,

1993). Simple gels are those with one type of gelling agent (for example, sodium alginate), while composite gels are those with two types of gelling agents (sodium alginate and silica gel or chitosan or carrageenan or any other agent).

Based on their gel-forming procedure, the encapsulation agents generally are categorized into three groups (Barbotin et al., 1993).

1. Gel formation by ionic cross-linking of a charged polymer (alginate, chitosan, carrageenan, etc.)
2. Gel formation by cooling a heated polymer (agar, agarose, carrageenan, etc.)
3. Gel formation by a chemical reaction (polyacrylamide)

Although several gelling agents have been tested for encapsulation, sodium alginate has become the most commonly used gel for the preparation of synthetic seeds. The characteristics of some of these agents are described in the following sections.

Agar and Agarose

This is a polysaccharide extracted from agarophytes (marine algae including *Gracilaria, Gelidium, Pterocladia*). Chemically agar is 1 agarobiose backbone alternating with 1,3-linked β-D-galactopyranose and 1,4-linked 3,6-anhydro-∝-L-galactopyranose. For encapsulation, agar is boiled at 100°C, cooled, and then, before solidification, cells or propagules are embedded. The advantage with agar is that after solidification it can be cut into the desired size and shape; however, there is possibility of heat damage to the entrapped propagules/cells during agar encapsulation, and this can limit its usage. Agarose is a neutral gelling fraction obtained from agar with high purity. It has a lower gelling temperature (about 30°C) and can be used in a similar way to agar for entrapment.

Alginates

Alginic acid is a phyco-colloid, discovered in 1883 by a British pharmacist, Stanford, who called it algin (Lewis et al., 1990). Alginate is one of the most abundant naturally occurring polymers, and thus availability of the raw material imposes no problems for application. Alginates are extracted with sodium hydroxide from giant brown seaweeds such as *Macrocystis pyrifera* and *Laminaria hyperborea*. They are linear 1,4-linked copolymers of β-D-mannuronic acid (M unit) and its C5 epimer, ∝-L-guluronic acid

(G unit). Alginate gel is formed by the replacement of monovalent ion of sodium with divalent ion of calcium (Ca^{2+}) and the ionic cross-linking results in the formation of polymeric network in the polysaccharide. The gels with the highest mechanical strength, lowest shrinkage, best stability toward monovalent cations, and the highest porosity are especially made from alginate with a content of L-guluronic acid higher than 70 percent and an average length of the G-blocks higher than fifteen (Martinsen et al., 1989).

The use of sodium alginate for the preparation of synthetic seeds has certain advantages, such as nontoxicity and the thermoindependence of the entire gelling process. The procedure involves mixing of propagules/somatic embryos and dropping the mixture with the help of a pasteur pipette or forceps into a calcium chloride solution (50 to 300 mM). This results in the formation of beads that can be hardened during 15 minutes to 12 hours in calcium chloride. The beads are not thermoreversible but can be dissolved in the presence of a chelating agent such as EDTA or in phosphate and citrate buffers. Alginate solution is generally autoclaved and used; however, UV sterilization is recommended, as autoclaving is known to affect viscosity and degradation of alginate polymer molecules (Sakamoto et al., 1992).

Carrageenan

Carrageenans are isolated from the cell walls of red sea weeds such as *Chondrus, Euchema,* and *Gigartina,* etc. There are three types: kappa (K), iota (ι) and lambda (λ), of which, K-carrageenan is more suitable for immobilization. It is a condensation product of 1,3 linked α-galactose-4-sulfate and 1,4-linked 3,6-β-anhydrogalactose. Alkali metal ions (K^+, Rb^+ , Cs^+), alkaline earth metal ions (Ca^{2+}) or trivalent ions (Al^{3+}), and also NH_4^+ and amines (hexamethylenediamine, histamine) promote gelation. The mechanism involves crossing over of the chains forming several helices. Ethylene glycol or glycerol also promotes gelation that is thermally reversible. The rigidity of the gel increases with the increase in potassium ion concentration. One of the disadvantages is its instability in the presence of ions and the use of temperature of 30 to 50°C for immobilization.

Chitosan

Chitosan is a partially deacetylated chitin consisting of a poly (1-4) 2-amino-β-D-glucose. Vorlop and Klein (1981) first demonstrated the use of a viscose chitosan acetate solution. The mixture is added drop-wise in 2 percent $K_4Fe(CN)_6$ solution (pH 5.7), and after 30 minutes, the chitosan beads are washed with phosphate buffer. One of the major drawbacks with this is the low viability of plant cells. The use of K-carrageenan/chitosan or alginate/chitosan mixture can enable the viability of cells. The addition of $CaCl_2$ to chitosan and glucose to alginate also can be used to increase the capsule strength.

Gellan Gum

Gellan gum, an industrial anionic polysaccharide, is produced in an extracellular form by *Pseudomonas elodea.* It is a linear homopolymer with a tetra saccharide-repeating unit consisting of two β-D-glucose molecules: one β-D-glucuronic acid and one ∝-L-rhamnose residue (Jansson et al., 1983). The characteristics are comparable to those of K-carrageenan. Buitelaar et al. (1988) reported using a gellan gum in which plant cells were mixed with 1 percent in $MgSO_4$ at 40°C to form in a mold after cooling and subsequent hardening in $MgSO_4$. The use of $CaCl_2$ or KCl has also been reported in encapsulation (Norton and Lacroix, 1990).

Pectate

Pectates are useful material for cell immobilization (Gemeiner et al., 1991). Pectin, an important constituent of the cell walls and soft tissues of higher plants, is extracted commercially from apple waste or from the peel of citrus fruits. It is composed of long, regular sequences of 1,4-linked ∝-D-galacturonate residues. Gel is formed by sodium pectate after cross-linking with calcium ions (Vanek et al., 1989).

Polyacrylamide

This is a synthetic polymer, formed by the polymerization of monomers (acrylamide and N,N,N-methylene bisacrylamide). Such preparation (in phosphate buffer) is mixed with a cell suspension, and N,N,N',N' tetramethylethylene diamine (TEMED) and ammonium persulfate are added as catalysts. The mixture is left at 4°C for 15 minutes. The resulting gel block is made into granules by pressing it through a sieve (Klein and Schara, 1980; Skryabin and Koshcheenko, 1987).

In addition to these, photo-cross-linkable and urethane resins and protein gels such as gelatin have also been used for immobilization (Barbotin et al., 1993).

Dupuis et al. (1994) reported the use of pharmaceutical capsules as a coating system for synthetic seed preparation. The inside body of the capsule was covered by a watertight film composed of polyvinyl chloride

(PVC), polyvinyl acetate (PVA), and bentone as thickener to ensure the nutrient supply and the development of somatic embryos. The PVC used in this study has many advantages, including its ability to produce waterproof films, its widespread use arising from a high degree of chemical resistance, and a truly unique ability to be mixed with many additives as plasticizers, thickeners, surfactants, heat stabilizers, pigments, salts, etc. Mixtures of these additives are highly reproducible and cover a greater range of physical, chemical, and biological properties than any other plastic material. The cap of the synthetic seed was sealed with polymeric films, fibers, paper, and even oils. The best results with a film were obtained with polyvinyl alcohol in water. This material resulted in germination of 80 percent and subsequent conversion of 50 percent. The oil as a cap was found to be suitable for the conversion of synthetic seeds, and the best results were obtained with mineral oil. The carrot somatic embryos encapsulated using this method developed with 90 percent conversion.

Calcium alginate hollow beads have also been successfully employed to encapsulate plant material (Patel et al., 2000). Shoot tips or callus cells were suspended in a solution containing carboxymethyl cellulose and calcium chloride and then dropped into a stirred sodium alginate solution. Using carrot cells/seeds, encapsulation in hollow beads was done, with 100 percent of the seeds germinating into plants. Hollow beads based on ionotropic gels can be useful for encapsulation of plant material to be applied in conservation and distribution of in vitro cultures. Compared to conventional alginate beads, the authors suggest that entrapment and coating are realized in one step and that material be free from pathogens.

During the development of the syn seed, the shoot and root break through the gel beads, followed by further growth into a plantlet. Because alginate beads are known to have lower oxygen permeability, Sakamoto et al. (1992, 1995) developed self-breaking gel beads to compensate for limited oxygen supply. After the alginate beads are made, these are washed with running tap water and immersed in potassium nitrate solution, followed by rinsing again with running tap water. Upon sowing in humid conditions, the gel beads swell gradually and then break open. Lettuce adventitious shoots encapsulated with this method showed 96 percent conversion in vitro as compared to no growth with the conventional gel beads.

In order to improve coating and ensure tolerance to drying and microbial contamination, Redenbaugh and Reyes (1987) proposed membrane coating of gel beads with wax and resin. Hard beads with layers of coating have also been proposed to avoid problems of drying and microbial infection.

Dry Artificial Seeds

Fujii et al. (1989, 1992) advocated the concept of noncoated somatic embryos as artificial seeds, because hydrated syn seeds have the disadvantage of requiring coating materials. Hydrated artificial seeds are difficult to store because they lack quiescence. If somatic embryos could maintain their viability after their water content is reduced to very low levels similar to seeds, they can overcome the limitations of hydrated synthetic seeds. Desiccated artificial seeds can be economical with decreased delivery costs and also useful for germplasm preservation. Desiccation tolerance of somatic embryos is induced by treatments with physical (chilling) and chemical factors (abscisic acid, sucrose, proline). Desiccation tolerance has been induced in somatic embryos of grape (Gray 1990), orchard grass, alfalfa (Senaratna et al., 1989), rapeseed, and broccoli (Takahata et al., 1993). In broccoli, dried embryos remained viable with ability for plant conversion after three months of storage under room-temperature conditions. Plants that developed grew normally without any morphological or ploidy changes (Takahata et al., 1993).

DIFFERENT PROPAGULES USED FOR SYNTHETIC SEEDS

Among the different propagules used for making synthetic seeds, somatic embryos have been used in number of cases (Ara et al., 2000) since these are bipolar with a radicle and plumule and can develop into complete plantlets in a single step. There are many plant systems in which somatic embryogenesis is well established, and the fact that somatic embryos tolerate desiccation similarly to zygotic embryos has even encouraged researchers to encapsulate somatic embryos and use them for plant development and germplasm storage. Organogenesis, on the other hand, is demonstrated in several plant species. Both in vivo and in vitro derived plant explants (EUs) have been employed for developing syn seeds (see Figure 9.2, Table 9.2).

Axillary Buds and Apical Shoot Tips

In some cases, where somatic embryogeny has not been established, shoot tips or axillary buds have become the choice for encapsulation to prepare the synthetic seeds. These have shoot meristems and do not have root meristems. In many instances the meristems are pretreated with auxins to induce roots and are then encapsulated. In some cases, without such treatment, shoot tips/axillary buds are encapsulated and the syn seeds convert into plantlets (Bapat and Rao, 1990; Ganapathi et al., 1992). In apple, shoot

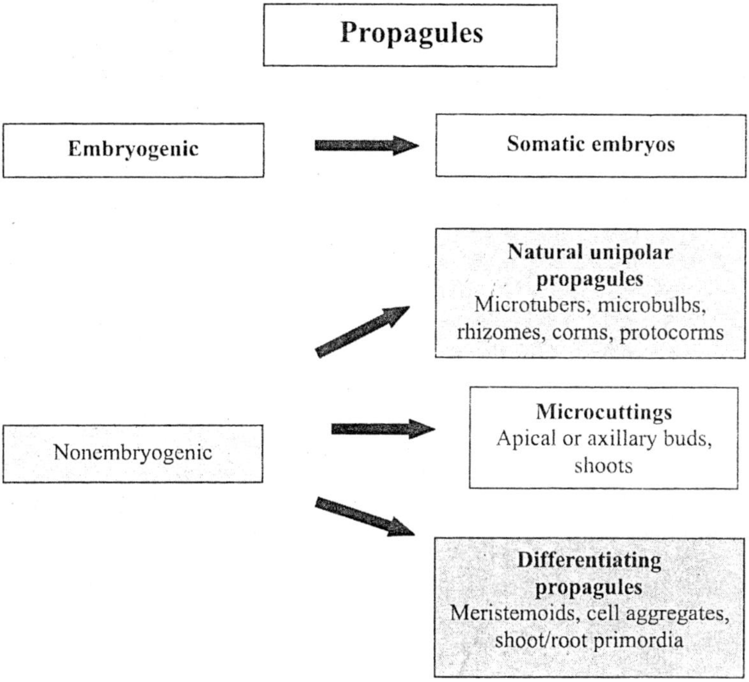

FIGURE 9.2. Different propagules used for preparing synthetic seeds.

buds are pretreated with IBA for three to six days prior to encapsulation (Capuano et al., 1998). This type of propagule encapsulation will be useful for germplasm transportation and exchange, and with a proper preservation technique it can be considered for germplasm conservation.

Embryogenic Masses

Highly regenerative embryogenic clumps are often the suitable choice for genetic manipulation studies. The encapsulation of embryogenic masses is primarily used for long-term maintenance to avoid the regular subcultures. This has the advantage of saving the labor involved in regular subculturing and the chances of contamination during subcultures. In addition, long-term culture of embryogenic callus can lead to accumulation of genetic variation. Embryogenic masses of *Pistacia vera* encapsulated in sodium alginate and stored at 4°C after treatment with BAP retained

TABLE 9.2. Reports of synthetic seeds in crop plants.

Crops	Part	Reference
Vegetables		
Carrot	Somatic embryos	Kitto and Janick, 1985
Tomato	Seeds	Garret et al., 1991
Solanum sp.	Shoot tips	Fabre and Dereuddre, 1990
Solanum melongena	Somatic embryos	Rao and Singh, 1991
Asparagus	Somatic embryos	Ghosh and Sen, 1994
Coriander	Axillary buds	Chen et al., 1995
Potato	Shoot tips	Patel et al., 2000
	Nodal segments	Sarkar and Naik, 1997
Celery	Somatic embryos	Redenbaugh et al., 1986
Ornamentals		
Dianthus	Nodal segments	Fukai et al., 1994
Dendrobium wardianum	Protocorm-like-bodies	Sharma et al., 1992
Cymbidium giganteum	Protocorms	Corrie and Tandon, 1993
Coelogyne oderatissima	Protocorms	Kamalakannan et al., 1999
Geodorum densiflorum	Protocorm-like-bodies	Gill et al., 1994
Spathoglottis pliccata	Protocorm-like-bodies	Singh, 1991
	Seeds	Tan et al., 1998
Geranium	Somatic embryos	Marsolais et al., 1991
Lillium	Somatic embryos	Piccioni et al., 1992
Syringa vulgaris	Axillary buds	Refouvelet et al., 1998
Phaius tankervillae	Protocorm-like-bodies	Malemngaba et al., 1996
Clitoria ternata	Somatic embryos	Malbadi and Nataraja, 2002a
Fruits and plantation crops		
Banana	Shoot apices	Ganapathi et al., 1992
	Somatic embryos	Ganapathi et al., 2001
Mango	Somatic embryos	Ara et al., 1999
Mulberry	Axillary buds	Bapat et al., 1987
		Patnaik and Chand, 2000
Mulberry	Vegetative buds	Patnaik et al., 1995
Apple	Shoot apices	Niino and Sakai, 1992
Pineapple	Shoot apices	Soneji et al., 2002
Grape	Somatic embryos	Gray and Purohit, 1991
Grape vine	Shoot tips	Plessis et al., 1991
Coffee	Somatic embryos	Hatanaka et al., 1994

Crops	Part	Reference
Tea (*Camellia sinensis* L.)	Axillary buds	Mandal et al., 2002
Kiwifruit, Blackberry, Raspberry	Axillary buds	Piccioni and Standardi, 1995
Rubus	Shoot buds	Piccioni and Standardi, 1995
Pistachio	Embryogenic	Onay et al., 1996
Camelia japonica	clumps	Janeiro et al., 1997
Papaya	Somatic embryos	Castillo et al., 1998
Cocoa	Zygotic embryos	Sudhakara et al., 2000
Citrus reticulata	Somatic embryos	Antonietta et al., 1998
Cleopatra tangerine	Zygotic and somatic embryos	Nieves et al., 1998
Spices, medicinal/aromatic plants		
Cardamom	Shoot apices	Ganapathi et al., 1994
Ginger	Shoot buds	Sharma et al., 1994
Ginseng	Somatic embryos	Sajina et al., 1997
Ginseng	Somatic embryos	Choi and Jeong, 2002
Black pepper	Shoot buds	Sajina et al., 1997
Turmeric	Adventitious buds	Sajina et al., 1997
Cinnamon	Shoot buds/somatic embryos	Sajina et al., 1997
Vanilla	Shoot buds	Sajina et al., 1997
Valiarana wallichi	Shoot tips and axillary buds	Mathur et al., 1989
Horse radish	Somatic embryos	Shigeta and Sato, 1994
Horse radish	Hairy roots with shoots	Uozumi et al., 1992
Caraway	Somatic embryos	Furamanowa et al., 1991
Coptis chinensis	Somatic embryos	Ke et al., 1995
Mentha arvensis	Axillary buds	Ahuja et al., 1989
Ocimum spp.	Axillary buds	Mandal et al., 2000
Forest trees		
Sandalwood	Somatic embryos	Bapat and Rao, 1988
Eucalyptus citriodora	Somatic embryos	Muralidharan and Mascarenhas, 1995
Norway spruce	Somatic embryos	Gupta et al., 1987
White spruce	Somatic embryos	Attree et al., 1994
Interior spruce	Somatic embryos	Lulsdorf et al., 1993
Black spruce	Somatic embryos	Lulsdorf et al., 1993
Hopea parviflora	Zygotic embryos	Sunilkumar et al., 2000
Guazuma crinata Mart.	Shoot tips/nodal segments	Maruyama and Kinoshita, 1997

TABLE 9.2 *(continued)*

Crops	Part	Reference
Cedera odorata L.		
Jacaranda mimosifolia D. Don.		
Cereals, legumes, and other crop plants		
Rice	Somatic embryos	Suprasanna et al., 1996
Finger millet	Embryogenic clumps	George and Eapen, 1995
Barley	Somatic embryos	Datta and Potrykus, 1989
Groundnut	Somatic embryos	Padmaja et al., 1995
Mustard	Shoot buds	Arya and Beg, 1998
Betula pandula	Shoot buds	Piccioni and Standardi, 1995
Crataegus oxycantha	Shoot buds	Piccioni and Standardi, 1995
Alfalfa	Somatic embryos	Redenbaugh et al., 1986
Sugarcane	Somatic embryos	Nieves et al., 2001
Mothbean *(Vigna aconitifolia)*	Somatic embryos	Malbadi and Nataraja, 2002b
Cassava	Nodal cuttings and shoot tips	Danso and Ford-Lloyd, 2003
Yam	Nodal segments	Hasan and Takagi, 1995

pro-liferative capacity for up to two months (Onay et al., 1996). Small clusters of somatic embryos of *Eleucine coracona* L. encapsulated in 3 percent sodium alginate showed signs of germination in 10 to 12 days. There was no visible sign of desiccation of the matrix and the plantlets grew normally, similar to plantlets developed from nonencapsulated embryo clusters (George and Eapen, 1995).

Protocorm or Protocorm-Like Bodies

These have been successfully demonstrated in *Cymbidium giganteum, Dendrobium, Geodorum densiflorum, Phaius tonkervillae,* and *Spathoglottis plicata* (reviewed in Ara et al., 2000; Rao et al., 2000). The encapsulated protocorms of *C. giganteum* converted into healthy plantlets under in vitro and in vivo conditions (Corrie and Tandon, 1993). The advantage with this type of propagule encapsulation is the direct transfer of protocorms to soil, eliminating the cost involved in raising the plantlets and subsequent hardening (Corrie and Tandon, 1993).

Hairy Root Encapsulation

Uozumi et al. (1992) reported an efficient method of producing plantlets from encapsulated hairy root fragments in *Armoracia rusticana* (horseradish). Though frequency of plantlet formation decreased due to encapsulation, dehydration treatment improved the frequency of plantlet development. This type of propagule can be used for large-scale production of transgenic plants, as hairy roots grow fast and much biomass can be produced within a short span of time. Small fragments from these hairy roots then can be encapsulated and plantlets regenerated. Use of encapsulated plantlets as artificial seeds can shorten the time required for the development of healthy plants (Nakashimada et al., 1995).

Advantages of Encapsulating Unipolar Propagules

The advantages with unipolar propagule encapsulation for synthetic seed preparation are similar to micropropagation. The main advantages are high production efficiency, sanitary plant conditions, and storage potentiality; their small size also eases mechanization and makes for easy for transport (Standardi and Piccioni, 1998). The incidence of somaclonal variation would be reduced, as mutations are known to occur more frequently when the de differentiation and redifferentiation processes are required to regenerate a plant. In alfalfa, Piccioni et al. (1997) used organized meristematic buds for encapsulation and found that incidence of mutations is less. However, buds from different parts of a plant vary in their response to regrowth, dormancy, and lack of tap root system (Capuano et al., 1998). Rao et al. (1993) opined that encapsulated shoot apices could offer convenient material for germplasm exchange. Recently, Danso and Ford-Lloyd (2003) reported encapsulation of nodal cuttings and shoot tips for storage and germplasm exchange, and suggested that the method of encapsulation of in vitro–derived nonembryogenic propagules can be useful for germplasm storage and exchange.

APPLICATIONS

Potential applications are generally based on the viable and efficient in vitro culture systems, need for improvement, and, last, the cost involved in comparison to seed-based methods and advantages, if any. Synthetic seeds should offer low-cost alternatives in vegetable crops, forage crops, cereals, and commercial crops, as the production costs of micropropagation (in ornamental and plantation crops) and hybrid seed production are very high

(Rao et al., 2000). Exhibit 9.1 presents the list of potential candidate crops for synthetic seed research. In the case of forest species, application of synthetic seeds will be valuable because of the need for vegetative propagation and lack of other cost-effective methods. This has significant potential, particularly for reducing the tree breeding cycle by 5 to 20 years by eliminating the need for seed production orchards. In addition, polyploids with elite characters, transgenic plants, and sterile plants can be propagated through synthetic seeds. The production of new lettuce varieties generally begins with an F1 cross followed by 8 to 12 generations of selfing and evaluation

EXHIBIT 9.1.
List of potential crops for synthetic seeds.

Category I: Strong Technological Basis[a]

Alfalfa	Oil palm, date palm
Caraway	Orchard grass
Carrot	Orange
Celery	*Panicum*
Coffee	Walnut
Eggplant	

Category II: Strong Commercial Basis[b]

Asparagus	Impatiens
Begonia	Lettuce
Broccoli	Lobolly pine
Cauliflower	Petunia
Corn	Potato
Cotton	Rice
Cucumber	Soybean
Cyclamen	Spinach
Douglas fir	Sugarcane
Garlic	Tobacco
Geranium	Tomato
Gerbera	Watermelon
Grape	Banana

Source: Modified from Redenbaugh et al., 1991.
[a]Good somatic embryogenesis is available.
[b]Economic importance based on value of vegetative propagation.

prior to varietal release. Using lettuce artificial seeds, single plant selections can be made from selected F1 plants, and evaluation and trial can then begin immediately; no breeding would be needed to fix the desirable genes (Redenbaugh et al., 1988). In alfalfa, which shows heavy inbreeding depression, syn seeds can be employed for cloning of selected parents, which otherwise may be self-sterile (McKersie and Bowley, 1998).

Micropropagation of many ornamental plants is in high demand because the maintenance of inbred lines for the production of F1 seed is expensive and laborious. Limited supply, clonal fidelity, seasonal variability, and poor germination are other problems. Although tissue culture propagation is practiced on a commercial scale in some ornamental crops, research on somatic embryogenesis with optimized frequencies of somatic embryo conversion and germination can have a significant impact on synthetic seed development. In orchids that are micropropagated generally, aseptically grown encapsulated protocorms can be directly transplanted in soil, thereby reducing the cost of raising and acclimatizing in vitro plantlets (Corrie and Tandon, 1993). Orchid seeds are minute and difficult to handle as tens of thousands of seeds are germinated in a container. This can be achieved using encapsulation of dustlike orchid seeds. Tan et al. (1998) encapsulated *Sphathoglottis plicata* seeds in capsules of alginate-chitosan or alginate-gelatin and infected with mycorhizal fungus. High-frequency germination (84 percent) was observed in the presence of fungus alone without any supplementation with sucrose. Delivery of syn seeds can obviate the routinely high cost of propagation methods.

For agronomic crops, commercial vegetative propagation can be done through somatic embryogenesis and syn seed can play an important role for the development and production of hybrids with alfalfa, cotton, rice, soybean, and sugar beet. Rice hybridization generally requires manual emasculation of anthers, and this can impede development and commercialization of hybrid rice. In this regard, artificial seeds can be used for the propagation of high-yielding, elite hybrids. High-efficiency somatic embryogenesis with good conversion into plants is critical, and such a system can be useful for developing synthetic seed in rice. Somatic embryogenesis through the formation of direct embryos from cultured explant offers a unique means to propagate male sterile and elite hybrids (Sahasrabudhe et al., 1999). In vitro cultured inflorescence has been induced to form somatic embryos without an intervening callus phase, and hence there can be less chances of somaclonal variation and the purity of cytoplasmic male sterile lines can be maintained (Sahasrabudhe et al., 1999).

In commercial crops such as sugarcane, somatic embryogenesis and artificial seed technology would have direct applications for propagation. Because labor is a major constituent of this micropropagation industry, savings can be realized by using artificial seeds. Nieves et al. (2001) demonstrated survival of encapsulated somatic embryos following desiccation with abscisic acid or jasmonic acid. These embryos, encapsulated and dehydrated to 30 to 60 percent water content, showed 52 percent survival as compared to 28 percent in the control. The authors suggested that encapsulated–hydrated embryos could be similar to seeds in handling, germination, conversion, and ex vitro conditions and seedling development in the field.

Micropropagation has been suggested as an alternative for the multiplication of selected vegetable crops to propagate parental genotypes for hybrid production and rapid production of asexually propagated crops. Being labor intensive, this method involves several in vitro steps and acclimatization problems and, therefore, extensive costs. A highly mechanized, direct greenhouse or field planting system that would enable production cost to be comparable to that of seeds can be an alternative (Redenbaugh et al., 1986). In some crops, such as tomato, single seeds are difficult to isolate for handling and planting, and hence encapsulated seeds can be useful for handling (Garret et al., 1991). Somatic embryos of celery and cauliflower have been encapsulated as single-embryo beads with subsequent in vitro germination (Redenbaugh et al., 1986). In celery, 10 percent of the beads converted into plants when transplanted into sand trays or transplant plugs. Onishi et al. (1994) reported 50 to 80 percent conversion on an artificial medium (horticubes) with celery encapsulatable units under greenhouse conditions without any supply of carbohydrate and other nutrients.

Micropropagation followed by synthetic seed technique can play a significant role in the propagation of hybrids. It is possible to attain sufficient number in a given period through micropropagation in asexually propagated crops. For example in asparagus and certain cucurbits, tissue culture techniques are being effectively utilized and syn seed protocols can be standardized for micropropagation of elite superior hybrids. In sweet potato, nursery space requirements (one acre of nursery space required for ten acres of field) can be minimized drastically if somatic embryogeny and a syn seed system are effectively utilized (Chee and Cantliffe, 1992).

Application of tissue culture technology for improvement of both temperate and tropical fruits has been extensively worked out, and commercially viable protocols have been established for banana, coffee, strawberry, pineapple, grapes, and apple. However, in many of the fruit and nut crops, although conventional techniques such as grafting, air layering, seed germination, or removal of suckers are routinely used, shortage of elite clonal material cannot meet the demand of planting material. Vegetatively propagated fruit and nut crops with long breeding cycles and those with self-incompatibilities can be good candidates for synthetic seed development. In cases such as grape, where grafting to a rootstock is necessary, syn seeds can be useful for germplasm preservation. Grape is an excellent candidate for testing applications of syn seed for germplasm preservation because embryogenic cultures can be readily established and potential exists for conservation of germplasm (Gray and Compton, 1993).

Mulberry is propagated conventionally by cuttings, and only 30 to 40 percent of cuttings survive the period between pruning, transportation, and final transplantation, whereas the encapsulated axillary buds could be easily packed in small containers and transported, thus limiting space while increasing viability and survival rate (Bapat, 1993).

New and effective means of propagating bananas would be advantageous over the conventional use of sucker material for germplasm maintenance, exchange, and transportation (Rao et al., 1993). Papaya is another valuable fruit crop normally propagated by seeds. Clonal propagation through uniform somatic embryo production and syn seed development can be useful for practical applications in agriculture (Castillo et al., 1998).

Developing protocols for direct in vivo development of encapsulated shoot tips of cardamom may have greater impact on cardamom cultivation (Ganapathi et al., 1994). In the conventional tissue culture, multiple shoots have to be separated singly and put for rooting to achieve root-shoot balance, and the plantlets have to be hardened in the greenhouse before field planting. Syn seeds will provide an ideal delivery system, enabling easy transport as compared to large parcels of seedlings or plantlets. Further, the encapsulated shoot tips can be stored in bottles and transported without loosing viability.

CASE STUDIES

Encapsulation of Shoot Tips and Somatic Embryos of Banana

Banana is an economically profitable fruit crop, and most of the edible triploid bananas are vegetatively propagated using suckers. Alternatively, in vitro cultures can be used for propagating bananas that can be superior to the conventional use of sucker material, for rapid propagation, germplasm maintenance, exchange, and transportation. The encapsulated shoot tips can be handled like a seed and could be useful in minimizing the cost of production, as 1 mL of medium is sufficient for encapsulation of a single shoot tip compared to 15 to 20 mL for conversion of shoot tips into plantlets. By directly sowing the encapsulated shoot tips in soil, the two-stage process such as rooting and hardening can be eliminated. As compared to suckers, encapsulated shoot tips can become inexpensive, easier, and safer material for germplasm exchange, maintenance, and transportation (Rao et al., 1993).

In the author's laboratory, shoot tips excised from the shoot cultures of banana cv. Basrai were encapsulated in 3 percent sodium alginate solution prepared either in distilled water or MS medium with 0.1 percent activated charcoal and an antibiotic mixture (Ganapathi et al., 1992). These encapsulated shoot tips developed into plants on a variety of substrates, including moist cotton supplemented with nutrients. Prior treatment with NAA to induce early rooting has also been found to be effective in improving germination of syn seeds (98 percent) compared to nontreated shoot apices (86.6 percent). Preculture treatment has also led to shoot emergence followed by faster growth of the shoots.

In recent experiments, we studied the effect of double coating on viability of encapsulated shoot apices. After the first coating with sodium alginate in nutrient medium, a second coating was given with gel consisting of an antibiotic (250 $mg \cdot L^{-1}$ timentin). Increased germination (36 percent) and survival up to ten days in case of cv. Basrai and Robusta were observed when the beads were placed under in vivo conditions (Srinivas, 2002). The addition of antibiotic has been suggested to be effective in preventing the microbial growth during the germination of syn seeds (Bapat and Rao, 1990). Encapsulated shoot apices of different banana and plantain cultivars belonging to genomic groups AAA, AAB, AB, ABB responded differently with respect to postencapsulation viability. Cultivars with B genome exhibited better response over those with only A genome (Suprasanna et al., 2001).

Somatic embryos encapsulated in 5 percent sodium alginate and cultured on different culture media showed the emergence of shoots in two to three weeks and complete plantlet development in four weeks (Ganapathi et al., 2001; see Figure 9.3, parts A and B). Although plantlet development was observed on all the culture media, the frequency varied considerably (see Figure 9.4). The inclusion of distilled water or tap water in encapsulation gel affected the plant conversion response. When distilled water was used, only 33 percent mean somatic embryo germination frequency was noted, while with the use of tap water, the frequency rose to 53.3 percent. Among the different media tested, MS medium with sucrose gave the maximum conversion (66 percent) of encapsulated somatic embryos into plantlets; transplantable plantlets developed in six weeks, and the plants were hardened in the greenhouse for about eight weeks. Encapsulated somatic embryos cultured on MS+BAP (4.44 μM) exhibited multiple shoot development (five to seven shoots/embryo), whereas only single shoot emergence was noted in the case of synthetic seeds cultured on NAA or GA_3 added medium.

FIGURE 9.3. Synthetic seeds of banana, rice, and cardamom. A: Encapsulated somatic embryos of banana showing emergence of shoot and root; B: Complete plantlet development from encapsulated somatic embryos of banana; C: Encapsulated somatic embryos of rice; D: Plantlet development from synthetic seeds of rice on MS medium; E: Encapsulated shoot tips of cardamom packed in a plastic bag; F: Developing encapsulated shoot tips on MS medium; G: Hardened plants of cardamom in the greenhouse.

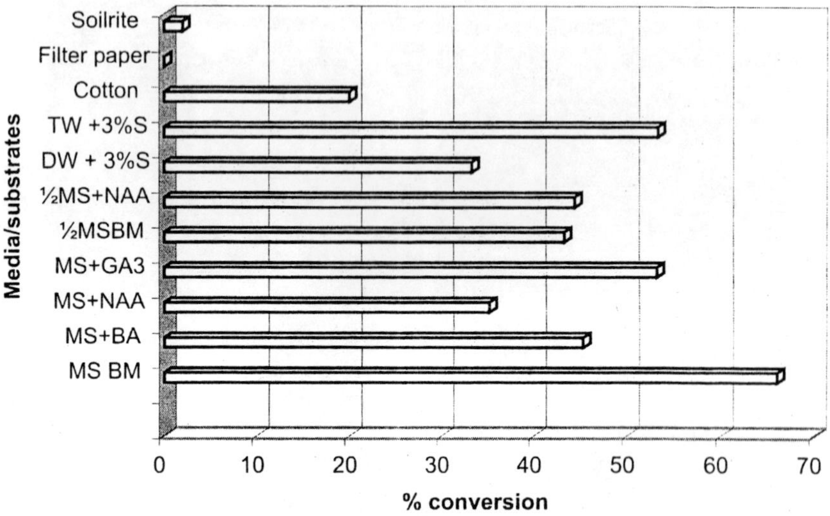

FIGURE 9.4. Effect of different substrates on the conversion of encapsulated somatic embryos of banana.

In a second experiment, encapsulated somatic embryos were cultured on different substrates, including cotton, soilrite, and blotting paper strips moistened with one-fourth strength MS liquid medium, for germination (Ganapathi et al., 2001). Plant conversion from encapsulated embryos was noted on cotton and soilrite (20 percent), but in general the frequency was low as compared to nutrient media (Figure 9.4). The encapsulated somatic embryos placed on blotting paper strips dried completely and did not develop into plantlets.

Inclusion of tap water in the encapsulation mixture exhibited better response over distilled water, presumably because the former has several minerals/salts, which might have contributed to growth in addition to the complexing of the gel. Media substrates, in general, gave higher plant conversion frequencies than cotton, soilrite, and blotting paper. The use of full- strength MS medium yielded the highest plant conversion response (66 percent) as well as shoot and root growth compared to other treatments (Figure 9.5), suggesting that ingredients in the encapsulation mixture contribute as an artificial endosperm to the developing encapsulated somatic embryo (Ganapathi et al., 2001).

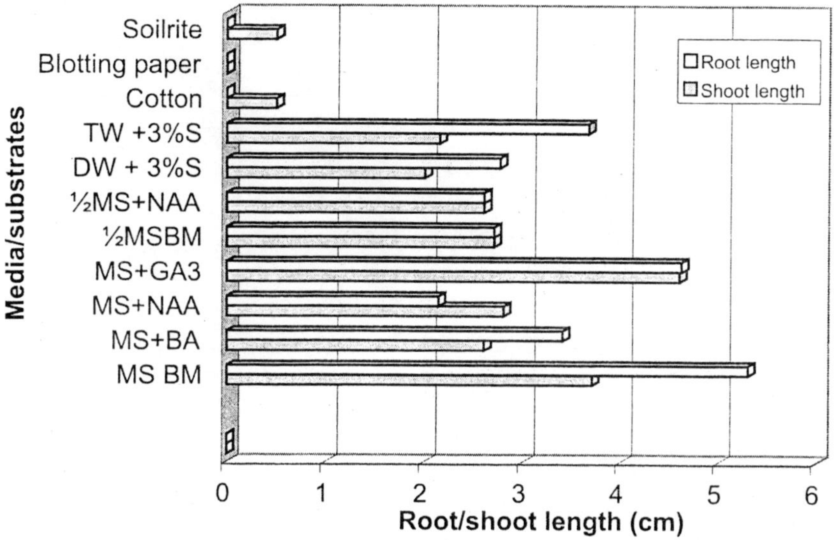

FIGURE 9.5. Development of encapsulated somatic embryos on different sub-strates.

Encapsulation of Shoot Tips of Cardamom

Cardamom is the "queen of spices" and is generally propagated vegetatively as well as through seeds. This crop is highly cross-pollinated, and seed-derived plants show extensive variation. Shoot tips isolated from multiple shoot cultures were encapsulated in 3 percent sodium alginate with different gel matrices (Ganapathi et al., 1994). Maximum conversion of the encapsulated shoot tips into plantlets was observed and the plantlets were grown successfully in soil (Figure 9.6, parts F and G).

Although a large number of plantlets can be produced in tissue culture through multiple shoot induction from shoot tip culture, their delivery is often tedious and cumbersome. By employing cardamom synthetic seeds, the plantlets can be regenerated on different substrates that can eliminate the phase of compulsory rooting of shoots. This will provide an ideal delivery system enabling easy transport as compared to large parcels of seedlings or plantlets. Further, the encapsulated shoot tips can be packed in sterile bags or containers for storage and transport (Figure 9.6, part E).

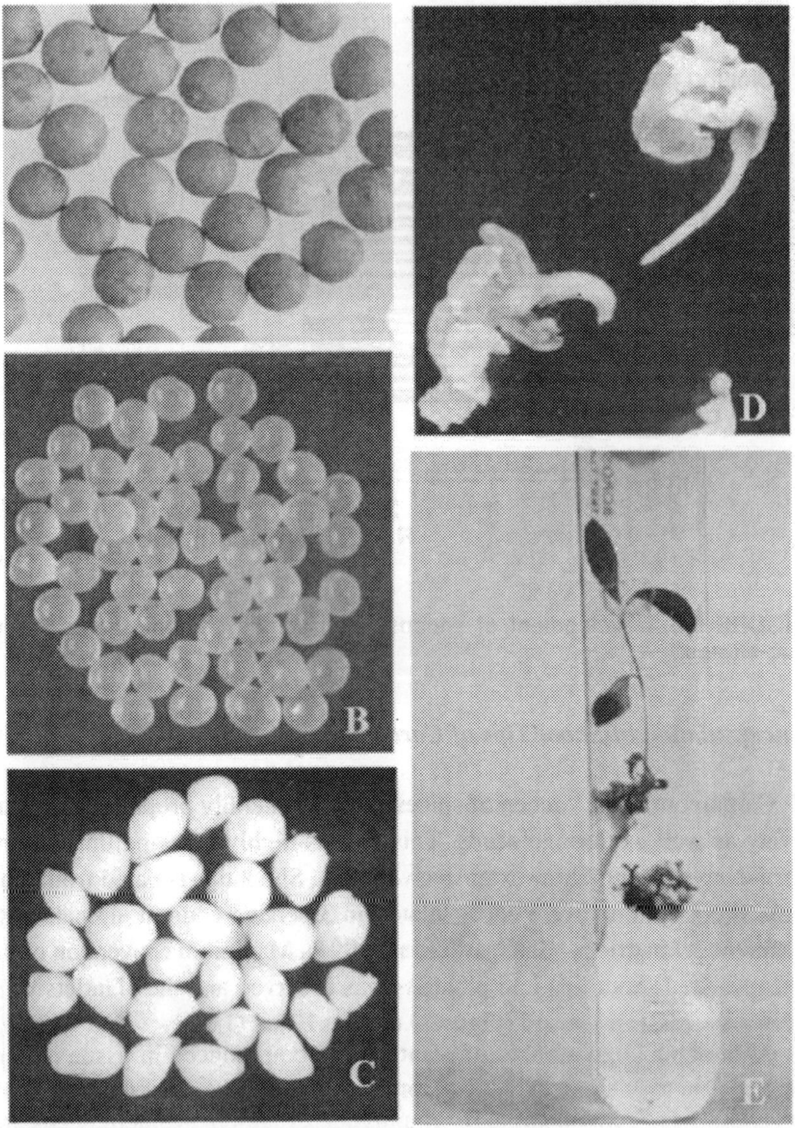

FIGURE 9.6. Plant regeneration for synthetic seeds of sandalwood (*Santalum album* L.). A: Natural seeds; B: Artificial seeds prepared in sodium alginate; C: Artificial seeds prepared in a composite gel; D: Artificial seeds growing on MS medium six weeks after germination; E: Plantlet growing on MS medium six weeks after germination.

Sandalwood Plantlets from Synthetic Seeds

In sandalwood, a commercially valuable forest tree in India, synthetic seeds were prepared by encapsulating somatic embryos in a matrix consisting of 3 percent sodium alginate (Bapat and Rao, 1988). Somatic embryogenesis was induced from internodal segments derived from a 20-plus-year-old tree. Somatic embryos were carefully isolated, blot dried, and then dipped in sodium alginate prepared in MS basal medium. After a brief incubation, the embryos were picked up and then dropped into a solution of calcium chloride. Each drop, consisting of a single somatic embryo, is referred to as a bead akin to normal seeds (see Figure 9.6, parts A and B). These germinated with root and shoot development (see Figure 9.6, parts D and E). The synthetic seeds could be stored at 4°C for 45 days and germinated into plants. The technique exemplifies the potential of the synseeds of sandalwood to revive after encapsulation and also that they can be useful for propagation and storage.

In further experiments on maturation, embryogenic callus and somatic embryos exposed to different levels of abscisic acid and sucrose were allowed to desiccate for 10 to 30 days. Even after drying for 30 days, dried callus showed viability (40 percent) and revival of regeneration (Bapat and Rao, 1992). In order to see if the gelation can contribute to increasing the germination response of synthetic seeds, Priya et al. (1994) studied the effect of simple (sodium alginate) and composite (sodium alginate + silica) gels (Figure 9.6, part C). Composite gel exhibited better germination (96 percent) and percent conversion to plants (27 percent) compared to controls (77 and 10 percent, respectively).

Encapsulation of Somatic Embryos of Rice

In cereals such as rice, syn seed technology has vast potential for exploitation of hybrid vigor, as it would permit multiplication and large-scale propagation of superior rice hybrids. Embryogenic rice callus with ability to regenerate into plantlets can be a suitable system for immobilization, as the callus can be stored at low temperatures. Moon et al. (1999) developed callus immobilization in polyurethane foam and found that immobilized callus maintained high regeneration because shear stress and hydrodynamic damage were avoided. Although embryogenic callus can be immobilized, problems exist for long-term culturability and/or treatments with growth regulators for further somatic embryo development. Somatic embryos with root and shoot axis are a better choice, as the encapsulated embryos can be placed directly on a germination medium for plant development without need for treatment with growth regulators.

In this direction, we proposed encapsulation of somatic embryos with a scutellum and coleoptile to develop artificial seeds in rice (Suprasanna et al., 1996). Embryogenic callus was established from mature seeds, and the resulting whitish, compact embryogenic callus was transferred to an embryogenic induction medium consisting of abscisic acid, kinetin, mannitol, and casein hydrolysate (Suprasanna et al., 1996). The embryos were isolated, were blot dried on a sterile filter sheet, mixed in a gel, and encapsulated in 3 percent sodium alginate prepared either in MS or White's medium (see Figure 9.3, part C). Their further development into plants was demonstrated on a variety of substrates including MS or White's media, cotton, and filter paper irrigated with liquid nutrient media (see Figure 9.3, part D). Encapsulated embryos germinated well into plants with better root and shoot system compared to nonencapsulated embryos. Further research is needed on optimization of large-scale production of somatic embryos in bioreactors and plant conversion and direct in vivo planting of rice synthetic seeds. This will have implications for delivery of the germplasm and for preservation of improved genotypes.

Encapsulation of Somatic Embryos of Groundnut

Groundnut (*Arachis hypogaea* L.) is an oil seed crop with high oil-yielding potential. Somatic embryogenesis has been established in several genotypes, and several laboratories have employed this system for genetic manipulation studies. Encapsulation of somatic embryos for the preparation of synthetic seeds has so far not been attempted. In a first report, Padmaja et al. (1995) demonstrated the potential of this technique. Synthetic seeds were developed by encapsulating 30-day-old cotyledonary-stage embryos in 2.5 percent sodium alginate (Figure 9.7, parts A and B). These beads were tested for germination and plant development on a variety of media with different concentrations of sucrose, maltose, BAP, and NAA. The highest germination response was noted on half-strength MS medium with sucrose and maltose at 1 percent (see Figure 9.7, parts C and D). The beads showed low conversion (8.2 percent) upon storage at 4°C for about 40 days. Germinated synthetic seeds developed into complete plants, and the plants were grown in the greenhouse with a survival rate of 54 percent.

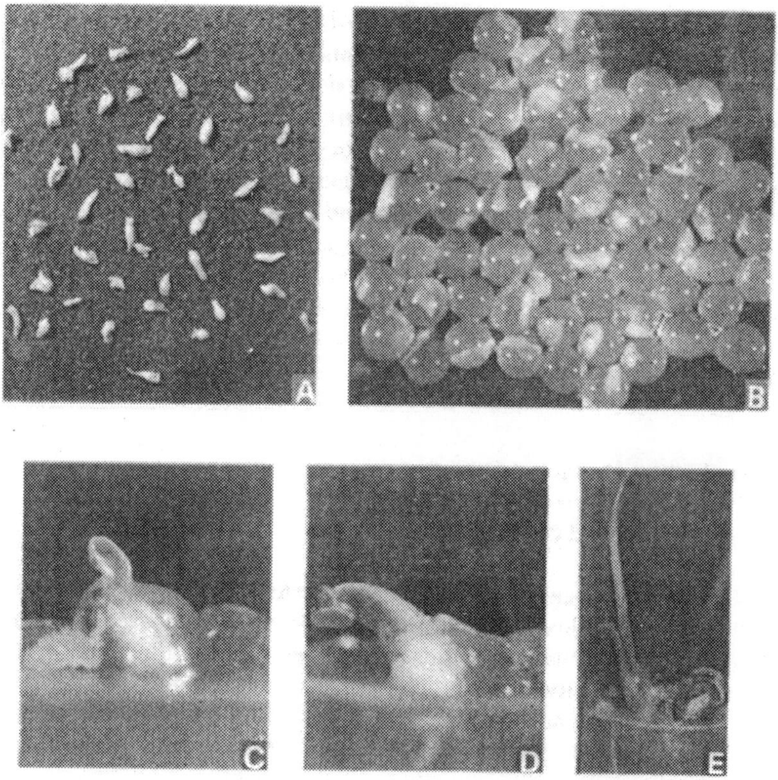

FIGURE 9.7. Plant regeneration for synthetic seeds of groundnut (*Arachis hypogaea* L.). A: Isolated immature embryo derived somatic embryos of groundnut cv. ICG221; B: Somatic embryos of groundnut encapsulated in alginate matrix prepared in distilled water; C: Emerging shoot and root from synthetic seeds on MS medium with 1 percent sucrose and 1 percent maltose 8 days after planting; D: Emerging shoot and root from synthetic seeds on MS medium with 1 percent sucrose and 1 percent maltose 20 days after planting; E: Developed plant originating from synthetic seed (60 days after planting).

CRITICAL CONSIDERATIONS

Somatic Embryogeny

Among the propagation methods practiced by tissue culturists, somatic embryogenesis occupies a prime position as it possesses advantages over others for automation that include growth of a large number of embryogenic cells, sorting of embryos, and encapsulation (Redenbaugh et al., 1988; Suprasanna and Rao, 1997). Embryogenesis is an important phase in plant development, representing a unique feature of plant cells in culture to develop into a complete plant through a process known as somatic embryogenesis. Somatic embryogenesis recapitulates zygotic embryogenesis with the production of embryos having shoot and root apices. Any somatic tissue when cultured under appropriate nutrient and hormonal conditions can be induced to undergo somatic embryogenesis to form somatic embryos which subsequently germinate into complete plants. Zygotic and somatic embryogenesis share several similarities in their developmental pattern at the morphological, biochemical, and molecular levels.

Somatic embryogenesis mainly involves three stages:

1. induction of embryogenic competence
2. somatic embryo development
3. maturation and conversion of embryos

In the induction process embryos can originate directly from the cultured explants or from callus. It is necessary in some cases to modify the medium and transfer of cultures to an auxin-free medium to trigger the process. Once embryo induction phase is achieved, several strategies are adopted for further proliferation and development. Developing embryos are taken out from the induction medium and good-quality embryos are selected for continued growth to plantlet regeneration. Regulation of embryogenic cultures is often exercised by using either physical separation of embryogenic tissues to yield uniformly sized cell mass that favors embryo development, or use of growth regulators to synchronize the growth of embryogenic cells (Bapat and Rao, 1996).

Desiccation Tolerance

The process of maturation and desiccation that is prevalent during zygotic embryogenesis is also responsible for the development of somatic embryos (Bapat and Rao, 1996). Zygotic embryos pass through a maturation phase in which sufficient nutrients are accumulated that are necessary for germination. This is followed by a desiccation period that invariably helps embryos to withstand the harsh environment until the seed gets the proper environment for germination. The poor conversion of somatic embryos to plants in many cases has been attributed to not subjecting embryos to maturation and desiccation process by proper manipulation of several factors that operate at this stage (Bornman, 1993). Desiccation tolerance in developing somatic embryos may be induced by such exogenous signals in the tissue culture medium such as addition of osmotically active substances, ABA, PEG, sucrose, and chilling stress which have been effective for the maturation and stimulation of embryo-plant conversion in several plant species (Lecoutex et al., 1993). The most effective factor that has been successfully tested is the incorporation abscisic acid into the medium. In zygotic embryos the concentration of ABA is always very high and then declines when the embryo embarks upon germination. Addition of abscisic acid followed by desiccation has increased frequency of germination of somatic embryos in several plants (Nieves et al., 2001).

According to Gray and Purohit (1991), there are two possible causes of dormancy in somatic embryos:

1. dormancy that is typical for zygotic embryos may also be expressed by somatic embryos of the same species in vitro and
2. dormancy may be artificially induced by aspects of the culture environment such as exogenously supplied growth hormones.

Induction of quiescence in rapidly growing somatic embryos by lowering water content is logical since dehydration and rehydration cause arrest and resumption of growth in natural seeds. Parrott et al. (1988) found that desiccation promoted rapid and uniform germination of somatic embryos if they were allowed to mature to a minimum age on a basal medium. Inclusion of a desiccation period after embryo maturation resulted in a fourfold increase in the percentage of embryos that converted to plants in peanut. Besides improvement in germination, desiccation of whole somatic embryos is also an alternative method of germplasm storage. Somatic embryos are produced continuously year-round, and these could therefore be dried and stored until the appropriate season or shipped to new locations.

Large-Scale Production and Handling

In order to be considered as a choice of material for mass production of somatic embryos, synthetic seeds should necessarily meet production and high conversion ability and scale up of culture system, and process integration between somatic embryogenesis and encapsulation. The process of somatic embryogenesis is very convenient for automation. All the phases of embryogenic stages have been automated (Cerevelli and Senaratna, 1995). Specific bioreactors have been designed for initiation and multiplication of the embryonic cells. The advanced technique of computer-aided image analysis has also been attempted for sorting out the right type of embryos (Kurata, 1995). Sieving and filtration of embryonic suspension is another way of isolating uniform-type embryos. The plant should have potentiality to generate a large number of embryos, and embryos should not loose their viability during scaling up and sustain the mechanization process.

Because bulk handling, automation, or other techniques are inherent in the high labor costs involved with somatic embryogeny, artificial seeds offering ease in handling and storage would be an advantage. However, constraints such as scaling up of embryo production, maturation, and conversion are still critical for the realization of syn seed technology. In lieu of artificial seed, a propagation system based on in vitro germination and delivery of bare-root emblings to soil was employed. In study with interior spruce, somatic embryos were allowed to germinate on substrates (agar, Oasis foam, Grow-sticks, Sorbrods) and the bare-root emblings were transplanted directly into the soil by the mechanized planting system.

In Vivo Conversion and Field Planting

One climax of synthetic seed research is the demonstration of plant conversion from encapsulated units ex vitro directly under field conditions. This is often a critical point of intense research. Syn seeds provide a novel method to deliver tissue-cultured plant material directly to the greenhouse or field. Poor development of encapsulated propagules under soil conditions has often been attributed to low conversion of embryos and improper maturation of the embryo. In alfalfa, encapsulated and naked somatic embryos were planted directly in the field with row coverings either with plastic or cloth to protect the somatic embryos during conversion. Compared to 60 percent obtained with naked embryos in a potting mix, encapsulated embryos converted with 40 percent. Under the field conditions, the frequencies were 9 (cloth) and 13 percent (plastic) with naked embryos compared to 14 and 5 percent, respectively, with the encapsulated embryos. Onishi et al. (1994) demonstrated successful sowing of encapsulated carrot somatic embryos on humid soil in the greenhouse. In their study, novel self-breaking gel beads were employed with sustained release of microcapsules for carrot embryos, at a rate of 80,000 per day. Encapsulated embryos showed 52 percent conversion frequency. Mamiya and Sakamoto (2001) developed a method to produce encapsulatable units in *Asparagus officianalis* L. Somatic embryos with a preencapsulation treatment with plant growth regulators (IAA) produced compact encapsulatable units, which converted with 72 percent frequency in nonsterile soil.

Artificial seed technology can provide one of the most valuable opportunities in case of forest crops because of the need for vegetative propagation and lack of other cost-effective methods for plantation. Maruyama and Kinoshita (1997) reported good conversion rates from encapsulated shoot tips of *C. odorata, G. crinata,* and *J. mimosaefolia* grown on nonsterilized soil. Compared to the single coating for the shoot tips that gave poor conversion rates (up to 6.7 percent), shoot tips with double-layered coatings gave about 30 percent conversion. Further, these authors suggested that growing of the encapsulated shoot tips on nutrient medium for one week before transfer to nonsterilized soil enhanced the plant conversion rates to 100 percent. It is possible that during the early period of growth the beads require nutrients for promoting swift growth process, and this can help them upon transfer to soil conditions.

Preencapsulation culture of somatic embryos or other propagules is often necessary to produce syn seeds that can convert or germinate at high frequency. Cold pretreatment of microcuttings of kiwifruit provided bud vigor and subsequent conversion (Adriani et al., 2000). Sugar has been found to be important during the conversion process in asparagus (Mamiya and Sakamoto, 2001). Sucrose increases during some steps of the protocol can have profound influence on conversion; however, sucrose addition in a nonsterile environment can be extremely crucial for plant survival in a competitive environment with microbes (Bapat and Rao, 1990; Standardi and Piccioni, 1998; Adriani et al., 2000). Root development is an essential growth phase to obtain a high survival rate in the greenhouse (Conner et al., 1992).

It is also possible to germinate syn seeds in a sugar-free media and its exclusion can have advantages of eliminating microbial contamination. In addition, sucrose in the inner alginate matrix could result in a high osmotic pressure, inhibiting the germination process by limiting water availability (Tan et al., 1998). Sodium alginate-chitosan encapsulated orchid *(Sphathoglottis pliccata)* seeds infected with mycorrhizal fungus *Rhizoctonia* AM9 developed into protocorms (Tan et al., 1998). Although the results did not show much growth in terms of plant development, the study is a first step in the direction toward growing syn seeds in a nonsterile environment.

LIMITATIONS AND PROSPECTS

Several factors influence the commercialization of somatic embryogenesis and synthetic seed technology. Murashige (1977) observed that to be applicable, a syn seed cloning method should be extremely rapid, capable of producing millions of plants, and competitive with the seed-based method. It is also often difficult to compare the seed-based and syn-seed-based methods, as economic viability still is not completely determined with the latter technology. Somatic embryogenesis can be compared with other methods of vegetative propagation in terms of production of a large number of propagules; however, the problems of reliability, somaclonal variation, and development of quality plant material remain to be resolved as somatic embryogenesis and consequently syn seed methods are employed in propagation and storage of germplasm (Cerevelli and Senaratna, 1995).

Practical implementation of syn seed technology is also constrained by factors intrinsic to the developmental systems for in vitro propagules including somatic embryos and processes leading to their growth and plant regeneration. These could be

1. production of viable propagules or encapsulatable units,
2. synchronization of the developmental process of somatic embryos,
3. maturation and germination of somatic embryos, and
4. lack of dormancy and stress tolerance associated with somatic embryos (Bapat and Rao, 1996; Ara et al., 2000).

Syn seed technique has immense potential for storage and exchange of germplasm. Further, syn seeds are useful for the clonal propagation of high-value transgenic plants or for the preservation of elite and unique plant material. Even in the absence of proper in vivo growth, growth under in vitro conditions before greenhouse planting and distribution can enable a syn-seed system to be an effective conduit for bulk transportation of germplasm. Syn-seed technology is almost two decades old, and there have been numerous reports in different crop plants. In the near future, it is hoped that limitations of production, viability, and conversion might be overcome and that the technology will eventually find a place in commercial micropropagation of important and high-value plant species.

REFERENCES

Adriani, M., E. Piccioni, and A. Standardi (2000). Effect of different treatments on the conversion of 'Hayward' kiwifruit synthetic seeds to whole plants following encapsulation of in vitro derived bulbs. *New Zealand Journal Crop and Horticulture Science* 28: 59-67.

Ahuja, P.S., J. Mathur, N. Lal, A. Mathur, and A.K. Kukreja (1989). Toward developing artificial seeds by shoot bud encapsulation. In A.K. Kukreja, A.K. Mathur P.S. Ahuja, and R.S. Thakur (eds.), *Tissue culture and biotechnology of medicinal and aromatic plants* (pp. 20-78). Lucknow, India: CIMAP.

Antonietta, G.M., P. Emanuelle, and S. Alvarro (1998). Effects of encapsulation on *Citrus reticulata* Blanco somatic embryo conversion. *Plant Cell, Tissue & Organ Culture* 55(3): 235-237.

Ara, H., U. Jaiswal, and V.S. Jaiswal (1999). Germination and plantlet regeneration from encapsulated somatic embryos of mango (*Mangifera indica* L.). *Plant Cell Reports* 19: 166-170.

Ara, H., U. Jaiswal, and V.S. Jaiswal (2000). Synthetic seed: Prospects and limitations. *Current Science* 78(12): 1438-1444.

Arditti, J., A.P. Oliva, and J.D. Michaud (1982). Practical germination of North American and related orchids: II. *Goodyera oblongifolia* and *G tesselata. American Orchid Society Bulletin* 51: 394-397.

Arya, K.R. and M.U. Beg (1998). Microcloning and propagation of endosulphan tolerant genotypes of mustard *(Brassica campestris)* through apical shoot bud encapsulation. *Indian Journal Experimental Biology* 36: 1162-1164.

Attree, S.M., M.K. Pomeroy, and L.C. Fowke (1994). Production of vigorous disiccation tolerant white spruce [*Picea glauca* (Moench.) Voss.] synthetic seeds in a bioreactor. *Plant Cell Reports* 13: 601-606.

Bapat, V.A. (1993). Studies on synthetic seeds of sandalwood and mulberry. In K. Redenbaugh (ed.), *Synthetic seeds: Applications of synthetic seeds to crop improvement* (pp. 381-384). Boca Raton, FL: CRC Press.

Bapat, V.A., M. Mhatre, and P.S. Rao (1987). Propagation of *Morus indica* L. (mulberry) by encapsulated shoot buds. *Plant Cell Reports* 6: 393-395.

Bapat, V.A. and P.S. Rao (1988). Sandalwood plantlets from synthetic seeds. *Plant Cell Reports* 7: 434-436.

Bapat, V.A. and P.S. Rao (1990). In vivo growth of encapsulated axillary buds of mulberry (*Morus indica* L.). *Plant Cell Tissue Organ Culture* 20: 69-70.

Bapat, V.A. and P.S. Rao (1992). Plantlet regeneration from encapsulated and nonencapsulated desiccated somatic embryos of a forest tree: Sandalwood. *Journal Plant Biochemistry & Biotechnology* 1: 109-113.

Bapat, V.A. and P.S. Rao (1996). Maturation and desiccation. In L.K. Pareek (ed.), *Trends in plant tissue culture and biotechnology* (pp. 53-64). Jaipur, India: Agrobotanikal Publishers.

Barbotin, J.N., J.E.N. Saucedo, C. Bazinet, A. Kersulec, V. Thomasset, and D. Thomas (1993). Immobilization of whole cells and somatic embryos: Coating process and cell matrix interactions. In K. Redenbaugh (ed.), *Synthetic seeds: Applications of synthetic seeds to crop improvement* (pp. 65-103). Boca Raton, FL: CRC Press.

Bornman, C.H. (1993). Maturation of somatic embryos. In K. Redenbaugh (ed.), *Synthetic seeds: Applications of synthetic seeds to crop improvement* (pp. 105-113). Boca Raton, FL: CRC Press.

Buitelaar, R.M., A.C. Hulst, and J. Tramper (1988). Immobilization of biocatalysts in thermogels using the resonance nozzle for rapid drop formation and an organic solvent for gelling. *Biotechnology Techniques* 2: 109.

Capuana, M. and P.C. Debergh (1997). Improvement of the maturation and germination of horse chestnut somatic embryos. *Plant Cell, Tissue Organ Culture* 48: 23-29.

Capuano, G., E. Piccioni, and A. Standardi (1998). Effect of different treatments on the comversion of M26 apple rootstock synthetic seeds obtained from encapsulated apical and axillary micropropagated buds. *Journal of Horticultural Science* 73: 299-305.

Castillo, B., M.A.L. Smith, and U.L. Yadava (1998). Plant regeneration from encapsulated somatic embryos of *Carica papaya* L. *Plant Cell Reports* 17(3): 172-176.

Cervelli, R. and T. Senaratna (1995). Economic aspects of somatic embryogenesis. In J. Atkin-Christie, T. Kozai, and M.P.L. Smith (eds.), *Automation and environmental control in plant tissue culture* (pp. 29-64). Dordrecht, the Netherlands: Kluwer Academic Publ.

Chee, R.P. and D.J. Cantliffe (1992). Improved production procedures for a somatic embryos of sweet potato for a synthetic seed system. *Horticulture Science* 27(12): 1314-1315.

Chen, R.R., J.T. Zhang, L.B. Ping, S.S. Guo, and J.P. Hao (1995). Somatic embryogenesis and synthetic seeds in coriander *(Corindrum sativum)*. In Y.P.S. Bajaj (ed.), *Biotechnology in agriculture and forestry*, Volume 31: *Somatic embryogenesis and synthetic seeds II* (pp. 334-342). Berlin: Springer Verlag.

Choi, Y.E. and J.H. Jeong (2002). Dormancy induction of somatic embryos of Siberain ginseng by high sucrose concentrations enhances the conservation of hydrated artificial seeds and dehydration resistance. *Plant Cell Reports* 20: 1112-1116.

Conner, A.J., D.J. Abernthy, and P.G. Falloon (1992). Importance of in vitro storage root development for the successful transfer of micropropagated asparagus plants to greenhouse conditions. *New Zealand Journal Crop Horticulture Science* 20: 477-481.

Corrie, S. and P. Tandon (1993). Propagation of *Cymbidium giganteum* Wall. through high frequency conversion of encapsulated protocorms under in vivo and in vitro conditions. *Indian Journal Experimental Biology* 31: 61-64.

Danso, K.E. and B.V. Ford-Lloyd (2003). Encapsulation of nodal cuttings and shoot tips for storage and exchange of cassava germplasm. *Plant Cell Reports* 21: 718-725.

Datta, S.K. and Potsykus (1989). Artificial seeds in barley: Encapsulation of microspore-derived embryos. *Theoretical and Applied Genetics* 77: 820-824.

Dupuis, J.M., C. Roffat, R.T. DeRose, and F. Molle (1994). Pharmaceutical capsules as coating system for artificial seeds. *Bio/Technology* 12: 385-389.

Fabre, J. and J. Dereuddre (1990). Encapsulation-dehydration, a new approach to cryopreservation of *Solanum* shoot tips. *CryoLetters* 11: 413-426.

Fujii, J., D. Slade, R. Olsen, S.E. Ruzin, and K. Redenbaugh (1992). Field planting of alalfa artificial seed. *In Vitro Cellular & Developmental Biology-Plant* 28: 73-80.

Fujii, J., D. Slade, and K. Redenbaugh (1989). Maturation and green house planting of alfalfa synthetic seeds. *In Vitro Cellular & Developmental Biology-Plant* 25: 1179-1182.

Fukai, S., M. Togashi, and M. Goi (1994). Cryopreservation of in vitro grown *Dianthus* by encapsulation-dehydration. *Technical Bulletin Faculty of Agriculture* Kagawa Univ. (Japan) 46(2): 101-107.

Furamanowa, M., D. Sowinska, and A. Pietrosiuk (1991). *Carum carvi* L. (caraway): In vitro culture, embryogenesis and the production of aromatic compounds. In Y.P.S. Bajaj (ed.), *Biotechnology in agriculture & forestry,* Volume 15: *Medicinal & aromatic plants III* (pp. 176-192). Berlin, Heidelberg: Springer Verlag.

Gamborg, O.L. (2002). Plant tissue culture, biotechnology, milestones. *In Vitro Cellular & Development Biology-Plant* 38(2): 84-92.

Ganapathi, T.R., V.A. Bapat, and P.S. Rao (1994). In vitro development of encapsulated shoot tips of cardamom. *Biotechnology Techniques* 8(4): 239-244.

Ganapathi, T.R., L. Srinivas, P. Suprasanna, and V.A. Bapat (2001). Regeneration of plants from alginate-encapsulated somatic embryos of banana cv. Rasthali (*Musa* spp. AAB group). *In Vitro Cellular & Developmental Biology.-Plant* 37: 178-181.

Ganapathi, T.R., P. Suprasanna, V.A. Bapat, and P.S. Rao (1992). Propagation of banana through encapsulated shoot tips. *Plant Cell Reports* 11: 571-575.

Garret, R.E., J.J. Mehlschau, N.E. Smith, and M.K. Redenbaugh (1991). Encapsulation of tomato seeds. *American Society Agriculture & Engineering* 7(1): 25-31.

Gemeiner, P., L. Kurillova, O. Markovic, A. Malovikova, D. Uhrin, M. Ilavsky, V. Stefuca, M.Polakovic, and V. Bales (1991). Calcium pectate gel beads for cell entrapment: 3. Physical properities of calcium pectate and calcium alginate gel beads. *Biotechnology Applied Biochemistry* 13(3): 335-345.

George, L. and S. Eapen (1995). Encapsulation of somatic somatic embryos of finger millet (*Eleusine coracana* Gaertn.). *Indian Journal Experimental Biology* 33: 291-293.

Ghosh, B. and S. Sen (1994). Plant regeneration from alginate encapsulated somatic embryos of *Asparagus cooperi* Baker. *Plant Cell Reports* 13: 381-385.

Gill, R., T. Senratna, and P.K. Saxena (1994). Thidiazuron induced somatic embryogenesis enhances viability of hydrogel encapsulated somatic embryos of geranium. *Journal Plant Physiology* 143(6): 726-729.

Gray, D.J. (1990). Synthetic seed technology for clonal production of crop plants. In R.D Taylorson (ed.), *Recent advances in development and germination of seeds* (pp. 29-45). New York: Plenum Press.

Gray, D.J. and M.E. Compton (1993). Grape somatic embryo dormancy and quiescence: Potential of dehydrated synthetic seeds for germplasm conservation. In K. Redenbaugh (ed.), *Synthetic seeds: Applications of synthetic seeds to crop improvement* (pp. 381-384). Boca Raton, FL: CRC Press.

Gray, D.J. and A. Purohit (1991). Somatic embryogenesis and development of synthetic seed technology. *Critical Review Plant Science* 10(1): 33-61.

Gupta, P.K., D. Shaw, and D.J. Durjan (1987). Lobolly pine: Micropropagation, somatic embryogenesis and encapsulation. In J.M. Donja and D.J. Durjan (eds.), *Cell and tissue culture in forestry* (pp. 101-108). Dordrecht, the Netherlands: Martinus Nijhoff Publ.

Hasan, S.M.Z. and H. Takagi (1995). Alginate-coated nodal segments of yam (*Dioscorea* spp.) for germplasm exchange and distribution. *Plant Genetic Resources Newsletter* 103: 32-35.

Hatanaka, T., T. Yasuda, T. Yamaguchi, and A. Sakai (1994). Direct regrowth of encapsulated somatic embryos of coffee (*Coffee canephora)* after cooling in liquid nitrogen. *CryoLetters* 15: 47-52.

Honda, H., C. Liu, and T. Kobayashi (2001). Large scale plant micropropagation. *Advances Biochemical/Bioengineering* 72: 158-182.

Janeiro, L.V., A. Ballester, and A.M. Vieitez (1997). In vitro response of encapsulated somatic embryos of *Camellia. Plant Cell Tissue Organ Culture* 51: 119-125.

Jansson, P.E., B. Lindberg, and P.A. Sanford (1983). Structural studies of gellan gum, an extra cellular polysaccharide elaborated by *Pseudomonas elodea. Carbohydrate Research* 124: 135.

Kamalakannan, R., B.V. Narmatha, and L. Jeyekudi (1999). Plantlet regeneration from encapsulated protocorms of the endemic orchid *Celogyne odeoratissima* var. *Angustifolia* Lindl. *Journal Phytology Research* 12(1-2): 21-23.

Ke, S., Y. Gui, and R.M. Skirvin (1995). Somatic embryogenesis and synthetic seeds in *Coptis chinensis*. In Y.P.S. Bajaj (ed.), *Biotechnology in agriculture and forestry*, Volume 31: *Somatic embryogenesis and synthetic seeds II* (pp. 323-333). Berlin: Springer Verlag.

Kitto, S. and J. Janick (1980). Water soluble resins as artificial seed coats. *Horticulture Science* 15: 439.

Kitto, S. and J. Janick (1985). A citrus embryo assay to screen water soluble resins as synthetic seed coats. *Horticulture Science* 20: 98.

Klein, J. and P. Schara (1980). Entrapment of living microbial cells in covalent polymeric networks: I. Preparation and properties of different networks. *Journal Solid-Phase Biochemistry* 5: 61.

Kurata, K. (1995). Automated systems for organogenesis. In J. Aitken-Chirstie, T. Kozai, and M.A.L. Smith (eds.), *Automation and environmental control in plants* (pp. 187-214). Dordrecht, the Netherlands: Kluwer.

Lai, F.M., C.G. Lecoutex, and B.D. McKersie (1995). Germination of alfalfa (*Medicago sativa*) seeds and somatic embryos: I. Mobilization of storage reserves. *Journal Plant Physiology* 145: 507-513.

Le coutex, C., F.M. Kai, and B.D. McKersie (1993). Maturation of somatic embryos by abscisic acid, sucrose and chilling. *Plant Science* 94: 207-213.

Lewis, J.G., N.F. Stanley, and G.G. Guist (1990). Commercial production and applications of algal hydrocolloids. In C.A. Lembai and J.R. Walland (eds.), *Algae and human affairs* (pp. 205-235). Cambridge, UK: Cambridge University Press.

Lulsdorf, M.M., T.E. Tautorus, S.I. Kikcio, T.D. Bethune, and D.I. Dunston (1993). Germination of encapsulated embryos of interior spruce (*Picea glauca engelmannii* complex) and black spruce (*Picea mariana* Mill.). *Plant Cell Reports* 12: 385-389.

Malbadi, R.B. and K. Nataraja (2002a). In vitro storage of synthetic seeds in *Clitoria ternatea* Linn. *Phytomorphology* 52(2-3): 231-237.

Malbadi, R.B. and K. Nataraja (2002b). Large scale production storability of encapsulated somatic embryos of mothbean (*Vigna aconitifolia* Jacq.). *Journal Plant Biochemistry and Biotechnology* 11(1): 61-64.

Malemngaba, H., B.K. Roy, S. Bhattacharya, and P.C. Deka (1996). Regeneration of encapsulated protocorms of *Phaius tankervillae* stored at low temperature. *Indian Journal Experimental Biology* 34: 801-805.

Mamiya, K. and Y. Sakamoto (2001). A method to produce encapsulatable units for synthetic seeds in *Asparagus officianalis*. *Plant Cell Tissue Organ Culture* 64: 27-32.

Mandal, J., S. Patnaik, and P.K. Chand (2000). Alginate encapsulation of axillary buds of *Occimum americanum* L. (hoary basil), *O bacillicum* L. (sweet basil), *O. gratissimum* L (shrubby basil) and *O. sanctum* L. (sacred basil). *In Vitro Cellular & Developmental Biology-Plant* 36(4): 287-292.

Marsolais, A.A., D.P.M. Wilson, M.J. Tsugita, and T. Senratna (1991). Somatic embryo and artificial seed production in zonal *(Pelargonium × Hortorum)* and regal *(Pelargonium × domesticum)* geranium. *Canadian Journal Botany* 69: 1188-1193.

Martinsen, A., G. Skjak-Braek, and O. Smidsrod (1989). Alginate as immobilization material: I. Correlation between chemical and physical properties of alginate gel beads, *Biotechnology and Bioengineering* 33: 79.

Maruyama, E. and I. Kinoshita (1997). High plant recovery from double layered synthetic seeds. *Silvae Genetica* 46(1): 17-23.

Mathur, A., P.S. Ahuja, N. Lal, and A.K. Mathur (1989). Propagation of *Valeriana wallichii* DC. Using encapsulated apical and axial shoot buds. *Plant Science* 60: 111-116.

McKersie, B.D. and S.R. Bowley (1998). Somatic embryogenesis: Forage improvement using synthetic seeds and plant transformation. In E.C. Brummer, N.S. Hill, and C. Roberts (eds.), *Molecular and cellular technologies for forage improvement,* Volume 26 (pp. 117-134). Madison, WI: Crop Science Society of America.

Moon, K.H., H. Honda, and T. Kobayashi (1999). Development of a bioreactor suitable for embryogenic rice callus culture. *Journal Bioscience Bioengineering.* 87(5): 661-665.

Muralidharan, E.M. and A.F. Mascarenhas (1995). In S. Jain, P.K. Gupta, and R.J. Newton (eds.), *Somatic embryogenesis in woody plants* (pp. 101-108). Dordrecht, the Netherlands: Kluwer Academic Publishers.

Murashige, T. (1977). Plant cell and organ cultures as horticultural practices. *Acta Hcrticulturae* 78: 17.

Nakashimada, Y., N. Uozumi, and T. Kobayashi (1995). Production of plantlets for use as artificial seeds from horseradish hairy roots fragmented in a blender. *Journal Fermentation Bioengineering* 79: 458-464.

Nieves, N., J.C. Lovenzo, A. Blanco Mania de Los, J. Gonalez, H. Perlta, M. Hernandez, R. Santos, O. Concecion, C.A. Borroto, E. Borrto, et al. (1998). Artificial endosperm of *Cleopatra tangerine* zygotic embryos—A model for somatic embryos encapsulation. *Plant Cell Tissue Organ Culture* 54(2): 77-83.

Nieves, N., M.E. Martinez, R. Castillo, M.A. Blanco, and J.L. Gonzalez-Olmedo (2001). Effect of abscisic acid and jasmonic acid on partial desiccation of encapsulated somatic embryos of sugarcane. *Plant Cell Tissue Organ Culture* 65: 15-21.

Niino, T. and A. Sakai (1992). Cryopreservation of alginate coated in vitro grown shoot tips of apple, pear and mulberry. *Plant Science* 87: 199-206.

Norton, S. and C. Lacroix (1990). Gellan gum as entrapment matrix for high temperature fermentation processes: A rheological study. *Biotechnology Techniques* 4: 351.

Onay, A., C. Jeffrey, and M.M. Yeoman (1996). Plant regeneration from encapsulated embryoids and embryogenic mass of pistachio, *Pistacia vera* L. *Plant Cell Reports* 15: 723-726.

Onishi, N., Y. Sakamoto, and T. Hirosawa (1994). Synthetic seeds as an application of mass production of somatic embryos. *Plant Cell Tissue Organ Culture* 39: 137-145.

Padmaja, G., L.R. Reddy, and G.M. Reddy (1995). Plant regeneration from synthetic seeds of groundnut, *Arachis hypogaea* L. *Indian Journal Experimental Biology* 33: 967-971.

Parrott, W.A., G. Dryden, S. Vogt, D.F. Hildebrand, B.G. Collins, and E.G. Williams (1988). Optimizing somatic enbryogenesis and embryo germination in soybeans *In Vitro Cell and Developmental Biology* 24: 817-820.

Patel, A., I. Pusch, G. Mix-Wagner, and K.D. Vorlop (2000). A novel encapsulation technique for the production of artificial seeds. *Plant Cell Reports* 19: 868-874.

Patnaik, S.K. and P.K. Chand (2000). Morphogenic response of the alginate encapsulated axillary buds from in vitro shoot cultures of six mulberries. *Plant Cell Tissue Organ Culture* 60(3): 177-185.

Patnaik, S.K., Y. Sahoo, and P.K. Chand (1995). Efficient plant retrieval from alginate encapsulated vegetative buds of mature mulberry trees. *Scientia Horticulture* 61: 227-239.

Piccioni, E., A. Barraccia, M. Falcanelli, and A. Standardi (1997). Estimating alfalfa somaclonal variation in axillary branching propagation and indirect somatic embeyrogenesis by RAPD fingerprinting. *International Journal Plant Science* 158: 556-562.

Piccioni, E., E. Gasbarro, and A. Standardi (1992). Preliminary experiments on encapsulation of in vitro propagules of *Lillium* and M.27. *Annals Facolta di Agraria* XLVI: 357-371.

Piccioni, E. and A. Standardi (1995). Encapsulation of micropropagated buds of six woody species. *Plant Cell Tissue Organ Culture* 42: 221-226.

Plessis, P., C. Leddet, and J. Dereuddreu (1991). Resistant to dehydration and to freezing in liquid nitrogen of alginate coated shoot tips of grapevine (*Vitis vinifera* L. cv. Chardonnay). *C R Academy Science, Paris* 313: 373-380.

Priya, C.F., V.A. Bapat, and P.S. Rao (1994). Investigations on the development of somatic seeds of *Santalum album* L. (sandalwood). *Proceedings National Academy of Science, (India)* LXIV: 1-8.

Rao, P.S., T.R. Ganapathi, P. Suprasanna, and V.A. Bapat (1993). Encapsulated shoot tips of banana: A novel propagation and delivery system. *InfoMusa* 3: 4-5.

Rao, P.S., T.R. Ganapathi, P. Suprasanna, and V.A. Bapat (1996). Synthetic seed technology as a method of plant propagation and delivery of tissue cultured plants. In L.K. Pareek (ed.), *Trends in plant tissue culture and biotechnology* (pp. 47-52). Jaipur, India: Agrobotanikal Publishers.

Rao, P.S. and P. Suprasanna (1999). Augmenting plant productivity through plant tissue culture and genetic engineering. *Journal Plant Biology* 26(2): 119-127.

Rao, P.S., P. Suprasanna, and T.R. Ganapathi (1997). Plant biotechnology and agriculture: Prospects for improving and increasing productivity. *Science & Culture* 62(7-8): 185-191.

Rao, P.S., P. Suprasanna, T.R. Ganapathi, and V.A. Bapat (1998). Synthetic seeds: Concept, methods and applications. In P.S. Srivastava (ed.), *Plant tissue culture and molecular biology: Applications & prospects* (pp. 607-619). New Delhi, India: Narosa Publishing House.

Rao, P.S., P. Suprasanna, T.R. Ganapathi, and V.A. Bapat (2000). Synthetic seed technology in horticultural crops. In K.L. Chadha (ed.), *Biotechnology in horticultural and plantation crops* (pp. 55-64). ICAR. New Delhi, India: Malhotra Publ.

Rao, P.V. and B. Singh (1991). Plant regeneration from encapsulated somatic embryos of *Solanum melongena* L. *Plant Cell Reports* 10: 7-11.

Redenbaugh, K. (1990). Application of synthetic seeds to tropical crops. *Horticulture Science* 25: 251-255.

Redenbaugh, K. (1993). *Synseeds: Applications of synthetic seeds to crop improvement.* Boca Raton, FL: CRC Press.

Redenbaugh, K., J. Fujii, and D. Slade (1988). Encapsulated plant embryos. In A. Mizrahi and A.R. Liss (eds.), *Biotechnology in agriculture* (pp. 225-248). New York.

Redenbaugh, K., J. Fujii, D. Slade, P. Viss, and M. Kossler (1991). Artificial seeds-encapsulated somatic embryos. In Y.P.S. Bajaj (ed.), *Biotechnology in agriculture and forestry,* Volume 17: *High tech and micropropagation I* (pp. 395-416). New York: Springer-Verlag.

Redenbaugh, K., J. Nichol, M. Kossler, and B. Paasch (1984). Encapsulation of somatic embryos for artificial seed production. *In Vitro Cellular & Developmental Biology* 20: 256.

Redenbaugh, K., B.D. Paash, J.W. Nichol, M.E. Kossler, P.R. Viss, and K.A. Walker (1986). Somatic seeds: Encapsulation of asexual embryos. *Bio/Technology* 4: 797-801.

Redenbaugh, K. and Z. Reyes (1987). Artificial seed coats for botanical seed analogs. United States Patent 4,715,143.

Redenbaugh, K., D. Slade, P. Viss, and J. Fijii (1987). Encapsulation of somatic embryos in synthetic seed coats. *Horticulture Science* 22: 803-809.

Refouvelet, E., S. Lenours, C. Tallor, and F. Daguin (1998). A new method for in vitro propagation of liliac (*Syringa vulgaris* L.): Regrowth and storage conditions for axillary buds encapsulated in alginate beads, development of a pre-acclimatization. *Scientia Horticulture* 74: 233-241.

Sahasrabudhe, N.A., A. Nandi, R.A. Bahulikar, P.S. Rao, and P. Suprasanna (1999). A two step approach to scale up green plant regeneration through somatic embryogenesis from in vitro cultured immature inflorescences of a male sterile line and a maintainer line of rice. *Journal New Seeds* 2(4): 1-11.

Sajina, A., D. Minoo, S.P. Geetha, K. Samsuddin, J. Rema, K. Nirmalbabu, and P.N. Ravindran (1997). Production of synthetic seeds in few spice crops. In S. Edison, K.V. Ramanna, B. Sasikumar, K. Nirmalbabu, and J. Santosh (eds.), *Biotechnology of spices, medicinal and aromatic plants* (pp. 65-69). Eapen, Kerala, India: Indian Society of Spices.

Sakamoto, Y., T. Mashiko, A. Suzuki, H. Kawata, and A. Iwasaki (1992). Development of encapsulation technology for synthetic seeds. *Acta Horticulture* 319: 71-76.

Sakamoto, Y., N. Onishi, and T. Hirosawa (1995). Delivery systems for tissue culture by encapsulation. In J. Atkin-Christie, T. Kozai, and M.A.L. Smith (eds.), *Automation and environmental control in plant tissue culture* (pp. 215-243). Dordrecht, the Netherlands: Kluwer Academic Publ.

Sanada, M., Y. Sakamoto, M. Hayashi, T. Mashiko, A. Okamoto, and N. Ohnishi (1993). Celery and lettuce. In K. Redenbaugh (ed.), *Synseeds : Applications of synthetic seeds to crop improvement* (pp. 305-327). Boca Raton, FL: CRC Press.

Sarkar, D. and P.S. Naik (1997). Synseeds in potato: An investigation using nutrient-encapsulated in vivo nodal cutting segments. *Scientia Horticulture* 73: 179-184.

Senaratna, T., B.D. McKersie, and S.R. Bowley (1989). Desiccation tolerance of alfalfa (*Medicago sativa* L.) somatic embryos, influence of abscisic acid, stress treatments and drying rates. *Plant Science* 65: 253-259.

Sharma, A., P. Tandon, and A. Kumar (1992). Regeneration of *Dendrobium wardianum* Warner (Orchodaceae) from synthetic seeds. *Indian Journal Experimental Biology* 30: 747-748.

Sharma, T.R., B.M. Singh, and R.S. Chouhan (1994). Production of disease free encapsulated buds of *Zingiber officinale* Rose. *Plant Cell Reports* 13: 300-302.

Shigeta, J. and K. Sato (1994). Plant regeneration and encapsulation of somatic embryos of horseradish. *Plant Science* 102: 109-115.

Singh, F. (1991). Encapsulation of *Sphathoglottis pliccata* L. protocorms. *Lindleyana* 6: 61-63.

Skryabin, G.K. and K.A. Koshcheenko (1987). Immobilization of living microbial cells in polyacrylamide gel. *Methods Enzymology* 135: 198.

Soneji, J.R., P.S. Rao, and M. Mhatre (2002). Germination of synthetic seeds of pineapple (*Ananas comosus* L.) Mer. *Plant Cell Reports* 20: 891-894.

Srinivas, L. (2002). Morphogenetic studies in vitro in banana (*Musa* spp.). Master's thesis, University of Mumbai, Mumbai, India.

Standardi, A., and E. Piccioni (1998). Recent perspectives on synthetic seed technology using non-embryogenic in vitro derived explants. *International Journal Plant Science* 159(6): 968-978.

Sudhakara, K., B.N. Nagaraj, A.V. Santoshkumar, K.K. Sunilkumar, and N.K. Vijayakumar (2000). Studies on the production and storage potential of synthetic seed in cocoa (*Theobroma cacao* L.). *Seed Research* 28(2): 119-125.

Sunilkumar, K.K., K. Sudhakara, and N.K.Vijayakumar (2000). An attempt to improve storage life of *Hopea parviflora* seeds through synthetic seed production. *Seed Research* 28(2): 126-130.

Suprasanna, P., S. Anupama, T.R. Ganapathi, and V.A. Bapat (2001). In vitro growth and development of encapsulated shoot tips of different banana and plantain cultivars. *Journal New Seeds* 3(1): 19-25.

Suprasanna, P., T.R. Ganapathi, and P.S. Rao (1996). Synthetic seeds in rice (*Oryza sativa* L.): Encapsulation of somatic embryos from mature embryo callus cultures. *Asia-Pacific Journal Molecular Biology & Biotechnology* 4: 90-93.

Suprasanna, P. and P.S. Rao (1997). Somatic embryogenesis in crop plants. In T.V. Ramana Rao and I.L. Kothari (eds.), *Plant structure & morphogenesis* (pp. 29-35). Vallabh Vidyanagar, India: SP University.

Takahata, Y., D.C.W. Brown, W.A. Keller, and N. Kaizuma (1993). Dry artificial seeds and desiccation tolerance inducation in microspore derived embryos of broccoli. *Plant Cell Tissue and Organ Culture.*

Tan, T.K., W.S. Loon, E. Jhor, and C.S. Loh (1998). Infection of *Sphathoglottis pliccata* (Orchidaceae) seeds by mycorrhizal fungus. *Plant Cell Reports* 18: 14-19.

Thorpe, T.A. and C. Stasolla (2001). Somatic embryogenesis. In S.S. Bhojwani and W.Y. Soh (eds.), *Current trends in the embryology of angiosperms* (pp. 279-336). Dordrecht, the Netherlands: Kluwer Academic Publishers.

Uozumi, N., Y. Nakashimada, Y. Kato, and T. Kobayashi (1992). Production of artificial seeds from horseradish hairyroot. *Journal Fermentation Bioengineering* 74: 21-26.

Vanek, T., T. Macek, K. Stransky, and K. Ubik (1989). Plant cells immobilized in pectate gel: Biotransformation of verberol isomera by *Solanum aviculare* free and immobilized cells. *Biotechnology Techniques* 3: 411.

Vorlop, K.D. and J. Klein (1981). Formation of spherical chitosan biocatalysts by ionotropic gelation. *Biotechnology Letters* 3: 9.

SECTION II: SEED DORMANCY AND GERMINATION

Chapter 10

Dormancy and Germination

Henk W. M. Hilhorst
Leonie Bentsink
Maarten Koornneef

INTRODUCTION

Dormancy is a term that is used to denote a state of a plant or plant organ that is generally characterized by the virtual absence of metabolic activity and/or a lack of further development and growth. It is found among most known forms of plant life and may occur in seeds, bulbs, tubers, buds, and whole plants. Dormancy is a trait that has been acquired during evolution by selection for the capacity to survive unfavorable environmental conditions, such as heat, cold, and drought. There are indications that the evolutionary origin of dormancy is related to climatic changes during Earth's history. This is supported by observations that the number of plant species with dormancy generally increases with the geographical distance from the equator and, hence, with the amount of variation in precipitation and temperature (Baskin and Baskin, 1998). There is, however, little doubt that the phenomenon of dormancy has significantly contributed to the development of new species and the successful dispersion of those already present (Baskin and Baskin, 1998).

Definitions of Dormancy and Germination

Very often seed dormancy is simply regarded as the absence of germination. However, absence of germination can have several causes. First, the seed can be nonviable. Second, the environment may be limiting to germination, and, third, the cause may be in the seed or dispersal unit itself. With this in mind, a satisfactory definition of dormancy can be formulated: "Dormancy is the absence of germination of a viable seed under conditions that are favorable to germination." The ecological relevance of seed dormancy can be defined as an "effective delay of germination to avoid germination and subsequent growth under unfavorable climatic conditions."

Different definitions for germination are in use as well. In fundamental seed science, germination begins with the uptake of water by the dry seed and is completed with protrusion of the radicle. However, from an agronomic viewpoint germination includes the emergence and establishment of the seedling.

Categories of Dormancy and Germination

The many studies of seed dormancy have yielded a substantial number of classifications. Most classifications that are in use (Baskin and Baskin, 1998) are derived from one that was published by Nikolaeva (1977; see Table 10.1). An additional classification of dormancy is based on the timing of its occurrence: primary dormancy indicates the type of dormancy that occurs prior to dispersal as part of the seed's developmental program, whereas secondary dormancy denotes the acquisition of dormancy in a mature hydrated seed as a result of the lack of proper conditions for germination (Amen, 1968). The essence of this latter classification is a distinction between absence of germination, and dormancy. Within this context, "absence of germination" is identical to the term *quiescence*, which is commonly used to denote the absence of germination because of a lack of water.

In many species dormancy is composed of different types. For example, many seeds possess a seed coat that poses a mechanical restraint to embryonic growth but may also contain chemical inhibitors, such as phenolic compounds (mechanical and chemical dormancy). In addition, in the same seed, endosperm tissue may restrict embryo growth until it is degraded by hydrolytic enzymes induced by embryonic factors (physiological/mechanical dormancy). There does not seem to be a "preference" among higher plant families or genera for a certain category or type of dormancy.

In order to understand mechanisms and regulation of seed dormancy it is mandatory to study all the possible aspects that may contribute to the overall effect of dormancy on seed behavior. Unfortunately, most (developmental) studies have neglected this and have predominantly focused on physiological dormancy.

Types of germination also exhibit wide variety. The first part that protrudes from the seed coat in dicots is usually the radicle, although exceptions are known in, e.g., *Salsola,* in which the root protrudes first, and in some Cactaceae, in which the cotyledons may emerge first (Werker, 1997). In the Gramineae the coleorhiza is generally the first embryo organ to protrude, followed by the radicle and plumule. However, in the monocots there is considerably more variation among taxa regarding which organ is first to protrude.

TABLE 10.1. Categorization and types of seed dormancy.

Categories	Groups	Caused by	Mechanism
Primary dormancy (predispersal)	Exogenous (outside embryo)	Maternal tissues and/or endosperm or perisperm	• Inhibition of water uptake (physical dormancy) • Mechanical restraint to embryo expansion and radicle protrusion (mechanical dormancy) • Modification of gas exchange • Prevention of leaching of inhibitors from embryo • Supplying inhibitors to the endosperm (chemical dormancy)
	Endogenous (inside embryo)	• Underdeveloped embryo (morphological dormancy) • Metabolic blocks (physiological dormancy) • Morphophysiological dormancy (1 and 2 combined)	• Embryo in mature seed has to complete development prior to germination • Physiological mechanisms largely unknown
	Combinational	Combination of exogenous and endogenous dormancy	
Secondary dormancy (postdispersal)		Metabolic blocks, induced in nondormant seeds when the germination environment is unfavorable	• Physiological mechanisms largely unknown

Source: Adapted from Nikolaeva, 1977, and Baskin and Baskin, 1998.

THE REGULATION OF DORMANCY AND GERMINATION: A MUTANT APPROACH

Over the past 20 years the use of mutants, particularly those of *Arabidopsis thaliana*, has made a major contribution to our understanding of developmental processes in seeds, including dormancy and germination

(Koornneef et al., 2002). A vast range of seed phenotypes has been described, and genes that are associated with dormancy and germination have been identified. As yet, we are still far from a comprehensive view of the regulation and mechanisms of dormancy and germination, but, gradually, outlines are becoming visible. Particularly, the hormonal signaling involved in the regulation of these processes has been intensively studied and in this field most progress has been made. For this reason the larger part of this chapter is devoted to discussing the role of hormones. However, one should be aware that the number of species studied is very limited. Conclusions about general validity of concepts and hypotheses can as yet not be drawn.

ABA Signaling Is Involved in the Induction of Primary Dormancy

Primary dormancy has long been associated with the presence of the plant hormone abscisic acid (ABA) during seed development. Typically, ABA levels rise during the first half of development and decline during late maturation, when seed water content decreases. ABA can be present in all seed and fruit tissues. Apart from dormancy, this hormone has been associated with a number of other developmental processes, including storage protein synthesis, suppression of precocious germination, induction of late embryogenesis abundant proteins, and induction of desiccation tolerance. Studies with ABA-deficient and -responsiveness mutants of *Arabidopsis thaliana* (Karssen et al., 1983; Koornneef et al., 1984) and tomato (Koornneef et al., 1985; Groot and Karssen, 1992) have clearly demonstrated that defects in ABA signaling during seed development may result in the formation of nondormant seeds. In tomato and *Arabidopsis* this occurred only when the embryo contained the dominant *ABA* allele. Crosses between wild-type and mutant plants proved that maternal ABA (i.e., located in testa and fruit tissues) had no influence on dormancy Thus (transient) increase in embryonic ABA content during seed development is required to induce dormancy. Manipulation of seed ABA content by genetic modification of tobacco has shown that overexpression of zeaxanthin oxidase, one of the enzymes of the ABA-synthetic pathway, resulted in more dormant phenotypes, whereas "knocking out" of the gene encoding for this enzyme yielded less dormant phenotypes (Frey et al., 1999).

Sensitivity to ABA plays an equally important role in the expression of dormancy or inhibition of germination. The ABA-insensitive mutants *abi1* to *abi3* display marked reductions in seed dormancy. Conversely, the ABA-supersensitive mutant *era1* mutation produces seeds with enhanced seed dormancy (Cutler et al., 1996). In wheat and maize, several cultivars exhibit vivipary, also known as preharvest sprouting, under humid conditions. These sprouting-susceptible cultivars have a reduced sensitivity to ABA. Analysis of mutants in *Arabidopsis* and maize displaying vivipary has led to the identification of ABA-responsive genes that are responsible for these phenotypic characteristics: *ABI3* and *VP1*, respectively. These orthologous genes encode for transcription factors of the B3 domain family, which can activate the transcription of ABA-inducible genes that play a role in the regulation of gene expression during seed development (Finkelstein et al., 2002). They are involved in maintaining the developmental state in seeds and suppressing transition to the vegetative or growth stage. In other words, seeds of mutants lacking this gene will show characteristics of (germinative) growth.

Defects in ABA-Signaling Do Not Always Result in Nondormant Phenotypes

The *abi3*, *abi4*, and *abi5* mutants exhibit reduced expression of various seed maturation genes, but only *abi3* mutants are nondormant, which coincides with desiccation intolerance (Nambara et al., 1992; Ooms et al., 1993, 1994; Bies et al., 1999) in strong alleles. Surprisingly, no dormancy or other seed maturation phenotype was observed in *abi4* and *abi5* mutants. This may indicate that other genes are redundant in function to these seed-specific transcription factors, which are members of the APETALA2 domain *(ABI4)* (Finkelstein et al., 1998; Söderman et al., 2000) and basic leucine zipper factor *(ABI5)* families (Finkelstein and Lynch, 2000; Lopez-Molina et al., 2001). Data obtained by Lopez-Molina et al. (2001) suggest that *ABI5* may play a role in early seedling development by conferring an ABA-inducible desiccation tolerance in arrested seedlings, which positions its role mainly after germination has been completed.

Genetic screens based on ABA-regulated reporter genes in vegetative tissues have revealed additional ABA-related mutants. The *ade1* (ABA-deregulated gene expression) mutation enhances gene expression in response to ABA but not to cold (Foster and Chua, 1999), and *hos5* (high expression of osmotically responsive genes) mutant displays an increased sensitivity of gene expression to ABA and osmotic stress but not to cold (Xiong et al., 1999). However, the *ade1* and *hoc5* mutations have little effect on seed germination. This is also true for the *aao3* mutant, which affects a step in ABA biosynthesis which differs from those in the *aba1* to *aba3* mutants (Seo et al., 2000), probably because other redundant genes with seed-specific expression compensate for the function of the mutated genes.

ABA Signaling Shows Cross Talk with Sugar and Ethylene Signal Transduction

Sugars, such as sucrose and various hexoses, inhibit seed germination independently of their osmotic effects (Pego et al., 2000). Mutants that were insensitive to the inhibiting effect of glucose and sucrose were isolated by several groups and appeared to be defected in ABA biosynthesis or are among the ABA-insensitive mutants. The sugar-signaling mutants, *gin1*, *sis4*, and *isi4*, are allelic to *aba2* (Cheng et al., 2002); *gin5* and *los5* are alleles of *aba3;* and *gin6*, *sis5*, and *sun6* are *abi4* alleles (Huijser et al., 2000; Laby et al., 2000; Cheng et al., 2002). These results indicate that germination is controlled in an ABA- and sugar-dependent way. Garciarrubio et al. (1997) have shown that addition of sugars and amino acids allowed seeds to germinate in otherwise inhibitory concentrations of ABA and suggested that ABA inhibits the mobilization of food reserves. However, it cannot be excluded that these sugar effects are mediated by sugar signaling effects (Smeekens, 1998; Gibson and Graham, 1999) or that sugars induce ABA synthesis as suggested by Cheng et al. (2002). That storage reserve mobilization is not required for germination as such is shown by the ability of *kat2* (ketoacyl CoA thiolase-2)–deficient mutants to germinate when hardly containing sucrose (Germain et al., 2001). However, these mutants require sucrose for seedling establisment, which is the main defect also of isocitrate lyase–deficient mutants (Eastmond and Graham, 2001). Furthermore, when germination is inhibited by ABA, reserve mobilization continues and such ABA-inhibited seeds can contain two times more sucrose than wild-type seeds after three days of imbibtion (Pritchard et al., 2002). A not fully understood role of lipid metabolism was revealed by the cloning of the *COMATOSE (CTS)* gene (Footitt et al., 2002), which encodes an ABC transporter located in the peroxisome, which, when mutated, fails to breakdown lipids, but also does not germinate. Footitt et al. (2002) suggested that a product of lipid metabolism, accumulating in the *cts* mutant, prevents the release from dormancy. The observation that the *abi3* and *lec1-1* and *aba1* mutants are epistatic to *cts* (Russell et al., 2000), together with the expression pattern of *CTS* during imbibition (Footitt et al., 2002), suggests that CTS is essential for the progress of germination when dormancy is not induced and/or when ABA is absent.

Ethylene signaling mutants are also affected in their germination response. Ethylene may mimic the action of gibberellins. Seeds of the GA-deficient *ga1* mutant of *Arabidopsis* (which normally germinate only in the presence of GA_{4+7}) germinated in the presence of ethylene (Karssen et al. 1989). Seeds of ethylene-insensitive mutants such as *etr* and *ein2* germinate less well or after a longer period of after-ripening than wild-type seeds

(Bleecker et al., 1988). These ethylene mutants show phenotypes that resemble ABA and sugar signaling mutants. The *ein2* and *etr* mutants appeared to be hypersensitive to ABA (Beaudoin et al., 2000; Ghassemian et al., 2000), which is in agreement with the observation that *ein2* mutants were isolated as *abi1-1* suppressors. These observations, in combination with the nondormant phenotype of the *ein2*, *abi3-4* double mutant indicate that ethylene influences seed dormancy by inhibiting ABA action (Beaudoin et al., 2000). The presence of crosstalk between sugar signaling and ethylene was suggested by the sugar-insensitive phenotype of *ctr1* (Gibson et al., 2001), the selection of *gin6*, which is alellic to *ctr* as a glucose-insensitive mutant (Cheng et al., 2002), and the sugar-hypersensitive phenotype of *etr* (Zhou et al., 1998). Apparently ABA, ethylene, and sugar signaling strongly interact and express this at the level of germination and early seedling growth.

Dormancy Phenotypes Are Not Always Caused by Alterations in ABA Signaling

Mutants are known with altered dormancy characteristics but with normal ABA content and sensitivity. Examples are the *aberrant testa shape (ats), transparent testa (tt),* and *transparent testa glabra (ttg)* mutants of *Arabidopsis thaliana*. The *ats* (seed shape) mutant has a reduced testa thickness and germinates faster and to higher levels than wild-type seeds. This mutant produces ovules in which the integuments do not develop properly (Léon-Kloosterziel et al., 1994). The *tt* mutants produce seeds with defects in the flavonoid pigmentation of the testa. Mutants so far identified are *tt1* to *tt17, ttg1 and ttg2,* and *banyuls (ban).* The color of the *tt* mutants ranges from yellow to pale brown (Debeaujon et al., 2000). The *ban* mutant accumulates pink flavonoid pigments in the endothelium of immature seeds and grayish-green, spotted mature seeds (Albert et al., 1997; Devic et al., 1999). These testa modifications may enhance the uptake of water and oxygen and the leaching of inhibitory substances from the seed. Also, changes in testa dimensions may reduce the mechanical restraint to embryo growth. These examples are clear cases of altered physical and chemical dormancy in which modification of seed coat properties reduces the restraint to germination. Papi et al. (2000) described another mutant of which the dormancy is determined by the maternal genotype. This is in agreement with the expression pattern of the *DAG1* gene, which encodes a DOF transcription factor in the vascular tissues. It is genetically derived from the mother plant and enters the developing seeds. The *dag1* was shown to affect both the light requirement for germination and the structure of the testa (Papi et al., 2002).

Recently, a related gene, named *DAG2,* with a similar expression pattern as *DAG1,* was isolated. However the germination phenotype of the *dag2* mutant seeds is opposite to that of *dag1* seeds (Gualberti et al., 2002) as it shows increased dormancy.

The *rdo1* to *rdo4* mutants represent a class of mutants that was directly selected on the basis of reduced dormancy (Léon-Kloosterziel et al., 1996; Peeters et al., 2002). These mutants show mild pleiotropic adult plant effects, which indicates that the genes are not specific for dormancy/germination but affect other processes as well.

The analysis of genetic variation for seed dormancy between accessions of wild plants and among varieties of cultivated plants reveals additional loci controlling this trait. The polygenetic nature of this type of genetic variation, together with the large environmental effects on the expression of germination characteristics and of the involvement of many genes, make dormancy genetically a typically quantitative trait. Such traits are now better amenable to genetic analysis because the position of individual quantitative trait loci (QTL) and the relative contribution of these loci can be determined. QTL analysis for seed dormancy has been reported for *Arabidopsis thaliana* (Koornneef et al., 2002, and references therein), barley, rice, and wheat. It appears that QTL identified for wheat, colocate with barley QTL but not with rice QTL (Kato et al., 2001). Wild species often show stronger dormancy than cultivated species, which makes crosses between them useful for QTL analysis. QTL analysis can be followed by the study of the individual genes (or chromosome regions containing specific dormancy QTL) and fine mapping. These studies have been initiated in barley (Romagosa et al., 1999) and *Arabidopsis* (Alonso-Blanco et al., 2003; Bentsink, 2002). It is expected that the study of such QTL will allow the molecular identification of the respective genes by map-based cloning. However, the cloning of such dormancy QTL has not been reported yet.

Dormancy Is a Developmental Event

Mutant analysis has revealed a number of genotypes with altered dormancy characteristics but which possess normal ABA contents throughout their development. Examples are the *abi3, leafy cotyledon (lec1, lec2),* and *fusca (fus3)* mutants in *Arabidopsis.* These mutants all possess phenotypes that are characteristic of the vegetative state, e.g., reduced tolerance to desiccation, active meristems, expression of germination-related genes, and absence of dormancy. It appears that these three genes have partially overlapping functions in the overall control of seed maturation, leading to mutants defective in many aspects of seed maturation (Parcy et al., 1997).

LEC1 and *FUS3* loci probably regulate developmental arrest, as mutations in these genes cause a continuation of growth in immature embryos. *ABI3* is also active during vegetative quiescence processes in other parts of the plant in which it suppresses meristematic activity (Rohde et al., 2000).

Because maturation is defective in the *abi3, lec1, lec2,* and *fus3* mutants, no dormancy is initiated and the seeds germinate (precociously) because this is the default state (Nambara et al., 2000). However, detailed analysis of the *fus3, lec1,* and *abi3* mutants has shown that they differ in the time that premature germination occurs and double mutants of these mutants with GA-deficient mutants also behave differently. Thus, ABA-controlled dormancy (via *ABA* and *ABI*) may represent different mechanisms preventing germination, which occurs later and is additive to the developmental arrest controlled by *LEC1* and *FUS3* (Raz et al., 2001). The dependency of germination on GA is maintained in the *fus3* mutant but not in *lec1,* which suggests that these mutants affect the germination potential of seeds in different ways.

More evidence for developmental arrest and dormancy being different phenomena has been provided by studies with the ABA-deficient *sitiens* (*sit^w*) mutant of tomato. After the completion of embryonic histodifferentiation both wild-type (cv. Moneymaker) and mutant seeds enter a state of developmental arrest. In the wild type this is followed by the induction of dormancy, concurrent with the occurrence of a transient rise in ABA content (Hilhorst, 1995; Hilhorst and Karssen, 1992, and references therein). In the *sit^w* mutant there is no induction of dormancy and, hence, developmental arrest is not maintained. This results in viviparous germination within overripe fruits. However, at this stage wild-type seeds also possess undetectable levels of ABA, yet germination is suppressed. The osmotic environment (locular tissues) of seeds of both genotypes are comparable (Berry and Bewley, 1992; Liu et al., 1996). Apparently, the mutant seeds possess a greater "growth potential." Also, the mature mutant seeds display a stronger resistance to osmotic inhibition in vitro (Groot and Karssen, 1992). These results could be explained only by the aberrations in the testa of the mutant. It was shown that the mutant testa contained only one cell layer against four to five in the wild type (Hilhorst and Downie, 1996). In this sense the *sit^w* mutant resembles the *ats* mutant of *Arabidopsis.*

Overlap of Dormancy and Stress Responses

Differential screening for gene expression of dormant and nondormant lines of *Avena fatua* resulted in the isolation of several cDNA clones that were expressed in the dormant seeds but not in the nondormant ones (Li and

Foley, 1995). However, when the nondormant seeds, and even seedlings, were stressed by exposure to elevated temperatures the "dormancy genes" were expressed (Li and Foley, 1995). Many of these dormancy-associated transcripts encode members of the family of late embryogenesis abundant (LEA) proteins. These proteins are speculated to play a protective role during seed desiccation (Philips et al., 2002, and references therein). Other similar studies have also yielded stress-related genes, e.g., the *Per1* and *AtPer1* genes from barley and *Arabidopsis,* respectively, which encode for peroxiredoxins (Prx) (Aalen, 1999, and references therein). Peroxiredoxins are enzymes that play a role in protecting seeds from desiccation damage by exposure to reactive oxygen species (ROS). *Per1* gene expression can be induced by osmotic stress and ABA in immature barley embryos (Aalen et al., 1994), and the closely related pBS128 transcript is up-regulated in both dormant and nondormant imbibed mature embryos of *Bromus secalinus* (Goldmark et al., 1992). Based on the expression patterns of pBS128 in *B. secalinus* and *AtPer1* in *Arabidopsis* it was suggested that peroxiredoxins play a role in dormancy (Goldmark et al., 1992; Haslekas et al., 1998). However, more recent evidence casts doubt on this hypothesis. *AtPer1* expression in the nondormant *Arabidopsis aba1* mutant was similar as in wild-type seeds, which suggests that the presence of the *AtPer1* transcript alone is not sufficient to maintain dormancy. Furthermore, transgenic tobacco plants, overexpressing the rice 1Cys-peroxiredoxin *(R1C-Prx)* gene, showed a germination frequency similar to that of control plants (Lee et al., 2000).

More information is required to draw firm conclusions on crosstalk between stress- and dormancy-related signaling. Evidently, ABA is a common denominator in these signaling networks (Finkelstein et al., 2002; Xiong et al., 1999; Xiong and Zhu, 2003). In view of the ecological relevance of dormancy, overlapping responses could be favorable. Seeds strive to maintain their dormancy during unfavorable periods for growth. Such periods are generally stressful: too cold, too hot, or too dry. It is during this period that seeds also have to be equipped to tolerate these stresses.

Is ABA Involved in the Maintenance of Dormancy?

The transient increase in ABA content in seeds halfway through development may contribute to the prevention of precocious germination. This was, for example, shown by the maternal ABA effects in the extreme *aba1-abi3-1* double mutant (Koornneef et al., 1989). However, as discussed for the tomato *sitiens* mutant, pleiotropic effects of the absence or diminishing of ABA signaling, e.g., on testa morphology, cannot be excluded. Indeed, the *aba* mutants shows aberrations in the seed coat, which possesses a mucilage layer of reduced thickness (Karssen et al., 1983). The rate of water uptake may be affected by the thickness of this layer. It is also feasible that this layer influences oxygen uptake. Because in wild-type seeds ABA levels decrease at the end of seed maturation, it was proposed that after the onset of dormancy endogenous ABA is not required for its maintenance (Karssen et al., 1983). However, recent studies have shown that inhibitors of ABA biosynthesis, such as norfluorazon, may promote germination in *Arabidopsis* (Debeaujon and Koornneef, 2000). In imbibing seeds of *Nicotiana plumbaginifolia* it was indeed shown that ABA is synthesized de novo (Grappin et al., 2000). A detailed comparison of ABA levels in dormant and nondormant barley seeds indicated that in dormant seeds ABA levels did not decrease upon imbibition, whereas this occurred rapidly in nondormant seeds (Jacobsen et al., 2002).

Also, in imbibing coffee *(Coffea arabica)* seeds de novo synthesis of ABA has been demonstrated (da Silva, 2002). In this species it was shown that ABA synthesis accounted for one day in delay of germination. These results suggest that the maintenance of dormancy in imbibed seeds may be an active process involving de novo synthesis of ABA. It remains enigmatic as to why seeds synthesize ABA when they have switched to the germination mode. As yet, ecological explanations for this phenomenon are lacking.

Gibberellins Are a Prerequisite for Germination

Gibberellins (GAs) play an important role in the stimulation of seed germination. Gibberellin-deficient mutants do not germinate unless GAs are added to the germination medium (Koornneef and van der Veen, 1980). In addition, inhibitors of GA biosynthesis, including tetcyclacis and paclobutrazol, inhibit germination. GAs can promote germination of dormant seeds by their ability to overcome or "short-circuit" the requirement for environmental factors that are required for germination, including after-ripening, light, and cold. This has led to the hypothesis that such environmental factors may induce GA biosynthesis during the early phases of germination (Hilhorst and Karssen, 1992). Indeed, it was shown that red light enhances GA_1 levels in photoblastic lettuce seeds (Toyomasu et al., 1993). It was later confirmed that phytochrome could induce one of two 3-β hydroxylase enzymes involved in GA biosynthesis, encoded by the *GA4H* gene (Yamaguchi et al., 1998). In addition, Toyomasu et al. (1994) showed that endogenous levels of ABA were substantially lowered by red-light irradiation and exogenous GA_3. The incubation medium contained only trace amounts of ABA. Thus, it was concluded that ABA was either catabolized or conjugated as a result of the action of red light or GAs. It was proposed that light-induced germination was mediated by a GA-controlled inactivation of ABA.

In *Arabidopsis thaliana* seeds levels of (physiologically active) GA_4 increased during chilling. However, the seeds were still light-requiring for germination, but light did not further increase GA levels. Thus, a clear correlation between GA levels and germination could not be found. Therefore, it was concluded that the sensitivity to GAs plays an equally important role in increasing germinability. (Derkx and Karssen, 1993). Indeed, in several species it has been shown that after-ripening and chilling are accompanied by an increase in sensitivity to (exogenous) GAs (Hilhorst and Karssen, 1992, and references therein).

The role of GA is further confirmed by the observation that GA signaling mutants (reviewed in Olszewski et al., 2002), such as sleepy *(sly)* (Steber et al., 1998), do not germinate. Negatively acting factors are the RGA/GAI proteins that all affect plant height when the genes are mutated, but some also affect germination (Olszewski et al., 2002). These proteins of the GRAS family are encoded by five related genes of which *RGL1* (Wen and Chang, 2002) and *RGL2* (Lee et al., 2002) are involved in germination, as concluded from the resistance of loss of function mutants of these genes to the germination-inhibiting compound paclobutrazol. The *SPINDLY (SPY)* gene, encoding a UDP-GlcNAc protein transferase-like protein (Olszewski et al., 2002), has a similar effect since the *spy ga1-2* double mutant does not require GA for germination (Jacobsen and Olszewski, 1993). Not all *RGA/RGL/GAI*-like genes affect germination since the *rga gai-t6 ga1* triple mutant still requires GA for germination.

A gene encoding for a putative G protein-coupled receptor, *GCR1*, for which no specific link with GA has been shown, was found to abolish dormancy when overexpressed (Colucci et al., 2002). Mutants defective in the alpha subunit of the G protein are also affected in their germination responses, and an interaction with brassinosteroid signaling was suggested (Ullah et al., 2002). The brassinosteroids (BRs), which are a group of over 40 naturally occurring plant steroid hormones found in a wide variety of plant species plants (Clouse and Sasse, 1998; Schumacher and Chory, 2000), are also involved in the control of germination in *Arabidopsis*. It has been suggested that the BR signal is required to reverse ABA-induced dormancy and stimulate germination (Steber and McCourt, 2001). In this sense BR action resembles that of GAs. BRs could overcome the lack of germination of the *sleepy1* mutant, probably by bypassing its GA requirement.

Two BR mutants, *det2* and *bri1*, displayed reduced germination, but eventually germinated without BR, indicating that, in contrast to GAs, BRs are not absolutely required for germination (Steber and McCourt, 2001).

DORMANCY AND GERMINATION: MECHANISMS

The final stage of seed maturation is generally marked by maturation drying, which facilitates seed dispersal and survival. Seed germination begins with the (re)imbibition by water of the "dry" seed and ends with the protrusion of the radicle through the surrounding tissues. Water uptake during germination displays a triphasic pattern (Bewley and Black, 1994). Phase I shows a sharp increase in water content of the seed, which is due to a large water potential ($\Delta\Psi$) gradient between the seed and the environment. Phase II is a period of variable duration in which little or no change in water content of the seeds is observed and $\Delta\Psi \sim 0$. During phase III, the radicle starts to take up extra water as it protrudes through the surrounding tissues, such as endosperm or pericarp. Obviously, radicle protrusion is the net result of the expansion force of the embryo and the opposing restraint of surrounding tissues, such as endosperm and testa. For a full account of the water relations during seed germination see Bradford (1995).

Weakening of Embryo-Surrounding Tissues During Seed Germination

Weakening of the tissues opposing the radicle tip has been proposed to be a prerequisite for radicle protrusion in tomato (Haigh and Barlow, 1987; Groot and Karssen, 1987), muskmelon (Welbaum et al., 1995), *Datura ferox* (de Miguel and Sanchez, 1992) and pepper (Watkins and Cantliffe, 1983). Here we will focus on the germination mechanism of the tomato seed which has served as the model species for this purpose for several decades (Hilhorst et al., 1998; Welbaum et al., 1998).

In tomato seeds the embryo is surrounded by a rigid endosperm and a thin (dead) testa. Water uptake of the embryo is inhibited by the endosperm, possibly via restriction of embryo swelling, which results in a lower embryonic water content as compared to the endosperm, and a $\Delta\Psi$ of -1.5 MPa or lower during phase II, whereas the whole seed is in equilibrium with that of the imbibing solution ($\Delta\Psi = 0$) (Haigh and Barlow, 1987; Liu et al., 1996). For the seeds to germinate the radicle has to protrude through the part of endosperm opposing the radicle tip: the endosperm cap. It is well established that the endosperm cap is weakened by hydrolytic enzyme activity during germination of tomato (Karssen et al., 1989; Nomaguchi et al., 1995;

Toorop et al., 1996). Due to the water potential gradient, weakening of the endosperm cap facilitates protrusion of the radicle. For protrusion of the radicle extra water uptake is required while cells in the radicle elongate. Possibly, the weakening of the endosperm cap also facilitates this uptake of extra water by the radicle (Haigh and Barlow, 1987; Liu et al., 1996).

GAs Regulate the Weakening of Tissues Surrounding the Embryo

Evidence for a role of GAs in decreasing the mechanical resistance of the endosperm to radicle protrusion has mainly been obtained from studies with the gibberellin-deficient tomato mutant *gib1*. During tomato seed germination the thick walls of the endosperm are degraded, initially in a region opposite the radicle tip (Groot and Karssen, 1987). The tissue of this micropylar endosperm (or endosperm cap) can be distinguished anatomically by having smaller cells and thinner walls than the rest of the endosperm (the lateral endosperm).

It was shown that concomitant with hydrolysis of the endosperm cell walls a decrease in the puncture force required to break through endosperm and testa occurred (Groot and Karssen, 1987; Toorop et al., 2000). This did not occur in the *gib1* mutant (Groot et al., 1988). Addition of exogenous GAs, however, completely reverted the mutant to the wild-type phenotype. Increasing evidence indicates that GAs promote the induction of the expression of genes encoding for a number of the hydrolytic enzymes, including those that degrade the hemicellulose fraction of the tomato endosperm cell walls. The hemicellulose fraction of tomato and other Solanaceous species typically consists of mannans and galactomannans. The endosperm cap cell walls of tomato contain 60 to 70 percent mannose, suggesting that they are primarily composed of mannan polymers, most likely galactomannans. These polysaccharides are hydrolyzed by the concerted action of (1,4)-β-mannan endohydrolases (EC 4.2.1.87), β-mannosidases (EC 3.2.1.25), and α-galactosidases (EC 3.2.1.22) (Reid, 1985; McCleary and Matheson, 1975).

Groot et al. (1988) have shown that endo-β-mannanase, β-mannosidase, and α-galactosidase activities were all present in GA-treated gibberellin-deficient (*gib-1*) tomato seeds, with endo-β-mannanase showing the most dramatic increase prior to radicle emergence. Multiple isoforms of endo-β-mannanase are present in tomato seeds and show temporal and spatial differences in expression during germination (Toorop et al., 1996; Nonogaki et al., 1998). An endo-β-mannanase cDNA (*LeMAN1*) has been isolated from germinated tomato seeds (Bewley et al., 1997), and a second cDNA

(*LeMAN2*) has been identified that is expressed exclusively in the endosperm cap prior to radicle emergence (Nonogaki et al., 2000). Thus, different mannanase genes are expressed in the different seed tissues. The pre-germinative endosperm cap isoform is likely to be associated with weakening processes, while the lateral endosperm isoform is involved with reserve mobilization.

However, other hydrolytic enzymes and related proteins of the tomato endosperm may also contribute to endosperm softening and may also be under the control of GA action (Table 10.2), including cellulase (β-1,4-endoglucanase) (Bradford et al., 2000), polygalacturonases (Sitrit et al., 1999), arabinosidase (Bradford et al., 2000), expansins (Chen and Bradford, 2000), and β-1,3-glucanase (Wu et al., 2000). Some of these enzymes are also under the control of ABA, which confirms the existence of a GA-ABA balance in germinating seeds. So far, a causal relationship between endosperm softening, hydrolytic enzyme activity, and germination in tomato is lacking. This is due to the absence of specific inhibitors of these enzymes and of studies using an antisense or or other gene knock-out approach. However, such an approach has been employed for the action of

TABLE 10.2. Hormonal regulation of genes presumably involved in the germination of tomato seeds.

		Tissue	Regulation of expression	
Gene	**cDNA**	**localization**	**ABA**	**GA**
Endo-β-mannanase	*LeMAN2*	CAP	+	o
	LeMAN1	LAT	+	–
Cellulase	*Cel55*	CAP, RT, ROS	+	o
Polygalacturonase	*LeXPG1*	CAP, RT		o
Arabinosidase	*LeARA1*	CAP, LAT	+	o
Xyloglucan endotransglycosylase	*LeXET*		+	
β-1,3-Glucanase	*GluB*	CAP	+	–
Chitinase	*Chi9*	CAP, RT	+	o
Expansin	*LeEXP4*	CAP	+	o

Source: After Bradford et al., 2000.
Note: Genes have been cloned from imbibed tomato seeds prior to radicle emergence. If known, the tissue localization of expression is indicated (CAP: endosperm cap; LAT: lateral endosperm; RT: radicle tip; ROS: rest of seed except micropylar tip). The qualitative effects of GA and ABA, if known, are also indicated (+: promotes expression; o: no effect on expression; –: inhibits expression; blank, information not available).

β-glucanase (βGlu) in the germination of tobacco *(Nicotiana tabacum)* seeds. β-1,3-Glucanases belong to the group of the pathogenesis-related (PR) proteins that are generally expressed in response to pathogens or wounding and in abscission zones or other places where interior plant tissues are (or will be) exposed. Together with chitinase, another PR protein, these enzymes can degrade the cell walls of many fungi and can enhance plant resistance to infection (for review, see Leubner-Metzger and Meins, 1999). It has been shown that β-1,3-glucanase mRNA, protein, and enzyme activity were present in the endosperm cap tissue of tobacco seeds prior to radicle emergence (Leubner-Metzger et al., 1996). ßGlu induction and germination are tightly linked in response to plant hormones and environmental factors, e.g., they are both promoted by gibberellins and inhibited by ABA. With the use of an ABA-inducible chimeric sense-transgene, resulting in overexpression of ßGlu in seeds, strong evidence was provided that ßGlu contributes to endosperm rupture (Leubner-Metzger and Meins, 2000). Surprisingly, antisense transformants with βGlu constructs containing endosperm-specific promoters were unaffected in endosperm rupture but were delayed in the onset of testa rupture in after-ripened seeds (Leubner-Metzger and Meins, 2001). Thus, two sites of ßGlu action were proposed: (1) after-ripening- mediated release of testa-imposed dormancy and (2) endosperm rupture during germination.

The Regulation of Embryo Expansion During Germination

Given the current knowledge of tomato seed germination, it is unlikely that the endosperm is just a simple physical barrier obstructing the radicle from water uptake and protrusion. The growth potential of the radicle or embryo is another important factor in germination (Ni and Bradford, 1993). When tomato embryos were dissected from fully imbibed seeds and placed on water, an instant increase of approximately 20 percent of the fresh weight was observed but actual growth of the radicle was apparent only after 30 h of imbibition, only a few hours before intact seeds started to germinate (Haigh and Barlow, 1987). This suggests the existence of an internal process, that controls outgrowth of the embryo during imbibition. Cell elongation is required for the radicle to protrude. For cells to elongate, the cell wall should become extensible. Once the cell walls in the radicle have become extensible and the restraint in the endosperm cap has been sufficiently weakened, the radicle will protrude through the endosperm, thereby taking up water.

GAs Are Involved in the Regulation of Embryo Expansion During Germination

Although our understanding of GA signaling in seed germination has undoubtedly increased over the past decade (for review, see Peng and Harberd, 2002, and this chapter) little is known yet about the possible targets of GA action in the expansion of the embryo during germination. Embryos respond to germination stimuli by an increase of the extensibility of the cell walls of the radicle in response to the internal turgor pressure. Although cell wall hydrolases are almost certainly involved in cell expansion, hydrolytic enzymes alone are generally unable to cause wall extension in in vitro assays of tissue stretching. Xyloglucan endotransglucosylase (XET) and expansins have been suggested to be involved in cell wall loosening, thus allowing cell expansion. XET breaks the xyloglucan chains and allows the cellulose microfibrils to move apart, driven by the internal cell turgor pressure (ψ_p) (Bewley, 1997). Nonenzymic proteins termed *expansins* have been identified that can cause extension of plant cell walls (Cosgrove, 1999). Expansins disrupt the noncovalent linkages (e.g., hydrogen bonds) at the cellulose-hemicellulose interface, allowing the polymers to creep under tension and the wall to expand. Cell wall hydrolases are proposed to work in conjunction with expansin to modify the hemicellulosic matrix by cleaving and reforming bonds to alter the strength or plasticity of the wall.

From their studies with ABA- and GA-deficient mutants of *Arabidopsis*, Karssen and Laçka (1986) proposed that GAs exert a direct effect on the growth potential of the embryo. This GA action was assumed to be counteracted by ABA, which is synthesized in the embryo. In seeds of *Brassica napus* an increase in the turgor and cell wall extensibility of the embryo is a prerequisite for radicle protrusion (Schopfer and Plachy, 1984, 1985). These processes are under the rigid control of ABA. ABA inhibits water uptake by preventing cell wall loosening of rape embryos. It was shown that the effect of ABA was to increase the minimum turgor for growth and to decrease cell wall extensibility. ABA did not influence hydraulic conductivity or solute accumulation. In coffee seed germination GAs are required to build up a temporal turgor pressure prior to radicle protrusion (da Silva, 2002). Inhibition of GA action by tetcyclacis or paclobutrazol inhibited the increase in turgor pressure and radicle protrusion. However, ABA allowed an increase of turgor pressure but inhibited germination of coffee seeds. It was shown that the inhibition of germination by ABA was through a suppression of one of the endosperm-degrading processes, in a similar fashion as in tomato (Toorop et al., 2000; da Silva, 2002).

Evidence is accumulating that expansins (EXP) and XETs are regulated by GA and ABA and, hence, are potential candidates for hormone-regulated cell expansion and embryo growth potential during germination. For example, in tomato seeds LeXET4 and LeEXP4 are specifically expressed in the endosperm cap whereas LeEXP8 and LeEXP10 are expressed in the radicle cortex or throughout the embryo, respectively (Chen et al., 2001).

The distinct temporal and spatial expression patterns of genes involved in the modification of cell walls may be a reflection of the multitude of interactions among these genes and their products and with distinct cell wall substrates. Thus, the inherent diversity of cell wall modifications may serve different stages of seed and plant development, including cell expansion and cell wall disassembly.

Engagement of the Cell Cycle in Seed Germination

Resumption of cell cycle activity is a specific feature of seed germination. In several species it has been shown that DNA synthesis and duplication occur prior to radicle emergence, as well as accumulation of components of the microtubular cytoskeleton, including tubulins (e.g., Górnik et al., 1997; de Castro et al., 2000; Vázquez-Ramos, 2000). Whereas DNA synthesis is clearly associated with DNA repair and cell division, accumulation of tubulins is required not only for the separation of cell organelles and daughter chromosomes during mitosis, but also for the preparation of cell enlargement (de Castro and Hilhorst, 2000; de Castro et al., 2000). The accumulation of β-tubulin has been observed during germination of several species, including tomato (de Castro et al., 1995), cucumber (Jing et al., 1999), cabbage (Górnik et al., 1997), and coffee (da Silva, 2002). In addition, in *Arabidopsis* seeds the accumulation of the cell-cycle-related proteins actin 7, α-2,4 tubulin, α-3,5 tubulin, β-tubulin, and a WD-40 repeat protein was associated with germination (Gallardo et al., 2002). Of these proteins, only α-2,4 tubulin was under control of GA. Accumulation of the other proteins in GA-deficient *ga1* mutant seeds and in wild-type seeds treated with paclobutrazol was similar to that of wild-type seeds (Gallardo et al., 2002). These results suggest that α-2,4 tubulin may be a key regulatory element in the sequence of processes leading to cell expansion and, hence, radicle protrusion.

The significance of DNA synthesis and duplication for radicle protrusion is doubtful. Inhibition of mitosis by hydroxyurea in cabbage seeds did not affect radicle protrusion (and β-tubulin accumulation) but resulted in severely stunted growth of the seedling (Górnik et al., 1997).

Germination, Dormancy, and the Environment

Depth of dormancy and the occurrence or absence of germination are generally responses to environmental factors, including light, temperature, and soil and atmospheric components. A constant stream of information about the suitability of the environment for successful seedling emergence and subsequent plant growth travels to the seed. This stream of information contains elements that are important for both immediate responses, e.g., start of germination, and for long-term responses, such as the seasonal regulation of dormancy of seeds in the soil seed bank, so-called dormancy cycling (Hilhorst et al., 1996). Although much is known about (hormonal) signaling inside the seed tissues, knowledge about the perception of these environmental factors by the seeds is still annoyingly limited.

So far, the perception of light by seeds through phytochrome-mediated signaling has received the most attention. It has been shown that light-dependent germination of *Arabidopsis* seeds proceeds through phyA and phyB, which mediate the very low and low fluence responses, respectively (for review, see Casal and Sánchez, 1998). More recently, the involvement of phyE in light-induced germination of *Arabidopsis* also has been demonstrated (Hennig et al., 2002). Phytochrome-mediated signaling interacts with the GA biosynthetic pathway to increase the content of physiologically active GAs that are utilized for the processes described previously, including embryo expansion and endosperm softening.

The primary targets of temperature perception in relation with dormancy changes or germination are largely unknown. Over the years a number of generic models involving cellular membranes have been described for temperature responses of seeds, mediated by changes in fluidity and order of membranes, but evidence for such a mechanism is largely circumstantial (Hilhorst, 1998; Hilhorst and Cohn, 2000; Hallett and Bewley, 2002).

Interactions of seeds with chemical components in the soil are crucial for germination (Hilhorst and Karssen, 2000). It has long been known that nitrate is an almost universal stimulator of germination (Roberts and Smith, 1977) that appears to exert a direct interaction with phytochrome signal transduction pathways, leading to germination (Hilhorst, 1990a,b). In addition, a vast range of organics, including alcohols and keto acids, are able to break dormancy in many species (Cohn and Hilhorst, 2000). It is as yet not known how these chemical signals are perceived and with which internal signaling pathways they may interfere.

The Termination of Dormancy and the Onset of Germination Are Distinct Events

Upon shedding, seeds of many species possess primary dormancy and may survive for extended periods without germination in the soil seed bank. Under natural conditions, the seed is exposed to conditions that release dormancy, often in a seasonal cycle. If the (nondormant) seed is not exposed to the right set of external signals to germinate, it will enter secondary dormancy. In many species, any inhibition of germination will ultimately lead to secondary dormancy. Induction and relief of secondary dormancy may occur in the same seed during successive seasons until conditions for germination for the particular seed become favorable. The most important ecological significance of this dormancy cycling is the prevention of germination during short spells of favorable conditions, when the prospects for seedling establishment and survival are poor. The main regulator of these annual dormancy cycles is temperature (Probert, 1992, and references therein), with other factors such as soil nitrate levels, light, temperature, and diurnal temperature fluctuations acting as stimuli for germination. Thus, dormancy and germination are considered separate phenomena. However, these stages are difficult to distinguish in many species and often appear as a continuum (Cohn, 1996).

In general, breaking of dormancy is characterized by a widening of the germination temperature range (temperature "window") and a reduction in the requirement for other germination promoters (e.g., light and nitrate) (Hilhorst et al., 1996). Induction of dormancy is characterized by the reverse of this process. Thus, temperature influences the state of dormancy in a seasonal cycle, but also regulates the germination process itself. The temperature optima for these two processes can differ widely (Bouwmeester and Karssen, 1992), again indicating that relief of dormancy and induction of germination are distinct processes. This has also been confirmed by more recent studies. For example, the *fus3* mutant of *Arabidopsis* is essentially nondormant but still has a GA requirement for germination (Raz et al., 2001). Similarly, the *dog1* (delay of germination) mutant is nondormant but requires GAs for germination (Bentsink, 2002).

PROSPECTS AND CHALLENGES

Over the past five years our understanding of the regulation of seed dormancy and germination has inreased substantially. It is expected that in the next five years major breakthroughs will occur, mainly through the utilization of high throughput "omic" technology in combination with advanced phenotyping of mutants. In addition, it is expected that the cloning of genes defective in mutants with dormancy/germination phenotypes and genes encoding QTL will reveal functional components of these processes. *Arabidopsis* is the species most widely used to unravel signaling networks in relation with germination and dormancy. However, to be able to generalize the results into realistic models, confirmation in other species is absolutely required. In our opinion, one of the greatest challenges in seed science is to explain the behavior of seeds in their natural environment, i.e., in the soil under highly variable conditions. Undoubtedly, the current state of molecular and genetic technology is such that a beginning can be made in this exciting field of science.

REFERENCES

Aalen, R.B. (1999). Peroxiredoxin antioxidants in seed physiology. *Seed Science Research* 9:285-295.

Aalen, R.B., H.-G. Opsahl-Ferstad, C. Linnestad, and O.-A. Olsen (1994). Transcripts encoding an oleosin and a dormancy-related protein are present in both the aleurone layer and the embryo of developing barley (*Hordeum vulgare* L.) seeds. *Plant Journal* 5:385-396.

Albert, S., M. Delseny, and M. Devic (1997). *BANYULS*, a novel negative regulator of flavonoid biosynthesis in the *Arabidopsis* seed coat. *Plant Journal* 11:289-299.

Alonso-Blanco, C., L. Bentsink, C.J. Hanhart, H. Blankestijn-de Vries, and M. Koornneef (2003). Analysis of natural variation at seed dormancy loci of *Arabidopsis thaliana. Genetics* 164:711-729.

Amen, R.D. (1968). A model of seed dormancy. *Botanical Review* 34:1-31.

Baskin, C.C. and J.M. Baskin (1998). *Seeds: Ecology, biogeography, and evolution of dormancy and germination.* San Diego: Academic Press.

Beaudoin, N., C. Serizet, F. Gosti, and J. Giraudat (2000). Interactions between abscisic acid and ethylene signaling cascades. *Plant Cell* 12:1103-1115.

Bentsink, L. (2002). Genetic analysis of seed dormancy and seed composition in *Arabidopsis thaliana* using natural variation. Thesis, Wageningen University, Wageningen, the Netherlands.

Berry, T. and J.D. Bewley (1992). A role for the surrounding fruit tissues in preventing the germination of tomato (*Lycopersicon esculentum*) seeds. *Plant Physiology* 100:951-957.

Bewley, J.D. (1997). Seed germination and dormancy. *The Plant Cell* 9:1055-1066.

Bewley, J.D. and M. Black (1994). *Seeds: Physiology of development and germination.* New York: Plenum Press.

Bewley, J.D., R.A. Burton, Y. Morohashi, and G.B. Fincher (1997). Molecular cloning of a cDNA encoding a (1-4)-β-mannan endohydrolase from the seeds of germinated tomato *(Lycopersicon esculentum). Planta* 203:454-459.

Bies, N., A. Da Silva Conceicao, J. Giraudat, M. Koornneef, K.M. Léon-Kloosterziel, C. Valon, and M. Delseny (1999). Importance of the B2 domain of the *Arabidopsis* ABI3 protein for EM and 2S albumin gene regulation. *Plant Molecular Biology* 40:1045-1054.

Bleecker, A.B., M.A. Estelle, C. Somerville, and H. Kende (1988). Insensitivity to ethylene conferred by a dominant mutation in *Arabidopsis thaliana*. *Science* 241:1086-1089.

Bouwmeester, H.J. and C.M. Karssen (1992). The dual role of temperature in the regulation of the seasonal changes in dormancy and germination of seeds of *Polygonum persicaria* L. *Oecologia* 90:88-94.

Bradford, K.J. (1995) Water relations in seed germination. In J. Kigel and G. Galili, (eds.), *Seed development and germination* (pp. 351-396). New York: Marcel Dekker Inc.

Bradford, K.J., M.B. Chen, P. Cooley, P. Dahal, B. Downie, K.K. Fukunaga, O.H. Gee, S. Gurusinghe, R.A. Mella, H. Nonogaki, et al. (2000). Gene expression prior to radicle emergence in imbibed tomato seeds. In M. Black, K.J. Bradford, and J. Vázquez-Ramos (eds.), *Seed biology: Advances and applications* (pp. 231-251). Wallingford, UK; CAB International.

Casal, J.J. and R.A. Sánchez (1998). Phytochromes and seed germination. *Seed Science Research* 8:317-329.

Chen, F. and K.J. Bradford (2000). Expression of an expansin is associated with endosperm weakening during tomato seed germination. *Plant Physiology* 124: 1265-1274.

Chen, F., P. Dahal, and K.J. Bradford (2001). Two tomato expansin genes show divergent expression and localization in embryos during seed development and germination. *Plant Physiology* 127:928-936.

Cheng, W-H., A. Endo, L. Zhou, J. Penney, H.-C. Chen, A. Arroyo, P. Leon, E. Nambara, T. Asami, M. Seo, et al. (2002). A unique short chain dehydrogenase/reductase in *Arabidopsis* glucose signalling and abscisic acid biosynthesis and functions. *Plant Cell* 14:2723-2743.

Clouse, S.D. and J.M. Sasse (1998). Brassinosteroids: Essential regulators of plant growth and development. *Annual Review of Plant Physiology and Plant Molecular Biology* 49:427-451.

Cohn, M.A. (1996). Operational and philosophical decisions in seed dormancy research. *Seed Science Research* 6:147-153.

Cohn, M.A. and H.W.M. Hilhorst (2000). Alcohols that break seed dormancy: The anesthetic hypothesis, dead or alive? In J. Crabbé and J.-D. Viemont (eds.), *Plan Dormancy* (pp. 259-274). Wallingford, UK: CAB International.

Colucci, G., F. Apone, N. Alyeshmerni, D. Chalmers, and M.J. Chrispeels (2002). *GCR1*, the putative *Arabidopsis* G protein-coupled receptor gene is cell cycle regulated, and its overexpression abolishes seed dormancy and shortens time to flowering. *Proceedings of the National Academy of Sciences USA* 99:4736-4741.

Cosgrove, D.J. (1999) Enzymes and other agents that enhance cell wall extensibility. *Annual Review of Plant Physiology* 50:391-417.

Cutler, S., M. Ghassemian, D. Bonetta, S. Cooney, and P. McCourt (1996). A protein farnesyl transferase involved in abscisic acid signal transduction in *Arabidopsis*. *Science* 273:1239-1241.

da Silva, E.A.A. (2002). Coffee (*Coffea arabica* cv. Rubi) seed germination: Mechanism and regulation. Thesis, Wageningen University, Wageningen, the Netherlands.

Debeaujon, I. and M. Koornneef (2000). Gibberelin requirement for *Arabidopsis thaliana* seed germination is determined both by testa characteristics and embryonic ABA. *Plant Physiology* 122:415-424.

Debeaujon, I., K.M. Léon-Kloosterziel, and M. Koornneef (2000). Influence of the testa on seed dormancy, germination and longevity in *Arabidopsis thaliana*. *Plant Physiol*ogy 122:403-414.

De Castro, R.D. and H.W.M. Hilhorst (2000). Dormancy, germination and the cell cycle in developing and imbibing tomato seeds. *Revista Brasileira de Fisiologia Vegetal, Special Edition* 12:105-136.

De Castro, R.D., A.A.M. van Lammeren, S.P.C. Groot, R.J. Bino, and H.W.M. Hilhorst (2000). Cell division and subsequent radicle protrusion in tomato seeds are inhibited by osmotic stress, but DNA synthesis and formation of microtubular cytoskeleton are not. *Plant Physiology* 122:327-335.

de Castro, R.D., X. Zheng, J.H.W. Bergervoet, C.H. Ric de Vos, and R.J. Bino (1995). β–tubulin accumulation and DNA replication in imbibing tomato seeds. *Plant Physiology* 109:499-504.

de Miguel, L. and R.A. Sanchez (1992). Phytochrome-induced germination, endosperm softening and embryo growth potential in *Datura ferox* seeds: Sensitivity to low water potential and time to escape to FR reversal. *Journal of Experimental Botany* 43:969-974.

Derkx, M.P.M. and C.M. Karssen (1993). Effects of light and temperature on seed dormancy and gibberellin-stimulated germination in *Arabidopsis thaliana*: Studies with gibberellin-deficient and -insensitive mutants. *Physiologia Plantarum* 89:360-368.

Devic, M., J. Guilleminot, I. Debeaujon, N. Bechtold, E. Bensaude, M. Koornneef, G. Pelletier, and M. Delseny (1999). The *BANYULS* gene encodes a DFR-like protein and is a marker of early seed coat development. *Plant Journal* 19:387-398.

Eastmond, P.J. and I.A. Graham (2001). Re-examining the role of the glyoxylate cycle in oil seeds. *Trends in Plant Science* 6:72-77.

Finkelstein, R.R., S.S.L. Campala, and C.D. Rock (2002). Abscisic acid signaling in seeds and seedlings. *Plant Cell* Supplement 2002: S15-S45.

Finkelstein, R.R. and T.J. Lynch (2000). The *Arabidopsis* abscisic acid response gene *ABI5* encodes a basic leucine zipper transcription factor. *Plant Cell* 12:599-609.

Finkelstein, R.R., M.L. Wang, T.J. Lynch, S. Rao, and H.M. Goodman (1998). The *Arabisopsis* abscisic acid response locus *ABI4* encodes an APETALA2 domain protein. *Plant Cell* 10:1043-1054.

Footitt, S., S.P. Slocombe, V. Larner, S. Kurup, Y.S. Wu, T. Larson, I. Graham, A. Baker, and M. Holdsworth, M. (2002). Control of germination and lipid mobilization by COMATOSE, the *Arabidopsis* homologue of human ALDP. *EMBO Journal* 21:2912-2922.

Foster, R. and N.-H. Chua (1999). An *Arabidopsis* mutant with deregulated ABA gene expression: Implication for negative regulator fuction. *Plant Journal* 17: 363-372.

Frey, A., C. Audran, E. Marin, B. Sotta, and A. Marion-Poll (1999). Engineering seed dormancy by the modification of zeaxanthin epoxidase gene expression. *Plant Molecular Biology* 39:1267-1274.

Gallardo, K., C. Job, S.P.C. Groot, M. Puype, H. Demol, J. Vandekerckhove, and D. Job (2002). Proteomics of *Arabidopsis* seed germination: A comparative study of wild-type and gibberellin-deficient seeds. *Plant Physiology* 129:823-837.

Garciarrubio, A., J.P. Legaria, and A.A. Covarrubias (1997). Abscisic acid inhibits germination of mature *Arabidopsis* seeds by limiting the availability of energy and nutrients. *Planta* 203:182-187.

Germain, V., E.L. Rylott, T.R. Larson, S.M. Sherson, N. Bechtold, J.P. Carde, J.H. Bryce, I.A. Graham, and S.M. Smith (2001). Requirement for 3-ketoacyl-CoA thiolase-2 in peroxisome development, fatty acid beta-oxidation and breakdown of triacylglycerol in lipid bodies of *Arabidopsis* seedlings. *Plant Journal* 28: 1-12.

Ghassemian, M., E. Nambara, S. Cutler, H. Kawaide, Y. Kamiya and P. McCourt (2000). Regulation of abscisic acid signaling by the ethylene response pathway in *Arabidopsis*. *Plant Cell* 12:1117-1126.

Gibson, S.I. and I.A. Graham (1999). Another player joins the complex field of sugar-regulated gene expression in plants. *Proceedings of the National Academy of Sciences USA* 96:4746-4748.

Gibson, S.I., R.J. Laby, and D. Kim (2001). The *sugar-insensitive1 (sis1)* mutant of *Arabidopsis* is allelic to *ctr1*. *Biochemical and Biophysical Research Communications* 280:196-203.

Goldmark, P.J., J. Curry, C.F. Morris, and M.K. Walker-Simmons (1992). Cloning and expression of an embryo-specific mRNA up-regulated in hydrated dormant seeds. *Plant Molecular Biology* 19:433-441.

Górnik, K., R.D. de Castro, Y. Liu, R.J. Bino, and S.P.C. Groot (1997). Inhibition of cell division during cabbage (*Brassica oleracea* L.) seed germination. *Seed Science Research* 7:333-340.

Grappin, P., D. Bouinot, B. Sotta, E. Miginiac, and M. Jullien (2000). Control of seed dormancy in *Nicotiana plumbaginifolia*: Post-imbibition abscisic acid synthesis imposes dormancy maintenance. *Planta* 210:279-285.

Groot, S.P.C. and C.M. Karssen (1987). Gibberellins regulate seed germination in tomato by endosperm weakening: A study with gibberellin-deficient mutants. *Planta* 171:525-531.

Groot, S.P.C. and C.M. Karssen (1992). Dormancy and germination of abscisic acid-deficient tomato seeds: Studies with the *sitiens* mutant. *Plant Physiology* 99:952-958.

Groot, S.P.C., B. Kieliszewska-Rokicka, E. Vermeer, and C.M. Karssen (1988). Gibberellin-induced hydrolysis of endosperm cell walls in gibberellin-deficient tomato seeds prior to radicle protrusion. *Planta* 174:500-504.

Gualberti, G., M. Papi, L. Bellucci, L. Ricci, D. Bouchez, C. Camilleri, P. Costantino, and P. Vittorioso (2002). Mutations in the Dof zinc finger genes DAG2 and DAG1 influence with opposite effects the germination of *Arabidopsis* seeds. *Plant Cell* 14:1253-1263.

Haigh, A.M. and E.W.R. Barlow (1987). Water relations of tomato seed germination. *Australian Journal of Plant Physiology* 14:485-492.

Hallett, B.P. and J.D. Bewley (2002). Membranes and seed dormancy: Beyond the anaesthetic hypothesis. *Seed Science Research* 12:69-82.

Haslekas, C., R.A.P. Stacy, V. Nygaard, F.A. Culianez-Macia, and R.B. Aalen (1998). The expression of a peroxiredoxin antioxidant gene, *AtPer1,* in *Arabidopsis thaliana* is seed specific and related to dormancy. *Plant Molecular Biology* 36:833-845.

Hennig, L., W.M. Stoddart, M. Dieterle, G.C. Whitelam, and E. Schäfer (2002). Phytochrome E controls light-induced germination of *Arabidopsis. Plant Physiology* 128:194-200.

Hilhorst, H.W.M. (1990a). Dose-response analysis of factors involved in germination and secondary dormancy of seeds of *Sisymbrium officinale:* 1. Phytochrome. *Plant Physiology* 94:1090-1095.

Hilhorst, H.W.M. (1990b). Dose-response analysis of factors involved in germination and secondary dormancy of seeds of *Sisymbrium offinicale:* 2. Nitrate. *Plant Physiology* 94:1096-1102.

Hilhorst, H.W.M. (1995). A critical update on seed dormancy: I. Primary dormancy. *Seed Science Research* 5:61-73.

Hilhorst, H.W.M. (1998). The regulation of secondary dormancy: The membrane hypothesis revisited. *Seed Science Research* 8:77-90.

Hilhorst, H.W.M. and M.A. Cohn (2000). Are cellular membranes involved in the control of seed dormancy? In J. Crabbé and J.-D. Viemont (eds.), *Plant dormancy* (pp. 275-289). Wallingford, UK: CAB International.

Hilhorst, H.W.M., M.P.M. Derkx, and C.M. Karssen (1996). An integrating model for seed dormancy cycling; characterization of reversible sensitivity. In G. Lang, (ed.), *Plant dormancy: Physiology, biochemistry and molecular biology* (pp. 341-360). Wallingford, UK: CAB International.

Hilhorst, H.W.M. and B. Downie (1996). Primary dormancy in tomato (*Lycopersicon esculentum* cv. Moneymaker)—Studies with the *sitiens* mutant. *Journal of Experimental Botany* 47:89-97.

Hilhorst, H.W.M., S.P.C. Groot, and R.J. Bino (1998). The tomato seed as a model system to study seed dormancy and germination. *Acta Botanica Neerlandica* 47:169-183.

Hilhorst, H.W.M. and C.M. Karssen (1992). Seed dormancy and germination: The role of abscisic acid and gibberellins and the importance of hormone mutants. *Plant Growth Regulation* 11:225-238.

Hilhorst, H.W.M. and C.M. Karssen (2000). Effect of chemical environment on seed germination. In M. Fenner (ed.), *Seeds: The ecology of regeneration in* plant communities, Second edition (pp. 293-309). Wallingford, UK: CAB International.

Huijser, C., A.J. Kortstee, J.V. Pego, P. Weisbeek, E. Wisman and S. Smeekens (2000). The *Arabidopsis sucrose uncoupled-6* gene is identical to *abscisic acid insensitive-4*: Involvement of abscisic acid in sugar responses. *Plant Journal* 23:577-585.

Jacobsen, J.V., D.W. Pearce, A.T. Poole, R.P. Pharis, and L.N. Mander (2002). Abscisic acid, phaseic acid and gibberellin contents associated with dormancy and germination in barley. *Physiologia Plantarum* 115:428-441.

Jacobsen, S.E. and N.E. Olszewski (1993). Mutations at the *SPINDLY* locus of *Arabidopsis* alter gibberellin signal transduction. *Plant Cell* 5:887-896.

Jing, H.-C., A.A.M. van Lammeren, R.D. de Castro, H.W.M. Hilhorst, R.J. Bino, and S.P.C. Groot (1999). ß-Tubulin accumulation and DNA synthesis are sequentially resumed in embryo organs of cucumber (*Cucumis sativus* L.) seeds during imbibition. *Protoplasma* 208:230-239.

Karssen, C.M., D.L.C. Brinkhorst-van der Swan, A.E. Breekland, and M. Koornneef (1983). Induction of dormancy during seed development by endogeneous abscisic acid: Studies on abscisic acid deficient genotypes of *Arabidopsis thaliana* (L.) Heynh. *Planta* 157:158-165.

Karssen, C.M. and E. Laçka (1986). A revision of the hormone balance theory of seed dormancy: Studies on gibberellin and/or abscisic acid-deficient mutants of *Arabidopsis thaliana*. In M. Bopp (ed.), *Plant growth substances 1985* (pp. 315-323). Berlin, Heidelberg: Springer-Verlag.

Karssen, C.M., S. Zagórski, J. Kepczynski, and S.P.C. Groot (1989). Key role for endogenous gibberellins in the control of seed germination. *Annals of Botany* 63:71-80.

Kato K., W. Nakamura, T. Tabiki, H. Miura, and S. Sawada (2001). Detection of loci controlling seed dormancy on group 4 chromosomes of wheat and comparative mapping with rice and barley genomes. *Theoretical & Applied Genetics* 102:980-985.

Koornneef, M., L. Bentsink, and H.W.M. Hilhorst (2002). Seed dormancy and germination. *Current Opinion in Plant Biology* 5:33-36.

Koornneef, M., J.W. Cone, C.M. Karssen, R.E. Kendrick, J.H. Van der Veen, and J.A.D. Zeevaart (1985). Plant hormone and photoreceptor mutants in *Arabidopsis* and tomato. In M. Freeling (ed.), *Plant genetics, UCLA symposia on molecular and cellular biology,* New series, Volume 35 (pp. 1-12). New York: Alan Liss Inc.

Koornneef, M., C.J. Hanhart, H.W.M. Hilhorst, and C.M. Karssen (1989). In vivo inhibition of seed development and reserve protein accumulation in recombinants of abscisic acid biosynthesis and responsiveness mutants in *Arabidopsis thaliana*. *Plant Physiology* 90:463-469.

Koornneef, M., G. Reuling, and C.M. Karssen (1984). The isolation and characterization of abscisic acid insensitive mutants of *Arabidopsis thaliana*. *Physiologia Plantarum* 61:377-383.

Koornneef, M. and J.H. van der Veen (1980). Induction and analysis of gibberellin-sensitive mutants in *Arabidopsis thaliana* (L.) Heynh. *Theoretical and Applied Genetics* 58:257-263.

Laby, R.J., S. Kincaid, D. Kim, and S.I. Gibson (2000). The *Arabidopsis sugar-insensitive mutants sis4* and *sis5* are defective in abscisic acid synthesis and response. *Plant Journal* 23:587-596.

Lee, K.O., H.H. Jang, B.G. Jung, Y.H. Chi, J.Y. Lee, Y.O. Choi, J.R. Lee, C.O. Lim, M.J. Cecho, and S.Y. Lee (2000). Rice 1Cys-peroxiredoxin over-expressed in transgenic tobacco does not maintain dormancy but enhances antioxidant activity. *FEBS-Letters* 486:103-106.

Lee, S., H. Cheng, K.E. King, W. Wang, Y. He, A. Hussain, J. Lo, N.P. Harberd, and J. Peng (2002). Gibberellin regulates *Arabidopsis* seed germination via *RGL2*, a *GAI/RGA*-like gene whose expression is up-regulated following imbibition. *Genes & Development* 16:646-658.

Léon-Kloosterziel, K.M., C.J. Keijzer, and M. Koornneef (1994). A seed shape mutant of *Arabidopsis* that is affected in integument development. *Plant Cell* 6:385-392.

Léon-Kloosterziel, K.M., G.A. van de Bunt, J.A.D. Zeevaart, and M. Koornneef (1996). *Arabidopsis* mutants with a reduced seed dormancy. *Plant Physiology* 110:233-240.

Leubner-Metzger, G., C. Fründt, and F. Meins Jr. (1996). Effects of gibberellins, darkness and osmotica on endosperm rupture and class I ß-1,3-glucanase induction in tobacco seed germination. *Planta* 199:282-288.

Leubner-Metzger, G. and F. Meins Jr. (1999). Functions and regulation of plant ß-1,3-glucanases (PR-2). In S.K. Datta and S. Muthukrishnan (eds.), *Pathogenesis-related proteins in plants* (pp. 49-76). Boca Raton, FL, CRC Press LLC.

Leubner-Metzger, G. and F. Meins Jr. (2000). Sense transformation reveals a novel role for class I ß-1,3-glucanase in tobacco seed germination. *Plant Journal* 23:215-221.

Leubner-Metzger, G. and F. Meins Jr. (2001). Antisense-transformation reveals novel roles for class I ß-1,3-glucanase in tobacco seed afterripening and photodormancy. *Journal of Experimental Botany* 52:1753-1759.

Li, B. and M. Foley (1995). Cloning and characterization of differentially expressed genes in imbibed dormant and afterripened *Avena fatua* embryos. *Plant Molecular Biology* 29:823-831.

Liu, Y., R.J. Bino, C.M. Karssen, and H.W.M. Hilhorst (1996). Water relations of GA- and ABA-deficient tomato mutants during seed and fruit development and their influence on germination. *Physiologia Plantarum* 96:425-432.

Lopez-Molina, L., S. Mongrand, and N.-H. Chua (2001). A postgermination developmental arrest checkpoint is mediated by abscisic acid and requires the *ABI5* transcription factor in *Arabidopsis*. *Proceedings of the National Academy of Sciences USA* 98:4782-4787.

McCleary, B.V. and N.K. Matheson (1975). Galactomannan structure and β-mannanase and β-mannosidase activity in germinating legume seeds. *Phytochemistry* 14:1177-1184.

Nambara, E., R. Hayama, Y. Tsuchiya, M. Nishimura, H. Kawaide, Y. Kamiya, and S. Naito (2000). The role of *ABI3* and *FUS3* loci in *Arabidopsis thaliana* on phase transition from late embryo development to germination. *Developmental Biology* 220:412-423.

Nambara, E., S. Naito, and P. McCourt (1992). A mutant of *Arabidopsis* which is defective in seed development and storage protein accumulation is a new *abi3* allele. *Plant Journal* 2:435-441.

Ni, B.R. and K.J. Bradford (1993). Germination and dormancy of abscisic acid and gibberellin-deficient mutant tomato *(Lycopersicon esculentum)* seeds: Sensitivity of germination to abscisic acid, gibberellin, and water potential. *Plant Physiology* 101: 607-617.

Nikolaeva, M.G. (1977) Factors controlling the seed dormancy pattern. In A.A. Khan (ed.), *The physiology and biochemistry of seed dormancy and germination* (pp. 51-74). Amsterdam, New York, Oxford: North-Holland Publishing Company.

Nomaguchi, M., H. Nonogaki, and Y. Morohashi (1995). Development of galactomannan-hydrolysing activity in the micropylar endosperm tip of tomato seed prior to germination. *Physiologia Plantarum* 94:105-109.

Nonogaki, H., O.H. Gee, and K.J. Bradford (2000). A germination-specific endo-ß-mannanase gene is expressed in the micropylar endosperm cap of tomato seeds. *Plant Physiology* 123:1235-1245.

Nonogaki, H., M. Nomaguchi, N. Okumoto, Y. Kaneko, H. Matsushima, and Y. Morohashi (1998). Temporal and spatial pattern of the biochemical activation of the endosperm during and following imbibition of tomato seeds. *Physiologia Plantarum* 102:236-242.

Olszewski, N., T.-P. Sun, and F. Gubler (2002). Gibberellin signaling: Biosynthesis, catabolism, and response pathways. *Plant Cell* Supplement 2002:S61-S80.

Ooms, J.J.J., K.M. Léon-Kloosterziel, D. Bartels, M. Koornneef, and C.M. Karssen (1993). Acquisition of desiccation tolerance and longevity in seeds of *Arabidopsis thaliana:* A comparative study using ABA-insensitive *abi3* mutants. *Plant Physiology* 102:1185-1191.

Ooms, J.J.J., R. van der Veen, and C.M. Karssen (1994). Abscisic acid and osmotic stress or slow drying independently induce desiccation tolerance in mutant seeds of *Arabidopsis thaliana. Physiologia Plantarum* 92:506-510.

Papi, M., S. Sabatini, M.M. Altamura, L. Hennig, E. Schafer, P. Costantino, and P. Vittorioso (2002). Inactivation of the phloem-specific dof zinc finger gene DAG1 affects response to light and integrity of the testa of *Arabidopsis* seeds. *Plant Physiology* 128:411-417.

Papi, M., S. Sabatini, D. Bouchez, C. Camilleri, P. Costantino, and P. Vittorioso (2000). Identification and disruption of an *Arabidopsis* zinc finger gene controlling seed germination. *Genes and Development* 14:28-33.

Parcy, F., C. Valon, A. Kohara, S. Miséra, and J. Giraudat (1997). The *ABSCISIC ACID-INSENSITIVE3, FUSCA3,* and *LEAFY COTOLEDON1* loci act in concert to control multiple aspects of *Arabidopsis* seed development. *Plant Cell* 9:1265-1277.

Peeters, A.J.M., H. Blankestijn-de Vries, C.J. Hanhart, K.M. Leon-Kloosterziel, J.A.D. Zeevaart, and M. Koornneef (2002). Characterization of mutants with reduced seed dormancy at two novel *rdo* loci and a further characterization of *rdo1* and *rdo2* in *Arabidopsis. Physiologia Plantarum* 115:604-612.

Pego, J.V., A.J. Kortstee, C. Huijser, and S. Smeekens (2000). Photosynthesis, sugars and the regulation of gene expression. *Journal of Experimental Botany* 51:407-416.

Peng, J.R. and N.P. Harberd (2002). The role of GA-mediated signaling in the control of seed germination. *Current Opinion in Plant Biology* 5:376-381.

Philips, J.R., M.J. Oliver, and D. Bartels (2002). Molecular genetics of desiccation and tolerant systems. In M. Black and H. Pritchard (eds.), *Desiccation and survival in plants: Drying without dying* (pp. 319-341). Wallingford, UK: CAB International.

Pritchard, S.L., W.L. Charlton, A. Baker, and I.A. Graham (2002). Germination and storage reserve mobilization are regulated independently in *Arabidopsis. Plant Journal* 31:639-647.

Probert, R.J. (1992). The role of temperature in germination ecophysiology. In M. Fenner (ed.), *Seeds: The ecology of regeneration in plant communities* (pp. 285-325). Wallingford, UK: CAB International.

Raz, V., J.H.W. Bergervoet, and M. Koornneef (2001). Sequential steps for developmental arrest in *Arabidopsis* seeds. *Development* 128:243-252.

Reid, J.S.G. (1985). Cell wall storage carbohydrate in seeds: Biochemistry of the seed 'gums and hemicelluloses'. *Advances in Botanical Research* 11:125-155.

Roberts, E.H. and R.D. Smith (1977). Dormancy and the pentose phosphate pathway. In A.A. Khan (ed.), *The physiology and biochemistry of seed dormancy and germination* (pp. 385-411). Amsterdam: Elsevier/North-Holland Biomedical Press.

Rohde, A., S. Kurup, and M.J. Holdsworth (2000). ABI3 emerges from the seed. *Trends in Plant Science* 5:418-419.

Romagosa I., F. Han, J.A. Clancy, and S.E. Ullrich (1999). Individual locus effects on dormancy during seed development and after ripening in barley. *Crop Science* 39:74-79.

Russell, L., V. Larner, S. Kurup, S. Bougourd, and M.J. Holdsworth (2000). The *Arabidopsis COMATOSE* locus regulates germination potential. *Development* 127:3759-3767.

Schopfer, P. and C. Plachy (1984). Control of seed germination by abscisic acid: II. Effect on embryo water uptake in *Brassica napus* L. *Plant Physiology* 76:155-160.

Schopfer, P. and C. Plachy (1985). Control of seed germination by abscisic acid. III. Effect on embryo growth potential (minimum turgor pressure) and growth coefficient (cell wall extensibility) in *Brassica napus* L. *Plant Physiology* 77:676-686.

Schumacher, K. and J. Chory (2000). Brassinosteroid signal transduction: Still casting the actors. *Current Opinion in Plant Biology* 3:79-84.

Seo, M., A.J.M. Peeters, H. Koiwai, M. Oritani, A. Marion-Poll, J.A.D. Zeevaart, M. Koornneef, Y. Kamiya, and T. Koshiba (2000). The *Arabidopsis aldehyde oxidase 3* (*AAO3*) gene product catalyzes the final step in abscisic acid biosynthesis in leaves. *Proceedings of the National Academy of Sciences USA* 97:12908-12913.

Sitrit, Y., K.A. Hadfield, A.B. Bennett, K.J. Bradford, and B. Downie (1999). Expression of a polygalacturonase associated with tomato seed germination. *Plant Physiology* 121:419-428.

Smeekens, S. (1998). Sugar regulation of gene expression in plants. *Current Opinion in Plant Biology* 1:230-234.

Söderman, E.M., I.M. Brocard, T.J. Lynch, and R.R. Finkelstein (2000). Regulation and function of the *Arabidopsis* ABA-insensitive gene in seed abscisic acid response signaling networks. *Plant Physiology* 124: 1752-1765.

Steber, C.M., S. Cooney, and P. McCourt (1998). Isolation of the GA-response mutant *sly1* as a suppressor of *ABI1-1* in *Arabidopsis thaliana. Genetics* 149:509-521.

Steber, C.M. and P. McCourt (2001). A role for brassinosteriods in germination in *Arabidopsis. Plant Physiology* 125:763-769.

Toorop, P.E., J.D. Bewley, and H.W.M. Hilhorst (1996). Endo-ß-mannanase isoforms are present in the endosperm and embryo of tomato seeds, but are not essentially linked to the completion of germination. *Planta* 200:153-158.

Toorop, P.E., A.C. van Aelst, and H.W.M. Hilhorst (2000). ABA controls the second step of the biphasic endosperm cap weakening that mediates tomato (*Lycopersicon esculentum*) seed germination. *Journal of Experimental Botany* 51: 1371-1379.

Toyomasu, T., H. Tsuji, H. Yamane, M. Nakayama, I. Yamaguchi, N. Murofushi, N. Takahashi, and Y. Inoue (1993). Light effects on endogenous levels of gibberellins in photoblastic lettuce seeds. *Journal of Plant Growth Regulation* 12:85-90.

Toyomasu, T., H. Yamane, N. Murofushi, and Y. Inoue (1994). Effects of exogenously applied gibberellin and red light on the endogenous levels of abscisic acid in photoblastic lettuce seeds. *Plant & Cell Physiology* 35:127-129.

Ullah, H., J.-G. Chen, S. Wang, and A.M. Jones (2002). Role of a heterotrimeric G protein in regulation of *Arabidopsis* seed germination *Plant Physiology* 129: 897-907.

Vázquez-Ramos, J.M. (2000). Cell cycle control during maize germination. In M. Black, K.J. Bradford, and J. Vázquez-Ramos (eds.), *Seed biology: Advances and applications* (pp. 261-269). Wallingford, UK: CAB International.

Watkins, J.T. and D.J. Cantiliffe (1983). Mechanical resistance of the seed coat and endosperm during germination of *Capsicum annum* at low temperature. *Plant Physiology* 72:146-150.

Welbaum, G.E., K.J. Bradford, K.-O. Yim, D.T. Booth, and M.O. Oluouch (1998). Biophysical, physiological and biochemical processes regulating seed germination. *Seed Science Research* 8:161-172.

Welbaum, G.E., W.J. Muthui, J.H. Wilson, R.L. Grayson, and R.D. Fell (1995). Weakening of muskmelon perisperm envelope tissue during germination. *Journal of Experimental Botany* 46:391-400.

Wen, C.-K. and C. Chang (2002). *Arabidopsis RGL1* encodes a negative regulator of gibberellin responses. *Plant Cell* 14:87-100.

Werker, E. (1997). *Seed anatomy: Encyclopedia of plant anatomy*, Spezieller Teil, Band X, Teil 3. Berlin, Stuttgart: Gebrüder Borntraeger.

Wu, C.-T., G. Leubner-Metzger, F. Meins Jr., and K.J. Bradford (2000). Class I ß-1,3-glucanase and chitinase are expressed in the micropylar endosperm of tomato seeds prior to radicle emergence. *Plant Physiology* 126:1299-1313.

Xiong, L., M. Ishitani, H. Lee, and J.-K. Zhu (1999). *HOS5*—a negative regulator of osmotic stress-induced gene expression in *Arabidopsis thaliana. Plant Journal* 19:569-578.

Xiong, L. and J.K. Zhu (2003). Regulation of abscisic acid biosynthesis. *Plant Physiology* 133: 29-36.

Yamaguchi, S., M.W. Smith, R.G. Brown, Y. Kamiya, and T. Sun (1998). Phytochrome regulation and differential expression of *gibberellin 3* β-*hydroxylase* genes in germinating *Arabidopsis* seeds. *Plant Cell* 10:2115-2126.

Zhou, L., J.-C. Jang, T.L. Jones, and J. Sheen (1998). Glucose and ethylene signal transduction crosstalk revealed by an *Arabidopsis* glucose-insensitive mutant. *Proceedings of the National Academy of Sciences USA* 95:10294-10299.

Chapter 11

Hormonal Interactions During Seed Dormancy Release and Germination

Gerhard Leubner-Metzger

INTRODUCTION

Plant hormones are of upmost importance in the regulation of the genetic, physiological, and biochemical properties of seeds. Seeds are propagation and dispersal structures usually containing fully developed plant embryos and functioning to ensure the establishment of a new plant generation. Only little is known about the interconnected molecular key processes regulating seed dormancy and germination in response to plant hormones and environmental cues. The dry dormant seed is well equipped to survive extended periods of unfavorable conditions.

Seed dormancy can be "coat-imposed" and/or determined by the embryo itself and is a temporary failure or block of a viable seed to complete germination under physical conditions that normally favor the process (for reviews see Hilhorst, 1995; Bewley, 1997b; Li and Foley, 1997; Koornneef et al., 2002). Considerable genetic variation of seed dormancy can be found within species, accessions of wild type, and varieties of cultivated plants. The substantial influence of environmental effects on the expression of germination characteristics and the involvement of many genes make dormancy a typical quantitative trait, which is subject to QTL (quantitative trait loci) analysis (Koornneef et al., 2002).

The process of germination commences with the uptake of water by imbibition of the dry seed, followed by embryo expansion growth. This usually culminates in rupture of the covering layers and emergence of the

I thank Eberhard Schäfer and Klaus Harter (Institut für Biologie II, Botanik, Albert-Ludwigs-Universität, Freiburg i. Br., Germany) for their critical comments. My research is supported by a grant from the Deutsche Forschungsgemeinschaft (LE 720/3), which is gratefully acknowledged.

radicle, generally considered to be the completion of germination. Radicle protrusion during seed germination depends on embryo expansion, which is a growth process driven by water uptake. It is generally believed that cell elongation is necessary and sufficient for the completion of radicle protrusion and that cell division is not necessary. In many plant species with coat-imposed dormancy the seed envelope imposes a physical constraint to radicle protrusion, which has to be overcome by the growth potential of the embryo.

In nonendospermic species endosperm assimilation occurs during seed development. In many mature nonendospermic seeds the cotyledons are the sole storage organs and the embryo is enclosed by the testa (seed coat) as the sole covering layer (e.g., Schopfer and Plachy, 1984, 1993; Kretsch et al., 1995). The testa imposes a restraint during the germination of radish (Schopfer and Plachy, 1993) and *Arabidopsis* (Debeaujon and Koornneef, 2000; Debeaujon et al., 2000), but the testa is no hindrance during the germination of rape (Schopfer and Plachy, 1984) and pea (Petruzzelli et al., 2000). In nonendospermic seeds, as well as in *Arabidopsis* with only a remainder of the endosperm, the testa characteristics are responsible for the degree of coat-imposed dormancy (Debeaujon and Koornneef, 2000; Debeaujon et al., 2000).

In addition to the testa, a diploid and entirely maternal tissue, in endospermic seeds the seed covering layers also include the endosperm, which is usually triploid in angiosperms, and two-thirds of its genome originates from the mother plant. In endospermic seeds the contributions of both the testa and the endosperm layers to the degree of coat-imposed dormancy have to be considered (for reviews see Hilhorst, 1995; Bewley, 1997a; Leubner-Metzger, 2003b). Endosperm rupture is the main germination-limiting process in members of the Asteraceae (e.g., lettuce) and Solanaceae (e.g., tomato and tobacco). In these cases of endosperm-limited germination, weakening of the micropylar endosperm surrounding the radicle tip seems to be required for radicle protrusion and is likely to involve cell wall hydrolysis by the action of hydrolytic enzymes.

Major recent contributions to the issue of endosperm-limited germination were achieved by using tomato and *Nicotiana* species as suitable model systems for endospermic seeds (for reviews see Hilhorst, 1995; Koornneef et al., 2002; Leubner-Metzger, 2003b). However, recent progress in the field of seed biology was especially achieved by molecular genetic studies with *Arabidopsis thaliana* hormone mutants. This review focuses mainly on dicot seeds and on the interactions among abscisic acid (ABA), gibberellins (GA), ethylene, and brassinosteroids (BR) in regulating the interconnected molecular key processes that determine dormancy and germination. I attempt to present an integrated view of the efforts made using approaches of molecular genetics, physiology, and biochemistry to unravel these fascinating processes of seed biology.

ABSCISIC ACID (ABA): A POSITIVE REGULATOR OF DORMANCY INDUCTION AND MAINTENANCE, A NEGATIVE REGULATOR OF GERMINATION

Seed development is completed by a period of maturation when water content decreases, ABA and storage proteins accumulate, and desiccation tolerance and primary dormancy are established. In many plant species endogenous ABA is involved in both the induction and maintenance of the dormant state (for reviews see Hilhorst, 1995; Bewley, 1997b; Li and Foley, 1997; Koornneef et al., 2002; Leubner-Metzger, 2003b).

ABA deficiency during seed development is associated with absence of primary dormancy of the mature seed. The formation of nondormant seeds and in some cases even precoccious germination on the mother plant (vivipary) have been reported for ABA-deficient biosynthesis mutants, e.g., for *aba1* and *aba2* of *Arabidopsis* (Karssen et al., 1983; Koornneef and Karssen, 1994), *sitiens* (*sit*^{w}) and *notabilis* (*not*) of tomato (Hilhorst, 1995; Thompson et al., 2000), *aba2* of *Nicotiana plumbaginifolia* (Marin et al., 1996), and for several of the *viviparous (vp)* mutants of maize (Liotenberg et al., 1999; White et al., 2000). For the *not* mutant of tomato this ABA deficiency is due to a mutation in the gene encoding a putative 9-*cis*-epoxycarotenoid dioxygenase that leads to decreased ABA levels during seed development and to reduced seed dormancy (Hilhorst, 1995; Thompson et al., 2000). Overexpression of (wild-type) NOT in transgenic tomato can cause enhanced ABA biosynthesis and enhanced seed dormancy. Similar results have been obtained for transgenic *Arabidopsis* seeds with enhanced ABA levels (Lindgren et al., 2003). In the *aba2* mutant of *N. plumbaginifolia* ABA deficiency is due to a mutation in the *ABA2* gene, encoding zeaxanthin epoxidase, a key step in ABA biosynthesis (Marin et al., 1996). Antisense- and sense-*ABA2* transformation of *N. plumbaginifolia* results in decreased and increased ABA biosynthesis and seed dormancy, respectively (Frey et al., 1999).

The onset of dormancy induction in *Nicotiana tabacum* is correlated with a peak in ABA content at approximately 15 to 20 days after pollination (DAP); a rapid decline in ABA content follows during further seed maturation; and dormancy has been established when seeds are harvested at DAP 25 (Yamaguchi-Shinozaki et al., 1990; Jiang et al., 1996; Phillips et al., 1997; Leubner-Metzger and Meins, 2000).

Seed dormancy is not established in transgenic tobacco expressing an anti-ABA antibody that causes deficiency in free ABA (Phillips et al., 1997). ABA production during seed development can be of dual origin; it occurs both in the embryo and in the maternal tissues. Reciprocal crosses and/or grafting experiments between wild-type and ABA-deficient mutants of *Arabidopsis* (Karssen et al., 1983; Koornneef and Karssen, 1994), tomato (Groot and Karssen, 1992; Hilhorst, 1995), and *N. plumbaginifolia* (Frey et al., 1999) have shown that only ABA produced by the embryo itself, and not maternal ABA, is necessary to impose a lasting dormancy. Exogenously applied ABA resembles maternal ABA and also fails to to induce seed dormancy. However, sensitivity of seeds to ABA is partially maternally controlled and embryonic, maternal, and applied ABA affect other aspects of seed development (Finkelstein, 1994; Koornneef and Karssen, 1994).

The *Arabidopsis ABA-insensitive (abi)* response mutants *abi1* to *abi5* have all been indentified by selecting for seeds capable of germination in the presence of ABA concentrations that are inhibitory to the wild type (Koornneef and Karssen, 1994; Leung and Giraudat, 1998). Several of these ABA-insensitive mutants also exhibit, similar to the ABA-deficient mutants, a marked reduction in seed dormancy (see Figure 11.1). The *ABI1* and *ABI2* genes encode homologous serine/threonine protein phosphatases 2C (PP2C) (Grill and Himmelbach, 1998; Leung and Giraudat, 1998; Beaudoin et al., 2000). The abi1-1 and abi2-1 mutations are dominant and lead to ABA-insensitive phenotypes that include reduced seed dormancy, ABA-insensitive germination, and vegetative responses. In contrast to the *abi1-1* response mutant, loss-of-function alleles of *ABI1* lead to ABA-hypersensitive response mutants, indicating that the ABI1 PP2C is a negative regulator of ABA responses. Seed-related PP2C have also been reported from other species, e.g., transcripts of a PP2C are synergistically up-regulated by ABA and calcium in dormant seeds of *Fagus sylvatica* (Lorenzo et al., 2002). The seed responses of the *Arabidopsis abi3* mutants are severe compared to the *abi1*, *abi2*, and the ABA-deficient mutants and are mostly limited to seeds (Leung and Giraudat, 1998; Schwechheimer and Bevan, 1998; Raz et al., 2001). Similarily, the ABA-insensitive *viviparous1* (*vp1*) mutant of maize is characterized by severe seed responses including reduced sensitivity of germination to exogenous ABA and vivipary (e.g., McCarty, 1995; Li and Foley, 1997; Schwechheimer and Bevan, 1998). The *Arabidopsis ABI3* and the maize *VP1* are orthologous genes that encode seed-specific transcription factors of the B3 domain class that are essential for ABA action. The VP1/ABI3-like proteins can function as activators and repressors of ABA-dependent and -independent gene expression in seeds (e.g., Ezcurra et al., 2000; Holdsworth et al., 2001; Suzuki et al., 2001;

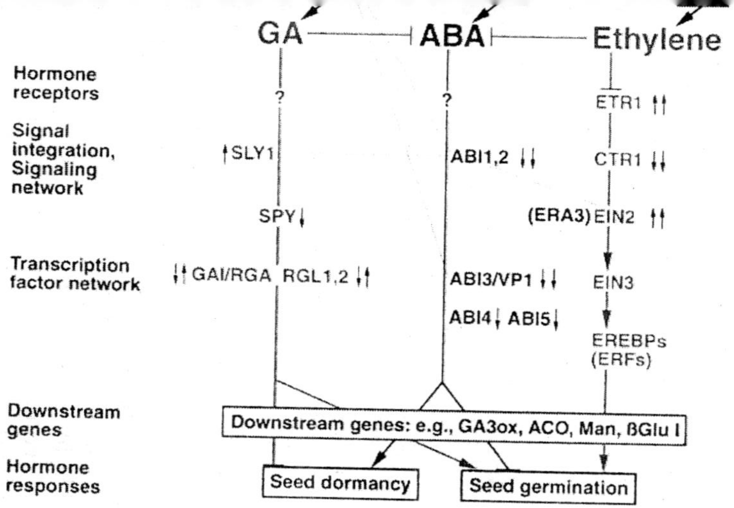

FIGURE 11.1. Schematic representation of the interactions between the gibberellin (GA), absisic acid (ABA), and ethylene signaling pathways in the regulation of seed dormancy and germination. The model is mainly based on *Arabidopsis* hormone mutant analyses, the positions of some components is speculative, and details are explained in the text. Promotion or inhibition of a process is indicated by thick arrows and blocks, respectively. Interactions based on extragenic suppressor or enhancer screens are indicated by thin gray lines. Small arrows indicate enhancement (up-arrow) or reduction (down-arrow) of seed dormancy or seed ABA sensitivity upon mutation of the corresponding protein. Corresponding hormone mutants of *A. thaliana*: *aba1*, *aba2* = *ABA-deficient1,2*; *abi1* to *abi5* = *ABA-insensitive1* to *ABA-insensitive5*; *ctr1* = *constitutive triple response1*; *ein2*, *ein3* = *ethylene insensitive2, 3*; *era3* = *enhanced response to ABA3*; *gai* = *GA-insensitive*; *sly1* = *sleepy1*; *spy* = *spindly*; *rga* = *repressor-of-ga1-3*; *rgl 1*, *rgl2* = *rga-like1, 2;* ACO = ACC oxidase; ACS = ACC synthase; EREBP = ethylene responsive element binding protein; ERF = ethylene responsive factor; GA3ox = GA 3-oxidase; Man = mannanase; βGlu I = class I β-1,3-glucanase; *vp1* = *viviparous1* (maize mutant).

Clerkx et al., 2003; Zeng et al., 2003). Ectopic expression of ABI3 or VP1 in *Arabidopsis* plantlets confers ABA- and seed-specific responses to vegetative tissues (Leung and Giraudat, 1998; Suzuki et al., 2001). These findings suggest that the VP1/ABI3-like proteins are multifunctional transcription factors that integrate ABA and other regulatory signals of seed maturation and dormancy. Factors that interact with VP1/ABI3 have been identified and it has been proposed that the VP1/ABI3-like transcription factors not only regulate seed responses, but also might function as general regulators for the timing of developmental transitions throughout the life cycle of plants (Hobo et al., 1999; Rohde et al., 2000; Holdsworth et al., 2001).

Arabidopsis ABI4 shows the greatest sequence homology with ethylene-responsive element binding protein (EREBP)-type transcription factors and contains an APETALA2 (AP2)-like DNA binding domain characteristic for AP2/EREBP family transcriptional regulators (Finkelstein et al., 1998; Ohta et al., 2000; Söderman et al., 2000). ABI4 is likely to be seed specific and may act downstream of ABI3 or in a parallel pathway. ABA signal transduction via ABI4 is also involved in responses of seeds to sugar (Huijser et al., 2000; Finkelstein and Gibson, 2002). The *Arabidopsis ABI5* gene encodes a basic leucine zipper (bZIP) transcription factor important in determining ABA responsiveness during embryogenesis and regulating the transition from germination to vegetative growth (Finkelstein and Lynch, 2000; Lopez-Molina et al., 2001, 2002). A postgermination developmental arrest checkpoint is mediated by ABA and requires the ABI5 transcription factor. In agreement with this, the transition from the initial imbibition phase (water uptake phases 1 and 2) during seed germination to the growth phase (water uptake phase 3) is inhibited by ABA, and ABA inhibits further embryo extension and seedling growth after radicle emergence. However, ABA appears not to inhibit initial imbibition of water, initial embryo extension growth, and testa rupture of many nonendospermic, nondormant seeds, e.g., of rape (Schopfer and Plachy, 1984). Work with *Arabidopsis* seeds showed that the ABA block on germination is not a consequence of inhibition of storage lipid mobilization (Pritchard et al., 2002). Two independent programs appear to operate, one that is blocked by ABA, governing developmental growth resulting in germination, and a second that governs storage lipid mobilization which is largely ABA independent. ABI3, ABI4, and ABI5 interact, and it is proposed that these transcription factors are key factors in a signaling network that regulates ABA-related seed responses (Finkelstein et al., 2002; Brocard-Gifford et al., 2003). The *Arabidopsis* ABA-hypersensitive response mutants, including *enhanced response to ABA1* (*era1*), exhibit enhanced seed dormancy (Cutler et al., 1996; Ghassemian et al., 2000). Their germination is already inhibited by low ABA concentrations that do not affect the wild type. The *ERA1* gene encodes the ß subunit of a farnesyl transferase, and it is proposed that a negative regulator of ABA sensitivity is modulated by protein farnesylation. The *Arabidopsis reduced dormancy* (*rdo*) mutants are not altered in ABA biosynthesis or sensitivity (Leon-Kloosterziel et al., 1996; Peeters et al., 2002). Finally, reduced dormancy of the *Arabidopsis dof affecting germination1-1 (dag1-1)* mutant is also not due to altered ABA sensitivity (Papi et al., 2000). The *DAG1* gene encodes a Dof ("DNA-binding with one finger") transcription factor involved in the maternal control of germination. Inactivation of the phloem-specific *DAG1* gene affects the seed response to light and the integrity of the testa (Papi et al., 2002). Dof-type transcription factors interact with phytochrome and GA signaling in seeds (e.g., Diaz et al., 2002; Hennig et al., 2002; Papi et al., 2002). DAG1 and DAG2 are proposed act on a maternal switch that controls seed germination (Gualberti et al., 2002). Taken together, factors determining spatial and temporal ABA content and sensitivity pattern in seeds positively regulate dormancy induction and maintenance, and negatively regulate dormancy release and germination.

ABA is a positive regulator of not only dormancy induction, but also dormancy maintenance. Dormancy can be released during after-ripening, i.e., a period of dry storage of freshly harvested, mature seeds (Bewley, 1997b; Li and Foley, 1997). Further decline in ABA content, decreased sensitivity to ABA, and increased sensitivity to GA are involved in the after-ripening-mediated transition from the dormant to the nondormant state of many species (e.g., Hilhorst, 1995; Li and Foley, 1997; Benech-Arnold et al., 1999; Debeaujon and Koornneef, 2000; Grappin et al., 2000; Romagosa et al., 2001; Koornneef et al., 2002; Schmitz et al., 2002). The work of Grappin et al. (2000) demonstrates this for *N. plumbaginifolia* and shows in addition de novo ABA biosynthesis in imbibed fresh (dormant), but not after-ripened (nondormant) seeds. Expression studies in *N. tabacum* seeds of ABA-regulated genes are in agreement with this finding and suggest that a common role of ABA during the after-ripening-mediated dormancy maintenance and release among *Nicotiana* species (Leubner-Metzger and Meins, 2000; Leubner-Metzger, 2002). Vegetation-derived ABA is also of ecological importance in the regulation of seed dormancy and germination. ABA leached from vegetation litter plays an important role in the germination control of the postfire annual *Nicotiana attenuata* (Krock et al., 2002). Rupture of the testa and the endosperm are distinct and temporally separate events during the germination of tobacco seeds (see Figure 11.2; Leubner-Metzger, 2003b; Web: <http:// www.seedbiology.de/>). The after-ripening-mediated promotion of tobacco germination is due to the promotion of both testa and subsequent endosperm rupture. Addition of ABA to the medium during imbibition resembles maternal ABA during seed development and

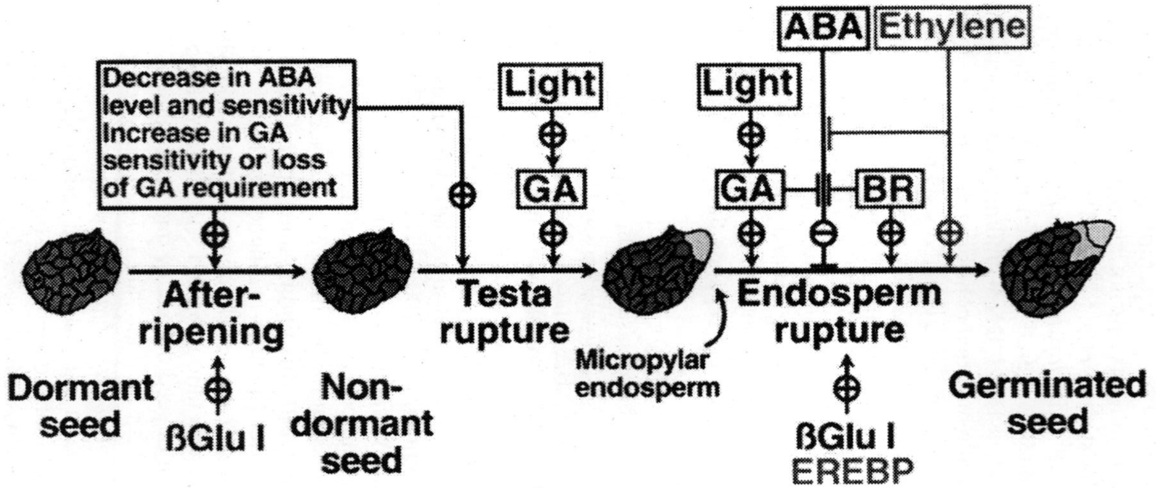

FIGURE 11.2. Hormonal interactions during tobacco seed after-ripening, dormancy release, and germination and their effects on testa rupture and endosperm rupture. Expression of the abscisic acid (ABA)-inhibited βGlu I genes contribute to the release of coat-imposed dormancy and the promotion of germination by acting at two sites. First, decrease in ABA level and sensitivity eventually permit βGlu I expression in seeds during after-ripening. This βGlu I contributes to the release of coat-imposed dormancy and promotes testa rupture in the light. Second, βGlu I is induced by the light/gibberellin (GA) pathway in the micropylar endosperm and facilitates endosperm rupture. Endosperm-specific βGlu I expression and endosperm rupture are inhibited by ABA and promoted by light, GA, and ethylene. The light/GA pathway also counteracts ABA effects by promoting ABA degradation. Ethylene and brassinosteroids (BR) counteract ABA effects and promote endosperm rupture, but do not affect testa rupture. EREBPs (ethylene responsive element binding proteins) are transcription factors that mediate hormonal regulation of βGlu I expression and endosperm rupture. BR and light/GA promote tobacco endosperm rupture by distinct signal transduction pathways. A plus (+) sign means promotion and a minus (−) sign inhibition of a process.

residual ABA in mature seeds. ABA treatment does not appreciably affect the kinetics of testa rupture of fresh or after-ripened tobacco seeds, but it delays endosperm rupture and results in the formation of a novel structure, consisting of the enlarging radicle with a sheath of greatly elongated endosperm tissue. The visible distinction between testa and endosperm rupture and the finding that ABA inhibits endosperm rupture, but not testa rupture, is typical for the seed germination of the Cestroideae (*Nicotiana* spp., *Petunia hybrida*) subgroup of Solanaceous species (Krock et al., 2002; Petruzzelli, Müller, et al., 2003). Class I β-1,3-glucanase (βGlu I) is induced after testa rupture and just prior to endosperm rupture of *Nicotiana* seeds (Leubner-Metzger, 2003b). This induction is exclusively localized in the micropylar endosperm at the site where the radicle will emerge. ABA inhibits the induction of the βGlu I genes during tobacco seed germination and specifically delays endosperm rupture (Leubner-Metzger et al., 1995). The close correlation between βGlu I induction and the onset of endosperm rupture under a variety of physiological conditions support the hypothesis that βGlu I contributes to endosperm rupture (Figure 11.2). This hypothesis is further supported by results from other species representing the Cestroideae *(N. plumbaginifolia, N. sylvestris, N. tabacum, Petunia hybrida)* and Solanoideae *(Capsicum annuum, Lycopersicon esculentum, Physalis peruviana)* subgroup of the Solanaceae (Wu et al., 2000; Leubner-Metzger, 2003b; Petruzzelli, Müller, et al., 2003). ABA inhibition of germination and ßGlu accumulation in the micropylar endosperm appears to be a widespread event during the seed germination of Solanaceous species. Direct evidence for a causal role of βGlu I during endosperm rupture comes from sense-transformation of *N. tabacum* with a chimeric ABA-inducible βGlu I transgene (Leubner-Metzger and Meins, 2000; Leubner-Metzger, 2003b; Web: <http://www.seedbiology.de/>). This has been achieved be transformation of tobacco with a sense-βGlu I construct consisting of the genomic DNA fragment of the tobacco β*GluIB* gene regulated by the castor bean *Cat1* gene promoter, which is known to confer ABA-inducible, endosperm-specific transgene expression in germinating tobacco seeds. Seeds have been harvested from independent sense-βGlu I lines (TKSG7) and, for the purpose of proper controls, from empty-vector lines (TCIB1). Sense-βGlu I transformation results in overexpression of βGlu I in TKSG7 seeds and promotes endosperm rupture of fresh, mature (dormant) seeds and ABA-treated after-ripened (nondormant) seeds. In contrast to fresh and ABA-treated after-ripened TKSG7 seeds, βGlu I overexpression does not promote endosperm rupture of after-ripened TKSG7 seeds imbibed in medium without ABA added. ABA down-regulates the βGlu I host genes in TCIB1 seeds, but due to the ABA-inducible βGlu I transgene it causes high-level βGlu I expression in TKSG7 seeds. ABA treatment delays endosperm rupture of after-ripened TCIB1 and TKSG7 seeds, but due to βGlu I overexpression this delay is significantly reduced in TKSG7 seeds. βGlu I overexpression reduces the ABA-mediated delay in endosperm rupture of fresh and after-ripened seeds, but ABA treatment does not affect the kinetics of testa rupture. Taken together, these results support the view that a threshold βGlu I content is necessary, but not sufficient, for endosperm rupture (Leubner-Metzger et al., 1995; Leubner-Metzger, 2003b). In the presence of ABA βGlu I becomes a limiting factor for endosperm rupture, and removal of this block due to expression of the ABA-inducible βGlu I-transgene in TKSG7 seeds promotes endosperm rupture until other ABA-sensitive processes become limiting (Leubner-Metzger and Meins, 2000). Although these results do not exactly show how βGlu I promote endosperm rupture, they directly show that βGlu I is causally involved and that it substantially contributes to endosperm rupture.

GIBBERELLINS RELEASE DORMANCY, PROMOTE GERMINATION, AND COUNTERACT ABA EFFECTS

According to the revised hormone-balance hypothesis for seed dormancy proposed by Karssen and Laçka (1986), ABA and GA act at different times and sites during "seed life." ABA induces dormancy during maturation, and GA plays a key role in the promotion of germination. GA biosynthesis in developing seeds of many species leads to the accumulation and storage of either bioinactive GA precursors or bioactive GA (e.g., Groot et al., 1987; Toyomasu et al., 1998; Kamiya and Garcia-Martinez, 1999; Yamaguchi et al., 2001). GA biosynthesis in developing seeds seems not to be involved in the induction of primary dormancy per se (Karssen and Laçka, 1986; Groot et al., 1987; Koornneef and Karssen, 1994; Bewley, 1997b), but GA biosynthesis seems to be involved in some aspects of seed development including fertilization, embryo growth, assimilate uptake, fruit growth, and the prevention of seed abortion of tomato, pea, and several species of the Brassicaceae (e.g., Groot et al., 1987; Swain et al., 1997; Batge et al., 1999; Hays et al., 2002; Koornneef et al., 2002; Singh et al., 2002). However, recent experiments with ABA-deficient and -insensitive mutants of maize and with GA biosynthesis inhibitors demonstrate that GA is a positive regulator of vivipary (White et al., 2000; White and Rivin, 2000). Bioactive GA accumulate prior to the ABA peak during embryo development of maize. GA biosynthesis inhibition mimicks the effects of exogenous ABA, e.g., in suppressing vivipary. Interestingly, the GA/ABA ratio, and not the absolute levels of these hormones, appears to control vivipary. Thus, it is possible that GA directly antagonizes ABA signaling during maize kernel development. Cereal endosperm development and postgerminational aleurone layer senescence are promoted by GA and ethylene, and are inhibited by ABA (e.g., Bethke and Jones, 2001), but this well-investigated field is beyond the scope of this review. One result from these studies that might also be important for dormancy and germination of cereal caryopses is the finding that the sensitivity of barley aleurone tissue is heterogeneous and may be spatially determined (Ritchie et al., 1999). An increase in GA sensitivity is probably the most important factor for seed dormancy release and germination of most species.

GA-deficient biosynthesis mutants of *Arabidopsis* (e.g., *ga1*) and tomato (e.g., *gib1*) have been isolated based on the assumption that GA is required for the release of dormancy and the promotion of germination (Karssen et al., 1989; Hilhorst and Karssen, 1992; Koornneef and Karssen, 1994; Richards et al., 2001). Seed germination of several of these GA-deficient mutants absolutely depends on the addition of GA to the medium during imbibition in the light or in darkness and germinated seedlings develop into dwarfed rosette plants. The *Arabidopsis GA1* gene encodes copalylpyrophosphate synthetase, which catalyzes a key cyclization step in early GA biosynthesis and has been cloned by genomic substraction based on a 5-kb deletion in the severe *ga1-3* mutant allele (Sun and Kamiya, 1994). The mechanisms imposing a GA requirement to promote the germination of dormant and nondormant *Arabidopsis* seeds have been analyzed using the GA-deficient mutant *ga1*, the ABA-deficient mutant *aba1*, and several testa mutants that exhibit reduced seed dormancy (Debeaujon and Koornneef, 2000; Debeaujon et al., 2000). Testa mutants are not resistant to GA biosynthesis inhibitors. However, in the presence of the inhibitors or when transferred to a GA-deficient background, they are more sensitive to exogenous GA than wild type. The germination capacity of the *ga1-1* mutant could be integrally restored, without the help of exogenous GA, by removing the envelopes or by transferring the mutation to a testa mutant background. The ABA biosynthesis inhibitor norflurazon is partially efficient in releasing the dormancy of wild-type and mutant seeds. Debeaujon and Koornneef (2000) conclude that dormancy and germination are probably the net result of a balance between many promoting and inhibiting factors including GA and ABA that have the embryo and the testa as targets. Their results support the view that the GA requirement for dormancy release and germination is determined by (1) ABA produced in the developing seeds and/or the state of dormancy set by ABA and (2) ABA produced upon imbibition especially in dormant seeds. Furthermore, when the restraint to radicle protrusion imposed by the seed envelopes was weakened by the testa mutations, the embryo growth potential threshold required for germination is lowered. Thus, the GA requirement for *Arabidopsis* seed germination is determined both by testa characterisitics and by embryonic ABA.

Among the GA-response mutants of *Arabidopsis*, the GA-insensitive *gai* mutant is characterized by a dwarf phenotype, increased GA levels, and complex seed effects that are consistent with severely decreased GA sensitivity of dormancy release and germination (Derkx and Karssen, 1993; Koornneef and Karssen, 1994; Richards et al., 2001). No appreciable seed germination of *gai* occurs in the dark, and only a combination of light with either chilling or dry after-ripening causes dormancy release and germination. The *Arabidopsis GAI* gene and its orthologues in other species seem to encode nucleus-localized proteins that may act as transcription factors and appear to be negative regulators of the GA-signal transduction pathway. The *GAI* gene belongs to the DELLA subfamily of GRAS regulatory genes, which includes several other negative regulators of GA responses, e.g., *RGA (repressor-of-ga1-3), RGL1 (RGA-like1), RGL2*, and *RGL3* (Richards et al., 2001; Peng and Harberd, 2002). The DELLA domain region is thought to be involved in modulating the GA response and its deletion in the *gai* mutant causes a gain-of-function mutation characterized by dominant GA-insensitive repression of GA responses. In agreement with this, the loss-of-function allele *gai-t6* confers increased paclobutrazol resistence of stem growth, wild-type like germination in the light, and *gai-t6* can not rescue the nongermination phenotype of *ga1-3* (Dill and Sun, 2001; Richards et al., 2001). The different negative regulators of the GRAS family appear to possess separate as well as overlapping roles in GA responses (Figure 11.1): RGL1 appears to play a greater role in seed germination than do GAI and RGA (Wen and Chang, 2002), but RGL2 appears to be the most important regulator of *Arabidopsis* seed germination in response to GA (Lee et al., 2002). Loss-of-function *rgl2* alleles suppress the GA-deficient seed germination phenotype conferred either by treatment with paclobutrazol or by *ga1-3*. *RGL2* transcript levels rise rapidly following seed imbibition and then decline rapidly as germination proceeds. *RGL2* mRNA expression in imbibed seeds is restricted to elongating regions of radicles. In addition, RGL2 may function as an integrator of environmental and endogenous cues to control seed germination (Lee et al., 2002).

The constitutive GA-response mutant *spindly (spy)* of *Arabidopsis* has been isolated in a screen for seeds able to germinate in the presence of the GA-biosynthesis inhibitor paclobutrazol (Izhaki et al., 2001; Richards et al., 2001; Swain et al., 2001). Thus, the GA requirement for seed dormancy release and germination is decreased in the *spy* mutant. The elongated growth phenotype of *spy* mutant plants resembles wild-type plants treated with GA. Furthermore, *spy* mutations suppress the effects of GA-deficiency on germination, and overexpression of (wild-type) SPY inhibits seed germination of *Arabidopsis* and petunia. The *Arabidopsis SPY* gene and its orthologues in other species encode tetratricopeptide repeat proteins that might function as *O*-linked *N*-acetylglucosamine (*O*-GlcNAc) transferases. SPY appears to act as a negative regulator of GA responses upstream of GAI. A possible function of SPY could be to regulate GAI function by *O*-GlcNAc modification and thereby influencing the nuclear localization of GAI (Richards et al., 2001). The *Arabidopsis sleepy1 (sly1)* and the *comatose (cts)* mutants exhibit marked seed germination effects that cannot be rescued by GA (Steber et al., 1998; Steber and McCourt, 2001; Russell et al., 2000). The SLY1 and CTS proteins are proposed to be key factors involved in GA signaling of seeds. The *sly1* mutant has been selected in a screen for suppressors of the ABA-insensitive *abi1-1* mutant, and the *cts* mutant is impaired in seed dormancy release in the dry or imbibed state by after-ripening or cold treatment, respectively. *CTS* encodes a peroxisomal protein of the ATP binding cassette (ABC) transporter class, and regulation of CTS function seems to be a major control point for the switch between dormancy and germination (Footitt et al., 2002). The *Arabidopsis SLY1* gene is a positive regulator of GA signaling and encodes a putative F-box subunit of a SCF E3 ubiquitin ligase (McGinnis et al., 2003). RGA is a putative substrate of SLY1 and suggests that SCF^{SLY1}-targeted degradation of RGA through the 26S proteasome pathway is involved in GA signaling. GA appears to positively regulate dormancy release and germination by a complex interaction with ABA and environmental conditions (Figure 11.1). This direct or indirect GA-ABA antagonism is supported by physiological and biochemical experiments and by screens for suppressor mutants. Several of the nondormant *Arabidopsis* mutants that originated from a screen for germination in the presence of a GA biosynthesis inhibitor turned out to be a new alleles of *abi3* and *aba2* (Nambara et al., 1992, 1998). Suppressor screens of the ABA-insensitive *abi1-1* mutation yielded not only *sly1,* but also *ga1* mutant alleles (Steber et al., 1998; Richards et al., 2001). The sly1 alleles may be loss-of-function mutations and SLY1 is postulated to be a key factor in the control of GA signaling.

ABA and after-ripening, but not GA, are the primary regulators of seed dormancy of *Avena fatua* (Fennimore and Foley, 1998). In contrast, in most species dormancy release during after-ripening involves a further decline in ABA content, a decrease in ABA sensitivity, and an increase in GA sensitivity or a loss of GA requirement (Hilhorst, 1995; Bewley, 1997b; Li and Foley, 1997; Beaudoin et al., 2000; Debeaujon and Koornneef, 2000; Grappin et al., 2000; Romagosa et al., 2001; Koornneef et al., 2002; Leubner-Metzger, 2002; Schmitz et al., 2002). Release of dormancy during after-ripening can have different targets in nonendospermic seeds, the embryo, and the testa. In *Arabidopsis* this involves hormone signaling to promote embryo expansion growth and testa characteristics to lower the constraints of the seed envelopes (Beaudoin et al., 2000; Debeaujon and Koornneef, 2000). In addition, in endospermic seeds such as from *Nicotiana* species, the contributions of both the testa and the endosperm layers have to be considered (Leubner-Metzger, 2003b; Petruzzelli, Müller, et al., 2003). A key step for predisposing *N. plumbaginifolia* seeds to germinate is the inhibition of the capacity for ABA biosynthesis and/or the stimulation of ABA degradation (Grappin et al., 2000). Experiments comparing seeds that were freshly harvested (dormant) or after-ripened (nondormant), ABA deficient or wild type, treated with GA, ABA, or the ABA-biosynthesis inhibitor fluridone have shown that this predisposition takes place during after-ripening. GA or fluridone are both able to release dormancy and to inhibit de novo ABA biosynthesis that occurs in imbibed dormant seeds. In contrast, nondormant seeds do not synthesize ABA upon imbibition, a feature therefore associated with the after-ripening-mediated release of dormancy. The after-ripening-mediated promotion of *N. tabacum* seed germination is due to a promotion of testa rupture and a similar promotion of subsequent endosperm rupture (Leubner-Metzger and Meins, 2000; Leubner-Metzger, 2002, 2003b). Reciprocal crosses between wild-type tobacco and sense-βGlu I transformant lines (TKSG7) have shown that βGlu I overexpression in the seed covering layers can replace the promoting effect of after-ripening on testa rupture in light, but only if the mother plant is a sense-βGlu I line. This maternal effect supports the model of two sites for βGlu I action (Figure 11.2): (1) βGlu I contribution to the after-ripening-mediated release of dormancy in the dry seed state, which is manifested in promotion and ABA insensitivity of testa rupture during imbibition, and (2) ABA-sensitive expression of βGlu I in the micropylar endosperm, which contributes to endosperm rupture. Promotion of ABA-delayed seed germination of *N. plumbaginifolia* by light or GA involves stimulation of ABA degradation and inhibition of ABA synthesis (Kraepiel et al., 1994; Grappin et al., 2000). Although light can induce GA biosynthesis and increase GA sensitivity of seeds, after-ripening and chilling appear to act primarily by increasing the GA sensitivity of seeds.

The release of photodormancy, a block in dark germination, and promotion of germination of light-requiring seeds of many species are regulated by the phytochrome system (Kamiya and Garcia-Martinez, 1999; Neff et al., 2000; Hennig et al., 2002). Red light has been shown to up-regulate the biosynthesis of bioactive GA_1 and GA_4 by inducing GA biosynthetic genes in germinating seeds of lettuce and *Arabidopsis* (Toyomasu et al., 1993, 1998; Yamaguchi et al., 1998, 2001). Gene induction of GA 3β-hydroxylases and GA 3-oxidases during seed imbibition seems to be controlled by light via the phytochrome system. Furthermore, GA biosynthesis during *Arabidopsis* seed germination seems to take place in two separate locations, with the early step occurring in the provasculature and the later steps in the cortex and endodermis. This implies that intercellular transport of an intermediate of the GA biosynthetic pathway is required to produce bioactive GA (Yamaguchi et al., 2001; Ogawa et al., 2003). Tobacco seed germination is also regulated by the phytochrome system, and GA can substitute for the red-light trigger needed to release photodormancy and to induce dark germination (Khalil, 1992; Kretsch et al., 1995; Emmler and Schäfer, 1997; Leubner-Metzger, 2003b). Saturating endogenous GA_1 and GA_4 levels present in light-imbibed tobacco seeds can explain why GA treatment does not promote germination in the light, while it promotes dark germination of nonphotodormant seeds (Leubner-Metzger, 2001, 2003b). Far less is known about the role of GA sensitivity during the after-ripening-mediated release of photodormancy. Fresh tobacco seeds are photodormant, i.e., they do not germinate during imbibition in darkness, and even prolonged incubation in the dark does not induce testa rupture, βGlu I accumulation, or endosperm rupture. After-ripening contributes to the release of photodormancy, and this effect varies greatly for different seed batches as reported for several tobacco cultivars (Leubner-Metzger and Meins, 2001; Leubner-Metzger, 2002). The GA requirements for photodormancy release of fresh and completely photodormant after-ripened seed batches are equal. Nonphotodormant tobacco seeds have lost the GA requirement for dark germination, which could be due to increased GA sensitivity and/or increased endogenous GA levels. Endospermic and nonendospermic seeds seem to share GA requirements and testa characteristics as common factors in the after-ripening-mediated release of coat-imposed dormancy. In addition, in imbibed endospermic seeds GA seems to directly counteract the germination-delaying effects of ABA on the micropylar endosperm.

In members of Solanaceae (e.g., tomato and tobacco) and Asteraceae (e.g., lettuce), the micropylar endosperm and testa tissues impose a constraint to radicle protrusion (Ni and Bradford, 1993; Hilhorst, 1995; Bewley, 1997a; Leubner-Metzger, 2003b). That the micropylar seed-covering layers impose a physical constraint to radicle protrusion is supported by puncture force measurements and surgical experiments. Removal of the micropylar testa and the endosperm tissues permits radicle growth under conditions that inhibit germination of intact seeds, e.g., of tobacco (Kincaid, 1935; Leubner-Metzger, 2003b), tomato (Liptay and Schopfer, 1983; Hilhorst, 1995), and lettuce (Bewley, 1997a; Dutta et al., 1997). Weakening of the micropylar endosperm appears to be a prerequisite for tomato germination and is likely to be achieved by cell wall hydrolysis by the collaborative or successive action of several GA-induced cell wall hydrolases. Numerous cell wall modifying proteins, e.g., endo-β-mannanase, β-mannosidase, α-galactosidase, cellulase, pectin methylesterase, polygalacturonase, xyloglucan endo-transglycosylase, β-1,3-glucanase, chitinase, peroxidase, and expansin, have been investigated for their possible role in endosperm weakening (e.g., Bewley, 1997a; Welbaum et al., 1998; Chen and Bradford, 2000; Nonogaki et al., 2000; Leubner-Metzger, 2003b; Petruzzelli, Müller, et al., 2003). Two phases prior to radicle protrusion can be distinguished in tomato:

1. The early phase is not inhibited by ABA and includes ABA-insensitive endosperm weakening associated with micropylar-endosperm specific, ABA-independent expression of endo-β-mannanase, expansin, and other proteins, but not of βGlu I expression (e.g., Bewley, 1997a; Toorop et al., 1998; Chen and Bradford, 2000; Nonogaki et al., 2000; Toorop et al., 2000; Wu et al., 2000; Mo and Bewley, 2002). Endo-β-mannanase, which can hydrolyze isolated micropylar endosperm cell walls in vitro, appears to be necessary for endosperm weakening, but is not sufficient for the completion of tomato germination and is not down-regulated by ABA.
2. The late phase is critical since it includes the final ABA-controlled step of radicle emergence. It is associated with ABA-sensitive βGlu I expression in the micropylar endosperm, and βGlu I therefore could contribute to radicle emergence of tomato (Wu et al., 2000).

It is proposed that the late phase includes a second, ABA-controlled step of endosperm weakening, which thereby is a biphasic process in tomato (Toorop et al., 2000). Tomato endosperm weakening is usually measured as the force required to puncture micropylar seed halves that include the endosperm plus the testa tissues (Toorop et al., 1998; Chen and Bradford, 2000; Toorop et al., 2000; Wu et al., 2000). The micropylar endosperm confers the major part of the mechanical resistance (Groot and Karssen, 1987). The testa accounts for approximately 20 percent of the mechanical resistance during the early phase of seed imbibition, and this declines just prior to radicle protrusion. ABA deficiency of the tomato *sit^w* mutant is correlated with a thinner micropylar testa and faster seed germination (Hilhorst and Downie, 1995). These authors conclude from their experiments that, although the testa resistance is smaller compared to the endosperm resistance, it is the micropylar testa that finally controls the completion of germination, i.e., radicle emergence of tomato (Hilhorst and Downie, 1995). Thus, testa rupture could be important in the late phase and could be achieved by an ABA-sensitive process that is characterized by wall breakage at preformed breaking points. ßGlu I could contribute to this process, and possible mechanisms have been proposed (Leubner-Metzger, 2003b). Reactive oxygen species (ROS; including $O_2^{\cdot-}$, H_2O_2, $^{\cdot}OH$) could also contribute to the process of radicle protrusion. ROS are released by germinating seeds, and ROS production is under hormonal and developmental control (Schopfer et al., 2001). ROS production by the seed coat and the embryo, as well as germination of radish, are both promoted by GA and inhibited by ABA. It is also proposed that promotion of *Zinnia elegans* seed germination by H_2O_2 is due to oxidative ABA degradation (Ogawa and Iwabuchi, 2001). On the other hand, H_2O_2 has been also shown to inactivate ABI1, a negative regulator of ABA responses, and thereby enhance the ABA responses in *Arabidopsis* seedlings (Meinhard and Grill, 2001). ROS are proposed to be rate-limiting second messengers in ABA signaling during seed germination (Kwak et al., 2003). Thus, the mechanism(s) by which the antagonists GA and ABA regulate the balance between the extension force of the embryo and the opposing force of the micropylar seed layers, i.e., how they control radicle protrusion, appears to be complex and remains to be elucidated.

ETHYLENE PROMOTES SEED GERMINATION AND COUNTERACTS ABA EFFECTS ON SEEDS

Ethylene is implicated in the promotion of germination of nondormant seeds of many species (for reviews see Esashi, 1991; Kepczynski and Kepczynska, 1997; Matilla, 2000). For several cases release by ethylene of primary (e.g., peanut, sunflower) and secondary (e.g., lettuce, sunflower) dormancy also has been reported. However, the precise molecular mechanism(s) of ethylene action, and of dormancy release itself, are unknown. Ethylene alone is not sufficient to release seed dormancy in many species, even if it promotes germination of nondormant seeds of this species. For example, ethylene treatment does not release photodormancy needed for dark germination of positively photoblastic lettuce or tobacco seeds, but endogenous ethylene is involved in germination responses of these species (e.g., Saini et al., 1989; Leubner-Metzger et al., 1998). Increasing ethylene evolution accompanies germination of most dicot seeds, e.g., *Cicer arietinum* (Matilla, 2000; Gómez-Jiménez et al., 2001), *Pisum sativum* (Gorecki et al., 1991; Petruzzelli et al., 1995), *Lycopersicon esculentum* (Lashbrook et al., 1998), *Nicotiana tabacum* (Khalil, 1992; Leubner-Metzger et al., 1998), *Lactuca sativa* (Saini et al., 1989; Matilla, 2000). Several of these studies utilize the competitive ethylene action inhibitor 2,5-norbornadiene (NBD) and thereby demonstrate that endogenous ethylene is also required for optimal germination of nondormant seeds. A higher amount of ethylene production is obvious in nondormant compared to dormant seeds (Esashi,

1991; Kepczynski and Kepczynska, 1997; Matilla, 2000). The amount of ethylene evolution is positively correlated with the germination vigor of nondormant pea seeds (Gorecki et al., 1991). A major peak of ethylene evolution coincides with the completion of germination by radicle protrusion, but ethylene production is already detectable very early during imbibition and prior to radicle protrusion through the covering layers. Several hypothesis have been proposed concerning the mechanism(s) of ethylene action in seeds (Esashi, 1991; Kepczynski and Kepczynska, 1997; Matilla, 2000). It has been suggested for lettuce that the primary action of ethylene is the promotion of radial cell expansion in the embryonic hypocotyl. A positive regulation by ethylene of seed respiration and water potential has also been proposed. High-level induction of ABA-sensitive class I β-1,3-glucanase in the micropylar endosperm of tobacco requires endogenous ethylene, which thereby promotes endosperm rupture (Leubner-Metzger et al., 1998). The molecular mechanisms of gene regulation by ethylene have been thoroughly studied in vegetative tissues, senescing flowers, and fruit ripening (Hall et al., 2001). In contrast, only a few reports present molecular data about ethylene regulated gene induction during seed germination. These include genes encoding three members of the *ETR* gene family encoding putative ethylene-receptors in tomato (Lashbrook et al., 1998), a cysteine proteinase in chickpea (Cervantes et al., 1994), a class I β-1,3-glucanase in tobacco and pea (Leubner-Metzger et al., 1998; Petruzzelli et al., 1999), and a 1-aminocyclopropane-1-carboxylic acid (ACC) oxidase (ACO) of pea (Petruzzelli et al., 2000; Petruzzelli, Sturaro, et al., 2003). Ethylene biosynthesis and sensitivity are important for the seed germination of *Arabidopsis* (Beaudoin et al., 2000; Ghassemian et al., 2000; Gallardo et al., 2002a,b). Poor germination is a feature of the ethylene-insensitive mutant *ethylene response1 (etr1)* (Bleecker et al., 1988), and seed dormancy and germination are also altered in other ethylene signal transduction mutants of this species (Beaudoin et al., 2000; Ghassemian et al., 2000). A major conclusion from the studies with these mutants is that ethylene appears to be a negative regulator of ABA action during germination and that it decreases the sensitivity to ABA. Ethylene and ABA signaling seem to interfere with each other on several levels of the signal transduction pathways, a finding which is in agreement with physiological results about the regulation of seed germination in other species (e.g., Finch-Savage and Clay, 1994; Leubner-Metzger et al., 1998; Kepczynski et al., 2003).

Ethylene is perceived by a family of receptors related to ETR1 of *Arabidopsis,* and ethylene binding inhibits the signaling activities of these receptors (Hall et al., 2001). An increase in responsiveness toward ethylene is correlated with the germination of nonendospermic and endospermic seeds, e.g., pea (Petruzzelli et al., 2000) and tomato (Lashbrook et al.

1998). The transcript levels of three genes of the tomato ethylene receptor family increase during seed germination. ABA inhibits the seed germination of *Arabidopsis*, and this inhibitory effect can be partially reversed by treatment with ACC (Ghassemian et al., 2000). Not only is the *etr1* mutant of *Arabidopsis* characterized by poor germination (Bleecker et al., 1988), but *etr1* seed germination is also hypersensitive to ABA (Beaudoin et al., 2000). ETR1 and the related ethylene receptors are hybrid-type histidine kinases, and it is likely that ethylene is signaling through two-component systems (Lohrmann and Harter, 2002). In the absence of ethylene, the receptors activate CTR1, which is a negative regulator of downstream signaling components. CTR1 is inactive in the presence of ethylene, and the *constitutive triple response1 (ctr1)* mutant of *Arabidopsis* is characterized by a constitutive ethylene response (Figure 11.1). Seed germination of *ctr1* is less sensitive to ABA, and freshly harvested *ctr1* seeds germinate slightly faster than wild-type seeds (Beaudoin et al., 2000). CTR1 is thought to function as a RAS-like mitogen-activated protein kinase kinase kinase (MAPKKK), and an ethylene receptor complex is proposed to consist of ETR1, CTR1, and phosphorelay intermediates (Hall et al., 2001; Lohrmann and Harter, 2002). EIN2 is a downstream signaling component which, in the absense of ethylene, is negatively down-regulated by CTR1. As *etr1*, the *ethylene insensitive2 (ein2)* mutants of *Arabidopsis* are characterized by higher seed dormancy and by hypersensitivity to ABA. Interestingly, *ctr1* and *ein2* mutants were recovered as enhancer and suppressor mutants, respectively, of the ABA-insensitive seed germination phenotype of the *abi1-1* mutant (Beaudoin et al., 2000). *Arabidopsis enhanced response to ABA3 (era3)* mutants are characterized by increased sensitivity of the seed to ABA and by overaccumulation of ABA (Ghassemian et al., 2000). Subsequent genetic analysis of the *era3* alleles has shown that they are new alleles of the *EIN2* locus (Figure 11.1). The *ein2-45* allele increases seed dormancy, but this effect is completely counteracted by severely ABA-insensitive mutants such as *abi3-4* (Beaudoin et al., 2000). The *ein2-45 abi3-4* double mutant is as nondormant as the *abi3-4* single mutant. The nondormant phenotype of the *ein2 abi3-4* double mutant indicates that ethylene may suppress seed dormancy by inhibiting ABA action. Thus, EIN2 is a negative regulator of ABA biosynthesis. The enhanced dormancy of *etr1* and *ein2* seeds suggests that endogenous ethylene is a negative regulator of *Arabidopsis* seed dormancy. The finding that *ein2* mutations can suppress and *ctr1* mutations can enhance the ABA-insensitive germination phenotype of *abi1* suggests that ethylene must also influence the sensitivity of the seed to ABA. These results indicate a strong interaction between the ethylene and ABA signal transduction pathways, and a key conclusion is that ethylene can promote germination by directly interfering with ABA signaling (Beaudoin et al., 2000; Ghassemian et al., 2000). In summary, it seems likely that ethylene alone cannot act as a positive regulator of germination but most likely acts by interfering with ABA.

Genetically downstream of EIN2 are transcription factors that are localized in the nucleus, bind to *cis*-regulatory elements in gene promoter regions, and thereby regulate gene expression in response to ethylene (Figure 11.1). Ethylene-ABA interactions are also manifested at this level of ethylene signaling, e.g., a mutation in the *Arabidopsis* EIN3 transcription factor confers a reduced ABA responsiveness in roots (Ghassemian et al., 2000). EIN3 binds to the promoter of the *Arabidopsis ethylene responsive factor 1* (*ERF1*) gene, which encodes an ethylene responsive element binding protein (EREBP)-type transcription factor, and thereby confers an hierarchy of transcription factors involved in ethylene signaling (Solano et al., 1998). The EREBP-type transcription factors mediate ethylene regulation of gene expression (Figures 11.1 and 11.2). The AP2-like DNA binding domain is characteristic for these transcriptional regulators of the AP2/EREBP family, e.g., *Arabidopsis* ERF1 and ABI4, and tobacco EREBP1 to EREBP4 (also named ERF1 to ERF4) (Finkelstein et al., 1998; Solano et al., 1998; Ohta et al., 2000). The EREBP bind to the GCC box within the positively acting ethylene-responsive element (ERE) of target promoters, and this *cis*-regulatory element is necessary and sufficient for the regulation of transcription by ethylene. Ethylene is involved in endosperm rupture and high-level βGlu I expression during tobacco seed gemination, but it does not affect the spatial and temporal pattern of βGlu I expression (Leubner-Metzger et al., 1998; Leubner-Metzger, 2003b). A promoter deletion analysis of a tobacco βGlu I gene in germinating tobacco seeds suggests that the distal region, which contains the ERE, is required for high-level, ethylene-sensitive expression; that the proximal region is necessary and sufficient for low-level micropylar-endosperm specific expression; and that both regions contribute to down-regulation by ABA. Enhancer activity and ethylene responsiveness of βGlu I depend on the integrity of the AGCCGCC sequence present as two copies in the ERE. Transcripts of the EREBPs show a novel pattern of expression during tobacco seed germination (Figure 11.2): light or gibberellin are required for EREBP-3 and EREBP-4 expression; EREBP-4 expression is constitutive and unaffected by ABA or ethylene; EREBP-3 shows transient induction just before endosperm rupture, which is earlier in ethylene-treated seeds and is inhibited by ABA. The results suggest that transcriptional regulation of βGlu I could depend on activation of ethylene signaling pathways acting via EREBP-3 with the ERE as the target, and ethylene-independent signaling pathways with targets in the proximal promoter region that are likely to determine spatial and temporal patterns of expression (Leubner-Metzger et al., 1998). Interestingly, sequences that are homologous to the GCC box are also present in the promoters of some ACO genes (see references in (Petruzzelli et al., 2000).

Ethylene promotes ethylene biosynthesis during pea seed germination by positive feedback regulation of ACO (Petruzzelli et al., 2000; Petruzzelli, Sturaro, et al. 2003). An early onset and sequential induction of ACC biosynthesis, *Ps-ACO1* mRNA, 36-kDa Ps-ACO1 protein, and ACO activity accumulation and ethylene production are localized almost exclusively in the embryonic axis but not in the cotyledons. Within the embryonic axis, ethylene biosynthesis and responsiveness were localized to the cell elongation and differentiation zones of the radicle. Maximal levels of ACC, Ps-ACO1, and ethylene evolution are found when radicle emergence is just complete. Treatment of germinating seeds with ethylene alone or in combination with NBD have shown that endogenous ethylene regulates its own biosynthesis through a positive feedback loop that enhances Ps-ACO1 expression. Accumulation of *Ps-ACO1* mRNA, protein, and ACO enzyme activity in the embryonic axis during the late phase of germination required ethylene, whereas *Ps-ACS1* mRNA levels for ACC synthase and overall ACC contents are not induced by ethylene treatment. Ethylene does not induce ACO in the embryonic axis during the early phase of germination. Ethylene-independent signaling pathways regulate the spatial and temporal pattern of ethylene biosynthesis, whereas the ethylene signaling pathway regulates high-level ACO expression in the embryonic axis, and thereby enhances ethylene evolution during seed germination. Similar findings have been made with chickpea seeds, and the ACO induced in the embryonic axis of chickpea shares 86 percent of amino acid sequence identity with the Ps-ACO1 of pea (Nicolas et al., 1998; Matilla, 2000; Gómez-Jiménez et al., 2001). ABA inhibits chickpea ACO expression, ACC accumulation, and ethylene production prior and during germination, but not after germination. This suggests that interference of ethylene and ABA signaling could confer an opposing feedback regulation of ethylene production by the embryonic axis. Ca^{2+} is required for the ethylene responses during pea and chickpea seed germination (Nicolas et al., 1998; Petruzzelli, Sturaro, et al., 2003). Ca^{2+} release from internal pools, the Ca^{2+}/calmodulin complex, and the PI cycle seem to be required for seed ethylene responses such as ACO induction. Ca^{2+} is also involved in ABA responses, and ABA inhibits the expression of calmodulin during germination. Taken together, this suggests that Ca^{2+} could mediate at least some of the interactions of ethylene and ABA signaling and at least some of the opposing effects of ethylene and ABA on seed germination. Although ABA and ethylene act antagonistically on seed dormancy and germination, they both inhibit root elongation of germinated pea seeds and of *Arabidopsis* seedlings (Beaudoin et al., 2000; Ghassemian et al., 2000; Petruzzelli, Sturaro, et al., 2003). Thus, seeds and vegetative tissues differ with regard to the ethylene-ABA interactions and integrated responses.

An excess of ethylene can bypass the GA requirement and induce full germination of *Arabidopsis ga1* mutant seed imbibed in the light, the effect being much weaker in darkness (Karssen et al., 1989; Koornneef and Karssen, 1994). The GA-ethylene interaction seems reciprocal, because high GA concentrations restore the germination of *etr1* mutant seed to wild-type level (Bleecker et al., 1988). By contrast, ethylene does not break dormancy and stimulate germination of the tomato *gib1* mutant, although it promotes the germination of tomato wild-type seeds (Nelson and Sharples, 1980; Groot and Karssen, 1987). Treatment with either GA, ethylene, or cytokinin alone is not able to overcome lettuce thermoinhibition in the dark (Saini et al., 1989; Matilla, 2000). In addition to exogenous ethylene, at least one other hormone or light is required. Ethylene biosynthesis is essential for the relief of thermoinhibition in the dark by applications of GA or a cytokinin, as well as for the light-induced relief of thermoinhibition. The cytokinin or ethylene requirements for overcoming thermoinhibition of lettuce can also be eliminated or diminished by removing or weakening the endosperm. Thus, ethylene biosynthesis is involved as a negative regulator of coat-imposed dormancy. Furthermore, thermoinhibition of lettuce is correlated with the accumulation of ABA (Yoshioka et al., 1998). Thus, an antagonistic action of ABA and ethylene on dormancy release and germination is obivous for the seeds of many species.

BRASSINOSTEROIDS PROMOTE SEED GERMINATION

Brassinosteroids (BR) and GA interact with light in regulating elongation growth of shoots and photomorphogenesis of seedlings by what appear to be independent pathways (Altmann, 1999; Neff et al., 2000). Endogenous BR have been identified in seed of several species (e.g., Adam and Marquardt, 1986; Schmidt et al., 1997). BR application has been reported to enhance germination of certain parasitic angiosperms (Takeuchi et al., 1991, 1995), cereals (Yamaguchi et al., 1987), *Arabidopsis* (Steber and McCourt, 2001), and tobacco (Leubner-Metzger, 2001), but not of non-photodormant, nonendospermic cress seeds imbibed in the dark (Jones-Held et al., 1996). Germination of the endospermic seeds of parasitic *Orobranche* and *Striga* species is, in contrast to *Arabidopsis* and tobacco, inhibited by light (Takeuchi et al., 1991, 1995). Neither BR, ethylene, nor GA can substitute for the conditioning treatment with strigol, which is needed for inducing germination of unconditioned (i.e., dormant) seed. Conditioning removes the restriction on the ethylene biosynthetic pathway and increases the capacity to produce ethylene (Babiker et al., 2000). Treatment with BR promotes the germination of conditioned (i.e., nondormant) *Orobranche* and *Striga* seeds imbibed in the light and in the dark.

BR promotes the germination of prechilled (i.e., nondormant) seeds of the BR-deficient biosynthesis mutant *det2-1* and the BR-insensitive response mutant *bri1-1* of *Arabidopsis* imbibed in the light (Steber and McCourt, 2001). Seed germination of *det2-1* and *bri1-1* is more strongly inhibited by ABA than is germination of the wild type, and BR is therefore able to partially overcome the inhibition of germination by ABA. BR treatment rescues the germination phenotype of the severe GA-deficient biosynthesis mutant *ga1-3*, which normally requires GA treatment for dormancy release and germination. BR treatment also partially rescues the germination phenotype of the severe GA-insensitive response mutant *sly1*, which can not be rescued by treatment with GA. Interestingly, a new allele for *sly1* was identified in a screen for BR-dependent germination and suggests interactions between BR and GA signaling in seeds (Steber et al., 1998; Steber and McCourt, 2001). These results point to a role for BR in stimulating germination of *Arabidopsis* seeds. This is further supported by the germination phenotype of the *gpa1* mutant of *Arabidopsis* (Ullah et al., 2002). The *GPA1* gene encodes the alpha subunit of a heterotrimeric G protein. Seeds with the *gpa1* null mutation are 100-fold less responsive to GA, and GPA1 overexpressing seeds are hypersensitive for GA. The *gpa1* mutant seeds are also completely insensitive to BR rescue of germination when the level of GA in seeds is reduced. These findings support the view that there is a complex interaction between GA and BR in regulating seed germination of *Arabidopsis*.

BR promotes seedling elongation and germination of nonphotodormant tobacco seeds, but does not appreciably affect testa rupture and the subsequent induction of βGlu I in the micropylar endosperm (Leubner-Metzger, 2001). Treatment with BR, but not GA, accelerates endosperm rupture of tobacco seeds imbibed in the light. BR and GA promote endosperm rupture of dark-imbibed nonphotodormant seeds, but only GA enhances βGlu I induction. Promotion of endosperm rupture by BR is dose dependent, and 0.01 μM brassinolide is most effective. BR and GA promote ABA-inhibited dark germination of nonphotodormant seeds, but only GA replaces light in inducing βGlu I. These results indicate that BR and GA promote tobacco seed germination by distinct signal transduction pathways and distinct mechanisms. GA and light act in a common pathway to release photodormancy, whereas BR does not release photodormancy. βGlu I induction in the micropylar endosperm and release of coat-imposed dormancy seem to be associated with the GA/light pathway, but not with BR signaling. Xyloglucan endo-transglycosylase (XET) enzyme activity accumulates in the embryo and the endosperm of germinating tobacco seeds, and this appears to be partially controlled by BR (Leubner-Metzger, 2003a). A GA-regulated XET mRNA is expressed exclusively in the micropylar endosperm of tomato seeds (Chen et al., 2002). It is therefore possible that the XET induction in the tobacco embryo is controlled by BR. These findings suggest a model for the endosperm-limited germination of tobacco (Figure 11.2):

1. Photodormancy is released exclusively by the GA/light-pathway.
2. Promotion of subsequent endosperm rupture by the BR and the GA/light signal transduction pathways is achieved by independent and distinct mechanisms.
3. ABA inhibits endosperm rupture by interfering with both pathways.
4. The GA/light pathway regulates βGlu I induction in the micropylar endosperm and seems to control endosperm weakening.
5. It is proposed that the BR pathway promotes endosperm rupture of nondormant seeds by enhancing the growth potential of the embryo (Leubner-Metzger, 2001, 2003b).

Taken together, these findings suggest that GA and BR act in parallel to promote cell elongation and germination and to counteract the inhibitory action of ABA on seeds. Because BR stimulates germination of the GA-insensitive mutant *sly1,* it is unlikely that BR acts by increasing GA sensitivity. It is possible that BR acts by stimulating GA biosynthesis in *Arabidopsis* seeds imbibed in the light (Steber and McCourt, 2001). BR action via stimulation of GA biosynthesis is, however, unlikely for tobacco, because BR does not promote the expression of ßGlu I, which is induced by GA in the dark (Leubner-Metzger, 2001). It is known that BR can stimulate ethylene production, and ethylene treatment can rescue the germination phenotype of the GA-deficient *Arabidopsis ga1-1* mutant (Karssen et al., 1989; Koornneef and Karssen, 1994; Steber and McCourt, 2001). However, there are several arguments against the hypothesis that BR acts via ethylene:

1. Ethylene levels are not increased in cress seedlings following BR treatment of seeds (Jones-Held et al., 1996).
2. Endogenous ethylene promotes βGlu I accumulation in the micropylar endosperm of tobacco, but BR treatment promotes endosperm rupture without enhancing βGlu I accumulation (Leubner-Metzger et al., 1998; Leubner-Metzger, 2001).

3. Ethylene rescue of *ga1-1* seed germination results in seedlings exhibiting triple response, but BR rescue of *ga1-3* seed germination results in seedlings that do not exhibit triple response (Steber and McCourt, 2001).

Another possibility would be BR action via auxin. Auxin also stimulates cell elongation, but it does not rescue germination of *ga1-3* (Koornneef and Karssen, 1994). Thus, if BR stimulates germination via embryo expansion, this effect is likely specific to seed germination. Finally, the *Arabidopsis sax1 (hypersensitive to abscisic acid and auxin)* dwarf mutant is impaired in BR biosynthesis and exhibits pleiotropic seedlings effects with respect to ABA, auxin, GA, ethylene, and BR, but seed germination of *sax1* and wild type is not differentially inhibited by ABA (Ephritikhine et al., 1999).

CYTOKININS AND AUXINS

Cytokinins are present in developing seeds and accumulate predominantly in the liquid endosperm (Emery et al., 2000; Fischer-Iglesias and Neuhaus, 2001; Mok and Mok, 2001). It has been proposed that the endosperm functions as a source for cytokinins needed for the promotion of cell devision in the embryo. They may have roles in embryogenesis, in embryonic pattern formation, in the early period of grain filling of cereals, and in enhancing sink strength. During germination of sorghum grains the cytokinin content is high in the embryo, low in endosperm, and declines during imbibition (Dewar et al., 1998). After radicle protrusion a postgerminational cytokinin peak is associated with α-amylase accumulation. This cytokinin might play a role in cell division and elongation of the emerged root. It may also be redistributed within the embryo to regions where it effectively concentrates and directs root growth. ABA content in the embryo prior germination is high, and it is proposed that the cytokinin/ABA interaction plays a significant role in controlling sorghum germination (Dewar et al., 1998). During the conditioning process of parasitic *Orobranche* and *Striga* species and the relief of lettuce thermoinhibition, cytokinins seem to contribute to the promotion of dormancy release and germination by enhancing ethylene biosynthesis (Saini et al., 1989; Babiker et al., 2000; Matilla, 2000). A cytokinin-ethylene connection is also supported by the discovery that the *Arabidopsis cytokinin-resistant1* (*ckr1*) mutant is also insensitive to ethylene and allelic to the ethylene-insensitive mutant *ein2* (Fischer-Iglesias and Neuhaus, 2001). Cytokinin-resistant mutants of *N. plumbaginifolia* have been isolated that exhibit reduced seed dormancy and pleiotropic seed effects that suggest cytokinin-ABA interactions (Rousselin et al., 1992).

Auxins seem to play a major role in embryogenesis, providing positional information for the coordination of correct cellular patterning from the globular stage onward (Fischer-Iglesias and Neuhaus, 2001; Hamann, 2001). Recent molecular and genetic data support an essential role of auxin for apical-basal pattern formation during embryogenesis, but very little is known on the molecular level about auxin during seed germination. Free indoleacetic acid (IAA) decreases during the imbibition of sorghum grains (Dewar et al., 1998), and auxin regulates catalase expression in the scutellum of germinating maize kernels (Guan and Scandalios, 2002). IAA is released from conjugates stored in seeds of Scots pine during germination (Ljung et al., 2001). A peak of free IAA occurs prior to the initiation of root elongation and coincides with initial seed swelling during imbibition. Free IAA contents decline dramatically during radicle emergence, and new IAA synthesis is established in the emerged seedling. An IAA-modified protein from bean seeds, IAP1, is associated with the developmental period of rapid growth during seed development (Walz et al., 2002). Moreover, this protein undergoes rapid degradation during germination. IAA-modified proteins represent a distinct class of conjugated phytohormones and appear to be the major form of auxin in bean seeds. The postgerminational acculumation of ACO and ethylene production of chickpea seedlings is promoted by IAA (Gómez-Jiménez et al., 2001). IAA has no appreciable effect on ACO expression prior or during germination. Several auxin-resistant mutants of *N. plumbaginifolia* have been isolated that exhibit reduced seed dormancy and pleiotropic seed effects that suggest auxin GA interactions (Rousselin et al., 1992). However, there appears to be no auxin effect on the seed germination of GA-deficient *Arabidopsis* and tomato mutants (Koornneef and Karssen, 1994).

CONCLUSIONS AND PERSPECTIVES

Dormancy and germination are complex traits that are controlled by a large number of genes, which are affected by both developmental and environmental factors (Koornneef et al., 2002). It is fascinating to see that many of the results on hormonal interactions in seeds obtained by using methods of physiology and biochemistry are now supported by molecular genetic data obtained with hormone mutants. Taken together, a glimpse of the complex network of hormonal interactions became visible and will be further unraveled by transcriptome and proteome approaches. A crucial role for ABA has been identified in inducing and maintaining seed dormancy. GA counteracts ABA effects in either a direct or indirect manner by releasing dormancy and promoting germination. Thus, the revised hormone-balance hypothesis for seed dormancy proposed by Karssen and Laçka (1986) appears to be valid, in that ABA action during seed dormancy induction or maintenance may be antagonized by GA action during dormancy release and germination. BR and ethylene also counteract the inhibitory effects of ABA on seed germination, but in most species they appear to act after dormancy has been released by GA. In *Arabidopsis* two-component signaling elements are of upmost importance for the perception and initial transduction of hormonal and environmental signals (e.g., Hall et al., 2001; Frankhauser, 2002; Haberer and Kieber, 2002; Lohrmann and Harter, 2002). This type of signaling involves protein phosphorylation and signal receptors for ethylene, cytokinin, and osmolarity are likely to function as histidine kinases. They interact directly or indirectly with phosphotransfer proteins and response regulators. During this interaction the phosphoryl group is transferred to an conserved aspartate residue within the receiver domain of the cognate response regulator. This "His-to-Asp phosphorelay" permits the conversion of hormonal and environmental signals to a biochemical reaction. The GA and ABA receptors are not known, but it is possible that they also function via phosphorelay. The phosphorelay not only allows linking of the different signals to form a complex signaling network, but also allows signal integration on the level of phosphotransfer proteins and response regulators. I therefore propose the intriguing hypothesis that regulation of dormancy and germination by plant hormones and environmental factors is achieved by convergency of different phosphorelay signals on the level of phosphotransfer proteins and response regulators. Several key components of hormone signal transduction pathways as well as downstream target genes have been identified. Some of these are involved in the regulation of coat-imposed dormancy, which can determine germination in response to hormones and environmental cues. The importance of GA signaling and testa characteristics appear to be a common feature in the after-ripening-mediated release of coat-imposed dormancy of endospermic and nonendospermic seeds. In addition, regulation of coat-imposed dormancy of endospermic seeds requires hormonal interactions between the embryo and the micropylar endosperm. Hormonal interactions and responses differ between vegetative tissues and seeds. The interaction of seed tissues is likely to depend on seed-tissue-specific signaling factors that are important for imposing, maintaining, and releasing embryo extension growth and the types of dormancy constraints confered by the different covering layers.

REFERENCES

Adam, G. and V. Marquardt (1986). Brassinosteroids. *Phytochemistry* 25:1787-1799.

Altmann, T. (1999). Molecular physiology of brassinosteroids revealed by the analysis of mutants. *Planta* 208:1-11.

Babiker, A.G.T., Y.Q. Ma, Y. Sugimoto, and S. Inanaga (2000). Conditioning period, CO_2 and GR24 influence ethylene biosynthesis and germination of *Striga hermonthica*. *Physiologia Plantarum* 109:75-80.

Batge, S.L., J.J. Ross, and J.B. Reid (1999). Abscisic acid levels in seeds of the gibberellin-deficient mutant lh-2 of pea *(Pisum sativum)*. *Physiologia Plantarum* 105:485-490.

Beaudoin, N., C. Serizet, F. Gosti, and J. Giraudat (2000). Interactions between abscisic acid and ethylene signaling cascades. *The Plant Cell* 12:1103-1115.

Benech-Arnold, R.L., M.C. Giallorenzi, J. Frank, and V. Rodriguez (1999). Termination of hull-imposed dormancy in developing barley grains is correlated with changes in embryonic ABA levels and sensitivity. *Seed Science Research* 9: 39-47.

Bethke, P.C. and R.L. Jones (2001). Cell death of barley aleurone protoplasts is mediated by reactive oxygen species. *The Plant Journal* 25:19-29.

Bewley, J.D. (1997a). Breaking down the walls—A role for endo-ß-mannanase in release from seed dormancy? *Trends in Plant Science* 2:464-469.

Bewley, J.D. (1997b). Seed germination and dormancy. *The Plant Cell* 9:1055-1066.

Bleecker, A.B., M.A. Estelle, C. Somerville, and H. Kende (1988). Insensitivity to ethylene conferred by a dominant mutation in *Arabidopsis thaliana*. *Science* 241:1086-1089.

Brocard-Gifford, I.M., T.J. Lynch, and R.R. Finkelstein (2003). Regulatory networks in seeds integrating developmental, abscisic acid, sugar, and light signaling. *Plant Physiology* 131:78-92.

Cervantes, E., A. Rodriques, and G. Nicolas (1994). Ethylene regulates the expression of a cysteine proteinase gene during germination of chickpea (*Cicer arietinum* L.). *Plant Molecular Biology* 25:207-215.

Chen, F. and K.J. Bradford (2000). Expression of an expansin is associated with endosperm weakening during tomato seed germination. *Plant Physiology* 124: 1265-1274.

Chen, F., H. Nonogaki, and K.J. Bradford (2002). A gibberellin-regulated xyloglucan endotransglycosylase gene is expressed in the endosperm cap during tomato seed germination. *Journal of Experimental Botany* 53:215-223.

Clerkx, E.J.M., H. Blankestijn-De Vries, G.J. Ruys, S.P.C. Groot, and M. Koorneef (2003). Characterization of *green seed*, an enhancer of *abi3-1* in *Arabidopsis* that affects seed longevity. *Plant Physiology* 132:1077-1084.

Cutler, S., M. Ghassemian, D. Bonetta, S. Cooney, and P. McCourt (1996). A protein farnesyl transferase involved in abscisic acid signal transduction in *Arabidopsis*. *Science* 273:1239-1241.

Debeaujon, I. and M. Koornneef (2000). Gibberellin requirement for *Arabidopsis* seed germination is determined both by testa characteristics and embryonic abscisic acid. *Plant Physiology* 122:415-424.

Debeaujon, I., K.M. Léon-Kloosterziel, and M. Koornneef (2000). Influence of the testa on seed dormancy, germination, and longevity in *Arabidopsis*. *Plant Physiology* 122:403-413.

Derkx, M. and C.M. Karssen (1993). Effects of light and temperature on seed dormancy and gibberellin-stimulated germination in *Arabidopsis thaliana*—Studies with gibberellin-deficient and gibberellin-insensitive mutants. *Physiologia Plantarum* 89:360-368.

Dewar, J., J.R.N. Taylor, and P. Berjak (1998). Changes in selected plant growth regulators during germination in sorghum. *Seed Science Research* 8:1-8.

Diaz, I., J. Vicente-Carbajosa, Z. Abraham, M. Martinez, I. Isabel La Moneda, and P. Carbonero (2002). The GAMYB protein from barley interacts with the DOF transcription factor BPBF and activates endosperm-specific genes during seed development. *The Plant Journal* 29:453-464.

Dill, A. and T.P. Sun (2001). Synergistic derepression of gibberellin signaling by removing RGA and GAI function in *Arabidopsis thaliana*. *Genetics* 159:777-785.

Dutta, S., K.J. Bradford, and D.J. Nevins (1997). Endo-β-mannanase activity present in cell wall extracts of lettuce endosperm prior to radicle emergence. *Plant Physiology* 113:155-161.

Emery, R.J.N., Q. Ma, and C.A. Atkins (2000). The forms and sources of cytokinins in developing white lupine seeds and fruits. *Plant Physiology* 123:1593-1604.

Emmler, K. and E. Schäfer (1997). Maternal effect on embryogenesis in tobacco overexpressing rice phytochrome A. *Botanica Acta* 110:1-8.

Ephritikhine, G., S. Pagant, S. Fujioka, S. Takatsuto, D. Lapous, M. Caboche, R.E. Kendrick, and H. Barbier-Brygoo (1999). The *sax1* mutation defines a new locus involved in the brassinosteroid biosynthesis pathway of *Arabidopsis thaliana*. *The Plant Journal* 18:315-320.

Esashi, Y. (1991). Ethylene and seed germination. In A.K. Mattoo and J.C. Suttle (eds.), *The plant hormone ethylene* (pp. 133-157). Boca Raton, FL: CRC Press.

Ezcurra, I., P. Wycliffe, L. Nehlin, M. Ellerstrom, and L. Rask (2000). Transactivation of the *Brassica napus* napin promoter by ABI3 requires interaction of the conserved B2 and B3 domains of ABI3 with different cis-elements: B2 mediates activation through an ABRE, whereas B3 interacts with an RY/G-box. *The Plant Journal* 24:57-66.

Fennimore, S.A. and M.E. Foley (1998). Genetic and physiological evidence for the role of gibberellic acid in the germination of dormant *Avena fatua* seeds. *Journal of Experimental Botany* 49:89-94.

Finch-Savage, W.E. and H.A. Clay (1994). Evidence that ethylene, light and abscisic acid interact to inhibit germination in the recalcitrant seeds of *Quercus robur* L. *Journal of Experimental Botany* 45:1295-1299.

Finkelstein, R.R. (1994). Maternal effects govern variable dominance of two abscisic acid response mutations in *Arabidopsis thaliana*. *Plant Physiology* 105:1203-1208.

Finkelstein, R.R., S.S.L. Gampala, and C.D. Rock (2002). Abscisic acid signaling in seeds and seedlings. *The Plant Cell* 14:S15-S45.

Finkelstein, R.R. and S.I. Gibson (2002). ABA and sugar interactions regulating development: Cross-talk or voices in a crowd? *Current Opinion in Plant Biology* 5:26-32.

Finkelstein, R.R. and T.J. Lynch (2000). The *Arabidopsis* abscisic acid response gene *ABI5* encodes a basic leucine zipper transcription factor. *The Plant Cell* 12:599-609.

Finkelstein, R.R., M.L. Wang, T.J. Lynch, S. Rao, and H.M. Goodman (1998). The *Arabidopsis* abscisic acid response locus ABI4 encodes an APETALA2 domain protein. *The Plant Cell* 10:1043-1054.

Fischer-Iglesias, C. and G. Neuhaus (2001). Zygotic embryogenesis—Hormonal control of embryo development. In S.S. Bhojwani and W.Y. Soh (eds.), *Current trends in the embryology of angiosperms* (pp. 223-247). Dordrecht, Boston, London: Kluwer Academic Publishers.

Footitt, S., S.P. Slocombe, V. Larner, S. Kurup, Y. Wu, T. Larson, I. Graham, A. Baker, and M. Holdsworth (2002). Control of germination and lipid mobilization by *COMATOSE*, the *Arabidopsis* homologue of human ALDP. *European Molecular Biology Organization Journal* 21:2912-2922.

Frankhauser, C. (2002). Light perception in plants: Cytokinins and red light join forces to keep phytochrome B active. *Trends in Plant Science* 7:143-145.

Frey, A., C. Audran, E. Marin, B. Sotta, and A. Marion-Poll (1999). Engineering seed dormancy by the modification of zeaxanthin epoxidase gene expression. *Plant Molecular Biology* 39:1267-1274.

Gallardo, K., C. Job, S.P.C. Groot, M. Puype, H. Demol, J. Vandekerckhove, and D. Job (2002a). Importance of methionine biosynthesis for *Arabidopsis* seed germination and seedling growth. *Physiologia Plantarum* 116:238-247.

Gallardo, K., C. Job, S.P.C. Groot, M. Puype, H. Demol, J. Vandekerckhove, and D. Job (2002b). Proteomics of *Arabidopsis* seed germination: A comparative study of wild-type and gibberellin-deficient seeds. *Plant Physiology* 129:823-837.

Ghassemian, M., E. Nambara, S. Cutler, H. Kawaide, Y. Kamiya, and P. McCourt (2000). Regulation of abscisic acid signaling by the ethylene response pathway in *Arabidopsis*. *The Plant Cell* 12:1117-1126.

Gómez-Jiménez, M.C., E. Garcia-Olivares, and A.J. Matilla (2001). 1-Aminocyclopropane-1-carboxylate oxidase from embryonic axes of germinating chick-pea (*Cicer arietinum* L.) seeds: Cellular immunolocalization and alterations in its expression by indole-3-acetic acid, abscisic acid and spermine. *Seed Science Research* 11:243-253.

Gorecki, R.J., H. Ashino, S. Satoh, and Y. Esahi (1991). Ethylene production in pea and cocklebur seeds of differing vigour. *Journal of Experimental Botany* 42: 407-414.

Grappin, P., D. Bouinot, B. Sotta, E. Miginiac, and M. Jullien (2000). Control of seed dormancy in *Nicotiana plumbaginifolia*: Post-imbibition abscisic acid synthesis imposes dormancy maintenance. *Planta* 210:279-285.

Grill, E. and A. Himmelbach (1998). ABA signal transduction. *Current Opinion in Plant Biology* 1:412-418.

Groot, S.P.C., J. Bruinsma, and C.M. Karssen (1987). The role of endogenous gibberellin in seed and fruit development of tomato: Studies with a gibberellin-deficient mutant. *Physiologia Plantarum* 71:184-190.

Groot, S.P.C. and C.M. Karssen (1987). Gibberellins regulate seed germination in tomato by endosperm weakening: A study with gibberellin-deficient mutants. *Planta* 171:525-531.

Groot, S.P.C. and C.M. Karssen (1992). Dormancy and germination of abscisic acid-deficient tomato seeds. *Plant Physiology* 99:952-958.

Gualberti, G., M. Papi, L. Bellucci, L. Ricci, D. Bouchez, C. Camilleri, P. Costantino, and P. Vittorioso (2002). Mutations in the Dof zinc finger genes DAG2 and DAG1 influence with opposite effects the germination of *Arabidopsis* seeds. *The Plant Cell* 14:1253-1263.

Guan, L.Q.M. and J.G. Scandalios (2002). Catalase gene expression in response to auxin-mediated developmental signals. *Physiologia Plantarum* 114:288-295.

Haberer, G. and J.J. Kieber (2002). Cytokinins: New insights into a classic phytohormone. *Plant Physiology* 128:354-362.

Hall, M.A., I.E., Moshkov, G.V. Novikova, L.A.J. Mur, and A.R. Smith (2001). Ethylene signal perception and transduction: Multiple paradigms? *Biological Reviews* 76:103-128.

Hamann, T. (2001). The role of auxin in apical-basal pattern formation during *Arabidopsis* embryogenesis. *Journal of Plant Growth Regulation* 20:292-299.

Hays, D.B., E.C. Yeung, and R.P. Pharis (2002). The role of gibberellins in embryo axis development. *Journal of Experimental Botany* 53:1747-1751.

Hennig, L., W.M. Stoddart, M. Dieterle, G.C. Whitelam, and E. Schäfer (2002). Phytochrome E controls light-induced germination of *Arabidopsis*. *Plant Physiology* 128:194-200.

Hilhorst, H.W.M. (1995). A critical update on seed dormancy: I. Primary dormancy. *Seed Science Research* 5:61-73.

Hilhorst, H.W.M. and B. Downie (1995). Primary dormancy in tomato (*Lycopersicon esculentum* cv. Moneymaker): Studies with the *sitiens* mutant. *Journal of Experimental Botany* 47:89-97.

Hilhorst, H.W.M. and C.M. Karssen (1992). Seed dormancy and germination: The role of abscisic acid and gibberellins and the importance of hormone mutants. *Plant Growth Regulation* 11:225-238.

Hobo, T., Y. Kowyama, and T. Hattori (1999). A bZIP factor, TRAB1, interacts with VP1 and mediates abscisic acid-induced transcription. *Proceedings National Academy of Sciences USA* 96:15348-15353.

Holdsworth, M., J. Lenton, J. Flintham, M. Gale, S. Kurup, R. McKibbin, P. Bailey, V. Larner, and L. Russell (2001). Genetic control mechanisms regulating the initiation of germination. *Journal of Plant Physiology* 158:439-445.

Huijser, C., A. Kortstee, J. Pego, P. Weisbeek, E. Wisman, and S. Smeekens (2000). The *Arabidopsis* SUCROSE UNCOUPLED-6 gene is identical to ABSCISIC ACID INSENSITIVE-4: Involvement of abscisic acid in sugar responses. *The Plant Journal* 23:577-585.

Izhaki, A., S.M. Swain, T.S. Tseng, A. Borochov, N.E. Olszewski, and D. Weiss (2001). The role of SPY and its TPR domain in the regulation of gibberellin action throughout the life cycle of *Petunia hybrida* plants. *The Plant Journal* 28:181-190.

Jiang, L., S.R. Abrams, and A.R. Kermode (1996). Vicilin and napin storage-protein gene promoters are responsive to abscisic acid in developing tobacco seed but lose sensitivity following premature desiccation. *Plant Physiology* 110: 1135-1144.

Jones-Held, S., M. Vandoren, and T. Lockwood (1996). Brassinolide application to *Lepidium sativum* seeds and the effects on seedling growth. *Journal of Plant Growth Regulation* 15:63-67.

Kamiya, Y. and J.L. Garcia-Martinez (1999). Regulation or gibberellin biosynthesis by light. *Current Opinion in Plant Biology* 2:398-403.

Karssen, C.M., D.L.C. Brinkhorst-van der Swan, A.E. Breekland, and M. Koornneef (1983). Induction of dormancy during seed development by endogenous abscisic acid: Studies on abscisic acid deficient genotypes of *Arabidopsis thaliana* (L.) Heynh. *Planta* 157:158-165.

Karssen, C.M. and E. Laçka (1986). A revision of the hormone balance theory of seed dormancy: Studies on gibberellin and/or abscisic acid-deficient mutants of *Arabidopsis thaliana*. In M. Bopp (ed.), *Plant growth substances 1985* (pp. 315-323). Berlin, Heidelberg: Springer-Verlag.

Karssen, C.M., S. Zagórsky, J. Kepczynski, and S.P.C. Groot (1989). Key role for endogenous gibberellins in the control of seed germination. *Annals of Botany* 63:71-80.

Kepczynski, J., M. Bihun, and E. Kepczynska (2003). The release of secondary dormancy by ethylene in *Amaranthus caudatus* L. seeds. *Seed Science Research* 13:69-74.

Kepczynski, J. and E. Kepczynska (1997). Ethylene in seed dormancy and germination. *Physiologia Plantarum* 101:720-726.

Khalil, M.K. (1992). Nature of growth regulators effects on *Nicotiana tabacum* seed germination. *Angewandte Botanik* 66:106-108.

Kincaid, R.R. (1935). The effects of certain environmental factors on the germination of Florida cigar-wrapper tobacco seeds. *Technical Bulletin, University of Florida, Agricultural Experiment Station* 277:5-47.

Koornneef, M., L. Bentsink, and H. Hilhorst (2002). Seed dormancy and germination. *Current Opinion in Plant Biology* 5:33-36.

Koornneef, M. and C.M. Karssen (1994). Seed dormancy and germination. In E.M. Meyerowitz and C.R. Somerville (eds.), *Arabidopsis* (pp. 313-334). New York: Cold Spring Harbor Laboratory Press.

Kraepiel, Y., P. Rousselin, B. Sotta, L. Kerhoas, J. Einhorn, M. Caboche, and E. Miginiac (1994). Analysis of phytochrome- and ABA-deficient mutants suggests that ABA degradation is controlled by light in *Nicotiana plumbaginifolia*. *The Plant Journal* 6:665-672.

Kretsch, T., K. Emmler, and E. Schäfer (1995). Spatial and temporal pattern of light-regulated gene expression during tobacco seedling development: The photosystem II-related genes *Lhcb (Cab)* and *PsbP (Oee2)*. *The Plant Journal* 7:715-729.

Krock, B., S. Schmidt, C. Hertweck, and I.T.M. Baldwin (2002). Vegetation-derived abscisic acid and four terpenes enforce dormancy in seeds of the post-fire annual, *Nicotiana attenuata*. *Seed Science Research* 12:239-252.

Kwak, J.M., I.C. Mori, Z.-M. Pei, N. Leonhardt, M.A. Torres, J.L. Dangl, R.E. Bloom, S. Bodde, J.D.G. Jones, and J.I. Schroeder (2003). NADPH oxidase *AtrbohD* and *AtrbohF* genes function in ROS-dependent ABA signaling in *Arabidopsis*. *European Molecular Biology Organization Journal* 22:2623-2633.

Lashbrook, C.C., D.M. Tieman, and H.J. Klee (1998). Differential regulation of the tomato *ETR* gene family throughout plant development. *The Plant Journal* 15:243-252.

Lee, S.C., H. Cheng, K.E. King, W.F. Wang, Y.W. He, A. Hussain, J. Lo, N.P. Harberd, and J.R. Peng (2002). Gibberellin regulates *Arabidopsis* seed germination via *RGL2*, a *GAI/RGA*-like gene whose expression is up-regulated following imbibition. *Genes & Development* 16:646-658.

Leon-Kloosterziel, K.M., G.A. van de Bunt, J.A.D. Zeevaart, and M. Koornneef (1996). *Arabidopsis* mutants with a reduced seed dormancy. *Plant Physiology* 110:233-240.

Leubner-Metzger, G. (2001). Brassinosteroids and gibberellins promote tobacco seed germination by distinct pathways. *Planta* 213:758-763.

Leubner-Metzger, G. (2002). Seed after-ripening and over-expression of class I β-1,3-glucanase confer maternal effects on tobacco testa rupture and dormancy release. *Planta* 215:659-698.

Leubner-Metzger, G. (2003a). Brassinsteroids promote seed germination. In S. Hayat and A. Ahmad (eds.), *Brassinosteroids: Bioactivity and crop productivity*. Dordrecht, The Netherlands: Kluwer Academic Publisher, pp. 119-128.

Leubner-Metzger, G. (2003b). Functions and regulation of β-1,3-glucanase during seed germination, dormancy release and after-ripening. *Seed Science Research* 13:17-34.

Leubner-Metzger, G., C. Fründt, R. Vögeli-Lange, and F. Meins, Jr. (1995). Class I β-1,3-glucanase in the endosperm of tobacco during germination. *Plant Physiology* 109:751-759.

Leubner-Metzger, G. and F. Meins, Jr. (2000). Sense transformation reveals a novel role for class I β-1,3-glucanase in tobacco seed germination. *The Plant Journal* 23:215-221.

Leubner-Metzger, G. and F. Meins Jr. (2001). Antisense-transformation reveals novel roles for class I β-1,3-glucanase in tobacco seed after-ripening and photodormancy. *Journal of Experimental Botany* 52:1753-1759.

Leubner-Metzger, G., L. Petruzzelli, R. Waldvogel, R. Vögeli-Lange, and F. Meins, Jr. (1998). Ethylene-responsive element binding protein (EREBP) expression and the transcriptional regulation of class I β-1,3-glucanase during tobacco seed germination. *Plant Molecular Biology* 38:785-795.

Leung, J. and J. Giraudat (1998). Abscisic acid signal transduction. *Annual Review of Plant Physiology and Plant Molecular Biology* 49:199-222.

Li, B.L. and M.E. Foley (1997). Genetic and molecular control of seed dormancy. *Trends in Plant Science* 2:384-389.

Lindgren, L.O., K.G. Stalberg, and A.-S. Höglund (2003). Seed-specific overexpression of an endogenous *Arabidopsis* phytoene synthase gene results in delayed germination and increased levels of carotenoids, chlorophyll, and abscisic acid. *Plant Physiology* 132:779-785.

Liotenberg, S., H. North, and A. Marion-Poll (1999). Molecular biology and regulation of abscisic acid biosynthesis in plants. *Plant Physiology and Biochemistry* 37:341-350.

Liptay, A. and P. Schopfer (1983). Effect of water stress, seed coat restraint, and abscisic acid upon different germination capabilities of two tomato lines at low temperature. *Plant Physiology* 73:935-938.

Ljung, K., A. Östin, L. Lioussanne, and G. Sandberg (2001). Developmental regulation of indol-3-acetic acid turnover in scots pine seedlings. *Plant Physiology* 125:464-475.

Lohrmann, J. and K. Harter (2002). Plant two-component signaling systems and the role of response regulators. *Plant Physiology* 128:363-369.

Lopez-Molina, L., B. Mongrand, D.T. McLachlin, B.T. Chait, and N.H. Chua (2002). ABI5 acts downstream of ABI3 to execute an ABA-dependent growth arrest during germination. *The Plant Journal* 32:317-328.

Lopez-Molina, L., S. Mongrand, and N.-H. Chua (2001). A postgermination developmental arrest checkpoint is mediated by abscisic acid and requires ABI5 transcription factor in *Arabidopsis*. *Proceedings of the National Academy of Sciences of the United States of America* 98:4782-4787.

Lorenzo, O., C. Nicolas, G. Nicolas, and D. Rodriguez (2002). Molecular cloning of a functional protein phosphatase 2C (FsPP2C2) with unusual features and synergistically up-regulated by ABA and calcium in dormant seeds of *Fagus sylvatica*. *Physiologia Plantarum* 114:482-490.

Marin, E., L. Nussaume, A. Quesada, M. Gonneau, B. Sotta, P. Hugueney, A. Frey, and A. Marion-Poll (1996). Molecular identification of zeaxanthin epoxidase of *Nicotiana plumbaginifolia*, a gene involved in abscisic acid biosynthesis and corresponding to the ABA locus of *Arabidopsis thaliana*. *European Molecular Biology Organization Journal* 15:2331-2342.

Matilla, A.J. (2000). Ethylene in seed formation and germination. *Seed Science Research* 10:111-126.

McCarty, D.R. (1995). Genetic control and integration of maturation and germination pathways in seed development. *Annual Review of Plant Physiology and Plant Molecular Biology* 46:71-93.

McGinnis, K.M., S.G. Thomas, J.D. Soulea, L.C. Straderc, J.M. Zalea, T.-P. Sunb, and C.M. Steber (2003). The *Arabidopsis SLEEPY1* gene encodes a putative F-box subunit of an SCF E3 ubiquitin ligase. *The Plant Cell* 15:1120-1130.

Meinhard, M. and E. Grill (2001). Hydrogen peroxide is a regulator of ABI1, a protein phosphatase 2C from *Arabidopsis*. *Federation European Biochemical Society Letters* 508:443-446.

Mo, B.X. and J.D. Bewley (2002). β-Mannosidase (EC 3.2.1.25) activity during and following germination of tomato (*Lycopersicon esculentum* Mill.) seeds: Purification, cloning and characterization. *Planta* 215:141-152.

Mok, D.W.S. and M.C. Mok (2001). Cytokinin metabolism and action. *Annual Review of Plant Physiology* 52:89-118.

Nambara, E., H. Kawaide, Y. Kamiya, and S. Naito (1998). Characterization of an *Arabidopsis thaliana* mutant that has a defect in ABA accumulation: ABA-dependent and ABA-independent accumulation of free amino acids during dehydration. *Plant and Cell Physiology* 39:853-858.

Nambara, E., N. Satoshi, and P. McCourt (1992). A mutant of *Arabidopsis* which is defective in seed development and storage protein accumulation is a new *abi3* allele. *The Plant Journal* 2:435-441.

Neff, M.M., C. Fankhauser, and J. Chory (2000). Light: An indicator of time and place. *Genes & Development* 14:257-271.

Nelson, J.M. and G.C. Sharples (1980). Stimulation of tomato, pepper and sugarbeet seed germination at low temperatures by growth regulators. *Journal of Seed Technology* 5:62-68.

Ni, B.R. and K.J. Bradford (1993). Germination and dormancy of abscisic acid-deficient and gibberellin-deficient mutant tomato (*Lycopersicon esculentum*) seeds—Sensitivity of germination to abscisic acid, gibberellin, and water potential. *Plant Physiology* 101:607-617.

Nicolas, C., J.M. Deprada, O. Lorenzo, G. Nicolas, and D. Rodriguez (1998). Abscisic acid and stress regulate the expression of calmodulin in germinating chick-pea seeds. *Physiologia Plantarum* 104:379-384.

Nonogaki, H., O.H. Gee, and K.J. Bradford (2000). A germination-specific endo-β-mannanase gene is expressed in the micropylar endosperm cap of tomato seeds. *Plant Physiology* 123:1235-1245.

Ogawa, K. and M. Iwabuchi (2001). A mechanism for promoting the germination of *Zinnia elegans* seeds by hydrogen peroxide. *Plant and Cell Physiology* 42:286-291.

Ogawa, M., A. Hanada, Y. Yamauchi, A. Kuwahara, Y. Kamiya, and S. Yamaguchi (2003). Gibberellin biosynthesis and response during *Arabidopsis* seed germination. *The Plant Cell* 15:1591-1604.

Ohta, M., M. Ohme-Takagi, and H. Shinshi (2000). Three ethylene-responsive transcription factors in tobacco with distinct transactivation functions. *The Plant Journal* 22:29-38.

Papi, M., S. Sabatini, M.M. Altamura, L. Hennig, E. Schäfer, P. Costantino, and P. Vittorioso (2002). Inactivation of the phloem-specific dof zinc finger gene *DAG1* affects response to light and integrity of the testa of *Arabidopsis* seeds. *Plant Physiology* 128:411-417.

Papi, M., S. Sabatini, D. Bouchez, C. Camilleri, P. Costantino, and P. Vittorioso (2000). Identification and disruption of an Arabidopsis zinc finger gene controlling seed germination. *Genes & Development* 14:28-33.

Peeters, A.J.M., H. Blankestijn de Vries, C.J. Hanhart, K.M. Leon-Kloosterziel, J.A.D. Zeevaart, and M. Koornneef (2002). Characterization of mutants with reduced seed dormancy at two novel *rdo* loci and a further characterization of *rdo1* and *rdo2* in *Arabidopsis*. *Physiologia Plantarum* 115:604-612.

Peng, J. and N.P. Harberd (2002). The role of GA-mediated signalling in the control of seed germination. *Current Opinion in Plant Biology* 5:376-381.

Petruzzelli, L., I. Coraggio, and G. Leubner-Metzger (2000). Ethylene promotes ethylene biosynthesis during pea seed germination by positive feedback regulation of 1-aminocyclopropane-1-carboxylic acid oxidase. *Planta* 211:144-149.

Petruzzelli, L., F. Harren, C. Perrone, and J. Reuss (1995). On the role of ethylene in seed germination and early growth of *Pisum sativum*. *Journal of Plant Physiology* 145:83-86.

Petruzzelli, L., C. Kunz, R. Waldvogel, F. Meins, Jr., and G. Leubner-Metzger (1999). Distinct ethylene- and tissue-specific regulation of β-1,3-glucanases and chitinases during pea seed germination. *Planta* 209:195-201.

Petruzzelli, L., K. Müller, K. Hermann, and G. Leubner-Metzger (2003). Distinct expression patterns of β-1,3-glucanases and chitinases during the germination of Solanaceous seeds. *Seed Science Research* 13:139-153.

Petruzzelli, L., M. Sturaro, D. Mainieri, and G. Leubner-Metzger (2003). Calcium requirement for ethylene-dependent responses involving 1-aminocyclopropane-1-carboxylic acid oxidase in radicle tissues of germinated pea seeds. *Plant, Cell and Environment* 26:661-671.

Phillips, J., O. Artsaenko, U. Fiedler, C. Horstmann, H.P. Mock, K. Müntz, and U. Conrad (1997). Seed-specific immunomodulation of abscisic acid activity induces a developmental switch. *European Molecular Biology Organization Journal* 16:4489-4496.

Pritchard, S.L., W.L. Charlton, A. Baker, and I.A. Graham (2002). Germination and storage reserve mobilization are regulated independently in *Arabidopsis*. *The Plant Journal* 31:639-647.

Raz, V., J.H.W. Bergervoet, and M. Koornneef (2001). Sequential steps for developmental arrest in *Arabidopsis* seeds. *Development* 128:243-252.

Richards, D.E., K.E. King, T. Aitali, and N.P. Harberd (2001). How gibberellin regulates plant growth and development: A molecular genetic analysis of gibberellin signaling. *Annual Review of Plant Physiology* 52:67-88.

Ritchie, S., A. McCubbin, G. Ambrose, T.H. Kao, and S. Gilroy (1999). The sensitivity of barley aleurone tissue to gibberellin is heterogeneous and may be spatially determined. *Plant Physiology* 120:361-370.

Rohde, A., S. Kurup, and M. Holdsworth (2000). ABI3 emerges from the seed. *Trends in Plant Science* 5:418-419.

Romagosa, I., D. Prada, M.A. Moralejo, A. Sopena, P. Munoz, A.M. Casas, J.S. Swanston, and J.L. MolinaCano (2001). Dormancy, ABA content and sensitivity of a barley mutant to ABA application during seed development and after ripening. *Journal of Experimental Botany* 52:1499-1506.

Rousselin, P., Y. Kraepiel, R. Maldiney, E. Miginiac, and M. Caboche (1992). Characterization of three hormone mutants of *Nicotiana plumbaginifolia*: Evidence for a common ABA deficiency. *Theoretical and Applied Genetics* 85:213-221.

Russel, L., V. Larner, S. Kurup, S. Bougourd, and M.J. Holdsworth (2000). The Arabidopsis *COMATOSE* locus regulates germination potential. *Development* 127: 3759-3767.

Saini, H.S., E.D. Consolacion, P.K. Bassi, and M.S. Spencer (1989). Control processes in the induction and relief of thermoinhibition of lettuce seed germination: Actions of phytochrome and endogenous ethylene. *Plant Physiology* 90:311-315.

Schmidt, J., T. Altmann, and G. Adam (1997). Brassinosteroids from seeds of *Arabidopsis thaliana*. *Phytochemistry* 45:1325-1327.

Schmitz, N., S.R. Abrams, and A.R. Kermode (2002). Changes in ABA turnover and sensitivity that accompany dormancy termination of yellow-cedar *(Chamaecyparis nootkatensis)* seeds. *Journal of Experimental Botany* 53:89-101.

Schopfer, P. and C. Plachy (1984). Control of seed germination by abscisic acid. II. Effect on embryo water uptake in *Brassica napus* L. *Plant Physiology* 76:155-160.

Schopfer, P. and C. Plachy (1993). Photoinhibition of radish (*Raphanus sativus* L.) seed germination—Control of growth potential by cell-wall yielding in the embryo. *Plant, Cell and Environment* 16:223-229.

Schopfer, P., C. Plachy, and G. Frahry (2001). Release of reactive oxygen intermediates (superoxide radicals, hydrogen peroxide, and hydroxyl radicals) and peroxidase in germinating radish seeds controlled by light, gibberellin, and abscisic acid. *Plant Physiology* 125:1591-1602.

Schwechheimer, C. and M.W. Bevan (1998). The regulation of transcription factor activity in plants. *Trends in Plant Science* 3:378-383.

Singh, D.P., A.M. Jermakow, and S.M. Swain (2002). Gibberellins are required for seed development and pollen tube growth in *Arabidopsis*. *The Plant Cell* 14: 3133-3147.

Söderman, E.M., I.M. Brocard, T.J. Lynch, and R.R. Finkelstein (2000). Regulation and function of the *Arabidopsis ABA-insensitive4* gene in seed and abscisic acid response signaling networks. *Plant Physiology* 124:1752-1765.

Solano, R., A. Stepanova, Q.M. Chao, and J.R. Ecker (1998). Nuclear events in ethylene signaling: A transcriptional cascade mediated by ETHYLENE-INSENSITIVE3 and ETHYLENE- RESPONSE-FACTOR1. *Genes & Development* 12:3703-3714.

Steber, C.M., S.E. Cooney, and P. McCourt (1998). Isolation of the GA-response mutant *sly1* as a suppressor of *ABI1-1* in *Arabidopsis thaliana*. *Genetics* 149: 509-521.

Steber, C.M. and P. McCourt (2001). A role for brassinosteroids in germination in *Arabidopsis*. *Plant Physiology* 125:763-769.

Sun, T. and Y. Kamiya (1994). The *Arabidopsis GA1* locus encodes the cyclase *ent*-kaurene synthetase A of gibberellin biosynthesis. *The Plant Cell* 6:1509-1518.

Suzuki, M., C.Y. Kao, S. Cocciolone, and D.R. McCarty (2001). Maize VP1 complements *Arabidopsis* abi3 and confers a novel ABA/auxin interaction in roots. *The Plant Journal* 28:409-418.

Swain, S.M., J.B. Reid, and Y. Kamiya (1997). Gibberellins are required for embryo growth and seed development in pea. *The Plant Journal* 12:1329-1338.

Swain, S.M., T.S. Tseng, and N.E. Olszewski (2001). Altered expression of SPINDLY affects gibberellin response and plant development. *Plant Physiology* 126:1174-1185.

Takeuchi, Y., Y. Omigawa, M. Ogasawara, K. Yoneyama, M. Konnai, and A.D. Worsham (1995). Effects of brassinosteroids on conditioning and germination of clover broomrape (*Orobanche minor*) seeds. *Plant Growth Regulation* 16:153-160.

Takeuchi, Y., A.D. Worsham, and A.E. Awad (1991). Effects of brassinolide on conditioning and germination of witchweed (*Striga asiatica*) seeds. In H.G. Cuttler, T. Yokota, and G. Adam (eds.), *Brassinosteroids: Chemistry, bioactivity and applications* (pp. 298-305). Washington, DC: ACS Symposium series.

Thompson, A.J., A.C. Jackson, R.C. Symonds, B.J. Mulholland, A.R. Dadswell, P.S. Blake, A. Burbidge, and I.B. Taylor (2000). Ectopic expression of a tomato 9-cis-epoxycarotenoid dioxygenase gene causes over-production of abscisic acid. *The Plant Journal* 23:363-374.

Toorop, P.E., A.C. van Aelst, and H.W.M. Hilhorst (1998). Endosperm cap weakening and endo-ß-mannanase activity during priming of tomato (*Lycopersicon esculentum* cv. Moneymaker) seeds are initiated upon crossing a threshold water potential. *Seed Science Research* 8:483-491.

Toorop, P.E., A.C. van Aelst, and H.W.M. Hilhorst (2000). The second step of the biphasic endosperm cap weakening that mediates tomato *(Lycopersicon esculentum)* seed germination is under control of ABA. *Journal of Experimental Botany* 51:1371-1379.

Toyomasu, T., H. Kawaide, W. Mitsuhashi, Y. Inoue, and Y. Kamiya (1998). Phytochrome regulates gibberellin biosynthesis during germination of photoblastic lettuce seeds. *Plant Physiology* 118:1517-1523.

Toyomasu, T., H. Tsuji, H. Yamane, M. Nakayama, I. Yamaguchi, N. Murofushi, N. Takahashi, and Y. Inoue (1993). Light effects on endogenous levels of gibberellins in photoblastic lettuce seeds. *Journal of Plant Growth Regulation* 12:85-90.

Ullah, H., J.G. Chen, S.C. Wang, and A.M. Jones (2002). Role of a heterotrimeric G protein in regulation of *Arabidopsis* seed germination. *Plant Physiology* 129: 897-907.

Walz, A., S. Park, J.P. Slovin, J. LudwigMuller, Y.S. Momonoki, and J.D. Cohen (2002). A gene encoding a protein modified by the phytohormone indoleacetic acid. *Proceedings National Academy of Sciences USA* 99:1718-1723.

Welbaum, G.E., K.J. Bradford, K.-O. Yim, D.T. Booth, and M.O. Oluoch (1998). Biophysical, physiogical and biochemical processes regulating seed germination. *Seed Science Research* 8:161-172.

Wen, C.K. and C. Chang (2002). *Arabidopsis* RGL1 encodes a negative regulator of gibberellin responses. *The Plant Cell* 14:87-100.

White, C.N., W.M. Proebsting, P. Hedden, and C.J. Rivin (2000). Gibberellins and seed development in maize: I. Evidence that gibberellin/abscisic acid balance governs germination versus maturation pathways. *Plant Physiology* 122:1081-1088.

White, C.N. and C.J. Rivin (2000). Gibberellins and seed development in maize: II. Gibberellin synthesis inhibition enhances abscisic acid signaling in cultured embryos. *Plant Physiology* 122:1089-1097.

Wu, C.-T., G. Leubner-Metzger, F. Meins, Jr., and K.J. Bradford (2000). Class I β-1,3-glucanase and chitinase are expressed in the micropylar endosperm of tomato seeds prior to radicle emergence. *Plant Physiology* 126:1299-1313.

Yamaguchi, S., Y. Kamiya, and T.P. Sun (2001). Distinct cell-specific expression patterns of early and late gibberellin biosynthetic genes during *Arabidopsis* seed germination. *The Plant Journal* 28:443-453.

Yamaguchi, S., M.W. Smith, R.G.S. Brown, Y. Kamiya, and T.P. Sun (1998). Phytochrome regulation and differential expression of gibberellin 3β-hydroxylase genes in germinating *Arabidopsis* seeds. *The Plant Cell* 10:2115-2126.

Yamaguchi, T., T. Wakizuka, K. Hirai, S. Fujii, and A. Fujita (1987). Stimulation of germination in aged rice seeds by pretreatment with brassinolide. *Proceedings of the Plant Growth Regulation Society of America* 14:26-27.

Yamaguchi-Shinozaki, K., M. Mino, J. Mundy, and N.-H. Chua (1990). Analysis of an ABA-responsive rice gene promoter in transgenic tobacco. *Plant Molecular Biology* 15:905-912.

Yoshioka, T., T. Endo, and S. Satoh (1998). Restoration of seed germination at supraoptimal temperatures by fluridone, an inhibitor of abscisic acid biosynthesis. *Plant and Cell Physiology* 39:307-312.

Zeng, Y., N. Raimondi, and A.R. Kermode (2003). Role of an ABI3 homologue in dormancy maintenance of yellow-cedar seeds and in the activation of storage protein and Em gene promoters. *Plant Molecular Biology* 51:39-49.

Chapter 12

Photoregulation of Seed Germination

Chizuko Shichijo
Osamu Tanaka
Tohru Hashimoto

INTRODUCTION

Adherent higher plants must grow and reproduce their offspring where their seeds germinate, and this necessitated evolution of the ability to sense environmental conditions such as light, temperature, water, oxygen, and other chemical substances suitable for the plants to finish their life cycle. Seeds sense light and respond to it in various ways modified by the other environmental factors.

The response of seeds to light varies from promotion to inhibition as well as indifference. Seeds sense light mostly through phytochromes (phy) among the plant photoreceptors known to be involved in plant photomorphogenesis. Phytochromes have two forms: one is a physiologically inactive, red-light-absorbing form, and the other is an active, far-red-light-absorbing form. A phy molecule is composed of a chromophore and a protein moiety. The protein moiety is encoded by several different genes, and thus differentiates phy into chemical species known as the phy family (Furuya, 1993; Pratt, 1995; Quail, 1994; Quail et al., 1995; Smith, 1994, 1995, 2000). Interestingly, the multiplicity of phy species arising from the difference in protein moiety and the difference of seed stages at which phy operates may explain the diverse photoresponses of seed germination.

Short Research History of Seed Germination Photoregulation—Recognition of Phytochromes As Photoreceptors

Since long ago, humans must have noted that light affects seed germination. However, only in the nineteenth century was it scientifically shown that light promotes germination. In 1860, Caspary observed that seeds of *Bulliardia aquatica* germinated better under full sunlight than in diffuse light (after Toole,

1973). In 1881, Stebler proved that seeds of some grass species germinated better in light than in darkness (after Rollin, 1972). In 1904, Heinricher and Remer demonstrated that light inhibited the germination of some plant species (after Rollin, 1972). In 1907, after testing seeds of 964 species, Kinzel found that 70 percent were promoted by light, 27 percent were inhibited, and 3 percent were indifferent. Thus, seed germination is affected in opposite ways. Seeds whose germination is promoted by light are defined as positive photoblastic seeds, while seeds whose germination is inhibited are termed negative photoblastic seeds. The nature of such seeds as affected by light is collectively termed photoblasticity or photoblastism.

Compared with seeds of wild plants, a smaller number of cultivated plant species demonstrate photoblasticity, probably due to artificial selection of species whose seeds germinate readily without light.

Effects of the wavelengths of light were examined as early as 1883 by Cieslar by filtering sunlight or artificial white light. He found that yellow light favored the germination of seeds of a certain species, while violet light retarded it. These findings, confirmed by subsequent researchers, were described in a textbook by Molisch (1930), explaining that yellow to red light promotes germination, while violet, blue, or green light inhibits it (after Flint and McAlister, 1935). Irradiating lettuce seeds (var. Arlington Fancy) with spectral monochromatic light, Flint and McAlister (Flint, 1936; Flint and McAlister, 1935, 1937) first constructed an action spectrum (see Figure 12.1) demonstrating precise wavebands for promotion and inhibition of germination thus far observed with filtered broadband light. Seeds were

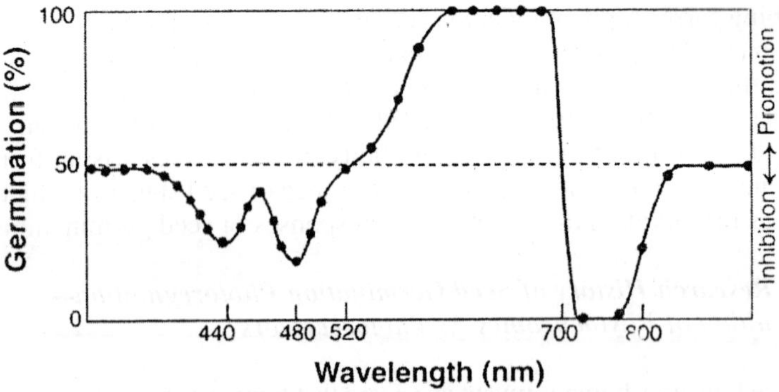

FIGURE 12.1. The historical first experiment to show the spectral bands for promotion and inhibition of seed germination. After red light was given to induce 50 percent germination, monochromatic light obtained through a prism was given for 24 h. Seed: *Lactuca sativa* var. Arlington Fancy. (*Source:* Adapted from Flint, 1936, and Flint and McAlister, 1937.)

preirradiated with red light (R) so as to show 50 percent germination and given spectral light for 24 h (note that it was not a pulse). The resulting action spectrum had a broad waveband of promotion from 600 nm to 700 nm, as well as three peaks of inhibition at 440, 480, and 760 nm. These researchers considered the possibility of the photoreceptor for promotion being chlorophyll based on some similarity of the action peak to the absorption peak of chlorophyll. Another great discovery in this study was that the effect of R (for 50 percent germination) was reversed by far-red light (FR), which later led Borthwick et al. (1952) to indicate R/FR reversibility.

Using the large, powerful spectrograph newly built at Beltsville, Maryland, Borthwick et al. (1952, 1954) determined action spectra in a true sense for germination promotion as well as for inhibition of inductive R action. For germination promotion, 'Grand Rapids' lettuce seeds imbibed for 16 h in darkness were irradiated with spectral light of 560 to 700 nm without preirradiation, and for germination inhibition, with wavelengths longer than 700 nm after R of 2.5 min sufficient for full germination. These researchers used a pulse irradiation of spectral light as short as 8 or 12 min, for example, although the irradiation period varied depending on the intensity of the spectral light available and the quantity of light to be given to seeds. For each wavelength the quantity of light required for 50 percent response of the full action was plotted against wavelengths. Thus, obtained action spectra (see Figure 12.2) were the ones which should reflect absorption spectra of the relevant photoreceptors, and they showed the maximum response, respectively, at 660 nm for germination promotion and at 720 to 750 nm for inhibition of R-induced germination, qualitatively confirming the action spectra determined by Flint and McAlister (1937) at the wavelength region longer than 525 nm. At shorter wavelengths, however, Borthwick et al. (1954) could not trace such inhibition by blue light (B) after R as shown by Flint and McAlister (1937), but they did observe promotion of germination of dark-imbibed seeds. In later years this discrepancy of the results resolved such that the inhibition observed by Flint and McAlister as well as other previous researchers was high-irradiance response (HIR) of phy due to a prolonged irradiation with B, and the promotion shown by Borthweick et al. was a low-fluence response (LFR) due to a short pulse irradiation.[1]

Success in causing maximum germination by a 1 min pulse of R and almost full suppression of the R effect by 4 min FR led Borthwick et al. (1952) to test the effects of alternate R and FR pulses repeatedly. They found that if R came last, seeds germinated, but if FR was last, seeds did not germinate (see Figure 12.3). These findings led these researchers to assume that two interchangeable forms of one photoreceptor may control seed germination, R converting the inactive form into the active form and FR

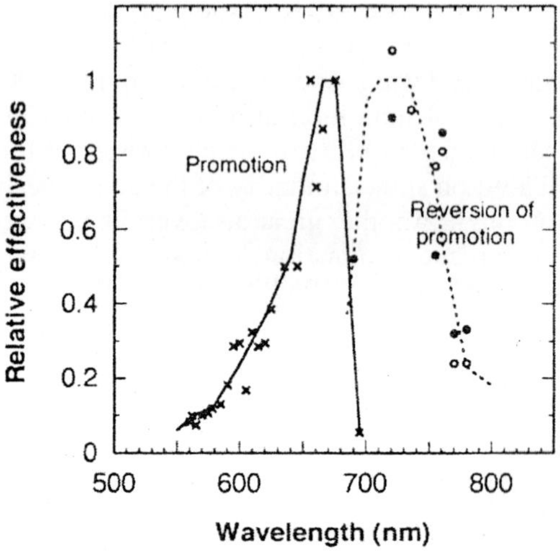

FIGURE 12.2. Action spectra for promotion and its reversion of seed germination by monochromatic pulse. For reversion of promotion, a saturating R pulse was given, followed by a pulse of monochromatic light. Each relative value of the reciprocal of the light energy required for 50 percent promotion or 50 percent reversion was plotted as relative effectiveness. Seeds in *Lactuca sativa* var. Grand Rapids. (*Source:* Adapted from Borthwick et al., 1952.)

reversing the active form to inactive form. Similar R/FR reversibility was found to occur in the seeds of other species.

Meanwhile, it was discovered that many light responses in plants, including the floral initiation inhibition of short-day plants and the floral initiation induction of long-day plants, as well as the unfolding of plumular hook and leaf expansion in dark-grown seedlings, show action spectra similar to that of lettuce seed germination promotion. The reversibility of R/FR was also confirmed in these responses. Thus, it was suggested that the same photoreversible pigment operates as a photoreceptor in a wide range of light response in plants.

The active form of the pigment arising on R was expected to have a larger absorbance at the FR region, and the inactive form, at the R region. Using a differential spectrophotometer, Butler et al. (1959) showed such changes in absorbance occurred first in etiolated maize shoots, implying the occurrence of the expected pigment. The pigment was then named phytochrome ("plant color" in Greek) and was publicized in 1960 at a Cold Spring Harbor Symposium. It did not take long until the pigment, phytochrome, was partially purified from etiolated pea seedlings (Bonner, 1960) and obtained in a

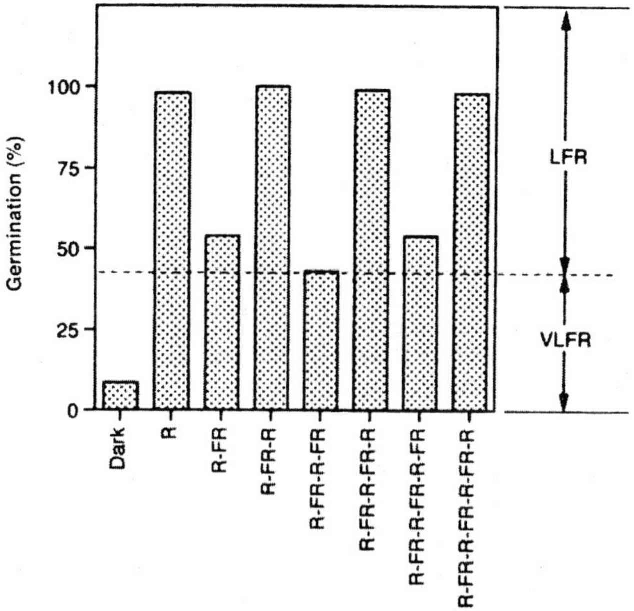

FIGURE 12.3. Photoreversibility of seed germination in Grand Rapids lettuce. After imbibition for 16 h in darkness, seeds were exposed to alternate R (1 min) and FR (4 min) and then incubated for 2 days in darkness at 20°C. (*Source:* Adapted from Borthwick et al., 1952.)

pure form from etiolated oat seedlings (Siegelman and Firer, 1964). The R/FR reversibility originally found in seed germination contributed to the discovery as well as the purification of the pigment, facilitating the detection during the purification process (Sage, 1992).

PHYTOCHROMES

Chemical Structure

Phytochromes found in seed plants are relatively well characterized, although other *PHY*-like genes were recently found to occur widely in plants, including gymnosperms, ferns, mosses, and algae, as well as in cyanobacteria and even in certain other bacteria. The PHY-like proteins in prokaryotes are now called bacteriophytochrome. Here we explain seed plant phy regulating germination.

The chemical structure and properties concern mostly those of phytochrome A (phyA), one of the phy-family members that was most extensively studied. Phytochromes are water-soluble conjugated protein consisting of an apoprotein of approximately 120 kD in monomer and the chromophore of an open-chain tetrapyrrole named phytochromobilin (see Figure 12.4), which combines covalently with a cysteine residue of the phy-apoprotein; they occur as dimers in the native state. Several molecular species are known, which differ in the amino acid sequence of the apoprotein but have the same chromophore in common. In *Arabidopsis* the known phytochromes are phyA, phyB, phyC, phyD, and phyE (Sharrock and Quail, 1989; Clack et al., 1994), and in tomato, phyA, phyB1, phyB2, phyE, and phyF (Hauser et al., 1995). In dark-grown seedlings of most species phyA exists in amounts 10 to 50 times as much as phyB or other phy species (Sharrock and Clack, 2002). Hence, phyA is a sole phy which is spectrometrically measurable in vivo and has thus far been isolated in pure form.

Photoconversion of Phytochromes

When synthesized in the cell, phys are in the R-absorbing form (Pr), which is physiologically inactive. On absorbing R it converts within 1 s (slower reaction if compared with photoconversion of rhodopsin) to the active, FR-absorbing form (Pfr) through several intermediates. Pfr reverses to Pr on absorbing FR through distinct intermediates. The change between Pr and Pfr can be repeated many times without loss. In the interconversion between Pr and Pfr the chromophore changes from one isomeric form to another (see

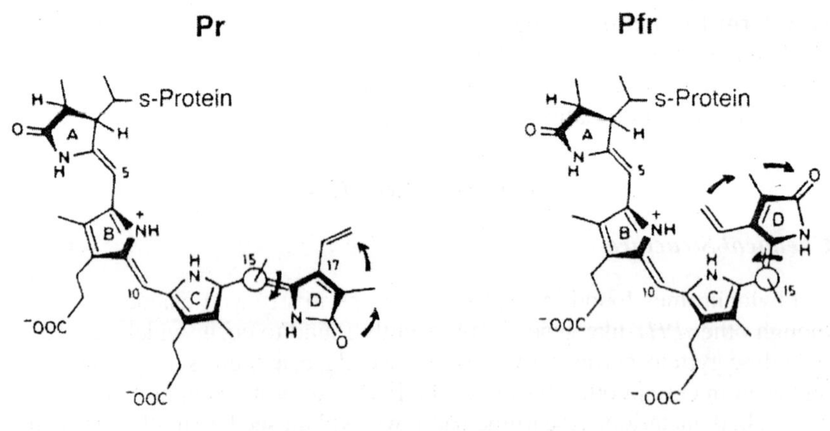

FIGURE 12.4. Structure of phytochrome chromophore. Left: Pr form (15 *cis* isomer), right: Pfr form (15 *trans* isomer). Arrows indicate the assumed movement of ring D upon photoconversion. (*Source:* Adapted from Rüdiger, 1992.)

Figure 12.4), and the conformation of the apoprotein also varies accordingly. Thus, the interconversion is a photochemical reaction, and its rate is not very sensitive to temperature. In darkness Pfr reverses directly to Pr at a temperature-dependent manner, at a slow rate without absorbing FR, because Pfr is at a higher free energy level than Pr (Sineshchekov, 1995).

Red (about 660 nm) and FR (about 730 nm) for the Pr→Pfr and Pfr→Pr conversions, respectively, are most effective, but other wavelengths are not without effect (see Figure 12.5). The efficiency for photoconversion at a given wavelength λ is expressed in Equation 12.1:

$$\sigma_{R\,\lambda} = 2.3 \cdot \varepsilon_{R\lambda} \cdot \Phi_R \qquad (12.1)$$

where $\sigma_{R\,\lambda}$ is a photoconversion cross-section (efficiency as represented by an action spectrum) for Pr→Pfr; $\varepsilon_{R\lambda}$, extinction coefficient of Pr, as represented by an absorption spectrum; Φ_R, quantum yield for Pr→Pfr. Similarly, Equation 12.2 holds for the photoreversion Pfr→Pr.

$$\sigma_{FR\,\lambda} = 2.3 \cdot \varepsilon_{FR\lambda} \cdot \Phi_{FR} \qquad (12.2)$$

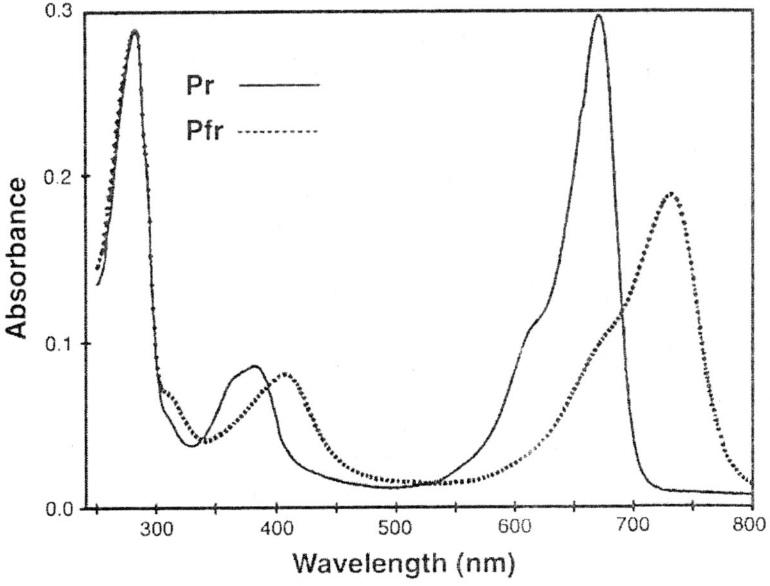

FIGURE 12.5. Absorption spectra of Pr and Pfr of 124 kDa *Avena* pytochrome. (*Source:* Adapted from Lagarias et al., 1987.)

The Φ_R and Φ_{FR} values determined with oat phyA, for example, are 0.152 and 0.069, respectively (Kelly and Lagarias, 1985). When irradiated by light of a given wavelength λ, Pr converts to Pfr according to Equation 12.1, and then Pfr reverses to Pr by absorbing the same light according to Equation 12.2, and finally an equilibrium between Pr and Pfr is reached, which is determined by $\sigma_{R\,\lambda}$ and $\sigma_{FR\,\lambda}$ (see Exhibit 12.1). Thus, Pfr/P ratio at photoequilibrium (ϕ_λ) depends solely on the wavelength, but not the fluence rate (Table 12.1).[2]

EXHIBIT 12.1.
Photoconversion of phytochrome and parameters.

$$\mathrm{Pr} \underset{K_{2\lambda}}{\overset{h\nu \downarrow \; K_{1\lambda}}{\rightleftarrows}} \mathrm{Pfr}$$

$K_{1\lambda}$: Rate constant of Pr to Pfr photoconversion at wavelength λ.
$K_{2\lambda}$: Rate constant of Pr to Pfr photoconversion at wavelength λ.
Pr + Pfr = P : total phytochrome

ϕ_λ: Pfr/P ratio photoequilibrium at λ.
$\phi\lambda = K_{1\lambda}/(K_{1\lambda} + K_{2\lambda}) = \sigma_{R\lambda} / (\sigma_{R\lambda} + \sigma_{FR\lambda})$

$K_{1\lambda} = N_\lambda \cdot \sigma_{R\lambda} = N_\lambda \cdot 2.3 \cdot \varepsilon_{R\lambda} \cdot \Phi_R$
$K_{2\lambda} = N_\lambda \cdot \sigma_{FR\lambda} = N_\lambda \cdot 2.3 \cdot \varepsilon_{FR\lambda} \cdot \Phi_{FR}$

N_λ : photon fluence rate ($\mathrm{mol \cdot m^{-2}}$ per second) at λ.
$\sigma_{R\lambda}$: photconversion cross-section of Pr at λ.
$\sigma_{FR\lambda}$: photoconversion cross-section of Pfr at λ.
$\varepsilon_{R\lambda}$: extinction coefficient of Pr at λ.
$2.3 \cdot \varepsilon_{R\lambda}$: molar absorption cross-section of Pr at λ. ($ln10 \fallingdotseq 2.3$)
$\varepsilon_{FR\lambda}$: extinction coefficient of Pfr at λ.
$2.3 \cdot \varepsilon_{FR\lambda}$: molar absorption cross-section of Pfr at λ. ($ln10 \fallingdotseq 2.3$)
Φ_R : quantum yield for Pr to Pfr photoconversion.
Φ_{FR} : quantum yield for Pr to Pfr photoconversion.

Definition:

$\sigma = \zeta \cdot \Phi$, ζ is molar absorption cross-section ($\mathrm{area \cdot mol^{-1}}$).
The relationship between ζ and ε is: $\zeta = \varepsilon \cdot ln10$.

TABLE 12.1. Photoconversion cross-sections of Pr (σ_R) and Pfr (σ_{FR}), and Pfr/P at photoequilibrium (ϕ) of phytochrome (phyA).

Wavelength (nm)	σ_R (m^2·mol^{-1})	σ_{FR} (m^2·mol^{-1})	ϕ
300	1404	728.3	0.66
350	981.2	328.0	0.75
400	845.5	650.6	0.57
440	188.7	280.7	0.40
460	114.7	146.1	0.44
480	76.80	91.95	0.45
500	67.12	71.52	0.48
520	101.0	64.71	0.61
550	287.5	71.76	0.80
600	1508.0	190.1	0.89
620	2069	296.5	0.87
640	3062	482.0	0.86
660	4963	743.9	0.87
680	3770	945.7	0.80
700	515.8	1223	0.30
710	160.2	1416	0.102
720	65.45	1602	0.039
730	35.53	1701	0.020
750	20.39	1237	0.016

Source: Data from Mancinelli, 1994.

Intermediates of Photoconversion

The conversion of phy involves intermediates as shown in Figure 12.5 (cf. Kendrick and Spruit, 1977). The conversion Pr→lumi-R requires R and Pfr→lumi-F, FR. The other steps proceed without light. Under the dehydrated conditions meta-Ra→meta-Rb→Pfr and meta-Fa→meta-Fb→Pr do not occur. When Pr absorbs R under the dehydrated conditions, Pr converts to meta-Ra via lumi-R. When R is turned off, lumi-R and meta-Ra reverse to Pr, but meta-Rb formed under hydrated conditions does not reverse. Hence, once Pr has reached meta-Rb under hydrated conditions and then dehydrated, the meta-Rb does not reverse and stays as it is. On being hydrated, it readily converts to Pfr. With the reversion steps of Pfr→Pr, similar phenomena occur.

Peculiar Feature of phyA

In darkness phyA is synthesized as Pr in long-imbibed seeds and dark-grown seedlings, and is contained in plenty. It is stable as long as it stays as Pr. Once it becomes Pfr, however, it is tagged by ubiquitin, a small (76 amino acid) protein ubiquitous to all plant cells, and rapidly destructed by 26S proteasome, a large specific proteolytic complex (Vierstra, 1994). Exposure to light also suppresses new synthesis of phyA (Sharrock and Quail, 1989). Thus, the level of phyA in plants grown under or transferred to the light conditions is as low as that of phyB or other phy species (Somers et al., 1991; Sharrock and Clack, 2002). This is unique to phyA and not found in other phys yet.

Phytochrome Signal Transduction

The phy molecule is composed of an N-terminal domain possessing chromophore-bearing site and a C-terminal domain housing dimerization sites. By overexpression of the *phy* gene, whose N-terminal domain-encoding part was exchanged between *phyA* and *phyB*, the N-terminal domain was shown to be responsible for the photosensory specificity, i.e., the *phy* having the *phyA* N-terminal domain showed HIR and the *phy* having *phyB* N-terminal domain exhibited LFR, and the C-terminal domain possessed regulatory sites which transfer perceived information to downstream transduction pathways (Wagner et al., 1996).

Phytochrome is very likely to translocate into the nucleus light specifically to regulate the gene expression. Sakamoto and Nagatani (1996) first succeeded in showing it by expressing DNA encoding the C-terminal fragment of phyB connected with β-glucuronidase *(GUS)* gene. Later the same result was confirmed by expression of genes encoding the intact molecule of phyA and phyB connected with green fluorescent proteins (Kircher et al., 1999; Yamaguchi et al., 1999). Interestingly, an FR pulse (FRp) caused translocation of phyA, but not phyB, and inhibited translocation of phyB to be observed after an R pulse (Rp). These findings support the view that phyA is responsible for VLFR, and phyB, for LFR.

The C-terminal region shows serine/threonine kinase activity (Yeh and Lagarias, 1998) and binds a phy-interacting factor, PIF3, which is found in the nucleus (Ni et al., 1998, 1999; Martínez-García et al., 2000). PIF3 is bound to a promoter of a gene, thus it is assumed that a phy molecule may attach to the promoter via PIF3 and convey a signal to start the transcription of the gene.

ACTION MODES (LFR, VLFR, AND HIR) OF PHYTOCHROMES IN SEED GERMINATION

On absorbing light (photons) Pr is converted to Pfr to varied extents, depending on the wavelengths and the fluence rate of the light as well as the duration of the irradiation, and finally reaches the photoequilibria, the conversion toward Pfr being balanced by the reversion to Pr caused by absorption of the same light by Pfr. The ratios of Pfr/P at equilibria depend solely on the relevant wavelengths (Table 12.1, Exhibit 12.1). A short pulse of light can often afford a Pfr/P ratio less than maximum, whereas long, continuous irradiation usually results in an equilibrium of Pr and Pfr. During long, continuous irradiation the total amount of phy may change due to its new synthesis and/or destruction. Moreover, different phy species also show different actions. Accordingly, phy actions are diverse and classified into three classes: low fluence response (LFR), very low fluence response (VLFR), and high irradiance response (HIR).

Low Fluence Response (LFR)

This class of responses is mostly caused by a short pulse of light, and appears at the fluence range from 1 to 1000 $\mu mol \cdot m^{-2}$, if R (660 nm) is used. The magnitude of response depends on the total fluence (fluence rate × duration), thus it obeys the reciprocity law (an equal value of intensity times duration gives an equal magnitude of response) before photoequilibrium is reached. It shows R/FR reversibility (Figure 12.3), which often serves as the most convenient criterion to distinguish LFR from the other classes of responses. In LFR, Pfr releases, at rather slow, temperature-dependent rates, an intracellular signal or signals for photomorphogenesis, including seed germination and growth. If given before Pfr delivers its signal(s), FR nullifies R effect, but if given after, FR is no more effective. This is called escape from FR effect. The rate of escape varies depending on seed species as well as temperature to which seeds are exposed (Table 12.2). For example, in an Rp-induced seed germination of 'Grand Rapids' lettuce, the half time of escape was about 10 h at 20°C, but 5 to 6 h at 25°C.

Most seeds showing LFR contain phyB and, in some cases, phyE. Seeds of phyA-deficient *Arabidopsis thaliana* show germination promotion by Rp as well as R/FR reversibility, whereas those of the phyB mutant lack such photoresponses (Shinomura et al., 1994, 1996). Similarly, the phyB mutant lacks the LFR part of the biphasic fluence-response curve or has a curve obscured at the part of LFR (Botto et al., 1995), compared with the wild type (WT). Compared with WT of *Arabidopsis,* seeds containing

TABLE 12.2. Escape of R effect from FR reversion in seed germination.

Seed species	Time (h) for 50 percent escape	Time (h) for full escape	Temperature (°C)
Amaranthus caudatus	3	7	25
Betula pubescens		12	20
Carnegiea gigantea		24	30
Chenopodium album	15	40	23
Lactuca sativa var. Grand Rapids	10	20	20
	5-6	8-9	25
Nicotiana tabacum var. Bicchu		5	25
Nicotiana tabacum var. Oodaruma		9	25
Paulownia tomentosa[a]	72	110	25
Pinus sylvestris		48	25
Pinus thunbergii		36	25
Rumex obtusifolius		30	20

Source: Data from Table V in Toole, 1973.
[a]FRp given immediately after prolonged R.

overexpressed phyB show a greater germination percentage than WT at the same Pfr/P ratio, indicating that the quantity of Pfr rather than Pfr/P determines the response (Shinomura et al., 1998).

Even in the *Arabidopsis* phyB-mutant, fresh seeds show an LFR comparable with that of WT (Botto et al., 1995). When imbibed for 24 h, phyA/phyB double mutant has R/FR reversibility (Poppe and Schäfer, 1997). Recently, Hennig et al. (2002) demonstrated that although a triple mutant phyA/phyB/phyD showed the reversibility, phyA/phyB/phyE did not. Thus, it turned out that LFR involves phyB and phyE.

Dry seeds are not sensitive to light but become sensitive after imbibed in water because dehydrated phy becomes ready to operate first after being rehydrated. If the stored phy is Pr, the seeds germinate in response to Rp, and if Pfr, they germinate in darkness, but their germination is inhibited by FRp. The dark germination is also induced by Pfr arising from a dehydrated phy-intermediate (meta-Rb) on rehydration (Kendrick and Spruit, 1977). It is thought that phyB is involved in the dark germination of *Arabidopsis,* because, in darkness, the phyB mutant does not germinate (Shinomura et al., 1994) and phyB-overexpressed seeds germinate much better than WT, but their germination is inhibited by FRp (McCormac et al., 1993).

Seeds of tomato germinate readily in darkness as well as under R, but it is inhibited by FRp and has reversibility. It is clear that Pfr stored in the seeds causes germination. Since both the phyA mutant and the phyB1 mutant also behave similarly to WT, it is not conclusive, but possible that a phy other than phyA and phyB1 may operate in the dark germination (Shichijo et al., 2001).

Very Low Fluence Response

Very low fluence response is a response 1000 to 10,000 times more sensitive than LFR. This response is elicited, if R, at 10^{-4} to 1 μmol·m^{-2} (Figure 12.7). Such a small fluence of light gives a Pfr/P as low as 10^{-2} or less, which still elicits observable photoresponses. Green safelight is not safe, being effective in causing VLFR. Far-red light is also effective at fluences larger than 10^{-3} μmol·m^{-2}, giving a Pfr/P high enough to cause a VLFR (cf. Table 12.1), and hence FR does not nullify the VLFR caused by an Rp (Figure 12.3). Accordingly, the phenomenon that FR causes the same sort of response as R, e.g., germination induction, serves as a criterion to distinguish VLFR from LFR. VLFR is widely found in photoresponses of not only seed germination, but also seedling growth.

Discovery of super-sensitive seed germination induction by extremely weak light developed this new class of photoresponse in plant photomorphogenesis, VLFR. When dark-germinating lettuce seeds, var. May Queen and var. Noran, were incubated in darkness at 37°C, they became positively photoblastic, i.e., not germinating in darkness (thermodormancy) but requiring light for germination. Testing a wide fluence range of R, Blaauw-Jansen and Blaauw (1975) found that the fluence-response curve of seed germination of these two varieties extended toward a very low fluence region and was biphasic as shown with *Arabidopsis* seed. Subsequently, similar shapes of curves were obtained with sensitized seeds of 'Grand Rapids' lettuce, *Arabidopsis, Rumex obtusifolius,* and other species (VanDerWoude, 1985; Cone et al., 1985; Kendrick and Heeringa, 1986) as well as with light-induced growth promotion and suppression of the *Avena* coleoptile and mesocotyl, respectively (Mandoli and Briggs, 1981). The lower phase of the curve which plateaus between 0.1 to 1 μmol·m^{-2} of R belongs to VLFR, while another phase found at higher fluence region belongs to LFR. The part of the response caused by the very low fluences of Rp is also induced by about ten times as intense FR (Figure 12.7).

For sensitizing seeds various methods are effective. Applying prechilling to seeds, VanDerWoude and Toole (1980) observed VLFR with Grand Rapids lettuce (typical LFR seed). After 1 h imbibition followed by an FRp and further 23 h dark incubation at 20°C, they chilled the seeds at 4°C for a few hours. With extension of chilling period the seeds became increasingly FRp-sensitive, and showed by 4 h a characteristic biphasic fluence-response curve, and by 24 h full germination at the VLFR fluence range (VanDerWoude, 1985). By a combined treatment of one week at 4°C and 10 mM KNO_3, Hartmann and Mollwo (2000) were able to determine an action spectrum at about ten-times lower fluence range (for example, 2 $nmol \cdot m^{-2}$ at a half-percent germination at 655 nm) than that of usual VLFR, indicating that a pulse exposure to moonlight or nightly skylight can induce germination of the seeds. It was described earlier that incubation at a high temperature also sensitizes seeds.

The photoreceptor for VLFR is phyA. This has been fully evidenced by action spectra determined with *Arabidopsis thaliana* seed (Shinomura et al., 1996) as well as germination test of *Arabidopsis* seed of phyA or phyB mutants (Johnson et al., 1994; Botto et al., 1996). In *Arabidopsis* seed, phyA is synthesized after a long imbibition of 4 to 24 h (Shinomura et al., 1996; Sharrock and Clack, 2002) to show VLFR. This finding is suggestive to understand that *Oryzopsis miliacea* (Negbi and Koller, 1964) seed shows germination only after 18 h dark imbibition at 25°C in the manner of VLFR.

In VLFR phyA seems to interact with some other phy species: phyB-mutant *Arabidopsis* seeds germinated better than WT, suggesting that phyA action is suppressed by phyB (Hennig et al., 2001). Similar suppression has been reported with phyD (Hennig et al., 2001). By contrast, phyE is needed for phyA to operate, because a phyE mutant failed to show VLFR (Hennig et al., 2002).

Ecological significance of VLFR is great. In the fields seeds may easily be sensitized by low or high temperature while they are in the soil for a long time, and may accomplish germination by very weak light coming through the gaps of soil or litters.

High Irradiance Response

High irradiance response is caused by irradiation of a long duration ranging from several hours to a few days. Since it is characterized by irradiance dependence, the response is called high irradiance response (HIR) (Shropshire, 1972). It does not plateau, but increases with increasing light intensity if compared at the same duration of irradiation (fluence-rate dependency), unless the conditions other than the photorecepting system become limiting. It does not obey the reciprocity law and does not show R/FR reversibility. The action spectrum is characteristic, having the main peak between 710 and 720 nm and the subpeak between 460 and 480 nm (Gwynn and Scheibe, 1972), which do not agree with the absorption spectrum of phy (see Figure 12.5). The positive effects of prolonged FR as well as B are peculiar to HIR, and the effect of prolonged FR, in particular, serves as criterion for recognizing HIR. HIR is found in the inhibition of seed germination of many plants as well as the suppression of growth and induction of anthocyanin synthesis in etiolated seedlings.

First, the action spectrum suggested the involvement of a photoreceptor other than phy, but Hartmann (1966) showed that HIR could be produced by combinations of R of a fixed wavelength and FR of various wavelengths, and the maximum effect resulted at a Pfr/P ratio of 0.03. This Pfr/P ratio accords with that given by 720 nm light (cf. Table 12.1). In prolonged irradiation Pr and Pfr are in photoequilibrium, interchanging between the two. The interchange is named cycling, and the rate of cycling is proportional to the fluence rate. Assuming that the phy is unstable in the form of Pfr and that fresh Pfr newly converted from Pr is functional, Hartmann explained that at a Pfr/P of 0.03 the product of the Pfr times the cycling rate is kept at the maximum. Later, computer calculation demonstrated that the maximum value of the product was realized between 710 and 720 nm in the case of unstable phy (Johnson and Tasker, 1979; Wall and Johnson, 1983)

The photoreceptor for HIR has been assumed to be a phy that is unstable in the form of Pfr (Hartmann, 1966). Conclusive identification of it as phyA was made by a phyA mutant of *Arabidopsis* (Smith, 1994, 1995; Quail et al., 1995; Casal et al., 1998). For HIR of seed germination, however, the same *Arabidopsis* mutant was useless, because *Arabidopsis* seed does not have HIR (Casal and Sánchez, 1998). PhyA mutant of tomato was also examined for HIR by placing seeds under continuous FR (FRc) from the start of imbibition, but the mutant still showed germination inhibition similar to WT (Van Tuinen et al., 1995), casting doubt that the photoreceptor for HIR of seed germination might be different from that for seedling growth. Using the same tomato mutant, however, Shichijo et al. (2001) gave the seed FRc, beginning 24 h after the start of imbibition, when LFR was finished and only HIR remained to occur (cf. Figure 12.6), and found that the germination of phyA mutant was not inhibited, whereas that of WT or phyB1 mutant was inhibited. Thus, the phy for HIR of seed germination was also identified as phyA.

In seed germination HIR results in inhibition (Frankland and Taylorson, 1983; Cone and Kendrick, 1986; Frankland, 1986; Casal and Sánchez, 1998), and it becomes apparent only in ready-to-germinate seeds, activated by the LFR Pfr which arises either from Pr or from the phy conversion intermediate(s). The phy action as HIR operates even after Pfr for LFR escaped from FR nullification. The seed exposed to FR for some hours or days enter

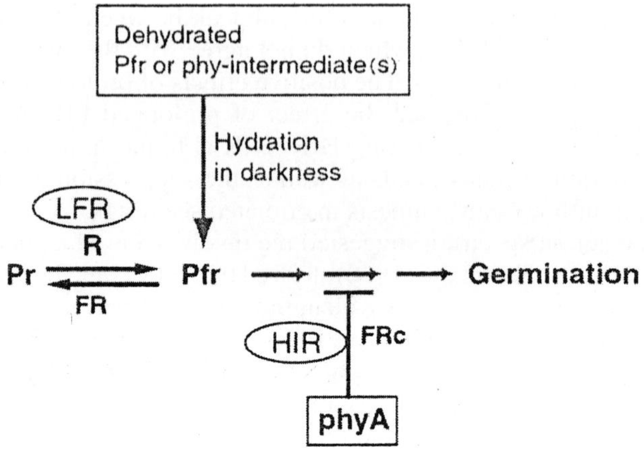

FIGURE 12.6. Proposed action scheme of LFR and HIR in seed germination. LFR and HIR are temporally separated in the germination process. (*Source:* Adapted from Shichijo et al., 2001.)

so-called far-red dormancy, but if an Rp is given, the seed germinates as LFR, showing that the LFR system is not affected by the HIR. Thus, the HIR is very likely to interfere with the signal transduction system of LFR (Figure 12.6).

Seed germination inhibition as HIR occurs not only by prolonged FR but also under sunlight (Górski and Górska, 1979). The inhibition was caused not only when the R component of sunlight is reduced (FR-rich sunlight), but under the complete spectrum. The percent inhibition increases to reach complete inhibition with increasing intensity. Seeds exposed on the soil surface are in danger of desiccation due to intense sunlight if they germinate. The germination inhibition by sunlight is believed to contribute to the survival of seeds.

MULTIPLE MODES OF SEED GERMINATION AND EXPLANATION BY ACTION MODES OF PHYTOCHROMES

Photoresponse of the seed of an individual species is not limited to one mode, but is mostly diversified. Amazingly, these various photoresponses are controlled by different modes of phy actions and their combination.

Phytochrome-Controlled Dark Germination

Seeds of many cultivated species such as *Phaseolus vulgaris, Pisum sativum,* and *Zea mays* germinate without any influence of light. These seeds may have lost the phy control mechanism in the long breeding processes, although some other cultivated seeds retain phy control, e.g., *Lactuca sativa, Lycopersicon esculentum (Solanum lycopersicum),* and *Nicotiana tabacum.*

Seeds of *Avena fatua* (nondormant), *Cucumis sativus,* and *Lycopersicon esculentum* as well as the lower seed of *Xanthium pennsylvanicum* readily germinate in darkness. *Dactylis glomerata* seed makes germination in darkness after a long storage. If decoated, *Datura ferox* seed germinates without light. The germination of these seeds is suppressed by an FRp or FRps and has R/FR reversibility, characteristic of LFR, and hence is under phy control.

Simple LFR

Most seeds that germinate in response to a single Rp have necessarily R/FR reversibility. The Rp-induced germination of seeds of this group occurs without treatment other than imbibition as in *Amaranthus blitoides, A. retroflexus,* 'Grand Rapids' lettuce, *Lepidium campestre,* and *L. densiflorum,* but is often promoted by pretreatment with low or high temperature or KNO_3, etc., as in *Arabidopsis thaliana, Betula papyrifera,* and *Digitalis purpurea.* Although some seeds, such as *Boenninghausenia albiflora,* Arlington Fancy lettuce, and *Nigella damascena* had no test of reversibility because of old experiments, their germination induced by a single Rp should be considered LFR. For *Barbarea orthoceras, Cardamine nipponica, Chamaecyparis obtusa, Clinopodium chinense, Erigeron canadensis, Lysimachia acroadenia, Panicum bisulcatum, Polygonum cuspidatum, Rumex japonicus, Silene armeria, Sisyrinchium angustifolium,* and *Veronica persica,* a white-light pulse (Wp) or an incandescent-light pulse (Ip) was used instead of Rp, but these germinations could also be regarded as phy LFR, because no photoreceptor other than phy for seed germination is known.

LFR Found After Prolonged/Intermittent Irradiation

Among seeds having reversibility, the germination of many seeds has been shown to be induced or promoted by prolonged or intermittent irradiation, e.g., *Aechmea coelestis, Avena fatua* (dormant), *Cecropia obtusifolia,*

Chenopodium album, Kalanchoe blossfeldiana, Lythrum salicaria, Paulownia tomentosa, Piper auritum, and *Ranunculus sceleratus*. For full germination *Paulownia tomentosa* seed, for example, requires continuous irradiation for more than three days (Borthwick et al., 1964), but the irradiation can be replaced by two Rps separated by a corresponding dark period, and the resulting percent germination depends on the fluence of the last pulse, suggesting the involvement of two distinct phy, respectively, operating for seed sensitization and germination induction (Grubišic and Konjevic, 1990). The latter phy action also shows R/FR reversibility. Thus, germination by prolonged/intermittent irradiation contains LFR as its essential part of light action. *Chenopodium album* and *Portulaca oleracea* seeds were also shown to require continuous R (Rc) of 48 h and 16 h, respectively, for full germination and had reversibility. The percent germination increases with the increasing duration of Rc, but it plateaus once at about 50 percent germination before resuming the increase, thus giving a two-phase curve (Karssen, 1970; Van Rooden et al., 1970). These two seed species may probably have consisted of two populations different in time requirement for seed sensitization, while the phy for germination induction is functional throughout the germination period.

LFR Found After Far-Red-Induced Dormancy (HIR)

Seeds of *Amaranthus caudatus, Lactuca sativa* var. Great Lakes, May Queen, Noran, Vanguard, and Cud Vorburgu, *Cucumis sativus,* and *Lycopersicon esculentum* germinate in darkness, and an obscure R/FR reversibility has been shown to occur with *C. sativus* and *L. esculentum,* hence LFR. However, striking actions of Rp and FRp are observed after FR dormancy (FD) imposed by prolonged FR irradiation (HIR). This LFR is presumably controlled by the same phy as that for dark germination or phy which is newly synthesized during imbibition.

Germination Inhibition, Exceptional Action of LFR

As an exception in LFR, seeds of *Aristida murina, Bromus sterilis,* and *Gossypium hirsutum* germinate in darkness, and the germination is inhibited by a single Rp and has R/FR reversibility. For *B. sterilis* daily intermittent Rps were also shown to be inhibitory (Hilton, 1982). The continuous-white-light (Wc)-induced inhibition for *G. hirsutum* may rather be placed in a prolonged/intermittent category of LFR, but at present no distinction is possible. These seeds possibly possess an LFR phy-controlled system to inhibit their germination.

HIR and Prolonged Irradiation-Induced Germination Inhibition

Germination of many seeds is inhibited by prolonged irradiation with FR and other kinds of light. Of these inhibitory effects, the ones having been proven to be HIR by testing for fluence-rate dependency and other characteristics, have been separated as HIR in Table 12.2. It includes the responses obtained with sunlight and incandescent light or under leaf canopy (FR-rich sunlight coming through tree leaves). Noteworthy is that besides FR and B, R is also effective in causing HIR in *Eschscholzia californica* and *Nemophila insignis*, and UV light as well as R, in *Phacelia tanacetifolia*. An example of R-fluence-rate dependency (R-HIR) involving an unidentified phy has been reported in anthocyanin synthesis (Shichijo and Hashimoto, 1997; Shichijo et al., 1999). Such a phy is activated by R and near UV rather than FR and B. The germination promotion by continuous FR found in *Artemisia monosperma*, *Billbergia pyramidalis*, and *Wittrockia superba* seed is interesting to note.

NOTES

1. It was verified that the inhibition observed by Flint and McAlister in FR region was also HIR, but the one by Borthwick et al. was LFR.
2. P: Total phytochrome, Pr + Pfr.

REFERENCES

Blaauw-Jansen, G. and O.H. Blaauw (1975). A shift of the response threshold to red irradiation in dormant lettuce seeds. *Acta Botanica Neerlandica* 24:199-202.

Bonner, B.A. (1960). Partial purification of the photomorphogenic pigment from pea seedlings. *Plant Physiology* 35 (suppl. xxxii).

Borthwick, H.A., S.B. Hendricks, M.W. Parker, E.H. Toole, and V.K. Toole (1952). A reversible photoreaction controlling seed germination. *Proceedings of the National Academy of Science, U.S.A.* 38:662-666.

Borthwick, H.A., S.B. Hendricks, E.H. Toole, and V.K. Toole (1954). Action of light on lettuce-seed germination. *Botanical Gazette* 115:205-225.

Borthwick, H.A., E.H. Toole, and V.K. Toole (1964). Phytochrome control of *Paulownia* seed germination. *Israel Journal of Botany* 13:122-133.

Botto, J.F., R.A. Sánchez, and J.J. Casal (1995). Role of phytochrome B in the induction of seed germination by light in *Arabidopsis thaliana*. *Journal of Plant Physiology* 146:307-312.

Botto, J.F., R.A. Sánchez, G.C. Whitelam, and J.J. Casal (1996). Phytochrome A mediates the promotion of seed germination by very low fluences of light and canopy shade light in *Arabidopsis*. *Plant Physiology* 110:439-444.

Butler, W.L., K.H. Norris, H.W. Siegelmann, and S.B. Hendricks (1959). Detection, assay, and preliminary purification of the pigment controlling photoresponsive development of plants. *Proceedings of the National Academy of Science, U.S.A.* 45:1703-1708.

Casal, J.J. and R.A. Sánchez (1998). Phytochromes and seed germination. *Seed Science Research* 8:317-329.

Casal, J.J., R.A. Sánchez, and J.F. Botto (1998). Mode of action of phytochromes. *Journal of Experimental Botany* 49:127-138.

Clack, T., S. Mathews, and R.A. Sharrock (1994). The phytochrome apoprotein family in *Arabidopsis* is encoded by five genes: The sequences and expression of *PHYD* and *PHYE*. *Plant Molecular Biology* 25:413-427.

Cone, J.W., P.A.P.M. Jaspers, and R.E. Kendrick (1985). Biphasic fluence-response curves for light induced germination of *Arabidopsis thaliana* seeds. *Plant, Cell and Environment* 8:605-612.

Cone, J.W. and R.E. Kendrick (1986). Photocontrol of seed germination. In R.E. Kendrick and G.H.M. Kronenberg (eds.), *Photomorphogenesis in plants* (pp. 443-465). Dordrecht, the Netherlands: Martinus Nijhoff/Dr W. Junk Publishers.

Flint, L.H. (1936). The action of radiation of specific wave-lengths in relation to the germination of light-sensitive lettuce seed. *Proceedings of the International Seed Testing Association* 8:1-4.

Flint, L.H. and E.D. McAlister (1935). Wave lengths of radiation in the visible spectrum inhibiting the germination of light-sensitive lettuce seed. *Smithsonian Miscellaneous Collections* 94:1-11.

Flint, L.H. and E.D. McAlister (1937). Wave lengths of radiation in the visible spectrum promoting the germination of light-sensitive lettuce seed. *Smithsonian Miscellaneous Collections* 96:1-8.

Frankland, B. (1986). Perception of light quantity. In R.E. Kendrick and G.H.M. Kronenberg (eds.), *Photomorphogenesis in plants* (pp. 219-235). Dordrecht, the Netherlands: Martinus Nijhoff/Dr W. Junk Publishers.

Frankland, B. and R. Taylorson (1983). Light control of seed germination. In W. Shropshire Jr. and H. Mohr (eds.), *Encyclopedia of plant physiology*, New series, Volume 16A: *Photomorphogenesis* (pp. 428-456). Berlin: Springer-Verlag.

Furuya, M. (1993). Phytochromes: Their molecular species, gene families, and functions. *Annual Review of Plant Physiology and Plant Molecular Biology* 44:617-645.

Górski, T. and K. Górska (1979). Inhibitory effects of full daylight on the germination of *Lactuca sativa* L. *Planta* 144:121-124.

Grubišic, D. and R. Konjevic (1990). Light and nitrate interaction in phytochrome-controlled germination of *Paulownia tomentosa* seeds. *Planta* 181:239-243.

Gwynn, D. and J. Scheibe (1972). An action spectrum in the blue for inhibition of germination of lettuce seed. *Planta* 106:247-257.

Hartmann, K.M. (1966). A general hypothesis to interpret 'high energy phenomena' of photomorphogenesis on the basis of phytochrome. *Photochemistry and Photobiology* 5:349-366.

Hartmann, K.M. and A. Mollwo (2000). The action spectrum for maximal photosensitivity of germination. *Naturwissenschaften* 87:398-403.

Hauser, B.A., M.-M. Cordonnier-Pratt, F. Daniel-Vedele, and L.H. Pratt (1995). The phytochrome gene family in tomato includes a novel subfamily. *Plant Molecular Biology* 29:1143-1155.

Hennig, L., C. Poppe, U. Sweere, A. Martin, and E. Schäfer (2001). Negative interference of endogenous phytochrome B with phytochrome A function in *Arabidopsis*. *Plant Physiology* 125:1036-1044.

Hennig, L., W.M. Stoddart, M. Dieterle, G.C. Whitelam, and E. Schäfer (2002). Phytochrome E controls light-induced germination of *Arabidopsis*. *Plant Physiology* 128:194-200.

Hilton J.R. (1982). An unusual effect of the far-red absorbing form of phytochrome: Photoinhibition of seed germination in *Bromus sterilis* L. *Planta* 155:524-528.

Johnson, C.B. and R. Tasker (1979). A scheme to account quantitatively for the action of phytochrome in etiolated and light-grown plants. *Plant, Cell and Environment* 2:259-265.

Johnson, E., M. Bradley, N.P. Harberd, and G.C. Whitelam (1994). Photoresponses of light-grown *phyA* mutants of *Arabidopsis*. *Plant Physiology* 105:141-149.

Karssen, C.M. (1970). The light promoted germination of the seeds of *Chenopodium album* L.: VI. Pfr requirement during different stages of the germination process. *Acta Botanica Neerlandica* 19:297-312.

Kelly, J.M. and J.C. Lagarias (1985). Phytochemistry of 124-kilodalton *Avena* phytochrome under constant illumination in vitro. *Biochemistry* 24:6003-6010.

Kendrick, R.E. and G.H. Heeringa (1986). Photosensitivity of *Rumex obtusifolius* seeds for stimulation of germination: Influence of light and temperature. *Physiologia Plantarum* 67:275-278.

Kendrick, R.E. and C.J.P. Spruit (1977). Phototransformations of phytochrome. *Photochemistry and Photobiology* 26:201-214.

Kircher, S., L. Kozma-Bognar, L. Kim, E. Adam, K. Harter, and E. Schäfer (1999). Light quality-dependent nuclear import of the plant photoreceptors phytochrome A and B. *The Plant Cell* 11:1445-1456.

Lagarias, J.C., J.M. Kelly, K.L. Cyr, and W.O. Smith Jr. (1987). Comparative photochemical analysis of highly purified 124 kilodalton oat and rye phytochromes in vitro. *Photochemistry and Photobilogy* 46:5-13.

Mancinelli, A.L. (1994). The physiology of phytochrome action. In R.E. Kendrick and G.H.M. Kronenberg (eds.), *Photomorphogenesis in plants,* Second edition (pp.211-269). Dordrecht, the Netherlands: Kluwer Academic Publishers.

Mandoli, D.F. and W.R. Briggs (1981). Phytochrome control of two low-irradiance responses in etiolated oat seedlings. *Plant Physiology* 67:733-739.

Martínez-García, J.F., E. Huq, and P.H. Quail (2000). Direct targeting of light signals to a promoter element-bound transcription factor. *Science* 288:859-863.

McCormac, A.C., H. Smith, and G.C. Whitelam (1993). Photoregulation of germination in seed of transgenic lines of tobacco and *Arabidopsis* which express an introduced cDNA encoding phytochrome A or phytochrome B. *Planta* 191:386-393.

Molisch, H. (1930). *Pflanzenphysiologie als Theorie der Gärtnerei.* Jena: Verlag von G. Fischer.

Negbi, M. and D. Koller (1964). Dual action of white light in the photocontrol of germination of *Oryzopsis miliacea. Plant Physiology* 39:247-253.

Ni, M., J.M. Tepperman, and P.H. Quail (1998). PIF3, a phytochrome-interacting factor necessary for normal photoinduced signal transduction, is a novel basic helix-loop-helix protein. *Cell* 95:657-667.

Ni, M., J.M. Tepperman, and P.H. Quail (1999). Binding of phytochrome B to its nuclear signalling partner PIF3 is reversibly induced by light. *Nature* 400: 781-784.

Poppe, C. and E. Schäfer (1997). Seed germination of *Arabidopsis thaliana phyA/phyB* double mutants is under phytochrome control. *Plant Physiology* 114:1487-1492.

Pratt, L.H. (1995). Phytochromes: Differential properties, expression patterns and molecular evolution. *Photochemistry and Photobiology* 61:10-21.

Quail, P.H. (1994). Phytochrome genes and their expression. In R.E. Kendrick and G.H.M. Kronenberg (eds.), *Photomorphogenesis in plants,* Second edition (pp. 71-104). Dordrecht, the Netherlands: Kluwer Academic Publishers.

Quail, P.H., M.T. Boylan, B.M. Parks, T.W. Short, Y. Xu, and D. Wagner (1995). Phytochromes: Photosensory perception and signal transduction. *Science* 268: 675-680.

Rollin, P. (1972). Phytochrome control of seed germination. In K. Mitrakos and W. Shropshire Jr. (eds.), *Phytochrome* (pp. 229-254). London and New York: Academic Press.

Rüdiger, W. (1992). Events in the phytochrome molecule after irradiation. *Photochemistry and Photobiology* 56:803-809.

Sage, L.C. (1992). *Pigment of the imagination: A history of phytochrome research.* San Diego: Academic Press.

Sakamoto, K. and A. Nagatani (1996). Nuclear localization activity of phytochrome B. *The Plant Journal* 10:859-868.

Sharrock, R.A. and T. Clack (2002). Patterns of expression and normalized levels of the five *Arabidopsis* phytochromes. *Plant Physiology* 130:442-456.

Sharrock, R.A. and P.H. Quail (1989). Novel phytochrome sequences in *Arabidopsis thaliana:* Structure, evolution, and differential expression of a plant regulatory photoreceptor family. *Genes and Development* 3:1745-1757.

Shichijo, C. and T. Hashimoto (1997). A red light signal distinct from the far-red-absorbing form of phytochrome an anthocyanin induction of *Sorghum bicolor. Journal of Photochemistry and Photobiology B: Biology* 38:70-75.

Shichijo, C., K. Katada, O. Tanaka, and T. Hashimoto (2001). Phytochrome A-mediated inhibition of seed germination in tomato. *Planta* 213:764-769.

Shichijo, C., S. Onda, R. Kawano, Y. Nishimura, and T. Hashimoto (1999). Phytochrome elicits the cryptic red-light signal which results in amplification of anthocyanin biosynthesis in sorghum. *Planta* 208:80-87.

Shinomura, T., H. Hanzawa, E. Schäfer, and M. Furuya (1998). Mode of phytochrome B action in the photoregulation of seed germination in *Arabidopsis thaliana*. *The Plant Journal* 13:583-590.

Shinomura, T., A. Nagatani, J. Chory, and M. Furuya (1994). The induction of seed germination in *Arabidopsis thaliana* is regulated principally by phytochrome B and secondarily by phytochrome A. *Plant Physiology* 104:363-371.

Shinomura, T., A. Nagatani, H. Hanzawa, M. Kubota, M. Watanabe, and M. Furuya (1996). Action spectra for phytochrome A- and B-specific photoinduction of seed germination in *Arabidopsis thaliana*. *Proceedings of the National Academy of Science, U.S.A.* 93:8129-8133.

Shropshire, W., Jr. (1972). Action spectroscopy. In K. Mitrakos and W. Shropshire Jr. (eds.), *Phytochrome* (pp. 161-181). London and New York: Academic Press.

Siegelman, H.W. and E.M. Firer (1964). Purification of phytochrome from oat seedlings. *Biochemistry* 3:418-423.

Sineshchekov, V.A. (1995). Photobiophysics and photobiochemistry of the heterogeneous phytochrome system. *Biochimica et Biophysica Acta* 1228:125-164.

Smith, H. (1994). Sensing the light environment: The functions of the phytochrome family. In R.E. Kendrick and G.H.M. Kronenberg (eds.), *Photomorphogenesis in plants*, Second edition (pp. 377-416). Dordrecht, the Netherlands: Kluwer Academic Publishers.

Smith, H. (1995). Physiological and ecological function within the phytochrome family. *Annual Review of Plant Physiology and Plant Molecular Biology* 46: 289-315.

Smith, H. (2000). Phytochromes and light signal perception by plants —An emerging synthesis. *Nature* 407:585-591.

Somers, D.E., R.A. Sharrock, J.M. Tepperman, and P.H. Quail (1991). The *hy3* long hypocotyl mutant of *Arabidopsis* is deficient in phytochrome B. *The Plant Cell* 3:1263-1274.

Toole, V.K. (1973). Effects of light, temperature and their interactions on the germination of seeds. *Seed Science and Technology* 1:339-396.

VanDerWoude, W.J. (1985). A dimeric mechanism for the action of phytochrome: Evidence from photothermal interactions in lettuce seed germination. *Photochemistry and Photobiology* 42:655-661.

VanDerWoude, W.J. and V.K. Toole (1980). Studies of the mechanism of enhancement of phytochrome-dependent lettuce seed germination by prechilling. *Plant Physiology* 66:220-224.

Van Rooden, J., L.M.A. Akkermans, and R. Van Der Veen (1970). A study on photoblastism in seeds of some tropical weeds. *Acta Botanica Neerlandica* 19:257-264.

Van Tuinen, A., L.H.J. Kerckhoffs, A. Nagatani, R.E. Kendrick, and M. Koornneef (1995). Far-red light-insensitive, phytochrome A-deficient mutants of tomato. *Molecular and General Genetics* 246:133-141.

Vierstra, R.D. (1994). Phytochrome degradation. In R.E. Kendrick and G.H.M. Kronenberg (eds.), *Photomorphogenesis in plants,* Second edition (pp.141-162). Dordrecht, the Netherlands: Kluwer Academic Publishers.

Wagner, D., C.D. Fairchild, R.M. Kuhn, and P.H. Quail (1996). Chromophore-bearing NH_2-terminal domains of phytochromes A and B determine their photosensory specificity and differential light lability. *Proceedings of the National Academy of Science, U.S.A.* 93:4011-4015.

Wall, J.K. and C.B. Johnson (1983). An analysis of phytochrome action in the "high-irradiance response." *Planta* 159:387-397.

Yamaguchi, R., M. Nakamura, N. Mochizuki, S.A. Kay, and A. Nagatani (1999). Light-dependent translocation of a phytochrome B-GFP fusion protein to the nucleus in transgenic *Arabidopsis. The Journal of Cell Biology* 145:437-445.

Yeh, K.-C. and J.C. Lagarias (1998). Eukaryotic phytochromes: Light-regulated serin/threonine protein kinases with histidine kinase ancestry. *Proceedings of the National Academy of Science, U.S.A.* 95:13976-13981.

SECTION III: SEED ECOLOGY

Chapter 13

Competition for Pollination and Seed Set

Beverly J. Brown

INTRODUCTION

Competition for pollinator service among plants will occur whenever there are too few visits by pollinators to provide flowers with sufficient conspecific pollen, or heterospecific pollen transfer reduces reproductive success (Waser, 1983b). Pollinator sharing may contribute to lower seed set in many natural populations. Maximizing seed set and producing seeds that will yield robust progeny are outcomes that should occur as the result of selection processes. High seed yield, with dispersal, allows the species to persist and even increase its distribution. High-quality seeds—i.e., seeds with few deleterious traits—and variation within the gene pool allow for adaptation to new habitats as well as the ability to adapt to a changing environment.

Evolution should select for characteristics and capabilities that allow a population to receive sufficient amounts of pollen (pollen quantity) and high-quality pollen (outcrossed and conspecific). If the species is heterostylous, pollen in addition to being outcrossed and conspecific should be from a compatible morph (legitimate). High pollen quality without sufficiently high pollen quantity will not allow for maximum seed set. Likewise, high pollen deposition may not lead to maximum seed set if the pollen is of poor quality (either heterospecific, illegitimate, or self rather than outcross pollen) and incapable of fertilizing ovules. Although other extrinsic factors can affect resource availability and limit seed set, they are not the focus of this chapter.

If a population is pollen limited and suffers reduced pollen quantity to the extent that seed set is not maximized to the limits of available resources, then natural selection should favor changes that enhance pollinator

I would like to thank John Bell, Jerry Denno, and Liz Tapp for their comments on the manuscript. This work benefited tremendously from their comments.

visitation. For example, natural selection should favor traits that increase attractiveness, including changes in floral color, plant and flower size or shape, plant and flower number and density, or increased pollinator rewards. If pollen quality is poor, then selection will be for characters that minimize receipt of heterospecific or illegitimate pollen. Such adaptations could involve changes in stigma and stamen height, floral architecture, flowering time, or other characteristics. Any of these changes may enhance pollinator visitation, but it is important to note that seed set may not always increase with increased visitation (Young, 1988; Young and Young, 1992).

Although maximum seed set with outcross pollen is the optimal goal, seed set alone does not ensure that a species will survive. Flowers need to form fruits, and the mature seeds they contain should have high germination rates and produce robust progeny. Species persist because they are capable of reproducing in a given location and can spread to additional locations. Sexual reproduction, with its inherent recombination of characteristics essential to maintaining polymorphism, is an advantage that makes receipt of outcross pollen desirable. However, if outcross pollen is unavailable, it may be advantageous for plants to self-pollinate, or even reproduce vegetatively. The strategy a particular species uses to allocate its resources reflects past selection pressures and is constrained by the genetic variation with which selection can work.

Many environmental factors determine what suite of characteristics will result in reproductive success for a given population. Are there many plants in one location or are they dispersed? Are there enough birds, bats, insects, other organisms, or abiotic resources to move sufficient quantities of pollen? How many flowers (total floral display) do you need to attract pollinators? What color and size should corollas be? Does floral height make a difference in attracting pollinators? What amount and type of reward (nectar, pollen, fragrance, etc.) is sufficient to entice, yet small enough to force the pollinator to move on? How does flowering phenology influence the composition of the pollinator community and their behavior throughout the season? Questions such as these are the core of research in pollination biology and the subject of this chapter.

Generalists versus Specialists: Pollinators and Flowers

Generalist flowers can be pollinated by a variety of pollinators and have a suite of characteristic morphological traits. They tend to be open, with multiple, brightly colored petals and easy access to stigmas and stamen. The direction from which pollinators approach a generalist flower is not critical, nor is their orientation while on the flower. They may offer pollen or nectar as rewards. Bumblebees are quintessential generalist pollinators, typically wandering over a flower or inflorescence, gathering pollen and distributing it indiscriminately.

Generalist flowers are advantageous if no one pollinator species is sufficiently abundant to be an adequate group of pollinators (Elam and Linhart, 1988), or if pollinators vary over a season (Motten, 1986; Rathcke, 1988a) or between years (Waser et al., 1996; Fleming et al., 2001; Hiei and Suzuki, 2001). For example, Hiei and Suzuki (2001) found that three species of bumblebees effectively pollinated *Melampyrum roseum* var. *japonica* and that two visits were sufficient to remove almost all pollen from the anthers as well as to ensure maximum seed set. Although there were differences in the ability of the three species to effectively remove or deposit pollen (as measured by seed set), the system as it exists is effective at achieving maximum seed set if at least one species of bumblebee is present. Because the dominant bumblebee species may vary from year to year, there is no selection pressure for flower shape to evolve such that it is adapted for a single pollinator species. If this situation exists throughout the species' distribution, it would explain why in this species of *Melampyrum* there is no modification of flower shape across its range (Hiei and Suzuki, 2001) and a generalist morphology exists.

Linum lewisii, another generalist flower, occurs at a variety of elevations in Colorado. At lower elevations, bumblebees are more effective as pollinators on a per visit basis, but as elevation increases bees are less common and flies become more common (Kearns and Inouye, 1994). At higher elevations, flies are more important as pollinators based on total pollen deposition due to increased visitation, rather than increased effectiveness (Kearns and Inouye, 1994). Having a floral morphology that allows multiple types of pollinators may help balance shifts in pollinators that are likely to occur from year to year (Fleming et al., 2001) or in areas, such as higher elevations, that are susceptible to wide variations in weather and pollinator availability (Kearns and Inouye, 1994).

Weberbauerocereus weberbauri, a columnar cactus, exhibits a wide variety of morphologies including floral size, color, and nectar production. Bats, hummingbirds, and diurnal insects can be important pollinators for this species, depending on the year. In this case, selection pressures shift from year to year and it may be unlikely that a "pure" pollination syndrome for either hummingbirds or bats will evolve (Sahley, 1996). Fleming et al. (2001) investigated the evolution of pollination systems where spatiotemporal timing of flowering and presence of pollinators (primarily bats) varied. They found that generalist pollination systems or shifts away from nocturnal bat pollination were favored in species without dependable pollination by bats.

Potential problems with generalist pollinators include self-pollination, geitonogamy (flowers receiving pollen from other flowers on the same plant), and heterospecific pollen deposition. These can be avoided or minimized in a variety of ways. For example, pollen for different plant species might be carried on different parts of the pollinator's body and deposited in relatively pure loads on the correct species. If heterospecific pollen is received, pollen incompatibility can limit pollen grain germination and arrest pollen tube growth. However, excessive heterospecific pollination can result in stigma clogging, which may prevent germination of conspecific pollen (Brown and Mitchell, 2001).

The concept of pollination systems or syndromes inevitably tending toward specialization and therefore ensuring the effective transport of conspecific pollen has recently been brought into question (Waser et al., 1996). Petterson's (1991) work with *Silene vulgaris* showed that the annual and seasonal variation in pollinators, none of which show marked advantages as pollinators, counters any trend toward specialization. Others have found advantages to generalist pollination when the species is a spring wildflower subject to great variation in pollinators (Motten, 1986) and that generalist pollinators can be quite effective as pollinators (Kearns and Inouye, 1994).

Pollination Syndromes

Although generalist pollination systems seem to be more common, it is useful to discuss apparent pollination syndromes as reference points along the continuum from generalists to obligate specialists with dependence on one pollinator species. Several researchers have proposed that plants typically exhibit syndromes associated with a specific type of pollination (Faegri and van der Pijl, 1979).

Plants pollinated abiotically, for example, via wind (anemophily) or water (hydrophily), have small, severely minimized flowers which produce copious amounts of light-weight (anemophilous) pollen or filamentous (hydrophilous) pollen (Ackerman, 2000). Pollen is released from anthers that extend beyond the flower, increasing the wind or water's ability to move pollen to a receptive stigma. In anemophilous plants the stigmas typically extend beyond the flower and have brush- or featherlike surfaces to receive pollen (Proctor et al. 1996). Although one might expect that wind-pollinated species might suffer from heterospecific pollen loads, the combination of fluid dynamics and biological structures appears to reduce heterospecific pollen deposition (Ackerman, 2000). Hydrophilous plants may send pollen, anthers, or detached flowers to float toward carpellate flowers (Ackerman, 2000).

Biotically pollinated flowers might show evidence of selection pressures if one type of pollinator is exclusively responsible for pollination. For instance, flowers pollinated by bats are typically large and sturdy in order to tolerate the forceful thrust of the bat's head into the corolla. They are generally light in color, with flowers presented in tall inflorescences that open in the evening. These exposed inflorescences are easily accessible to bats. They typically produce a fragrance of fermenting or rotting fruit, and large amounts of nectar and pollen (Faegri and Van der Pijl, 1979; Slauson, 2000). Pollen from bat-pollinated plants tends to be larger than pollen of plants with other pollination syndromes (Stroo, 2000). Bats may be better able to access flowers along and under branches, while birds may be more likely to pollinate flowers at the tips of branches (Fleming et al., 1996). These preferences could influence inflorescence structure over time.

Beetle-pollinated species have often been characterized as having strong and malodorous scents, but this does not hold true for all beetle-pollinated plants (Bernhardt, 2000). The flowers are usually large enough to allow several beetles into the flower at the same time. Flowers are arranged in one of two ways: the first has large individual and usually bisexual flowers whose petals form bowls or urnlike shapes, and the second type has many tiny florets which are often unisexual and either displayed on an exposed shoot or branch or surrounded by a bract (Bernhardt, 2000). Beetles typically arrive at an inflorescence and stay there for 24 hours, wandering over the inflorescence, mating, feeding, and walking around the stigmas (Young, 1988).

For some time, the flowers of hummingbird-pollinated plants have been characterized as tubular, red, diurnal, and odorless, with large amounts of nectar and pollen (Faegri and van der Pijl, 1979). However, research is showing that the color red is not necessarily a preferential attractant (Elam and Linhart, 1988). Moth-pollinated flowers tend to be white or light colored, strongly scented, have more nectar than bee or butterfly plants (frequently deeply hidden in long tubes), and open at night (Faegri and van der Pijl, 1979).

Effective Pollinators

A wide variety of animals, such as bees, butterflies, beetles, and other insects, birds, and mammals, are known to act as pollinators. Baker and Cruden (1991) found that even thrips and aphids can play an active roll in pollination, contributing approximately one-third of fruit set in *Potentilla rivalis*. Effectiveness of pollinators is a major factor in maximizing seed set and can be evaluated using two broad criteria:

1. Do they deposit sufficient pollen to allow maximum seed set either by leaving sufficient pollen in one visit or through repeated visitation?
2. Do they deposit high-quality (conspecific, outcross) pollen?

Pollinators affect male function by removing pollen and delivering it to conspecific stigmas (male function) that leads to eventual seed set (a measure of female function).

Factors that lead to effective pollination may vary based on the pollinator. For instance, Mitchell and Waser (1992) increased nectar content in *Ipomopsis aggregata* and found that increasing the time hummingbirds spent at a flower did not increase male or female function, but that repeated visits by hummingbirds did. They attribute this to the difference in bee and hummingbird behavior. While bees tend to remove most pollen on their initial probe, hummingbirds remove only small portions of pollen with each probe, giving much greater advantage to repeat visitation in hummingbird-pollinated plants (Mitchell and Waser, 1992). Therefore, selection pressures in a predominately hummingbird-pollinated population could lead to locating nectar in a location requiring multiple probes, or other adaptation to induce similar behavior.

The following sections will look at factors that increase both pollen removal and deposition and the quantity of pollen received, as well as factors that increase the likelihood of receiving high-quality pollen loads.

POLLEN QUANTITY

Without pollinator visitation many animal-pollinated species have reduced to very low seed or fruit set. Ensuring adequate visitation is the first step toward maximizing seed or fruit set. When a population has ovules that are not fertilized, it is referred to as pollen limited. Pollen limitation has been frequently demonstrated (for instance Zimmerman and Pyke, 1988; Young and Young, 1992) and may be attributed to a variety of factors including lack of appropriate pollinators (Motten, 1986; Zimmerman and Pyke, 1988; Kearns et al., 1998), inability to compete with coflowering species for pollination (Brown et al., 2002), and habitat fragmentation (Kearns et al., 1998; Moody-Weis and Heywood, 2001). Larson and Barrett's (1999) study of *Rhexia virginica* is a good example of pollen limitation. It is a buzz-pollinated species that releases the lowest recorded amount of pollen per visit. It suffers pollen limitation due to infrequent visitation combined with low pollen dispersal (10.2 percent of pollen) when it is visited. Typically, 42 percent of pollen produced remains in the anthers at floral senescence.

Zimmerman and Pyke (1988) used a long-term study of *Polemonium* to address factors that can potentially confound determination of pollen limitation. These factors include reallocation of resources within a plant (seed set per plant was not increased, but seed set in the hand-pollinated portions was), reallocation of resources affecting future seed set (potential reduction in seed set in future years due to allocation of resources in the current year), and differences in seed set within one season due to variation in pollinator or other resource availability over a season. Addressing all these issues is not always possible and should not prevent studies from being undertaken. However, research needs to be interpreted with these issues in mind.

Testing for pollen limitation typically involves randomly selecting flowers and applying excess conspecific pollen followed by bagging the flowers. At the same time, flowers are randomly marked and left to be open pollinated. After fruits are mature, seed set in each treatment is determined. If seed set is reduced in the open-pollinated flowers, then the population is regarded as pollen limited.

Ackerman and Montalvo (1990) studied fruit set in *Epidendrum ciliare*, a neotropical orchid, and found that hand pollination increased fruit production. However, as fruit set increased, seed crop mass decreased. Experimental plants (that experienced increased fruit set) also suffered reduced plant size and vegetative proliferation in the following year, suggesting that resource constraints might be more of a factor in reproduction than pollen limitation even though hand pollination increased seed set.

Even if stigmas receive sufficient conspecific pollen to maximize seed set, receiving excess pollen is advantageous for enhanced female function. High loads of outcross pollen can lead to gametophytic selection and can affect phenotypic characteristics of the next generation (Snow, 1986; Winsor et al., 2000). Winsor et al. (2000) studied gametophytic selection in a wild population of *Cucurbita foetidissima*. Although only approximately 900 pollen grains were required for full seed set and 650 were deposited in a single visit, they typically found about 4000 pollen grains on stigmas, suggesting multiple visits from pollinators and the potential for strong competition between pollen grains. They found that plants with multiple visits, and therefore probably pollen from multiple outcross sources, had higher seed set compared to plants with single visits. Progeny were also more vigorous based on five vegetative criteria. Herrera (2000) exposed natural populations of *Lavandula latifolia* to either predominantly butterflies and small bees or predominantly large bees and flies. Not only did he find that higher visitation and seed set were associated with butterflies and small bees, but the chance of an ovule producing a third-year seedling was significantly greater for those plants pollinated predominantly by butterflies and small bees. He attributes this advantage to increased outcrossing and gametophytic competition.

Factors affecting visitation rates include display (number and grouping of plants and flowers, color, height), other attractants (fragrance), and rewards offered by the flower/plant (amount of pollen, nectar, fragrance, waxes, oils). The morphology of the flower/inflorescence with the structure and behavior of the pollinator determine pollinator effectiveness. We will examine each of these areas in detail in the following sections.

Attracting Pollinators

Display

Display can refer to the number of flowers on a plant, relative appearance of flowers (a few vs. many, large vs. smaller flowers, and various combinations) as well as the total number of plants present (ignoring distance between conspecific plants, "population size" or "patch size") or the number of conspecifics in a given area and their proximity to one another ("local density") (Kunin, 1997). Although increased display can be linked to increased pollination (Sih and Baltus, 1987; Feinsinger et al., 1991) or seed set (Roll et al., 1997), it is not always the case (Horvitz and Schemske, 1988; Sabat and Ackerman, 1996; Goulson et al., 1998; Caruso 1999). Increased display may vary within a season (O'Neil, 1999) or between years (Fleming et al., 2001), may affect pollinators differently (Goulson et al., 1998), or may become a factor in combination with other characteristics such as rewards (Rathcke, 1988a).

Size of individual flowers may play a role in attracting pollinators (Young and Stanton, 1990). However, whether flower size affects male or female function is debatable. Stanton and Preston (1988) studied corolla size and its impact on female and male function in *Raphanus sativus*. They found no increase in estimates of female success with increase in corolla size. However, Young and Stanton (1990) found that honeybees showed a preference for large flowers and either visited them more frequently or visited them before visiting smaller flowers. They suggest that selection through variation in male success would lead to increased corolla size.

High numbers of flowers require a significant investment of plant resources and may limit the amount of energy available for manufacture of rewards or vegetative growth (Kay, 1987) or maturing fertilized ovules. If a single species is not present in sufficient numbers to attract pollinators, it may benefit from being near a species that does attract pollinators (Rathcke, 1983). Alternatively, two or more species may mimic each other (Müllerian mimicry), or a nonrewarding species may mimic a rewarding one (Batesian mimicry) (Roy and Widmer, 1999). These adaptations can result in heterospecific pollination, which can be avoided if pollinators carry pollen on different portions of their bodies (Waser, 1983b) or in pollinia that are relatively specific to species (Johnson and Edwards, 2000).

Display may work in conjunction with other aspects of attractiveness to pollinators. For example, Rathcke (1988a) studied four species of coflowering shrubs and found that sucrose production per flower did not explain visitation rates. However, the number of flowers present per plant and total average sucrose production did parallel visitation patterns.

Diurnal versus Nocturnal Display

Attracting pollinators during the day or evening requires different strategies. During the day, colors are clearly visible and sight is an effective way for pollinators to locate flowers. In this case, the issues of floral display discussed previously are central to a species' ability to attract an adequate number of pollinators. In the evening, sight is less reliable. Nocturnally pollinated flowers tend to be white or light yellow to allow for maximum recognition in the dark. They also are frequently scented, allowing location of the population even from a distance beyond the limits of vision.

Seasonal Display

Plants that occur sympatrically with overlapping phenologies can logically be expected to compete for pollination if pollinators are a limited resource. Although there is substantial evidence that plants flowering at the same time compete for pollinators (Macior, 1970; Waser, 1978; Motten, 1982; Campbell, 1985; Petanidou et al., 1998; Brown and Mitchell, 2001), this may not always be the case (Armbruster and McGuire, 199)1. To avoid competition, species could be expected to diverge in flowering time, to become self-compatible, or to coevolve with one effective pollinator (Rathcke, 1988a).

Color

For years color has been regarded as central to pollinator visitation. However, while color can enhance visitation (Odell et al., 1999), color can also have no impact on visitation (Elam and Linhart, 1988). Nocturnally visited species might benefit from light-colored flowers which are more easily spotted in reduced light. Red-flowered species were thought to be likely candidates for hummingbird pollination, but this association is by no means universal (Waser, 1983a). Elam and Linhart (1988) studied red, white, and pink morphs of *Ipomopsis aggregata*. Although hummingbirds did visit more red flowers, hummingbirds and moths visited all color morphs. In general, innate color preferences are not pronounced and are more likely governed by association with rewards (Waser, 1983a).

Scent

Floral scent can play a key role in attracting pollinators (Galen and Kevan, 1980), especially over great distances. Scent is an advantage if plants are rare or nocturnal and use scent to help pollinators locate flowers. Scent may also play a role on a smaller scale. Floral scents are complex combinations of volatiles that can be very distinct. The extent to which pollinators can discern differences in species based on scent is unclear, but some evidence does suggest that pollinators can be fairly selective on the basis of scent (Kunze and Gumbert, 2001).

Floral scent may enhance recognition of subtle color differences. Kunze and Gumbert (2001) found that naive bumblebees could more quickly identify rewarding and nonrewarding flowers whose colors differed slightly when both flowers had identical scents compared to when both flowers were unscented. They postulate that color discrimination may be limited by attention and that scent may increase attention. Furthermore, scent may be the first attractant recognized, rather than color (Kunze and Gumbert, 2001).

Populations of *Polemonium viscosum* are polymorphic for scent and grow at 3,540 m to 4,000 m in Colorado. Flowers can have either a skunky or sweet scent. Galen and Newport (1988) found that at the higher elevation, where flies were more common as pollinators, the skunky-scented flowers received more outcross pollen grains, while at the lower elevation, where bumblebees were more common as pollinators, the sweet-scented flowers had higher levels of outcross pollen. Preliminary studies of pollen scents show that the chemical constituents of pollenkit are species specific and could serve as unique chemical cues (Dobson, 1988).

Height

There is some conjecture that pollinators may prefer to maintain their flight at a stable height, minimizing energy expended to gather rewards (optimal foraging theory, MacArthur and Pianka, 1966). However, this aspect of pollinator preference has not been extensively studied (but see Levin, 1973; Donnelly et al., 1998), possibly due to the difficulty of isolating the sole effect of height. Donnelly et al. (1998) looked at the effect of height on pollination in *Verbascum thapsis* and found that plants that were visited were taller than their neighbors. However, their results were confounded because taller plants also had a greater number of flowers. Further studies need to be conducted before generalizations, if any, can be made.

Nectar and Other Pollinator Rewards

The amount and quality of a reward that can be offered to pollinators is critical and can contribute to increased visitation (Rathcke, 1988b; Mitchell and Waser, 1992; Fleming et al., 1996) and outcrossing (Devlin and Stephenson, 1985). If a flower or plant offers resources that satiate the pollinator's need, then it will not venture beyond that plant to deposit pollen on other flowers. However, if rewards are minimal or nonexistent, then the pollinator may not visit that particular species (Motten, 1982, 1986). Rewards can include perfumes, waxes, and oils, but most commonly pollen and nectar act as rewards. Increased nectar reward may improve male function (Cresswell and Galen, 1991; Mitchell and Waser, 1992; Mitchell, 1993) and/or female function (Mitchell and Waser, 1992; Ladio and Aizen, 1999). However, even if increased nectar does increase visitation time, male and/or female function may not be affected (Cresswell, 1999). There are also no assurances that rewards will go only to pollinators. Nectar robbing is common (Arizmendi et al., 1996) and may reduce seed or fruit set (Irwin and Brody, 1998; Navarro, 2001), although this result is not universal (Arizmendi et al., 1996, and others).

The presence of nectar can also affect outcrossing rates. Devlin and Stephenson (1985) found that staminate flowers produced consistently and significantly more nectar (total sugar) than pistillate flowers. Since these protandrous (anthers mature before stigmas) inflorescences mature acropetally (flowers mature from the base of the inflorescence toward the top), the top third of the inflorescence contain the most nectar. The primary pollinators (hummingbirds) appeared to recognize this difference and would arrive mid-inflorescence and proceed up the inflorescence. Flowers in the middle of the inflorescence could be in the pistillate or staminate stage so that outcrossing was encouraged by the pollinator behavior.

Pollen and Ovule Production

In general, pollen size is negatively correlated with pollen number (Cruden, 2000) and pollen grain number is positively related to ovule number (Cruden, 2000). The ratio of pollen grains to number of ovules is strongly correlated with the type of breeding system, with the ratio decreasing in value from xenogamous (outcrossing) to facultative xenogamous to autogamous (selfing) species (Cruden, 2000).

Increasing male fitness can be accomplished in several ways: increasing the production of pollen, and improving the effectiveness of pollen packaging or pollen removal. For instance, Young and Stanton (1990) found that flowers of *Raphanus sativus* that produced more pollen were rewarded with a greater percentage of pollen removal than those that produced less pollen. Harder and Thomson (1989) investigated pollen removal and subsequent deposition on stigmas. They found that only 0.6 percent of the pollen removed from *Erythonium grandiflorum* reached conspecific stigmas. In this case, limiting the amount of pollen removed with each visit may increase male fitness. However, this conclusion cannot be uniformly applied to other species because evolutionary constraints on inflorescence size, flower morphology, anthesis patterns, nectar production, and dichogamy (separation of male and female function in time) combined with the pollinators involved may lead to other solutions to the issue of maximizing male function (Harder and Thomson, 1989). Pollen wastage has been minimized in systems where pollen is packaged or clumped. The Onagraceae, Caesalpinioideae, Asclepidaceae, and Orchiadeae have pollen-to-ovule ratios much lower than would be expected in xenogamous species (Johnson and Edwards, 2000).

Selfing

Self-compatibility is a mechanism that increases reproductive assurance in the absence of pollinators, as is the case with *Hepatica americana* which avoids reduced seed set due to competition for pollination by vigorously selfing (86 percent of ovules, Motten, 1982). Selfing is can be advantageous for spreading over distances and is an advantage when pollination service may be poor or unreliable (as summarized in Fishman and Wyatt, 1999). It does, however, usually result in abortion of ovules due to inbreeding depression. This is a postzygotic process that wastes maternal resources and usurps ovules that might otherwise be available for cross-pollination.

POLLEN QUALITY

When a species is rare within a community or when more than one species is plentiful, pollinators may visit more than one species (Kunin, 1993; Stout et al., 1998). Plants may benefit from this situation (facilitation, Rathcke, 1983), especially if it enhances visitation to rare flowers

(Oostermeijer et al., 1998). However, flowers may receive foreign pollen from these indiscriminate pollinators (McLernon et al., 1996), decreasing their chances for high levels of seed set. Thus, even if a particular flower is visited frequently, it may not receive sufficient conspecific pollen (Petanidou et al., 1998; McClernon et al., 1996; Ramsey and Vaughton, 2000) if the pollinator is inconstant.

Pollinators that visit multiple species typically carry mixed pollen loads (Feinsinger and Tiebout, 1991) and are likely to leave pollen on heterospecific stigmas, resulting in pollen wastage (Waser, 1983b) and reducing male fitness (Campbell, 1985; Campbell and Motten, 1985; Feinsinger and Tiebout, 1991; Murcia and Feinsinger, 1996). Female fitness may be reduced by stigma or stylar clogging (Galen and Gregory, 1989; Randall and Hilu, 1990, but see Kohn and Waser, 1985), pollen allelopathy (Sukhada and Jayachandra, 1980; Murphy and Aarssen, 1995a,b), or premature closing of the stigma (Waser and Fugate, 1986) or if ovule usurpation occurs (Fishman and Wyatt, 1999).

Heterospecific pollen does not always affect seed set (Kohn and Waser, 1985; Caruso, 1999). Even if a pollinator is inconstant, heterospecific pollen deposition can be avoided if the pollinator carries pollen from different species on different locations on its body. Armbruster et al. (1994) found that species of *Stylidium* explosively place pollen in specific locations on insects when the insects touch a sensitive area at the base of the fused staminal and pistillate tissue (gynostemium). The flowers are protandrous and deliver pollen for one to two days. At this time the stigma becomes receptive and its usually bristly stigma picks up pollen from the exact location on the pollinator's body, thereby avoiding heterospecific pollen acquisition.

The impact of heterospecific pollen on pollination and seed set may vary depending on the breeding systems of the species involved. For instance, Feinsinger et al. (1991) found that self-incompatible *Palicourea lasiorrachis* suffered stronger competition for visitation and seed set with increasing density of heterospecific plants than the self-compatible *Besleria triflora*. Even with conspecific pollen deposition, pollen quality can be improved with increased outcrossing, which results in higher seed and fruit set, possibly the result of increased gametophytic competition (Herrara, 2000).

Problems of Heterospecific Pollen

Stigma Clogging, Stylar Clogging, Ovule Usurpation

Heterospecific pollen, if received in sufficient quantities, can block conspecific pollen's access to the stigmatic surface (Rathcke, 1983). In most cases, without contact with the chemicals on the stigmatic surface, pollen germination cannot occur (Raghavan, 1997). Clogging can be avoided or reduced by not allowing pollen to attach to stigma. A non-papillate stigma may deter heterospecific pollen attachment. Campbell and Motten (1985) proposed that species with small papillae may receive less heterospecific pollen if the diameter of the heterospecific pollen is large in comparison to the size of the papillae, while small pollen grains could adhere to and even clog stigmas with large papillae (as proposed by Galen and Gregory, 1989).

A more definitive example of heterospecific pollen impact is that of *Lythrum salicaria* and *L. alatum*. Brown and Mitchell (2001) demonstrated that when stigmas of *L. alatum* receive mixed loads of pollen, seed set is reduced even if adequate conspecific pollen is present. Because few pollen tubes were present in styles when mixed pollen loads were present and no pollen tubes were present with pure *L. salicaria* pollen loads, some type of stigma clogging or inhibition is indicated. Pollen tubes in pure *L. salicaria* treatments exhibited corkscrew shapes characteristic of stigmatic inhibition of pollen tube growth.

If pollen tubes start to grow, they may block conspecific access to the stigma (Brown and Mitchell, 2001) or clog the style, preventing conspecific pollen tubes from growing to the ovules. If the heterospecific pollen tubes do grow into the ovules, they can usurp ovules and make them unavailable to conspecific pollen (Fishman and Wyatt, 1999). If fertilization with heterospecific pollen occurs, chromosomal mismatches may prevent the seeds from maturing or may allow development of a sterile seeds, thus wasting resources on seed development. In plants with few ovules, the impact on seed set could be significant.

Given these difficulties, it is not surprising that selection pressures have created ways to circumvent or reduce the impact of heterospecific pollen deposition. For example, some species close the stigma when sufficient pollen has been received. *Campsis radicans* flowers that eventually set fruit closed in 15 hrs, while those that did not set fruit closed in 42 hrs (Bertin, 1982). However, heterospecific pollen may induce premature stigma closing (Waser and Fugate, 1986).

Randall and Hilu (1990) studied heterospecific pollen movement between *Impatiens capensis* and *I. pallida*. *Impatiens capensis* suffered reduced fruit set when mixed pollen was applied to the stigma, but the reverse was not the case. *Impatiens pallida* pollen will adhere to *I. capensis* stigmas, will germinate, and will grow down the style to the ovule, but will not fertilize the ovule. *Impatiens capensis* pollen does not adhere well to *I. pallida* stigmas and will not germinate. Thus, it is possible for *I. capensis* to suffer from stigma and/or stylar clogging while *I. pallida* does not. *Impatiens capensis* counters this assault by relying on pollinators that do not visit

I. pallida, maintaining a receptive stigma throughout the female phase, being able to accomplish full seed set with one pollination event and producing cleistogamous flowers (flowers that self-pollinate and the perianth never opens), especially late in the flowering season (Randall and Hilu, 1990).

Allelopathy

Heterospecific pollen can prevent the germination of conspecific pollen on stigmas (Sukhada and Jayachandra, 1980). Where pollen allelopathy has been demonstrated, just a few heterospecific pollen grains can be enough to prevent conspecific germination (Murphy and Aarssen, 1995a,b). If pollen grains are allelopathic, then adequate conspecific pollen deposition can be totally undermined. Because pollen allelopathy has been demonstrated in the wind-pollinated grasses, allelopathy may be an issue even if pollinator visitation is not. The prevalence of this effect remains unknown and warrants further study.

Pollinator Behavior

Pollinators should visit rewarding flowers more frequently than unrewarding flowers (MacArthur and Pianka, 1966) when given a choice, and there is some evidence that this is the case (Pleasants, 1981). Pollination biologists frequently observe pollinators moving between only a few species when other species are nearby, or switching species only over time, and not when flowers become immediately available. A pollinator that visits only a few species of plants when others equally or more rewarding are present is described as being constant. Pollinators that appear to have a few specific, unalterable search images are said to have fixed preference, while others that change plant species as some plant species wane in presence and others increase are said to have labile preference (Waser, 1986). Individuals of some pollinator species may exhibit different preferences.

Waser (1986) proposed that much of the behavior of pollinators is due to memory constraints. If pollinators could learn many flowers and recognize the more efficient means of removing rewards, then recognition of each flower would lead to immediate access to memory which would reduce handling times. If pollinators can remember only a few search images, this could explain why some pollinators would continue to visit the same species, even if other more rewarding species were present.

The order of visitation may make a difference in the impact of heterospecific pollen transfer. Caruso and Alfaro (2000) found no effect on pollen receipt or seed set in *Ipomopsis aggregata* when *Castilleja linariaefolia* pollen arrived at the same time as conspecific pollen. However, if *C. linariaefolia* pollen arrived first, then both pollen receipt and seed set were significantly decreased.

Floral Shape

The shape and arrangement of flowers can have a major impact on increase or decrease in heterospecific pollen deposition (McLernon et al., 1996). Although many have hypothesized about the effect of floral shape and pollination efficacy, there are few conclusions (Neal et al., 1998). The variation in pollinator behavior, both between and within species, when combined with possible variations in floral morphology, arrangement, and density make it unlikely that simple patterns will easily emerge. For instance, Stout et al. (1998) found that bumblebee behavior was affected by plant floral complexity and the identity of the species in the arrays. The three bumblebee species also differed in their response to the experimental conditions. Neal et al. (1998) have provided an excellent review of research to date and have neatly laid the groundwork for future research in this area.

Color Differentiation

If pollinators selectively visit flowers based on color, then clear color differentiation can reduce heterospecific pollen deposition. Flower color has been the subject of discussion for some time (Proctor et al., 1996, for example), and color preferences based on type of pollinator were considered likely (Faegri and van der Pijl, 1979). However, recent research has not supported the existence of pollinator-specific colors (Elam and Linhart, 1988; Galen and Kevan, 1980). Color can reduce heterospecific pollen receipt if the pollinator species responds differentially to color. Levin (1985) noted that allopatric populations of *Phlox drummondii* and *P. cuspidata* had pink corollas, while when the two were sympatric *P. drummondii* was red. He introduced red and pink morphs of *P. drummondii* to a *P. cuspidata* population and found that the 38 percent of the seed progeny of the pink morph were hybridized, while only 13 percent of the red morph's progeny were hybrids, a significant difference. Thus, character displacement of corolla color appears to reduce heterospecific pollen transfer and potentially reduce pollen wastage. Color change after pollination may also reduce heterospecific pollen deposition (Proctor and Harder, 1995).

Dichogamy and Herkgogamy

Separation of anthers and stigmas in space (herkogamy) or in time (dichogamy) can reduce heterospecific pollen deposition or interflower pollination on an individual plant (geitonogamy). Whether flowers are self-incompatible or self-compatible, it is usually more valuable to receive outcross pollen than self pollen (Helenurm and Schal, 1996; Herrera, 2000). To minimize self-pollination, the stigma or stamen mature first (protogynous or protandrous, respectively) (Wyatt, 1983). Typically, the stamen will mature first, with the stigma becoming receptive toward the end of anthesis (Galen and Kevan, 1980), or, in some cases, the two periods do not overlap at all (Bertin, 1982), forcing the plant to rely on outcross pollen. *Ranunculus scleratus* is protandrous but takes additional steps to ensure seed set. Anthesis begins as the bud opens, but self-pollination does not occur until the second day when the stamens bend in and contact the opposing stigmas. This process does not allow for all stigmas to receive self-pollen but does allow for fruit set, although it is reduced from 85 to 90 percent to 50 to 60 percent (Baker and Curden, 1991).

Arum italicum is protogynous, with the female flowers surrounded by a spadix that effectively traps pollinators (deceptive pollination attracted by foul odors) until the second day. Male flowers dehisce at this time, and escaping pollinators are covered with pollen on their exodus (Méndez and Díaz, 2001), effectively ensuring that only outcross pollen reaches the stigmas. *Veronica cusickii*, while being weakly protogynous, has widely separated anthers and stigma and rarely self-pollinates (Campbell, 1987). Character displacement in timing of anthesis is also known (Stone et al., 1998).

Another solution to the problem of heterospecific pollen deposition is to make it mechanically impossible for pollen deposition to occur unless the pollen is species specific. The idea that divergent shapes of flowers with varying placements of stigma and stamen should prevent interspecific pollen flow has been termed the "sexual architecture hypothesis" (Murcia and Feinsinger, 1996). Subsequently, many pollinators may carry pollen at different places on their bodies, preventing interspecific pollen transfer (Waser, 1983b; Feinsinger et al., 1986; Murcia and Feinsinger, 1996; Armbruster et al., 1994).

In a heterostylous flower, anthers and stigmas are at different heights, and pollen from same-height stamens is required for fertilization of same-height stigmas. Plants are di- or trimorphic and typically self-incompatible. Pollination by an incompatible morph is referred to as illegitimate. The heterostyly reduces self-fertilization and increases proper pollen transfer between plants (Barrett et al., 2000), although this is not always the case

(Stone, 1995). Pollinators may carry pollen from each stamen height on different portions of their bodies, causing minimal illegitimate pollen deposition (Lloyd and Webb, 1992; Murcia and Feinsinger, 1996).

Murcia and Feinsinger (1996) propose the "pollen scraping hypothesis" to explain why flowers with the greatest difference in architecture, i.e., the short flowers of *Palicourea* and the long flowers of *Hansteinia*, are most impacted by intervening heterospeciific visitation. These hummingbird-pollinated flowers allow placement of pollen on species specific areas of the hummingbirds' bodies. However, entry into the long flowers of *Hansteinia* scrapes off pollen from the short-flowered *Palicourea*. Heterospecific pollen is not deposited on *Hansteinia,* and it does not incur loss of male or female function. However, *Palicourea* has suffered loss of male function through pollen wastage. The pollen scraping hypothesis suggests another set of characters and processes for natural selection to act upon.

The Asclepidaceae and Orchidaceae have developed a packaging strategy of using pollinia or other masses of pollen to move large amounts of pollen and ensure conspecific deposition. The pollinia are carried by a pollinator who must approach flowers in a specific orientation to receive its reward. The pollinia attach via sticky viscin threads and can be removed when they approach another flower designed to catch and hold the pollinia (Roubik, 2000).

Timing of Flowering

If several species are blooming at the same time, then the chances of receiving heterospecific pollen are greatly increased. Selection pressures would seem to select for divergence in blooming time, either within a day or a season, to minimize the chances of heterospecific pollen transfer (Waser, 1978; Armbruster and McGuire, 1991). Multiple factors (i.e., time of maximum nectar secretion, length of flowering) combined with timing of flowering may also increase receipt of compatible pollen (Fleming et al., 1996). Timing of pollen release within a 24-hour period may also minimize pollen wastage (Stone et al., 1998).

Reproductive success can vary over the blooming period, depending on other species that are blooming simultaneously (Jennersten et al., 1988). When plant species bloom at different times during a season, or their blooming periods overlap only briefly, heterospecific pollen transfer can be reduced or prevented (Waser, 1983b; Feinsinger and Tiebout, 1991) if such variation is within the genetic capability of the species (Kochmer and Handel, 1986). If the pollinator pool were relatively constant, this would allow adequate conspecific visitation without jeopardizing pollen quality.

Armbruster and McGuire (1991) investigated reproductive competition between *Erigeron glabellus* and *Aster sibiricus*, two sympatric species in interior Alaska. They artificially manipulated blooming time in *E. glabellus* to coincide with that of *A. sibiricus* and found no significant effect on seed set in either species and no significant effect of heterospecific pollen on seed set. They suggest that selection for sequential blooming would result from reduction in male fitness through pollen loss during interspecific visitation. McGuire (1993) found neither facilitation nor competition between two synchronously flowering species of *Hedysarum* with similar color and morphology. In this instance most pollinators exhibited constancy, leaving little opportunity for natural selection based on competition for pollination and interspecific movement.

Timing of Rewards Within a Day

To a pollinator, the act of pollination is incidental. Pollinators might actually be called pollen gatherers, nectar gatherers, and wax or scent gatherers (Faegri and van der Pijl, 1979). The plant merely uses availability of pollen, nectar, wax, or scents to lure the would-be pollinator to the plant. If multiple species are blooming at the same time, varying the timing and amount of rewards may regulate when pollinators will visit. If heightened rewards do not occur at the same time of day, pollinators may tend to visit within a given species and heterospecific pollen receipt may be minimized (Stone et al., 1998).

Developing Specialist Relationships: Coevolution with Pollinators

Coevolution of plant species with their pollinators leads to dependent relationships between plants and pollinators. Effective pollinators of plant species may have anatomically similar body dimensions highly correlated with the corolla length of the species they visit (Macior, 1969). Flowers may mimic their pollinators, leading to pseudocopulation such as occurs in the orchids (Faegri and van der Pijl, 1979).

Another example of coevolution is the interdependence between *Yucca* and yucca moths. The plant relies on the yucca moth for pollination and the moth relies on the plant for suitable environment and food to feed its larvae (Marr et al., 2000). Relying on one pollinator reduces the issue of poor pollen quality but puts the population at risk for low pollen quantity.

Self-Pollination As a Pollen Quality Issue

Receiving self-pollen affects species differently. In self-incompatible species it may cause many of the problems associated with heterospecific pollen deposition. In a primarily outcrossing species capable of self-fertilization, reliance on pollinators for seed set may be reduced, but issues of inbreeding depression may arise. All of the previously mentioned factors can be used to reduce self-pollination.

Some systems exhibit cryptic self-incompatibility. This occurs when self- and cross-pollen on different flowers will be successful, but when self- and cross-pollen are applied to the same stigma, self-pollen does not fare as well (Becerra and Lloyd, 1992). Becerra and Lloyd (1992) found this type of incompatibility may occur at the whole flower level with *Phormium tenax*. The likelihood of fruit abortion increased with the proximity of a selfed flower to an outcrossed flower. Likewise, in *Yucca* the proportion of pollen analog (dye) that was outcrossed was 35 percent, while genetic evidence of outcrossing in fruit was 94 percent (Marr et al., 2000).

Fishman and Wyatt (1999) explored the selection pressures for selfing in populations of *Arenaria uniflora* that occur with and without its congener *A. glabra*. Populations of *A. uniflora* growing with *A. glabra* are selfers, while those not near *A. glabra* populations are not. When outcrossing *A. uniflora* was grown with *A. glabra, A. uniflora* suffered reduced fruit set and total seed production; thus, selection pressure for selfing in this species would be strong. Self-compatibility could be selected for in sympatric populations of *Phlox* to discourage hybridization. Whereas self-progeny were 65 percent as fit as outcross progeny, they were more fit than hybrid progeny, which rarely reproduced (Levin, 1985).

Agriculture

Pollinators are essentially, either directly or indirectly, responsible for about one-third of the food humans consume (Buchman and Nabham, 1996). Pollination affects seed production, seed quality, fruit production, and fruit quality, and creates hybrid seed. Disappearance of honeybees due to infectious mites, reduction of wild habitat, grazing, and pesticide and herbicide use all reduce the likelihood of adequate pollination of crops (Kearns et al., 1998). Studies of competition for pollination in crops are almost non-existent and merit more attention.

Disturbed Areas

Fragmentation of habitat can have a negative effect on seed set due to reduction in pollinators (Kearns et al., 1998). In a study of *Dianthus deltoides*, Jennersten (1988) found that plants in a fragmented habitat were pollen limited, and the site had lower plant and animal diversity than an unfragmented site. Since standing crop nectar, ovule production, seed set per hand-pollinated flower, and seed set per bagged flower were not different between the fragmented and unfragmented sites, fragmentation and the resulting reduction in pollinator service are strongly implicated in reduced female function. Those involved in conservation efforts would do well to prevent fragmentation whenever possible if preserving pollinated plants is a project goal.

REFERENCES

Ackerman, J.D. (2000). Abiotic pollen and pollination: Ecological, functional and evolutionary perspectives. In A. Dafni, M. Hesse, and E. Pacini (eds.), *Pollen and pollination* (pp. 167-185). Wien, Austria: Springer-Verlag.

Ackerman, J.D. and A.M. Montalvo (1990). Short-and long-term limitations to fruit production in a tropical orchid. *Ecology* 71(1):263-272.

Arizmendi, M.C., C.A. Domínguez, and R. Dirzo (1996). The role of an avian nectar robber and of hummingbird pollinators in the reproduction of two plant species. *Functional Ecology* 10:119-127.

Armbruster, W.S., M.E. Edwards, and E.M. Debevec (1994). Floral character displacement generates assemblage structure of Western Australian triggerplants *(Stylidium). Ecology* 75(2):315-329.

Armbruster, W.S. and A.D. McGuire (1991). Experimental assessment of reproductive interactions between sympatric *Aster* and *Erigeron* (Asteraceae) in interior Alaska. *American Journal of Botany* 78(10):1449-1457.

Baker, J.D. and R.W. Cruden (1991). Thrips-mediated self-pollination of two facultatively xenogamous wetland species. *American Journal of Botany* 78(7): 959-963.

Barrett, S.C.H., L.K. Jesson, and A.M. Baker (2000). The evolution and function of stylar polymorphism in flowering plants. *Annals of Botany* 85:253-265.

Becerra, J.X. and D.G. Lloyd (1992). Competition-dependent abscission of self-pollinated flowers of *Phormium tenax* (Agavaceae): A second action of self-incompatibility at the whole flower level? *Evolution* 46:458-469.

Bernhardt, P. (2000). Convergent evolution and adaptive radiation of beetle-pollinated angiosperms In A. Dafnie, M. Hesse, and E. Pacini (eds.), *Pollen and pollination* (pp. 243-269). Wien: Springer-Verlag.

Bertin, R.I (1982). Floral biology, hummingbird pollination and fruit production of trumpet creeper (*Campsis radicans*, Bignoniaceae). *American Journal of Botany* 69(1):122-134.

Brown, B.J. and R.J. Mitchell (2001). The impact of foreign pollen from an invasive plant on seed set in a native congener. *Oecologia* 129:43-49.

Brown, B.J., R.J. Mitchell, and S.A. Graham (2002). Competition for pollination between an invasive species (purple loosestrife) and a native congener. *Ecology* 83:2328-2336.

Buchman, S.L. and G.P. Nabhan (1996). *The forgotten pollinators*. Washington, DC: Island Press.

Campbell, D.R. (1985). Pollinator sharing and seed set in *Stellaria pubera:* Field experiments on competition for pollination. *Ecology* 66:544-553.

Campbell, D.R. (1987). Interpopulational variation in fruit production: The role of pollination-limitation in the Olympic Mountains. *American Journal of Botany* 74(2):269-273.

Campbell, D.R. and A.F. Motten (1985). The mechanism of competition for pollination between two forest herbs. *Ecology* 66:554-563.

Caruso, C.M. (1999). Pollination of *Ipomopsis aggregata* (Polemoniaceae): Effects of intra-vs. interspecific competition. *American Journal of Botany* 86(5):663-668.

Caruso, C.M. and M. Alfaro (2000). Interspecific pollen transfer as a mechanism of competition: Effect of *Catilleja linariaefolia* pollen on seed set of *Ipomopsis aggregata*. *Canadian Journal of Botany* 78:600-606.

Cresswell, J.E. (1999). The influence of nectar and pollen availability on pollen transfer by individual flowers of oil-seed rape (*Brassica napus*) when pollinated by bumblebees *(Bombus lapidarius). Journal of Ecology* 87:670-677.

Cresswell, J.E. and C. Galen (1991). Frequency-dependent selection and adaptive survades for floral character combinations: The pollination of *Polemonium viscosum. The American Naturalist* 138(6):1342-1353.

Cruden, R.W. (2000). Pollen grains: Why so many? In A. Dafni, M. Hesse, and E. Pacini (eds.), *Pollen and pollination* (pp. 143-165). Wien: Springer-Verlag.

Devlin, B. and A.G. Stephenson (1985). Sex differential floral longevity, nectar secretion, and pollinator foraging in a protandrous species. *American Journal of Botany* 72(2):303-310.

Dobson, H.E.M. (1988). Survey of pollen and pollenkitt lipids—Chemical cues to flower visitors? *American Journal of Botany* 75:170-182.

Donnelly, S.E., C.J. Lortie, and L.W. Aarssen (1998). Pollination in *Verbascum thapsus* (Scrophulariaceae): The advantage of being tall. *American Journal of Botany* 85(11):1618-1625.

Elam, D.R. and Y.B. Linhart (1988). Pollination and seed production in *Ipomopsis aggregata*: Differences among and within flower color morphs. *American Journal of Botany* 75(9):1262-1274.

Faegri, D. and L. Van der Pijl (1979). *The principles of pollination biology*. Oxford: Pergamon Press.

Feinsinger, P., K.G. Murray, S. Kinsman, and W.H. Busby (1986). Floral neighborhood and pollination success in four hummingbird-pollinated cloud forest plant species. *Ecology* 67(2):449-464.

Feinsinger, P. and H.M. Tiebout III (1991). Competition among plants sharing hummingbird pollinators: Laboratory experiments on a mechanism. *Ecology* 72(6):1946-1952.

Feinsinger, P., H.M. Tiebout III, and B.E. Young (1991). Do tropical bird-pollinated plants exhibit density-dependent interactions? Field experiments. *Ecology* 72(6):1953-1963.

Fishman, L. and R. Wyatt (1999). Pollinator-mediated competition, reproductive character displacement, and the evolution of selfing in *Areniaria uniflora* (Caryophyllaceae). *Evolution* 53(6):1723-1733.

Fleming, T.H., C.T. Sahley, J.N. Holland, J.D. Nason, and J.L. Hamrick (2001). Sonoran desert columnar cacti and the evolution of generalized pollination systems. *Ecological Monographs* 7(14):511-530.

Fleming, T.H., M.D. Tuttle, and M. A. Horner (1996). Pollination biology and the relative importance of nocturnal, and diurnal pollinators in three species of Sonoran desert columnar cacti. *Southwestern Naturalist* 41(3):257-269.

Galen, C. and T. Gregory (1989). Interspecific pollen transfer as a mechanism of competition: Consequences of foreign pollen contamination for seed set in the alpine wildflower, *Polemonium viscosum*. *Oecologia* 81:120-123.

Galen, C. and P.G. Kevan (1980). Scent and color, floral polymorphisms and pollination biology in *Polemonium viscosum* Nutt. *American Midland Naturalist* 104(2):291-289.

Galen, C. and M.E.A. Newport (1988). Pollination quality, seed set, and flower traits in *Polemonium viscosum:* Complementary effects of variation in flower scent and size. *American Journal of Botany* 65(6):900-905.

Goulson, D., J.C. Stout, S.A. Howson, and J.A. Allen (1998). Floral display size in comfrey, *Symphytum officinale* L. (Boraginaceae). *Oecologia* 113:502-508.

Harder, L.D. and J.D. Thomson (1989). Evolutionary options for maximizing pollen dispersal of animal-pollinated plants. *American Naturalist* 133(3):323-344.

Helenurm, K. and B.A. Schal (1996). Genetic load, nutrient limitation, and seed production in *Lupinus texensis* (Fabaceae). *American Journal of Botany* 83(12): 1585-1595.

Herrera, C.M. (2000). Flower-to-seedling consequences of different pollination regimes in an insect-pollinated shrub. *Ecology* 81(1):15-29.

Hiei, K. and K. Suzuki (2001). Visitation frequency of *Melampyrum roseum* var. *japonicum* (Scrophulariaceae) by three bumblebee species and its relation to pollination efficiency. *Canadian Journal of Botany* 79:1167-1174.

Horvitz, C.C. and D.W. Schemske (1988). A test of the pollinator limitation hypothesis for a neotropical herb. *Ecology* 69(1):200-206.

Irwin, R.E. and A.K. Brody (1998). Nectar robbing in *Ipomopsis aggregata*: Effects on pollinator behavior and plant fitness. *Oecologia* 116:519-527.

Jennersten, O. (1988). Pollination in *Dianthus deltoides* (Caryophyllaceae): Effects of habitat fragmentation on visitation and seed set. *Conservation Biology* 2:359-366.

Jennersten, O., L. Berg, and C. Lehman (1988). Phenological differences in pollinator visitation, pollen deposition and seed set in the sticky catchfly, *Viscaria vulgaris*. *Journal of Ecology* 76:1111-1132.

Johnson, S.D. and T.J. Edwards (2000). The structure and function of orchid pollinaria. In A. Dafnie, M. Hesse, and E. Pacini (eds.), *Pollen and pollination* (pp. 243-269). Wien: Springer-Verlag.

Kay, Q.O.N. (1987). The comparative ecology of flowering. *New Phytologist* 106(1):265-281.

Kearns, C.A. and D.W. Inouye (1994). Fly pollination of *Linum lewsii*. *American Journal of Botany* 81(9):1091-1095.

Kearns, C.A., D.W. Inouye, and N.M. Waser (1998). Endangered mutualisms: The conservation of plant-pollinator interactions. *Annual Review of Ecology and Systematics* 29:83-112.

Kochmer, J.P. and S.N. Handel (1986). Constraints and competition in the evolution of flowering phenology. *Ecological Monographs* 56(4):303-325.

Kohn, J.R. and N.M. Waser (1985). The effect of *Delphinium nelsonii* pollen on seed set in *Ipomopsis aggregata*, a competitor for hummingbird pollination. *American Journal of Botany* 72(7):1144-1148.

Kunin, W.E. (1993). Sex and the single mustard: Population density and pollinator behavior effects on seed-set. *Ecology* 74(7):2145-2160.

Kunin, W.E. (1997). Population size and density effects in pollination: Pollinator foraging and plant reproductive success in experimental arrays of *Brassica kaber*. *Journal of Ecology* 85:225-234.

Kunze, J. and A. Gumbert (2001). The combined effect of color and odor on flower choice behavior of bumble bees in flower mimicry systems. *Behavioral Ecology* 12(4):447-456.

Ladio, A.H. and M.A. Aizen (1999). Early reproductive failure increases nectar production and pollination success of late flowers in south Andean *Alstroemeria aurea*. *Oecologia* 120:235-241.

Larson, B.M.H. and S.C.H. Barrett (1999). The ecology of pollen limitation in buzz-pollinated *Rhexia virginica* (Melastomataceae). *Journal of Ecology* 87:371-381.

Levin, D.A. (1973). Assortative pollination for stature in *Lythrum salicaria*. *Evolution* 27:144-152.

Levin, D.A. (1985). Reproductive character displacement in *Phlox*. *Evolution* 39(6):1275-1281.

Lloyd, D.G. and C.J. Webb (1992). The selection of heterostyly. In S.C.H. Barrett (ed.), *Evolution and function of heterostyly* (pp. 179-207). Berlin: Springer-Verlag.

MacArthur, R.H. and E.R. Pianka (1966). On optimal use of a patchy environment. *American Naturalist* 100:603-609.

Macior, L.W. (1969). Pollination adaptation in *Pedicularis lanceolata*. *American Journal of Botany* 56(8):853-859.

Macior, L.W. (1970). The pollination of *Pedicularis* in Colorado. *American Journal of Botany* 57(6):716-728.

Marr, D.L., J. Leebens-Mack, L. Elms, and O. Pellmyr (2000). Pollen dispersal in *Yucca filamentosa* (Agavaceae): The paradox of self-pollination behavior by *Tegeticula yuccasella* (Prodoxidae). *American Journal of Botany* 87(5):670-677.

McGuire, A.D. (1993). Interactions for pollination between two synchronously blooming *Hedysarum* species (Fabaceae) in Alaska. *American Journal of Botany* 80(2):147-152.

McLernon, S.M., S.D. Murphy, and L.W. Aarssen (1996). Heterospecific pollen transfer between sympatric species in a midsuccessional old-field community. *American Journal of Botany* 83(9):1168-1174.

Méndez, M. and A. Díaz (2001). Flowering dynamics in *Arum italicum* (Araceae): Relative role of inflorescence traits, flowering synchrony, and pollination context on fruit initiation. *American Journal of Botany* 88(10):1774-1780.

Mitchell, R.J. (1993). Adaptive significance of *Ipomopsis aggregata* nectar production: Observation and experiment in the field. *Evolution* 47(1):25-35.

Mitchell, R.J. and N.M. Waser (1992). Adaptive significance of *Ipomopsis aggregata* nectar production: Pollination success of single flowers. *Ecology* 72(2): 633-638.

Moody-Weis, J.M. and J.S. Heywood (2001). Pollination limitation to reproductive success in the Missouri evening primrose, *Oenothera macrocarpa* (Onagraceae). *American Journal of Botany* 88(9):1615-1622.

Motten, A.F. (1982). Autogamy and competition for pollinators in *Hepatica americana* (Ranunculaceae). *American Journal of Botany* 69(8):1296-1305.

Motten, A.F. (1986). Pollination ecology of the spring wildflower community of a temperate deciduous forest. *Ecological Monographs* 56(1):21-42.

Murcia, C. and P. Feinsinger (1996). Interspecific pollen loss by hummingbirds visiting flower mixtures: Effects of floral architecture. *Ecology* 77(2):550-560.

Murphy, S.D. and L.W. Aarssen (1995a). Allelopathic extract from *Phleum pratense* L. (Poaceae) reduces seed set in sympatric species. *International Journal of Plant Science* 156(4):435-444.

Murphy, S.D. and L.W. Aarssen (1995b). In vitro allelopathic effects of pollen from three *Hieracium* species (Asteraceae) and pollen transfer to sympatric Fabaceae. *American Journal of Botany* 82(1):37-45.

Navarro, L. (2001). Reproductive biology and effect of nectar robbing on fruit production in *Macleania bullata* (Ericaceae). *Plant Ecology* 152:59-65.

Neal, P.R., A. Dafni, and M. Giurfa (1998). Floral symmetry and its role in plant-pollinator systems: Terminology, distribution and hypotheses. *Annual Review of Ecology and Systematics* 29:345-373.

Odell, E., R.A. Raguso, and K.N Jones (1999). Bumblebee foraging responses to variation in floral scent and color in snapdragons. *American Midland Naturalist* 142(2):257-264.

O'Neil, P. (1999). Selection of flowering time: An adaptive fitness surface for nonexistent character combinations. *Ecology* 80(3):806-820.

Oostermeijer, J.G.B., S.H. Luijten, Z.V. Krenová, and H.C.M. Den Nijs (1998). Relationships between population and habitat characteristics and reproduction of the rare *Gentiana pneumonanthe* L. *Conservation Biology* 12(5):1042-1053.

Petanidou, T., A.C. Ellis-Adam, J.C.M. den Nijs, and J.G.B. Oostermeijer (1998). Pollination ecology of *Gentianella uliginosa*, a rare annual of the Dutch coastal dunes. *Nordic Journal of Botany* 18(5):537-548.

Petterson, M.W. (1991). Pollination by a guild of fluctuating moth populations: Options for unspecialization in *Silene vulgaris*. *Journal of Ecology* 79:591-604.

Pleasants, J.M. (1981). Bumblebee response to variation in nectar availability. *Ecology* 62:1648-1661.

Proctor, H.C. and L.D. Harder (1995). Effect of pollination success on floral longevity in the orchid *Calypso bulbosa* (Orchidaceae). *American Journal of Botany* 82(9):1131-1136.

Proctor, M., P. Yeo, and A. Lack (1996). *The natural history of pollination.* Portland, OR: Timber Press.

Raghavan, V. (1997). *Molecular embryology of flowering plants.* Cambridge, UK: Cambridge University Press.

Ramsey, M. and G. Vaughton (2000). Pollen quality limits seed set in *Burchardia umbellata* (Colchicaceae). *American Journal of Botany* 87(6):845-852.

Randall, J.L. and K.W. Hilu (1990). Interference through improper pollen transfer in mixed stands of *Impatiens capensis* and *I. pallida* (Balsaminaceae). *American Journal of Botany* 77(7):939-944.

Rathcke, B. (1983). Competition and facilitation among plants for pollination. In L. Real (ed.), *Pollination biology* (pp. 305-329). Orlando, FL: Academic Press, Inc.

Rathcke, B. (1988a). Flowering phenologies in a shrub community: Competition and constraints. *Journal of Ecology* 76:975-994.

Rathcke, B. (1988b). Interactions for pollination among coflowering shrubs. *Ecology* 69(2):446-457.

Roll, J., R.J. Mitchell, R.J. Cabin, and D.L. Marshall (1997). Reproductive success increases with local density of conspecifics in a desert mustard (*Lesquerella fendleri*). *Conservation Biology* 11(3):738-746.

Roubik, D.W. (2000). Deceptive orchids with Meliponini as pollinators. In A. Dafni, M. Hesse, and E. Pacini (eds.), *Pollen and pollination* (pp. 271-279). Wien: Springer-Verlag.

Roy, B. and A. Widmer (1999). Floral mimicry: A fascinating yet poorly understood phenomenon. *Trends in Plant Science* 4(8):414-426.

Sabat, A.M. and J.D. Ackerman (1996). Fruit set in a deceptive orchid: The effect of flowering phenology, display size, and local floral abundance. *American Journal of Botany* 83(9):1181-1186.

Sahley, C.T. (1996). Bat and hummingbird polliantion of an autotetraploid columnar cactus, *Weberbauerocereus weberbaueri* (Cactaceae). *American Journal of Botany* 83(10):1329-1336.

Sih, A. and M.-S. Baltus (1987). Patch size, pollinator behavior, and pollinator limitation in catnip. *Ecology* 68(6):1679-1690.

Snow, A.A. (1986). Pollination dynamics in *Epilobium canum* (Onagraceae): Consequences for gametophytic selection. *American Journal of Botany* 73:139-151.

Slauson, L.A. (2000). Pollination biology of two chiropterophilous Agaves in Arizona. *American Journal of Botany* 87(6):825-836.

Stanton, M.L. and R.E. Preston (1988). Ecological consequences and phenotypic correlates of petal size variation in wild radish, *Raphanus sativus* (Brassicaceae). *American Journal of Botany* 75:528-529.

Stone, G.N., P. Willmer, and J.A. Rowe (1998). Partitioning of pollinators during flowering in and African acacia community. *Ecology* 79(8):2808-2826.

Stone, J.L. (1995). Pollen donation patterns in a tropical distylous shrub (*Psychotria suerrensis*: Rubiaceae). *American Journal of Botany* 82(11):1390-1398.

Stout, J.C., J.A. Allen, and D. Goulson (1998). The influence of relative plant density and floral morphological complexity on the behaviour of bumblebees. *Oecologia* 117:543-550.

Stroo, A. (2000). Pollen morphological evolution in bat-pollinated plants. In A. Dafni, M. Hesse, and E. Pacini (eds.), *Pollen and pollination* (pp. 226-242). Wien: Springer-Verlag.

Sukhada, K. and Jayachandra (1980). Pollen allelopathy—A new phenomenon. *New Phytologist* 84(4):739-746.

Waser, N.M. (1978). Competition for hummingbird pollination and sequential flowering in two Colorado wildflowers. *Ecology* 59:934-944.

Waser, N.M. (1983a). The adaptive nature of floral traits: Ideas and evidence. In L. Real (ed.), *Pollination biology* (pp. 242-285). Orlando, FL: Academic Press, Inc.

Waser, N.M. (1983b). Competition for pollination and floral character differences among sympatric plant species: A review of evidence. In C.E. Jones and R.J. Little (eds.), *Handbook of experimental pollination biology*. New York: Scientific and Academic Editions, Van Nostrand Reinhold Company.

Waser, N.M. (1986). Flower constancy: Definition, cause, and measurement. *American Naturalist* 127(5):593-603.

Waser, N.M., L. Chittka, M.V. Price, N.M. Williams, and J. Ollerton (1996). Generalization in pollination systems, and why it matters. *Ecology* 77(4): 1043-1060.

Waser, N.M. and M.L. Fugate (1986). Pollen precedence and stigma closure: A mechanism of competition for pollination between *Delphinium nelsonii* and *Ipomopsis aggregata*. *Oecologia* 70(4):573-577.

Winsor, J.A., S. Peretz, and A.G. Stephenson (2000). Pollen competition in a natural population of *Cucurbita foetidissima* (Cucurbitaceae). *American Journal of Botany* 87(4):527-532.

Wyatt, R. (1983). Pollinator—Plant interactions and the evolution of breeding systems. In C.E. Jones and R.J. Little (eds.), *Handbook of experimental pollination biology*. New York: Scientific and Academic Editions, Van Nostrand Reinhold Company.

Young, H.J. (1988). Differential importance of beetle species pollinating *Dieffenbachia longispatha* (Araceae). *Ecology* 69(3):832-844.

Young, H.J. and M.L. Stanton (1990). Influences of floral variation on pollen removal and seed production in wild radish. *Ecology* 71(2):536-547.

Young, H.J. and T.P. Young (1992). Alternative outcomes of natural and experimental high pollen loads. *Ecology* 73(2):639-647.

Zimmerman, M. and G.H. Pyke (1988). Reproduction in *Polemonium*: Assessing the factors limiting seed set. *American Naturalist* 131(5):723-738.

Chapter 14

Seed Size

Jorge Castro
José A. Hódar
José M. Gómez

INTRODUCTION

Seed size plays a pivotal role in plant life history. It is a result of the interrelationship of a wide array of proximate and ultimate factors—many of these acting in conflicting directions—and at the same time is the starting point for a new generation of plants. Seed size partly determines the number of seeds that can be produced and, on the other hand, bears far-reaching consequences for the fate of the seeds themselves and for the ensuing seedlings, including dispersal, predation, germination, emergence, and seedling performance as well as competitive ability. Furthermore, seeds are fundamental in human nutrition, either directly as food or indirectly as feed for domestic animals. In any case, most crops require seeds of adequate quality, often in terms of size, for propagation. Knowledge of the factors that determine seed size and of the consequences of seed size for the successive phases of the plant's life cycle are thus of prime interest in many spheres of plant biology and ecology, and have broad implications for forestry, agriculture, and ecosystem management. In this chapter, we summarize recent investigation concerning different aspects of seed size. Although information on seed size is copious, many topics are still poorly understood, and knowledge is fragmented into different fields, mostly agriculture, forestry, and ecology, where information is poorly shared despite having the same biological basis. Here, we attempt to condense and bring together knowledge from all these different fields of research, in an effort to provide a fuller understanding of the causes and consequences of seed size.

We thank Regino Zamora, Carmelo Ruiz-Rejón, and Manuel Ruiz-Rejón for helpful comments on the manuscript. This work was supported by postdoctoral grants from Granada University and from the Spanish Ministry of Education and Science to Jorge Castro.

SETTING TERMINOLOGY

For practical reasons, it is common to refer indifferently to seed size, seed weight, or seed mass, "seed size" being the term indexed in most textbooks. Rigorous differentiation among these terms is not normally required, but it is worth considering certain terminology that may refer to different facets of seed biology. Seed weight and mass refer to the same trait, although the second term is strictly speaking more accurate. Seed mass and seed size, however, are different traits. The relationship between size (volume) and the mass (thus density) may differ according to the structure and composition of the coats, the embryo, and the nutritional reserves (e.g., Arditti and Ghani, 2000). This matter is of interest, because both seed mass and seed dimensions are basic parameters in seed ecology, related to dispersal patterns, predation, or seed distribution in the soil (e.g., Thompson et al., 1993; Cerdà and García-Fayos, 2002).

Seed size or mass is in many cases linked (even intimately) to some part of the fruit that is generally considered together with the seed, such as in the caryopsids of Gramineae, in the achenes of Compositae, or in the acorns of Fagaceae. In such cases, the rule is that larger diaspores (whenever unisperms) bear larger seeds. Therefore, these can be considered true seeds for the purpose of this chapter, and we will refer to them as such. However, it is important to be aware that the relationship between seed mass and coat, embryo or endosperm mass may change with the size of the diaspore, thus altering the effect attributable to seed mass. For instance, Tweney and Mogie (1999) found that in larger achenes the seeds had a greater proportion of total achene mass in relation to smaller achenes; Lacey et al. (1997) reported changes in seed mass attributable to greater coat mass without changes in the embryo mass; while Malcolm et al. (2003) similarly reported a nonlinear relationship between seed weight and endocarp weight in peach.

Finally, it should be borne in mind that either seed mass or size may fail to explain many aspects of seed ecology if seed shape is not considered, as this last parameter can similarly govern major features of seed biology such as dispersal, predation, distribution in the soil seed bank, or removal by surface runoff (e.g., Bekker et al., 1998; Cerdà and García-Fayos, 2002). All these possibilities complicate generalizations concerning the ecological consequences of seed mass. In general, however, when the context is specified, this does not create confusion. Thus, for the sake of simplicity, throughout this chapter we will refer alternately to seed size, mass (or weight), and shape, given that, for a given species that maintains a constant seed shape, a seed with larger size or mass usually has a larger embryo and/or greater reserves.

SEED MASS VARIABILITY

Seed-mass variation in the plant kingdom is enormous. The smallest known seeds, from some Orchidaceae, measure to 0.00031 mg, while several other families with species that grow in association with mychorrizal fungus or that parasitize other higher plants (e.g., Pyrolaceae, Orobanchaceae, Scrophulariaceae, Balanophoraceae, Sarraceniaceae) also have dustlike seeds weighing micrograms (Tweddle et al., 2002). By contrast, the largest known seed, the double coconut *(Lodoicea maldivica),* weigh around 20 kg. Apart from these exceptions, seeds around 20 g are common in tropical rain forest trees (Murali, 1997; Tweddle et al., 2002). This results in a difference of 10^{11} between the largest and the smallest seed. Within species, the range of seed-size variation is usually up to one order of magnitude (heaviest/lightest), though larger intraspecific variations (at the population level) are not rare, as for instance up to 13.7-fold in apomictic *Taraxacum* achenes within the same capitulum (thus reducing size variation due to factors other than genetic differences between seeds; Tweney and Mogie, 1999), 16-fold in *Cryptotaenia canadensis* (Hendrix and Sun, 1989), 29-fold in *Coreopsis lanceolata* (Banovetz and Scheiner, 1994), or 30-fold in a *Chrysophyllum* species (Green, 1999). The variation range is expected to be higher at the population level than at the individual plant level (Haase et al., 1995; Susko and Lovett-Doust, 2000), reflecting maternal constraints on seed mass. Within genera and families, seed-size variation again becomes huge, as the 78-fold difference in Orchidaceae (Arditti and Ghani, 2000), or the 212-fold difference within species of the tribe Genisteae from southwest Spain (López et al., 2000; and see Tweddle et al., 2002, for other genera and families having large differences). Despite this range, seed-size variation has traditionally been considered to be one of the least variable plant traits (e.g., Harper et al., 1970). One reason for this apparent constancy is that seed mass is often determined using bulk samples, usually 100 or 1000, this parameter being one of the least variable yield components. This does not impede, however, that individual seed mass maintains high variability at all taxonomical levels.

SEED-SIZE DETERMINATION

Genetic Factors

The genetic control of seed size is potentially determined by three genetic components: the pollinating plant, the seed-bearing plant, and the genetic constitution of the embryo (Silvertown and Lovett-Doust, 1993). As pollen quality may affect seed size, mechanisms reducing the gene flow, such as reduction in number of pollen donors (Bañuelos and Obeso, 2003), or self-pollination (Pellmyr et al., 1997; Gigord et al., 1998; Affre and Thompson, 1999), can trigger an inbreeding depression in terms of lower seed mass. In this regard, hermaphroditic individuals of gynodioecious plants consistently produce smaller seeds than do female individuals (reviewed in Shykoff et al., 2003). This might be related to selfing avoidance in females, although other mechanisms such as reallocation of resources saved from pollen production cannot be discarded. Similarly, in some species, self-pollinated cleistogamous flowers reportedly produce smaller seeds than do cross-pollinated chasmogamous flowers (Baskin and Baskin, 1998; Berg and Redbo-Torstensson, 1999), although this trend is found only occasionally and may be influenced by the self-compatibility of the system and by differences in seed set per flower type or by flower position in the inflorescence (Baskin and Baskin, 1998). Overall, the effect of the sire component is minor in relation to the dame component in determining seed weight (e.g., Biere, 1991; Platenkamp and Shaw, 1993; Lipow and Wyatt, 1999) and may be modulated by the nutritional status of the pollinating parent (see next section). Similarly, genetic constitution of the embryo appears to have a weak influence on seed size compared to the genotype of the maternal parent, and thus the main genetic control of seed size is exerted by the seed-bearing plant (Nakamura and Stanton, 1989; Mojonnier, 1998; Lipow and Wyatt, 1999). This means that, under similar environmental conditions, variation in seed size from different mothers is due mainly to genetic differences among maternal plants (Biere, 1991; Castro, 1999).

A positive correlation between nuclear-DNA content and seed size (as well as with other agronomically important traits) has been observed in several species, although opposite results have also been reported (see Chung et al., 1998). Nevertheless, where the relationship holds positive, it can be useful for breeders seeking the selection of strains with larger seed size. Similarly, polyploidy generally results in greater seed size (e.g., Bretagnolle et al., 1995; Villar et al., 1998; Pegtel, 1999)—often also accompanied of reduced seed set—and the induction of polyploids has served as an important tool in developing larger seeds in crops and in improving the seedling vigor of small-seeded pastures species (Agrawal, 1998; Kelman et al., 1999).

Despite the enormous agronomic significance of seed weight, very little is still known about the genes involved in its regulation. It is clear that seed weight is under complex genetic control, being determined by polygenes placed through different regions of the genome (also different chromosomes) and in addition being modulated by several genetic interactions (Cho and Scott, 2000; Doganlar et al., 2000; Rahman and Saad, 2000). This makes impractical the classical approaches for gene identification. Nevertheless, seed weight has been linked to specific quantitative trait loci (QTLs) in several crops (Maughan et al., 1996; Mian et al., 1996; Vaz Patto et al., 1999; Doganlar et al., 2000), which is allowing the identification of genes responsible of seed weight and eventually might translate to the creation of genetically modified crops with modifications in seed size (Doganlar et al., 2000). Given the transcendental relevance of seed size for human nutrition and for the markets, either directly as a food product or as a secondary product to be eliminated to increase marketability, we can expect concerted efforts on that topic and possible drastic changes in the seed mass of crops in the future.

Environmental Factors

The environmental conditions, including not only abiotic but also biotic effects related to other organisms and to the physiological status of the seed-bearing plant, are a main determinant of seed size. Ultimately, given that environmental conditions determine plant performance, and that the maternal plant and the seeds form part of the same organism, conditions that benefit the plant development generally increase seed yield and weight. Thus, as a general framework, greater resource availability augments mass concomitantly with seed yield and plant performance, as reflected by nutrient fertilization (Grewal et al., 1997; Cookson et al., 2000; Cheema et al., 2001; Tungate et al., 2002), stronger light intensity (Talwar et al., 1996; Warringa et al., 1998), or *Rhizobium* or mychorrizal inoculation (Shumway and Koide, 1994; El Hadi and Elsheikh, 1999). A clear example of this simultaneous variation emerges when comparing seed production of branches with different resource availability within the same maternal plant, eliminating annual variability and reducing genetic interactions. For instance, the needles of Scots pine in Sierra Nevada, southern Spain, may be consumed by a phytophagous insect, the pine processionary caterpillar *(Thaumetopoea pytiocampa).* If the larvae reach a tree late in the season, they may completely defoliate some branches but lack time to consume other branches before pupating. Defoliated branches, poorer in photosynthates, produce smaller cones, fewer seeds per cone, and smaller seeds than do undefoliated branches (Hódar et al., 2003) (see Figure 14.1). The nutritional status (including light intensity) of the pollinating plant (sire component) may also have impact on seed mass, although the few studies available have reported contrasting effects that in addition may be modulated by the nutritional environment of the maternal plant (Galloway, 2001; Etterson and Galloway, 2002).

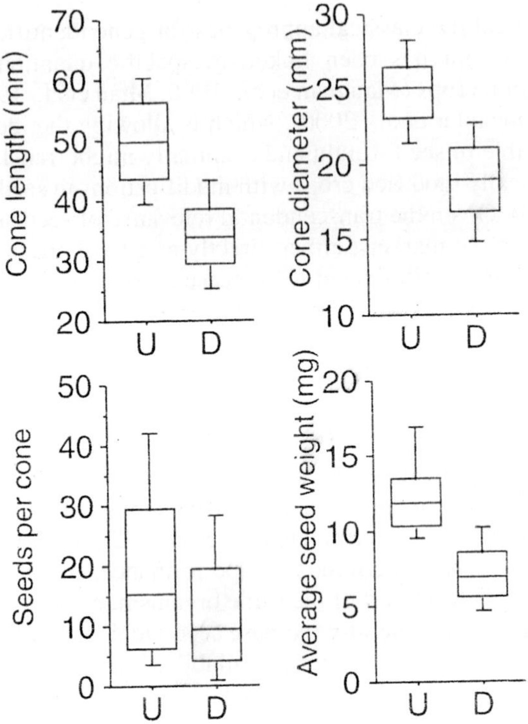

FIGURE 14.1. Differences in cone size (length and diameter), number of seeds per cone, and average seed weight per cone, from undefoliated (U) and defoliated (D) branches in Scots pine in Sierra Nevada, southeastern Spain. Defoliation is produced by the pine processionary caterpillar *(Thaumetopoea pytiocampa).*

Water also affects seed size. Drought is a major factor reducing seed size and/or yield in wild plants as well as crops (Egli, 1998; Sinniah et al., 1998; Leport et al., 1999). The impact of moisture stress on yield depends on timing, duration, and severity of stress, so that at the seed-maturation stage (e.g., drought stress late in reproductive phase), yield loss is related primarily to reduced seed size, whereas drought stress earlier in the reproductive phase tends to reduce the number of seeds (ovules) without reducing seed size (Larson and Mullen, 1995; Sweeney and Granade, 2002). This is likely a consequence of a trade-off between the capacity of the plant to respond to environmental variability and the quality of the offspring to be produced, and might apply similarly for other nutrients provided at contrasting periods during seed maturation. For large, perennial plants such as trees, rainfall might be considered as a main resource controlling seed size (although interaction with nutrient uptake cannot be disregarded), as in principle it is not likely that nutrient availability changes too much from one year to another for plants with large root systems. For example, seed size in Scots pine in the Mediterranean mountains is related to annual rainfall, with all maternal plants producing consistently heavier seeds when cones ripen in rainy years (Castro, 1999). In general, yearly variation in environmental conditions *sensu lato* triggers annual seed-mass variations (Vaughton and Ramsey, 1998; Martin and Puech, 2001).

Environmental conditions provoking physiological stress in plants may also be detrimental to seed size, as in the case of salinity (Abdullah et al., 2001), inappropriate soil pH (Michail et al., 1995), herbicides (L. Andersson, 1996), or temperatures above or below the optimal at flowering or during seed maturation (McKee and Richards, 1998; Prasad et al., 2002). In turn, any factor reducing the performance of the plant or of the fruit usually exerts a negative impact on seed size, even if environmental conditions hold constant, as in delayed flowering (even if it is within the same individual) (Warringa et al., 1998; Zhang, 1998; Susko and Lovett-Doust, 2000), delayed emergence, or, analogously, delayed sowing or transplanting in the case of cropping systems (Baskin and Baskin, 1998; Keeve et al., 2000). Within the same individual plant, situations altering a source-sink balance may alter seed size, so that seeds closer to the source (e.g., lower parts of the inflorescence or basal portion of the fruit) may grow larger than others farther away (Obeso, 1993; Méndez, 1997; Susko and Lovett-Doust, 2000). Similarly, biotic interactions weakening plant performance also shrink seed size, as in competition (Platenkamp and Shaw, 1993; Karlsson and Örlander, 2002), or disease and herbivory (De Clerck-Floate, 1999; Mwase and Kapooria, 2000; Hódar et al., 2003).

In the allocation of resources to different plant compartments a trade-off between vegetative growth and reproduction as well as between the yield components is expected. Thus, if resources remain constant, the reduction in the number of seeds per fruit (Ericksson, 1999; Simons and Johnston, 2000), per inflorescence (Vaughton and Ramsey, 1998; Hiei and Ohara, 2002), or per plant (Pilson and Decker, 2002) usually results in greater seed mass. All these possibilities have obvious consequences in cropping systems, and not only determine the quality of the crop in terms of seed size, but also may be managed in order to achieve the desired balance between seed size and yield characteristics. In this way, fertilizer application, sowing density, reduction of flower or fruit number, and many other agricultural practices may be mechanisms for controlling the size in crops where the seed is the product of interest. An illustrative case is reported by the work of Spurr et al. (2002) with onions cultivated as seed sources. An early harvest of the seed crop ensures a high seed yield, while a late harvest implies a loss of seeds because of the dehiscence of mature capsules. However, the remaining seeds after these losses reach greater seed mass than those maturing earlier and render seedlings with better performance. Thus, early harvest may compromise the quality of the seeds and future seedling vigor, and the timing of the harvest represents therefore a trade-off between a higher number of seeds with lower seed mass or fewer but heavier seeds.

In any event, cases in which changes in environmental conditions (e.g., competition, delayed sowing, fertilization) alter seed yield but not seed mass are not rare (e.g., Myers, 1998; López-Bellido et al., 2000), either because of species-specific characteristics or because of particularities attributable by environmental conditions. In addition, the expected change in seed mass may be mitigated by other factors acting in the opposite direction, as for example when distal fruits of the inflorescence hold fewer seeds, thus giving rise to seeds with the same mass as fruits closer to the stem (Simons and Johnson, 2000). In short, although the processes described are common as general trends, the study of the response of a particular species is necessary in order to ascertain the relationship between environmental factors and seed mass.

Thus far, we have seen that proximate factors such as environmental conditions *sensu lato* have critical consequences for seed mass. But environmental conditions may be also a main factor ultimately determining seed size within the same species, giving rise to ecotypes with genetically fixed differences in seed mass. In this sense, Kobayashi et al. (2001) found that shaded populations of *Plantago asiatica* produced larger seeds (advantageous for seedling establishment under shaded conditions) than did populations from more sunny habitats. Seed size has also shown quite consistently to be smaller for populations growing at higher latitudes or altitudes (Aizen and Woodcock, 1992; Li et al., 1998; Oleksyn et al., 1998), despite that under lower temperatures a greater seed mass favors seedling establishment. This geographical variation may result as a response to lower temperature and shorter growing season, reducing seed mass, and also could be the consequence of past size-selective dispersal of smaller seeds (e.g., birds dispersing smaller acorns) that accompanied the postglacial range of expansion (Aizen and Woodcock, 1992). Geographical patterns of seed-mass variation attributable to other factors such as dryness have been also suggested (Dangasuk et al., 1997). Such geographical differences in seed mass among populations are also clearly reflected by the usually pronounced differences in seed mass among crop cultivars of different origins.

DISPERSAL

Seed-dispersal distances have profound consequences for many aspects of plant ecology. The dispersal ability is closely linked to the weight of the seed not only for wind-dispersed plants but also in the case of animal-dispersed species. For a particular wind-dispersed species, dispersal ability decreases with increased seed mass (Greene and Johnson, 1993). If the seed bears a samara (or other dispersal structure), the flight structure may compensate for the increased seed weight by enlarging the samara surface. However, even when this compensation exists, large samaras provide poorer dispersal, mainly because for exact compensation, a change in samara shape is needed in addition to the samara size (Greene and Johnson, 1993; Sipe and Linnerooth, 1995). In *Acer rubrum*, Peroni (1994) found that samaras coming from early successional environments had a lower wing loading than did late successional stands, thus favoring their dispersal ability. In plants with heterocarpy (the production of more than one type of fruit in one plant), seeds commonly have contrasting dispersal ability not only because of differences in seed mass, but also in the size and type of dispersal structures (pappus, bracts, etc.), as exemplified by members of Compositae (e.g., Imbert, 2001). This may be important for coping with spatiotemporal variability of habitats for establishment, smaller and well-dispersed seeds acting as "colonizers" of new habitats, and larger, poorly dispersed seeds as "maintainers" of occupied habitats, and might lead to microevolutionary changes in natural populations (Imbert, 2001).

Across species, there are constraints for the enlargement of dispersal structures with increasing seed mass, and within particular taxonomic groups larger-seeded species tend to have lower dispersal capacity. For instance, among pine species with small seeds, seed mass is isometrically related to samara shape, but among the largest pines, seed disc loading increases and dispersal capacity declines at an accelerated rate as seed mass increases, which might arise in part from the energetic cost of constructing long-scaled cones to house the seeds and their wings (Benkman, 1995).

For animal-dispersed plants, the selection on seed size may also depend on the selection on fruit size, which nevertheless often correlates positively with seed mass. In the case of frugivorous animals, the relationship between seed size and dispersal ability depends on the preferences of the dispersers, which may select either smaller fruits (thus smaller seeds) or larger ones (e.g., Riera et al., 2002), although especially for birds an important constraint against consuming big fruits is imposed by gape size (Wheelwright, 1985; Jordano, 1987). Among species, variations in fruit size may condition the disperser identity: big fruits are usually dispersed by mammals, while birds prefer smaller fruits (Herrera, 1989). Seeds may also be dispersed by hoarding animals (e.g., rodents, birds) that, regardless of predators, may act as dispersers of a considerable proportion of unrecovered seeds, smaller seeds being generally dispersed over longer distances (Aizen and Woodcock, 1992; Brewer, 2001) as expected from the effort done by the animal. In the case of mammalian herbivores that consume a range of seed sizes of different species (e.g., browsing and grazing ungulates) the size of the seeds have an upper threshold ensuring that seeds are not destroyed by chewing. In this case, smaller and rounded seeds are advantageous (Pakeman et al., 2002), and most forbs and grasses dispersed by this way are small-seeded species (Malo and Suárez, 1995; Leishman and Westoby, 1994; Pakeman et al., 2002).

DEPREDATION

Larger seeds usually have a higher probability of being depredated than smaller ones, both within (Moegenburg, 1996; Willot et al., 2000; Jackai et al., 2001; Parciak, 2002) and across species (Díaz, 1990; Reader, 1993; Szentesi and Jermi, 1995; Hulme, 1998; Blaney and Kotanen, 2001). This preference occurs for many seed predators irrespective of their taxonomical affiliation. For instance, Gómez and Zamora (1994) found that the curculionid beetle *Ceutorhynchus* sp. oviposits on the crucifer *Hormathophylla spinosa* flowers that produce the largest fruits and which, in turn, bear the larger seeds. In the same way, the acorns of Holm oak *Quercus ilex* in southeast Spain are preyed by two vertebrate predators, the wood mouse *Apodemus sylvaticus* and the wild boar *Sus scrofa*. Both mammals highly prefer to consume bigger acorns (see Figure 14.2), presumably because larger seeds are more nutritious since proportion of proteins and carbohydrate reserves increases with increasing seed mass in acorns, hence providing more available energy to seed consumers. Larger seeds are also more easily located when buried (Hulme, 1998; Hulme and Borelli, 1999; Brewer, 2001), therefore suffering higher losses from the soil seed bank. Because of this selection for seed size, granivory may exert a profound influence on the rate and direction of vegetation change, by modulating colonization, succession, and alien plant invasion (Brown and Heske, 1990; Reader, 1993; Blaney and Kotanen, 2001; Katz et al., 2001).

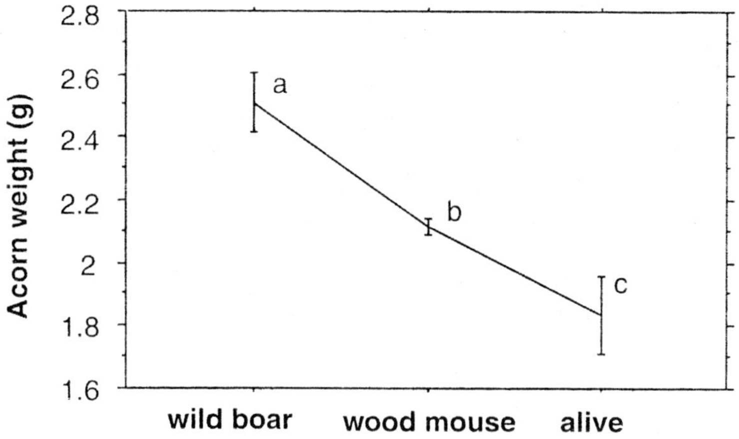

FIGURE 14.2. Average weight of *Quercus ilex* acorns consumed by wild boar *Sus scrofa*, wood mouse *Apodemus sylvaticus*, or surviving to postdispersal predation in Sierra Nevada, southeast Spain. (*Source:* J.M. Gómez, in prep.)

GERMINATION

The relationship between seed mass and germination rate within a species has received contrasting reports. Larger seeds have been shown to render higher germination percentages (S. Andersson, 1996; Milberg et al., 1996; Vera, 1997; Simons and Johnston, 2000; Humara et al., 2002) and faster germination (Vera, 1997; Galloway, 2001; Malcolm et al., 2003), results that may be regarded as a consequence of higher reserves (often also with higher mineral-nutrient concentration), thereby increasing the chances for embryo development and radicle protrusion through the seed coat. Larger seeds may also have better germination probabilities than do smaller seeds after pathogen infection or predator damage (Hare et al., 1999), or in fire-prone species after thermal shock (Escudero et al., 2000; Delgado et al., 2001; but see Hanley et al., 2003). This may simply be a result of having more resistance to environmental stress, i.e., more nutrients to overcome pathogens and predators, or thicker coats and lower surface:volume ratio to resist heating. On the contrary, however, larger seeds may also have slower germination rates than do smaller seeds (Khan et al., 1999; Susko and Lovett-Doust, 2000; Tungate et al., 2002), perhaps due to thicker coats that may retard imbibition and gas exchange, or to slower embryo growth rates. These differences between small and large seeds could also result from factors other than those related to coats or reserves, as for instance differences in the physiological status (degree of dormancy) related either to embryo maturity or to different concentrations of growth regulators. All this may have important consequences for offspring performance, as fast germination can result in a priority effect in which the first seedlings to establish have greater access to resources and therefore a higher probability of survival, which is particularly relevant in crowded (competitive) habitats such as cropping systems. In any case, many studies show no relationship between seed mass and germination, and thus the relationship between seed mass and germination must be considered for each species of interest.

SEEDLING PERFORMANCE

It is widely accepted that higher seed mass usually correlates with seedling performance, with larger seeds being able to emerge from greater depths (Ruiz de Clavijo, 2001), and rendering larger seedlings (Vaughton and Ramsey, 1998; Castro, 1999; Simons and Johnston, 2000; Chacón and Bustamante, 2001; Wennström et al., 2002), having higher survival rates (Bonfil, 1998; Simons and Johnston, 2000), and competitive ability (Ericksson, 1999). These trends similarly hold when comparing populations (ecotypes) differing in seed mass (van Rijn et al., 2000), as well as across species—that is, larger-seeded species can emerge from greater depths (Bond et al., 1999; Pearson et al., 2002), render higher emergence rates (Dalling and Hubbell, 2002), produce larger seedlings which have higher survival rates (Long and Jones, 1996; Milberg et al., 1998; Walters and Reich, 2000; Green and Newbery, 2001), as well as stronger competitive ability (Turnbull et al., 1999). This is logical in terms of the starting capital in the form of energy, given that larger seeds can supply more nutrients to the developing embryo and to the recently emerged seedling, which in turn gain greater access to soil nutrients, water, and sunlight. Thus, higher mass is therefore particularly advantageous to seedling performance under resource scarcity. In fact, the implications of seed-size variation depend on the environmental conditions undergone by the progeny (e.g., competitive environment, water stress, light intensity, nutrient availability, sowing depth), with performance differences between large and small seeds generally being greatest under adverse conditions, while less pronounced or nil in more favorable conditions, both within (Houssard and Escarré, 1991; Meyer and Carlson, 2001; Seiwa et al., 2002; Wennström et al., 2002; Paz and Martínez-Ramos, 2003) and across species (Milberg et al., 1998). The positive relationship between seed mass and seedling growth is, however, usually restricted to the first weeks or months after emergence, while seedling performance depends on or is influenced by initial seed reserves, but disappears later in the life of the seedling (Castro, 1999; Smart and Moser, 1999; Meyer and Carlson, 2001). This relationship is nevertheless still crucial given that early stage of establishment may be decisive between survival and death.

Despite the generally better seedling performance offered by larger seeds, the relative growth rate of seedlings is nevertheless negatively correlated to seed mass across species (Wright and Westoby, 1999; Green and Newbery, 2001; Meyer and Carlson, 2001). This confers a certain competitive advantage to small-seeded species in nutrient-rich environments such as agricultural systems, in which rapid growth fuelled by intensive resource uptake allows small-seeded weeds to overtake the generally larger-seeded, slower-growing crops (e.g., Liebman and Davis, 2000). In addition, the relative growth rate of smaller-seeded species may respond proportionally more to light than in larger-seeded species (Green and Newbery, 2001). However, this competitive advantage comes at the cost of growth reduction under lower nutrient availability (Milberg et al., 1998; Liebman and Davis, 2000). This may have broad implications for weed management. Given that crop seeds are usually between one to three orders of magnitude larger than weed seeds with which they compete, farming practices that minimize nutrient availability early in the growing season may leave small-seeded weeds at a disadvantage with respect to better-provisioned crop seeds (Liebman and Davis, 2000).

The effect of seed size upon seedling performance can be easily confused with the effect exerted by the genetic determination of the maternal plant when experiments are performed with mixed groups of genotypes. For example, let us consider a given population having (1) minor differences in seed mass within the maternal plants, attributable to a lack of genetic embryonic effect, and (2) clear differences in seed mass between maternal plants, attributable to genetic maternal effects. If from this population we collect seeds mostly from maternal plants that produce similar-sized seeds, we might conclude that seed mass is not related to seedling performance, given the low seed-size variation within maternal plants (see Figure 14.3, left). On the contrary, if we collect seeds mostly from plants that differ substantially in the size of seeds produced, we might positively correlate seed mass and plant performance, a relationship that in reality was owed to maternal plants (see Figure 14.3, right). Similarly, if we perceive a positive relationship, we might not know whether to attribute it to a maternal effect on seed mass or an effect exerted by the embryo on seed size. This situation can be illustrated with the Scots pine. Larger seeds produce larger seedlings when subdividing the seeds in category classes (Reich et al., 1994). However, for a population from Sierra Nevada (southeast Spain), it was shown that seed size was maternally determined (Castro, 1999), and thus larger seedlings arose from the maternal plants producing the heavier seeds. The

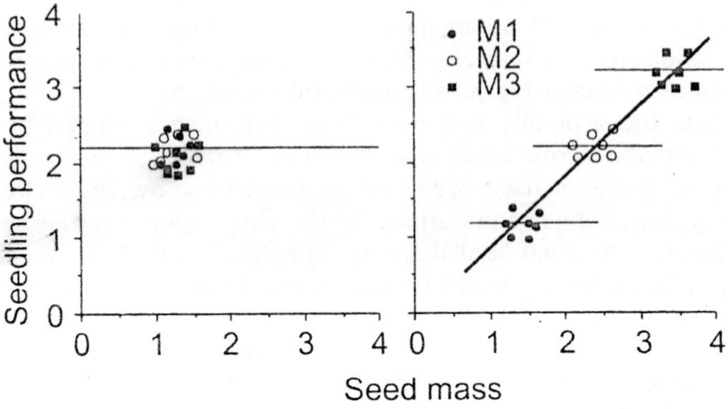

FIGURE 14.3. Theoretical relationship between seed mass and seedling performance for three maternal plants (M1, M2, and M3) having low within-plant seed mass differences. On the left side, three maternal plants with similar seed size give no seed mass-seedling performance relationship. On the right side, maternal plants with different seed size give a significant seed mass-seedling performance relationship, despite that the relationship is not significant within maternal plants.

conclusion that larger seeds give rise to larger seedlings is thus drawn similarly when pooling seeds or considering the maternal origin (Castro, 1999), but knowledge of maternal origin is crucial when seeking an accurate interpretation of the effect of seed size on seedling performance, indicating, for example, that the evolution of seed size may be driven by selection on maternal plants.

In summary, as a general hypothesis it can be stated that seedlings arising from heavier seeds not only have greater survival and growth, but also become more resistant to nutrient stress, allelopathic compounds, diseases, and herbivory. This is an elemental point for seed producers intending to market seeds that promise high emergence and establishment success. In fact, seed quality is often expressed in terms of the thousand-seed weight, with the larger the value the better the seed. Nevertheless, it is essential to consider that the effect of seed mass may be related to factors other than the provisioning of nutrients. For example, the advantage of heavier seeds in dry environments may be the result of higher production of osmotic substances that reduce water stress in the seedlings, or simply the result of smaller area relative to volume from which to lose water in larger seedlings (White et al., 2001); heavier, basal seeds of a fruit could render larger seedlings because of higher concentrations of growth regulators (Susko and Lovett-Doust, 2000); or the effect of seed mass could be mediated by the temperature of origin (Oleksyn et al., 1998) or by the maternal plant (Castro, 1999). On the other hand, exceptions to the positive relationship between seed mass and seedling performance are not rare, caused either by particularities of the system being studied or by characteristics of the species, so caution is advisable when predicting trends and management strategies concerning the relationship between seed size and seedling performance, either in natural ecosystems or in agriculture and forestry.

HABITAT TYPE AND PLANT TRAITS

Seed size in a broad sense is related to the type of habitat occupied by the species as well as with other life-history traits of the individual plant (Westoby et al., 1992; Rees et al., 2001; Coomes and Grubb, 2003). Seed size is linked to many other plant features, and thus cannot be segregated from a myriad of factors conditioning plant traits that, in turn, suit a species to a particular habitat. Furthermore, factors that appear to control seed size simultaneously govern other life traits, and, if selective pressures for traits other than seed size are stronger, seed size might evolve under the constraints of other traits, even maladaptively. One of the clearest and most accepted relationships between seed mass and habitat type concerns light requirements, so that light-demanding species have significantly lower seed mass than shade-tolerant ones (Metcalfe and Grubb, 1995; Hodkinson et al., 1998). This is logical in terms of higher reserves needed in order to reach soil nutrients and light under crowded canopies, bettering chances of reaching suitable growing conditions. However, species living in closed canopies are often trees, which, being long-lived, persistently pursue resources, can reproduce over a long time span, and thus may invest in a larger seed mass at a cost of reduced seed production. It is thus not possible to determine whether higher seed mass is a causal condition for being a typical shade-tolerant plant or is a consequence of other plant traits. Similarly, colonizer species (which usually have lower seed mass) may be evolutionarily forced to produce a high number of seeds of small size either because of a trade-off in size/number or because other selective pressures (e.g., dispersal) conducive to low seed mass. Nevertheless, correlations between seed mass, habitat/type, and plant traits are useful in studying plant communities, helping to identify plant strategies and answer ecological questions. Thus, as a general trend,

1. early successional species (thus also colonizers) generally produce smaller seeds than late successional ones;

2. seed mass positively correlates with species longevity (as well as with plant size);
3. seed mass correlates negatively with seed predation (at least when comparing the range of sizes accessible to particular predator groups);
4. seed mass also correlates negatively with seed longevity, as large seeds can be more easily located and depredated, resulting in selective pressure for fast germination;
5. therefore, small seeded species tend to have dormancy mechanisms and form large, persistent seed banks, whereas seed banks are usually absent in large-seeded species; and
6. species from drier habitats normally produce larger seeds (Rees, 1997; Hodkinson et al., 1998; Onipchenko et al., 1998; Rees et al., 2001).

SEED-SIZE EVOLUTION

The study of the evolution of propagule size has traditionally engaged the attention of many evolutionary biologists (Clutton-Brock, 1991; Roff, 2002; Stearns, 1992). In a seminal paper, Smith and Fretwell (1974) proposed a model that predicted the evolution toward an optimal propagule size that maximizes the fitness of the mother. This model was based on a main assumption, the occurrence of a simple and direct effect of investment per offspring and offspring quality, which results in a positive association between offspring mass and fitness (Smith and Fretwell, 1974). Most theoretical treatments about the evolutionary ecology of offspring size developed later have included this precondition (Clutton-Brock, 1991; Rees, 1997; Rees and Westoby, 1997; Sakai and Harada, 2001; Roff, 2002; and references therein). However, cumulative evidence is demonstrating that seed size can affect other fitness components in addition to seedling establishment fate in many plant species. It is important to consider that seed size is a compromise between conflicting selective pressures, i.e., larger seeds have lower dispersal ability, higher risk of predation, lower longevity, etc., and the relationship appearing between seed size and several fitness components may be negative rather than positive, the smaller seeds being favored against the larger ones (Moegenburg, 1996; Ganeshaiah and Uma Shaanker, 1991; Hegde et al., 1991). This suggests that the relationship between seed size and fitness seems to be much more complex than previously accepted (Geritz, 1995, 1998; Sakai et al., 1998; Eriksson and Jakobsson, 1999). To obtain an accurate understanding of the association between seed size and offspring fitness, it is necessary to quantify the phenotypic selection acting on seed size during different fitness components of the plants. When a given phenotypic trait covaries with more than one fitness component, the appearance of trade-offs due to conflictive selection pressures is possible, and several biologists have already found conflicting selection regimes for seed size. For example, larger acorns of Holm oak *Quercus ilex* in southeast Spain render higher seedling performance but at the same time suffer higher predation rates, the overall effect being a negative relationship between seed mass and seedling recruitment (see Figure 14.4). In the same way, several authors have suggested that seed-size evolution can respond to the opposite pressures exerted by seedling establishment and wind-dispersal efficiency (Hegde et al., 1991; Ganeshaiah and Uma Shaanker, 1991; Eriksson and Jakobsson, 1999). Thus, it is very convenient to consider the entire life cycle of the organism, as well as the complete ecological scenario in which the interactions take place, to obtain an accurate view of the selective pressures acting on seed size.

These ideas also suggest that the general assumption made by most theoretical models examining the evolution of offspring size, that success of each offspring is a fixed positive function of seed size (Smith and Fretwell, 1974; Rees, 1997; Rees and Westoby, 1997; Sakai and Harada, 2001), needs to be reconsidered for some systems. In this sense, Geritz (1998) extended Smith and Fretwell's basic model to incorporate the effect of seed predators on seed size using an evolutionary stable strategy approach,

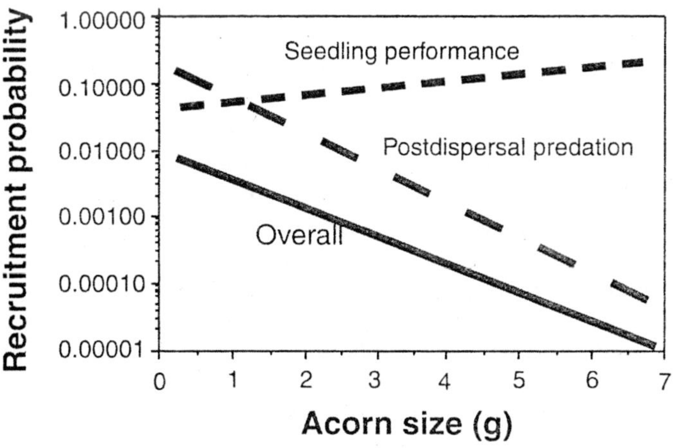

FIGURE 14.4. Probability of recruitment (in log) of *Quercus ilex* throughout different postdispersal demographic phases depending on acorn size (Sierra Nevada, southeast Spain). "Overall" refers to the stages from dispersed seeds to two-year seedlings. (*Source:* J.M. Gómez, in prep.)

finding that the action of seed predators can sometimes favor high intraplant variability in seed mass. Eriksson and Jakobsson (1999) predicted that selection should favor higher fecundity and smaller seeds when seed size-recruitment relationship is absent or negative. Irrespective of the specific predictions, it is necessary to incorporate the effect of other selective agents acting on seed size during different fitness components in order to improve the accuracy and reliability of the theoretical models studying seed size evolution.

SEED SIZE AND GLOBAL CHANGE

Atmospheric CO_2 concentrations doubling or tripling the current level of around 355 ppm may have an effect comparable to fertilization by increasing resource availability (carbon), thereby augmenting the mass of the seeds produced (Fordham et al., 1997; Monje and Bugbee, 1998; Dijkstra et al., 1999; Smith et al., 2000). However, opposite results has been also obtained (Andalo et al., 1998; Huxman et al., 1998; Smith et al., 2000; Wagner et al., 2001). On the other hand, the reduction in nitrogen concentration (as well as other nutrients) in plant tissues under conditions of elevated CO_2 (e.g., Cotrufo et al., 1998; Loladze, 2002) may similarly translate in a reduction of the nitrogen content of seeds (Andalo et al., 1998; Huxman et al., 1998; Steinger et al., 2000; Loladze, 2002). More complex still, seed-mass reduction under these circumstances could also be swayed by other life-history characteristics. For example, with more CO_2, nitrogen-fixing plants could increase seed mass because of a certain control of their nitrogen budget, whereas plants that are not nitrogen fixers would more likely reduce seed size. It should be also noted that CO_2 concentrations above 1,000 ppm (common in growth chambers and greenhouses) diminish seed size without affecting growth parameters in wheat (Reuveni and Bugbee, 1997), which may have implications for crop cultivation under such conditions or for studying seed-size parameters in growth-chamber experiments.

Higher CO_2 concentrations may also indirectly alter seed-size parameters. For instance, CO_2 boosts net photosynthesis more in plants with the C_3 pathway than in plants with the C_4 pathway (Bowes, 1996). This may change the competitive relationship of plants in the same community and alter the size of their seeds, in turn possibly magnifying crop-yield losses to competing C_3 weeds (Ziska, 2000). Rising temperatures during climatic change may also affect seed size (McKee and Richards, 1998), whereby potential benefits of higher CO_2 concentrations on seed mass may be cancelled by negative effects of higher temperatures (Prasad et al., 2002). On the other hand, UV-B radiation showed no effect on *Silene vulgaris* (van de

Staaij et al., 1997) but increased seed number and size in *Cistus creticus* (Stephanou and Manetas, 1998), while elevated ozone concentration has similarly produced contrasting results (Black et al., 2000). In short, although much is needed to provide a complete picture of the effect of global change on seed mass, seed mass will almost certainly be involved, potentially altering the performance of the species, their relationship with the environment, crop production and quality, and future community composition (Stephanou and Manetas, 1998; Huxman et al., 1998; Ziska, 2000; Smith et al., 2000; Loladze, 2002; Thürig et al., 2003).

HUMANS AND SEED SIZE

Since the origins of civilization, human lives have being tightly linked to seed size, attempts to boost crop production by increasing seed mass being continuous throughout history. In many cases, however, the production of crops with smaller seeds is desirable in order to improve acceptance among consumers. On the other hand, agriculture has acted as a selective pressure for the evolution of the seed size of weeds, so that the closer in size and shape to the seed of the crop in which the weed grows, the more possibilities of survival (Zimdahl, 1999). Humans have therefore shaped the seed mass of crops and, unintentionally, of weeds.

The foregoing sections clearly demonstrate the profound implications of seed mass on many aspects of agriculture and forestry, such as predation, germination, emergence, or competitive interactions, all of which may be of interest for planning effective agricultural or forestry practices. For example, pathogen infestation may reduce plant performance and seed size in weeds (De Clerck-Floate, 1999), which may in turn weaken the ability of those seeds to germinate and become established, such interactions being of potential use as a tool for biological control. Seed predators also may be useful for the biological control of weeds having seeds within the size range consumed by the predator (Hartke et al., 1998), and of course the size range of seeds consumed may similarly help in decision concerning the crop and the cultivar to be sown. The timing of seedling emergence may be critical when trying to reduce competition with weeds, and cultivars with larger seeds and faster emergence may be preferable (Ntanos and Koutroubas, 2000). Seeds in the soil are mostly concentrated in the aggregate fraction most closely matching their sizes. This regulates seed germination and seedling establishment by creating microsites with different moisture, temperature, and aeration regimes (Reuss et al., 2001). Consequently, either for weed control or for crop cultivation, the relationship between seed size and microtopography may be of interest. The differences in seed mass between crops and weeds, by influencing different traits such as susceptibility to predation, depth of emergence, resistance to allelopathic compounds, growth rate, or susceptibility to low nutrient availability, are thus a primary biological basis for an integrated control of weeds (Liebman and Davis, 2000).

Seed mass is also a fixed trait in cultivars and thus may be used for cultivar characterization. The weight of the seeds can be correlated to specific traits of interest such as crop protein and oil content (Önder and Babaoglu, 2001), sugar content (Tang and Tigerstedt, 2001), accumulation of toxins during pathogen infestation (Bai et al., 2001), or insect damage (Lopes et al., 1997). This is useful for an inexpensive assessment of preliminary screening for traits of interest, for selecting the cultivars to be used, and even for planning selection programs for traits of interest. Similarly, the fact that seed size may differ and be genetically fixed according to their origin (ecotypes), together with the correlation to seedling emergence and performance, may be important criteria for planning seed-source selection and reforestation not only for trees (Surles et al., 1993; Dangasuk et al., 1997), but also for grasses or other plants used for restoration purposes (Greipsson and Davy, 1995). Furthermore, seed size may be a criteria to select the appropriate species for planning successful vegetation recovery under particular circumstances, as for instances in semiarid, poorly vegetated areas with high water erosion (Cerdà and García-Fayos, 2002).

REFERENCES

Abdullah, Z., M.A. Khan, and T.J. Flowers (2001). Causes of sterility in seed set of rice under salinity stress. *Journal of Agronomy and Crop Science* 187:25-32.

Affre, L. and J.D. Thompson (1999). Variation in self-fertility, inbreeding depression and levels of inbreeding in four *Cyclamen* species. *Journal of Evolutionary Biology* 12:113-122.

Agrawal, R.L. (1998). *Fundamentals of plant breeding and hybrid seed production*. Enfield, NH: Science Publishers.

Aizen, M.A. and H. Woodcock (1992). Latitudinal trends in acorn size in eastern North American species of *Quercus*. *Canadian Journal of Botany* 70:1218-1222.

Andalo, C., C. Raquin, N. Machon, B. Godelle, and M. Mousseau (1998). Direct and maternal effects of elevated CO_2 on early root growth of germinating *Arabidopsis thaliana* seedlings. *Annals of Botany* 81:405-411.

Andersson, L. (1996). Characteristics of seeds and seedlings from weeds treated with sublethal herbicide doses. *Weed Research* 36:55-64.

Andersson, S. (1996). Seed size as determinant of germination rate in *Crepis tectorum* (Asteraceae): Evidence from a seed burial experiment. *Canadian Journal of Botany* 74:568-572.

Arditti, J. and A.K.A. Ghani (2000). Numerical and physical properties of orchid seeds and their biological implications. Tansley Review No. 110. *New Phytologist* 145:367-421.

Bai, G.-H., R. Plattner, A. Desjardins, and F. Kolb (2001). Resistance to *Fusarium* head blight and deoxynivalenol accumulation in wheat. *Plant Breeding* 120:1-6.

Banovetz, S.J. and S.M. Scheiner (1994). The effects of seed mass on the seed ecology of *Coreopsis lanceolata. The American Midland Naturalist* 131:65-74.

Bañuelos, M.J. and J.R. Obeso (2003). Maternal provisioning, sibling rivalry and seed mass variability in the dioecious shrub *Rhamnus alpinus. Evolutionary Ecology* 17:19-31.

Baskin, C.C. and J.M. Baskin (1998). *Seeds: Ecology, biogeography, and evolution of dormancy and germination.* San Diego: Academic Press.

Bekker, R.M., J.P. Bakker, U. Grandin, R. Kalamees, P. Milberg, P. Poschlod, K. Thompson, and J.H. Willems (1998). Seed size, shape and vertical distribution in the soil: Indicators of seed longevity. *Functional Ecology* 12:834-842.

Benkman, C.W. (1995). Wind dispersal capacity of pine seeds and the evolution of different seed dispersal modes in pines. *Oikos* 73:221-224.

Berg, H. and P. Redbo-Torstensson (1999). Offspring performance in three cleistogamous *Viola* species. *Plant Ecology* 145:49-58.

Biere, A. (1991). Parental effects in *Lychnis flos-cuculi:* I. Seed size, germination and seedling performance in a controlled environment. *Journal of Evolutionary Biology* 3:447-465.

Black, V.J., C.R. Black, J.A. Roberts, and C.A. Stewart (2000). Impact of ozone on the reproductive development of plants. Tansley Review No. 115. *New Phytologist* 147:421-447.

Blaney, C.S. and P.M. Kotanen (2001). Post-dispersal losses to seed predators: An experimental comparison of native and exotic old field plants. *Canadian Journal of Botany* 79:284-292.

Bond, W.J., M. Honig, and K.E. Maze (1999). Seed size and seedling emergence: An allometric relationship and some ecological implications. *Oecologia* 120: 132-136.

Bonfil, C. (1998). The effects of seed size, cotyledon reserves, and herbivory on seedling survival and growth in *Quercus rugosa* and *Q. laurina* (Fagaceae). *American Journal of Botany* 85:79-87.

Bowes, G. (1996). Photosynthetic responses to changing atmospheric carbon dioxide concentration. In N.R. Baker (ed.), *Photosynthesis and the environment* (pp. 397-407). Dordrecht, the Netherlands: Kluwer.

Bretagnolle, F., J.D. Thompson, and R. Lumaret (1995). The influence of seed size variation on seed germination and seedling vigour in diploid and tetraploid *Dactylis glomerata* L. *Annals of Botany* 76:607-615.

Brewer, S.W. (2001). Predation and dispersal of large and small seeds of a tropical palm. *Oikos* 92:245-255.

Brown, J.H. and E.J. Heske (1990). Control of a desert-grassland transition by a keystone rodent guild. *Science* 250:1705-1707.

Castro, J. (1999). Seed mass versus seedling performance in Scots pine: A maternally dependent trait. *New Phytologist* 144:153-161.

Cerdà, A. and P. García-Fayos (2002). The influence of seed size and shape on their removal by water erosion. *Catena* 48:293-301.

Chacón, P. and R.O. Bustamante (2001). The effects of seed size and pericarp on seedling recruitment and biomass in *Cryptocarya alba* (Lauraceae) under two contrasting moisture regimes. *Plant Ecology* 152:137-144.

Cheema, M.A., M.A. Malik, A. Hussain, S.H. Shah, and S.M.A. Basra (2001). Effects of time and rate of nitrogen and phosphorous application on the growth and the seed and oil yields of canola (*Brassica napus* L.). *Journal of Agronomy and Crop Science* 186:103-110.

Cho, Y. and R.A. Scott (2000). Combining ability of seed vigor and seed yield in soybean. *Euphytica* 112:145-150.

Chung, J., J.-H. Lee, K. Arumuganathan, G.L. Graef, and J.E. Specht (1998). Relationships between nuclear DNA content and seed and leaf size in soybean. *Theoretical and Applied Genetics* 96:1064-1068.

Clutton-Brock, T.H. (1991). *The evolution of parental care*. Princeton, NJ: Princeton University Press.

Cookson, W.R., J.S. Rowarth, and K.C. Cameron (2000). The response of a perennial rygrass (*Lolium perenne* L.) seed crop to nitrogen fertilizer application in the absence of moisture stress. *Grass and Forage Science* 55:314-325.

Coomes, D.A. and P.J. Grubb (2003). Colonization, tolerance, competition and seed-size variation within functional groups. *Trends in Ecology and Evolution* 18:283-291.

Cotrufo, M.F., P. Ineson, and A. Scott (1998). Elevated CO_2 reduces the nitrogen concentration of plant tissues. *Global Change Biology* 4:43-54.

Dalling, J.W. and S.P. Hubbell (2002). Seed size, growth rate and gap microsite conditions as determinants of recruitment success for pioneer species. *Journal of Ecology* 90:557-568.

Dangasuk, O.G., P. Seurei, and S. Gudu (1997). Genetic variation in seed and seedling traits in 12 African provenances of *Faidherbia albida* (Del.) A. Chev. at Lodwar, Kenya. *Agroforestry Systems* 37:133-141.

De Clerck-Floate, R. (1999). Impact of *Erysiphe cynoglossi* on the growth and reproduction of the rangeland weed *Cynoglossum officinale*. *Biological Control* 15:107-112.

Delgado, J.A., J.M. Serrano, F. López, and F.J. Acosta (2001). Heat shock, mass-dependent germination, and seed yield as related components of fitness in *Cistus ladanifer*. *Environmental and Experimental Botany* 46:11-20.

Díaz, M. (1990). Interspecific patterns of seed selection among granivorous passerines: Effect of seed size, seed nutritive value and bird morphology. *Ibis* 132:467-476.

Dijkstra, P., AD H.M.C. Schapendonk, K. Groenwold, M. Jansen, and S.C. van de Geijn (1999). Seasonal changes in the response of winter wheat to elevated atmospheric CO_2 concentration grown in open-top chambers and field tracking enclosures. *Global Change Biology* 5:563-576.

Doganlar, S., A. Fray, and S.D. Tanksley (2000). The genetic basis of seed-weight variation: Tomato as a model system. *Theoretical and Applied Genetics* 100: 1267-1273.

Egli, D.B. (1998). *Seed biology and the yield of grain crops.* Wallingford, UK: CAB International.

El Hadi, E.A. and E.A.E. Elsheikh (1999). Effects of *Rhizobium* inoculation and nitrogen fertilization on yield and protein content of six chickpea (*Cicer arietinum* L.) cultivars in marginal soils under irrigation. *Nutrient Cycling in Agroecosystems* 54:57-63.

Eriksson, O. (1999). Seed size variation and its effect on germination and seedling performance in the clonal herb *Convallaria majalis*. *Acta Oecologica* 20:61-66.

Eriksson, O. and A. Jakobsson (1999). Recruitment trade-offs and the evolution of dispersal mechanisms in plants. *Evolutionary Ecology* 13:411-423.

Escudero, A., Y. Núñez, and F. Pérez-García (2000). Is fire a selective force of seed size in pine species? *Acta Oecologica* 21:245-256.

Etterson, J.R. and L.F. Galloway (2002). The influence of light on paternal plants in *Campanula americana* (Campanulaceae): Pollen characteristics and offspring traits. *American Journal of Botany* 89:1899-1906.

Fordham, M., J.D. Barnes, I. Bettarini, A. Polle, N. Slee, C. Raines, F. Miglietta, and A. Raschi (1997). The impact of elevated CO_2 on growth and photosynthesis in *Agrostis canina* L. ssp. *monteluccii* adapted to contrasting atmospheric CO_2 concentrations. *Oecologia* 110:169-178.

Galloway, L.F. (2001). The effect of maternal and paternal environments on seed characters in the herbaceous plant *Campanula americana* (Campanulaceae). *American Journal of Botany* 88:832-840.

Ganeshaiah, K.N. and R. Uma Shaanker (1991). Seed size optimization in a wind dispersed tree *Butea monosperma*: A trade-off between seedling establishment and pod dispersal efficiency. *Oikos* 60:3-6.

Geritz, S.A.H. (1995). Evolutionary stable seed polymorphism and small-scale spatial variation in seedling density. *The American Naturalist* 146:685-707.

Geritz, S.A.H. (1998). Co-evolution of seed size and seed predation. *Evolutionary Ecology* 12:891-911.

Gigord, L., C. Lavigne, and J.A. Shykoff (1998). Partial self-incompatibility and inbreeding depression in a native tree species of La Réunion (Indian Ocean). *Oecologia* 117:342-352.

Gómez, J.M. and R. Zamora (1994). Top-down effects in a tritrophic system: Parasitoids enhance plant fitness. *Ecology* 75:1023-1030.

Green J.J. and D.M. Newbery (2001). Light and seed size affect establishment of grove-forming ectomycorrhizal rain forest tree species. *New Phytologist* 151:271-289.

Green, P.T. (1999). Seed germination in *Chrysophyllum* sp. nov., a large-seeded rainforest species in north Queensland: Effect of seed size, litter depth and seed position. *Australian Journal of Ecology* 24:608-613.

Greene, D.F. and E.A. Johnson (1993). Seed mass and dispersal capacity in wind-dispersed diaspores. *Oikos* 67:69-74.

Greipsson, S. and A.J. Davy (1995). Seed mass and germination behaviour in populations of the dune-building grass *Leymus arenarius*. *Annals of Botany* 76:493-501.

Grewal, H.S., L. Zhonggu, and R.D. Graham (1997). Influence of subsoil zinc on dry matter production, seed yield and distribution of zinc in oilseed rape genotypes differing in zinc efficiency. *Plant and Soil* 192:181-189.

Haase, P., F.I. Pugnaire, and L.D. Incoll (1995). Seed production and dispersal in the semi-arid tussock grass *Stipa tenacissima* L. during masting. *Journal of Arid Environments* 31:55-65.

Hanley, M.E., J.E. Unna, and B. Darvill (2003). Seed size and germination response: A relationship for fire-following plant species exposed to thermal shock. *Oecologia* 134:18-22.

Hare, M.C., D.W. Parry, and M.D. Baker (1999). The relationship between wheat seed weight, infection by *Fusarium culmorum* or *Microdochium nivale*, germination and seedling disease. *European Journal of Plant Pathology* 105:859-866.

Harper, J.L., P.H. Lovell, and K.G. Moore (1970). The shapes and sizes of seeds. *Annual Review of Ecology and Systematics* 1:327-356.

Hartke, A., F.A. Drummond, and M. Liebman (1998). Seed feeding, seed caching, and burrowing behaviors of *Harpalus rufipes* De Geer larvae (Coleoptera: Carabidae) in the Maine potato agroecosystem. *Biological Control* 13:91-100.

Hegde, S.G., R. Uma Shaanker, and K.N. Ganeshaiah (1991). Evolution of seed size in the bird-dispersed tree *Santalum album* L.: A trade-off between seedling establishment and dispersal efficiency. *Evolutionary Trends in Plants* 5:131-135.

Hendrix, S.D. and I.-F. Sun (1989). Inter- and intraespecific variation in seed mass in seven species of umbellifer. *New Phytologist* 112:445-451.

Herrera, C.M. (1989). Frugivory and seed dispersal by carnivorous mammals, and associated fruit characteristics, in undisturbed Mediterranean habitats. *Oikos* 55:250-262.

Hiei, K. and M. Ohara (2002). Variation in fruit- and seed set among and within inflorescences of *Melampyrum roseum* var. *japonicum* (Scrôphulariaceae). *Plant Species Biology* 17:13-23.

Hódar, J.A, J. Castro, and R. Zamora (2003). Pine processionary caterpillar *Thaumetopoea pityocampa* as a new threat for relict Mediterranean Scots pine forests under climatic warming. *Biological Conservation* 110:123-129.

Hodkinson, D.J., A.P. Askew, K. Thompson, J.G. Hodgson, J.P. Bakker, and R.M. Bekker (1998). Ecological correlates of seed size in the British flora. *Functional Ecology* 12:762-766.

Houssard, C. and J. Escarré (1991). The effect of seed mass on growth and competitive ability of *Rumex acetosella* from two successional old-fields. *Oecologia* 86:236-242.

Hulme, P.E. (1998). Post-dispersal seed predation and seed bank persistence. *Seed Science Research* 8:513-519.

Hulme, P.E. and T. Borelli (1999). Variability in post-dispersal seed predation in deciduous woodland: Relative importance of location, seed species, burial and density. *Plant Ecology* 145:149-156.

Humara, J.M., A. Casares, and J. Majada (2002). Effect of seed size and growing media water availability on early seedling growth in *Eucalyptus globulus*. *Forest Ecology and Management* 167:1-11.

Huxman, T.E., E.P. Hamerlynck, D.N. Jordan, K.J. Salsman, and S.D. Smith (1998). The effects of parental CO_2 environment on seed quality and subsequent seedling performance in *Bromus rubens*. *Oecologia* 114:202-208.

Imbert, E. (2001). Capitulum characters in a seed heteromorphic plant, *Crepis sancta* (Asteraceae): Variance partitioning and inference for the evolution of dispersal rate. *Heredity* 86:78-86.

Jackai, L.E.N., S. Nokoe, B.O. Tayo, and P. Koona (2001). Inferences on pod wall and seed defences against the brown cowpea coreid bug, *Clavigralla tomentosicollis* Stäl. (Hem., Coreidae) in wild and cultivated *Vigna* species. *Journal of Applied Entomology* 125:277-286.

Jordano, P. (1987). Frugivory, external morphology and digestive system in Mediterranean sylviid warblers *Sylvia* spp. *Ibis* 129:175-189.

Karlsson, C. and G. Örlander (2002). Mineral nutrients in needles of *Pinus sylvestris* seed trees after release cutting and their correlations with cone production and seed weight. *Forest Ecology and Management* 166:183-191.

Katz, G.L., J.L. Friedman, and S.W. Beatty (2001). Effects of physical disturbance and granivory on establishment of native and alien riparian trees in Colorado, U.S.A. *Diversity and Distributions* 7:1-14.

Keeve, R., H.L. Loubser, and G.H.J. Krüger (2000). Effects of temperature and photoperiod on days to flowering, yield and yield components of *Lupinus albus* (L.) under field conditions. *Journal of Agronomy and Crop Science* 184:187-196.

Kelman, W.M., R.N. Oram, and J.E. Hayes (1999). Characterization, establishment and persistence under grazing of nitrous oxide-induced octoploid *Phalaris aquatica* (L.). *Grass and Forage Science* 54:62-68.

Khan, M.L., P. Bhuyan, U. Shankar, and N.P. Todaria (1999). Seed germination and seedling fitness in *Mesua ferrea* L. in relation to fruit size and seed number per fruit. *Acta Oecologica* 20:599-606.

Kobayashi, T., K. Okamoto, and Y. Hori (2001). Variation in size structure, growth and reproduction in Japanese platain (*Plantago asiatica* L.) between exposed and shaded populations. *Plant Species Biology* 16:13-28.

Lacey, E.P., S. Smith, and A.L. Case (1997). Parental effects on seed mass: Seed coat but not embryo/endosperm effects. *American Journal of Botany* 84:1617-1620.

Larson, K.L. and R.E. Mullen (1995). Drought injury and resistance of crop plants. In U.S. Gupta (ed.), *Production and improvement of crops for drylands* (pp. 167-191). Enfield, NH: Science Publishers.

Leishman, M.R. and M. Westoby (1994). Hypothesis on seed size: Tests using the semiarid flora of New South Wales, Australia. *The American Naturalist* 143:890-906.

Leport, L., N.C. Turner, R.J. French, M.D. Barr, R. Duda, S.L. Davies, D. Tennant, and K.H.M. Siddique (1999). Physiological responses of chickpea genotypes to terminal drought in a Mediterranean-type environment. *European Journal of Agronomy* 11:279-291.

Li, B., J.-I. Suzuki, and T. Hara (1998). Latitudinal variation in plant size and relative growth rate in *Arabidopsis thaliana*. *Oecologia* 115:293-301.

Liebman, M. and A.S. Davis (2000). Integration of soil, crop and weed management in low-external-input farming systems. *Weed Research* 40:27-47.

Lipow, S.R. and R. Wyatt (1999). Diallel crosses reveal patterns of variation in fruit-set, seed mass, and seed number in *Asclepias incarnata*. *Heredity* 83:310-318.

Loladze, I. (2002). Rising atmospheric CO_2 and human nutrition: Toward globally imbalanced plant stoichiometry? *Trends in Ecology and Evolution* 17:457-461.

Long, T.J. and R.H. Jones (1996). Seedling growth strategies and seed size effects in fourteen oak species native to different soil moisture habitats. *Trees* 11:1-8.

Lopes, E.C.A., D. Destro, R. Montalván, M.U. Ventura, and E.P. Guerra (1997). Genetic gain and correlations among traits for stink bug resistance in soybeans. *Euphytica* 97:161-166.

López, J., J.A. Devesa, A. Ortega-Olivencia, and T. Ruiz (2000). Production and morphology of fruit and seeds in Genisteae (Fabaceae) of southwest Spain. *Botanical Journal of the Linnean Society* 132:97-120.

López-Bellido, L., M. Fuentes, and J.E. Castillo (2000). Growth and yield of white lupin under Mediterranean conditions: Effect of plant density. *Agronomy Journal* 92:200-205.

Malcolm, P.J., P. Holford, W.B. McGlasson, and S. Newman (2003). Temperature and seed weight affect the germination of peach rootstock seeds and the growth of rootstock seedlings. *Scientia Horticulturae* 98:247-256.

Malo, J.E. and F. Suárez (1995). Herbivorous mammals as seed dispersers in a Mediterranean dehesa. *Oecologia* 104:246-255.

Martin, A. and S. Puech (2001). Interannual and interpopulation variation in *Helichrysum stoechas* (Asteraceae), a species of disturbed habitats in the Mediterranean region. *Plant Species Biology* 16:29-37.

Maughan, P.J., M.A.S. Maroof, and G.R. Buss (1996). Molecular-marker analysis of seed weight: Genomic locations, gene action, and evidence for orthologous evolution among three legume species. *Theoretical and Applied Genetics* 93: 574-579.

McKee, J. and A.J. Richards (1998). The effect of temperature on reproduction in five *Primula* species. *Annals of Botany* 82:359-374.

Méndez, M. (1997). Sources of variation in seed mass in *Arum italicum*. *International Journal of Plant Sciences* 158:298-305.

Metcalfe, D.J. and P.J. Grubb (1995). Seed mass and light requirements for regeneration in Southeast Asian rain forest. *Canadian Journal of Botany* 73:817-826.

Meyer, S.E. and S.L. Carlson (2001). Achene mass variation in *Ericameria nauseosus* (Asteraceae) in relation to dispersal ability and seedling fitness. *Functional Ecology* 15:274-281.

Mian, M.A.R., M.A. Bailey, J.P. Tamulonis, E.R. Shipe, T.E. Carter Jr., W.A. Parrott, D.A. Ashley, R.S. Hussey, and H.R. Boerma (1996). Molecular markers associated with seed weight in two soybean populations. *Theoretical and Applied Genetics* 93:1011-1016.

Michail, N.N., F.S. Faris, M.W.A. Hassan, and R.G. Kerlous (1995). Improvement of leguminous vegetables production in calcareous soil by addition of some acidifying material: I. Effects on yield and yield components. *Fertilizer Research* 43:87-91.

Milberg, P., L. Andersson, C. Elfverson, and S. Regnér (1996). Germination characteristics of seeds differing in mass. *Seed Science Research* 6:191-197.

Milberg, P., M.A. Pérez-Fernández, and B.B. Lamont (1998). Seedling growth response to added nutrients depends on seed size in three woody genera. *Journal of Ecology* 86:624-632.

Moegenburg, S.M. (1996). *Sabal palmetto* seed size: Causes of variation, choices of predators, and consequences for seedlings. *Oecologia* 106:539-543.

Mojonnier, L. (1998). Natural selection on two seed-size traits in the common Morning glory *Ipomoea purpurea* (Convolvulaceae): Patterns and evolutionary consequences. *The American Naturalist* 152:188-203.

Monje O. and B. Bugbee (1998). Adaptation to high CO_2 concentration in an optimal environment: Radiation capture, canopy quantum yield and carbon use efficiency. *Plant, Cell and Environment* 21:315-324.

Murali, K.S. (1997). Patterns of seed size, germination and seed viability of tropical tree species in southern India. *Biotropica* 29:271-279.

Mwase, W.F. and R.G. Kapooria (2000). Incidence and severity of frogeye leaf spot and associated yield losses in soybeans in agroecological zone II of Zambia. *Mycopathologia* 149:73-78.

Myers, R.L. (1998). Nitrogen fertilizer effect on grain amaranth. *Agronomy Journal* 90:597-602.

Nakamura, R.R. and M.L. Stanton (1989). Embryo growth and seed size in *Raphanus sativus*: Maternal and paternal effects in vivo and in vitro. *Evolution* 43:1435-1443.

Ntanos, D.A. and S.D. Koutroubas (2000). Competition of barnyardgrass with rice varieties. *Journal of Agronomy and Crop Science* 184:241-246.

Obeso, J.R. (1993). Seed mass variation in the perennial herb *Asphodelus albus*: Sources of variation and position effect. *Oecologia* 93:571-575.

Oleksyn, J., J. Modrzynski, M.G. Tjoelker, R. Zytkowiak, P.B. Reich, and P. Karolewski (1998). Growth and physiology of *Picea abies* populations from elevational transects: Common garden evidence for altitudinal ecotypes and cold adaptation. *Functional Ecology* 12:573-590.

Önder, M. and M. Babaoglu (2001). Interactions amongst grain variables in various dwarf dry bean (*Phaseolus vulgaris* L.) cultivars. *Journal of Agronomy and Crop Science* 187:19-23.

Onipchenko, V.G., G.V. Semenova, and E. van der Maarel (1998). Population strategies in severe environments: Alpine plants in the northwestern Caucasus. *Journal of Vegetation Science* 9:27-40.

Pakeman, R.J., G. Digneffe, and J.L. Small (2002). Ecological correlates of endozoochory by herbivores. *Functional Ecology* 16:296-304.

Parciak, W. (2002). Seed size, number, and habitat of a fleshy-fruited plant: Consequences for seedling establishment. *Ecology* 83:794-808.

Paz, H. and M. Martínez-Ramos (2003). Seed mass and seedling performance within eight species of *Psychotria* (Rubiaceae). *Ecology* 84:439-450.

Pearson, T.R.H., D.F.R.P. Burslem, C.E. Mullins, and J.W. Dalling (2002). Germination ecology of neotropical pioneers: Interacting effects of environmental conditions and seed size. *Ecology* 83:2798-2807.

Pegtel, D.M. (1999). Effect of ploidy level on fruit morphology, seed germination and juvenile growth in scurvy grass (*Cochlearia officinalis* L. s.l., Brassicaceae). *Plant Species Biology* 14:201-215.

Pellmyr, O., L.K. Massey, J.L. Hamrick, and M.A. Feist (1997). Genetic consequences of specialization: Yucca moth behavior and self-pollination in yuccas. *Oecologia* 109:273-278.

Peroni, P.A. (1994). Seed size and dispersal potential of *Acer rubrum* (Aceraceae) samaras produced by populations in early and late successional environments. *American Journal of Botany* 81:1428-1434.

Pilson, D. and K.L. Decker (2002). Compensation for herbivory in wild sunflower: Response to simulated damage by the head-clipping weevil. *Ecology* 83:3097-3107.

Platenkamp, G.A.J. and R.G. Shaw (1993). Environmental and genetic maternal effects on seed characters in *Nemophila menziesii*. *Evolution* 47:540-555.

Prasad, P.V.V., K.J. Boote, L.H. Allen Jr., and J.M.G. Thomas (2002). Effect of elevated temperature and carbon dioxide on seed-set and yield of kidney bean (*Phaseolus vulgaris* L.). *Global Change Biology* 8:710-721.

Rahman, M.A. and M.S. Saad (2000). Estimation of additive, dominance and digenic epistatic effects for certain yield characters in *Vigna sesquipedalis* Fruw. *Euphytica* 114:61-66.

Reader, R.J. (1993). Control of seedling emergence by ground cover and seed predation in relation to seed size for some old-field species. *Journal of Ecology* 81:169-175.

Rees, M. (1997). Evolutionary ecology of seed dormancy and seed size. In J. Silvertown, M. Franco, and J.L. Harper (eds), *Plant life histories: Ecology, phylogeny and evolution* (pp. 121-142). Cambridge, UK: Cambridge University Press.

Rees, M., R. Condit, M. Crawley, S. Pacala, and D. Tilman (2001). Long-term studies of vegetation dynamics. *Science* 293:650-655.

Rees, M. and M. Westoby (1997). Game-theoretical evolution of seed mass in multi-species ecological models. *Oikos* 78:116-126.

Reich, P.B., J. Oleksyn, and M.G. Tjoelker (1994). Seed mass effects on germination and growth of diverse European Scots pine populations. *Canadian Journal of Forest Research* 24:306-320.

Reuss, S.A., D.D. Buhler, and J.L. Gunsolus (2001). Effects of soil depth and aggregate size on weed seed distribution and viability in a silt loam soil. *Applied Soil Ecology* 16:209-217.

Reuveni, J. and B. Bugbee (1997). Very high CO_2 reduces photosynthesis, dark respiration and yield in wheat. *Annals of Botany* 80:539-546.

Riera, N., A. Traveset, and O. García (2002). Breakage of mutualisms by exotic species: The case of *Cneorum tricoccon* L. in the Balearic Islands (western Mediterranean sea). *Journal of Biogeography* 29:713-714.

Roff, D.A. (2002). *Life history evolution*. New York: Chapman and Hall.

Ruiz de Clavijo, E. (2001). The role of dimorphic achenes in the biology of the annual weed *Leontodon longirrostris*. *Weed Research* 41:275-286.

Sakai, S. and Y. Harada (2001). Sink-limitation and the size-number trade-off of organs: Production of organs using a fixed amount of reserves. *Evolution* 55:467-476.

Sakai, S., K. Kikuzawa, and K. Umeki (1998). Evolutionary stable resource allocation for production of wind-dispersed seeds. *Evolutionary Ecology* 12:477-485.

Seiwa, K., A. Watanabe, T. Saitoh, H. Kanno, and S. Akasaka (2002). Effects of burying depth and seed size on seedling establishment of Japanese chestnuts, *Castanea crenata*. *Forest Ecology and Management* 164:149-156.

Shumway, D.L. and R.T. Koide (1994). Within-season variability in mycorrhizal benefit to reproduction in *Abutilon theophrasti* Medic. *Plant, Cell and Environment* 17:821-827.

Shykoff, J.A., S.-O. Kolokotronis, C.L. Collin, and M. López-Villavicencio (2003). Effects of male sterility on reproductive traits in gynodioecious plants: A meta-analysis. *Oecologia* 135:1-9.

Silvertown, J.W. and J. Lovett-Doust (1993). *Introduction to plant population biology*. London, UK: Blackwell Science.

Simons, A.M. and M.O. Johnston (2000). Variation in seed traits of *Lobelia inflata* (Campanulaceae): Sources and fitness consequences. *American Journal of Botany* 87:124-132.

Sinniah, U.R., R.H. Ellis, and P. John (1998). Irrigation and seed quality development in rapid-cycling *Brassica:* Seed germination and longevity. *Annals of Botany* 82:309-314.

Sipe, T.W. and A.R. Linnerooth (1995). Intraspecific variation in samara morphology and flight behavior in *Acer saccharinum* (Aceraceae). *American Journal of Botany* 82:1412-1419.

Smart, A. and L.E. Moser (1999). Switchgrass seedling development as affected by seed size. *Agronomy Journal* 91:335-338.

Smith, C.C. and S.D. Fretwell (1974). The optimal balance between size and number of offspring. *The American Naturalist* 108:499-506.

Smith, S.D., T.E. Huxman, S.F. Zitzer, T.N. Charlet, D.C. Housman, J.S. Coleman, L.K. Fenstermaker, J.R. Seeman, and R.S. Nowak (2000). Elevated CO_2 increases productivity and invasive species success in an arid ecosystem. *Nature* 408:79-82.

Spurr, C.J., D.A. Fulton, P.H. Brown, and R.J. Clark (2002). Changes in seed yield and quality with maturity in onion (*Allium cepa* L., cv. "early cream goald"). *Journal of Agronomy and Crop Science* 188:275-280.

Stearns, C.C. (1992). *The evolution of life history*. Oxford, UK: Oxford University Press.

Steinger, T., R. Gall, and B. Schmid (2000). Maternal and direct effects of elevated CO_2 on seed provisioning, germination and seedling growth in *Bromus erectus*. *Oecologia* 123:475-480.

Stephanou, M. and Y. Manetas (1998). Enhanced UV-B radiation increases the reproductive effort in the Mediterranean shrub *Cistus creticus* under field conditions. *Plant Ecology* 134:91-96.

Surles, S.E., T.L. White, G.R. Hodge, and M.L. Duryea (1993). Relationship among seed weight components, seedling growth traits, and predicted field breeding values in slash pine. *Canadian Journal of Forest Research* 23:1550-1556.

Susko, D.J. and L. Lovett-Doust (2000). Patterns of seed mass variation and their effects on seedling traits in *Alliaria petiolata* (Brassicaceae). *American Journal of Botany* 87:56-66.

Sweeney, D.W. and G.V. Granade (2002). Effect of a single irrigation at different reproductive growth stages on soybean planted in early and late June. *Irrigation Science* 21:69-73.

Szentesi, Á. and T. Jermi (1995). Predispersal seed predation in leguminous species: Seed morphology and bruchid distribution. *Oikos* 73:23-32.

Talwar, G., A. Dua, and R. Singh (1996). CO_2 exchange, primary photochemical reactions and certain enzymes of photosynthetic carbon reduction cycle of *Brassica* pods under reduced irradiance. *Photosynthetica* 32:221-229.

Tang, X. and P.M.A. Tigerstedt (2001). Variation of physical and chemical characters within an elite sea buckthorn (*Hippophae rhamnoides* L.) breeding population. *Scienta Horticulturae* 88:203-214.

Thompson K., S.R. Band, and J.G. Hodgson (1993). Seed size and shape predict persistence in soil. *Functional Ecology* 7:263-241.

Thürig, B., C. Körner, and J. Stöcklin (2003). Seed production and seed quality in a calcareous grassland in elevated CO_2. *Global Change Biology* 9:873-884.

Tungate, K.D., D.J. Susko, and T.W. Rufty (2002). Reproduction and offspring competitiveness of *Senna obtusifolia* are influenced by nutrient availability. *New Phytologist* 154:661-669.

Turnbull, L.A., M. Rees, and M.J. Crawley (1999). Seed mass and the competition/colonization trade-off: A sowing experiment. *Journal of Ecology* 87:899-912.

Tweddle, J.C., R.M. Turner, and J.B. Dickie (2002). Seed Information Database. Release 3.0. July 2002. Available online at <http://www.rbgkew.org.uk/data/sid>.

Tweney, J. and M. Mogie (1999). The relationship between achene weight, embryo weight and germination in *Taraxacum* apomicts. *Annals of Botany* 83:45-50.

van de Staaij, J.W.M., E. Bolink, J. Rozema, and W.H.O. Ernst (1997). The impact of elevated UV-B (280-320) radiation levels on the reproduction biology of a highland and a lowland population of *Silene vulgaris*. *Plant Ecology* 128:173-179.

van Rijn, C.P.E., I. Heersche, Y.E.M. van Berkel, E. Nevo, H. Lambers, and H. Poorter (2000). Growth characteristics in *Hordeum spontaneum* populations from different habitats. *New Phytologist* 146:471-481.

Vaughton, G. and M. Ramsey (1998). Sources and consequences of seed mass variation in *Banksia marginata* (Proteaceae). *Journal of Ecology* 86:563-573.

Vaz Patto, M.C., A.M. Torres, A. Koblizkova, J. Macas, and J.I. Cubero (1999). Development of a genetic composite map of *Vicia faba* using F_2 populations derived from trisomic plants. *Theoretical and Applied Genetics* 98:736-743.

Vera, M.L. (1997). Effects of altitude and seed size on germination and seedling survival on heathland plants in north Spain. *Plant Ecology* 133:101-106.

Villar, R., E.J. Veneklas, P. Jordano, and H. Lambers (1998). Relative growth rate and biomass allocation in 20 *Aegilops* (Poaceae) species. *New Phytologist* 140:425-437.

Wagner J., A. Lüscher, C. Hillebrand, B. Kobald, N. Spitaler, and W. Larcher (2001). Sexual reproduction of *Lolium perenne* L. and *Trifolium repens* L. under free air CO_2 enrichment (FACE) at two levels of nitrogen application. *Plant, Cell and Environment* 24:957-965.

Walters, M.B. and P.B. Reich (2000). Seed size, nitrogen supply, and growth rate affect tree seedling survival in deep shade. *Ecology* 81:1887-1901.

Warringa, J.W., R. de Visser, and A.D.H. Kreuzer (1998). Seed weight in *Lolium perenne* as affected by interactions among seeds within the inflorescence. *Annals of Botany* 82:835-841.

Wennström, U., U. Bergsten, and J.E. Nilsson (2002). Effects of seed weight and seed type on early seedling growth of *Pinus sylvestris* under harsh and optimal conditions. *Scandinavian Journal of Forest Research* 17:118-130.

Westoby, M., E. Jurado, and M. Leishman (1992). Comparative evolutionary ecology of seed size. *Trends in Ecology and Evolution* 7:368-372.

Wheelwright, N.T. (1985). Fruit size, gape width, and the diets of fruit-eating birds. *Ecology* 66:808-818.

White, T.A., B.D. Campbell, and P.D. Kemp (2001). Laboratory screening of the juvenile responses of grassland species to warm temperature pulses and water deficits to predict invasiveness. *Functional Ecology* 15:103-112.

Willot, S.J., S.G. Compton, and L.D. Incoll (2000). Foraging, food selection and worker size in the seed harvesting and *Messor bouvieri*. *Oecologia* 125:35-44.

Wright, I.J. and M. Westoby (1999). Differences in seedling growth behaviour among species: Trait correlations across species, and trait shifts along nutrient compared to rainfall gradients. *Journal of Ecology* 87:85-97.

Zhang, J. (1998). Variation and allometry of seed weight in *Aeschynomene americana*. *Annals of Botany* 82:843-847.

Zimdahl, R.L. (1999). *Fundamentals of weed science* (Second edition). San Diego: Academic Press.

Ziska, L.H. (2000). The impact of elevated CO_2 on yield loss from C_3 and C_4 weed in field-grown soybean. *Global Change Biology* 6:899-905.

Chapter 15

Seed Predation

Jose M. Serrano
Juan A. Delgado

INTRODUCTION

Plants produce seeds for sexual reproduction and to enable the establishment of new individuals beyond parental influence. These seeds are developed inside structures which, regardless of whether they come from an enlarged ovary or whether they present another anatomic origin, are called *fruits* in ecological literature (see for example, Herrera, 1992). The mission of fruits is to protect seeds and, in some cases, to favor their dispersion. The use of the word *seed* is also common in a more general sense than the strictly anatomic one, to refer to the fertilized ovule and its associated structures (Harper et al., 1970).

Seeds may be attacked by fungi and bacteria or, more usually, by animals. Animals can destroy seeds accidentally when they are feeding on vegetative structures or attempting to obtain nutritive resources from the fruits. In all these cases plants lose seeds and coevolutive processes can be set in motion. Nevertheless, it could be of great interest to distinguish these situations from those in which seeds are the main feeding item (granivory).

Seeds have an important nutritive value and are hierarchically aggregated into fruits, later into branches, and finally into individual plants, which constitute a valuable source of nutrition for animals. Predation actually begins at the earliest stage of seed formation and continues once they have been dispersed (Thompson, 1985; Youti and Miller, 1986; Traveset, 1990; Forget et al., 1999; Holland and Fleming, 1999).

Seeds are usually consumed completely, although seed predation sometimes results in several seeds being partially eaten. These damaged seeds will be destroyed if reservoir materials have been depleted (and seedlings cannot survive) or if the embryo has been damaged. In other cases, seeds may present opportunities for germination. The effects of this partial consumption on seed viability should be considered in order to estimate seed predation rates and the consequent real effect of seed predators on plant fitness (Koptur, 1989; Mack, 1998).

Seed predation could be approached from a plant-ecology perspective, where granivory is another factor decreasing plant fertility. Seed predation has often been reported as an important selective force that accounts for the existence of several plant traits in a plant-animal coevolution process. In these studies seed predation could be estimated as a whole, without partitioning total predation values among the different species of seed predators. However, the same plant species, the same individual, the same fruit, and even the same seed could be predated simultaneously or consecutively by two or more species (Green and Palmbald, 1975; Cunningham, 1997; Hulme, 1997; Holbrook et al., 2000). Sharing total predation scores between different predator species appears to be of interest because the same level of seed predation could be achieved through different combinations of seed predation rates exerted for different predator species, which can present spatial and temporal variations. These processes are of great interest in order to understand plant-animal coevolution. This interest increases on considering the great variety of taxonomic origins of seed predators (see Table 15.1) which could have different evolution rates.

Seed predators can individually attack a plant randomly or in nonrandom patterns when seed predators feed cooperatively (Serrano et al., 2001). The more characteristic example is that of granivorous ants which build trunk

TABLE 15.1. Taxonomic affiliation of common seed predators.

Taxonomic group	Reference
Invertebrates	
Insects: Mainly Diptera, Lepidoptera, Coleoptera, and Hymenoptera (ants and chalcid wasps)	See extensive review in Crawley, 1992
Mollusca: Snails and slugs	Godnan, 1983; Kollman and Bassin, 2001
Crustacea: Crabs and isopods	Fishman and Orth, 1996; Holbrook et al., 2000; Nakaoka, 2002
Vertebrates	
Mammals: Ungulates, rodents, primates, etc.	Kinzey, 1992; Anapol and Lee, 1994; Happel, 1998; Forget et al., 1999; Hulme and Borelli, 1999; Alcántara et al., 2000; Curran and Webb, 2000; Kollmann and Bassin, 2001
Birds: Finches, pigeons, sparrows, etc.	Fuentes and Shupp, 1998; Linzey and Washok 2000
Fishes	Kubitzki and Ziburski, 1994

trail systems to facilitate seed transport to the nest. These systems are distributed in space not only as a response to food abundance (López, Acosta, and Serrano, 1994) but also to the decrease in intracolonial competitive interactions (Acosta et al., 1995). Interestingly, these trunk trail systems have been considered to be modular structures similar to those of plant anatomy (López, Serrano, and Acosta, 1994).

DEFENDING SEEDS

Seed predation rates could be very high (see reviews in Janzen, 1969, 1971; Crawley, 1992), and plants have developed many strategies to attenuate such high loses. An overview of theses strategies would appear to show that seeds, fruits, and individual plants are involved in defensive tasks.

In the defense of seeds from predators, plants have developed different strategies that could be grouped into three basic types: mechanical, chemical, and phenological. Mechanical defenses have been reported not only in seeds but also in the fruits and stems that bear them. The presence of spiny trunks in some species has been interpreted as an adaptation to prevent climbing rodents from reaching seeds (Janzen, 1969; Herrera, 1984).

Fruits are, however, the most obvious structures of mechanical protection of seeds. The existence of fleshy or lignified tissues around seeds could be an efficient barrier that prevents predator access to seeds. It has been suggested that pulp in vertebrate-dispersed fruits might have originated as a defense from insect seed predators (Mack, 2000). It has been reported that hard, spiny, hairy fruits and those that produce sticky gum when damaged can protect seeds, reducing the frequency of attack by predators (Janzen, 1969, 1971; Green and Palmbald, 1975).

Fruit compartmentation constitutes a peculiar form of mechanical protection. Loculate fruits have several cavities that partition the set of seeds in isolated groups. In this way a seed predator must bore the fruit wall several times in order to gain access to the seeds of more than one locule. Furthermore, seeds within intact locules are safe from secondary predators, which are unable to bore the fruit wall (Janzen, 1971; Bradford and Smith, 1977; Serrano et al., 2001). Fruit compartmentation also results in the division of space and food resources, which could reduce inter- and intraspecific competition between seed predators (see Figure 15.1).

Seeds also present mechanical defenses, the presence of hard coats reducing the probability of predation (Janzen, 1969, 1971; Blate et al., 1998; Rodgerson, 1998; Alcántara et al., 2000). In the same way, a smooth surface could be of some use because seeds are harder to manage, jaws slip on biting, and the attachment of eggs is hindered (Janzen, 1969). Furthermore,

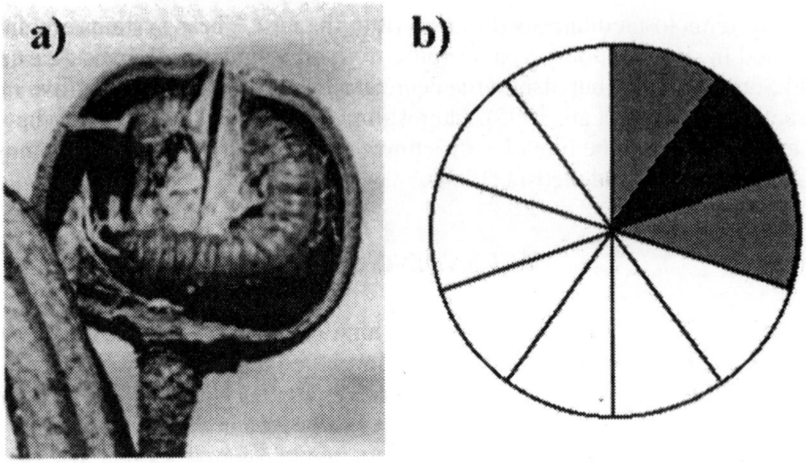

FIGURE 15.1. Fruit compartmentation and competition between seed predators.

the likelihood of being damaged when a mammal feeds on the fruit is reduced because hard and smooth seeds are easier to swallow.

There is also an interesting relationship between seed size and seed predation because

1. large seeds are usually preferred by predators (Janzen, 1969; Bradford and Smith, 1977; Westoby et al., 1992; Howe and Brown, 2001);
2. mechanical defenses of large seeds are more effective (Blate at al., 1998; Alcántara et al., 2000); and
3. they are more likely to survive attack by insects (Mack, 1998).

Other factors must also be considered, regardless of seed predation, which might also affect the fitness of different sized seeds. First, once the reproductive investment has been established, the larger the seeds the lower the number of them that can be produced (Smith and Fretwell, 1974). Second, the dispersal ability of seeds is usually inversely related to their size (Howe and Westley, 1986; Willson, 1992). Third, seedlings from large seeds are more likely to become successfully established (Gross and Kromer, 1986; Tripathi and Kahn, 1990; Leishman and Westoby, 1994).

Plants have numerous chemical substances that can work as defensive agents against herbivores (Harborne, 1977). The presence of these products in fruits and seeds may constitute an effective mechanism for reducing seed loss due to predation (Janzen et al., 1986; Singh and Singh, 1998; Krishnaveni et al., 1999; Ignacimuthu et al., 1999). The toxicity of these defensive substances must be considered to be a relative parameter. Toxicity depends on the amount of toxin ingested by a given animal weight unit in a given unit of time, the age and sanitary status of the subject, and the way the toxin is absorbed and excreted, relevant inter- and intraspecific differences in toxic tolerance also existing (Harborne, 1977).

It must be pointed out that, in order to constitute a valuable defense, a chemical does not need to kill a predator or cause severe poisoning. An unpleasant taste in seeds or a low nutritional value may also be a good defense system. This kind of plant-animal interaction could propitiate at least two different evolutionary processes:

Plants generate chemical substances, and seed predators acquire mechanisms that allow for detoxification.

Plants develop traits featuring fruit and seed toxicity (color, odor, taste, or a combination of these), and predators evolve efficient sensors to avoid feeding on toxic seeds.

The development of chemical substances has resulted in a great variety of products with different kind of effects on seed predators. Some chemicals alter insect hormonal balance (edycsone and juvenile hormone). Other products are toxic with different actuation mechanisms (cyanogenic glycosides, alkaloids, proteins), and others simply reduce fruit and seed palatability (curcubitacins) or diminish their nutritional quality (tannins). Three different chemical pathways are usually recognized:

1. Nitrogen-based compounds
2. Phenolic-derived substances
3. Terpenoids (Harborne, 1977)

In the nitrogen pathway there are nonprotein amino acids such as L-dopa and L-cyanoalanine, proteins such as abrin and phytohaemogglutinin, alkaloids such as ricinine, and cyanogenic compounds. These latter compounds are not toxic in themselves, but during digestive processes they produce cianhidric acid (HCN), which could result in the death of predators.

Among terpenoids, the more relevant ones are fitohormones, which are analogous to the juvenile hormone (interfering with insect development) and curcubitacins, which are triterpenoids with a bitter taste and strong deterrent effects (Harborne, 1977). Tannins are relevant phenolic compounds of astringent taste whose effects on predators are based on a decrease in the digestibility of proteins (Swain, 1979). Tannins combine with proteins at low pH, protecting them from the action of trypsin and other digestive enzymes. At high densities they can produce damage in the digestive epithelium (Hudson et al., 1971).

The evolutionary race to obtain toxins by plants and detoxification mechanisms by animals has resulted in two distinct patterns. Some plant families are specialized in the production of just one kind of chemical, for example crucifers with mustard oils (Pedras et al., 2002). However, other families have greatly diversified their toxic defenses, for instance legumes, which present amino acids, proteins, alkaloids, cyanogenic compounds, and isoflavones (Janzen, 1969; Oboh et al., 1998). Various defensive chemicals are sometimes present in seeds of the same species, which increase their defensive capabilities (Castellanos and Espinosa-García, 1997). This is the case of castor beans *(Ricinus communis)*. Seeds are protected inside dehiscent red-bright spiny capsules. The seed surface is smooth and shiny, and the color is grayish brown mottled with reddish brown—traits making seeds quite conspicuous. The defensive system of the seed is composed of irritant oil (40 to 50 percent of seed weight), a toxic alkaloid (ricinine), and a toxic protein (ricin). If a seed is swallowed after the coat has been broken, intoxication produces severe gastrointestinal irritation, anaphylaxis, and shock. The brightly colored pattern of fruits and seeds should facilitate recognition by predators.

The presence of bright colors and/or aposematic color patterns in conjunction with toxins is a well-documented phenomenon in animals (Mallet and Gilbert, 1995; Willson, 2000; Golding and Edmunds, 2000). In plant seeds and fruits protected by toxic compounds, this combination of traits is not uncommon either. For example, the black and red seeds of *Abrus precatorius* (Fabaceae) and *Argemone mexicana* (Papaveraceae) present highly toxic compounds to both vertebrates and invertebrates (Harborne, 1977; Singh and Singh, 1998). Furthermore, the relationship between the content of tannins and the color of the seed in wild legumes has also been reported (Oboh et al., 1998). Nevertheless, seed toxicity does not appear to be clearly related to aposematic coloration, and müllerian and batesian mimicry have never aroused much interest in the context of seed predation. Perhaps because insects are the main seed predators (Crawley, 1992) and their interpretation of color patterns is probably very different from that of vertebrates.

Phenologic defenses form the last group of the systems against seed predation. Specifically, they consist of variations in the temporal distribution of seed production within a reproductive season or among different years. For example, producing seeds before (or after) the best environmental conditions exist for predators may result in a fraction of the seed crop not being exposed to predation (Biere and Honders, 1996; Brody, 1997; Forget et al., 1999; Albrectsen, 2000).

The production of huge seed crops could decrease seed loss due to predator satiation. Predispersal seed predators are unable to consume all seeds before dispersal, and postdispersal seed predators are unable to consume all seeds before these germinate or are buried (De Steven, 1981; O'Dowd and Gill, 1984; Sperens, 1997; Curran and Web, 2000). In order to establish an effective system of defense, the growth of populations of seed predators must be sufficiently restrained in order to respond in a density-dependent manner. Large interannual variations in seed production have been reported in several species. Years of huge production in all individuals (masting), which satiate predators, alternate with years of low seed production, which starve predators. Although the existence of masting is acknowledged as an evolutive response to seed predation, several authors have suggested alternative explanations such as climatic conditions, pollination, dispersal by animals, or the need to obtain large resource reservoirs (Fenner, 1991; Haase et al., 1995; Isagi et al., 1997; Satake and Iwasa, 1999).

To conclude, seed defense consists of mechanical structures, chemical compounds, and phenological responses which, as seed production and seed predation, are structured along the hierarchical organization of plant anatomy and also across supra-individual levels (population, community).

A HIERARCHICAL PERSPECTIVE

Seed predation is a complex process generated at very different scales, both spatial and temporal, which originates a very heterogeneous pattern. Aspects such as the influence of ecological factors in seed production, selection of the reproductive material by the predator, pre- and postdispersal, or loss of plant fitness can be considered across a spectrum of spatial and temporal scales. Study of seed predation using the hierarchical paradigm enables us to better understand this type of process.

The great complexity and organization of biological systems has led us to study them over the past few decades from a more functional perspective, considering them as hierarchically structured systems. We are all familiar with the hierarchical conception of ecological structure (biosphere, biogeographical region, biome, landscape, ecosystem, community, population, organism) in which each level is comprised of a set of lower levels (Allen and Starr, 1982).

It is precisely in the consideration of a hierarchically structured organization of biological systems where the concept of the modular organism arises. One of the most characteristic representatives and, at the same time, one of the most abundant in modular organisms in the biosphere is found in green plants, although these can be represented in many other groups of heterotrophic organisms (sponges, hydrozoans, corals, bryozoans, ascidians, and other colonial organisms).

The relatively recent application of the concept of modular organisms to studies of vegetation has revealed the importance of modular organization as a decisive factor in the understanding of the biology of this type of organism, providing us with a greater interpretative capacity regarding their demography, growth, and evolution (Harper and White, 1974; Harper and Bell, 1979; White, 1979; Schmid, 1990).

Autotrophic organisms are comprised of a varied series of basic elements (modules), both vegetative (the leaf with its axillar bud, the cladode) and reproductive (flowers, fruits). The iterativity between these modules in the structure of the plant and the demographic dynamics themselves established in the subpopulation of modules enables these organisms to show great morphological and functional plasticity. This modularity allows the structural and reproductive characteristics among modular individuals belonging to one population to be incredibly different, their growth, reproduction, and evolution programs being unpredictable and very dependent on their interaction with the environment.

An interesting aspect of the modular structures is that the modular nature is frequently seen at more than one infra-individual level. Examples of this are the fruit, the branch upon which the fruits are arranged infrutescence, the systems of branches which repeat a characteristic pattern, the vegetative-reproductive fraction, shoot-root. Thus, levels of organization are seen below that of the individual, each one of these becoming new objects of study for disciplines such as botany, zoology, or ecology, without forgetting, of course, that all of these form part of a hierarchical structure. We could say that the plant is a modular organism made up of a population which is iterative with its own demography, which in turn is hierarchically linked to the individual demography and to the dynamics of the population, which is made up of modular individuals.

Indeed, other hierarchical levels, in this case, supra-individual ones, can be based on this modular structure. An example of this is the result of the asexual reproduction taking place in clonal plants in which the parent individual produces subunits with a certain autonomy called clonal elements ("ramet"). These subunits can become physically disconnected from the parental unit or remain connected, at least temporally, by means of underground rhizomes or aerial stolons. In the latter case, grouping together all the clonal units along with the parental one forms the so-called genetic unit ("genet"), leading to a new level of organization (van Groenendael and de Kroon, 1990; Huber, 1997; de Kroon and van Groenendael, 1997; Stuefer, 1997). In the higher plants, a similar type of organization is very common, the difference being that the individuals that remain grouped together are the result of sexual reproduction; that is, their origin is based on seeds of the same or of a different parental plant, the fusion of micorhyzal hyphae present in their root systems one of their most generalized connecting mechanisms.

It must be pointed out that, when the study is approached from the different hierarchical levels, we must consider that although each one of them has different attributes, their dynamics may be the result of their interactions, with both higher and lower levels. Within this concept, a functional approach is provided with regard to analyzing the processes by which the organisms present strategies for optimizing growth and reproduction, enabling basic rules to be established in order to determine the structure and development of the individuals and their flexibility to respond to external factors.

The relationship existing between the scale of observation and the hierarchical structure is obvious. In a given spatiotemporal scenario, what we see at one level and the processes involved therein may be different from the situation and the processes at another. For instance, seed predation in one plant species may be less variable (it may even remain constant in time, predation rate = 0) than that existing in individual plants of the same species or even between two branches of an individual. In this case, the higher hierarchical levels, such as the populations, show greater stability in their dynamics than the lower hierarchical levels, such as the individual components.

A very common case, which can be extrapolated to other trophic groups, is that herbivorous animals rarely feed on the whole plant. They usually select specific parts; an example of this is the high rate of consumption of seeds due to their high nutritional value. Study of the dynamics of the reproductive modules, such as fruits, should be approached by taking into consideration that they are individual units whose individual contribution to fitness ought to be evaluated, while acknowledging that seed production is hierarchically structured (fruits, infrutescence, individual, population), and, as a consequence, any process affecting this, such as predation by herbivores, should also be evaluated. Naturally included in this consideration is one of the processes that has the most influence on the fitness and population dynamics of plant species, such as predispersal seed predation to which much of the flora existing on the planet is subjected.

The density-dependent phenomena of seed predation have been demonstrated on numerous occasions. (Bernstein, 1975; Green and Palmbald, 1975; Rissing and Wheeler, 1976). The spatial and/or temporal distribution of the seeds has important implications in the foraging activities for predators, which exploit an aggregated resource, divided into discrete parcels. Evidently, the number of seeds produced by a population of individuals of a determined plant species can present great variability in space and in time, but this variability can also be expressed at very different levels of organization, both above and below the individual.

On one hand, internal ecological factors pertaining to the plant, such as age, size, or the distribution of resources among the different fractions it is composed of (photosynthetic structure and those of conduction, positioning, reproduction, and absorption), can lead to differential seed production at the different levels of organization existing. Modular growth is a clear expression of resource distribution strategies among structures of the organism. Translocations of nutrients among individuals (such as in clonal plants by means of rhyzomes or stolons, or in higher plants by means of hyphae of micorrhyzal fungi) or among parts of the individual (shoot-root, vegetative-reproductive fraction, only among reproductive fraction) can originate a differential production of reproductive material at different hierarchical levels. Seen from another point of view, the distribution can be established among structures with a different functionality or among units or structures with the same function within the hierarchical structure. A very generalized situation is that in which the modules that originate only the vegetative fraction or only the reproductive one produce more biomass of these fractions than those modules which originate both structures simultaneously; this suggests the existence of a partial limitation of resources for modular growth and the consequent differential distribution of these in the structures produced. Distribution of resources appears to bring about phenomena of inter- and intra-individual interactions.

On the other hand, external factors could lead to a differential production of reproductive material which could be encompassed within the spatiotemporal distribution of the different resources in the environment. An example of this are the processes of structural optimization, such as the different forms of growth, called "phalanx" and "guerilla," observed in clonal plants (Lovett Doust, 1981; Lovett Doust and Lovett Doust, 1982; Slade and Hutchings, 1987a,b; Hutchings, 1988) or in the systems of resource exploitation developed by granivorous ants (López, Acosta, and Serrano, 1994; López, Serrano, and Acosta, 1994; Acosta et al., 1995), where the modular demographic structure was presented as a consequence of the functional expression of their foraging strategies in respect to the abundance and distribution of resources and the inter- and intra-individual competitive and co-operative interactions.

Furthermore, the seed predation rate is also affected by a wide range of factors (Fenner, 1992) such as density of the plant population, density of the predator population, size and spatial distribution of seed production, availability of alternative food sources for generalist seed predators, pollination rate, phenology, or meteorological conditions.

It could be said that the different possible scenarios originated by differential seed production among the different levels of organization could have other consequences for the seed predation process itself at each of these levels, and for the process of phenotypical selection or for predispersal fitness, thus affecting the dynamics of the predator and plant populations.

The hierarchical consideration of the predator-plant interactions provides other interesting possibilities for analysis. One case which is very much related to the predispersal production and predation of seeds is the importance of predator satiation and the consequences of this for the predispersal fitness of the plant. The existence of different levels of organization implies different seed availability for the predator (Janzen, 1971), so that this defense mechanism of the plant and its inference in the spatiotemporal dynamics of the animal-plant interactions are expressed at different scales of analysis. The "satiation strategy" established in a plant population could be the consequence of different distributions of total seed production among the existing hierarchical levels. Thus, a situation of satiation can be based on different combinations:

1. a large number of plants producing fruits,
2. a large number of fruits produced per plant, and
3. a large number of seeds produced per fruit.

The simultaneous analysis of predation at different hierarchical levels within the plant population is a useful way to ascertain the relative importance of each of these components.

Although, as we have mentioned, the predation process of reproductive material in plants can be expressed at different spatial scales, this process could also be expressed in a temporal hierarchical structure.

Analysis of the modular dynamics and the effects of predation in a cycle of annual development of the plant (from the growth bud which will produce reproductive material to the recruitment of individuals, at the different stages of flower, fruit, and seed bank) reveals different hierarchical units or subpopulations of modules which contribute to the reproductive performance of the organism. Each subpopulation is therefore considered to descend from the preceding hierarchical level, that is, it contributes to the following level by means of its survival and its production. In this case, study of predation and its effects on reproductive modular demography, considering the different hierarchical levels, provides a valuable methodology for determining those states of development that are critical in order for the plant to be reproductively successful. Losses of reproductive material at one hierarchical level can be compensated for by means of an increase in the number of modules produced at other levels without causing any change in final reproductive performance. An example of this can be seen in the genus *Cistus* whose seeds have a low survival rate and whose plants must have a high production rate in order to compensate for the losses of the soil seed bank (Troumbis and Trabaud, 1986, 1987).

When the same hierarchy of organizational levels is considered, different functional responses in the seed production/predation process can be directly compared among years in a population, among several populations, or even among populations of different species. This functional analysis can be carried out by means of a contrast between the magnitude of heterogeneity in predation pressure and that of seed production across the levels. This provides an integrated picture of the interacting processes that shape the predispersal fitness outcome.

THE CASE OF CISTUS LADANIFER *L.*

Cistus ladanifer is a Mediterranean woody perennial species with two varieties: var. *albiflorus* Dun. and var. *maculatus* Dun., following an obligatory seeder strategy under the influence of recurrent fires. The fruits are lignified globular capsules with seven to ten valvae, which delimit the same number of internal compartments containing the seeds (26 to 173 seeds/compartment). Let us see in this case the occurrence of a differential seed production at different levels of organization and the interest of carrying out a hierarchical analysis.

In this species, Acosta et al. (1993) differentiate among five hierarchical levels with respect to reproductive material: subpopulations of varieties (white flower = var. *albiflorus* Dun., and spotted flower = var. *maculatus* Dun.), plants (structural or genetic individuals), branches (inter-related shoot modules), fruits, and compartments of a fruit (internal compartments containing the seeds), using the number of seeds as a unit of fitness (see Figure 15.2). These authors detect a phenotypical variability, expressed as different seed production at three levels of organization: among individuals, among branches, and among fruits, but not between varieties. The importance of this heterogeneity is that it establishes two essential conditions in order for phenotypical selection to operate: (1) a property—seed production—exists which varies among units at certain levels, which could originate differential predispersal seed predation, and (2) there is a relationship between this property and some aspects of fitness, the consequences of which would be that the genetic units could be differentially conveyed to successive generations.

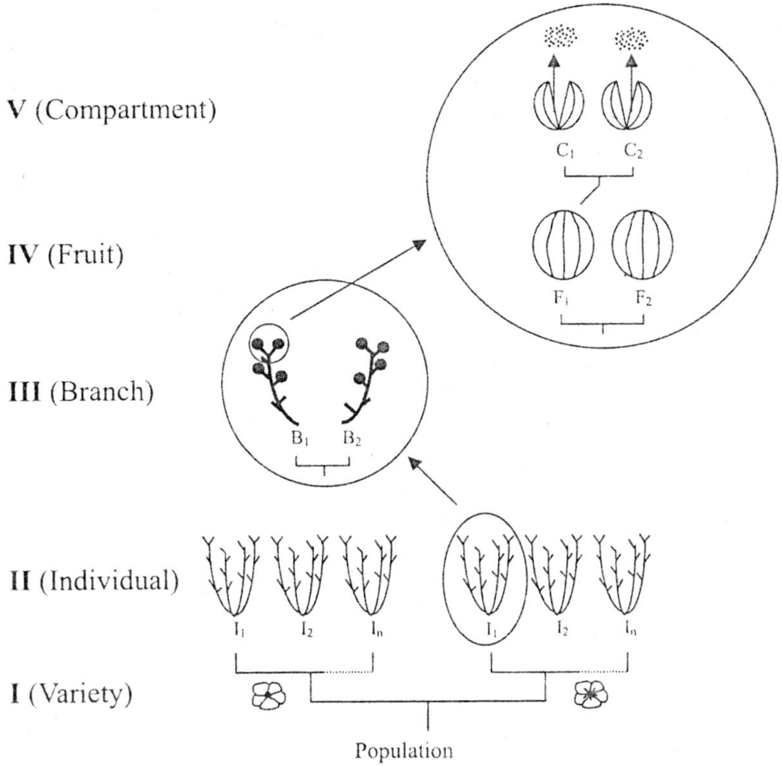

FIGURE 15.2. Hierarchical structure of the different levels considered in the seed production in a population of *C. ladanifer.* A nested analysis of variance enabled us to highlight three levels of heterogeneity of the reproductive output: inter-individual, inter-branch, and inter-capsule variation. I: Individuals, B: Branches, C: Capsules, V: Compartments separated by valvae.

Therefore, in modular organisms, and by extension, at supra-individual levels, the concept of natural selection reaches beyond the classical concepts of individual selection and group selection. It might very well be appropriate to consider the natural selection in this type of organism as a multilevel process (Heisler and Damuth, 1987) or as hierarchical selection (Tuomi and Vuorisalo, 1989; Eriksson and Jerling, 1990; Pedersen and Tuomi, 1995; Vuorisalo et al., 1997; among others), which encompasses components of fitness within and among groups, as well as individual and group effects which act upon the fitness of the individuals. Along the same lines of argument, it is obviously of interest to approach the study of seed predation within a hierarchical biological structure, which adds a new functional and explanatory dimension to our knowledge of this type of process. Within this perspective, let us analyze in this same species predispersal seed predation. In a population of this species, Serrano et al. (data unpublished) analyze the process of predispersal seed predation at three levels of organization (see Figure 15.3):

1. the population, where the proportion of individuals predated upon in the population is measured;
2. the individual, where the proportion of fruits predated upon in the individual is measured; and
3. the fruit, where the proportion of seeds predated upon in the fruit is measured.

Considering this hierarchical distinction, we can analyze the pressure exercised by predation through the different levels and through the degree of heterogeneity in the distribution of this pressure, which is a measurement of its selective effects. Furthermore, analysis by levels enables us to integrate the effects of predation at each level into a single process (a new function resulting from the multiplication of the partial functions on being independent estimates of the predation rate), which shows changes in fitness in the population as a whole. If we deal only with the final predation rate and not with its temporal changes, the hypothetical case could arise in which a population of this species suffers a predispersal seed predation of 100 percent of the fruits of each individual and, to the contrary, that the proportion of seeds predated upon per fruit is 10 percent; this situation appears to imply that the population as a whole saves a high number of seeds, and that there are two levels of organization which are very vulnerable to this type of predator. But the same percentage of seeds could be saved if 100 percent of the seeds of the fruit were predated upon, 100 percent of the fruits of the individual, and only 10 percent of the individuals of the population. In this example, we can see how in the population of *Cistus* a determined number of seeds can be saved by different intensities of predation to which the different levels of organization considered are subjected.

As we can see, a final predispersal fitness value can be reached with many different combinations of predation rates and also with many different temporal developments (faster or slower increases in predation rates, reaching identical values at the end of the fruiting season).

Analysis by levels provides greater explanatory power and operativity, in terms of both the temporal dynamics of the infestation process and the hierarchical structure itself in relation to appraisal of seed predation and its effects on the fitness of the population as a whole (in this case, predispersal).

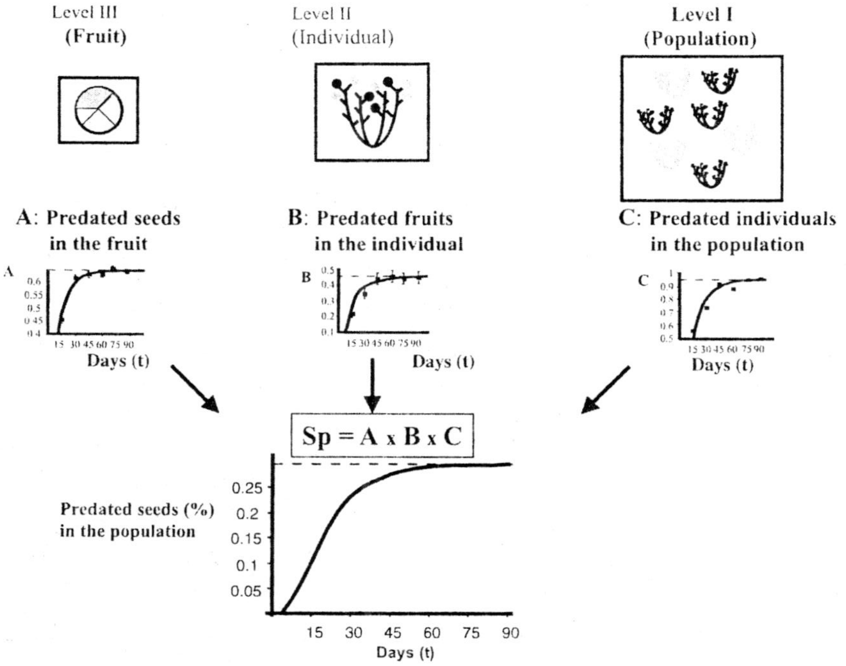

FIGURE 15.3. Seed predation at the different levels considered in the population of *C. ladanifer*. Level I: Individual plants in the population. Level II: Fruits in an individual plant. Level III: Compartments in a fruit. Effects of insect predation along the fruiting season, in terms of proportion of predated seeds in the population, is expressed as a single function (obtained by multiplying the functions at each level) which summarizes the simultaneous changes in predation of the different hierarchical levels in the population.

Application of the concept of modularity enables us to see how seed production and any process influencing this, such as predispersal seed predation, affects the fitness of the genetic individual. But consideration of a hierarchical structure opens up new possibilities for broader knowledge of reproduction and the phenomena related to fitness in modular organisms.

REFERENCES

Acosta, F.J., F. López, and J.M. Serrano (1995). Dispersed versus central-place foraging: Intra- and intercolonial competition in the strategy of trunk trail arrangement of a harvester ant. *American Naturalist* 145:389-411.

Acosta, F.J., J.M. Serrano, C. Pastor, and F. López (1993). Significant potential levels of hierarchical phenotypic selection in a woody perennial plant, *Cistus ladanifer*. *Oikos* 68:267-272.

Albrectsen, B.R. (2000). Flowering phenology and seed predation by a tephritid fly: Escape of seeds in time and space. *Ecoscience* 7:433-438.

Alcántara, J.M., P.J. Rey, A.M. Sánchez-Lafuente, and F. Valera (2000). Early effects of rodent post-dispersal seed predation on the outcome of the plant-seed disperser interaction. *Oikos* 88:362-370.

Allen, T.F.H. and T.B. Starr (1982). *Hierarchy: Perspectives for ecological complexity*. Chicago: Chicago University Press.

Anapol, F. and S. Lee (1994). Morphological adaptation to diet in platyrrhine primates. *American Journal of Physical Anthropology* 94:239-261.

Bernstein, R.A. (1975). Foraging strategies of ants in response to variable food density. *Ecology* 56:213-219.

Biere, A. and S.J. Honders (1996). Impact of flowering phenology of *Silene alba* and *S. dioica* on susceptibility to fungal infection and seed predation. *Oikos* 77:467-480.

Blate, G.M., D.R. Peart, and M. Leighton (1998). Post-dispersal predation on isolated seeds: A comparative study of 40 tree species in a Southeast Asian rainforest. *Oikos* 82:522-538.

Bradford, D.F. and C.C. Smith (1977). Seed predation and seed number in *Scheelea* palm fruits. *Ecology* 58:667-673.

Brody, A.K. (1997). Effects of pollinators, herbivores, and seed predators on flowering phenology. *Ecology* 78:1624-1631.

Castellanos, I. and F.J. Espinosa-García (1997). Plant secondary metabolite diversity as a resistance trait against insects: A test with *Sitophilus granarius* (Coleoptera: Curculionidae) and seed secondary metabolites. *Biochemical and Systematics Ecology* 25:591-602.

Crawley, M.J. (1992). Seed predators and plant population dynamics. In M. Fenner (ed.), *Seeds: The ecology of regeneration in plant communities* (pp. 157-191). Wallingford, UK: C.A.B. International.

Cunningham, S.A. (1997). Predator control of seed production by a rain forest understory palm. *Oikos* 79:282-290.

Curran, L.M. and C.O. Webb (2000). Experimental tests of the spatiotemporal scale of seed predation in mast-fruiting Dipterocarpaceae. *Ecological Monographs* 70:129-148.

de Kroon, H. and J. Van Groenendael (eds.) (1997). *The ecology and evolution of clonal plants*. Leiden, the Netherlands: Backhuys Publishers.

De Steven, D. (1981). Predispersal seed predation in a tropical shrub (*Mabea occidentalis*, Euphorbiaceae). *Biotropica* 13:146-150.

Eriksson, O. and L. Jerling (1990). Hierarchical selection and risk spreading in clonal plants. In J. van Groenendael and H. de Kroon (eds.), *Clonal growth in plants: Regulation and function* (pp. 79-94). The Hague, the Netherlands: SPB Academic Publishing.

Fenner, M. (1991). Irregular seed crops in forest trees. *Quarterly Journal of Forestry* 85:166-172.

Fenner, M. (ed.) (1992). *Seeds: The ecology of regeneration in plant communities*. Wallingford, UK: C.A.B. International.

Fishman, J.R. and R.J. Orth (1996). Effects of predation on *Zostera marina* L. seed abundance. *Journal of Experimental Marine Biology and Ecology* 34:11-26.

Forget, P.M., K. Kitajima, and R.B. Foster (1999). Pre- and post-dispersal seed predation in *Tachigali versicolor* (Caesalpiniaceae): Effects of timing of fruiting and variation among trees. *Journal of Tropical Ecology* 15:61-81.

Fuentes, M. and E.W. Schupp (1998). Empty seeds reduce seed predation by birds in *Juniperus osteosperma*. *Evolutionary Ecology* 12:823-827.

Godnan, D. (1983). *Pest slugs and snails*. Berlin: Springer-Verlag.

Golding, T.C. and M. Edmunds (2000). Behavioural mimicry of honeybees (*Apis mellifera*) by droneflies (Diptera: Syrphidae: *Eristalis* spp). *Proceedings of the Royal Society of London, Series B: Biological Sciences* 267:903-909.

Green, T.W. and I.G. Palmbald (1975). Effects of insect seed predators on *Astragalus utahensis* (Leguminosae). *Ecology* 56:1435-1440.

Gross, K.L. and M.L. Kromer (1986). Seed weight effects on growth and reproduction in *Oenothera biennis* L. *Bulletin of the Torrey Botanical Club* 113:252-258.

Haase, P., F.I. Pugnaire, and L.D. Lincoll (1995). Seed production and dispersal in the semi-arid tussock grass *Stipa tenacissima* L. during masting. *Journal of Arid Environment* 31:55-65.

Happel, R. (1988). Seed-eating by West African cercopithecines, with reference to the possible evolution of bilophodont molars. *American Journal of Physical Anthropology* 75:303-307.

Harborne, J.B. (1977). *Introduction to ecological biochemistry*. London: Academic Press.

Harper, J.L. and A.D. Bell (1979). The population dynamics of growth form in organisms with modular construction. In R.M. Anderson (ed.), *Population dynamics* (pp. 29-52). 20th Symposium of the British Ecological Society. Oxford: Blackwell.

Harper, J.L., P.H. Lovell, and K.G. Moore (1970). The shapes and sizes of seeds. *Annual Review of Ecology and Systematics* 1:327-356.

Harper, J.L., and J. White (1974). The demography of plants. *Annual Review of Ecology and Systematics* 5:419-463.

Heisler, I.L. and J. Damuth (1987). A method for analyzing selection in hierarchically structured populations. *American Naturalist* 130:582-602.

Herrera, C.M. (1984). Seed dispersal and fitness determination in wild rose: Combined effects of hawthorn, birds, mice and browsing ungulates. *Oecologia* 63:386-393.

Herrera, C.M. (1992). Interespecific variation in fruit shape: Allometry, phylogeny, and adaptation to dispersal agents. *Ecology* 73:1832-1841.

Holbrook, S.J., D.C. Reed, K. Hansen, and C.A. Blanchette (2000). Spatial and temporal patterns of predation on seeds of the surfgrass *Phyllospadix torreyi*. *Marine Biology* 136:739-747.

Holland, J.N. and T.H. Fleming (1999). Mutualistic interactions between *Upiga virescens* (Pyralidae), a pollinating seed-consumer, and *Lophocerus schottii* (Cactaceae). *Ecology* 80:2074-2084.

Howe, H.F. and J.S. Brown (2001). The ghost of granivory past. *Ecology Letters* 4:371-378.

Howe, H.F. and L.C. Westley (1986). Ecology of pollination and seed dispersal. In M.J. Crawley (ed.), *Plant ecology* (pp. 185-215). Oxford: Blackwell Scientific Publications.

Huber, H. (1997). Architectural plasticity of stoloniferous and erect herbs in response to light climate. Ph D thesis, Utrecht University, the Netherlands.

Hudson, S.N., R.L. Levin, and D.H. Smith (1971). Absorption from the alimentary tract. In D.J Bell and B.M. Freeman (eds.), *Physiology and biochemistry of the domestic fowl*, Volume 1 (pp. 51-71). London: Academic Press.

Hulme, P.E. (1997). Post-dispersal seed predation and the establishment of vertebrate dispersed plants in Mediterranean scrublands. *Oecologia* 111:91-98.

Hulme, P.E. and T. Borelli (1999). Variability in post-dispersal seed predation in deciduous woodland: Relative importance of location, seed species, burial and density. *Plant Ecology* 145:149-156.

Hutchings, M.J. (1988). Differential foraging for resources and structural plasticity in plants. *Trends in Ecology and Evolution* 3:200-204.

Ignacimuthu, S., S. Janarthanan, and B. Balachandran (1999). Chemical basis of resistance in pulses to *Callosobruchus maculatus* (F.) (Coleoptera: Bruchidae) *Journal of Stored Products Research* 36:89-99.

Isagi, Y., K. Sugimura, A. Sumida, and H. Ito (1997). How does masting happen and synchronize? *Journal of Theoretical Biology* 187:231-239.

Janzen, D.H. (1969). Seed eaters vs. seed size, number, toxicity, and dispersal. *Evolution* 23:1-27.

Janzen, D.H. (1971). Seed predation by animals. *Annual Review of Ecology and Systematics* 2:465-492.

Janzen, D.H., C.A. Ryan, I.E. Liener, and G. Pearce (1986). Potentially defensive proteins in mature seeds of 59 species of tropical Leguminosae. *Journal of Chemical Ecology* 12:1469-1480.

Kinzey, W.G. (1992). Dietary and dental adaptations in the Pitheciinae. *American Journal of Physical Anthropology* 88:499-514.

Kollmann, J. and S. Bassin (2001). Effects of management on seed predation in wildflower strips in northern Switzerland. *Agriculture Ecosystems and Environment* 83:285-296.

Koptur, S. (1989). Effect of seed damage on germination in the common vetch (*Vicia sativa* L.). *American Midland Naturalist* 140:393-396.

Krishnaveni, S., G.H. Liang, S. Muthukrishnan, and A. Manickam (1999). Purification and partial characterization of chitinases from sorghum seeds. *Plant Science* 144:1-7.

Kubitzki, K. and A. Ziburski (1994). Seed dispersal in flood plain forests of Amazonia. *Biotropica* 26:30-43.

Leishman, M.R. and M. Westoby (1994). The role of seed size in seedling establishment in dry soil conditions: Experimental evidence from semi-arid species. *Journal of Ecology* 82:249-258.

Linzey, A.V. and K.A. Washok (2000). Seed removal by ants, birds and rodents in a woodland savanna habitat in Zimbabwe. *African Zoology* 35:295-299.

López, F., F.J. Acosta, and J.M. Serrano (1994). Guerilla vs. phalanx strategies of resource capture: Growth and structural plasticity in the trunk trail system of the harvester ant *Messor barbarus*. *Journal of Animal Ecology* 63:127-138.

López, F., J.M. Serrano, and F.J. Acosta (1994). Parallels between the foraging strategies of ants and plants. *Trends in Ecology and Evolution* 9:150-153.

Lovett Doust, L. (1981). Intraclonal variation and competition in *Ranunculus repens*. *New Phytologist* 89:495-502.

Lovett Doust, L. and J. Lovett Doust (1982). The battle strategies of plants. *New Scientist* 91:81-84.

Mack, A.L. (1998). An advantage of large seed size: Tolerating rather than succumbing to seed predators. *Biotropica* 30:604-608.

Mack, A.L. (2000). Did fleshy fruit pulp evolve as a defence against seed loss rather than as a dispersal mechanism? *Journal of Biosciences* 25:93-97.

Mallet, J. and L.E. Gilbert Jr. (1995). Why are there so many mimicry rings? Correlations between behaviour and mimicry in *Heliconius* butterflies. *Biological Journal of the Linnean Society* 55:159-180.

Nakaoka, M. (2002). Predation on seeds of seagrasses *Zostera marina* and *Zostera caulescens* by a tanaid crustacean, *Zeuxo* sp. *Aquatic Botany* 72:99-106.

Oboh, H.A., M. Muzquiz, C. Burbano, C. Cuadrado, M.M. Pedrosa, G. Ayet, and A.V. Osagie (1998). Antinutritional constituents of six underutilized legumes grown in Nigeria. *Journal of Chromatography A* 823:307-312.

O'Dowd, D.J. and A.M. Gill (1984). Predator satiation and site alteration following fire: Mass reproduction of alpine ash (*Eucalyptus delegatensis*) in Southern Australia. *Ecology* 65:1052-1066.

Pedersen, B. and J. Tuomi (1995). Hierarchical selection and fitness in modular and clonal organisms. *Oikos* 73:167-180.

Pedras, M.S.C., C.M. Nycholat, S. Montaut, Y. Xu, and A.Q. Kahn (2002). Chemical defenses of crucifers: Elicitation and metabolism of phytoalexins and indle-3-acetonitrile in brown mustard and turnip. *Phytochemistry* 59:611-625.

Rissing, S.W. and J. Wheeler (1976). Foraging responses of *Veromessor pergandei* to changes in seed production. *Pan Pacific Entomologist* 52:63-72.

Rodgerson, L. (1998). Mechanical defense in seeds adapted for ant dispersal. *Ecology* 79:1669-1677.

Satake, A. and Y. Iwasa (1999). Pollen coupling of forest trees: Forming synchronized and periodic reproduction out of chaos. *Journal of Theoretical Biology* 203:63-84.

Schmid, B. (1990). Some ecological and evolutionary consequences of modular organization and clonal growth in plants. *Evolutionary Trends in Plants* 4:25-34.

Serrano, J.M., J.A. Delgado, F. López, F.J. Acosta, and S.G. Fungairiño (2001). Multiple infestation by seed predators: The effect of loculate fruits on intraespecific insect larval competition. *Acta Oecologica* 22:153-160.

Singh, S. and D.K. Singh (1998). Mulluscicidal activity of *Abrus precatorius* Linn. and *Argemone mexicana* Linn. *Chemosphere* 38:3319-3328.

Slade, A.J. and M.J. Hutchings (1987a). Clonal integration and plasticity in foraging behaviour in *Glechoma hederacea*. *Journal of Ecology* 75:1023-1036.

Slade, A.J. and M.J. Hutchings (1987b). The effects of nutrient availability on foraging in the clonal herb *Glechoma hederacea*. *Journal of Ecology* 75:95-112.

Smith, C.C. and S.D. Fretwell (1974). The optimal balance between size and number of offspring. *American Naturalist* 108:499-506.

Sperens, U. (1997). Fruit production in *Sorbus aucuparia* L (Rosaceae) and pre-dispersal seed predation by the apple fruit moth (*Argyresthia conjugella* Zell). *Oecologia* 110:368-373.

Stuefer, J.F. (1997). Division of labour in clonal plants? On the response of stoloniferous herbs to environmental heterogeneity. PhD thesis, Utrecht University, the Netherlands.

Swain, T. (1979). Tannins and lignins. In G.A. Rosenthaland and D.H. Janzen (eds.), *Herbivores: Their interaction with secondary plant metabolites* (pp. 657-682). New York: Academic Press.

Thompson, J.N. (1985). Postdispersal seed predation in *Lomatium* spp. (Umbeliferae): Variation among individuals and species. *Ecology* 66:1608-1616.

Traveset, A. (1990). Post-dispersal predation of *Acacia farnesiana* seeds by *Stator vachelliae* (Bruchidae) in Central America. *Oecologia* 84:506-512.

Tripathi, R.S. and M.L. Kahn (1990). Effects of seed weight and microsite characteristics on germination and seedling fitness in two species of *Quercus* in a subtropical wet hill forest. *Oikos* 57:289-296.

Troumbis, A. and L. Trabaud (1986). Comparison of reproductive biological attributes of two *Cistus* species. *Oecologia Plantarum* 7:235-250.

Troumbis, A. and L. Trabaud (1987). Dynamique de la banque de graines de deux espècies de Cistes dans les maquis grecs. *Oecologia Plantarum* 8:167-179.

Tuomi, J. and T. Vuorisalo (1989). Hierarchical selection in modular organisms. *Trends in Ecology and Evolution* 4:209-213.

van Groenendael, J. and H. de Kroon (eds.) (1990). *Clonal growth in plants: Regulation and function*. The Hague, the Netherlands: SPB Academic Publishing.

Vuorisalo, T., J. Tuomi, B. Pedersen, and P. Kaar (1997). Hierarchical selection in clonal plants. In H. de Kroon and J. van Groenendael (eds.), *The ecology and evolution of clonal plants* (pp. 243-261). Leiden, the Netherlands: Backhuys Publishers.

Westoby, M., E. Jurado, and M. Leishman (1992). Comparative ecology of seed size. *Trends in Ecology and Evolution* 11:368-372.

White, J. (1979). The plant as a metapopulation. *Annual Review of Ecology and Systematics* 10:109-145.

Willson, M.F. (1992). The ecology of seed dispersal. In M. Fenner (ed.), *Seeds: The ecology of regeneration in plant communities* (pp. 61-85). Wallingford, UK: C.A.B. International.

Willson, K. (2000). How the locust got its stripes: The evolution of density-dependent aposematism. *Trends in Ecology and Evolution* 3:88-90.

Youti, B.A. and R.F. Miller (1986). Insect predation on *Astragallus filipes* and *A. purshii* seeds. *Northwest Science* 60:42-46.

Chapter 16

Natural Defense Mechanisms in Seeds

Gregory E. Welbaum

INTRODUCTION

Seeds contain stored protein, lipids, and carbohydrates to support seedling growth. These materials are attractive to a wide range of pests and pathogens, including specialized fungal pathogens, which require a specific host, to generalist feeders such bacteria, mammals, birds, and insects. Although the ecology of seed survival has been well studied (Baskin and Baskin, 1998), only recently have we begun to fully understand some of the natural mechanisms seeds employ to protect themselves against insects, fungi, bacteria, birds, and other predators. There has been a perception that seeds are extremely vulnerable, particularly to fungal attack, and for agricultural use chemical treatment is necessary to ensure successful establishment. Although the diversity and severity of disease pressure on agricultural seeds is not questioned, what is becoming increasingly apparent is that seeds are protected by elaborate and sometimes complex natural defense mechanisms. Defense strategies may take many forms. Some seed structures, such as the seed coat, provide physical protection, similar to a coat of armor, to protect against a wide range of predators. Other seeds have chemical defenses that actively protect against specific predators. In general, a plant molecule is considered to be part of a defensive mechanism against pathogenic microorganisms when (1) its synthesis is induced in response to pathogen challenge; (2) its expression level is dependent on specific race-cultivar interactions; and/or (3) it shows antimicrobial activity either in vivo or, more often, in fungal or bacterial growth inhibition assays (Kombrink and Somssich, 1995). An evolving field in seed biology is the study of organisms that provide biological control. These organisms are naturally associated with or sometimes applied to protect seeds. It is common for the

I thank Maura Wood, Donald Mullins, Tony Sturz, and Jerry Nowak for their help in preparing this manuscript.

seeds of a single species to combine a battery of defense responses to provide protection against broad range of pests and pathogens (Shewry, 2000). Barley, for example, contains thionin, endochitinases, ribosomal-inactivating proteins, β-glucanase, nonspecific lipid transfer protein, lectin, peroxidase, thaumatin-related antifungal proteins and inhibitors, or α-amylase and proteinases (Shewry and Lucas, 1997). Despite the presence of these defense compounds, barley is palatable to mammals and other animals and does not require processing prior to human consumption. Cases of defense proteins having a negative effect on human nutrition are fortunately the exception rather than the rule, despite the growing recognition that defense compounds are present in most seeds. Natural seed defenses have been manipulated and exploited commercially to protect seeds and improve their establishment. This trend will only continue as the knowledge of seed defense mechanisms increases. The following are some of the more widely recognized defense mechanisms that seeds have evolved.

SEED DEFENSE MECHANISMS

Avoidance

Seeds vary greatly in their rate of germination both among species and within individual seed lots. The longer seeds remain in the soil ungerminated, the greater their vulnerability to pathogenic attack or predation by animals or insects becomes. Seeds that germinate rapidly and establish healthy seedlings pass through vulnerable stages quickly, minimizing exposure to attack by seed scavengers and also from certain seed and seedling diseases. Seed priming, a controlled hydration process followed by redrying which often increases germination rate, is credited with reducing susceptibility to seed diseases, because the seeds are vulnerable to attack for a shorter period of time compared to slower germinating unprimed seeds (Welbaum et al., 1998). From an agricultural perspective, using seeds of the highest possible vigor level not only results in more rapid establishment, but also decreases losses due to soilborne diseases or seed predators by reducing the time the ungerminated seeds are susceptible to attack in the soil.

Physical Defense

In some seeds, specialized tissues provide a physical barrier to protect against pathogenic attack. Thick seed coats, some with lignin or trichomes for example, provide formidable barriers that inhibit predation by insects and animals (Bewley and Black, 1994; Baskin and Baskin, 1998). The structure and architecture of some seed coats likely delays or inhibits the growth of hyphae as well.In muskmelon (*Cucumis melo* L.) seeds, microbial growth around the seed during imbibition is inhibited because nutrient leakage from the embryo is confined by callow deposits that create a semi-permeable endosperm cell wall (Welbaum and Bradford, 1990; Yim and Bradford, 1998). It is also likely that the thick cell walls of the endosperm tissue provide a physical barrier to slow the penetration of fungal hyphae (Welbaum and Bradford, 1998). Endosperm layers with characteristics similar to muskmelon have been reported in lettuce *(Lactuca sativa)* (Hill and Taylor, 1989) and exist in other cucurbits (G. E. Welbaum, unpublished results).

Biological Control

Biological control uses other living organisms to modify the agricultural ecosystem to control disease or to prevent the establishment of a pest (Dowling and O'Gara, 1994). Some of the organisms that are used for biological control of plants are free living bacteria or fungi in the rhizosphere (e.g., the root surface and the soil directly influenced by the root system). Other biological control organisms are endophytes that exist inside the plant that may or may not be seed transmitted. In some circles, the term endophyte has been applied almost exclusively to fungi, including mycorrhizal fungi (Sturz and Nowak, 2000). A more comprehensive definition is one which includes fungi or bacteria that for part or all of their life cycle invade the tissues of living plants and cause unapparent and asymptomatic infections entirely within plant tissues, but cause no disease symptoms (Wilson, 1995). Bacteria isolated from the internal plant tissues of healthy plants comprise over 129 species representing more than 54 genera, with *Pseudomonas, Bacillus, Enterobacter,* and *Agrobacterium* being the most commonly isolated genera (Hallmann et al., 1997). Early reports regarded endophytes as contaminants resulting from incomplete surface sterilization or latent pathogens. However, recent research has demonstrated that bacterial endophytes can improve growth and reduce disease symptoms caused by several plant pathogens (Hallmann et al., 1997).

A thriving and stable microbial endoplant community may benefit host plants in several ways. Beneficial microbes may increase disease resistance, de novo synthesis of fungitoxic metabolites at sites of attempted fungal penetration is increased (Benhamou et al., 1996), systemic acquired resistance (SAR) is induced (Görlach et al., 1996; Richards, 1997; Sticher et al., 1997; Nowak, 1998), or other organisms are excluded by niche competition (Sturz and Nowak, 2000). Studies have shown that bacterialized plantlets grow faster than unbacterialized ones (Chanway, 1997; Bensalim et al., 1998), but are also sturdier, have a better developed root system (Nowak, 1998), and better withstand biotic stress and low-level disease pressures (Sharma and Nowak, 1998). Because microbes are amenable to genetic engineering, it is possible to optimize the beneficial effects of biological control organisms through genetic enhancement of the microbe or the plant (Sturz and Nowak, 2000).

Use of microbes and microbe-based agents to control the spread and severity of crop diseases in agriculture is incresing. However, many biological control mechanisms are involved in plant disease prevention, and many of these are only now becoming recognized and understood. The bacteria *Pseudomonus*, for example, produces metabolites that are implicated in the control of damping-off, flax wilt, and take-all diseases (Dowling and O'Gara, 1994). Although exoroot bacteria or mycorrhiza are often applied to plants as spray treatments, seeds can be used as a delivery system as well.

Less is known about the interaction of organisms that provide biological control and seeds. Applying bacterial seed treatments prior to planting does not ensure the establishment of a beneficial endo or exorhizal flora (Frommel et al., 1993), nor does it always enhance yield (Volkmar and Bremer, 1998). Introductions of nonlocal microfloras must compete with established microbial communities in the soil, in the rhizosphere, and within the plant in the case of endophytes. Both true seed and vegetative plant material are likely to carry a wide range of adapted endophytes, a portion of which will transferred to the subsequent progeny (Mundt and Hinkle, 1976; Holland and Polacco, 1994; Sturz and Nowak, 2000). Selections of endophytes can be introduced into the host plant through tissue culture (Varga et al., 1994; Nowak, 1998). The importance of seeds as a source of endophytic bacteria remains controversial (Hallman et al., 1997). Although endophytic bacteria have been microscopically detected in seeds, Mundt and Hinkle (1976) obtained endopytes only from seeds that had been physically damaged. Seed endophytes are mainly associated with protected sites near the seed coat, and they may have colonized this region through small openings in the coat. This hypothesis is supported by differences in the number of endophytic bacteria isolated from surface-disinfested seeds of cotton ranging from 1×10^3 to 1×10^5 CFU/g, while sweet corn populations were below 1×10 CFU/g (McInroy and Kloepper, 1995). These differences were explained by the rough grooves in the seed coat of cotton which are damaged during the ginning process to remove fiber from the seed, thus allowing access to bacteria. The sweet corn seed coat is smooth and intact and provides a more complete enclosure. Extremely low population densities may be present deeper in seed tissues, explaining the common observation that endophyte populations increase from nondetectable to detectable levels in seeds within

48 h after placing surface-disinfected seeds on water agar (Hallmann et al., 1997). Such bacterial endophyte populations were either below detectable limits or in a physiological state that was not culturable (Adams and Kloepper, 1996). Additional cytological research is needed to conclusively prove whether endophytes are seed transmissible and to determine more closely the specific sites of colonization within the seed.

Bacterial endophytes introduced as seed treatments on the seed coat colonized internal tissues of radicles shortly after emergence (Hallmann et al., 1997). In these studies, endophytic migration apparently began as the bacteria moved through the germination slit into the starchy endosperm. From the endosperm, the bacteria colonized the radicle and coleoptile before spreading systemically through the entire plant (Hinton and Bacon, 1995).

Exobacteria are often added to seeds in coating materials or during seed priming (Welbaum et al., 1998). Biopriming (a combination of preplant seed hydration and biological seed treatment) is especially useful for crops that are susceptible to imbibitional chilling injury such as lima bean (*Phaseolus lunatus* L.) and sweet corn (*Zea mays* L.) (Bennett et al., 1992; Khan, 1992). Biological agents may also be combined with seed during a priming process using aerated, dilute salt solutions. Tomato (*Lycopersicon esculentum* Mill.) seeds were bio-osmoprimed in aerated –0.8 MPa $NaNO_3$ at 20°C for 7 days with nutrient broth and *Pseudomonas aureofaciens* (bacterial strain AB254) added on day four (Warren, 1997). Bio-osmopriming allows rapid seed colonization by the beneficial organism(s) and more uniform coverage of the seed surface compared to other application techniques (Smith, 1996; Warren, 1997). Drum priming technology (Rowse, 1996; Warren and Bennett, 1997) also allows concurrent seed hydration and application/colonization of beneficial microbes as well. Biological seed treatments are apparently best suited to augment established consortia of microbial organisms (fungal, bacterial, and mycorrhizal) created as part of a long-term strategy of harmonizing crop (cultivar) selection and management practices (Sturz and Nowak, 2000).

Systemic Acquired Resistance

Induced systemic acquired resistance (SAR) is the ability of a bacterium, chemical, insect, or virus to induce broad plant defense mechanisms that lead to the systemic resistance of a number of pathogens (Sticher et al., 1997). Salicylic acid and its derivatives have been used extensively in agriculture to induce SAR (Görlach et al., 1996; Sticher et al., 1997). Attempts to induce SAR in ungerminated seeds by priming or treatment with SAR-inducing agents has produced mixed results (Shen and Welbaum, 1999).

Generally, it does not appear that adding salicylic acid or its derivatives to priming solutions can induce SAR in ungerminated seeds. However, SAR can be induced in seedlings after radicle protrusion (Shen and Welbaum, 1999). It remains to be determined whether SAR-inducing compounds added to seed coating material can effectively induce SAR in small seedlings after radicle emergence.

Chemical Defense: Secondary Metabolites

Seeds contain many minor constituents that cannot be regarded as storage components because they are not used during germination or seedling growth. Many of these compounds are inhibitors of insect and animal predation. Alkaloids, such as theobromine, caffeine, strychnine, brucine, and morphine, are natural compounds that protect seeds from insect or animal predation, and many of these compounds have been utilized as a source of drugs and stimulants for humans (Bewley and Black, 1994). Phytosterols such as sitosterols and stimasterols are present in some seeds such as soybean. Stimasterols are interesting because they are broken to produce the animal steroid hormone progesterone (Bewley and Black, 1994). Glucosides are bitter-tasting components in some seeds, e.g., amygdalin in almonds, peaches, and plums, as well as esculin from horse chestnut. Some of these compounds, such as saponin in tung seeds, may be deadly to humans and animals.

Phenoiic compounds such as coumarin and chlorogenic acid and their derivatives as well as compounds such as ferulic, caffeic, and sinapic acids occur in the coats of many seeds. These compounds are germination inhibitors that maintain quiescence in the seeds that contain them and leach out into the soil to inhibit germination of neighboring seeds (Bewley and Black, 1994).

Protective Proteins: Enzyme Inhibitors

Proteinase inhibitors are probably the most common group of defense proteins found in seeds. Over 20,000 papers have been published on proteinase inhibitors (Shewry, 2000). Inhibitors of serine proteases are widely distributed. Of the 12 families of plant proteinases, all except three have been detected in seeds, and all but one of these are serine proteinases (Shewry, 2000). Proteinase inhibitors are particularly abundant in legume seeds. In soybean seeds, for example, two classes of serine proteinase inhibitors have antinutritional effects in feeding trials unless inactivated by processing (Domoney, 1999). Cysteine proteinases are sometimes found in seeds, but their quantities are generally lower than serine proteinases and it is unclear whether they play a protective role in vivo. In vitro tests and expression in transgenic plants show that cysteine proteinases provide at least limited resistance to insects, nematodes, and mollusks that use cysteine endoproteinases for digestion (Shewry, 1999). Cereal seeds contain a different type of proteinase inhibitor than legumes (Carbonero and Garcia-Olmedo, 1999). Prolamins are a class of proteinases that include forms that inhibit trypsin, nonplant α-amylases, or both enzymes at separate inhibitory sites. Other families of proteinase inhibitors inhibit α-amylases, and some α-amylase inhibitors apparently constitute separate families (Shrewry, 1999). A major α-amylase inhibitor from bean confers insect resistance when expressed in peas (Domoney, 1999). In general, plant inhibitors of α-amylases and proteinases are low M_r proteins (<20,000) with tightly folded structures that are stabilized by infra-chain disulfide bonds, but exceptions to this general rule also occur (Shewry, 2000).

Lectins

Lectins are widely distributed in plant tissues including seeds and may have antinutritional effects unless inactivated by processing (Peumans and Van Damme, 1999). Seed lectins have been shown to kill or inhibit certain insect pests either in feeding studies or in transgenic plants. The mode of action for lectins is not fully understood but may involve impaired function of membranes in the digestive tract (Shewry, 2000).

Chitinase and Plant Defense

Another group of proteins that have received much research attention recently for their possible role in seed defense are endochitinases and β-1,3-glucanases which often, but not always, occur together (Gomez et al., 2002; Shewry, 2000). Chitin is an insoluble polysaccharide in many organisms. It is a component of fungal cell walls and also the cuticles of insects and nematodes. Chitin has not been found in higher plants or mammals, but plants and mammals produce a wide array of chitinases (EC 3.2.1.14) that are homologous to the ones in chitin-containing organisms. Some plant chitinases are expressed only during specific developmental stages such as seed maturation or flower and fruit development. These so-called constitutive chitinases have received less research attention compared with their stress-inducible homologues, even though constitutive chitinases are typically expressed in nonvegetative organs such as seeds (Neale et al., 1990; Leah et al., 1991; Collada et al., 1992; Cordero et al., 1994; Takakura et al., 2000).

Without excluding other functions, the major natural role for plant chitinases from vegetative tissues is defense against pathogens, either directly through their antifungal properties or indirectly through the release of chitin oligomers capable of eliciting plant defensive responses. Chitinase expression is often induced by microbial attack, causing many chitinases to be classified as pathogenesis-related proteins of the PR-3, PR-4, PR-8, and PR-11 families (Bol et al., 1990; Neuhaus et al., 1996; Fritig et al., 1998; Neuhaus, 1999). The in vitro antifungal activity of chitinases, either alone or in combination with other PR proteins, has been known for many years (Schlumbaum et al., 1986; Broekaert et al., 1988; Mauch et al., 1988; Boller, 1993). Chitinases are also synthesized in response to oligosaccharide elictors from microbial or plant cell walls and after treatments with defense-initiating compounds such as ethylene and salicylic acid as part of a SAR response as described previously (Boller, 1988; Fritig et al., 1998). The creation of transgenic plants expressing chitinase has demonstrated their antifungal role in plant defense (Broglie et al., 1991; Vierheilig et al., 1993; Zhu et al., 1994; Grison et al., 1996). Coexpression of barley chitinase and other defense proteins in transgenic tobacco increased quantitative resistance to fungal disease (Jach et al., 1995).

Classification and Characterization of Seed Chitinases

Compared to their vegetative counterparts, seed chitinases have been less widely investigated because vegetative chitinases have been associated with defensive responses against pathogen attack or other types of stress. However, many studies carried out with seed-specific chitinases have contributed to our knowledge of this protein family. Seed chitinase isoforms have been described in distantly related families such as Apiaceae, Fabaceae, Poaceae, Fagaceae, and Pinaceae (Gomez et al., 2002). All known seed chitinases are encoded by either *Chia* or *Chib* genes, and all seed chitinases encoded by *Chib* genes are members of class III. Class III chitinases are significantly different in structure and action from the other chitinases because they lack a chitin binding domain. (Gomez et al., 2002). Many chitinases are expressed during seed development, with mRNA synthesis typically beginning at early- or mid-maturation stages (Leah et al., 1991; Huynh et al., 1992; Yeboah et al., 1998). Some chitinase isoforms are also induced by microbial infection in developing and germinating seeds (Cordero et al., 1994; Ignatius et al., 1994; Wu et al., 1994; Caruso et al., 1999; Ji et al., 2000). Pathogen-responsive chitinases are often different from those described in leaves or roots (Majeau et al., 1990; Kragh et al., 1993;

Cordero et al., 1994; Ignatius et al., 1994; Krishnaveni et al., 1999). Seed chitinases have been reported to accumulate in different locations, including storage tissues (cotyledons and endosperm), aleurone, embryo, radicles, and seed coats (Swegle et al., 1989, 1992; Jensen, 1994; Leah et al., 1994; Wu et al., 1994). Genes encoding enzymes present in mature seeds may remain active during the first two or three days of germination when the transcription of new chitinase genes begins (Leah et al., 1991; Huynh et al., 1992; Swegle et al., 1992).

The production of two chitinases, *Cmchi1* and *Cmchi2*, of class III and II, respectively, was pinpointed to the onset of radicle emergence even though expression continued throughout muskmelon root development (Zou et al, 2002). These germination-specific chitinases were not encoded by the same genes responsible for chitinase activity detected in developing muskmelon seeds beginning 40 days after anthesis following the accumulation of maximum seed dry weight (Whitmer et a1., 2003). Seed chitinases are apparently not under the control of ABA or GA (Leah et al., 1991; Wu et al., 1994), but both ethylene-dependent and -independent pathways regulate chitinase expression during pea seed germination (Petruzzelli et al., 1999).

The first well-characterized plant chitinase was isolated from wheat germ in 1979 (Molano et al.). Since then, many chitinases have been purified from seeds, particularly from cereal crops. When comparing enzyme classes across species, no clear patterns emerge even when only major isoforms are considered. For example, the most abundant chitinases in barley and rye belong to classes I and II, while in maize they belong to classes I and IV (Leah et al., 1991; Kragh et al., 1991, 1993; Huynh et al., 1992; Swegle et al., 1992; Yamagami and Funatsu, 1993a,b, 1994; Wu et al., 1994). The major enzyme types can also differ within the same genus. This is evident when *Castanea sativa,* with one class I and two class II enzymes, is compared with *C. crenata* with only class II chitinases (Collada et al., 1992, 1993). However, SDS-PAGE analyses of seed chitinases from different cultivars of maize, sorghum, and wheat showed little intraspecific variability (Darnetty et al., 1993). In addition to these species, chitinases have also been purified from the seeds of *Benincasa hispida* (Shih et al., 2001), *Canavalia ensiforrnis* (Hahn et al., 2000), *Cucumis sativus* (Majeau et al., 1990), *Glycine max* (Wadsworth and Zikakis, 1984; Yeboah et al., 1998), and *Sorghum bicolor* (Krishnaveni et al., 1999). In addition to their hydrolytic action on chitin polymers, some of these enzymes also have lysozyme activity in vitro (Roberts and Selitrennikoff, 1988; Majeau et al., 1990; Swegle et al.. 1992; Shih et al., 2001). Chitinase from *Coix lachryma jobi* seeds (the Job's tears plant, Poaceae) also shows α-amylase inhibitory activity, which appears to be a special case among plant chitinases (Ary et al., 1989). The native structure of the Job's tear chitinase, unique among plants, is a dimeric protein with two similar or identical subunits of about 26 kDa linked by inter-chain disulfide bonds.

Evidence for Antifungal Chitinase Activity in Seeds

The highest accumulation of chitinase mRNA has been consistently observed during mid and late maturation stages in the seeds of different plant species when water content is highest and they are most sensitive to microbial attack (Leah et al., 1991; Swegle et al., 1992; Huynh et al., 1992; Garcia-Casado et al., 2000). The induction of seed chitinases by fungal attack occurs in some monocot species during both maturation and germination. For example, two genes encoding class I chitinases are induced in developing maize kernels after inoculation with *Aspergillus flavus.* These genes were activated in the embryo and the aleurone layer, but not in the endosperm (Wu et al., 1994). Ji et al. (2000) have shown that *A. flavus* and *Fusarium moniliforme* induce the expression of different chitinase isoforms, suggesting a specificity in the response to different pathogens. During germination, the latter fungus induced three chitinases and a β-1,3-glucanase in maize embryos. When seedlings were later analyzed, the accumulation patterns of glucanase and chitinase isoforms in radicles and coleoptiles were distinctive from embryos (Cordero et al., 1994). Different chitinase isoforms and other PR proteins were induced in germinating wheat seeds inoculated with *Fusarium culmorum* (Caruso et al., 1999). Less attention has been given to specific race-cultivar interactions in the response of seed tissues to pathogen infection, but analysis of different resistant and susceptible sorghum lines has shown a correlation between grain mold incidence and the levels of expression of chitinase and other antifungal proteins (Bueso et al., 2000).

The antifungal inhibitory effects of purified class I, II, and IV enzymes from monocots (wheat, barley, sorghum, and maize) and dicot (chestnut) seeds have been extensively tested against agricultural fungal pathogens and nonpathogenic model species (Roberts and Selitrennikoff, 1988; Collada et al., 1992; Huynh et al., 1992; Swegle et al., 1992; Allona et al., 1996; Garcia-Casado et al., 1998; Krishnaveni et al., 1499). Hyphal growth or fungal spore germination was monitored using microtiter well plate or disc-agar plate assays. In microtiter assay, the amount of purified protein that inhibited fungal growth by 50 percent (IC_{50}) varied between 0.1 and 2 μM. For the disc-agar plate assay, growth inhibition occurred with as little as 0.5 μg of protein per disc. Some of the purified enzymes were effective against phytopathogenic fungi such as *Fusarium sporotrichoides, F. oxysporum,*

F. moniliforme, *Alternaria solani*, or *Rhizoctonia solani*, but not against *Sclerotinia sclerotiorum* or *Gaeumannomyces graminis*. No inhibition has been observed when the pathogenic oomycetes *Phytophtora infestans* or *Pythium myrtotylum*, which do not contain chitin in their cell walls, were assayed (Huynh et al., 1992; Swegle et al., 1992). *Pythiam* may be such a devastating seedling disease for many species because it cannot be controlled by plant chitinases that apparently protect germinating seedlings from fungal attack.

Synergistic antifungal effects have been reported for combinations of chitinases and other seed defensive proteins. Hejgaard et al. (1991) and Leah et al. (1991) showed that mixtures of a barley seed chitinase with a β-1,3 glucanase, two thaumatin-like proteins, and a ribosome-inactivating protein from the same organ in all cases showed synergistic effects against *Trichoderma viride* and *F. sporotrichoides*. Analogous results have also been reported for a class I chitinase and a thaumatin-like protein purified from chestnut cotyledons (Garcia-Casado et al., 2000). The antifungal activity of seed chitinases has been confirmed in planta by Jach et al. (1995), who obtained transgenic tobacco plants expressing a class II chitinase (CHI), a β-1,3-glucanase (GLU), and a ribosome-inactivating protein (RIP) from barley seeds, as well as plants transformed with tandemly arranged CHI and GLU or CHI and RIP constructs. The latter plants performed significantly better after infection with *Rhizoctonia solani* than those expressing a single barley protein at similar levels of expression.

Despite the abundant literature on the antifungal properties of plant chitinases and their possible role in plant defense, the mode of chitinase action is not well understood. Both the inhibition of chitin synthesis due to the breakdown of newly formed chains, and the release of chitin fragments from isolated cell walls have been demonstrated in vitro using purified enzymes (Molano et al., 1979; Boller et al., 1983; Mauch et al., 1988; Brunner et al., 1998). Furthermore, the breakdown of chitin occurs when a bean chitinase was applied to actively growing mycelial cells of *R. solani* (Benhamou et al., 1993). Antifungal activity has been reported for chitinases that contain an N-terminal chitin binding domain (hevein domain; classes I and IV) and also for enzymes that lack this domain. Mature hevein, stinging nettle agglutinin (UDA), consisting of two heveinlike domains, and structurally related, nonenzymatic peptides from amaranth also inhibit fungal growth (Broekaert et al., 1989, 1992; Chrispeels and Raikhel, 1991; Van Parijs et al., 1991; Raikhel et al., 1993).

To determine the role of the hevein domain in antifungal activity, Iseli et al. (1993) expressed a wild-type tobacco class I chitinase and a truncated form lacking the hevein domain in transgenie *Nicotiana sylvestris* plants. By comparing both proteins, the authors concluded that a hevein domain is essential for chitin binding but not for catalytic or antifungal activity. Other studies with seed chitinases, however, support the hypothesis that the chitin binding domain itself can interfere with hyphal growth, as occurs with certain nonenzymatic lectins. Using site-directed mutagenesis, different forms of a class I chestnut chitinase (CsCh3) were generated with impaired chitinolytic activity (Gomez et al., 2002). The effects of these variants, wild- type CsCh3, and the homologous class II enzyme CsChl on the fungus *Trichoderma viride* showed that catalysis is not required for antifungal activity (Garcia-Casado et al., 1998). Moreover, the morphological changes caused by the wild-type enzyme and all its mutated forms in the hyphal tips were different from those caused by the class II enzyme. Apparently, the two domains present in class I chitinases, the chitin-binding domain, and the chitinolytic domain can alter apical growth, although through different mechanisms (Garcia-Casado et al., 1998). In a study conducted with a class II chitinase from barley seeds, mutants lacking chitinolytic activity retained some ability to inhibit the growth of *T. viride*. However, heat-inactivated wild-type chitinase was not inhibited (Andersen et al., 1997). The residual antifungal activity of the inactive mutant forms of the enzyme may reflect their ability to bind chitin. Recent analyses of a class I tobacco chitinase, including recombinant forms, suggests that the hevein domain has intrinsic antifungal activity (Suarez et al., 2001). Another comparison was made between two purified class IV enzymes sharing 87 percent identity from maize seeds, which showed a tenfold difference in antifungal activity. The enzyme with greater antimicrobial effects had threefold higher chitinolytic activity and a tenfold lower substrate binding constant (Huynh et al., 1992).

Chitinase Inhibition of Insects

Wounding by insects elicits the classic plant defense response in higher plants (Kramer and Muthukrishnan, 1997). Products of wounding produced by plants, such as oligoglacturonides, beta glucans, and oligo-N-acetylglucosamines will also initiate production of defense compounds. Plants that constitutively express chitinase would be anticipated to be resistant to insects and fungi because this exposure to chitinase might digest chitin in the gut lining of herbivores of plant pathogenic bacteria (Kramer and Muthukrishnan, 1997).

Chitinases isolated from cowpea seeds apparently have the ability to inhibit the growth of *Colletotrichum lindemuthianum* and *Colletotrichum musae* and to negatively affect the development of the cowpea weevil (*Callosobruchus maculatus*) in an artificial seed system (Gomes et al.,

1996). However, few additional data show that seed chitinases actually inhibit insect predation (Kramer and Muthukrishnan, 1997).

Chitinase genes cloned from bacteria, fungi, and insects also have been transferred to plants, although the expression has not been specifically targeted to seeds. Moreover, transgenic tobacco expressing an insect chitinase gene showed improved insect resistance (Ding et al., 1998). The growth of tobacco budworm *(Heliothis virescens)* larvae was reduced after feeding on transgenic plants. The feeding damage the budworm caused on the transgenic plants was also reduced. This indicates that transgenic seeds expressing an insect chitinase gene may have agronomic potential for improved control of certain insect pests.

Functions of Chitinase Not Related to Defense

It is likely that constitutive chitinases and other PR proteins are synthesized primarily to protect seeds and other nonvegetative organs from microbial attack. However, increasing evidence suggests that at least some seed chitinases may also be involved in nondefensive functions. For example, De Jong et al. (1992) demonstrated that a mutant cell line of carrot unable to form embryos was rescued by EP-3, a glycosylated class IV chitinase from wild-type carrot cells. The mutant phenotype did not result from structural differences in the enzyme but from a transient decrease in its amount compared to the wild type (De Jong et al., 1995). Further studies revealed at least four *EP-3* genes in carrot and that the proteins encoded by two of them had different effects on embryo formation (Kragh et al., 1996). In addition, a class I chitinase able to rescue the mutant line was also identified in this work. *EP-3* genes are expressed in the inner tegument of young carrot fruits and in a specific subset of endosperm cells, but not in zygotic embryos (Van Hengel et al., 1998). The function of EP-3 in embryogenesis involves the modification of GlcNAc/GlcN-containing arabinogalactan proteins. Indeed, pretreatment of such proteins with EP-3 resulted in increased activity to restore embryogenesis in cell-culture-derived protoplasts (Van Hengel et al., 2001). Interestingly, Nod-like factors might also be involved in this process, since several Nod factors have been shown to rescue the mutant line as well (De Jong et al., 1993; Dénarié and Cullimore, 1993). Somatic embryogenesis-related chitinases have been described in other plant species such as *Picea glassca, Linum usitatissimum, Pinus caribaea,* and a *Cichorium* hybrid (Dong and Dunstan, 1997; Roger et al., 1998; Domon et al., 2000; Helleboid et al., 2000). In *Pinus caribaea*, a 48 kDa chitinase-like protein secreted by embryogenic tissues acts on arabinogalactan proteins extracted from the same cells (Domon et al., 2000).

The finding that certain chitinases and β-1,3-glucanases are specifically expressed in tomato endosperm prior to radicle emergence has led to speculation regarding a possible role in seed germination (Wu et al., 2001). However, direct evidence that these enzymes participate in cell wall modification or tissue weakening is still lacking. In addition, ABA does not seem to affect chitinase expression during tomato seed germination (Wu et al., 2001).

Finally, some cold-inducible chitinases from winter rye leaves have been shown to possess antifreeze activity (Hon et al., 1995; Hiilovaara-Teijo et al., 1999). Such activity could be important to protect seed tissues from frost damage in temperate and boreal regions. Indeed, chitinases and other PR proteins accumulate at unusually high levels in recalcitrant chestnut seeds, which at abscission have one of the highest water contents known among seeds (Collada et al., 1992; Allona et al., 1996; Garcia-Casado et al., 1998, 2000). However, transgenic *Arabidopsis* plants constitutively expressing the major class I chitinase from chestnut seeds are not more tolerant to freezing temperatures than the control plants (Gomez et al., 2002).

Other Antifungal Seed Proteins

Other seed proteins also have antifungal activity. These include thionins, glucanases, nonspecific lipid-transfer proteins, osmotins, ribosome-inactivating proteins, polygalacturonase-inhibiting proteins, 2S albumin storage proteins, chitin binding proteins, which includes lectins, and several other low-M_r cysteine-rich antimicrobial peptides (Bol et al., 1990; Garcia-Olmedo et al., 1992; Broekaert et al., 1997; Yun et al., 1997; Shewry, 2000). Many diverse inhibitory mechanisms are responsible for the antifungal properties of these compounds; however, some apparently bind to hyphal membranes causing damage and leakage.

SUMMARY

The study of natural seed defense mechanisms has entered the mainstream of seed biology during the past decade. Our understanding and appreciation of natural defense mechanisms in seeds has grown tremendously as research in this area has increased dramatically. Environmental concerns and organic production will drive increased use of biocontrol measures to protect seeds. Defense gene manipulation will be another tool available for improving defense responses of seeds. Seed-specific promoters will allow the expression of defense genes to be targeted specifically to seeds. Once discovered and characterized, genes from other organisms will likely be introduced and expressed in plants to improve their disease and insect resistance.

REFERENCES

Adams, P.D. and J.W. Kloepper (1996). Seed-borne bacterial endophytes in different cotton cultivars. *Phytopathology* 86: S97 (abstract).

Allona, I., C. Collada, R. Casado, J. Paz-Ares, and C. Aragoncillo (1996). Bacterial expression of an active class Ib chitinase from *Castanea sativa* cotyledons. *Plant Molecular Biology* 32: 1171-1176.

Andersen, M.D., A. Jensen, J.D. Robertus, R. Leah, and K. Skriver (1997). Heterologous expression and characterization of wild-type and mutant forms of a 26 kDa endochitinase from barley (*Hordeum vulgare* L.). *Biochemical Journal* 322: 815-822.

Ary, M.B., M. Richardson, and P.R. Shewry (1989). Purification and characterization of an insect (α-amylase inhibitor/endochitinase from seeds of Job's Tears *(Coix lachryma jobi). Biochimica et Biophysica Acta* 993: 260-266.

Baskin, C.C. and J.M. Baskin (1998). *Seeds: Ecology, biogeography, and evolution of dormancy and germination.* San Diego, CA: Academic Press.

Benhamou, N., K. Broglie, R. Broglie, and L. Chet (1993). Antifungal effect of bean endochitinase on *Rhizoctonia solani:* Ultrastructural changes and cytochemical aspects of chitin breakdown. *Canadian Journal of Microbiology* 39: 318-328.

Benhamou, N., J.W. Kloepper, A. Quadt-Hallman, and S. Tuzun (1996). Induction of defence-related ultrastructural modifications in pea root tissues inoculated with endophytic bacteria. *Plant Physiology* 112: 919-929.

Bennett, M.A., V.A. Fritz, and N.W. Callan (1992). Impact of seed treatments on crop stand establishment. *HortTechnology 2:* 345-349.

Bensalim, S., J. Nowak, and S. Asiedu (1998). A plant growth promoting rhizobacterium and temperature effects on performance of 18 clones of potato. *American Potato Journal* 75: 145-152.

Bewley, J.D. and M. Black (1994). *Seeds: Physiology of development and germination,* Second edition. New York: Plenum Press.

Bol, J.F., H.J.M. Linthorst, and B.J.C. Cornelissen (1990). Plant pathogenesis related proteins induced by virus infection. *Annual Review of Phytopathology* 28: 113-138.

Boller, T. (1988). Ethylene and the regulation of antifungal hydrolases in plants. In B.J. Miflin (ed.), *Surveys of plant molecular and cell biology,* Volume 5 (pp. 145-174). Oxford, UK: Oxford University Press.

Boller, T. (1993). Antimicrobial functions of the plant hydrolases, chitinases and β-1,3 glucanase. In B. Fritig and L.M. Dordrecht (eds.), *Mechanism of plant defense response* (pp. 391-400). Dordrecht, the Netherlands: Kluwer Academic Press.

Boller, T., A. Gehri, F. Mauch, and U. Vögeli (1983). Chitinase in bean leaves: Induction by ethylene, purification, properties and possible function. *Planta* 157: 22-31.

Broekaert, W.F., B.P.A. Cammue, M.F.C. De Bolle, K. Thevissen, G.W. De Samblanx, and R.W. Osborn (1997). Antimicrobial peptides from plants. *Critical Reviews in Plant Sciences* 16: 297-323.

Broekaert, W.F., W. Mariën, F.R.G. Terras, M.F.C. De Bolle, P. Proost, J. Van Damme, L. Dillen, M. Claeys, S.B. Rees, J. Van der Leyden, and B.P.A. Cammue (1992). Antimicrobial peptides from *Amaranthus caudatus* seeds with sequence homology to the cysteine/glycine-rich domain of chitin-binding proteins. *Biochemistry* 31: 4308-4314.

Broekaert, W.F., J. Van Parijs, A.K. Allen, and W.J. Peumans (1988). Comparison of some molecular, enzymatic and antifungal properties of chitinases from thorn-apple, tobacco and wheat. *Physiological and Molecular Plant Pathology* 33: 319-331.

Broekaert, W.F., J. Van Parijs, F. Leyns, H. Joos, and W.J. Peumans (1989). A chitin-binding lectin from stinging nettle rhizomes with antifungal properties. *Science* 245: 1100-1102.

Broglie, K., L. Chet, M. Holliday, R. Cressman, P. Biddle, S. Knowlton, C.J. Mauvais, and R. Broglie (1991). Transgenic plants with enhanced resistance to the fungal pathogen *Rhizoctonia solani*. *Science* 254: 1194-1197.

Brunner, F., A. Stintzi, B. Fritig, and M. Legrand (1998). Substrate specificities of tobacco chitinases. *The Plant Journal* 14: 225-234.

Bueso, F.J., R.D. Waniska, W.L. Rooney, and F.P. Bejosano (2000). Activity of antifungal proteins against mold in sorghum caryopses in the field. *Journal of Agricultural and Food Chemistry* 48: 810-816.

Carbonero, P. and F. Garcia-Olmedo (1999). A multigene family of trypsin/α-amylase inhibitors from cereals. In P.R. Shewry and R. Casey (eds.), *Seed proteins* (pp. 159-169). Dordrecht, the Netherlands: Kluwer Academic Publishers.

Caruso, C., G. Chilosi, C. Caporale, L. Leonardi, L. Bertini, P. Magro, and V. Buonocore (1999). Induction of pathogenesis-related proteins in germinating wheat seeds infected with *Fusarium culmorum*. *Plant Science* 140: 107-120.

Chanway, C.P. (1997). Inoculation of tree roots with plant growth promoting soil bacteria: An emerging technology for reforestation. *Forestry Science* 43: 99-112.

Chrispeels, M.J. and N.V. Raikhel (1991). Lectins, lectin genes, and their role in plant defense. *Plant Cell* 3: 1-9.

Collada, C., R. Casado, and C. Aragoncillo (1993). Endochitinases from *Castanea crenata* cotyledons. *Journal of Agricultural and Food Chemistry* 41: 1716-1718.

Collada, C., R. Casado, A. Fraile, and C. Aragoncillo (1992). Basic endochitinases are major proteins in *Castanea sativa* cotyledons. *Plant Physiology* 100: 778-783.

Cordero, M.J., D. Raventos, and B. San Segundo (1994). Differential expression and induction of chitinases and β-1,3-glucanases in response to fungal infection during germination of maize seeds. *Molecular Plant-Microbe Interactions* 7: 23-31.

Darnetty, J.F., L. Subbaratnam Muthukrishnan, M. Swegle, A.J. Vigers, and C.P. Selitrennikoff (1993). Variability in antifungal proteins in the grains of maize, sorghum and wheat. *Physiologia Plantarum* 88: 339-349.

De Jong, A.J., J. Cordewener, F. Lo Schiavo, M. Terzi, J. Vandekerckhove, A. Van Kammen, and S.C. De Vries (1992). A carrot somatic embryo mutant is rescued by chitinase. *Plant Cell* 4: 425-433.

De Jong, A.J., R. Heidstra, H.P. Spaink, M.V. Hartog, E.A. Meijer, T. Hendriks, F. Lo Schiavo, M. Terzi, T. Bisseling, A. Van Kammen, and S.C. De Vries

(1993). *Rhizobium* lipooligosaccharides rescue a carrot somatic embryo mutant. *Plant Cell* 5: 615-620.

De Jong, A.J., T. Hendriks, E.A. Meijer, M. Penning, F. Lo Schiavo, M. Terzi, A. Van Kammen, and S.C. De Vries (1995). Transient reduction in secreted 32 kD chitinase prevents somatic embryogenesis in the carrot (*Daucus carota* L.) variant is 11. *Developmental Genetics* 16: 332-343.

Dénarié, J. and J. Cullimore (1993). Lipo-oligosaccharide nodulation factors: A new class of signalling molecules mediating recognition and morphogenesis. *Cell* 74: 951-954.

Ding, X., B. Gopalakrishnan, L.B. Johnson, F.F. White, X. Wang, T.D. Morgan, K.J. Kramer, and S. Muthukrishnan (1998). Insect resistance of transgenic tobacco expressing an insect chitinase gene. *Transgenic Research* 7: 77-84.

Domon, J.M., G. Neutelings, D. Roger, A. David, and H. David (2000). A basic chitinase-like protein secreted by embryogenic tissues of *Pinus caribaea* acts on arabinogalactan proteins extracted from the same cell lines. *Journal of Plant Physiology* 156: 33-39.

Domoney, C. (1999). Inhibitors of legume seed. In P.R. Shewry and R. Casey (eds.), *Seed proteins* (pp. 635-655). Dordrecht, the Netherlands: Kluwer Academic Publishers.

Dong, J.Z. and D.L. Dunstan (1997). Endochitinase and β-1,3-glucanase genes are developmentally regulated during somatic embryogenesis in *Picea glauca*. *Planta* 201: 189-194.

Dowling, D.N. and F. O'Gara (1994). Metabolites of *Pseudomonas* involved in biochemical control of plant disease. *Trends in Biotechnology* 12: 133-141.

Fritig, B., T. Heitz, and M. Legrand (1998). Antimicrobial proteins in induced plant defense. *Current Opinion in Immunology* 10: 16-22.

Frommel, M.I. J. Nowak, and G. Lazarovits (1993). Treatment of potato tubers with a growth promoting *Pseudomonas* sp.: Bacterium distribution in the rhizosphere and plant growth responses. *Plant and Soil* 150: 51-60.

Garcia-Casado, G., C. Collada, L. Allona, R. Casado, L.F. Pacios, C. Aragoncillo, and L. Gomez (1998). Site-directed mutagenesis of active site residues in a class I endochitinase from chestnut seeds. *Glycobiology* 8: 1021-1028.

Garcia-Casado, G., C. Collada, L. Allona, A. Soto, R. Casado, E. RodriguezCerezo, L. Gomez, and C. Aragoncillo (2000). Characterization of an apoplastic basic thaumatin-like protein from recalcitrant chestnut seeds. *Physiologia Plantarum* 110: 172-180.

Garcia-Olmedo, F., G. Salcedo, R. Sanchez-Monge, C. Hernandez-Lucas, M.J. Carmona, J.J. Lopez-Fando, J.A. Fernandez, L. Gomez, J. Royo, F. Garcia-Maroto, et al. (1992). Trypsin/a-amylase inhibitors and thionins: Possible defense proteins from barley. In P. Shewry (ed.), *Barley: Genetics, biochemistry, molecular biology and biotechnology* (pp. 335-350). Wallingford, UK: CAB International.

Gomes, V.M., A.E.A. Oliveira, and J.X. Filho (1996). A chitinase and a β-1, 3-glucanase isolated from the seeds of cowpea (*Vigna unguiculata* L. *Walp*) inhibit the growth of fungi and insect pests of the seed. *Journal of Science Food and Agriculture* 72: 86-90.

Gomez, L., L. Allona, R. Casado, and C. Aragoncillo (2002). Seed chitinases. *Seed Science Research* 12: 217-230.

Görlach, J., S. Volrath, G. Knauf-Beiter, G. Hengy, U. Beckhove, K.H. Kogel, M. Oostendorp, T. Staub, E. Ward, H. Kessmann, and J. Ryas (1996). Benzothiadiazole, a novel class of inducers of systemic acquired resistance, activates gene expression and disease resistance in wheat. *Plant Cell* 8: 629-643.

Grison, R., B. Grezes-Besset, M. Schneider, N. Lucante, L. Olsen, J.J. Leguay, and A. Toppan (1996). Field tolerance to fungal pathogens of *Brassica napus* constitutively expressing chimeric chitinase gene. *Nature Biotechnology* 14: 643-646.

Hahn, M., M. Henning, B. Schlesier, and W. Hohne (2000). Structure of jack bean chitinase. *Acta Crystallographica* 56: 1096-1099.

Hallman, J., J. Quadt-Hallmann, W.F. Mahaffee, and J.W. Kloepper (1997). Bacterial endophytes in agricultural crops. *Canadian Journal of Microbiology* 43: 895-914.

Hejgaard, J., S. Jacobsen, and L. Svendsen (1991). Two antifungal thaumatin-like proteins from barley grains. *FEBS Letters* 291: 127-131.

Helleboid, S., T. Hendriks, G. Bauw, D. Inze, J. Vasseur, and J.L. Hilbert (2000). Three major somatic embryogenesis related proteins in *Cichorium* identified as PR proteins. *Journal of Experimental Botany* 51: 1189-1200.

Hiilovaara-Teijo, M., A. Hannukkala, M. Griffith, X.M. Yu, and K. Pihakaski-Maunsbach (1999). Snow-mold-induced apoplastic proteins in winter rye leaves lack antifreeze activity. *Plant Physiology* 121: 665-673.

Hill, H.J. and A.G. Taylor (1989). Relationship between viability, endosperm integrity, and imbibed lettuce seed density and leakage. *HortScience* 24: 814-816.

Hinton, D.M. and C.W. Bacon (1995). *Enterobacter cloacae* is an endophytic symbiont of corn. *Mycopathologia* 129: 117-125.

Holland, M.A. and J.C. Polacco (1994) PPFM's and other covert contaminants: Is there more to plant physiology than just plant? *Annual Review Plant Physiology and Plant Molecular Biology* 45: 197-209.

Hon, W.C., M. Griffith, A. Mlynarz, Y.C. Kwok, and D.S.C. Yang (1995). Antifreeze proteins in winter rye are similar to pathogenesis-related proteins. *Plant Physiology* 109: 879-889.

Huynh, Q.K., C.M. Hironaka, E.B. Levine, C.E. Smith, J.R. Borgmeyer, and D.M. Shah (1992). Antifungal proteins from plants: Purification, molecular cloning, and antifungal properties of chitinases from maize seed. *Journal of Biological Chemistry* 267: 6635-6640.

Ignatius, S.M.J., J.K. Huang, R.K. Chopra, and S. Muthukrishnan (1994). Isolation and characterization of a barley chitinase genomic clone: Expression in powdery mildew infected barley. *Journal of Plant Biochemistry and Biotechnology* 3: 91-95.

Iseli, B., T. Boller, and J.M. Neuhaus (1993). The N-terminal cysteine-rich domain of tobacco class I chitinase is essential for chitin but not for catalytic or antifungal activity. *Plant Physiology* 103: 221-226.

Jach, G., B. Gornhardt, J. Mundy, J. Logemann, E. Pinsdorf, R. Leah, J. Schell, and C. Maas (1995). Enhanced quantitative resistance against fungal disease by combinatorial expression of different barley antifungal proteins in transgenic tobacco. *Plant Journal* 8: 97-109.

Jensen, L.G. (1994). Developmental patterns of enzymes and proteins during mobilization of endosperm stores in germinating barley grains. *Hereditas* 121: 53-72.

Ji, C., R.A. Norton, D.T. Wicklow, and P.F. Dowd (2000). Isoform patterns of chitinase and β-1,3-glueanase in maturing corn kernels (*Zea mays* L.) associated with *Aspergillus flavus* milk stage infection. *Journal of Agricultural and Food Chemistry* 48: 507-511.

Khan, A.A. (1992). Preplant physiological seed conditioning. *Horticultural Reviews* 14: 131-181.

Kombrink, E. and I.E. Somssich (1995). Defense responses of plants to pathogens. *Advances in Botanical Research* 21: 1-34.

Kragh, K.M., T. Hendriks, A.J. De Jong, F. Lo Schiavo, N. Bucherha, P. Hojrup, J.D. Mikkelsen, and S.C. De Vries (1996). Characterization of chitinases able to rescue somatic embryos of the temperature-sensible carrot variant *tsll*. *Plant Molecular Biology* 31: 631-645.

Kragh, K.M., S. Jacobsen, J.D. Mikkelsen, and K.A. Nielsen (1991). Purification and characterization of three chitinases and one β-1,3-glueanase accumulating in the medium of cell suspension cultures of barley (*Hordeum vulgare* L.). *Plant Science* 76: 65-77.

Kragh, K.M., S. Jacobsen, J.D. Mikkelsen, and K.A. Nielsen (1993). Tissue specificity and induction of class I, II and III chitinases in barley *(Hordeum vulgare)*. *Physiologia Plantarum* 89: 490-498.

Kramer, K.J. and S. Muthukrishnan (1997). Insect chitinases: Molecular biology and potential use as biopesticides. *Insect Biochemistry and Molecular Biology* 27: 887-900.

Krishnaveni, S., G.H. Liang, S. Muthukrishnan, and A. Manickam (1999). Purification and partial characterization of chitinases from sorghum seeds. *Plant Science* 144: 1-7.

Leah, R., K. Skriver, S. Knudsen, J. Ruud-Hamsen, N.V. Raikhel, and J. Mundy (1994). Identification of an enhancer/silencer sequence directing the aleurone-specific expression of a barley chitinase gene. *Plant Journal* 6: 579-589.

Leah, R., H. Tommerup, I. Svendsen, and J. Mundy (1991). Biochemical and molecular characterization of three barley seed proteins with antifungal properties. *Journal of Biological Chemistry* 266: 1564-1573.

Majeau, N., J. Trudel, and A. Asselin (1990). Diversity of cucumber chitinase isoforms and characterization of one seed basic chitinase with lysozyme activity. *Plant Science* 68: 9-16.

Mauch, F., B. Mauch-Mani, and T. Boller (1988). Antifungal hydrolases in pea tissue: II Inhibition of fungal growth by combinations of chitinase and β-1,3-glucanase. *Plant Physiology* 88: 936-942.

McInroy, J.A. and J.W. Kloepper (1995). Population dynamics of endophytic bacteria in freldgrown sweet corn and cotton. *Canadian Journal of Microbiology* 41: 895-901.

Molano, J., I. Polacheck, A. Duran, and E. Cabib (1979). An endochitinase from wheat germ: Activity on nascent and preformed chitin. *Journal of Biological Chemistry* 254: 4901-4907.

Mundt, J. O. and N.F. Hinkle (1976). Bacteria within ovules and seeds. *Applied Environmental Microbiology* 32: 694-698.

Neale, A.D., J.A. Wahleithner, M. Lund, H.T. Bonnett, A. Kelly, D.R. MeeksWagner, W.J. Peacok, and E.S. Dennis (1990). Chitinase, β-1,3-glucanase, osmotin, and extensin are expressed in tobacco explants during flower formation. *Plant Cell* 2: 673-684.

Neuhaus, J.M. (1999). Plant chitinases (PR-3, PR-4, PR-8, PR-11). In S.K. Datta and S. Muthukrishnan (eds.), *Pathogenesis-related proteins in plants* (pp. 77-105). Boca Raton, FL: CRC Press.

Neuhaus, J.M., B. Fritig, H.J.M. Linthorst, F. Meins Jr., J.D. Mikkelsen, and J. Ryals (1996). A revised nomenclature for chitinase genes. *Plant Molecular Biology Reporter* 14: 102-104.

Nowak, J. (1998). Benefits of in vitro "biotization" of plant tissue cultures with microbial inoculants. *In Vitro Cellular and Development Biology of Plants* 34: 122-130.

Petruzzelli, L., C. Kunz, R. Waldvogel, F. Meins Jr., and G. Leubner-Metzger (1999). Distinct ethylene and tissue-specific regulation of β-1,3-glucanases and ehitinases during pea seed germination. *Planta* 209: 195-201.

Peumans, W.J. and E.J. Van Damme (1999). Seed lectins, In P.R. Shewry and R. Casey (eds.), *Seed proteins* (pp. 657-683). Dordrecht, the Netherlands: Kluwer Academic Publishers.

Raikhel, N.V., H.I. Lee, and W.F. Broekaert (1993). Structure and functions of chitin-binding proteins. *Annual Review of Plant Physiology and Plant Molecular Biology* 44: 591-615.

Richards, J. (1997). Induced resistance responses in potato inoculated in vitro with a plant growth promoting pseudomonad. Master's thesis, Dalhousie University, Halifax, Nova Scotia, Canada.

Roberts, W.K. and C.P. Selitrennikoff (1988). Plant and bacterial chitinases differ in antifungal activity. *Journal of General Microbiology* 134: 169-176.

Roger, D., P. Gallusci, Y. Meyer, A. David, and H. David (1998). Basic chitinases are correlated with the morphogenic response of flax cells. *Physiologia Plantarum* 103: 271-279.

Rowse, H.R. (1996). Drum priming—A non-osmotic method of priming seeds. *Seed Science and Technology* 24: 281-294.

Schlumbaum, A., F. Mauch, U. Vogeli, and T. Boller (1986). Plant chitinases are potent inhibitors of fungal growth. *Nature* 324: 365-367.

Sharma, V.K. and J. Nowak (1998). Enhancement of verticillium wilt resistance in tomato transplants by in vitro co-culture of seedlings with a plant growth promoting rhizobaeterium (*Pseudomonas* sp. strain. PsJN). *Canadian Journal of Microbiology* 44: 528-536.

Shen, Z.-X. and G.E. Welbaum (1999). Comparison of fungicide, SAR inducers, and monopotassium phosphate as control measures for seedling diseases of muskmelon, cucumber, and tomato. In A. Liptay, C.S. Vavrina, and G.E. Welbaum (eds.), VI Symposium on Stand Establishment and ISHS Seed Symposium. May 1999. *Acta Horticulturae* 504: 113-120.

Shewry, P.R. (1999). Enzyme inhibitors of seeds: Types and properties, In P.R. Shewry and R. Casey (eds.), *Seed proteins* (pp. 587-615). Dordrecht, the Netherlands: Kluwer Academic Publishers.

Shewry, P.R. (2000). Seed proteins. In M. Black and J.D. Bewley (eds.), *Seed technology and its biological basis* (pp. 42-84). Boca Raton, FL: CRC Press.

Shewry, P.R. and J.A. Lucas (1997). Plant proteins that confer resistance to pests and pathogens. *Advances in Biological Research* 26: 135-192.

Shih, C., A.A. Khan, S. Jia, J. Wu, and D.S. Shih (2001). Purification, characterization, and molecular cloning of a chitinase from the seeds of *Benincasa hispida*. *Bioscience Biotechnology and Biochemistry* 65: 501-509.

Smith, V.L. (1996). Enhancement of snap bean emergence by *Giocladium virens*. *HortScience* 31: 984-985.

Sticher, L., B. Mauch-Mani, and J.P. Métraux (1997). Systemic acquired resistance. *Annual Review of Phytopathology* 35: 235-270.

Sturz, A.V. and J. Nowak (2000). Endophytic communities of rhizobacteria and the strategies required to create yield enhancing associations with crops. *Applied Soil Ecology* 15: 183-190.

Suarez, V., C. Staehelin, R. Arango, H. Holtorf, J. Hofsteenge, and F. Meins Jr. (2001). Substrate specificity and antifungal activity of recombinant tobacco class I chitinases. *Plant Molecular Biology* 45: 609-618.

Swegle, M., J.K. Huang, G. Lee, and S. Muthukrishnan (1989). Identification of an endochitinase cDNA clone from barley aleurone cells. *Plant Molecular Biology* 12: 403-412.

Swegle, M., K.J. Kramer, and S. Muthukrishnan (1992). Properties of barley seed chitinases and release of embryo-associated isoforms during early stages of imbibition. *Plant Physiology* 99: 1009-1014.

Takakura, Y., T. Ito, H. Saito, T. Inoue, T. Komari, and S. Kuwata (2000). Flower-predominant expression of a gene encoding a novel class I chitinase in rice (*Oryza sativa* L.). *Plant Molecular Biology* 42: 883-897.

Van Hengel, A.J., F. Guzzo, A. Van Kammen, and S.C. De Vries (1998). Expression pattern of the carrot EP3 endochitinase genes in suspension cultures and in developing seeds. *Plant Physiology* 117: 43-53.

Van Hengel, A.J., Z. Tadesse, P. Immerzeel, H. Schols, and A. van Kammen (2001). N-acetylglucosamine-containing arabinogalactan proteins control somatic embryogenesis. *Plant Physiology* 125: 1880-1890.

Van Parijs, J., W.F. Broekaert, I.J. Goldstein, and W.J. Peumans (1991). Hevein: An antifungal protein from rubber-tree *(Hevea brasilensis)* latex. *Planta* 183: 258-264.

Varga, S.S., P.A. Korányi, É. Preininger, and I. Gyrurán (1994). Artificial associations between *Daucus* and nitrogen-fixing *Azotobacter* cells in vitro. *Physiologia Plantarum* 90: 789-790.

Vierheilig, H., M. Alt, J.M. Neuhaus, T. Boller, and A. Wiemken (1993). Effect of chitinase overexpression on the colonization of roots by pathogenic and symbiotic fungi. *Molecular Plant-Microbe Interactions* 6: 261-264.

Volkmar, K.M. and E. Bremer (1998). Effects of seed inoculation with a strain of *Pseudomonas fluorescens* on root growth and activity of wheat in well-watered and drought stressed glass-fronted rhizotrons. *Canadian Journal of Plant Science* 78: 545-551.

Wadsworth, S.A. and J.P. Zikakis (1984). Chitinase from soybean seeds: Purification and some properties of the enzyme system. *Journal of Agricultural and Food Chemistry* 32: 1284-1288.

Warren, J.E. (1997). Bio-osmopriming tomato (*Lycoperricnn esculentum* Mill.) seeds for improved stand establishment. Master's thesis, The Ohio State University, Columbus, Ohio.

Warren, J.E., and M.A. Bennett (1997). Seed hydration using the drum priming system. *HortScience* 32: 1220-1221.

Welbaum, G.E. and K.J. Bradford (1990). Water relations of seed development and germination in muskmelon (*Cucumis melo* L.): IV. Characteristics of the perisperm envelope. *Plant Physiology* 92: 1038-1045.

Welbaum, G.E., Z.X. Shen, M.O. Oluoch, and L.W. Jett (1998). The evolution and effects of priming vegetable seeds. *Seed Technology* 20: 209-235.

Whitmer, X., H. Nonogaki, E.P. Beers, K.J. Bradford, and G.E. Welbaum (2003). Characterization of chitinase activity and gene expression in muskmelon seeds. *Seed Science Research* 13: 167-178.

Wilson, D. (1995). Endophyte—The evolution of a term, and clarification of its use and definition. *Oikos* 73: 274-276.

Wu, C.T., G. Leubner-Metzger, F. Meins, and K.J. Bradford (2001). Class I β-1,3 glucanase and chitinase are expressed in the micropylar endosperm of tomato seeds prior to radicle emergence. *Plant Physiology* 126: 1299-1313.

Wu, S., A.L. Kritz, and J.M. Widholm (1994). Molecular analysis of two cDNA clones encoding acidic class I chitinase in maize. *Plant Physiology* 105: 1097-1105.

Yamagami, T. and G. Funatsu (1993a). The complete amino acid sequence of chitinase-c from the seeds of rye *(Secale cereale). Bioscience Biotechnology and Biochemistry* 57: 1854-1861.

Yamagami, T. and G. Funatsu (1993b). Purification and some properties of three chitinases from the seeds of rye *(Secale cereale). Bioscience Biotechnology and Biochemistry* 57: 643-647.

Yamagami, T. and G. Funatsu (1994). The complete amino acid sequence of chitinase-a from the seeds of rye *(Secale cereale). Bioscience Biotechnology and Biochemistry* 58: 322-329.

Yeboah, N.A., M. Arahira, V.H. Nong, D. Zhang, K. Kadokura, A. Watanabe, and C. Fukazawa (1998). A class III acidic endochitinase is specifically expressed in the developing seeds of soybean (*Glycine max* [L.] Merr.). *Plant Molecular Biology* 36: 407-415.

Yim, K.O. and K.J. Bradford (1998). Callose deposition is responsible for apoplastic semipermeability of the endosperm envelope of muskmelon seeds. *Plant Physiology* 118: 83-90.

Yun, D.J., R.A. Bressan, and P.M. Hasegawa (1997). Plant antifungal proteins. *Plant Breeding Reviews* 14: 39-88.

Zhu, Q., E.A. Maher, S. Masoud, R.A. Dixon, and C.J. Lamb (1994). Enhanced protection against attack by constitutive co-expression of chitinase and glucanase genes in transgenic tobacco. *Bio-Technology* 12: 807-812.

Chapter 17

Seed Protease Inhibitors

A. M. Harsulkar
A. P. Giri
V. V. Deshpande
V. S. Gupta
M. N. Sainani
P. K. Ranjekar

INTRODUCTION

The roots of human civilization lie in the discovery that seeds of food plants could be collected, stored, and sown for assured harvest. This marked the advent of agrarian culture, which eventually culminated in the present civilized world. Besides food, seeds are used as feed for livestock, are an important object of trade, and also used as a raw material for industries. Plant seeds are thus instrumental for the evolution of human society, culture, and civilization. Biologically, seeds are designed for propagation and dispersal of plants. Aided with dormancy, they have the capacity to survive adverse climatic conditions. Seeds contain adequate stores of food/nutrients that support germination and early growth of seedlings. A true seed is defined as a fertilized mature ovule, consisting of an embryo, stored nutrients, and a protective coat. However, in the popular sense the term *seed* is also applied to single-seeded fruits such as caryopsis of cereals and cypsela of family Compositae.

Osborne (1924) proposed a systematic and largely accepted classification of seed proteins based on their solubility. Four major groups of proteins, commonly called "Osborne fractions," are albumins, globulines, prolamines, and glutelins. The modern classification, however, arranges proteins according to their function and categorizes them as storage proteins, structural proteins, metabolic proteins, or defense proteins. Among these, the defense proteins have attracted much attention, primarily because of their role in protecting plants from biotic and abiotic stresses. Recent advances in molecular biology, especially in the fields of genomics and

McKnight Foundation USA is gratefully acknowledged for the financial support to our chickpea research program.

functional genomics, are not only improving our knowledge about structure, function, expression, and inheritance of defense proteins but also instrumental in discovery of new genes by dissecting the complex responses of plants to pests and pathogens. Plant transformation, on the other hand, has opened avenues of transferring genes of defense proteins to bolster plants' inherent defense mechanisms and improve their tolerance. Moreover, transgenic plants promise bountiful harvest without over-reliance on external means of controlling pests and pathogens.

The rich reserves of proteins, carbohydrates, and lipids in seeds are attractive sources of food for humankind, pests, and pathogens alike. Insect attack is a major biotic stress to almost all crop plants, causing serious economic losses in cereals and legumes, in field and during storage. Interestingly, seeds are inherently equipped with an array of defense mechanisms. Hard testa made up of stone cells and the presence of cuticle or hairy/spiny coverings are examples of physical barriers, while secondary metabolites, toxins, and antinutritional compounds such as tannins, cyanogenic glucosides, and nonprotein amino acids accumulated in seeds constitute the biochemical defenses. Biologically active proteins/peptides that are potential deterrents for herbivores and grainivores are of significant biotechnological value since their genes can be isolated and expressed in target plants to find immediate application in agriculture. Lectins, amylase inhibitors, and protease inhibitors are the most important plant defense proteins and are extensively studied (Jouanin et al., 1998; Schuler et al., 1998). Lectins are proteins that have affinity to certain carbohydrate moieties. They bind to specific glycoproteins of the peritrophic membrane lining the insect midgut and disrupt osmotic balance and assimilation of nutrients (Brousseau et al., 1999). A lectin from snowdrop *(Galanthus nivalis)* has been found to be toxic to several insects (Foissac et al., 2000). Transgenic expressions of wheat germ agglutinin, pea lectin, and rice lectin in plants such as tobacco, maize, and potato have worked against aphids (Schuler et al., 1998). Amylase inhibitors (AIs) inactivate insect gut amylases and limit energy availability. α-AI from common bean *(Phaseolus vulgaris)* has been used in transgenic pea (Shade et al., 1994) and azuki bean (Ishimoto et al., 1996) to offer protection against bruchid beetles (Shade et al., 1994) and also a field pest, pea weevil (Morton et al., 2000). Proteinacious protease inhibitors that form complexes with proteases and inhibit their proteolytic activity are widespread in nature. Apart from their regulatory role in plant tissues, they are known to be important defense agents against pathogenic microbes (Joshi et al., 1998; Vernekar et al., 1999) and herbivorous insects. Plant PIs are the most documented defense proteins and are described in detail in the following sections.

PROTEASE INHIBITORS

Protease inhibitors (PIs) occur widely in the plant kingdom—especially in seeds and tubers. They may constitute 0 to 20 percent of total seed protein content and have molecular weights in the range of 4 to 20 kD. Tertiary structures of PI proteins are typically secured by the presence of one or more disulfide bonds, making them tolerant to heat, variations in pH, and enzymatic degradation. With a few exceptions, PIs are "double headed" since they can inhibit proteases in the ratio of 1:2 by virtue of two active domains. PIs are called bifunctional if they inhibit two or more enzymes of different classes, at either the same or different reactive sites. The most common examples of double-headed and bifunctional inhibitors are trypsin-chymotrypsin inhibitors from legume seeds, and trypsin-amylase inhibitors from cereal grains, respectively.

PIs from soybean were the first to be isolated and characterized by Kunitz (1945) and Bowman (1946) and were named Kunitz or Bowman-Birk type inhibitors. Since then a range of nonhomologous PIs have been isolated from microbial, plant, and animal sources and classified in eight to ten families (Birk, 1985, 1987; Garcia-Olmedo et al., 1987). Four classes of inhibitors are named according to the class of protease they inhibit, viz., serine, cysteine, aspartic, and metallo proteases (see Exhibit 17.1). Many excellent reviews have been published on PIs dealing with their classification and characterization (Garcia-Olmedo et al., 1987; Ryan, 1990; Valueva and Mosolov, 1999a,b; Shewry, 1999). Of all the classes of PIs, the serine PIs are the most abundant and also the most studied proteins; they have been categorized into seven families according to their sequence homology (Garcia-Olmedo et al., 1987; Ryan, 1990). Here we intend to give only a brief outline of major families of serine and cysteine PIs.

Bowman-Birk Family Inhibitors

These are the most abundant family of PIs, especially among legume seeds (Birk, 1985; 1987). They are "seed inhibitors" in true sense because they are found exclusively in seeds. The corresponding mRNAs accumulate early during seed development and are maintained until maturation (Foard et al., 1982). They are small proteins of approximately 8 to 9 kD with stable tertiary structures owing to the presence of seven disulfide bonds. They are double-headed inhibitors, usually showing inhibition against trypsin and chymotrypsin. Bowman-Birk inhibitors are encoded by low-copy, simple gene families that have no introns (Hammond et al., 1984).

EXHIBIT 17.1.
Various families of proteinase inhibitors.

Serine proteinase inhibitors

Kunitz family
Bowman-Birk family
Potato inhibitor I
Potato inhibitor II
Squash family inhibitor
Cereal trypsin and amylase inhibitor family
Ragi 1-2/Maize bifunctional inhibitor family

Cysteine proteinases

Family I or Steffins
Family II or Cystatins
Family III or Kininogen

Carboxypeptidase A and B inhibitors

Aspartyl proteinase inhibitors

Kunitz Family Inhibitors

Inhibitors belonging to Kunitz family are abundant in legume seeds and are also known to occur in cereals. Soybean Kunitz trypsin inhibitor (SKTI) was the first PI to be isolated and extensively characterized. Isolation and assay procedures of SKTI paved the way for studying other PIs from different plant sources. These are seed-specific inhibitors expressed in high amounts during seed development (Vodkin, 1981). A wide range of diversity is reported in Kunitz-type proteins. The majority of inhibitors from this family are single headed, although a very similar inhibitor was found in winged bean, which could inhibit bovine α-chymotrypsin in an inhibitor to a protease ratio of 1:2 (Kortt, 1980). Winged bean albumin-1 is a Kunitz-like storage protein, that accumulates up to 15 percent of total seed protein but has no inhibitory activity (Kortt et al., 1989). Kunitz inhibitors from cereals are bifunctional for proteases and α-amylases. The Kunitz family PIs have molecular weights of about 20 to 21 kD and contain four cystine residues.

Potato I and II Inhibitor Family

These are the best-known wound inducible-inhibitors from the family Solanaceae. They were first reported from potato tubers, where they constituted about 25 percent of the total soluble protein. Two nonhomologous inhibitors were found to be coexpressed in a systemic response to wounding in potato (Ryan, 1984) and hence were designated as potato-I and potato-II type inhibitors. Potato-I is an 8 kD protein which forms a pentamer, where each protomer has one disulfide bond and one reactive site. It exhibits a variety of inhibitory activities against serine endopeptidases, metallo-carboxypeptidases, papain, microbial proteases, and kallikreins. Homologous inhibitors are also known from *Vicia faba* (Svendsen et al., 1984), barley (Svendsen et al., 1980, 1982), and leech, *Hirudo medicinalis* (Seemuller et al., 1980). Potato inhibitor II exists as a dimer of two homologous 12 kD polypeptides where the protomers have five disulfide bonds and two reactive sites. Genes of inhibitor I from potato and tomato are reported to be interrupted by two introns, while inhibitor II genes have one intron (Lee et al., 1986; Cleveland et al., 1987; Keil et al., 1986).

Squash Family Inhibitors

These are the smallest known PIs of about 3 to 4 kD found in the family Cucurbitaceae; they are known to form the strongest complexes with their cognate proteases (Wieczorek et al., 1985). They are powerful inhibitors of trypsin and can also inhibit elastase, subtilisin, kallikrein, and thrombin (Otlewski and Krowarsch, 1996). The characteristic feature of these inhibitors is that they lack inhibitory activity toward human serine proteases (Hojima et al., 1980). In addition, their small size has made them attractive candidates for protein modeling studies. They have only one reactive site and three disulfide bridges. Similar to many storage proteins, squash family inhibitors show presence of "N" terminal leader peptide, suggesting that they are synthesized on endoplasmic reticulum and then transported to the protein bodies for storage (Ling et al., 1993). These inhibitors show clear signs of posttranslational modification with variations in "N" terminal residues (Garcia-Olmedo et al., 1987).

Cereal Trypsin-α-Amylase Inhibitor Family

Bifunctional inhibitors that can inhibit trypsin and α-amylase are known from barley, rice, wheat, and related cereals. The first trypsin inhibitor from this family was isolated from barley endosperm by Mikola and Soulinna

(1969), although their amylase-inhibitory activity was discovered only after 13 years (Odani et al., 1982, 1983). Members of this family have molecular weights that range from 12 to 16 kD and are expressed exclusively in endosperm of the kernels. They accumulate prior to the synthesis of storage proteins in developing seeds and are rapidly degraded during germination. Their potential to inhibit α-amylases is of much interest for controlling losses due to storage pests, which is a major problem in cereals (Gutierrez et al., 1990). It is interesting to note that several inhibitors are reported to have more affinity toward insect amylase over human salivary or pancreatic amylases (Garcia-Olmedo et al., 1987).

Cysteine Protease Inhibitors

Inhibitors of cysteine proteases are ubiquitous in seeds of all plant families of gymnosperms and angiosperms (Rele et al., 1980; Fernandes et al., 1991). Cysteine PIs are also known to occur in large amounts in tissues such as pineapple stem and potato tubers (Heinrikson and Kezdy, 1976; Hildmann et al., 1992). Homologous inhibitor proteins are of common occurrence in animals (Barret, 1987). Cysteine PIs are divided into three distinct families: family I, called "Steffins," includes proteins of 11 kD with no disulfide bonds; family II proteins, or "cystatins," are of 13 kD with two sulfide bridges; and family III, or "kininogem" family proteins, are of high molecular weight inhibitors (<40 kD) with six disulfide bonds (Barrett, 1987; Xavier Filho, 1992). Unlike serine PIs, cysteine PIs accumulate in metabolically active plant tissues in amounts comparable to endogenous proteases and are less diverse, suggesting their metabolic role (Fernendes et al., 1991). However, they are as important as serine PIs in plant defense and also in agrobiotechnology since several insects of Hemiptera and Coleoptera secrete cysteine proteases as major digestive proteases in their midguts. Beetles, which form the largest insect family, belong to Coleoptera and are important pests of field plants and stored seeds. Several reports show retardation in growth of coleopteran larvae when fed on cysteine PIs from cowpea, soybean, or rice.

EXPRESSION OF PROTEASE INHIBITORS IN SEEDS

Developing Seeds

Seed development is an intensive process at the molecular level, although the anatomical developments are relatively simple. About 20,000 different genes are expressed at any given stage of developing seed

(Kamalay and Goldberg, 1980). Many of them are nonhomologous multigene families of seed proteins, which vary in size, organization, and chromosomal locations (Kreis et al., 1985; Casey et al., 1986). Unique sets of seed-specific proteins are expressed almost exclusively during embryogenesis and are temporally and spatially regulated in the seed organs (Gatehouse et al., 1986; Goldberg et al., 1989). The seed-specific proteins include embryo-axis-specific proteins (expressed throughout embryogenesis) and seed-storage proteins. PIs are expressed from mid-maturation to late-maturation stages of embryogenesis (Goldberg et al., 1989). Presence of PIs during seed/tuber maturation might suggest that they facilitate accumulation of seed-storage proteins (Koiwa et al., 1997).

Accumulation of PIs during seed development in chickpea was studied in detail by Harsulkar et al. (1997) and Giri et al. (1998). They have shown that in developing chickpea seeds, synthesis of PIs begins at 24 days after flowering (DAF), and they accumulate steadily as the seed progresses toward maturity. Accumulation of seed proteins is relatively low between 12 and 24 DAF and dramatically increases between 36 and 48 days. On the other hand, inhibitor activity against trypsin, chymotrypsin, and bacterial protease is considerably high at 24 DAF and increases steadily until maturity. This phenomenon is a significant response of plants against feeding insects, since the bulk of seed proteins are synthesized only after PIs are produced (Giri et al., 1998). Another interesting feature of chickpea TIs is their differential expression. Chickpea produces seven electrophoretically separable TI isoforms, where the lower forms are early-stage specific and the higher forms are expressed latter during the seed development (see Figure 17.1). Interestingly, at mid-stage, where *Helicoverpa armigera* attack is at its maximum, all the isoforms are present (Harsulkar et al., 1997; Giri et al., 1998).

Mature Seeds

Legume seeds can be separated into three major tissues, viz., the seed coat, endosperm, and embryo, which can be further divided into a pair of cotyledons and the embryonic axis (EA). The testa is a maternal tissue that surrounds the embryo and attaches the seed to the pod through the funicle. The endosperm is a triploid tissue designated to nurture the developing embryo during the early stages of development. In several seeds, such as in legumes, the endosperm is consumed totally during development and does not exist in mature seeds, which are therefore called *nonendospermic*. As the seed matures and begins to loose moisture, the layers of testa become compressed and accumulate compounds that include lignin, suberin, cutin,

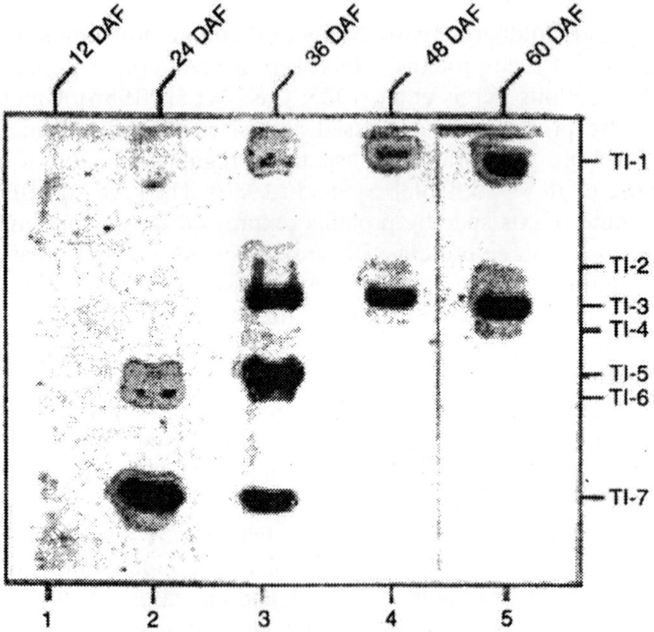

FIGURE 17.1. Differential expression of chickpea trypsin inhibitors during seed development. Equal trypsin inhibitor unit (0.03) of extracts of seeds harvested 24 DAF (lane 2), 36 DAF (lane 3), 48 DAF (lane 4), 60 DAF (lane 5) were loaded. For 12 DAF (lane 1) 60 μg of protein was loaded since no TI activity was detected in spectrophotometric assays. [*Source:* A.P. Giri, A.M. Harsulkar, V.S. Gupta, M.N. Sainani, V.V. Deshpande, and P.K. Ranjekar (1998). Chickpea defensive proteinase inhibitors can be inactivated by pod borer gut proteinases. *Plant Physiology* 116: 393-401. Copyright American Society of Plant Biologists, 1998. Reprinted with permission.]

callose, and tannins. In a mature seed, the testa is a dead tissue and serves as an excellent mechanical barrier to protect the embryo from biotic and abiotic aggression.

Patankar et al. (1999) demonstrated patterns of distribution and accumulation of TI activity in different seed organs of mature and developing chickpea seeds. The embryo-axis possesses about ninefold more TI activity even in immature seeds as compared to the cotyledons (see Table 17.1). The EA has considerably high protein as compared to the cotyledons at the mid-maturation stage. However, the rate of accumulation of TI activity in the cotyledons is considerably higher in the developing seed. The EA shows higher specific activity for PI than do the cotyledons, indicating a higher

TABLE 17.1. Distribution of trypsin inhibitor activity and storage proteins in different organs of developing and mature seeds of chickpea.

	Protein ($mg \cdot g^{-1}$)		Trypsin inhibitor ($Units \cdot g^{-1}$)	
Seed organ	**Midmature**	**Mature**	**Midmature**	**Mature**
Pod cover	0.24	0.18	ND	ND
Seed coat	0.21	0.37	ND	ND
Cotyledon	11.25	31.87	21.54	237.5
Embryo axis	52.15	42.85	181.62	336.68

Source: A.G. Patankar, A.M. Harsulkar, A.P. Giri, V.S. Gupta, M.N. Sainani, P.K. Ranjekar, and V.V. Deshpande (1999). Diversity in inhibitors of trypsin and *Helicoverpa armigera* gut proteinases in chickpea *(Cicer arietinum)* and its wild relatives. *Theoretical and Applied Genetics* 99: 719-726, Table 2. Copyright Springer-Verlag, 1999. Reprinted with permission.
ND: not detectable.

deposition of PIs in EA. At mid-maturation stage, 52 percent of TI activity is localized in the cotyledons and 48 percent in the EA. However, in the mature seed, the cotyledon contributes 91 percent of the TI activity versus only 9 percent by the embryo. This can be explained, since more than 80 percent of the 3×10^6 cells in a mature embryo are contained in cotyledons (Goldberg et al., 1981). Recently, Welham et al. (1998) demonstrated the immunolocalization of TI activity in the EA in developing as well as germinating seeds of *Pisum*. The high density of PI activity in the EA may be attributed to their defensive properties, which are utilized for protection of the embryo from insect pests. The PI activity of the seed increases with maturity; however, the increase is greater in the cotyledons than in other tissues. The cotyledons, which contribute up to 68 percent of the seed weight at mid-maturation, show increase in weight to 84 percent at maturity. Because the embryo shows very high specific activity of TIs (see Figure 17.2), use of their promoters to express genes of other antifeedent proteins for developing resistance to insects such as *H. armigera* would be of considerable interest.

Germinating Seeds

Proteolysis is a key event for mobilization of stored proteins in germinating seeds for making essential amino acids available to the growing embryo and young seedling before the onset of autotrophic growth. Some amount of proteolytic activity can be detected in most of the resting seeds, and the

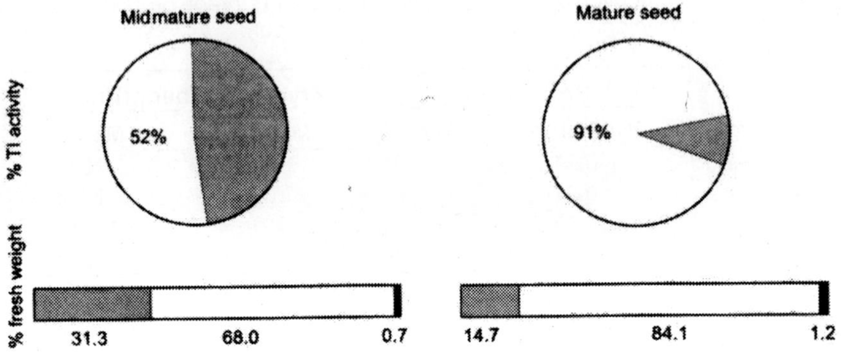

FIGURE 17.2. Distribution of trypsin inhibitor activity in seed organs of chickpea, cotyledon (□) and embryo axis (□), of midmature and mature seed of chickpea. The percent values of TIs were calculated as per the fresh weight contributed by the respective seed organs. The bar graphs show the percent distribution of the fresh weight of seed organs. The seed coat (□), cotyledon (□), and embryo (■) of midmature and mature seeds were separated and the percent fresh weight contributed by each of them was calculated. [*Source:* A.G. Patankar, A.M. Harsulkar, A.P. Giri, V.S. Gupta, M.N. Sainani, P.K. Ranjekar, and V.V. Deshpande (1999). Diversity in inhibitors of trypsin and *Helicoverpa armigera* gut proteinases in chickpea *(Cicer arietinum)* and its wild relatives. *Theoretical and Applied Genetics* 99: 719-726, Figure 2. Copyright Springer-Verlag, 1999. Reprinted with permission.]

activity tends to be higher in EA than in storage tissue. However, imbibition triggers a number of metabolic processes, one of which is the synthesis of proteases. Several studies suggest active proteolysis of PIs during seed germination (Garcia-Olmeda et al., 1987; Godbole et al., 1994; Ambekar et al., 1996). Giri et al. (1998) reported conversion of TIs into different forms before they totally disappear at 25 days after germination in chickpea. Wilson and Chen (1983) observed a rapid modification of TIs by limited proteolysis during early stages of seedling growth in mung bean. It has been documented by Yoshikawa et al. (1979) in adzuki bean that Bowman-Birk type inhibitors undergo extensive proteolysis during early germination. Proteolysis of inhibitors during germination suggests their role as storage proteins and as a depot of sulfur-rich amino acids. However, other reports of their release during imbibition attribute this to a defense function where they diffuse out from the seed and form a protective microenvironment around the germinating seed (Hwang et al., 1978; Wilson, 1980; Tan-Wilson and Wilson, 1982; Horisberger and Tacchini, 1982).

Ultrastructural Localization of PIs

Fractionation studies of seed proteins in pea have revealed that PIs are cytoplasmic in origin and are found outside the protein bodies (Hobday et al. 1973). Horisberger and Tacchini (1982) confirmed these findings with a detailed ultrastructural localization of the Kunitz and Bowman-Birk inhibitors in soybean. They have further shown that PIs are associated primarily with cytosol, but in some instances they are localized in protein bodies. They have also observed that unlike Kunitz type inhibitors, the Bowman-Birk type inhibitors are not present in the cell wall but are localized in the intercellular space, facilitating their rapid release in germinating soybean seeds (Hwang et al., 1978).

BIOLOGICAL ROLE OF PROTEASE INHIBITORS

The relative abundance of PIs in plant tissues that are common food and feed and their stability against heat, enzymatic degradation, and pH extremes have raised concerns about their antinutritive effects. Presence of PIs decreases the apparent nutritional quality of food by compromising the ability of proteases to digest dietary proteins, thus limiting biological availability of essential amino acids. Feeding studies in lower mammals have shown their adverse effects on growth, and prolonged exposure has resulted in pancreatic hypertrophy (Liener and Kakade, 1980). Therefore, PIs have been thought to be antinutritional compounds, especially in legume seeds, and have served as targets for seed quality improvement. In fact, breeders have a preference for genotypes with low seed inhibitor activity for breeding programs (Muel et al., 1998) and have even attempted to reduce levels of PIs in seeds through mutation breeding (Kothekar et al., 1996).

The biological role of PIs in plants has remained a mystery for quite some time, mainly because their target proteases, viz., trypsinlike serine proteases, are often not detected in plant tissues, or even if detected, these inhibitors do not seem to have an effect on them (Nishikata, 1984). The wide range of their occurrence in seeds, from 0 to 20 percent, suggests that their presence is physiologically irrelevant (Fernandes et al., 1991). Seeds of one soybean accession have been reported to be totally devoid of Kunitz type inhibitors and can still complete their entire life cycle without any noticeable difference (Orf and Hymowitz, 1979; Xavier Filho, 1992). The notion that they are storage proteins and serve as depots of sulfur-containing amino acids is supported by their proteolysis during seed germination but contradicts their presence in nonstorage tissues such as leaves, phloem, root tips, and floral parts. However, discovery of inhibitors of cysteine proteases that are common in metabolically active plant tissues has suggested a regulatory role for them. Pioneering work of Ryan and co-workers has further revealed that their expression is controlled by "systemin," a polypeptide hormone released by plants in response to mechanical or herbivoral damage, and has established unequivocally their role in plant defense (Ryan, 1990). Genetically engineered plants expressing PIs genes and their demonstrable improvement in pest resistance clear all doubts about their role in plant defense. Table 17.2 summarizes points in favor of their role as defense proteins, storage proteins, and regulatory agents of endogenous proteases.

TABLE 17.2. Various features supporting different roles of proteinase inhibitors in plants.

Features	Roles
Defense protein	• Transformed plants show enhanced levels of protection against herbivores and insect pests
	• Inducible by mechanical and herbivoral damage, by pathogen attack, and due to abiotic stress in local and remote tissues
	• Strong inhibitory activity against insect gut proteinases as demonstrated by in vitro and in vivo studies
	• Accumulation in amounts far exceeding than required to inhibit endogenous proteinases
	• Occurrence in nonstorage tissue such as leaves and floral parts
	• Demonstrated antimicrobial activity
	• Leached out from seeds during imbibition forming protective microenvironment
Storage protein	• Accumulation of high amounts in storage tissue
	• High content of "sulfur"-containing amino acids
	• Localized in protein bodies
	• Wide level of expression
	• Rapid proteolysis during germination
	• Leader sequences in nascent peptide chains typical to storage proteins
	• Evidence of posttranslational modifications
Regulation of endogenous proteinases	• Activity against endogenous proteinases
	• Presence in metabolically active tissue as in germinating seeds

Plant PIs Against Insect Herbivory

Green and Ryan (1972) showed for the first time that the PIs in potato and tomato were induced upon wounding due to herbivory. Subsequently, Gatehouse and Boulter (1983) demonstrated that the resistance of a cowpea variety to the bruchid beetle was due to elevated TI levels in the seeds. PIs are known to be induced under various stress-prone conditions such as insect attack (Jongsma et al., 1994; Korth and Dixon, 1997; Casaretto and Corcuera, 1998; Giri et al., 1998; Tamayo et al., 2000), mechanical wounding (Peña-Cortés et al., 1995; Botella et al., 1996; Zhao et al., 1996), pathogen attack (Pautọt et al., 1991; Cordero et al., 1994; Jongsma et al., 1994), and UV exposure (Conconi et al., 1996). Moreover, PIs are the end products of several defense cascades activated by numerous systemic and nonsystemic elicitors. These include systemin (Schaller and Ryan, 1995), ethylene (Botella et al., 1996), methyl jasmonate (Farmer and Ryan, 1990; Boulter, 1993), abscisic acid (Peña-Cortés et al., 1989; Hildmann et al., 1992), fungal cell wall oligomers (Doares et al., 1995), larval oral secretions (Korth and Dixon, 1997), and electrical and hydraulic signals (Wildon et al., 1992; Stankovic and Davies, 1997) leading to increased accumulation of PIs in local as well as remote tissues. All these studies have been carried out predominantly for the serine PIs except in a few cases where cysteine PIs have been studied (Bolter, 1993; Botella et al., 1996; Zhao et al., 1996). Ryan and co-workers have extensively characterized the wound-signalling pathways in tomato and have shown that the wound signal is an 18 amino acid polypeptide hormone, systemin (Pearce et al., 1991). Systemin travels from wounded parts of the plant to distal organs and induces the synthesis of defense proteins including PIs via the octadecanoid pathway (Schaller and Ryan, 1995).

PIs bring about physiological changes in insect pests by inhibiting their gut proteases and thereby adversely affecting protein digestion. The insect resorts to overproduction of proteases to compensate for the inhibited activity, leading to deficiency of essential amino acids (Broadway and Duffey, 1986a,b). When the induced activity is also inhibited by PIs, the insect is not able to compensate for the inhibited activity. This exerts additional physiological stress for the insect, resulting in inhibition of its growth. PIs also affect a number of vital metabolic processes in the insect including proteolytic activation of enzymes as well as molting and water balance (Hilder et al., 1993; Boulter, 1993). Lower mortality of insects minimizes the possibility of developing resistance in the insects and still ensures less crop damage. Another advantage is that PIs can be used along with many other components of the ecosystem for increased effectiveness. In tritrophic interactions (plants, pests, and their predators), the retarded insects become easy targets for greater parasitism by natural enemies (Lewis et al., 1997). For example, Heath et al. (1997) have reported that *H. armigera* and *Telleogryllus commodus* (black field cricket), which were inhibited by *Nicotiana alata* PI, showed increased lethargy and incapability to react to predators, thus increasing their susceptibility to predation. Thus, an outbreak of insect attacks can be controlled by retarding the development of the insect pest while sustaining their natural enemies (parasitoids) in the ecosystem. On the contrary, if "therapeutic" strategy involving Bt is resorted to, it results in immediate kill (and apparent success) but destroys a reservoir of natural enemies, leading to pest resurgence.

Dynamics of Plant-Pest Interaction: In Context to Seed PIs and Gut Proteases

Though PIs inhibit insect growth and development upon feeding, the actual mechanism is not completely understood and has been a focus of a number of studies. The effect of PIs as antinutritional factors on the regulation of gut proteolytic activity has been studied earlier in animals including rat, chicken, and dog (Ryan, 1990). In insects, Broadway and Duffey (1986a, b) first studied the effect of PIs on the gut proteolytic activity and insect growth in *Spodoptera exigua* and *Heliothis zea* and found an increased secretion of trypsin activity upon feeding of PIs. They also observed insect growth inhibition at high protein levels in the diet and concluded that there was overproduction of trypsinlike activity to digest the additional protein. This hyperproduction led to the depletion of essential amino acids, resulting in inhibited growth. Similar results were observed in *Ostrinia nubilalis* (Larque and Houseman, 1990) and *Spodoptera litura* (McManus and Burgess, 1995) upon feeding soybean TIs. However, Johnston et al. (1993) found a decrease in the trypsinlike activity and retarded growth of *H. armigera* fed on soybean TI. This was attributed to possible overproduction of protease activity in response to ingested PIs and masking of the increased tryptic activity by the excess inhibitor. Similarly, Burgess et al. (1994) observed 70 to 90 percent reduction of gut proteolytic activity accompanying mortality of black field cricket *(Telleogryllus commodus)* fed on potato PI-I and PI-II and attributed it to the in vivo inhibition of the gut enzymes by the ingested inhibitors. Broadway (1995) also observed retarded growth accompanied by a significant reduction (70 percent) in the gut proteolytic activity in *Agrotis ipsilon* fed on soybean TI. However, in case of *H. zea* fed on soybean TI, the retarded growth was accompanied by synthesis of some "inhibitor-resistant" activity (Broadway, 1995).

Concurrently, Jongsma et al. (1995) showed that the insect could overcome the effects of PIs by synthesis of inhibitor-insensitive activity. They observed a rise in the inhibitor-insensitive activity in *S. exigua* fed on potato PI-2. Bolter and Latoszek-Green (1997) reported that chronic feeding of the cysteine protease inhibitor, viz. E-64, to Colorado potato beetle had an adverse effect on its growth and adult fecundity. They found a two- to three-fold rise in the inhibitor-insensitive activity upon ingestion of the PI. This was, however, also accompanied by highly decreased gut proteolytic activity. These reports suggest that insects resort to increased production of both inhibitor-sensitive and -insensitive proteases in an attempt to adapt to the ingested PI. Depending upon the efficacy of the ingested PI, the induced protease activity may or may not compensate for the inhibited activity. Ultimately, the PIs that effect a significant reduction of gut proteolytic activity are able to inhibit insect growth efficiently.

PROTEASE INHIBITOR TRANSGENIC PLANTS FOR PEST CONTROL

Plant genetic engineering offers a tool to transfer insect-resistant genes across the barriers of conventional plant breeding and provides an environmentally friendly alternative to chemical pesticides. Although several different genes are proposed, protease inhibitors bear a distinct promise toward a range of economically important insects. Hilder et al. (1987) transferred the first PI gene (a cowpea TI) into tobacco plant and showed enhanced protection against tobacco budworm and corn earworm. Encouraged by these results, various other PI genes have been introduced into a variety of different plants (Johnson et al., 1989; Urwin et al., 1995; Xu et al., 1996; DeLeo et al., 1998; Lee et al., 1999; Cloutier et al., 2000). However, even after a series of serious attempts, commercialization of PI transgenic plants have not been successful even today and demands further research in this area.

Plant-pest interaction is a complex phenomenon. Studies carried out in our laboratory have revealed the complexity of gut protease composition of *H. armigera* fed on different host and nonhost plants. *Helicoverpa armigera* gut proteases showed predominance of serine proteases (Harsulkar et al., 1998; Patankar et al., 2001), although the larvae fed on different host plants revealed presence of metallo-proteases, aspartic proteases, and cysteine proteases (Patankar et al., 2001). On the other hand, the nonhost plant PIs were effective in reducing the level of gut proteases and retarding the larval growth (Harsulkar et al., 1999). Although pests necessarily adapt to counter the defensive measures of their host plants, they might be susceptible to parallel mechanisms in nonhost plants. Polyphagous insects such as

H. armigera may prove to be the exceptions to this rule since they are likely to be able to adapt as compared to specialist feeders.

In view of the complexity of gut proteases, it is clear that introduction of a single PI gene will not be successful in producing effective and sustainable transgenic plants. The current research on PIs is mainly focused on the expression of a single PI gene under a universal promoter (Jouanin et al., 1998; Schuler et al., 1998). However, several workers have proposed use of multiple PIs to inhibit full spectrum of gut proteases (Jongsma and Bolter, 1997; Michaud, 1997; Girard, Le Metayer, Bonade-Bottino, et al., 1998; Girard, Le Metayer, Zaccomer, et al., 1998). Combinations of PIs targeted to different proteinses were reported to act synergistically (Jongsma and Bolter, 1997). Combination of PIs increases their stability in the highly proteolytic environment of the gut, by preventing unwarranted proteolysis. It is known that insects synthesize insensitive proteases in response to the dietary PIs to counter their inhibitory effects (Broadway, 1995; Jongsma et al., 1995). Gut proteases insensitive to one PI are, however, shown to be inhibited by other PIs (Harsulkar et al., 1999). Because larvae of *H. armigera* feed on leaves for the first two instars and later shift on the developing seeds, their successive exposure to two different PIs can be engineered by expressing genes of PIs under leaf-specific and seed-specific promoters (Harsulkar et al., 1999). Many such combinatorial approaches using specific promoters for guided expression mark the beginning of the next generation of transgenic technology and open new avenues to tackle problems in modern agriculture. Resistance in transgenic plants could be even better if pyramiding of genes of different mechanistic classes can be achieved. The first example of gene pyramiding was demonstrated by Boulter (1993), who crossed two lines of transgenic tobacco expressing cowpea TI and pea lectin and reported additive effect of these two genes. Insect weight reduced to 90 percent in double expressing plant as compared to 50 percent reduction in plants expressing one of the genes. Macintosh et al. (1990) showed that PIs could enhance activity of Bt genes against their target insects. Expression of a truncated PI gene together with the Cry1 Ac gene results in a fusion protein which shows sixfold increase in Bt activity.

CONCLUSIONS

PIs are of much interest due to their abundance, antinutritive effects, and potential against insect pests. Moreover, they provide a unique system for understanding fundamental processes in plant defense and are instrumental in studies on the effects of environmental and developmental factors on natural defense mechanisms in plants. They are especially effective against insect pests that are highly dependent on proteases for digesting their food. Several studies including in vivo and in vitro trials have cleared all doubts about their role as defense proteins. In fact, genetic engineering of plants with PI genes provides an opportunity to enhance plant tolerance to a variety of field and storage pests. The choice of genes is critical. In view of ubiquitous nature of PIs, a large pool of genes with varied molecular properties is available which will enable selection of candidate inhibitors of maximum efficacy against proteases of the target pest. PIs have a high biotechnological value by virtue of their inherent mechanisms of action; they can improve protection against insect pests in transgenic plants without increasing insect mortality, thus reducing chances of developing resistance in insect populations. They might prove to be protective companions to many other defense proteins as they can limit untimely proteolysis. Innovative use of tissue-specific and -inducible promoters opens the second generation of transgenic technology for tailored expression of PI genes for durable resistance against insect pests.

REFERENCES

Ambekar, S. S., S. C. Patil, A. P. Giri, and M. S. Kachole (1996). Proteinaceous inhibitors of trypsin and amylase in developing and germinating seeds of pigeonpea (*Cajanus cajan* L. Millsp). *J. Sci. Food Agric.* 72: 57-62.

Barret, A. J. (1987). The cystatins: A new class of peptidase inhibitors. *Trends Biochem. Sci.* 12: 193-196.

Birk, Y. (1985). The Bowman-Birk inhibitor: Trypsin and chymotrypsin-inhibitor from soybeans. *Int. J. Pept. Prot. Res.* 25: 113-131.

Birk, Y. (1987). Proteinase inhibitors. In A. Neuberger and K. Brocklehurst (eds.), *Hydrolytic enzymes* (pp. 257-305). Amsterdam: Elsevier Science Publishers BV Biomedical Division.

Bolter, C. J. and M. Latoszek-Green (1997). Effect of chronic ingestion of the cysteine proteinase inhibitor, E-64, on Colorado potato beetle gut proteinases. *Entomol. Exper. Appli.* 82: 295-303.

Botella, M. A., Y. Xu, T. N. Prabha, Y. Zhao, M. L. Narasimhan, K. A. Wilson, S. S. Nielsen, R. A. Bressan, and P. M. Hasegawa (1996). Differential expression of soybean cysteine proteinase inhibitor genes during development and in response to wounding and methyl jasmonate. *Plant Physiol.* 112: 1201-1210.

Boulter, D. (1993). Insect pest control by copying nature using genetically engineered crops. *Phytochemistry* 34: 1453-1466.

Bowman, D. E. (1946). Differentiation of soybean antitrypsin factors. *Proc. Soc. Exp. Biol. Med.* 63: 547-550.

Broadway, R. M. (1995). Are insects resistant to plant proteinase inhibitors? *J. Insect Physiol.* 41: 107-116.

Broadway, R. M. and S. S. Duffey (1986a). The effect of dietary protein on the growth and digestive physiology of larval *Heliothis zea* and *Spodoptera exigua*. *J. Insect Physiol.* 32: 673-680.

Broadway, R. M. and S. S. Duffey (1986b). Plant proteinase inhibitors: Mechanism of action and effect on the growth and digestive physiology of larval *Heliothis zea* and *Spodoptera exigua*. *J. Insect Physiol.* 32: 827-833.

Brousseau, R., L. Masson, and D. Hegedus (1999). Insecticidal transgenic plants: Are they irresistible? *Review AgBiotechNet* July, ABN022.

Burgess, E. P. J., C. A. Main, P. S. Stevens, J. T. Christeller, A. M. R. Gatehouse, and W. A. Laing (1994). Effects of protease inhibitor concentration and combinations on the survival, growth and gut enzyme activities of the black field cricket, *Telleogryllus commodus*. *J. Insect Physiol.* 40: 803-811.

Casaretto, J. A. and L. J. Corcuera (1998). Proteinase inhibitor accumulation in aphid-infested barley leaves. *Phytochemistry* 49: 2279-2286.

Casey, R., C. Domoney, and N. Ellis (1986). Legume storage proteins and their genes. *Oxford Surv. Plant Molc. Cell Biol.* 3: 2-95.

Cleveland, T. E., R. W. Thornburg, and C. A. Ryan (1987). Molecular characterization of wound inducible inhibitor I gene from potato and the processing of its mRNA and protein. *Plant Molc. Biol.* 8: 199-207.

Cloutier, C., C. Jean, M. Fournier, S. Yelle, and D. Michaud (2000). Adult Colorado potato beetles, *Leptinotarsa decemlineata* compensate for nutritional stress on oryzacystatin I-transgenic potato plants by hypertrophic behaviour and overproduction of insensitive proteases. *Arch. Insect Biochem. Physiol.* 44: 69-81.

Conconi, A., M. J. Smerdon, G. A. Howe, and C. A. Ryan (1996). The octadecanoid signaling pathway in plants mediates a response to ultraviolet radiation. *Nature* 383: 826-829.

Cordero, M. J., D. Raventos, and B. sanSegundo (1994). Expression of a maize proteinase inhibitor gene is induced in response to wounding and fungal infection: Systemic wound response of a monocot gene. *Plant J.* 6: 141-150.

DeLeo, F., M. A. Bonade-Bottino, L. R. Ceci, R. Gallerani, and L. Jouanin (1998). Opposite effects on *Spodoptera littoralis* larvae of high expression level of a trypsin proteinase inhibitor in transgenic plants. *Plant Physiol.* 118: 997-1004.

Doares, S. H., T. Syrovets, E. W. Weiler, and C. A. Ryan (1995). Oligogalacturonides and chitosan activate plant defensive genes through the octadecanoid pathway. *Proc. Natl. Acad. Sci. USA* 92: 4095-4098.

Farmer, E. E. and C. A. Ryan (1990). Interplant communication: Airborne methyl jasmonate induces synthesis of proteinase inhibitors in plant leaves. *Proc. Natl. Acad. Sci. USA* 87: 7713-7716.

Fernandes, K. V. S., F. A. P. Campos, R. R. Doval, and J. Xavier-Filho (1991). The expression of papain inhibitors during development of cowpea seeds. *Plant Sci.* 74: 179-184.

Foard, D., P. Gutay, B. Ladin, R. Beachy, and B. Larkins (1982). In vitro synthesis of the Bowman-Birk and related soybean protease inhibitors. *Plant Molc. Biol.* 1: 227-243.

Foissac, X., N. T. Loc, P. Christou, A. M. R. Gatehouse, and J. A. Gatehouse (2000). Resistance to green leafhopper *(Nephotettix virescens)* and brown planthopper

(Nilaparvata lugens) in transgenic rice expressing snowdrop lectin *Galanthus nivalis* agglutinin; (GNA). *J. Insect Physiol.* 46: 573-583.

Garcia-Olmedo, F., G. Salcedo, R. Sanchez-Monge, L. Gomez, J. Royo, and P. Carbenero (1987). Plant proteinaceous inhibitors of proteinases and α-amylases. *Oxford Surv. Plant Molc. Cell Biol.* 4: 275-334.

Gatehouse, A. M. R. and D. Boulter (1983). Assessment of the antimetabolic effects of trypsin inhibitors from cowpea *(Vigna unguiculata)* and other legumes on development of the bruchid beetle *Callosobruchus maculatus. J. Sci. Food Agric.* 34: 345-350.

Gatehouse, J. A., I. M. Evans, R. R. D. Croy, and D. Boulter (1986). Differential expression of genes during legume seed development. *Phil. Trans R. Soc. Lond. B.* 314: 367-384.

Girard, C., M. Le Metayer, M. Bonade-Bottino, M. H. Pham-Delegue, and L. Jouanin (1998). High level of resistance to proteinase inhibitors may be conferred by proteolytic cleavage in beetle larvae. *Insect Biochem. Molc. Biol.* 28: 229-237.

Girard, C., M. Le Metayer, B. Zaccomer, E. Bartlet, I. Williams, M. Bonade-Bottino, M. H. Pham-Delegue, and L. Jouanin (1998). Growth stimulation of beetle larvae reared on trangenic oilseed rape expressing a cysteine proteinase inhibitor. *J. Insect Physiol.* 44: 263-270.

Giri A. P., A. M. Harsulkar, V. V. Deshpande, M. N. Sainani, V. S. Gupta, and P. K. Ranjekar (1998). Chickpea defensive proteinase inhibitors can be inactivated by podborer gut proteinases. *Plant Physiol.* 116: 393-401.

Godbole, S. A., T. G. Krishna, and C. R. Bhatia (1994). Changes in protease inhibitory activity from pigeonpea (*Cajanus cajan* (L.) Millsp) during seed development and germination. *J. Sci. Food Agric.* 66: 497-501.

Goldberg, R. B., S. J. Barker, and L. Perez-Grau (1989). Regulation of gene expression during plant embryogenesis. *Cell* 56: 149-160.

Goldberg, R. B., G. Hoschek, S. H. Tam, G. S. Ditta, and R. W. Breidenbach (1981). Abundance, diversity and regulation of mRNA sequence sets in soybean embryogenesis. *Dev. Biol.* 83: 201-217.

Green, T. R. and C. A. Ryan (1972). Wound induced proteinase inhibitors in plant leaves: A possible defence mechanism against insects. *Science* 175: 776-777.

Gutierrez, C., R. Sanchez-Monge, L. Gomez, M. Ruiz-Tapiador, P. Castanera, and G. Salcedo (1990). α-Amylase activities of agricultural insect pests are specifically affected by different inhibitor preparations from wheat and barley endosperm. *Plant Sci.* 72: 37-44.

Hammond, R., D. Foard, and B. Larkins (1984). Molecular cloning and analysis of a gene coding for the Bowman-Birk protease inhibitor in soybean. *J. Biol. Chem.* 259: 9883-9890.

Harsulkar, A. M., A. P. Giri, V. S. Gupta, M. N. Sainani, V. V. Deshpande, A. G. Patankar, and P. K. Ranjekar (1998). Characterization of *Helicoverpa armigera* gut proteinases and their interaction with proteinase inhibitors using gel-X-ray film contact print technique. *Electrophoresis* 19: 1397-1402.

Harsulkar, A. M., A. P. Giri, and V. S. Kothekar (1997). Protease inhibitors of chickpea (*Cicer arietinum* L.) during seed development. *J. Sci. Food Agric.* 74: 509-512.

Harsulkar, A. M., A. P. Giri, A. G. Patankar, V. S. Gupta, M. N. Sainani, P. K. Ranjekar, and V. V. Deshpande (1999). Successive use of non-host plant proteinase inhibitors required for effective inhibition of gut proteinases and larval growth of *Helicoverpa armigera. Plant Physiol.* 121: 497-506.

Heath, R. L., G. McDonald, J. T. Christeller, M. Lee, K. Bateman, J. West, R. van Heeswijck, and M. A. Anderson (1997). Proteinase inhibitors from *Nicotiana alata* enhance plant resistance to insect pests. *J. Insect Physiol.* 43: 833-842.

Heinrikson, R. L. and F. J. Kezdy (1976). Acid cysteine proteinase inhibitors from pineapple stem. *Method. Enzymol.* 45: 740-751.

Hilder, V. A., A. M. R. Gatehouse, and D. Boulter (1993). Transgenic plants conferring insect tolerance: Proteinase inhibitor approach. In S. D. Kung and R. Wu (eds.), *Transgenic plants engineering and utilization,* Volume 1 (pp. 317-338). San Diego, CA: Academic Press Inc.

Hilder, V. A., A. M. R. Gatehouse, S. E. Sheerman, R. F. Barker, and D. Boulter (1987). A novel mechanism of insect resistance engineered into tobacco. *Nature* 330: 160-163.

Hildmann, T., M. Ebneth, H. Pena-Cortes, J. Sanchez-Serrano, L. Willmitzer, and S. Prat (1992). General roles of abscisic and jasmonic acids in gene activation as a result of mechanical wounding. *Plant Cell* 4: 1157-1170.

Hobday, S. M., D. A. Thurman, and D. J. Barber (1973). Proteolytic and trypsin inhibitory activities in extracts of germinating *Pisum sativum* seeds. *Phytochemistry* 12: 1041-1046.

Hojima, Y., J. V. Pearce, and J. J. Pisano (1980). Plant inhibitors of serine proteinases: Hageman factor fragment, kallireins, plasmin, thrombin factor Xa, trypsin and chymotrypsin. *Thromb. Res.* 20: 163-171.

Horisberger, M. and M. Tacchini (1982). Ultrastructural localization of Kunitz trypsin inhibitor in soybeans using gold granules labeled with protein A. *Experientia* 38: 726.

Hwang, D. L., W. K. Yang, and D. E. Foaard (1978). Rapid release of protease inhibitors from soybeans: Immunochemical quantitation and parallels with lectin. *Plant Physiol.* 61: 30-34.

Ishimoto, M., T. Sato, M. J. Chrispeels, and K. Kitamura (1996). Bruchid resistance of transgenic azuki bean expressing seed α-amylase inhibitor of common bean. *Entomol. Experi. Appli.* 79: 309-315.

Johnson, R., J. Narvaez, G. An, and C. A. Ryan (1989). Expression of proteinase inhibitors I and II in transgenic tobacco plants: Effects on natural defense against *Manduca sexta. Proc. Natl. Acad. Sci. USA* 86: 9871-9875.

Johnston, K. A., J. A. Gatehouse, and J. H. Anstee (1993). Effect of soybean protease inhibitors on the growth and development of larval *Helicoverpa armigera. J. Insect Physiol.* 39: 657-664.

Jongsma, M. A., P. L. Bakker, J. Peters, D. Bosch, and W. J. Stiekema (1995). Adaptation of *Spodoptera exigua* larvae to plant proteinase inhibitors by induction of gut proteinase activity insensitive to inhibition. *Proc. Natl. Acad. Sci. USA* 92: 8041-8045.

Jongsma, M. A., P. L. Bakker, B. Visser, and W. J. Stiekema (1994). Trypsin inhibitor activity in mature tobacco and tomato plants is mainly induced locally in response to insect attack, wounding and virus infection. *Planta* 195: 29-35.

Jongsma, M. A. and C. J. Bolter (1997). The adaptation of insects to plant protease inhibitors. *J. Insect Physiol.* 43: 885-895.

Jongsma, M. A., J. Peters, W. J. Stiekema, and D. Bosch (1996). Characterization and partial purification of gut proteinases of *Spodoptera exigua* Hilbner (Lepidoptera: Noctuidae). *Insect Biochem. Molc. Biol.* 26: 185-193.

Joshi, B.N., M. N. Sainani, K. B. Bastawade, V. S. Gupta, and P. K. Ranjekar (1998). Cysteine protease inhibitor from pearl millet: A new class of antifungal protein. *Biochem. Biophys. Res. Commun.* 246: 382-387.

Jouanin L., M. Bonade-Bottino, C. Girard, G. Morrot, and M. Giband (1998). Transgenic plants for insect resistance. *Plant Sci.* 131: 1-11.

Kamalay, J. C. and R. B. Goldberg (1980). Regulation of structural gene expression in tobacco. *Cell* 19: 935-946.

Keil, M., J. Sanchez-Serrano, J. Schell, and L. Willmitzer (1986). Primary structure of proteinase inhibitor II gene from potato *(Solanum tuberosum). Nucl. Res.* 14: 5641-5650.

Koiwa, H., R. A. Bressan, and P. M. Hasegawa (1997). Regulation of protease inhibitors and plant defense. *Trends Plant Sci.* 2: 379-384.

Korth, K. L. and R. A. Dixon (1997). Evidence for chewing insect-specific molecular events distinct from a general wound response in leaves. *Plant Physiol.* 11: 1299-1305.

Kortt, A. A. (1980). Isolation and properties of a chymotrypsin inhibitor from winged bean (*Psophocarpus tetragonolobus* (L) DC). *Biochem. Biophys. Acta* 577: 237-248.

Kortt, A. A., P. M. Strike, and J. De Jersey (1989). Amino acid sequence of a crystalline seed albumin (winged bean albumin-1) from *Psophocarpus tetragonolobus* (L.) DC.: Sequence similarity with Kunitz-type seed inhibitors and 7S storage globulins. *Eur. J. Biochem.* 181: 403-408.

Kothekar, V. S., A. M. Harsulkar, and A. R. Khandelwal (1996). Low trypsin and chymotrypsin inhibitor mutants in winged bean (*Psophocarpus tetragonolobus* (L.) D.C.). *J. Sci. Food Agric.* 71: 137-140.

Kreis, M., P. R. Shewry, B. G. Forde, J. Forde, and B. J. Miflin (1985). Structure and evolution of seed storage proteins and their genes with particular reference to those of wheat, barley and rye. *Oxford Surv. Plant Molc. Cell Biol.* 2: 253-317.

Kunitz M. (1945). Crystallization of a soybean trypsin inhibitor from soybean. *Science* 101: 668-669.

Larque, A. M. and J. G. Houseman (1990). Effect of ingested soybean ovomucoid and corn protease inhibitors on digestive processes of the European corn borer, *Ostrinia nubilalis* (Lepidoptera: Pyralidae). *J. Insect Physiol.* 36: 691-697.

Lee, J. S., W. E. Brown, J. S. Graham, G. Pearce, E. A. Fox, T. W. Dreher, K. G. Ahern, G. D. Pearson, and C. A. Ryan (1986). Molecular characterization and phylogenetic studies of a wound-inducible proteinase inhibitor I gene in *Lycopersicon* species. *Proc. Natl. Acad Sci. USA* 83: 7277-7281.

Lee, S. I., S. H. Lee, J. C. Koo, H. J. Chun, C. O. Lim, J. H. Mun, Y. H. Song, and M. J. Cho (1999). Soybean Kunitz trypsin inhibitor (SKTI) confers resistance to the brown planthopper (*Nilaparvata lugens* Stal) in transgenic rice. *Mol. Breed.* 5: 1-9.

Lewis, W. J., J. C. van Lenteren, S. C. Phatak, and J. H. Tumlinson III (1997). A total system approach to sustainable pest management. *Proc. Natl. Acad Sci. USA* 94: 12243-12248.

Liener, I. E. and M. L. Kakade (1980). Proteinase inhibitors. In I. E. Liener (ed.), *Toxic constituents of plant foodstuffs,* Second edition. New York: Academic Press.

Ling, M. H., H. Qi, and C. Chi (1993). Protein, cDNA and genomic DNA sequences of the towel gourd trypsin inhibitor. *J. Biol. Chem.* 268: 810-814.

Macintosh, S. C., G. M. Kishore, F. J. Perlak, P. G. Marrone, T. B. Stone, S. R. Sims, and R. L. Fuchs (1990). Potentiation of *Bacillus thuringiensis* insecticidal activity by serine proteinase inhibitors. *J. Agric. Food Chem.* 38: 1145-1152.

McManus, M. T. and E. P. J. Burgess (1995). Effects of the soybean (Kunitz) trypsin inhibitor on growth and digestive proteases of larvae of *Spodoptera litura. J. Insect Physiol.* 41: 731-738.

Michaud, D. (1997). Avoiding protease-mediated resistance in herbivorous pests. *Trends Biotechnol.* 15: 4-6.

Mikola, J. and E. M. Soulinna (1969). Purification and properties of a trypsin inhibitor in barley. *Eur. J. Biochem.* 9: 555-557.

Morton, R. L., H. E. Schroeder, K. S. Bateman, M. J. Chrispeels, E. Armstrong, and T. J. V. Higgins (2000). Bean α-amylase inhibitor 1 in transgenic peas *(Pisum sativum)* provides complete protection from pea weevil *(Bruchus pisorum)* under field conditions. *Proc. Natl. Acad Sci. USA* 97: 3820-3825.

Muel, F., B. Carrouee, and F. Grosjean (1998). Trypsin inhibitor activity of pea cultivars: New data and a proposal strategy for breeding programs. In AEP (Association Europeenne de recherche sur les Proteagineux) (ed.), *Proceedings of the Third European Conference on Grain Legumes* (pp. 164-165). Valladolid, Spain:AEP, Paris.

Nishikata, M. (1984). Trypsin-like protease from soybean seeds: Purification and some properties. *J. Biochem.* 95: 1169-1177.

Odani, S., T. Koide, and T. Ono (1982). Sequence homology between barley trypsin inhibitor and wheat α-amylase inhibitors. *FEBS Lett.* 141: 279-282.

Odani, S., T. Koide, and T. Ono (1983). A possible evolutionary relationship between plant trypsin inhibitor, α-amylase inhibitor and mammalian pancreatic secretory trypsin inhibitor (Kazal). *J. Biochem.* 93: 1701-1704.

Orf, J. H. and T. Hymowitz (1979). Inheritance of the absence of the Kunitz trypsin inhibitor in seed protein of soybeans. *Crop Sci.* 19: 107-109.

Osborne, T. B. (1924). *The vegetable proteins,* Second edition. London: Longmans, Green and Co.

Otlewski, J. and D. Krowarsch (1996). Squash inhibitor family of serine proteinases. *Acta Biochem. Polonica* 43: 431-444.

Patankar, A. G., A. P. Giri, A. M. Harsulkar, M. N. Sainani, V. V. Deshpande, P. K. Ranjekar, and V. S. Gupta (2001). Complexity in specificities and expression of *Helicoverpa armigera* gut proteinases explains polyphagous nature of the insect pest. *Insect Biochem. Molec. Biol.* 31: 453-464.

Patankar, A. G., A. M. Harsulkar, A. P. Giri, V. S. Gupta, M. N. Sainani, P. K. Ranjekar, and V. V. Deshpande (1999). Diversity in inhibitors of trypsin and *Helicoverpa armigera* gut proteinases in chickpea *(Cicer arietinum)* and its wild relatives. *Theor. Appl. Genet.* 99: 719-726.

Pautot, V., F. M. Holzer, and L. L. Walling (1991). Differential expression of tomato proteinase inhibitor I and II genes during bacterial pathogen invasion and wounding. *Molc. Plant Microbe Interact.* 4: 284-292.

Pearce, G., D. Strydom, H. Johnson, and C. A. Ryan (1991). A polypeptide from tomato leaves induces wound-inducible proteinase inhibitor proteins. *Science* 253: 895-898.

Peña-Cortés, H., J. Fisahn, and L. Willmitzer (1995). Signals involved in wound-induced proteinase inhibitor II gene expression in potato and tomato plants. *Proc. Natl. Acad. Sci. USA* 92: 4106-4113.

Peña-Cortés, H., J. J. Sanchez-Serrano, R. Mertens, L. Willmitzer, and S. Prat (1989). Abscisic acid is involved in the wound-induced expression of the proteinase inhibitor II gene in potato and tomato. *Proc. Natl. Acad. Sci. USA* 86: 9851-9855.

Rele, M. V., H. G. Vartak, and V. Jagannathan (1980). Proteinase inhibitors from *Vigna ungiculata* subsp. *cylindrica:* Occurrence of thiol proteinase inhibitors in plants and purification from *Vigna ungiculata* subsp. *cylindrical. Arch. Biochem. Biophys.* 202: 117-128.

Ryan, C. A. (1984). Defence responses of plants. In D. P. S. Verma and T. H. Hohn (eds.), *Plant gene research: Genes involved in microbe plant interactions* (pp. 375-386). Wien, New York: Springer Verlag.

Ryan, C. A. (1990). Proteinase inhibitors in plants: Genes for improving defenses against insects and pathogens. *Annu. Rev. Phytopathol.* 28: 245-449.

Schaller, A. and C. A. Ryan (1995). Systemin—A polypeptide defense signal in plants. *BioEssays* 18: 27-33.

Schuler, T. H., G. M. Poppy, B. R. Kerry, and I. Denholm (1998). Insect-resistant transgenic plants. *Trends Biotechnol.* 16: 168-175.

Seemuller, U., M. Eulitz, H. Fritz, and A. Strobl (1980). Structure of the elastase-cathepsin G inhibitor of the leech *Hirudo medicinalis. Hoppe-Seyler's Z. Physiol. Chem.* 361: 1841-1846.

Shade, R. E., H. E. Schroeder, J. J. Pueyo, L. M. Tabe, L. L. Murdock, T. J. V. Higgins, and M. J. Chrispeels (1994). Transgenic pea seeds expressing α-amylase inhibitor of the common bean are resistant to bruchid beetles. *Bio/Technol.* 12: 793-796.

Shewry, P. R. (1999). Enzyme inhibitors of seeds: Types and properties. In P. R. Shewry and R. Casey (eds.), *Seed protein* (pp. 587-615). Amsterdam, the Netherlands: Kluwer Academic Publishers.

Stankovic, B. and E. Davies (1997). Intercellular communication in plants: Electrical stimulation of proteinase inhibitor gene expression in tomato. *Planta* 202: 402-406.

Svendsen, I., S. Boisen, and J. Hejgaard (1982). Amino acid sequence of serine protease inhibitor CI-1 from barley: Homology with barley inhibitor CI-2, potato inhibitor I and leech eglin. *Carlsberg Res. Commun.* 47: 45-53.

Svendsen, I., J. Hejgaard, and J. K. Chavan (1984). Subtilisin inhibitor from seeds of broad bean *(Vicia faba)*, purification, amino acid sequence and specificity of inhibition. *Carlsberg Res. Commun.* 49: 493-502.

Svendsen, I., I. Jonassen, J. Hejgaard, and S. Boisen (1980). Amino acid sequence homology between a serine protease inhibitor from barley and potato inhibitor I. *Carlsberg Res. Commun.* 45: 389-395.

Tamayo, M. C., M. Rufat, J. M. Bravo, and B. San Segundo (2000). Accumulation of a maize proteinase inhibitor in response to wounding and insect feeding, and characterization of its activity toward digestive proteinases of *Spodoptera littoralis* larvae. *Planta* 211: 62-71.

Tan-Wilson, A. L. and K. A. Wilson (1982). Nature of proteinase inhibitors related from soybeans during imbibition and germination. *Phytochemistry* 21: 1547-1551.

Urwin, P. E., H. J. Atkinson, D. A. Walter, and M. J. McPherson (1995). Engineered oryzacystatin I expressed in transgenic hairy roots confers resistance to *Globodera pallida. Plant. J.* 8: 121-131.

Valueva, T. A. and V. V. Mosolov (1999a). Protein inhibitors of proteinases in seeds: 1. Classification, distribution structure and properties. *Russian J. Plant Physiol.* 46: 362-378.

Valueva, T. A. and V. V. Mosolov (1999b). Protein inhibitors of proteinases in seeds: 2. Physiological functions. *Russian J. Plant Physiol.* 46: 379-387.

Vernekar, J. V., M. S. Ghatge, and V. V. Deshpande (1999). Alkaline protease inhibitor: A novel class of antifungal proteins against phytopathogenic fungi. *Biochem. Biophys. Res. Commun.* 262: 702-707.

Vodkin, L. O. (1981). Isolation and characterization of messenger RNAs for seed lectin and Kunitz inhibitor in soybeans. *Plant Physiol.* 68: 766-771.

Welham, T., M. O'Neill, S. Johnson, T. L. Wang, and C. W. Domoney (1998). Expression patterns of genes encoding seed trypsin inhibitors in *Pisum sativum. Plant Sci.* 131: 13-24.

Wieczorek, M., J. Otlewski, J. Cook, K. Parks, J. Leluk, A. Wilimowskapelc, A. Polanowski, T. Wilusz, and M. Laskowski Jr. (1985). The squash family of serine inhibitors: Amino acid sequences and association equilibrium constants of inhibitors from squash, summer squash, zucchini and cucumber seeds. *Biochem. Biophys. Res. Comm.* 126: 646-565.

Wilson, D. C., J. F. Thain, P. E. H. Minchin, I. R. Gub, A. J. Reily, Y. D. Skipper, H. M. Doherty, P. J. O'Donnell, and D. J. Bowles (1992). Electrical signaling and systemic proteinase inhibitor induction in the wounded plant. *Nature* 360: 62-65.

Wilson, K. A. (1980). The release of proteinase inhibitors from legume seeds during germination. *Phytochemistry* 19: 2517-2519.

Wilson, K. A. and J. C. Chen (1983). Amino acid sequences of mung bean trypsin inhibitors and its modified form appearing during germination. *Plant Physiol.* 71: 341-349.

Xavier Filho, J. (1992). The biological roles of serine and cystine proteinase inhibitors in plants. *R. Bras. Fisiol. Veg.* 4: 1-6.

Xu, D., Q. Xue, D. McElroy, Y. Mawal, V. A. Hilder, and R. Wu (1996). Constitutive expression of a cowpea trypsin inhibitor gene, *CpTi,* in transgenic rice plants confers resistance to two major rice insect pests. *Mol. Breed.* 2: 167-173.

Yoshikawa, M., T. Kiyohara, T. Iwasaki, and I. Yoshida (1979). Modification of proteinase inhibitor II in azuki beans during germination. *Agric. Biol. Chem.* 43: 1989-1990.

Zhao, Y., M. A. Botella, L. Subramanian, X. Niu, S. S. Nielsen, R. A. Bressan, and P. M. Hasegawa (1996). Two wound-inducible soybean cysteine proteinase inhibitors have greater insect digestive proteinase inhibitory activities than a constitutive homologue. *Plant Physiol.* 111: 1299-1306.

Chapter 18

Soil Seed Banks

A. J. Murdoch

WHAT IS THE SOIL SEED BANK?

The soil seed bank comprises viable seeds in the soil. Seed banks form in many environments ranging from the freshwater aquatic to the arid, from the tropical to the arctic (and antarctic), and from lowlands to highlands. Roberts (1981) and Leck et al. (1989) have reviewed seed banks in various ecosystems.

Longevity can be a useful criterion for classifying seed banks. Thompson and Grime (1979) first classified transient and persistent seed banks, defining them as those which persist for less or more than one year, respectively. The period of survival is not, however, an absolute characteristic of species. Thompson et al. (1997) utilized a modification of this classification, finding it useful to distinguish "short-term" and "long-term" persistency. The short-term persistent seeds survive for up to five years, while the long-term seeds are vialbe for more than five years. The reality is that persistence due to endogenous physiological factors is likely to comprise a continuum.

Popular literature inevitably highlights the remarkable record breakers in the seed survival game. There is a danger of these classifications drawing attention to the small fraction of long-lived individuals in some seed populations. Not only are survival periods not an absolute characteristic of species, but they vary widely within seed lots and even between seeds maturing on the same mother plant. One extreme example is the capsule of *Xanthium pensylvanicum,* which contains two seeds. The upper seed is more dormant than the lower, and at least one year usually separates the germination of the two seeds in the soil (Esashi and Leopold, 1968; Harper, 1977).

Seed survival curves lead to a prediction that small numbers of individuals may indeed survive for long periods. I therefore argue that practical applications of seed bank data for restoring degraded habitats and weed management strategies must be based on quantitative parameters of seed longevity derived from studies of the overall seed population rather than the behavior of a few extreme individuals.

In this chapter, I will consider the functions of and prerequisites for soil seed banks from both ecological and human perspectives. Alternative means of achieving these functions will also be mentioned. The size, composition, source, and destinies of soil seed banks will then be described.

FUNCTIONS OF SOIL SEED BANKS

The term *bank* implies a reservoir of viable seeds in the soil from which regeneration may take place. The soil seed bank is, therefore, a means of achieving survival and the temporal dispersal of a species. Although the periods for which the seeds of some species can survive are impressive, the banking analogy highlights a second function. The seed bank is like an account which can be drawn on especially when the environmental conditions may favor seedling establishment and successful regeneration. The seed bank is often a reserve that is drawn on in response to or immediately after an environmental disturbance. These may be large-scale disturbances such as fire, flooding, or the cultivation of a field, or smaller-scale effects such as a worm cast or a mole hill.

In essence, the function of the soil seed bank for the plant is to achieve a regeneration strategy to maximize the probability that seeds will be able to germinate in the right place at the right time (Murdoch and Ellis, 2000)—that is, when the chances of successful regeneration are high.

For other organisms in the soil, the function is of course completely different—it is often a food source. Humankind often perceives the seed bank as having a nuisance function, being the source of many weed infestations; but depending on the species present, it may also be the simplest means of regenerating degraded habitats with native genotypes.

Alternatives to Soil Seed Banks As Strategies to Promote Survival

Many species fail to form persistent seed banks, so it is important to emphasize that survival is achieved by diverse reproductive strategies. For example, some seeds are adapted for spatial dispersal by wind, water, or animals. Not all seed banks are in soil: some species form an aerial seed bank on the mother plant. Others achieve survival and the capacity to regenerate in appropriate circumstances through long-lived adults (e.g., many trees), seedling banks, bud banks, and vegetative organs such as tubers, corms, and bulbs.

PREREQUISITES FOR A SEED BANK

There are four prerequisites for seed banks, the first three of which are essential to achieve the first purpose of seed survival.

Viability Must Be Preserved

Many seeds have a remarkable capacity for longevity in the soil, but this is possible only if the viability can be maintained. Some species have such short intrinsic viability periods that they form only transient seed banks. Examples include species showing recalcitrant seed storage behavior (Roberts, 1973). Recalcitrant seeds rapidly lose viability at water potentials lower than approximately –1.5 to –5.0 MPa (Probert and Longley, 1989; Roberts and Ellis, 1989; Pritchard, 1991) and so have any potential for survival only in continually moist soil. These species comprised 7.4 percent of 6,919 species from 251 families (Hong and Ellis, 1996). Temperate species with recalcitrant seeds include oak (*Quercus* spp.), chestnut (*Castanea* spp.), and sycamore *(Acer pseudoplatanus)*. Seeds of these species are shed in the autumn, but their persistence in the soil is short and at most they may survive until they germinate in early spring after chilling of the seeds during the winter. Tropical species in this category include some members of the Dipterocarpaceae and Araucariaceae, and the formation of seedling rather than seed banks is more likely (Murdoch and Ellis, 2000).

At the other extreme, 88.6 percent of the 6,919 species probably exhibit orthodox seed storage behavior (Hong and Ellis, 1996). The longevity of orthodox seeds generally increases with decrease in temperature and in seed water potential to around –350 MPa in accordance with the viability equation (Roberts, 1973; Ellis and Roberts, 1980). The viability equation ceases to apply at seed water potentials above –10 to –20 MPa, and fully imbibed seeds survive for long periods provided air is present (Murdoch and Ellis, 2000). The metabolism occurring in imbibed seeds facilitates the preservation of viability since damage is repaired and microbial decay may be resisted (Murdoch and Ellis, 2000). Provided that the other prerequisites of seed banks are met, it is suggested that species with orthodox seed storage behavior are the only ones capable of forming persistent seed banks due to the variable soil moisture status of most soils.

About 1.9 percent of species were classified by Hong and Ellis (1996) as showing seed storage behavior that is intermediate between the orthodox and recalcitrant extremes.

Germination Must Be Prevented

No matter how spectacular the potential longevity, and it can be quantified for orthodox seeds by the parameter K_E in the viability equation, nothing approaching the potential will ever be realized if a seed germinates shortly after entering the soil seed bank. Dormancy or quiescence is therefore essential.

Dormancy and Quiescence

Dormancy can be defined as the failure of a viable seed to germinate given moisture, air and a suitable constant temperature for radicle emergence and seedling growth (Murdoch, 2003).

If one or more of the three prerequisites for germination of a nondormant seed are lacking in the environment, the seed is quiescent because metabolism is likely to be reduced. Quiescence may therefore be defined as the failure of a viable seed to germinate due to shortage of water, poor aeration or an unsuitable temperature for radicle emergence and seedling growth (Murdoch, 2003).

A seed in dry soil is therefore in an environmentally enforced state of quiescence. Arguably, it is not dormant, although many include it in their definition of dormancy (e.g., Baskin and Baskin, 1998). Commonly called hardseededness, this condition represents a type of endogenous quiescence.

Are Dormancy or Quiescence Linked to Longevity?

Some of the longest-lived seeds in soil seed banks are hardseeded and will survive for as long as the seed coat is impermeable to water. Longevity and dormancy are not, however, linked physiologically, though they often are ecologically (Roberts, 1973). Viable seeds in dry soil will gradually lose viability irrespective of their dormancy. However, as soon as sufficient moisture and a suitable temperature for seedling growth are provided, the nondormant seeds will, by definition, germinate, so persistence becomes dependent on dormancy or quiescence.

Predation Must Be Avoided

In this respect, small seeds have an immediate advantage over large seeds and buried seeds over those at the soil surface. For example, when the relatively large wild oat seeds (mature florets) were placed in polyester mesh packets at the soil surface in December 1974, very few intact, viable seeds survived beyond April or May 1975. The depleted packets gave evidence of predation: the packets had been torn open and the caryopses extracted, leaving the lemma and palea intact. By contrast, losses of seeds packaged identically but buried at depths of 25, 75, and 230 mm were mostly accounted for by in situ germination, amounting to about 40 percent after 19 months at each depth (Murdoch and Roberts, 1982; Murdoch, 1983).

Not surprisingly, it has therefore been suggested that the ability to avoid predation and hence to persist in the soil seed bank is associated with seed size and shape. The formulation of this hypothesis was first made by Thompson et al. (1993) based on 97 British species in which they noted that small, round-shaped seeds tended to be persistent. Several other studies have been carried out in a wide range of habitats and countries. Several of these studies support the hypothesis that small seeds tend to form long-term persistent seed banks (e.g., Thompson et al., 2001; Cerabolini et al., 2003; Peco et al., 2003). A notable exception appeared to be a survey of 101 species from a range of Australian habitats in which there was no correlation between seed mass (0.217 to 648.9 mg) and persistence (Leishman and Westoby, 1998). For 47 species in New Zealand forests, Moles et al. (2000) reported some tendency from small, rounded seeds to show greater persistence, but it was also not uncommon for large and/or elongated seeds to be persistent. The persistence of small seeds has been explained by the largely untested hypothesis that such seeds are more easily buried and are better able to avoid predation. It is also clear that other factors are important; in arable agroecosystems, for example, cultivation buries the relatively large seeds of *Avena fatua* and *Galium aparine* and permits them to avoid the early seed losses associated with predation. It is also emphasised that avoidance of predation is only one prerequisite for the formation of a persistent seed bank, so it is not surprising that exceptions do occur.

There is one final prerequiste without which a seed bank can never be said to be "successful." It relates to the second purpose of the soil seed bank, namely regeneration.

Some Seeds Must Germinate

The ultimate purpose of the seed bank is not, however, to maintain viability, avoid predation, or prevent germination. Perhaps the most important prerequiste without which a seed bank can never be said to be "successful" relates to the second purpose of regeneration. Seeds must eventually germinate. Using the banking analogy, some seeds need to be transferred from a deeply dormant "savings" account to a less or nondormant "current" account. These latter seeds need to be at a low enough level of dormancy so as to be capable of responding to environmental triggers associated with suitable conditions when the chances of successful regeneration are high. In transient seed banks, most if not all seeds will germinate in a given year, whereas in persistent seed banks, only a small fraction of the seed bank may become capable of germinating each year. Such quantitative or qualitative differences in dormancy within seed populations mean that even if some seeds germinate and then fail to establish due to adverse conditions after germination, others may survive for another day (cf. Sharifzadeh and Murdoch, 2001). In persistent seed banks, the proportion of seeds in a non-dormant condition depends on the annual dormancy cycle and the dormancy-relieving stimuli present in the soil (Murdoch, 1998).

HOW MANY SEEDS ARE IN THE SOIL?

Numbers vary widely within and among habitats (Leck et al., 1989), as well as with methods of assessment. Estimates based on the physical extraction of seeds from soil tend to give much higher estimates than the methods based on seedling emergence from soil samples. Many authors have carried out surveys, some extensive and others focusing on one or a few fields. I have limited this discussion to three surveys covering a large number of fields under fairly similar management, being arable or vegetable fields in the Midlands of England. Median seed banks in the plowed horizon of soils subject to frequent tillage and cropping in these surveys were 5-10,000 seeds·m^{-2} (see Table 18.1). Variation around these median values is high (see Figure 18.1) and some fields have very high populations of buried seeds (Table 18.1). Nevertheless, these seed densities are much lower than those reported by Jensen (1969) from his studies where the mean seed bank in 37 Danish cereal fields was 62,700 seeds·m^{-2} (see Table 18.2). Jensen's selection of fields was said to be biased toward heavily infested fields, but very high populations of *Juncus bufonis* in some fields and assessment by laboratory seed germination tests also account for some of the difference.

The Roberts and Stokes (1966) study was carried out in fields in which most had never been exposed to herbicides. The two later studies were for fields that had exposure to herbicides. The inference may be made that the use of herbicides may have caused a decline in the soil seed bank of about 50 percent based on the difference in median observations (Table 18.1). Caution must be exercised in making this comparison, however, as different fields were assessed in each case.

The practical importance of the soil seed bank is highlighted by comparing typical seed rates for crops with these observations. For example,

TABLE 18.1. Size and composition of some soil seed banks of commercial vegetable/arable fields in England and Wales.

Years of sampling	Number of fields	Number of seeds· m^{-2} Median	Min.	Max.	Number of species per field Mean	Min.	Max.	Species	Percentage of fields in which found	Number of fields with >n seeds·m^{-2}
1. Commercial vegetables with minimal herbicide use										n = *1,000*
1958-1962	58	10,200	1,600	86,000	14	4	30	*Stellaria media*	100	28
								Poa annua	98	39
								Urtica urens	93	30
								Senecio vulgaris	90	8
								Capsella bursa-pastoris	86	18
								Chenopodium album	84	18
								Veronica persica	71	10
2. Commercial vegetables with considerable herbicide use										n = *250*
1968-1975	89	4,120	0	24,330				*Poa annua*	90	68
								Stellaria media	74	42
								Matricaria matricarioides	69	41
								Polygonum aviculare	53	27
								Urtica urens	49	22
								Chenopodium album	49	18

TABLE 18.1 *(continued)*

Years of sampling	Number of fields	Number of seeds· m^{-2}			Number of species per field			Species	Percentage of fields in which found	Number of fields with >n seeds·m^{-2}
		Median	Min.	Max.	Mean	Min.	Max.			
								Capsella bursa-pastoris	48	18
								Veronica persica	39	18
								Senecio vulgaris	39	13
3. Arable fields sampled three times										n = *625*
1972-1973	64	4,360	1,500	67,000	17[a]	8[a]	32[a]	*Poa annua*	100	35
								Polygonum aviculare	92	17
1974-1975	64	4,930						*Stellaria media*	90	20
								Bilderdykia convolvulus	70	2
1976-1977	64	2,900						*Aethusa cynapium*	68	7
								Veronica persica	67	19
								Aloepecurus myosuroides	67	18
								Chenopodium album	66	7

Source: Collated from 1. Roberts and Stokes, 1966; 2. Roberts and Nielson, 1982; 3. Roberts and Chancellor, 1986, and Chancellor, 1979.
Note: All sampled to depth of 15 cm. All assays were based on seedling emergence from the soil samples over approximately two years.
[a]In 32 of the fields.

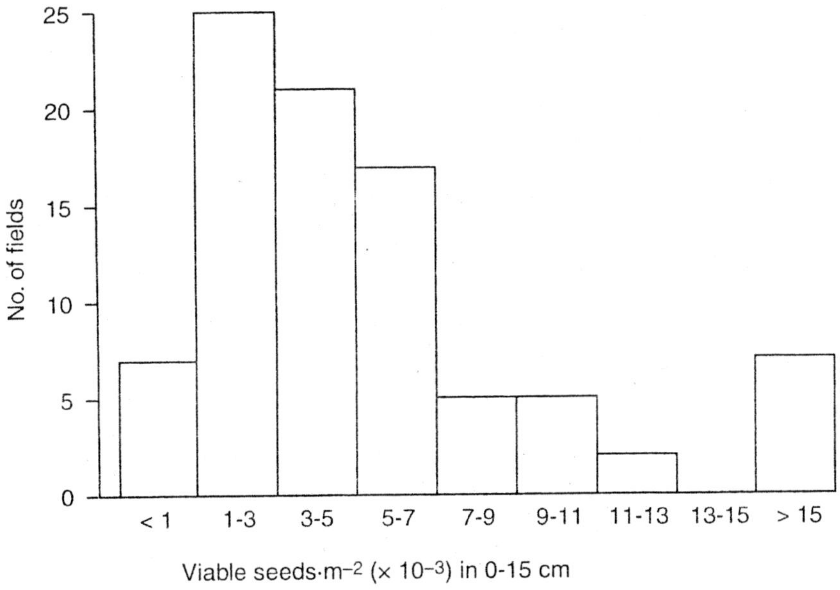

FIGURE 18.1. Frequency distribution of the size of soil seed banks in the top 15 cm among 89 fields in the United Kingdom. (*Source:* From H.A. Roberts and J.E. Neilson, Seed banks of soils under vegetable cropping in England, *Weed Research,* vol. 22, pp. 13-16. © 1982 Blackwell Publishing. Used with permission.)

TABLE 18.2 Viable seeds (× $10^3 \cdot m^{-2}$) in arable fields in Denmark (soils were sampled).

Cropping	Number of fields	All species Mean	All species Max.	All species Min.	All species except *Juncus bufonius* Mean	All species except *Juncus bufonius* Max.	All species except *Juncus bufonius* Min.
Roots	20	27.2	125.5	0.6	14.9	44.8	0.6
Cereals	37	62.7	496.2	2.2	28.5	100.7	2.2

Source: Created from data in Jensen, 1969.

sowing densities for lettuce, wheat, and *Phleum pratense* are approximately 100, 300, and, at the other extreme, 5,900 seeds·m^{-2}. This potential weed infestation is seldom realized on account of dormancy, quiescence, and microsite-specific factors.

From a grower's or farmer's perspective, the relative abundance of the various species in the soil seed bank is also important. In Roberts and Stokes's (1966) research, four or five species in each field individually comprised at least 5 percent, and one or two species >20 percent (cf. Table 18.1). The mean contribution of the principal species in each field is especially important and exceeded half of the total seed bank in about four out of ten fields. Its average contribution was 46 percent, with a range from 17 to 87 percent. Using the results for 32 fields in the third survey (Table 18.1), Chancellor (1979) noted a slightly lower dominance by the first species with a mean contribution of 38 percent and a range of 18 to 78 percent.

WHAT TYPES OF SEEDS ARE IN THE SOIL?

The soil seed bank of arable soils largely consists of those species that reproduce between cultivation treatments. Inspection of the names of the predominant species listed in Table 18.1 confirms that most are annuals. In a comprehensive survey of arable fields in Denmark, annuals, perennials with vegetative reproduction, and perennials without vegetative reproduction comprised 95 percent, 0.4 percent, and 5 percent of the seed bank, and 86 percent, 10 percent, and 4 percent of the plant population, respectively. Annuals and seed-producing perennials were therefore slightly over-represented in the seed bank, compared to their incidence in the vegetation.

Interestingly, as noted when discussing recalcitrant seeds, the climax vegetation is some ecosystems often fails to produce persistent seed banks. A classic example is *Quercus* spp. which produces recalcitrant seeds which are both short lived and vulnerable to predation.

WHERE DOES THE SEED BANK COME FROM?

Most seeds are found in the soil seed bank fairly close to where they were produced. An impression of seed production may be inferred from Table 18.3. Influx from elsewhere is not believed to be numerically unimportant in most circumstances compared to the numbers of seeds being produced per unit area (Table 18.3). Such introductions are highly significant if they lead to the introduction of a new species. Consider, for example, the impact of introductions with crop seeds or irrigation water.

TABLE 18.3. Seed outputs per unit area and per plant per year of selected species.

Species	Seeds per plant	Seeds·m^{-2}
Alopecurus myosuroides	43	2,500
Avena fatua	22 (range: 16-184)	1,000 (range: 393-4,784)
Sinapis arvensis	219	9,198
Quercus petraea	50,000 (0-90,000)	50 (0-175)
Senecio jacobaea	63,000 (4,760-174,230)	630,000

Source: Created from data in Sagar and Mortimer, 1976.

Seed Introduction in Certified Crop Seed

Taking the worst case scenario for standard certified cereal seed in the United Kingdom, the maximum permissible contamination is 20 weed seeds·kg^{-1}. Assuming a seed rate of 150·kg^{-1} then up to 3,000 weed seeds could be introduced per hectare. Even with the higher voluntary standard, where the allowable tolerance is only 8 weed seeds·kg^{-1}, up to 1,200 weed seeds could still be introduced per hectare.

Although these numbers may seem large, the important point is that they are respectively equivalent to only 0.3 and 0.12 weed seeds per m^2. Comparing these seed densities with the 5,000 to 10,000 viable seeds per m^2 in seed banks of arable soils, it is clear that contamination of crop seed will not contribute greatly to the seed bank in arable fields.

Does that mean such seeds can be ignored or that crop seed purity is a trivial concern? Certainly, new species may be introduced, which over time, may multiply.

Similar precautions must therefore be exercised when growers buy crop seeds or other products such as manure, nonsterile compost, or top soil that may introduce new species to the farm. Similar arguments apply to irrigation water. The central message is that it is always important to be on the lookout for new species, which may in time become serious problems.

WHAT HAPPENS TO THE SOIL SEED BANK?

Two questions need to be considered here. First, how long can seeds survive in the soil? Second, what are the main means of depletion?

How Long Do Seeds Survive in Soil?

Circumstantial Evidence 1: Apparent Seed Survival Through Successional Series

Livingston and Allessio (1968) studied the soil seed bank over a successional series including fields abandoned for different periods and white pine *(Pinus strobus)* stands of different ages at the Harvard Forest, Massachusetts. Extreme care was taken to avoid contamination when soil sampling, so that all leaf litter and humus was removed down to the mineral soil, which was then sampled to a depth of 11.4 cm. The results clearly suggested that many species and seeds were surviving for several decades without replenishment of the seed bank. In another survey of the seed bank and vegetation in ancient and recent forests in central Belgium, Bossuyt et al. (2002) likewise reported that the soil seed bank mainly contained species of early successional series or those which occur in short-term disturbances or clearings. In contrast to Livingston and Allessio (1968), however, they suggested that the seed bank of these Belgian forests was being maintained by seeds produced in such short-term disturbances. The corollary of both studies is, however, that the climax and even the normal understory vegetation is generally absent from the seed bank. From the point of view of conservation and regeneration of degraded habitats, therefore, it should be noted that disturbing forest soils is likely to lead to the emergence of "highly competitive and ruderal" species from the soil seed bank (Honnay et al., 2002). In other habitats, this caveat does not apply. For example, Jutila (2002) noted a close similarity between the seed bank and the aboveground vegetation in the meadows of river deltas in Western Finland, so it would therefore be possible to use the seed bank for habitat restoration. Kalamees and Zobel (2002) made a similar observation regarding regeneration of gaps in an Estonian calcareous grassland, where there was a close conformity between the seed bank and the vegetation. The probability of lack of seed of desirable species should, however, always be a criterion before advocating use of the soil seed bank for land restoration, and indeed will often be the key factor in limiting its use (Pywell et al., 2002).

Circumstantial Evidence 2: Archaeologically Dated Soil Samples

Odum (1978) has compiled a large collection of evidence supporting the extreme longevity of some seeds based on analyses of the seed banks of archaeologically dated soil samples in Denmark. One example will suffice here.

Hans Stiesdal (National Museum of Denmark) carried out an archaeological investigation under the floor of a church in Uggeløse, North Sjælland, Denmark. In 1971, a soil sample of the sandy filling of a medieval grave was taken at a depth of 1.5 meters under the floor in the tower. The grave was dated to the eleventh century and has not been disturbed. One seedling of *Verbascum thapsiforme* emerged from the soil. Odum (1978) concluded that the seedling came from an 850-year-old seed in the soil.

Experimental Evidence 1: Long-Term Experiments

The first real attempt to confirm the longevity of seeds in soil was started by Beal in 1879. He buried 50 seeds each of 20 species in bottles with moist sand at a depth of 46 cm. The bottles were slanting downward to prevent seed influx during storage. Seeds of *Rumex crispus* and *Oenothera biennis* survived for 80 years (Darlington and Steinbauer, 1961). After 100 years, 22 viable seeds of *Verbascum* and one of *Malva rotundifolia* were recovered (Kivilaan and Bandurski, 1981).

Duvel commenced a much larger experiment in 1902, employing more natural burial conditions. Batches of 100 or 200 seeds of 107 species were mixed with sterilized soil and buried in porous flower pots covered with inverted porous saucers at three depths (20, 56, and 107 cm). Thirty-six species survived for 39 years (Toole and Brown, 1946), with over 80 percent survival of *Datura stramonium*, *Solanum nigrum*, and *Phytolacca americana*. Among the crop species, 22 species were transient, but some seeds of tobacco *(Nicotiana tabacum)*, celery *(Apium graveolens)*, red clover *(Trifolium pratense)*, and Kentucky blue grass *(Poa pratensis)* persisted to the end of the experiment (Toole and Brown, 1946). The short longevity of most crop species in soil seed banks, other than some forage/pasture grasses and legumes, has been confirmed in many other studies and is of course the basis of crop production.

Experimental Evidence 2: Short-Term Seed Survival in Soil

Many experiments have been conducted to examine seed survival over periods of up to five years. Such research gives farmers a much more useful measure of longevity, and indeed the survival curves can be monitored.

Depletion in the absence of new influx often approximates the negative exponential decay curve

$$S = S_0 \, e^{-gt} \text{ or } \log_e S = \log_e S_0 - gt$$

where S is the number of surviving seeds from an initial population S_0 after t years and g is the annual rate of depletion.

This model is valid only on a year-to-year basis and in the absence of seed influx. Where applicable, implications of this model are twofold: seed banks have (1) a constant half life and (2) a constant annual probability of depletion.

These parameters immediately allow the effects of different treatments to be quantified and compared. For example, Roberts (1970) found that depletion of a mixed population mostly comprising annual broad-leaved weeds increased with an increase in the frequency of tillage over a period of five years. Annual probabilities of depletion were 22 percent, 30 percent, 36 percent, and 45 percent pa for untilled soil, and soil tilled twice a year, four times and five to six times a year, respectively.

The negative exponential model was accepted by Egley and Chandler (1983) only for the depletion of buried seeds of 10 out of 20 persistent weed species. Some deviation from this model was found for seven persistent species (*Abutilon theophrasti, Anoda cristata, Cassia obtusifolia, Ipomoea lacunosa, I. turbinata, Sesbania exaltata,* and *Sorghum halepense*). All apart from *Sorghum halepense* have hard seeds and depletion was generally quite low.

What Happens to Depleted Seeds?: The Means of Depletion

The principal means of depletion of buried seeds is usually germination. Not all germinated seeds emerge. For example, although germination of seeds of *Bidens pilosa* and *Acyranthes aspera* in Kenya decreased with depth below 1 or 2 cm, many still germinated at lower depths, but almost all seeds germinating at depths of 8, 16, and 32 cm failed to emerge (Figure 18.2). Seeds in the soil show a remarkable periodicity of emergence such that they tend to emerge at similar times in each year (Roberts, 1972). The control of this periodicity is in part due to an annual cycle of dormancy within the buried seed population and in part due to the environment. The combination of these factors leads to an annual germinability cycle within the seed population (Murdoch, 1998).

Within the soil seed bank this seed-to seed variation may include genetic variation, especially for species with a significant degree of out-crossing and for seeds produced in different years. In connection with desert annuals, Gutterman (2002) has also emphasized something that applies equally to species from other habitats. He argued that the maturation environment

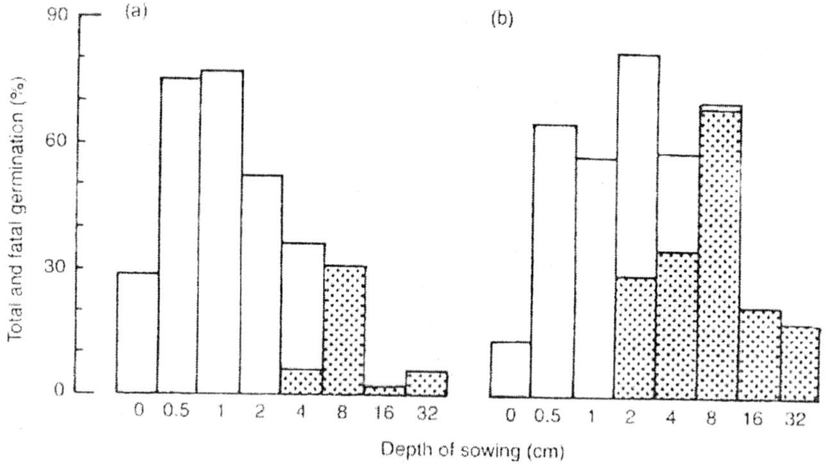

FIGURE 18.2. Seeds of (a) *Bidens pilosa* and (b) *Achyranthes aspera,* two local Kenyan ruderals. The experiment was carried out in murram soil at Nairobi, Kenya, using seeds sown at various depths. Seedlings were counted two weeks after sowing. (*Source:* From M. Fenner, *Seed ecology.* London: Chapman and Hall. © 1985, with kind permission of Kluwer Academic Publishers.)

on the mother plant could affect germination behavior for many years after seed shedding.

It is most significant that seeds vary in the degree to which they lose dormancy. Dormancy thereby helps to achieve temporal dispersal of seeds in the soil seed bank. For example, some *A. fatua* seeds (Murdoch, 1998)

- lost all dormancy as tested at 10°C (though not 20°C);
- required one (darkness or nitrate) and some both factors in order to germinate; and/or
- were too dormant to respond to anything (but they were still viable and germinated in response to gibberellic acid).

The control of seedling emergence is reflected in what may be called an annual germinability cycle under joint endogenous control of the dormancy cycle and exogenous control by the physical environment. These responses also vary with the depth of seed burial (Murdoch, 1998). In *Chenopodium album,* for example, light responses of seeds exhumed from the soil surface were, not surprisingly, rare. Responses to light and nitrate after periods of burial in the soil were typically additive. Interestingly, germinability increased with depth of burial within the plowed horizon.

Benvenuti et al. (2001) suggested that burial below the depth from which emergence was possible generally induces dormancy. That is, buried seeds do not germinate. Interestingly, the responsiveness of surviving seeds at depths from which emergence may not be possible is much greater than seeds at the soil surface. Thus, deeply buried seeds may not be as dormant as seeds at the soil surface, but the environment at those depths may not be sufficiently stimulatory for seed germination (Murdoch, 1998).

IMPACTS ON WEED MANAGEMENT

Annual dormancy cycle and periodicity of emergence are both fairly consistent, but the proportion of the seed bank that will germinate remains difficult to predict accurately. However, it is useful to be aware of the seasonal responses to nitrate and light, since these may be utilized to avoid cultivation.

An alternative to avoiding or minimizing cultivation is that of cultivating at night and/or shrouding the equipment. In 23 experiments, Juroszek et al. (2002) reported a mean reduction of 28 percent in weed infestations after dark cultivation. The species most affected were not surprisingly those that are positively photoblastic. Results from such treatments are, however, variable, and it should also be emphasized that the benefits do not always lead to yield improvements: the smaller number of weed seedlings may be subject to less intraspecific competition and density-dependent mortality.

A use of depletion information for the species present in seed banks is to determine those species that are transient and are therefore likely to be eliminated. Conversely, persistent seed banks are more likely to need to be contained. For example, taking a very simple case of an initial seed bank of 500 seeds; m^{-2}, assuming 33 percent depletion per annum, and somewhat unrealistic 100 percent effective weed control, it is predicted that the infestation would take 23 years to reach 10 plants per ha.

REFERENCES

Baskin, C.C. and J.M. Baskin (1998). *Seeds: Ecology, biogeography and evolution of dormancy and germination.* San Diego, CA: Academic Press.

Benvenuti, S., M. Macchia, and S. Miele (2001). Quantitative analysis of emergence of seedlings from buried weed seeds with increasing soil depth. *Weed Science* 49: 528-535.

Bossuyt, B., M. Heyn, and M. Hermy (2002). Seed bank and vegetation composition of forest stands of varying age in central Belgium: Consequences for regeneration of ancient forest vegetation. *Plant Ecology* 162: 33-48.

Cerabolini, B., R.M. Ceriani, M. Caccianiga, R. De Andreas, and B. Raimondi (2003). Seed size, shape and persistence in soil: A test on Italian flora from Alps to Mediterranean coasts. *Seed Science Research* 13: 75-85.

Chancellor, R.J. (1979). The long-term effects of herbicides on weed populations. *Annals of Applied Biology* 91: 141-144.

Darlington, H.T. and G.P. Steinbauer (1961). The eighty-year period for Dr. Beal's seed viability experiment. *American Journal of Botany* 48: 321-325.

Egley, G.H. and J.M. Chandler (1983). Longevity of weed seeds after 5.5 years in the Stoneville 50-year buried-seed study. *Weed Science* 32: 264-270.

Ellis, R.H. and E.H. Roberts (1980). Improved equations for the prediction of seed longevity. *Annals of Botany* 45: 13-30.

Esashi, Y. and C. Leopold (1968). Physical forces in dormancy and germination of *Xanthium* seeds. *Plant Physiology* 43: 871-876.

Fenner, M. (1985). *Seed ecology*. London: Chapman and Hall.

Gutterman, Y. (2002). Minireview: Survival adaptations and strategies of annuals occurring in the Judean and Negev Deserts of Israel. *Israel Journal of Plant Science* 50: 165-175.

Harper, J.L. (1977). *Population biology of plants*. London: Academic Press.

Hong, T.D. and R.H. Ellis (1996). *A protocol to determine seed storage behaviour*. Rome: International Plant Genetic Resources Institute.

Honnay, O., B. Bossuyt, K. Verheyen, J. Butaye, H. Jacquemyn, and M. Hermy (2002). Ecological perspectives for the restoration of plant communities in European temperate forests. *Biodiversity Conservation* 11: 213-242.

Jensen, H.A. (1969). Content of buried seeds in arable soil in Denmark and its relation to the weed population. *Dansk Botanisk Arkiv* 27(2): 1-56.

Juroszek, P., R. Gerhards, and W. Kuhbauch (2002). Photocontrol of weed germination of arable annual weeds. *Journal of Agronomy and Crop Science* 188: 389-397.

Jutila, H.M. (2002). Seed banks of river delta meadows on the west coast of Finland. *Annales Botanici Fennici* 39: 49-61.

Kalamees, R. and M. Zobel (2002). The role of the seed bank in gap regeneration in a calcareous grassland community. *Ecology* 83: 1017-1025.

Kivilaan, A. and R.S. Bandurski (1981). The one hundred-year period for Dr. Beal's seed viability experiment. *American Journal of Botany* 68: 1290-1292.

Leck, M.A., V.T. Parker, and R.L. Simpson (1989). *Ecology of soil seed banks*. London: Academic Press.

Leishman, M.R. and M. Westoby (1998). Seed size and shape are not related to persistence in soil in Australia in the same way as in Britain. *Functional Ecology* 12: 480-485.

Livingston, R.B. and M.L. Allessio (1968). Buried viable seed in successional field and forest stands, Harvard Forest, Massachusetts. *Bulletin Torrey Botanical Club* 95: 58-69.

Moles, A.T., D.W. Hodson, and C.J. Webb (2000). Seed size and shape and persistence in the soil in the New Zealand flora. *Oikos* 89: 541-545.

Murdoch, A.J. (1983). Environmental control of germination and emergence in *Avena fatua*. *Aspects of Applied Biology* 4: 63-69.

Murdoch, A.J. (1998). Dormancy cycles of weed seeds in soil. *Aspects of Applied Biology* 51: 119-126.

Murdoch, A.J. (2003). Seed dormancy. In R.M. Groodman (ed.), *Encyclopaedia of plant and crop science*. New York: Marcel Dekker.

Murdoch, A.J. and R.H. Ellis. (2000). Dormancy, viability and longevity. In M. Fenner (ed.), *Seeds: The ecology of regeneration and plant communities*, Second edition (pp. 183-214). Wallingford, UK: CABI.

Murdoch, A.J. and E.H. Roberts (1982). Biological and financial criteria of long-term control strategies for annual weeds. *Proceedings 1982 British Crop Protection Conference—Weeds*, 741-748.

Odum, S. (1978). *Dormant seeds in Danish ruderal soils*. KVL Horsholm Arboretum, Horsholm, Denmark.

Peco, B., J. Traba, C. Levassor, A.M. Sánchez, and F.M. Azcárate (2003). Seed size, shape and persistence in dry Mediterranean grass and scrublands. *Seed Science Research* 13: 87-95.

Pritchard, H.W. (1991). Water potential and embryonic axis viability in recalcitrant seeds of *Quercus rubra*. *Annals of Botany* 67: 43-49.

Probert, R.J. and P.L. Longley (1989). Recalcitrant seed storage physiology in three aquatic grasses (*Zizania palustris Spartina anglica* and *Porteresia coarctata*). *Annals of Botany* 63: 53-63.

Pywell, R.F., J.M. Bullock, A. Hopkins, K.J. Walker, T.H. Sparks, M.J.W. Burke, and S. Peel (2002). Restoration of species-rich grassland on arable land: Assessing the limiting processes using a multi-site experiment. *Journal of Applied Ecology* 39: 294-309.

Roberts, E.H. (1972). Dormancy: A factor affecting seed survival in the soil. In E.H. Roberts (ed.), *Viability of seeds* (pp. 321-359). London: Chapman and Hall.

Roberts, E.H. (1973). Predicting the storage life of seeds. *Seed Science and Technology* 1: 499-514.

Roberts, E.H. and R.H. Ellis (1989). Water and seed survival. *Annals of Botany* 63: 39-52.

Roberts, H.A. (1970). Viable weed seeds in cultivated soils. Report National Vegetable Reseach Station for 1969, pp. 25-38.

Roberts, H.A. (1981). Seed banks in soils. *Advances in Applied Biology* 6: 1-55.

Roberts, H.A. and R.J. Chancellor (1986). Seed banks of some arable soils in the English midlands. *Weed Research* 26: 251-257.

Roberts, H.A. and J.E. Neilson (1982). Seed banks of soils under vegetable cropping in England. *Weed Research* 22: 13-16.

Roberts, H.A. and F.G. Stokes (1966). Studies on the weeds of vegetable crops: VI. Seed populations of soil under commercial cropping. *Journal of Applied Ecology* 3: 181-190.

Sagar, G.R. and A.M. Mortimer (1976). An approach to the study of the population dynamics of plants with special reference to weeds. *Applied Biology* 1: 1-47.

Sharifzadeh, F. and A.J. Murdoch (2001). The effects of temperature and moisture on after-ripening *Cenchrus ciliaris* seeds. *Journal of Arid Environments* 49: 823-836.

Thompson, K., J.P. Bakker, and R.M. Bekker (1997). *Soil seed banks of North West Europe: Methodology, density, longevity*. Cambridge, UK: University Press.

Thompson, K., S.R. Band, and J.G. Hodgson (1993). Seed size and shape predict persistence in soil. *Functional Ecology* 7: 236-241.

Thompson, K. and J.P. Grime (1979). Seasonal variation in the seed banks of herbaceous species in ten contrasting habitats. *Journal of Ecology* 67: 893-921.

Thompson, K., A. Jalili, J.G. Hodgson, B. Hamzeh'ee, Y. Asri, S. Shaw, A. Shirvany, S. Yazdani, M. Khoshnevis, F. Zarrinkamar, et al. (2001). Seed size, shape and persistence in the soil in an Iranian flora. *Seed Science Research* 11: 345-355.

Toole, E.H. and E. Brown (1946). Final results of the Duvel buried seed experiment. *Journal of Agricultural Research* 72: 201-210.

Chapter 19

The Ecophysiological Basis of Weed Seed Longevity in the Soil

R. S. Gallagher
E. P. Fuerst

INTRODUCTION

Weed seeds are present in large numbers in agroecosystems, commonly ranging from 10^3 to $10^5 \cdot m^{-2}$, and weed seeds commonly persist in the soil for several years or decades (Cavers and Benoit, 1989; Cook, 1980; Yenish et al., 1992). Considerable research has been conducted on the longevity and persistence of the soil seed bank under a range of environments and management regimes. Although certain aspects of seed bank dynamics, such as mortality, dormancy, and recruitment, have been characterized to some extent, mechanisms regulating longevity of the soil seed bank remain poorly understood. A mechanistic understanding of the processes regulating weed seed longevity is needed to allow robust predictions of weed population dynamics and to optimize management decisions. In addition, the development of active seed bank management strategies, which are aimed at either depleting the seed bank pools or suppressing seed bank expression, requires an understanding of the interacting processes that drive seed bank dynamics. In this chapter, we will present a conceptual model that outlines the plant, soil, and climatic factors that we believe contribute to weed seed longevity in the soil. The purpose of this model is to provide a template in which to consider potential factors that may contribute to the variation in seed longevity within and among populations, and to identify potential avenues for seed bank management. The goal of this chapter is to summarize the interacting processes regulating weed seed longevity (excluding predation); this chapter is not intended to be an exhaustive review.

CONCEPTUAL MODEL FOR SEED LONGEVITY

We hypothesize that three primary, interacting factors regulate seed longevity: (1) plant genetics, (2) the environment in which a seed develops and matures, and (3) the chemical, physical, and microbial characteristics of the soil (see Figure 19.1). Each of these regulating factors ultimately impacts one or more of the three specific seed quality characteristics, which can be considered mechanisms of longevity:

1. *seed dormancy,* which maintains seed in a physiological stasis, reducing metabolic activity and thereby reducing the likelihood of fatal germination;
2. *seed vigor,* which enables seed to remain viable and to produce healthy seedlings over an extended time period under variable climatic conditions; and
3. *seed resistance to microbial decay,* which allows survival in a soil environment that can be teeming with pathogenic and opportunistic microorganisms.

Thus, genetics, weed maturation environment, and soil characteristics all have the potential of affecting the longevity mechanisms of seed dormancy, vigor, and resistance to decay. Genetics and maturation environment will affect these longevity mechanisms via their effects on seed quality, while soil characteristics will affect these longevity mechanisms via their effects on germination stimulants, allelochemicals, and microbial ecology. Soil processes, of course, depend on moisture and temperature, conditions which also directly influence seed dormancy. Germination stimulants and allelochemicals may be released directly by substrates in the soil or be the product of microbial activity. Seed quality, and thus longevity, can potentially be manipulated by modifying aspects of the competitive environment under which weed seed matures (e.g., crop densities, planting arrangements, and cultivars). Likewise, seed longevity may be manipulated by modifying the soil physical and chemical characteristics through tillage, crop rotations, and soil fertility management. Finally, natural selection will favor the individuals best suited to survive and reproduce in a particular environment. Therefore, shifts in the genetic composition of weed populations may be expected in response to strong selection pressures associated with weed, crop, and soil management.

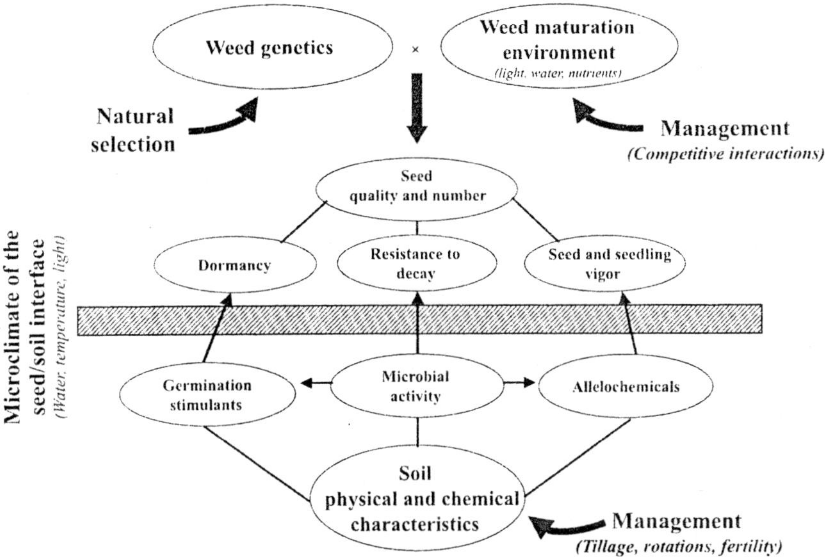

FIGURE 19.1. A hypothetical relationship outlining the effects of plant genetics and maturation environment on the number and quality of weed seed produced, and the relationships among seed quality, the physical and chemical characteristics of the soil, and microclimatic conditions of the soil/seed interface. The interaction between the plant genetics and the maturation environment (i.e., light, nutrients, water) regulates the number of seeds produced, and the quality of the seed with respect to dormancy, vigor, and resistance to decay. Soil physical and chemical characteristics influence the production of germination stimulants (i.e., nitrate and ethylene) alleochemicals and microbial activity in the soil. Germination stimulants and allelochemicals are released directly from soil substrates or are the product of microbial activity. Germination stimulants initiate the germination of conditionally dormant seeds, whereas allelochemicals tend to reduce seed and seedling vigor. Microbial populations directly degrade seed structures as well as feed on seed exudates. Management can be directed at manipulating the weed maturation environment by bolstering crop competitiveness, thereby reducing the number and the quality the weed seed produced. Less dormant and low vigor seed, and seed with compromised defense mechanisms will be more prone to rapid loss from the seed bank. Management can also be directed at manipulating the soil environment to favor the production of germination stimulants, allelochemicals, and enhancing microbial activity. Over the long term, weed populations may adapt to strong selection pressures in their local environment.

RESOURCE ALLOCATION TO SEED

Reproductive output (i.e., seed production per plant) is determined by the ability of a plant to acquire light, water, and nutrients for the production of plant resources, and by the allocation of plant resources to the seed (Sugiyama and Bazzaz, 1997; reviewed by Bazzaz et al., 2000). Reproductive allocation is the quantity of plant resources allocated to reproductive structures relative to other plant structures. Reproductive allocation is determined by both the genetics and maturation environment of a plant. The interactive effects of resource availability and plant genetics have been elegantly illustrated in ladysthumb smart weed (*Polygonum persicaria* L.). Ten individuals or genotypes (cloned for each treatment by replicate combination) each from two populations were grown under light, water, and nutrient (N, P, K) gradients in a common garden (data summarized in Table 19.1) (Sultan and Bazzaz, 1993a,b,c). Total seed biomass, total seed number, and mean seed weight all decreased as light levels were reduced. Under all light levels, there was evidence of significant genotypic variation in total seed number and mean seed weight, but not total seed weight. Drought stress reduced total seed weight and total seed number, but increased mean seed weight. Although there was significant genotypic variation among the

TABLE 19.1 A summary of the effects of environmental stress (i.e., suboptimal light, water, and nutrients), plant genetics, and their interaction (G × E) on reproductive allocation in *Polygonum persicaria.*

Plant characteristic	Resource deficiency	Genetic effect (G)	Environmental effect (E)	G × E
Total seed biomass	Light	No	Decrease	No
	Water	Yes	Decrease	Yes
	Nutrients	No	Decrease	No
Total seed number	Light	Yes[a]	Decrease	No
	Water	Yes	Decrease	Yes
	Nutrients	Yes[a]	Decrease	No
Mean seed weight	Light	Yes	Decrease	No
	Water	Yes	Increase	Yes[a]
	Nutrients	Yes	Decrease[a]	Yes[a]

Source: Data adapted from Sultan and Bazzaz, 1993a,b,c.
Note: Individual genotypes (i.e., genetic effects) from two populations were grown under a resource gradient (i.e., environmental effects) in a common garden. Data for the two populations have been pooled except where specified.[a]
[a]Occurred only in one of the two populations tested.

individuals, the degree of this variation depended on the water stress environment. Decreased nutrients generally reduced seed weight, total seed number, and mean seed weight. Within nutrient treatments, total seed number and mean seed weight varied among genotypes, but the degree of this variation in mean seed weight depended on the degree of nutrient stress. Total seed mass was similar among all the genotypes. Reduced light levels caused a relative shift in allocation from the seed to the leaves, whereas water and nutrient stress caused a relative shift in allocation from the stems to the seeds (data not shown). In a similar study with velvetleaf (*Abutilon theophrasti* Medisus), seed output and plant size declined as nutrients became limited, but genetic variability in seed output and plant size was considerably more pronounced under nutrient-rich versus nutrient-poor conditions (Sugiyama and Bazzaz, 1997). In six common annual weed species, seed production per gram of aboveground biomass was 1.4 to 8.0 times higher when plants were grown in monoculture without interspecific competition than when plants were grown in competition with a maize crop (see Table 19.2) (R. S. Gallagher, unpublished). Mean seed weight was 10 to 25 percent higher in the monoculture than when grown in competition (data not

TABLE 19.2. The effect of maturation environment on reproductive allocation in six common weed species (measured in seed number per g of dry aboveground biomass).

	Maturation environment	
Species	**Monoculture**	**In maize**
Abutilon theoprasti	47	33
Echinochola crus-gali	196	61
Chenopodium album	628	77
Amaranthus retroflexus	399	119
Setaria faberi	43	8
Ambrosia artemisiifolia	39	8

Note: Plants were grown under field conditions in monoculture or in competition with maize. Within each maturation environment, plants were established at the time of maize emergence, and at two and four weeks after maize emergence. Data for the established times were pooled within maturation environment. Seed were harvested as they matured, and aboveground biomass was harvested after all the seed had been removed. Reproductive allocation is based number of seed produced per g of above ground biomass. The experiment was conducted at Aurora, New York, 1997 by R.S.Gallagher (unpublished). All treatments were replicated four times.

shown). Unlike the studies in which specific resource limitations were isolated (described previously), weeds grown in competition with a crop would simultaneously experience limitations in light, water, and nutrients, in addition to experiencing a significantly lower red to far-red environment which tends to promote stem elongation and may alter resource allocation within the plant (Smith, 1995). Data suggest that the net effect of limiting more than one resource will be lower reproductive allocation and lower mean seed weight (Table 19.2).

The available data illustrate that resource limitations can have a profound effect on reproductive output and reproductive allocation, and that these effects can be further compounded by genetic variability within and among populations. Limitations in any resource are likely to reduce total reproductive output. Allocation to reproductive parts relative to other plant parts may increase or decrease, depending on the stress or suite of stress factors that a weed experiences. The effects of resource limitations on fitness have been primarily measured by reproductive output. Reproductive output, however, is just one aspect of fitness. Seed must also survive in the soil, germinate, and develop into reproductively fit individuals. Seed quality traits, including size, dormancy, vigor, and resistance to microbial decay, will also influence fitness (Figure 19.1). Seed size often decreases as resources become limited. Larger-seeded species may have a competitive advantage, particularly when establishing under shade, moisture, or nutrient stress, or from greater soil depth (Leishman, Wright, et al., 2000). The effect of reduced seed size on fitness within species has not been extensively characterized (Bazzaz et al., 2000). Seed dormancy is affected by both genetic and environmental factors (see Seed Dormancy section). A decrease in primary dormancy can reduce longevity (Peters, 1982) and may increase the incidence of fatal germination, due to precocious germination or germination from positions that do not favor seedling establishment, such as greater soil depths or in shade. In crop seed, less-vigorous seed often results in less-competitive seedlings, as well as seeds and seedlings that are more susceptible to soilborne diseases and decay (see Retention of Seed Vigor and Microbial Seed Decay sections). Although we would expect similar relationships among seed vigor, seedling vigor, and resistance to decay in weed species, insufficient data are available to draw any conclusions. Clearly, stresses during maturation alter reproductive allocation. There is a need to determine the extent to which these stresses affect subsequent survival in the seed bank and the likelihood that seedlings will develop into reproductive individuals.

SEED DORMANCY

Seed dormancy is an internal condition of the seed that impedes its germination under otherwise adequate hydric, thermal and gaseous conditions (Benech-Arnold et al., 2000). Dormancy protects seed against germination in environments that do not favor establishment and reproduction, such as germination that may occur out of season, at unfavorable soil depths, or in the shade. Dormancy also distributes germination and establishment within seasons and among years, ensuring that a seed pool is available for germination over a range of establishment opportunities. The three major categories or types of dormancy include (1) physical, (2) morphological, and (3) physiological (reviewed by Baskin and Baskin, 1998). Seed may also experience any combination of these dormancy types. Physical dormancy is due to the presence of a hard seed coat or fruit structures that prevent imbibition or physically restrict embryo expansion. Morphological dormancy is due the presence of an underdeveloped embryo. Physiological dormancy is due to the presence of endogenous inhibitors. Dormancy (all types) that occurs at physiological maturity is called primary dormancy, whereas dormancy that is induced by environmental conditions after physiological maturity is called secondary dormancy (physiological dormancy only).

Release of dormancy may be either a one-way or cyclic process, depending on the type of dormancy. For example, physical dormancy is overcome when the seed structure imposing the dormancy is sufficiently degraded to allow imbibition or embryo expansion. Likewise, morphological dormancy is overcome once embryo maturation has occurred. When physical or morphological dormancy is overcome, the seed will remain prone to germination if no other dormancy mechanisms are in place. In contrast, seed can cycle in and out of secondary (physiological) dormancy.

Physiological dormancy is not simply a toggle switch by which nondormant seed germinate and dormant seed do not. Rather, physiological dormancy is a continuum, with levels ranging from *dormant* to *conditionally dormant* to *nondormant* states (Baskin and Baskin, 1985, 1998). Dormant seed will not germinate under any conditions that would be reasonable for a particular species. Such dormant seed may sometimes be induced to germinate by artificial procedures such as wounding, scarification, or with hormones. Conditionally dormant seed require a somewhat specific suite of environmental conditions (i.e., temperature, moisture, chemicals, and/or light) to overcome dormancy. The environmental constraint on germination of conditionally dormant seed depends upon the degree of dormancy. Environmental conditions required for germination will be more stringent in more dormant seed, and less stringent in less dormant seed. Nondormant seed will germinate under the widest range of environmental conditions that are reasonable for a particular species. The level of primary seed dormancy depends on genetics and the environment during seed development and maturation (see Figure 19.1). Physiologically dormant seed can cycle in and out of dormant, conditionally dormant, and nondormant states in response to seasonal climatic fluctuations and changes in their chemical and physical microclimate. Seed response to germination cues, such as temperature, light, and chemical stimulants, will depend on the type and degree of seed dormancy (Baskin and Baskin, 1980; Bouwmeester and Karssen, 1993; Gallagher and Cardina, 1998a,c,d).

Regulation of Dormancy in Developing Seed

Genetic Factors

Seed dormancy for many species can be controlled to a large degree by the genetics of the plant (reviewed by Foley and Fennimore, 1998). Genetic diversity with respect to seed dormancy helps to ensure survival and reproduction in diverse environments. A number of studies have shown that plant species can have locally adapted dormancy strategies that protect against precocious germination under unfavorable environments, and that these adaptations are genetically based. For example in *Penstemon* spp., populations from regions with more severe winter conditions required longer chilling periods to overcome dormancy and had slower germination rates than populations from more mild regions (Meyers 1992; Meyers and Kitchen, 1994). Both of these strategies would help prevent premature germination when frost damage is most likely. In downy brome (*Bromus tectorum* L.), seed populations from intermountain regions had greater dormancy than seed populations from mountain regions (Meyers et al., 1997; Allen and Meyers, 2002). Greater dormancy in the intermountain population may protect seed from the sporadic summer rainfall events that are sufficient to initiate germination but not support plant establishment and reproduction. Similar dormancy strategies have been reported in slender oat (*Avena barbata* Brot.), soft brome (*Bromus mollis* L.) (Jain 1982), and rubberbush (*Calotropis procera* Aiton) (Amritphale et al., 1984). Naylor and Jana (1976) demonstrated a wide range of genetic variability for primary dormancy in wild oat. Studies cited earlier demonstrate that seed dormancy can vary immensely among plant populations within a species. In many cases, dormancy strategies appear to be locally adapted to prevent seed germination under unfavorable environmental conditions that are specific to a region.

Environmental Factors

As invasive species, weeds tend to be highly plastic with respect to establishment and reproduction in diverse environments. Because weeds may emerge over a wide time period, they must respond to a range of available heat units needed to bring a plant to reproduction and maturation and a range of photoperiods during key growth stages. Because many weeds are day-length sensitive, changes in day length may accelerate or delay reproduction. Weeds may also establish on resource-poor sites or under highly competitive situations where resources such as water, light, and nutrients are limited. Considerable evidence indicates that availability of water, light, and nutrients during seed development and maturation, and physiological signals that occur through photoperiod or light quality, can influence the dormancy of the seed produced.

Photoperiod. In temperate zones, day length can shift by 5 hr or more within a growing season. In weed species that are day-length sensitive, a shift from longer to shorter, or shorter to longer days initiates flowering. For example, winter annual weeds tend to emerge in the fall as days are getting shorter, and flowering is initiated in the spring as days become longer. In contrast, summer annual weeds germinate in the spring as day length is increasing, and flowering is generally initiated later in the summer as days become shorter. In regions where seasonal shifts in day length are pronounced, the timing of weed emergence responds not only to the degree days available to complete a growth cycle, but also to the duration of vegetative growth preceding the critical day length necessary to signal flower induction.

Several studies have been conducted evaluating the effect of day length on seed dormancy of summer annual weeds. In common lambsquarters (*Chenopodium album* L.), plants grown in a continuous 16 h photoperiod (LD) produced more dormant seeds with thicker seed coats than plants grown in a continuous 8 h photoperiod (SD) (Karssen, 1970). The SD seeds, however, had nearly twice the mass as the LD seeds. Similar results were reported for redroot pigweed (*Amarathus retroflexus* L.) (Kigel et al., 1977) and common purslane (*Portulaca oleracea* L.) (Gutterman, 1982). When redroot pigweed plants were shifted from a 16 to an 8 h photoperiod at the time of flowering, the dormancy of the seed produced was intermediate between the SD and LD seeds (Kigel et al., 1977). In common purslane, dormancy was positively correlated with the day length during the last eight days of seed maturation (Gutterman, 1982). These data suggest that in summer annual weeds, plants that emerge late in the season and mature under shorter days will produce fewer seed as well as seed that is less dormant than those that emerge earlier in the season and mature under longer days. To our knowledge, data are not available on the effect of day length on seed dormancy in winter or spring annual weeds.

Drought. Drought conditions can be related to regional or seasonal climatic patterns or be the result of competition. Sawhney and Naylor (1982) reported that several lines of wild oat (*Avena fatua* L.) seed from drought-stressed plants lost dormancy more quickly than seed from well-watered plants. In wild oat, Peters (1982) conducted an elegant set of experiments demonstrating that drought stress decreased total seed production, seed dormancy, and seed longevity in the soil. Three dormant lines of wild oat were grown in the greenhouse and subjected to drought stress starting at panicle emergence. Drought stress reduced the number of viable seeds approximately 50 percent. Seed from drought-stressed environments were consistently 12 to 20 percent less dormant at the time of harvest than seed from the well-watered plants. These differences in dormancy in the drought-stressed and well-watered seed resulted in quite pronounced differences in field emergence and longevity. For example, in the autumn in which the experiment was initiated, 4 percent of unstressed seed emerged, versus 66 percent of drought-stressed seed. In the second spring (19 months after initiation), 67 percent of unstressed seed emerged, versus 27 percent of drought-stressed seed. At the end of the experiment (40 months after initiation), 8 percent of unstressed seed remained viable, versus none of the stressed seed. The research by Sawhney and Naylor (1982) and Peters (1982) suggest that drought during seed development and maturation can reduce seed dormancy and longevity in the seed bank. Data from other species are needed to determine if this effect is unique to wild oat or generally true among weed species.

Shade. Seed production is generally reduced by shade (see Resource Allocation section). The reported effects of shade on seed dormancy have been mixed. For example, in eastern black nightshade (*Solanum ptycanthum* Dun.), seed from shade-grown plants were more dormant than seed from full-sun plants (Stoller and Myers, 1989). Similar results were reported with buckhorn plantain (*Plantaga lanceolata* L.) (Van Hinsberg, 1998). In velvetleaf, however, there were 10 percent fewer hard seeds from shade-grown velvetleaf plants compared with the full-sun control (Bello et al., 1995). Similarly, velvetleaf grown in monoculture had 36 percent hard seed compared to 16 percent hard seed in velvetleaf grown in competition with maize (R. S. Gallagher, unpublished research). Benvenuti et al. (1994), however, found that dormancy in velvetleaf, johnsongrass (*Sorghum halepense* L.), and jimsonweed (*Datura stramonium* L.) seed from plants grown under commercial shade cloth was similar to the dormancy in the plants grown in full sun. In all, the available data suggest that shade during seed development and maturation may increase, decrease, or have no effect on dormancy, depending on the weed species and the conditions of the experiment. It is likely that different weed species have different adaptations in response to shade. It should be noted, however, that shade provided by commercial shade cloth might produce a different physiological response than shade provided by a vegetative canopy. For example, the red to far-red (R:FR) ratio of plant canopy shade will be substantially lower than in the shade-cloth environment. This change in the R:FR ratio below the vegetative canopy is due to differential absorption of R compared to FR by chlorophyll (Holmes, 1981). A decrease in the R:FR ratio can result in an increase in internode elongation (Ballaré et al., 1988; Smith, 1995), which may also change resource allocation to seed. There is little information, however, comparing the effects of natural versus artificial shade during seed development and maturation on seed production, dormancy, or quality.

Interactions Between Seed Dormancy and External Germination Cues

Several environmental cues can regulate physiological dormancy, including temperature, seed water content, light, soil chemicals, fire, and smoke. Seeds can cycle in and out of physiological dormancy in response to seasonal climatic fluctuations (Baskin and Baskin, 1998). In general, summer annual weeds tend to lose dormancy over the fall and winter months, and may show an increase in secondary dormancy in response to hot and/or dry conditions in the summer. In contrast, winter annual species tend to lose dormancy over the summer months and may show an increase in secondary dormancy in response to cool/wet conditions over the fall and winter. Temperature and seed water content often play a critical role regulating the release of physiological dormancy (Bradford, 2002). In the field, regional and seasonal fluctuations in temperature and rainfall can influence that rate of dormancy release. For example, in annual ryegrass dormancy release in buried seed was faster at a high versus low summer rainfall location (R. Gallagher, unpublished data). This effect was attributed to the greater seed hydration that likely occurred at the high-rainfall location. In the same study, dormancy release in seed on the soil surface was faster at the location with the greatest thermal accumulation. In this case, dormancy release was probably regulated primarily by temperature, since extended hydration of seed on the soil surface was unlikely. Warm, moist conditions have been reported to promote the release of dormancy in other winter annual species (Baskin and Baskin, 1977; Roberts and Locke, 1978).

Light

Light can play a role in either promoting or inhibiting germination in many weed species (reviewed by Casal and Sánchez, 1998; Gallagher and Cardina, 1998b; Pons, 2000). Weed seed germination is usually stimulated in response to full sunlight. This response is generally mediated by the phytochrome family of photoreceptors that respond primarily to light in the red (600 to 700 nm) and far-red (FR) (700 to 800 nm) regions of the light spectrum (reviewed by Smith, 1995). Frequently in freshly imbibed seeds, red light in the range of 10 to 1,000 μmol·quanta·m^{-2} will initiate germination, whereas similar far-red fluences will reverse this effect. Prolonged dark imbibition, however, can induce an extreme light sensitivity (i.e., a very low fluence response) where seed germination is initiated by fluences well below 1 μmol quanta·m^{-2} of light in the 400 to 800 nm range. Extended exposure of the FR-rich light, such as occurs in plant canopy shade, often inhibits germination.

Researchers have attempted to use seed sensitivity to light to help manage weeds. For example, a number of studies have looked at conducting tillage operations at night as a means to reduce weed emergence (Scopel et al., 1994; Buhler, 1997; Gallagher and Cardina, 1998b). The basis for this approach is that weed seeds are exposed to light and reburied with daytime tillage operations. Conducting tillage at night prevents this exposure to light, assuming weed seeds are subsequently reburied after the tillage operation. In these experiments, weed emergence was often significantly lower following the night versus the day tillage operations. It is unclear, however, if the weed control achieved with night tillage is sufficient to be agronomically meaningful. For example, we found that weed densities in the night tillage treatments were still far in excess of the carrying capacity for adult plants, suggesting that substantial self-thinning of the weed stands would occur following both the day and night tillage treatments (Gallagher and Cardina, 1998b). Although initial weed emergence was reduced with night tillage, standing weed biomass after several weeks was similar in the day and night tillage treatments.

Alternatively, several studies have demonstrated that tillage can be an effective means to hasten seed bank decline (Chepil, 1946; Roberts and Feast, 1973; Mulugeta and Stoltenberg, 1997; Jensen, 1999). In these studies, seed bank populations dropped more quickly when the soil was disturbed on a regular basis compared to the undisturbed controls. Although the use of extensive tillage to deplete soil seed banks may often be in conflict with conservation tillage programs aimed at preventing soil erosion and soil organic matter degradation, innovative designs in tillage equipment may enable moderate soil disturbance while preserving a majority of crop residues on the soil surface. Row cultivators, rotary hoes, and various styles of harrows are commercially available for use in high-residue conservation tillage systems. These tillage tools may allow mechanical stimulation of seed germination from the soil without seriously compromising the benefits of conservation tillage systems.

Soil Chemicals

Nitrate and ethylene are probably the most common soil chemicals that act as germination stimulants for many weed species. Nitrate and ethylene directly stimulate weed seed germination, as well as sensitize weed seeds to other environmental stimuli such as light and suboptimal germination temperatures. Compared to seed germinated in water alone, nitrate concentrations of 70 ppm or lower have been reported to increase the germination of buckhorn plantain in the dark by 20 to 35 percent (Pons, 1989), wild oat by 30 percent (Saini et al., 1985a), and redroot pigweed in the dark and in response to very low fluences of red light (0.3 to 3.0 μmol quanta·m^{-2}) by 20 to 40 percent (Gallagher and Cardina, 1998d). A wide range of common plant species also show germination sensitivity to ethylene at concentrations as low as 0.01 to 10 ppb. These species include redroot pigweed (Egley, 1980; Schonbeck and Egley, 1988), common cocklebur (*Xanthium strumarium* L.) (Egley, 1990; Katoh and Esashi, 1975), subterranean clover (*Trifolium subterraneum* L.) (Eshashi and Leopold, 1969), common lambsquarters (Goudey et al., 1987; Saini et al., 1985b), love-lies bleeding *(Amaranthus caudatus),* common purslane (Egley, 1984), curly dock (*Rumex crispus* L.) (Corbineau and Côme, 1995), and wild oat (Saini et al., 1985a).

Attempts to use seed sensitivity to nitrate and ethylene to deplete soil seed banks have been accomplished primarily with exogenous additions of nitrogen fertilizers or ethylene to the soil. Field-scale applications of ammonium nitrate had no effect on weed emergence in one study (Fawcett and Slife, 1978), whereas applications of ammonium nitrate (34 to 67 kg NO_3^- ·ha^{-1}) and sodium azide (NaN_3) increased broadleaf emergence by 25 to 80 percent compared to the untreated controls (Hurtt and Taylorson, 1986). Similarly, field application of KNO_3 (200 kg·ha^{-1}) decreased primary dormancy in redroot pigweed seed and increased germination sensitivity to light and ethylene (Egley, 1989). In wild radish (*Raphanus raphanistrum* L.), 50 kg·ha^{-1} of added nitrogen resulted in a 15 percent increase in total seed bank emergence compared with the unfertilized controls (C. Murphy and J. Mathews, unpublished research). Soil applications of ethylene, either in the gaseous form or by the ethylene-evolving compound ethephon (2-chloroethane phosphoric acid), have been reported to significantly reduce soil seed bank populations of redroot pigweed and common cocklebur (Egley, 1980, 1990), common ragweed (*Ambrosia artemisiifolia* L.) (Samimy and Khan, 1983), and *Striga* spp. (Egley, 1986).

Applying chemical nitrogen fertilizers to reduce soil seed bank populations is unlikely to be an acceptable management strategy due to the negative environmental implications associated with off-site movement of nitrate. Likewise, the cost of applying exogenous ethylene may be prohibitive in many cases. Manipulating soil nitrate and ethylene levels through the use of cover crops or green manures, however, may be an alternative way to target weed seed sensitivity to these chemicals. For example, ethylene biosynthesis by soil microorganisms is positively correlated with soil organic matter and the addition of crop residue to the to the soil (Arshad and Frankenberger, 2002). We found that ethylene accumulation in a closed system was four to five times higher in soil amended with pea residues (1 percent dry wt basis) than in either nonamended soil or the moistened residues alone (see Figure 19.2) (R. S. Gallagher and A. D. Crawford, unpublished).

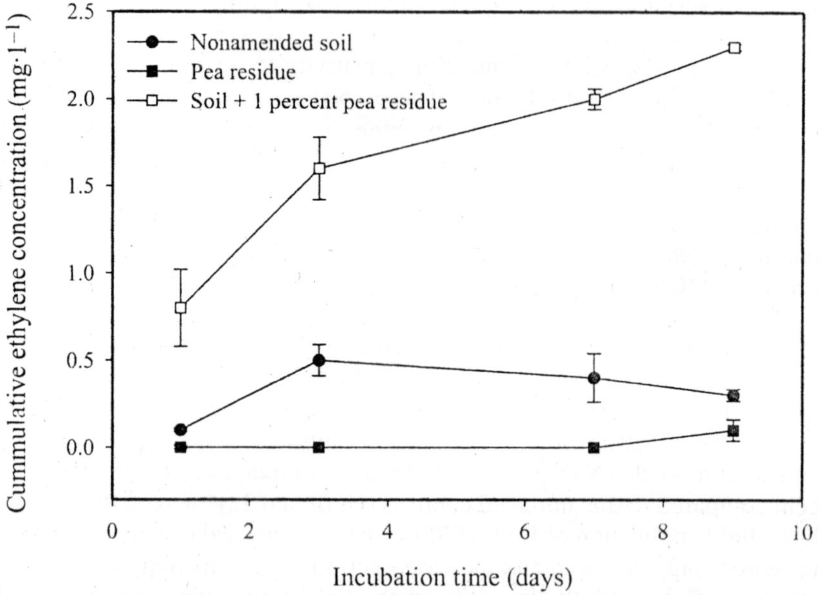

FIGURE 19.2. The evolution of ethylene from pea residues, soil amended with pea residues, and nonamended soil. Moistened residues, soil, and a mixture of the two were incubated in a sealed 500 mL jar. Ethylene was sampled from the headspace from one to nine days after the initiation of the experiment. (*Source:* Adapted from R. S. Gallagher and A. D Crawford, unpublished research.)

In addition, crop residues with a low carbon:nitrogen ratio, such as pea, would be expected to result in net nitrogen mineralization, thereby raising the nitrate concentration in the soil (Paul and Clark, 1996). Although largely untested, cover crops may alter the ethylene and nitrate environment of the soil to favor weed seed germination, allowing depletion of the soil seed bank via fatal germination and/or control of emerged plants.

The seed banks of the parasitic weeds *Striga* and *Orobanche* spp. have been successfully managed with the use of trap and catch crops (reviewed by Sauerborn, 1991; Parker and Riches, 1993). These weeds germinate in response to specific root exudates from the host plants. Trap crops produce stimulatory root exudates but are not susceptible to infection. Catch crops, on the other hand, also produce stimulatory root exudates but are susceptible infection. For catch crops to successfully deplete parasitic seed bank populations, they must be killed before the parasitic weeds set seed. In addition to promoting fatal germination of parasitic weeds, catch and trap crops that are used for green manures can improve the nitrogen fertility of the soil. Improved nitrogen fertility is thought to reduce the level and the impact of *Striga* and *Orobanche* infestations by reducing the host plant production of stimulatory root exudates, promoting shoot over root growth, and by increasing the host plant tolerance to infection (Parker and Riches, 1993).

Fire and Smoke

Fire and smoke can play a major role in stimulating seed germination in some habitats (reviewed by Van Staden, 2000). Heat from fire may overcome physical dormancy mechanisms by fracturing or desiccating the seed coat or other secondary structures. Fire will also change the microclimate of the soil surface, thereby influencing soil temperature fluctuations and the quantity and quality of light perceived by seeds. Fire will also induce chemical cues, such as smoke, nutrients, and gases which may act as germination stimulants. In Mediterranean-type ecosystems, smoke has been found to be an important germination cue for many native species. The smoke can be administered either directly by fire or through a smoke-water solution. Because fire is common in these ecosystems, smoke-induced germination appears to be an adaptation to initiate secondary succession. Smoke-water solutions have proven to be useful for habitat restoration efforts, enabling the establishment of nursery stocks of native plant species. In addition to the effects on native species, smoke-water solutions have also been found to stimulate the germination of a wide range of monocot and dicot weed species (Adkins and Peters, 2001). Although the specific mechanisms and compounds causing smoke-induced germination are not known, smoke has been shown to either replace or interact with other environmental germination cues such as light, nitrate, ethylene, and seed burial (Van Staden et al., 1995; Jager et al., 1996; Plummer et al., 2001; Tieu et al., 2001). The potential for weed seed bank management with smoke remains largely unexplored.

Summary of Seed Dormancy

Dormancy allows germination and establishment of many invasive weed species over an extended time period within seasons and among years. This is clearly a crucial adaptation for the survival of such species and for their perpetuation in agroecosystems. The dormancy of seeds is a remarkably plastic trait, regulated by endogenous factors and strongly influenced by several environmental cues, both during seed development and after dispersal. Endogenous factors include different mechanisms of dormancy, cycling of physiological dormancy, and genetic variability. Dormancy level is influenced by environmental factors during seed maturation, including drought, shade, photoperiod, and perhaps other factors. Environmental factors that influence dormancy of mature seeds include temperature, light, soil nitrate, soil ethylene, fire, and smoke. The numerous potential interactions among these endogenous and environmental factors are a major obstacle when attempting to forecast or control weed seed germination in the field.

ROLE OF SEED VIGOR

A key characteristic of most weeds is the ability to persist in the soil seed bank for many years. Although seed dormancy protects the seed against untimely germination, seed must also remain viable over time. Viability in itself, however, is not sufficient to ensure weed establishment and reproduction. Seed vigor, which enables germination and establishment over a wide range of environmental conditions, must also be maintained. Seed vigor differs from seed viability in that vigor is measured as germination or emergence under (or following) stress, and is considered a bett :r measure of potential seedling establishment in the field. Seed aging is the gradual process that leads to the loss of seed vigor, and ultimately, seed viability, and has been reviewed extensively (Priestly, 1986; Smith and Berjak, 1995; Walters, 1998; McDonald, 1999). These reviews focus primarily on crop seeds but do include some information on weedy species. Because relatively few studies are available on seed aging in weedy species, we will assume for the purpose of this discussion that the mechanisms associated with loss of vigor and viability are similar in crops and weeds.

Consequences of a Decline in Seed Vigor

Seed vigor is at a peak at physiological maturity (McDonald, 1999). As seeds age, vigor declines due to an impaired ability to synthesize key cell constituents, such as proteins and lipids, repair DNA, and convert starch reserves. As seed vigor declines, germination rate slows, seedling establishment is compromised, and eventually seed viability is lost (Priestly, 1986). In addition, a decrease in membrane integrity generally accompanies a loss in seed vigor, resulting in poor turgor regulation and increased leakage of solutes, such as ions and sugars, from the seed (Harmon and Stasz, 1986). Increased solute leakage from the seed is thought to make the seed more vulnerable to microbial attack (see Microbial Seed Decay).

A few studies with weedy species demonstrate that a loss in seed vigor may contribute to a decline in fitness or the ability to reproduce. For example in perennial ryegrass (*Lolium perenne* L.) seed, artificial aging under high humidity resulted in slower germination and lower seed viability when compared to nonaged seed (Naylor, 1989). In downy brome, time to 50 percent germination (T_{50}) was delayed by approximately three days in four-year-old seed compared to freshly harvested seed (Rice and Dyer, 2001). When downy brome was grown in competition with soft brome (*B. hordeaceaus* L.), the delayed germination in the four-year-old seed put the subsequent seedlings at a competitive disadvantage, resulting in lower biomass accumulation in the aged seed versus the nonaged seed. The detrimental effects of delayed emergence on competitiveness were compounded by poorer vigor in the adult plants from the aged seed. These data suggest that the age structure of the soil seed bank has the potential to influence the competitiveness of the weeds that emerge and ultimately their ability to return seed to the seed bank. To our knowledge, differential fitness among seed bank populations has not been accounted for in current models predicting weed emergence and competition.

Mechanisms of Seed Deterioration

In stored seed, aging is a function of temperature, seed water content, and time. There is a general rule that for every 5°C decrease in temperature or 1 percent decrease in seed moisture content that seed longevity will be doubled. This is a useful guideline, though more sophisticated viability equations have been developed (reviewed by Walters, 1998; Murdoch and

Ellis, 2000). Because weed seeds in the soil experience a wide range of temperature and moisture fluctuations and often exist in the imbibed state for extended periods, it is not clear that seed storage viability models can be applied to seed in situ.

Seed deterioration is largely attributed to damage caused by free radicals (McDonald, 1999). Free radicals cause direct damage to mitochondria, enzyme systems, and chromosomes, resulting in impaired metabolic function and the inability to repair genetic damage. Free radicals also cause lipid peroxidation, resulting in poor membrane integrity, reduced cellular and sub-cellular compartmentation, and leakage of solutes from the cell. Leakage of phosphate and other ions from the seed gives an early indication of membrane deterioration, whereas the leakage of carbohydrates indicates more extensive membrane damage (Ouyang et al., 2002). Antioxidants, such as carotenoids (Pinzino et al., 1999), phenolics, and other secondary metabolites (Vertucci and Farrant, 1995), may play a role in preserving seed vigor by scavenging free radicals.

Postdispersal desiccation/rehydration and freeze/thaw cycles also reduce seed vigor by causing damage to cell membranes and other cell constituents (reviewed by Horbowicz and Obendorf, 1994; Obendorf, 1997). Carbohydrates, such as the raffinose series of oligosacharides (sucrose, raffinose, stachyose, and verbascose), have been found to play a role in desiccation tolerance. These carbohydrates are thought to contribute to desiccation tolerance by assisting in the transition from the dry (gel-like) state to the imbibed (liquid-crystalline) state by stabilizing proteins and membranes, and minimizing cell leakage (Bewley and Black, 1994; Vertucci and Farrant, 1995). In addition, carbohydrates help in the formation of a superviscous fluid or glassy state during drying, which protects the cell constituents from deleterious oxidative reactions. Although temperate weed and crop seed are generally considered to be quite tolerant to desiccation, there is evidence that desiccation may negatively affect fitness. For example, in birdsfoot trefoil (*Lotus corniculatus* L.) seed, 18 h of dehydration over silica gel resulted in approximately a 20 percent decrease in seed viability, a 50 percent increase in seed leakage, and a 45 percent decrease in seedling length in seed imbibed for 12 h or more when compared to the seed that remained hydrated (McKersie and Tomes, 1979). To our knowledge, little information is available on how desiccation/rehydration and freeze/thaw cycles affect seed longevity in the soil seed bank and how tolerance of desiccation and freezing in weedy species is affected by genetic and environmental factors.

Effects of Maturation Environment on Seed Vigor

Many factors during seed development and maturation can contribute to seed vigor (reviewed by McDonald, 1999). In general, plant stress due to drought, nutrient deficiencies, temperature extremes, and competition negatively impact seed vigor. In wild oat seed, ion leakage, as measured by the electrical conductivity of the seed leachate, was approximately 20 percent higher in seed that matured within the wheat canopy than in either those seed that matured above the wheat canopy or in monoculture (see Table 19.3) (R. S. Gallagher, S. Higgins, and E. P. Fuerst, unpublished). This effect was exacerbated when the seed were subjected to two weeks of accelerated aging (37°C and 85 percent RH), resulting in seed leakage that was up to 47 percent higher in the seed collected from within the wheat canopy compared to the above canopy and monoculture seed. In wheat, postanthesis shading of 98 percent inhibited starch remobilization from the stems and leaves to the developing grain, resulting in numerous unfilled kernels (Kiniry, 1993). This study illustrated that sufficient light was required not only to build starch reserves, but also to remobilize starch reserves in times of stress. In triticale, postanthesis drought resulted in a 24 percent reduction in the soluble carbohydrates in the seed compared with well-watered plants (Zalewski et al., 2001). Specifically, there was a decrease in the raffinose:sucrose ratio which has been associated with a decrease in desiccation tolerance (Horbowicz and Obendorf, 1994; Vertucci and Farrant, 1995; Obendorf, 1997). In summary, these studies suggest that plant stress during seed development and maturation may alter the allocation of plant resources to carbohydrates associated with seed vigor. Seed from stressful environments are likely to be less vigorous at physiological maturity and be more prone to deterioration in response to aging and desiccation/rehydration and freeze/thaw cycles.

TABLE 19.3. The effect of seed maturation environment and seed aging on seed leakage as measured by the electrical conductivity of the seed leachate (measured in microamps·mg^{-1}).

Maturation environment	Nonaged seed	Aged seed (2 wk)
Within the wheat canopy	2.84[a]	3.53[a]
Above and wheat canopy	2.37[b]	2.59[b]
Wild oat monoculture	2.29[b]	2.39[b]

Note: Means followed by the same letter within a column are not significantly different.

Evaluating Seed Vigor

Seed vigor can be evaluated in many ways, though no one method is considered to be universally standard (reviewed by McDonald, 1999). Measuring seed viability under optimal germination conditions is the simplest indicator of seedlot vigor but may not reveal subtle differences in seed vigor that can affect performance of the seed in situ. Therefore, many seed vigor tests put the seed under one or more stress conditions to reveal potential differences in field performance. The following section is a summary of the seed vigor tests commonly used. Detailed methodologies can be found in the *Handbook of Vigour Test Methods* (Hampton and TeKrony, 1995). Unfortunately, weed seed dormancy is often a complicating factor in any of these tests, and it may be necessary to use natural or artificial procedures that overcome dormancy and/or to evaluate viability of ungerminated seed with a tetrazolium assay (Grabe, 1970; McDonald et al., 1993).

- *Germination rate:* Germination rate is commonly measured by determining the time to germination of 50 percent of the seed (i.e., T_{50}), and comparing T_{50} values among seedlots. This method requires frequent monitoring of germination and is difficult to conduct if seed germination is being studied in the dark. In addition, T_{50} values are seriously confounded by differences in dormancy among seedlots.
- *Cold test:* The cold test evaluates the germination potential of a seedlot at suboptimal temperatures for a particular species and is commonly used to measure vigor in crop seed. When working with less common species, preliminary experiments need to be conducted to determine the appropriate range of temperatures at which to conduct the test. The cold test may have particular merit in testing seed vigor in temperate spring annual species that must germinate and establish in cold, wet conditions.
- *Accelerated aging:* Accelerated aging tests have been designed to hasten seed deterioration by placing the seed in either warm/moist or oxygen-rich environments. The high temperature/humidity method typically places seed at 35 to 45°C and 85 to 100 percent relative humidity (RH). This method often achieves rapid seed deterioration and can be a useful indicator of field performance of a seedlot. However, in smooth pigweed (*Amaranthus hybridus* L.) and velvetleaf, accelerated aging at 35°C and 85 percent RH induced secondary dormancy, thereby hindering our ability to test for any differences in germination or vigor (R. S. Gallagher, personal experience). As in the cold test, temperature and humidity conditions for artificial aging can be modified to suit a particular species or situation. In stored seed, the high

temperature/humidity method has proven useful to screen out differences in seed vigor but has been criticized for inducing seed deterioration that is not characteristic of the natural seed aging process (Priestly, 1986). In contrast, aging seed in near pure oxygen conditions may hasten the oxidative reactions more typical of natural aging (R. Hannan, personal communication). For example, onion seed aged in a 100 percent oxygen environment at 40°C showed similar lipid degradation products as naturally aged seed (Hannan and Hill, 1991). Weed seed in the soil, however, generally experience repeated thermal and wetting/drying cycles. Insufficient data are available to indicate which accelerated aging method is most appropriate for detecting differences in potential longevity in weedy species.

- *Seed leakage tests:* Measuring the conductivity of the seed leachate solution is a common way to measure seed leakage (Pandey, 1992). Chloride, phosphate, sulfate, potassium, and magnesium are among the ions that leak from seeds. Alternatively, measuring optical density of the seed leachate at 260 nm gives an estimation of nucleic acid leakage from the seed (Deswal and Sheoran, 1993). This method correlated well with conductivity readings of the seed leachate and was found to be particularly well suited to testing small or single seeds.

Summary of Seed Vigor

Seed within a soil seed bank population are often assumed to have uniform potential to germinate and develop into reproductively fit individuals. However, germination and plant fitness will vary widely, depending upon seed vigor. Seed vigor is the net result of seed maturation conditions and aging processes in the soil. Seed that develop and mature under resource-poor conditions may have decreased vigor compared to seed that mature under resource-rich conditions, and may be prone to a more rapid loss of vigor when exposed to harsh environmental conditions in the soil. Limitation of light, water, or nutrients during seed development may constrain the plant's ability to allocate resources to the seed chemistries associated with seed vigor. For example, the raffinose series of carbohydrates appears to play a role in promoting membrane stability in the seed under wetting/drying and freeze/thaw cycles. Also, phenolics, which play roles in seed dormancy, resistance to decay, and as antioxidants, may be produced at lower levels in seed produced under resource-poor conditions. Once in the soil, seed vigor will decrease in response to aging processes and environmental conditions. As seed vigor declines, germination slows and seedling vigor decreases. Slower germination and less-vigorous seedlings may be less competitive with crops and other weeds. A key indicator of decreased seed vigor is increased leakage of cell constituents and solutes. Increased seed leakage will not only accelerate aging but also likely make seeds more prone to attack by soil microbes.

MICROBIAL SEED DECAY

Decay As a Cause of Seed Mortality

Most field studies of seed longevity have simply determined the number of seed surviving after various periods of time, without determining the causes of mortality. Seed mortality due to microbial decay is suspected in many cases but is quite difficult to demonstrate. Seed bank losses due to microbial decay have occasionally been estimated by indirect means, i.e., by subtraction of mortality from other causes, such as fatal germination, predation, environmental stress, and seed aging. An alternative indirect means to determine losses due to decay is to use treatments that inhibit microorganisms. Methodological hurdles to quantifying seed decay are further complicated by the fact that microbially induced seed mortality is not likely to be the result of a single event or mechanism. We hypothesize that multiple mechanisms are involved in determining seed vulnerability to microbial attack, including seed vigor, seed defenses (secondary defense compounds, seed coverings), aging, environmental stress, predation, and microbial ecology (Figure 19.1). Therefore, several gradual processes may precede the onset of decay.

The study of seed decay can be simplified in the laboratory, where bacteria and fungi can be isolated and tested under controlled conditions. Several studies have indicated that microbial isolates can inhibit seed germination under laboratory conditions. For example, several species of fungi isolated from seeds inhibited the germination of morning glory (*Ipomea hederacea* L.) and velvetleaf (Kirkpatrick and Bazzaz, 1979), and several rhizobacteria isolated from wheat and downy brome inhibited germination of downy brome (Kennedy et al., 1991). However, the study of seed decay in situ is very difficult due to the great diversity of microorganisms, the multiple causes of seed mortality, and the possibility that seed decay is secondary, i.e., decay follows weakening or injury from other causes (Kremer, 1993). These interactions make it quite difficult to directly apply laboratory information to the field situation.

A number of studies have provided indirect evidence for microbial seed decay by the use of fungicides, antibiotics, or soil sterilization to inhibit certain microorganisms and shift microbial populations. In these studies, any increases in seed longevity have been attributed to the assumed suppression of microbial decay. For example, when the fungicides carboxin plus thiram were used as a seed treatment, shattercane *Sorghum bicolor* (L.) Moench caryopsis survival increased from 0.5 to 17 percent over a five-month period (Fellows and Roeth, 1992). Seed treatment with the fungicide benomyl significantly increased survival of catclaw mimosa (*Mimosa pigra* L.) seeds by 7 to 10 percent over a seven-month period; however, fluctuating temperatures at the soil surface contributed more to mortality than did microorganisms (Lonsdale, 1993). In both of these studies, most of the lost viability was due to seed germination, but the number of germinated or emerged seeds was not directly measured. In a more recent report, a combination of captan, metalaxyl, and thiram fungicides was used as a seed treatment (Gallandt et al., 2002). In one experiment, in one out of two years, fungicide seed treatment significantly reduced the number of decayed seed by 4 percent and total dead seed by 7 percent over a ten-month period. However, in a second experiment, fungicide seed treatment had no effect on seed mortality (Gallandt et al. 2002). Leishman, Masters, et al. (2000) conducted a two-year study evaluating the effects of a combination of seed- and soil-applied fungicides on germination of recovered seeds; fungicide treatment significantly increased germination of black medic (*Medicago lupulina* L.), field bindweed (*Convolvulus arvensis* L.), and birdsfoot trefoil, but not blackberry (*Rubus fruticosus* L.), at various retrieval times up to two years. Using a related approach, surface-sterilized wild oat seeds had higher germination (64 percent) than non-surface sterilized seeds (34 percent) over a three-year period (Kiewnick, 1964). This was interpreted as implying that organisms on the seed surface contributed to seed decay (Kiewnick, 1964; Thurston and Cussans, 1976). However, much of the increased germination was due to a dormancy-breaking effect of surface sterilization (Thurston and Cussans, 1976); we have similarly observed that surface sterilization treatments often break wild oat dormancy (E. P. Fuerst, unpublished).

Soil sterilization techniques have also been used to evaluate a possible role of seed decay in seed mortality. The work of Kiewnick (1964), further analyzed by Thurston and Cussans (1976), demonstrated that when wild oat seeds were sown outdoors emergence was greater in steam-sterilized soil (60 percent) than in nonsterile soil (34 percent). The remaining seed were presumed dead after the three-year period. This implied that more seed decay occurred in the nonsterile soil. The lack of statistical analysis and replication in this study indicate the need to pursue this approach further. Gallandt et al. (2002) observed no change in wild oat mortality over 10 months in response to chemical fumigation of the soil.

Taken together, these studies suggest that microbial decay may be a modest or, in some cases, negligible component of seed mortality. However, this is probably not a safe generalization for a number of reasons:

1. Seeds used in the studies cited may have been of higher vigor than many of the seeds normally present in the soil. For example, Gallandt et al. (2002) used fresh wild oat seed that were grown in monoculture with adequate nutrients, light, and water. In contrast, seed produced under the environmental stresses caused by crop competition may result in production of seeds with reduced vigor, reduced resistance to decay, and/or reduced dormancy.
2. The freshly harvested seed used in studies cited are also likely to be more vigorous, and thus more resistant to decay, than seed that has been in the seed bank for some time (see Role of Seed Vigor).
3. It is possible that the fungicide and soil sterilization treatments used were only briefly and/or partially effective in preventing the activity of decay organisms, and substantially greater increase in seed survival (indicative of greater mortality due to seed decay in untreated plots/seeds) would have been seen with more effective treatments. Currently, there is no established test for efficacy of treatments that prevent seed decay over a period of several months or years.
4. Only the studies by Kiewnick (1964) and Leishman, Masters, et al. (2000) were longer than one year. This indicates a deficiency of longer-term studies evaluating the significance of seed decay.

We conclude that, at present, it is not possible to generalize to accurately assess the impact of microbial decay in the survival and mortality of seeds in the field.

Factors Regulating Decay

The seed itself probably contributes to the initiation of seed decay. As dry seeds imbibe, they exude carbohydrates, amino acids, fatty acids, nucleic acids, other organic acids, and inorganic ions into the spermosphere (Deswal and Sheoran, 1993: Harmon and Stasz, 1986). These chemicals serve as nutritional sources for soil microorganisms (Halloin, 1983). As seeds age, seed leakage usually increases (see Role of Seed Vigor). Field peas with high rates of carbohydrate leakage had high rates of *Fusarium solani* chlamydospore germination, resulting in greater vulnerability to seed and seedling rot (Short and Lacy, 1976). Harmon and Stasz (1986) reported similar relationships among seed exudation, fungal spore germination, seed decay, and seedling emergence in peas, beans, and soybean. Loss of seed vigor to environmental stress, aging, and microbial colonization may similarly promote leakage in weed seeds in the soil and consequently promote microbial invasion.

Seeds obviously must resist colonization and invasion by both pathogenic and deleterious saprotrophic organisms in order to survive for extended periods in the soil. Seeds modify and regulate microbial colonization and decay through two fundamental mechanisms: (1) physical barriers, represented by seed coverings, and (2) chemistry or antimicrobial chemicals of the seed, the chemistry of seed exudates, and the chemical characteristics of insoluble substrata (Kremer, 1993).

The Role of Seed Coverings

The seed coat (testa or pericarp) and other seed-covering tissues present physical barriers to microbial invasion and often contain antimicrobial chemicals that restrict or modify microbial colonization (Halloin, 1983; Kremer et al., 1984; Kremer, 1993). Many broadleaf weeds, such as velvetleaf (Kremer, 1993) and wild cotton (*Gossipium hirsutum* L.) (Halloin, 1983), have a physical dormancy trait called "hard-seededness," in which the testa has extremely dense palisade cells that resist penetration by soil fungi and also prevent imbibition and radicle protrusion. In these two species, the only discontinuity in the seed coat is the chalaza, which is also the only apparent site for microbial invasion of intact seeds (Kremer, 1993; Halloin, 1983). The importance of hulls in grasses in resisting decay is suggested by their ability to persist in the soil following seed germination or seed death. For example, the lemma of wild oat remains relatively unaltered for several months in the field (E. P. Fuerst, unpublished observation); this can be attributed to the dense and highly lignified cells of the wild oat lemma (Morrison and Dushnicky, 1982).

The Role of Seed Chemistry

Defense chemistries, such as seed phenolics, are thought to serve as a primary regulator of microbial growth on and in the seed as well as in the adjacent spermosphere (Putnam and Duke, 1978; Halloin, 1986). The testa of dicots often contains antimicrobial chemicals, including tannins and phenolics, that inhibit microbial colonization (Halloin, 1983; Kremer, 1993). In velvetleaf, compounds, including flavonoids, that diffused from surface-sterilized seeds inhibited 58 percent of the bacteria and 100 percent of the fungi evaluated (Kremer, 1986a; Paszkowski and Kremer, 1988). In another study, peas with dark seed coats were less susceptible to fungal attack than peas with light-colored seed coats (Halloin, 1983). Seeds of crops such as castor bean, soybean, and wheat often contain lectins, which may inhibit microbial colonization (Halloin, 1983; Mirelman et al., 1975). Seed lectins from a diversity of plant species can disrupt the growth of germinating fungal spores (Brambl and Glade, 1985). Lignans, tannins, and soluble phenolics can be found in the seed coat of dicots (Debeaujon et al., 2000; Morrison et al., 1995) and are often localized in the hulls, pericarp, and aleurone of monocots (Hatcher and Kruger, 1997; Fellows and Roeth, 1992). Resistance to microbial invasion in oat (*Avena sativa* L.) was attributed to the physical barrier of the hulls and the antifungal properties of phenolics and other compounds in the hulls; fungi colonized wild oat hulls more rapidly when the acetone-soluble secondary metabolites were removed (Picman et al., 1984). Several phenolic compounds have been extracted and identified from wild oat hulls (Chen et al., 1982). Shattercane seed survival and resistance to microbial colonization was attributed to high tannin and lignin content of glumes and the caryopsis (Fellows and Roeth, 1992). In barley seeds, mutants that did not express condensed tannins had significantly higher levels of *Fusarium* sp. infection than the wild-type or tannin over-expressing seeds (Skadhauge et al., 1997). Nelson et al. (1997) found that total phenolics was the most useful indicator of inhibition of microbial growth on seeds. The level of seed dormancy is often related to flavonoids and other phenolics in the seed coat and aleurone (Debeaujon et al., 2000; Shirley, 1998; Warner et al., 2000). Interestingly, Mortensen and Hsiao (1987) found that seeds of dormant wild oat lines generally had lower pre-existing levels of microbial colonization than nondormant lines, and that dormant lines tended to be less susceptible to fungal pathogens than non-dormant lines. As a group, therefore, phenolics may both enforce dormancy and inhibit microbial growth.

Stratification of Microbial Communities Around the Seed

The chemistry of the seed surfaces, including both soluble and insoluble components, may favor colonization by certain microorganisms, which in turn may serve to prevent the establishment of decay organisms. Colonization during seed development appears to play an important role in subsequent microbial colonization in the soil. For instance, microorganisms present on velvetleaf seeds at harvest became quickly established when seeds came in contact with soil, and prevented colonization by several species established in the soil; however, soilborne microorganisms became quickly established if seeds were surface sterilized (Kremer, 1986b).

Several studies suggest a stratification of microbial communities on and in seed retrieved from the soil, including the inhibition of external fungi by internal bacteria. In the case of velvetleaf, about 80 percent of bacteria isolated from inside the seed inhibited external seedborne fungi (Kremer et al., 1984). Also, about 80 percent of the bacteria isolated from five weed species exhibited antifungal activity (Kremer, 1987). Similarly, *Pseudomonas* spp. isolated from caryopses of eastern gamagrass (*Tripsacum dactyloides* L.) inhibited the growth of several fungi (Anderson et al., 1980). We observed differences in the types of bacteria inhabiting the wild oat hull, caryopsis surface, and interior of the caryopsis (Fuerst et al., 2003). Organisms from the caryopsis surface and interior potentially have roles of either resisting or promoting decay. Certain species were found only internally, others only on the caryopsis surface, and others both internally and on the caryopsis surface, but not on the hull. Substrate utilization analysis revealed that the soil microbial community structure was altered in soil adhering to the seed compared to the adjacent soil, probably due to the diversity of seed exudates (Fuerst et al., 2003). These studies suggest that the microbial community on and in the seed is highly stratified.

Summary of Resistance to Seed Decay

Seed coverings and seed chemicals, including inhibitors, exudates, and insoluble substrata, probably regulate colonization of the seed both during seed development and during incubation in the soil. Consequently, there is thought to be a stratification of microbial communities going from the interior of the seed, to the seed surface, and into the soil. This stratification may create a biological "shell" in which relatively innocuous or defensive organisms colonize the seed. These organisms act in concert with the chemicals and physical barriers of the seed to resist colonization and invasion by deleterious or pathogenic microorganisms.

IMPLICATIONS FOR SEED BANK MANAGEMENT

An understanding of the ecophysiological processes associated with weed seed longevity in the soil will be an important factor in determining our ability to accurately predict weed cycles under a range of environmental conditions and management regimes. In addition, active seed bank management strategies that aim to deplete seed bank populations or suppress weed emergence at key times will require a mechanistic understanding of seed bank dynamics. Seed longevity is regulated by (1) seed qualities, including dormancy, vigor, and resistance to microbial decay, and (2) the physical, chemical, and microbial environment of the soil seed bank Although the genetics of a weed are difficult to manage, manipulation of the environment in which a weed develops and matures may influence the number and the longevity of seed produced. For example, enhancing competitiveness of a crop by optimizing the date, rate, and arrangement of sowing may cause the weeds that escape control to produce fewer seed, and these fewer seed may be of poorer quality than seed produced with less-severe competition. Weeds maturing under resource-poor environments, such as beneath the crop canopy and/or when development is delayed relative to the crop, would be less of a management concern than weeds that have a stronger foothold in the crop. Weed management decisions should be based on the likelihood that weeds will have both an effect on short-term productivity of the crop or system and a long-term effect on the number and quality of seed in the soil. In addition to manipulating the maturation environment of weeds, manipulation of the seed bank environment may hasten or hinder seed bank depletion. For example, soil management through tillage and/or green manures can be directed at a seed population's sensitivity to light, chemicals, and other environmental germination cues. Green manures will also have a large effect on microbial activity and microbial community structure in the soil, which may influence the dynamics of seed decay. Researchers have only begun to understand the complexity of the interactions among plant genetics, plant maturation environment and weed seed quality, as well as the relationships among seed quality, soil quality, and seed longevity. In our opinion, these are valuable areas of research that will help elucidate the mechanisms regulating weed seed longevity in the soil, and move scientists beyond the current "black box" approach to seed bank dynamics.

IMPLICATIONS FOR FUTURE RESEARCH

In discussing our conceptual model for seed longevity, we hypothesized that seed longevity is regulated by plant genetics, the seed maturation environment, and the chemical, physical, and microbial characteristics of the soil (Figure 19.1). According to this model, each of these regulating factors ultimately impact one or more of the three specific seed quality characteristics, which can also be considered mechanisms of longevity: *seed dormancy, seed vigor,* and *seed resistance to microbial decay.* Within this framework, and based on discussions earlier in this chapter, we note many areas in which information is scarce or absent. We consider the following to be among the most crucial research needs:

- Evaluate the effect of specific resource limitations (light, nutrients, and water) as well as crop competition (and combined resource limitations) on seed quality ("quality" = dormancy, vigor, and resistance to decay) and on longevity in situ.
- Monitor actual seed moisture and temperature cycles in situ, and evaluate the effect of moisture and thermal cycles on seed quality and longevity, both in the laboratory and in situ.
- Evaluate the effect of aging in the seed bank on weed seed quality in the laboratory and on weed competitiveness in the greenhouse and in situ. Compare the similarities and differences that occur with natural versus artificial seed aging.
- Evaluate the effect of delayed weed maturation on seed quality and longevity.
- Characterize further the potential of smoke as a germination stimulant and identify specific compounds involved as well as mechanisms. Characterize the effects of heat and fire on seed quality.
- Evaluate cover crops and soil amendments as potential seed bank management options and determine their effects on soil nitrogen, ethylene, seed germination, seed quality, and seed decay.
- Assess the significance of microbial decay in seed mortality.
- Characterize the structure and function of microbial communities on and in weed seeds in the soil. Identify organisms that accelerate decay and determine ways of enhancing their activities while protecting crop seed.

In conclusion, improving our understanding of these key areas will better enable the development of ecologically based approaches to managing the soil seed bank.

REFERENCES

Adkins, S.W. and N.C.B. Peters (2001). Smoke derived from burnt vegetation stimulates germination of arable weeds. *Seed Science Research* 11:213-222.

Allen, P.S. and S.E. Meyers (2002). Ecology and ecological genetics of seed dormancy in downy brome. *Weed Science* 50:241-247.

Amritphale, D., J.C. Gupta, and S. Iyengar (1984). Germination polymorphism in sympatric populations of *Calotropis procera*. *Oikos* 42:220-224.

Anderson, R.C., A.E. Liberta, J. Packheiser, and M.E. Neville (1980). Inhibition of selected fungi by bacterial isolates from *Tripsacum dactyloides* L. *Plant and Soil* 56:149-152.

Arshad, M. and W.T. Frankenberger (2002). *Ethylene: Agricultural sources and applications*. New York: Kluwer Academic/Plenum Publishers.

Ballaré, C.L., R.A. Sánchez, A.L. Scopel, and C.M. Ghersa (1988). Morphological responses of *Datura ferox* L. seedlings to the presence of neighbours. *Oecologia* 76:288-293.

Baskin, C.C. and J.M. Baskin (1998). *Seeds: Ecology, biogeography, and evolution of dormancy and germination.* San Diego: Academic Press.

Baskin, J.M. and C.C. Baskin (1977). High temperature requirement for after-ripening in seeds of winter annuals. *New Phytologist* 77:619-624.

Baskin, J.M. and C.C. Baskin (1980). Ecophysiology of secondary dormancy in seeds of *Ambrosia artemisiifolia. Ecology* 61:475-480.

Baskin, J.M. and C.C. Baskin (1985). Annual dormancy in buried weed seeds: A continuum. *BioScience* 35:492-498.

Bazzaz, F.A., D.D. Ackerly, and E.G Reekie (2000). Reproductive allocation in plants. In M. Fenner (ed.), *Seeds: The ecology of regeneration in plant communities,* Second edition (pp. 1-29). Wallingford, UK: CABI Publishing.

Bello, I.A., M. Owen, and H.M. Hatterman-Valenti (1995). Effect of shade on velvetleaf *(Abutilon theophrasti)* growth, seed production, and dormancy. *Weed Technology* 9:452-455.

Benech-Arnold, R.L., R.A Sánchez., F. Forcella, B.C. Kruk, and C.M. Ghersa (2000). Environmental control of dormancy in weed seed banks in soil. *Field Crops Research* 67:105-122.

Benvenuti, S., M. Macchia, and A. Stefani (1994). Effects of shade on reproduction and some morphological characteristics of *Abutilon theophrasti* Medicus, *Datura stramonium* L. and *Sorghum halepense* L. *Weed Research* 34:283-288.

Bewley, J.D. and M. Black (1994). Metobolic consequences and causes of viability loss. In J.D. Bewley and M. Black, *Seeds: Physiology of development and germination* (pp. 402-420). New York: Plenum Press.

Bouwmeester, H.J. and C.M. Karssen (1993). Seasonal periodicity in germination of seeds of *Chenopodium album* L. *Annals of Botany* 72:462-473.

Bradford, K.J. 2002. Applications of hydrothermal time to quantifying and modeling seed germination and dormancy. *Weed Science* 50:248-260.

Brambl, R. and W. Glade (1985). Plant seed lectins disrupt growth of germinating fungal spores. *Physiologia Plantarum* 64:402-408.

Buhler, D. (1997). Effects of tillage and light environment on the emergence of 13 annual weeds. *Weed Technology* 11:496-501.

Casal, J.J. and R.A. Sánchez (1998). Phytochromes and seed germination. *Seed Science Research* 8:317-329.

Cavers, P.B. and D.L. Benoit (1989). Seed banks in arable land. In M.A. Leck, V.T. Parker, and R.L. Simpson, (eds.), *Ecology of soil seed banks* (pp. 309-328). San Diego, CA: Academic Press.

Chen, F.S., J.M. MacTaggart, and R.M. Elofson (1982). Chemical constituents in wild oat *(Avena fatua)* hulls and their effects on seed germination. *Canadian Journal of Plant Science* 62:155-161.

Chepil, W.S. (1946). Germination of weed seeds: II. The influence of tillage treatments on germination. *Scientific Agriculture* 26:307-346.

Cook, R.E. (1980). The biology of seeds in soil. In O. T. Solbrig (ed.), *Demography and evolution in plant populations* (pp. 107-129). Berkeley: University of California Press.

Corbineau, F. and D. Côme (1995). Control of seed germination and dormancy by the gaseous environment. In J. Kigel and G. Galili (eds.), *Seed development and germination* (pp. 397-424). New York: Marcel Dekker Inc.

Debeaujon, I., K.M. Leon-Kloosterziel, and M. Koorneef (2000). Influence of the testa on seed dormancy, germination, and longevity in *Arabidopsis. Plant Physiology* 122:403-413.

Deswal, D.P. and I.S Sheoran (1993). A simple method for seed leakage measurement: Applicable to seeds of any size. *Seed Science and Technology* 21:179-185.

Egley, G.H. (1980). Stimulation of common cocklebur and redroot pigweed seed germination by injection of ethylene into the soil. *Weed Science* 28:510-514.

Egley, G.H. (1984). Ethylene, nitrate and nitrite interactions in the promotion of dark germination of common purslane seeds. *Annals of Botany* 53:833-840.

Egley, G.H. (1986). Stimulation of weed seed germination in soil. *Reviews of Weed Science* 2:67-89.

Egley, G.H. (1989). Some effects of nitrate-treated soil upon the sensitivity of buried redroot pigweed (*Amaranthus retroflexus* L.) seeds to ethylene, temperature, light and carbon dioxide. *Plant, Cell and Environment* 12:581-588.

Egley, G.H. (1990). Ethephon reduction of redroot pigweed *(Amaranthus retroflexus)* seed population. *Weed Technology* 4:803-813.

Eshashi, Y. and A.C. Leopold (1969). Dormancy regulation in subterranean clover seeds by ethylene. *Plant Physoiology* 44:1470-1472.

Fawcett, R.S. and R.W. Slife (1978). Effects of field applications of nitrate on weed seed germination and dormancy. *Weed Science* 26:594-596.

Fellows, G.M. and F.W. Roeth (1992). Factors influencing shattercane *(Sorghum bicolor)* seed survival. *Weed Science* 40:434-440.

Foley, M.E. and S.A. Fennimore (1998). Genetic basis for seed dormancy. *Seed Science Research* 8:173-182.

Gallagher, R.S. and J. Cardina (1998a). Ecophysiological factors regulating phytochrome-mediated germination in soil seed banks. *Aspects of Applied Biology* 51:165-171.

Gallagher, R.S. and J. Cardina (1998b). The effect of light environment during tillage on the recruitment of various summer annuals. *Weed Science* 46:214-216.

Gallagher, R.S. and J. Cardina (1998c). Phytochrome-mediated *Amaranthus* germination: I. Effect of seed burial and germination temperature. *Weed Science* 46:48-52.

Gallagher, R.S. and J. Cardina (1998d). Phytochrome-mediated *Amaranthus* germination: II. Development of very low fluence sensitivity. *Weed Science* 46:53-58.

Gallandt, E.R., E.P. Fuerst, and A.C. Kennedy (2002). Tillage system effects on over-winter mortality of *Avena fatua* seed in soil. *Abstracts of the Weed Science Society of America* 42:84.

Goudey, J.S., H.S. Saini, and M.S. Spencer (1987). Uptake and fate of ethephon ([2-chloroethyl]phosphonic acid) in dormant weed seeds. *Plant Physiology* 85:155-157.

Grabe, D. F. (1970). *Tetrazolium testing handbook for agricultural seeds.* Contrib. No. 29, Handbook on Seed Testing. Las Cruces, NM: Association of Official Seed Analysts.

Gutterman, Y. (1982). Phenotypic maternal effect of photoperiod on seed germination. In A.A. Khan (ed.), *The physiology and biochemistry of seed development, dormancy and germination* (pp. 67-69). New York: Elsevier Biomedical Press.

Halloin, J.M. (1983). Deterioration resistance mechanisms in seeds. *Phytopathology* 73:335-339.

Halloin, J.M. (1986). Microorganisms and seed deterioration. In M.B. McDonald Jr. and C.J. Nelson (eds.), *Physiology of seed deterioration* (pp. 89-99). Madison, WI: CSSA.

Hampton, J.G. and D.M. TeKrony (1995). *Handbook of vigour test methods,* Third edition. Zurich, Switzerland: International Seed Testing Association.

Hannan, R.M. and H.H. Hill (1991). Analysis of lipids in aging seed using capillary supercritical fluid chromatography. *Journal of Chromotography* 547:393-401.

Harmon, G.E. and T.E. Stasz (1986). Influence of seed quality on soil microbes and seed rots. In S.H West (ed.), *Physiological-pathological interactions affecting seed deterioration* (pp. 11-37). Madison, WI: CSSA.

Hatcher, D.W. and J.E Kruger (1997). Simple phenolic acids in flours prepared from Canadian wheat: Relationship to ash content, color, and polyphenol oxidase activity. *Cereal Chemistry* 74:337-343.

Holmes, M.G. (1981). Spectral distribution of radiation within plant canopies. In H. Smith (ed.), *Plants and the daylight spectrum* (pp. 147-158). London: Academic Press.

Horbowicz, M. and R.L. Obendorf (1994). Seed desiccation tolerance and storability: Dependence on flatulence-producing oligosaccharides and cyclitols—Review and survey. *Seed Science Research* 4:385-404.

Hurtt, W. and R.B. Taylorson (1986). Chemical manipulation of weed emergence. *Weed Search* 26:259-267.

Jager, A.K., A. Strydom, and J. Van-Staden (1996). The effect of ethylene, octanoic acid and a plant-derived smoke extract on the germination of light-sensitive lettuce seeds. *Plant Growth Regulation* 19:197-201.

Jain, S.K. (1982). Variation and adaptive roles of seed dormancy in some annual grassland species. *Botanical Gazette* 143:101-106.

Jensen, P.K. (1999). The effect of timing and frequency of soil cultivation on emergence and depletion of the soil seed bank of volunteer grass seeds and dicotyledonous weeds. *Journal of Applied Seed Production* 17:27-34.

Karssen, C.M. (1970). The light promoted germination of the seeds of *Chenopodium album* L.: III. Effect of the photoperiod during growth and development of the plants on the dormancy of the produced seeds. *Acta Botanica Neerlandica* 19:81-93.

Katoh, H. and Y. Esashi (1975). Dormancy and impotency of cockleburr seeds: I. CO_2, C_2H_4, O_2 and high temperature. *Plant and Cell Physiology* 16:687-696.

Kennedy, A.C., L.F. Elliott, F.L. Young, and C.L. Douglas. (1991). Rhizobacteria suppressive to the weed downy brome. *Soil Science Society of America Journal* 55:722-727.

Kiewnick, L. (1964). Experiments on the influence of seedborne and soilborne microflora on the viability of wild oat seeds (*Avena fatua* L.): II. Experiments on the influence of microflora on the viability of seeds in the soil. *Weed Research* 4:31-43.

Kigel, J., M. Ofir, and D. Koller (1977). Control of the germination response of *Amaranthus retroflexus* L. seeds by the parental environment. *Journal of Experimental Botany* 28:1125-1136.

Kiniry, J.R. (1993). Non-structural carbohydrate utilization by wheat shaded during grain growth. *Agronomy Journal* 85:844-849.

Kirkpatrick, B.L. and F.A. Bazzaz (1979). Influence of certain fungi on seed germination and seedling survival of four colonizing annuals. *Journal of Applied Ecology* 16:515-527.

Kremer, R.J. (1986a). Antimicrobial activity of velvetleaf (*Abutilon theophrasti*) seeds. *Weed Science* 34:617-622.

Kremer, R.J. (1986b). Microorganisms associated with velvetleaf (*Abutilon theophrasti*) seeds on the soil surface. *Weed Science* 34:233-236.

Kremer, R.J. (1987). Identity and properties of bacteria inhabiting seeds of selected broadleaf weed species. *Microbial Ecology* 14:29-37.

Kremer, R.J. (1993). Management of weed seed banks with microorganisms. *Ecological Applications* 3:42-52.

Kremer, R.J., L.B. Hughes, and R.J. Aldrich (1984). Examination of microorganisms and deterioration resistance mechanisms associated with velvetleaf seed. *Agronomy Journal* 76:745-749.

Leishman, M.R., G.J. Masters, I.P. Clarke, and V.K. Brown (2000). Seed bank dynamics: The role of fungal pathogens and climate change. *Functional Ecology* 14:293-299.

Leishman, M.R., I.J. Wright, T.A. Moles, and M. Westoby (2000). The evolutionary ecology of seed size. In M. Fenner (ed.), *Seeds: The ecology of regeneration in plant communities,* Second edition (pp. 31-57). Wallingford, UK: CABI Publishing.

Lonsdale, W.M. (1993). Losses from the seed bank of *Mimosa pigra:* Soil microorganisms vs. temperature fluctuations. *Journal of Applied Ecology* 30:654-660.

McDonald, M.B. (1999). Seed deterioration: Physiology, repair and assessment. *Seed Science and Technology* 27:177-237.

McDonald, M., R. Danielson, and T. Gutormson (eds.) (1993). *Seed analyst training manual.* Las Cruces, NM: Association of Official Seed Analysts.

McKersie, B.D. and D.T. Tomes. (1979). Effects of dehydration treatments on germination, seedling vigour, and cytoplasmic leakage in wild oats and birdsfoot trefoil. *Canadian Journal of Botany* 58:471-476.

Meyers, S.E. (1992). Habitat-correlated variation in firecracker penstemon (*Penstemon eatonii:* Scrophulariaceae) seed germination response. *Bulletin of the Torrey Botanical Club* 119:268-279.

Meyers, S.E., P.S. Allen. and J. Beckstead (1997). Seed germination regulation in *Bromus tectorum* (Poaceae) and its ecological significant. *Oikos* 78:475-485.

Meyers, S.E. and S.G. Kitchen (1994). Habitat-correlated variation in seed germination response to chilling in *Penstemon* section Glabri *(Scrophulariaceae). American Midland Naturalist* 132:349-365.

Mirelman, D., E. Galun, N. Sharon, and R. Lotan (1975). Inhibition of fungal growth by wheat germ agglutinin. *Nature* 256:414-416.

Morrison, I.M., E.A. Asiedu, T. Stuchbury, and A.A. Powell (1995). Determination of lignin and tannin contents of cowpea seed coats. *Annals of Botany* 76:287-290

Morrison, I.N. and L. Dushnicky (1982). Structure of the covering layers of the wild oat *(Avena fatua)* caryopsis. *Weed Science* 30:352-359.

Mortensen, K. and A.I. Hsiao (1987). Fungal infestation of seeds from seven populations of wild oats (*Avena fatua* L.) with different dormancy and viability characteristics. *Weed Research* 27:297-304.

Mulugeta, D. and D.E. Stoltenberg (1997). Increased weed emergence and seed bank depletion by soil disturbance in a no-tillage system. *Weed Science* 45:234-241.

Murdoch, A.J. and R.H. Ellis (2000). Dormancy, viability and longevity. In M. Fenner (ed.). *Seeds: The ecology of regeneration in plant communities*, Second edition (pp. 183-214). Wallingford, UK: CABI International 2000.

Naylor, J.M. and S. Jana (1976). Genetic adaptation for seed dormancy in *Avena fatua. Canadian Journal of Botany* 54:306-312.

Naylor, R.E.L. (1989). Analysis of the differences in germination of seed lots of perennial ryegrass in response to artificial ageing. *The Journal of Agricultural. Science* 112:351-357.

Nelson, K.E., A.N. Pell, P.H. Doane, B.I. Gine-Chavez, and P. Schofield (1997). Chemical and biological assays to evaluate bacterial inhibition by tannins. *Journal of Chemical Ecology* 23:1175-1194

Obendorf, R.L. (1997). Oligosaccharides and galactosyl cyclitols in seed desiccation tolerance. *Seed Science Research* 7:63-74.

Ouyang, X., T. van Voorthuysen, P.E. Toorop, and H.W.M. Hilhorst (2002). Seed vigor, aging, and osmopriming affect anion and sugar leakage during imbibition of maize (*Zea mays* L.) caryopses. *International Journal of Plant Science* 163:107-112.

Pandey, D.K. (1992). Conductivity testing of seeds. In H. Lenskens and J. Jackson (eds.), *Seed analysis* (pp. 273-304). Berlin: Springer-Verlag.

Parker, C. and C. R. Riches (1993). *Parasitic weeds of the world: Biology and control.* Wallingford, UK: CAB International.

Paszkowski, W.L. and R.J. Kremer (1988). Biological activity and tentative identification of flavonoid components in velvetleaf (*Abutilon theophrasti* Medik.) seed coats. *Journal of Chemical Ecology* 14:1573-1582.

Paul, E.A. and F.E. Clark (1996). *Soil microbiology and biochemistry*, Second edition. Academic Press, San Diego.

Peters, N.C.B. (1982). Production and dormancy of wild oat *(Avena fatua)* seed from plants grown under waterstress. *Annals of Applied Biology* 100:189-196.

Picman, A.K., R. Giaccone, K.C. Ivarson, and I. Altosaar (1984). Antifungal properties of oat hulls. *Phytoprotection* 65:9-15.

Pinzino, C., A. Capocchi, L. Galleschi, and F. Saviozzi (1999). Aging, free radicals and antioxidants in wheat seeds. *Journal of Agricultural and Food Chemistry* 47:1333-1339.

Plummer, J.A., A.D. Rogers, D.W. Turner, and D.T. Bell (2001). Light, nitrogenous compounds, smoke and GA3 break dormancy and enhance germination in the Australian everlasting daisy, *Shoenia filifolia* subsp. subulifolia. *Seed Science and Technology* 29:321-330.

Pons, T.L. (1989). Breaking of seed dormancy by nitrate as a gap detection mechanism. *Annals of Botany* 63:139-143.

Pons, T.L. (2000). Seed responses to light. In M. Fenner (ed.), *Seeds: The ecology of regeneration in plant communities*, Second edition (pp. 237-260). Wallingford, UK: CABI Publishing.

Priestly, D.A. (1986). *Seed aging*. Ithaca, NY: Comstock Publishing Associates.

Putnam, A.R. and W.B. Duke (1978). Allelopathy in agroecomonic systems. *Annual Reviews of Phytopathology* 16:431-451.

Rice, K.J. and A.R. Dyer (2001). Seed aging, delayed germination and reduced competitive ability in *Bromus tectorum*. *Plant Ecology* 155:237-243.

Roberts, H.A. and P.M Feast (1973). Changes in the number of viable weed seeds in soil under different regimes. *Weed Research* 13:298-303.

Roberts, H.A. and H. Locke (1978). Seed dormancy and periodicity of seedling emergence in *Veronica hederifolia* L. *Weed Research* 18: 41-48.

Saini, H.S., P.K. Bassi, and M.S. Spencer (1985a). Interaction among ethephon, nitrate, and after-ripening in the release of dormancy of wild oat seed. *Weed Science* 34:43-47.

Saini, H.S, P.K. Bassi, and M.S. Spencer (1985b). Use of ethylene and nitrate to break seed dormancy of common lambsquarter. *Weed Science* 34:502-506.

Samimy, C. and A.A. Khan (1983). Effect of field application of growth regulators on secondary dormancy of common ragweed seeds. *Weed Science* 31:299-303.

Sauerborn, J. (1991). *Parasitic flowering plants: Ecology and management.* Verlag Josef Margraf, Scientific Books. FR Germany.

Sawhney, R. and J.M. Naylor (1982). Dormancy Studies in Seed of *Avena fatua* [paper no. 13]. Influence of drought stress during seed development on the duration of seed dormancy. *Canadian Journal of Botany* 60:1016-1020.

Schonbeck, M.W. and G.H. Egley (1988). Effects of ethylene and some other environmental factors on different stages of germination in redroot pigweed seeds. *Plant Cell and Environment* 11:189-197.

Scopel, A.L., C.L. Ballaré, and S.R. Radosevich (1994). Photostimulation of seed germination during soil tillage. *New Phytologist* 126:145-152.

Shirley, B.W. (1998). Flavonoids in seeds and grains: Physiological function, agronomic importance and the genetics of biosynthesis. *Seed Science Research* 8:415-422.

Short, G.E. and M.L. Lacy (1976). Carbohydrate exudation from pea seeds: Effect of cultivar, seed age, seed color, and temperature. *Phytopathology* 66:182-187.

Skadhauge, B., T.K. Kristian, and D. Von-Wettstein (1997). The role of the barley testa layer and its flavonoid content in resistance to *Fusarium* infections. *Hereditas-Lund* 126:147-160.

Smith, H. (1995). Physiological and ecological function within the phytochrome family. *Annual Reviews in Plant Physiology and Plant Molecular Biology* 46:289-315.

Smith, M.T. and P. Berjak (1995). Deteriorative changes associated with the loss of viability of stored desiccation tolerant and desiccation sensitive seeds. In J. Kigel and G. Galili (eds.), *Seed development and germination* (pp. 701-746). New York: Marcel Dekker.

Stoller, E.W. and R.A. Myers (1989). Effects of shading and soybean *Glycine max* (L.) interference on *Solanum ptycanthum* (Dun.) (eastern black nightshade) growth and development. *Weed Research* 29:307-316.

Sugiyama, S. and F.A. Bazzaz (1997). Plasticity of seed output in response to soil nutrients and density in *Abutilon theophrasti:* Implications for maintenance of genetic variation. *Oecologia* 112:33-41.

Sultan, S.E. and F.A. Bazzaz (1993a). Phenotypic plasticity in *Polygonum pericaria* I. Diversity and uniformity in genotypic norms in reactions to light. *Evolution* 47:1009-1031.

Sultan, S.E. and F.A. Bazzaz (1993b). Phenotypic plasticity in *Polygonum pericaria:* II. Norms of reaction to soil moisture and the maintenance of genetic diversity. *Evolution* 47:1032-1049.

Sultan, S.E. and F.A. Bazzaz (1993c). Phenotypic plasticity in *Polygonum pericaria:* III. The evolution of ecological breadth for nutrient environment. *Evolution* 47:1050-1071.

Thurston, J.M. and G.W. Cussans (1976). Seed behavior. In D.P. Jones (ed.), *Wild oats in world agriculture* (pp. 65-88). London: Agricultural Research Council.

Tieu, A., K.W. Dixon, K.A. Meney, and K. Sivasithamparam (2001). Interaction of soil burial and smoke on germination patterns in seeds of selected Australian native plants. *Seed Science Research* 11:69-76.

Van Hinsberg, A. (1998). Maternal and ambient environmental effects of light on germination in *Plantaga lanceolata:* Correlated responses to selection on leaf length. *Functional Ecology* 12:825-833.

Van Staden, J. (2000). Smoke as a germination cue. *Plant Species Biology* 15:167-178.

Van Staden, J., A.K. Jager, and A. Strydom (1995). Interaction between a plant-derived smoke extract, light and phytohormones on the germination of light-sensitive lettuce seeds. *Plant Growth Regulation* 17:213-218.

Vertucci, C.W. and J.M. Farrant (1995). Acquisition and loss of desiccation tolerance. In J. Kigel and G. Galili (eds.), *Seed germination and development* (pp. 237-271). New York: Marcel Dekker.

Walters, C. (1998). Understanding the mechanisms and kinetics of seed aging. *Seed Science Research* 8:223-244.

Warner, R.L., D.A. Kudrna, S.C. Spaeth, and S.S Jones (2000). Dormancy in white-grain mutants of Chinese spring wheat (*Triticum aestivum* L.). *Seed Science Research* 10:51-60.

Yenish, J.P., J.D. Doll, and D.D. Buhler (1992). Effects of tillage on vertical distribution and viability of weed seed in soil. *Weed Science* 40:429-433.

Zalewski, K., L.B. Lahuta, and M. Horbowicz (2001). The effect of soil drought on the composition of carbohydrates in yellow lupin seeds and triticale kernels. *Acta Physiologiae Plantarum* 23:73-78.

SECTION IV: SEED TECHNOLOGY

Chapter 20

Seed Quality Testing

Sabry Elias

INTRODUCTION

Increasing agricultural production through the use of high-quality seed, among other agricultural inputs, has become essential for providing enough food for the rising number of people in the world today. Selecting high-yielding varieties with disease, insect, lodging, and shattering resistance, along with other desirable characteristics that fit the area and the purpose of production, are basic keys to satisfactory crop performance, competitive yield, and better crop production systems (Copeland, 1977). Production of high-quality seed is the cornerstone of any successful agricultural program and is also a good marketing tool for increasing the potential sale of a crop seed, especially in today's competitive market.

High-quality seed does not happen by chance. Each stage in seed production from planting through weed control, fertility program, harvest, cleaning, storage, and shipping is critical for achieving high-quality seed.

WHAT IS SEED QUALITY?

The practical definition of seed quality can differ according to the end user. For example, a farmer may desire high-quality seed that produces rapid, uniform plants with high yielding capacity under a wide range of field conditions. On the other hand, a producer of oil seed crop for margarine or soap or similar industrial purposes may desire seed with a particular stable fatty acid profile as a measure of high quality

In general, the quality of seed is measured in many ways, including genetic and physical purity, germination, vigor, uniformity in size, freedom

Thanks to Drs. Larry Copeland, Bill Young, Lee Schweitzer, and Oscar Gutbrod for reviewing the manuscript and providing valuable suggestions.

from seedborne diseases, and other factors affecting seed performance in the field. Examples of these factors are heat or mechanical damage (Shah et al., 2001) and preharvest sprouting (Elias and Copeland, 1991) (see Figure 20.1). In other words, seed quality is a collective term for the condition of seed including genetic homogeneity, physical appearance, viability, vigor, and uniformity (Maguire, 1977). Other characteristics such as specific chemical composition or resistance to certain diseases or insects also contribute to the quality of seed.

Factors Influencing Seed Qualify

Complex interactions of genetic, environmental, pathological, physiological, biochemical, and cytological factors influence the expression of seed quality. Factors under genetic control may include, but are not limited to, seed size and color, chemical composition, hard-seededness, hybrid vigor, susceptibility to mechanical damage, and disease resistance (Dickson, 1980; Kneebone, 1976; McDaniel, 1973).

Success in producing quality seed of a particular crop in one area and failure in another illustrates the importance of environmental influence on seed development and maturation. Environmental conditions during seed development and maturation, including temperature, water stress or

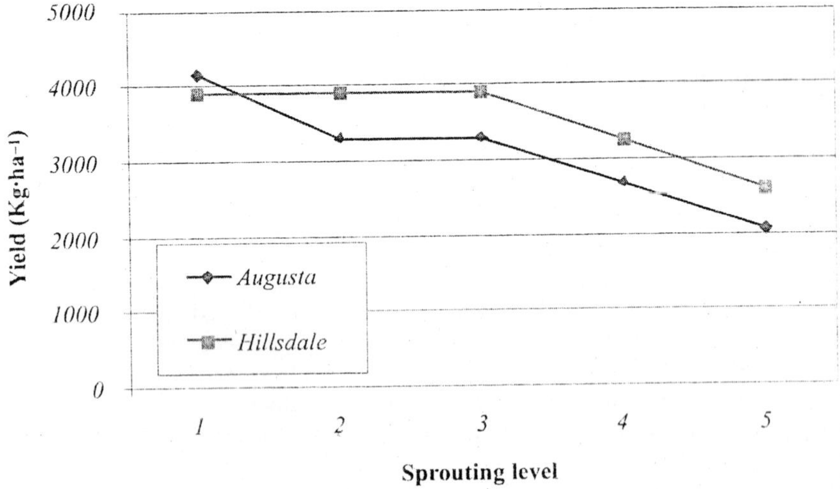

FIGURE 20.1. Effect of different sprouting levels on yield of one white (Augusta) and one red (Hillsdale) wheat cultivar. (*Source:* Adapted from Elias and Copeland, 1991.)

excessive rain, lack of nutrients, disease infestation, and insect pressure, influence seed quality (Delouche, 1980). The stage of seed development at harvest also affects seed quality. Harvesting too early or too late may reduce the quality of seed (Elias and Copeland, 2001). During prolonged and less than optimum storage conditions, physiological, biochemical, and cytological changes occur in seeds, leading to the deterioration of their quality. Slower growth rate, abnormal seedlings, and loss of vigor and viability are among the physiological changes (Edje and Burris, 1970; Harrington, 1977; Roos, 1980). Deterioration of cell membranes and reduction in enzyme and respiration activity are among the biochemical changes (Abdul-Baki and Anderson, 1972; Ching, 1973; Edje and Burris, 1970; Roos, 1980; Gill and Delouche, 1973; Likhatchev et al., 1984). Change in cell constituents and chromosomal aberrations are among the cytological changes (Roberts, 1973; Roos, 1980). Mechanical injuries due to harvesting, conditioning, and handling can also affect seed quality (Copeland and McDonald, 2001).

Storing seed under poor conditions of high temperature and relative humidity or at high seed moisture content accelerates seed deterioration (see Figure 20.2) and reduces its quality (Delouche, 1963). The extent of seed deterioration depends on species, storage environment, length of storage period, and initial quality of stored seed (Elias and Copeland, 1994).

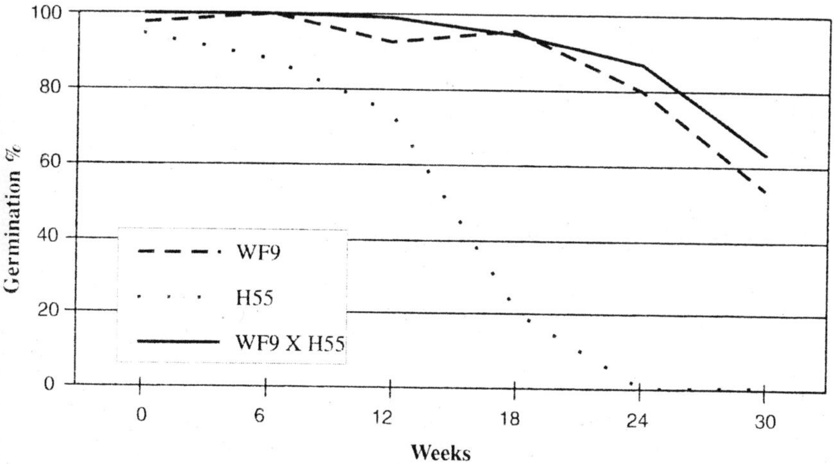

FIGURE 20.2. Effect of storage conditions on seed quality of three lots of corn during 30 weeks storage at 30° C, 75 percent RH. (*Source:* Adapted from Delouche, 1971.)

Therefore, testing the quality of seed stored for different lengths of time is important to determine the effect of aging on seed quality.

Seed Quality Attributes

Seed is a biological unit that is comprised of genetic (heritable) components contributed by the male and female parents. These components are born during the fertilization and the egg-sperm fusion and continue to develop and differentiate to form an embryo, storage tissues, seed coat, and other seed structures according to genetic codes. It is affected by the surrounding environment. It ages, loses vigor and viability, and eventually dies. The genetic components determine the biological and phenotypic characteristic of each individual seed. Environmental factors during seed formation, development, and maturation can affect gene expression and modify seed vigor, longevity, and other biological processes in the seed. The interaction between genetic characteristics and environmental conditions determines the quality of the end product seed.

Although it is widely recognized that physical purity (also called mechanical purity) and germination testing provide a baseline value of seed quality (Meyer, 1998), other characteristics such as seed health (i.e., freedom from diseases), vigor, uniformity, and freedom from mechanical damage are also attributes of seed quality. Physical purity is the freedom from contamination by any seed other than the stated species, and freedom from inert matter. Seed with excessive moisture content deteriorate rapidly and must be dried to a safe level to maintain quality during storage. Seed with high levels of unbroken dormancy affect field emergence, even though such dormant seed is viable. In such cases, moisture content and dormancy are considered to be factors in determining seed quality.

In general, the basic elements of seed quality can be summarized in Figure 20.3. However, seed quality is not limited to these elements; other characteristics can be added to the chart depending on the purpose of producing the seed.

IS TESTING THE QUALITY OF SEED REALLY IMPORTANT?

There is no doubt that using quality seed of known top-performing varieties reduces the risk of crop failure. Planting seed from unknown sources may result in poor stand establishment, field performance, and yield.

The main objective of evaluating the quality of seeds is to make a reasonable prediction of its performance in the field in order to determine the value for planting. High-quality seeds germinate better and faster under a

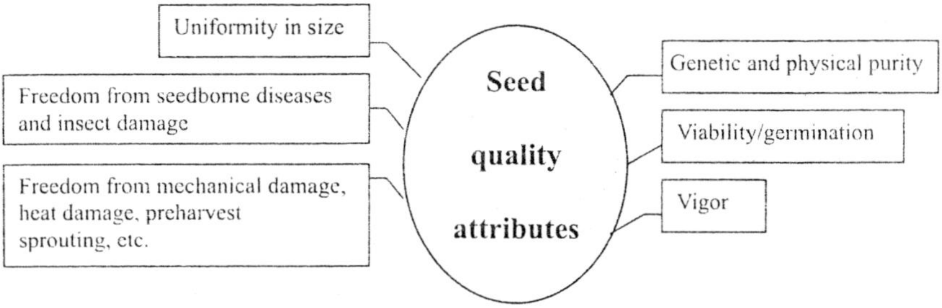

FIGURE 20.3. Diagrammatic chart for seed quality attributes.

wide range of field conditions and have a higher potential of satisfactory yield than poor quality seeds.

Seed testing is an important tool for providing assurance to farmers of getting the quality of seed they want based on the principle of "truth-in-labeling." For example, all seed in commercial trade is required to be labeled with the following information: variety and kind, purity and germination percent, and net weight. This label provides growers with some information about the quality of seed they are buying.

Seed testing is also used by seed law enforcement to protect buyers from fraudulent sales and to provide technical, professional opinions in case of litigation on the difference between seed labels and actual test results.

A question that puzzles some farmers from one planting season to the next is whether they can reduce the cost of production by planting seed of unknown or doubtful quality. Is buying certified seed or high-quality seed from a recognized source a good investment? In fact, if farmers prepared an excellent seed bed, used optimum cultural practices, and had ideal weather conditions throughout the growing season, but planted poor-quality seed, they may well expect poor stand establishment, inferior plants, and unsatisfactory yield. The selection of a good variety adaptable to the area of production and the use of high-quality seed are among the basic keys to successful field performance and satisfactory yield. Of course, following good cultural practices, including proper seeding date, rate, planting space and depth, fertilization, irrigation, weed and pest control, and pre- and postharvest management, will optimize the quality of seed produced and improve their potential value for planting and marketing.

In some cases, the use of poor-quality seed can result in stand failure and the need to reseed the whole field. The use of extra herbicides or pesticides may be necessary if the seed planted is contaminated with undesirable species or infected with pathogens. This ultimately increases the cost of production and/or results in poor yield and financial loss to the farmer.

With the development of the genetically modified crops, the need for varietal identification and testing the genetic purity became even more important than before. Some seed producers and consumers are interested in transgenic seed with specific traits, whereas others want conventional seed varieties free from any outside inserted genes.

Some countries prohibit planting any transgenic seed, while others take advantage of the useful traits that can be added to conventional varieties through molecular genetic techniques (e.g., tolerance to herbicides or resistance to certain bacteria and viruses). In such cases, testing for genetic purity of seed may be a key process in supplying such seed to consumers.

In general, testing seed assures that it meets minimum quality standards of genetic and physical purity as well as germination percentage. It minimizes the risk of crop failure and avoids problems that may result from using seed contaminated with noxious weeds, infested with disease or insects, and having a low viability level. Figure 20.4 demonstrates the typical seed testing processes in most seed laboratories.

SEED PURITY TESTING

Testing seeds for genetic integrity and physical purity is important for supplying high-quality seeds. It ensures trueness-to-type and confidence in the quality of seed purchased. There are two types of seed purity testing: genetic purity and physical purity.

Genetic Purity

For optimal crop production and marketing, it is important to know the genetic identity, uniformity, and purity of the cultivar you choose. When a plant breeder develops a new, improved cultivar, it usually has one or more specific characteristics that distinguish it from existing cultivars. Such characteristics may include disease resistance, high yield, lodging resistance, herbicide resistance, market quality, or other characteristics which may be critical for the marketed product.

Morphological tests, growth chamber tests, chemical varietal identification tests, electrophoresis techniques (*Cultivar Purity Testing Handbook*, AOSA, 1991), and DNA fingerprint are among the tests used to distinguish among cultivars and assure their genetic integrity.

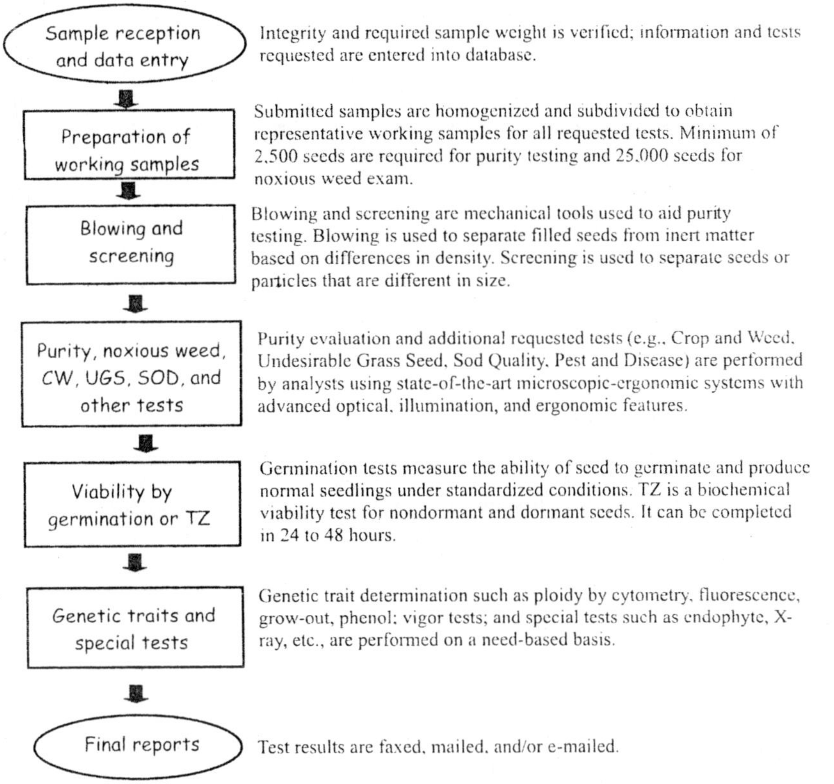

FIGURE 20.4. Flow chart of typical seed testing procedures. (*Source:* Modified from Sabry Elias and Adriel Garay, Oregon State University Seed Laboratory.)

Quick Tests to Determine Cultivar Purity

Morphological characteristics such as seed coat color, seed size, shape, and hilum color can be used to differentiate between closely related cultivars. However, differences in maturity and the effect of environmental conditions can result in differences in size and color of seeds within the same cultivar from year to year and from location to location. Chemical tests have been developed to offer rapid determination of cultivar purity of individual seeds based on the reaction between specific chemicals and certain components of the seed, e.g., enzymes or chemicals.

The following are some of the quick cultivar purity tests. They are described in detail in the *Cultivar Purity Testing Handbook* (AOSA, 1991):

- *Copper sulfate-ammonia (Cu SO and NH_4OH) test* for sweet clover: to distinguish between yellow sweet clover, *Melilotus officinalis* L. Lam, and white sweet clover, *M. alba* Medik
- *Peroxidase test* for soybean (*Glycine max* L. Merrill)
- *Amylose test* for rice *(Oryza sativa)*
- *Phenol test* for wheat *(Triticum aestivum)*, barley *(Hordeum vulgare)*, oat *(Avena sativa)*, annual ryegrass *(Lolium multiflorum)*, and Kentucky blue grass *(Poa pratensis)*
- *Potassium hydroxide test* for rice *(Oryza sativa)* and sorghum *(Sorghum halapense)*
- *Sodium hydroxide (NaOH) test* to distinguish between white and red wheat *(Triticum aestivum)*
- *Fluorescence test* for oat *(Avena sativa)*; hard *(Festuca longifoli)* and red *(Festuca rubra)* fescue; and annual *(Lolium multiflorum)* and perennial *(Lolium perenne)* ryegrass

Physical Purity

These tests are designed to identify and quantify any contaminants present in a seed sample that may reduce the value of the seed. Such contaminants are weed seed, other crop seeds, and inert materials. The specific objectives of purity analysis are as follows:

1. Determine the percentage by weight ofthe following sample components:
   - Pure seed—of the kind or cultivar under consideration (e.g., *Triticum aestivum*, variety specified by grower). To verify the variety, varietal identification test can be done separately.
   - Other seed—of any species present in the sample (e.g., *Avena sativa* or *Agrostis capliaris*, etc.) other than the one under consideration (e.g., *Triticum aestivum*). In North America, seed of other species are classified into two categories: weed seed and other crop seed according to the Rules for Testing Seeds of the Association of Official Seed Analysts (AOSA, Rules for Testing Seeds, 2001). This separation is made using to the *Uniform Classification of Weed and Crop Seeds Handbook* (AOSA, 2001). On the other hand, the International Seed Testing Association (ISTA) does not recognize this separation, but includes other crop and weed seeds

together as "other seeds" (International Rules for Seed Testing, ISTA, 2001). The function of AOSA and ISTA is to develop standardized rules and procedures for seed testing to North America (AOSA) and to the international community (ISTA).

- Inert matter—such as chaff, broken seed, soil particles, etc.

2. Determine the rate of occurrence of noxious weed seed per unit of weight (e.g., two seeds of *Cirsium arvense* per kg). Noxious weeds are species defined by state or federal seed law as being noxious (State Noxious-Weed Seed Requirements Recognized in the Administration of the Federal Seed Act, USDA, 2001). Noxious weed seed are usually highly objectionable when found in crop seed lots. ISTA lists noxious weed seeds under the "seeds of other species" category, whereas AOSA lists them as a separate category.
3. Identify the various species of seeds found in the sample.

Figure 20.5 demonstrates the various components of purity analysis.

The higher the percentage of the pure seed the better the physical quality of the seed lot. Contamination with weed seeds and/or other crop seeds will increase the cost of weed control and decrease crop yield. Higher contamination of inert matter can plug planting equipment and increase the volume and weight of seed lots, thus increasing storage and shipping costs.

To improve the physical quality of a lot, seeds have to be thoroughly cleaned after harvest. The quality of seed is improved during cleaning by removing impurities, separating seeds of other crops and weeds, and by getting rid of poor-quality (e.g., shriveled, immature, diseased, mechanically

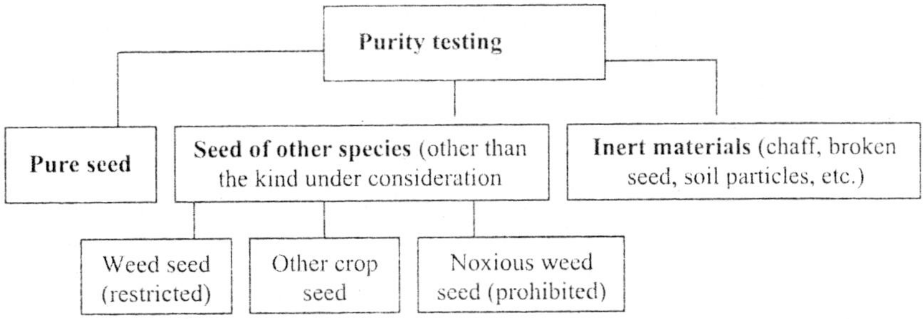

FIGURE 20.5. The main objective of mechanical purity testing is to separate sample components into pure seed, seed of other species, and intert materials. Further classification to seed of other species category may be made as in the AOSA Rules for Testing Seeds.

damaged, etc.) seeds. Air screen and gravity separators are among the equipment used for cleaning seeds.

The following are important processes and equipment used in purity and other seed quality testing.

Sampling

Since we cannot test each and every seed harvested from a field, testing a representative sample that is drawn from a population (seed lot) is an acceptable procedure in seed testing. It is worthy to note that no matter how accurately a seed testing is made to measure the quality of a seed lot, it can show the quality only of the sample submitted. Therefore, every effort should be made to ensure that the seed sample submitted for testing represents the lot of the seed to be tested (Rules for Testing Seeds, AOSA, 2002). If samples do not represent the actual seed lot, the analysis will not reflect the true quality of that lot.

Although a seed lot, by definition, should contain only one species of seed that is assumed to be uniform, this is almost never the case. Variation in seed maturity, occurrence of weed, disease infestation, lodging, and/or other factors within the same field may contribute to the lack of homogeneity in a seed lot. Lack of uniformity within a seed lot can also happen because of variation in adjustment of harvesting, and cleaning and blending equipment. Distribution of heavy and light seed within a bag is not always uniform. Heavy seed tend to migrate to the bottom and lighter seed stay on the top because of gravity.Therefore, seed within each bag must be mixed well, and samples must be collected from different parts of the bag. In general, samples must be taken at various locations of the seed lot for better representation (McDonald et al., 1993).

Sample Size

Since it is impossible to test each individual seed or particle in a lot (e.g., 10,000 kg), statistical procedures have been developed to determine a sample size that should achieve an acceptable, prescribed degree of precision for measuring seed purity (or any other characteristics of interest). An appropriate sample size is the one that is feasible to test and provides a test value as close as possible to what would have been obtained had all seeds in the lot been tested (Gomez and Gomez, 1984). According to ISTA standards, the maximum allowable seed lot size ranges from 1,000 to 10,000 kg depending on the species, but this is approximately 24,000 kg according to

AOSA standards. Regardless of seed lot size, the sample must truthfully represent the quality of the seed lot from which it is drawn.

Type of Samples for Seed Quality Tests

Submitted sample. The sample submitted to a seed laboratory for testing is called a *submitted sample.* It is assumed to be collected in a precise way, to be representative of the seed lot from which it is drawn, and of a minimum size. Working sample size of different species is prescribed in Table 1, Weight for working samples, in the AOSA, Rules for Testing Seeds (2002) or in Table 2A, Lot and sample weights, in the International Rules for Seed Testing (ISTA, 2001). The submitted sample must contain a minimum of 25,000 seeds (Usually ten times as much as the working sample size). The weight of submitted and working samples is affected by seed size of each species.

Obtaining a submitted sample. For sampling from bulk seed or bags, there are two recommended methods:

1. A trier or probe (see Photo 20.1) is used for free-flowing seeds (e.g., wheat and clover). The trier should be long enough to reach all portion of the bag, and samples should be taken from various locations in bags or bulk.
2. The hand method is recommended for nonflowing seed (e.g., bentgrass and bluegrass). The sample is taken by inserting the hand flat with fingers together to withdraw seeds from different locations in bags or in bulk (AOSA, Rules for Testing Seeds, 2002).

If samples are drawn during seed cleaning, a mechanical sampler can be used. It draws samples at specified intervals. Official samples for seed regulatory uses should be drawn by official samplers, properly sealed, identified, dated, and weighted, then forwarded to a seed lab for analysis.

Working sample. The sample on which purity analysis and other seed quality tests are performed is called the *working sample.* It is taken from the submitted sample using prescribed procedures and equipment to ensure that it accurately represents the submitted sample. It should include a minimum of 2,500 seeds.

Obtaining a working sample. Mechanical dividers such as Gamet Precision (centrifugal) (see Photo 20.2) and Boerner (conical) (Photo 20.3) are commonly used for obtaining working samples from submitted samples. Each time a sample passes through the divider, it is mixed and divided into one-half of the original weight. These dividers, however, are not suitable for

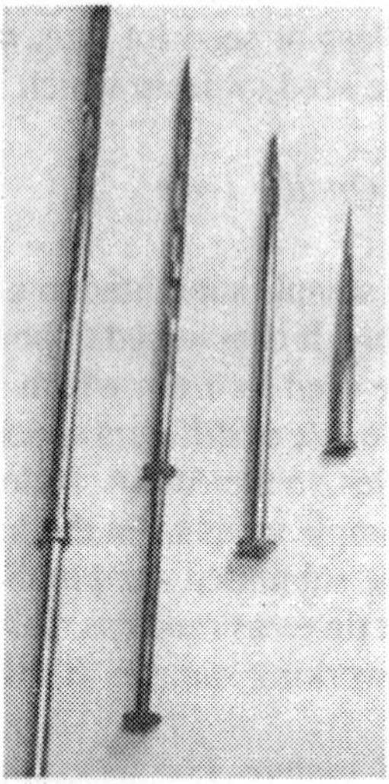

PHOTO 20.1. Triers (probes) for sampling seeds.

PHOTO 20.2. A Gamet (centrifugal) precision divider.

PHOTO 20.3. A Boerner (conical) seed divider.

chaffy seeds. A Riffle divider (Photo 20.4), which does not contain any moving parts, can be used for chaffy seed and for a wide range of seed species. If a suitable mechanical divider is not available, a hand method can be used where the sample is mixed thoroughly by hand and placed in a pile. The pile is divided into halves, then one half again divided into halves, and so on, until the minimum working sample weight is obtained.

Purity Testing Equipment

The traditional purity testing station usually includes the following equipment: a purity testing board or table, magnifying desk lens and forceps.

Scientists at Oregon State University have developed a microscopic inspection station with a mechanical vibrator to move the seed (see Photo 20.5). This type of equipment speeds up the purity analysis increasing efficiency in seed testing, and make it less tiring (ergonomic) for analysts, especially for small seeds such as bentgrass (*Agrostis* spp.) and orchardgrass

PHOTO 20.4. Riffle divider for subsampling of chaffy seeds.

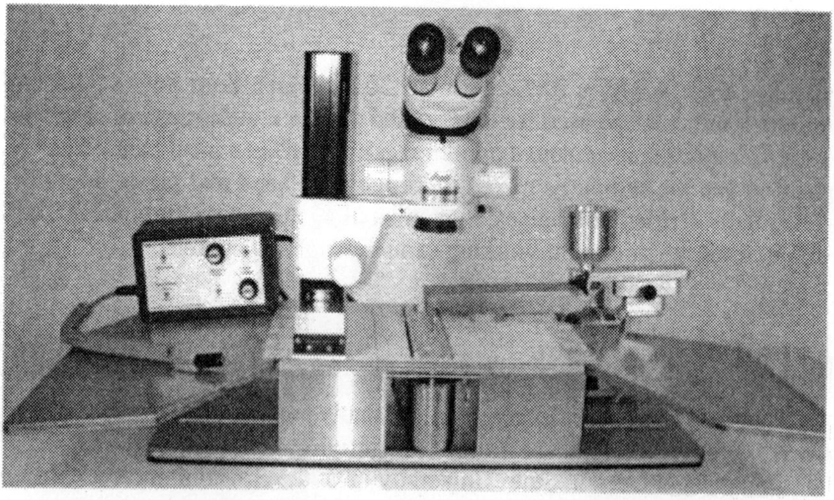

PHOTO 20.5. Mat/OSU 2002 Ergovision Purity Testing Station with mechanical seed feeder and ergonomic microscope. Developed by Oregon State University and Mater International, Corvallis, Oregon.

(Dactylis glomerata). Many seed laboratories, especially those dealing with small size seeds are currently using such equipment.

Blowing and Screening

Blowers and hand screens are used in purity testing to facilitate the separation of particles from the seed samples that are different in density (blowers) and size (screens) than the kind of seeds being tested. The General blower (see Photo 20.6), the Ottawa blower, and South Dakota blower are among the blowers that are used in seed testing laboratories. All separate seed on the basis of specific gravity and resistance to airflow.

Perforated metal or wire screens with different hole shapes (e.g., round, oblong, square, triangle) and sizes are used to separate differem seed species. Blowers and screens contribute to speeding up and improving uniformity of purity testing among seed laboratories.

PHOTO 20.6. The general seed blower.

Purity Tolerances

Variation in test results from different samples of the same seed lot or even from the same submitted sample is not unusual. This is because of the variability in distribution of components (e.g., contaminants, seed of different species, etc.) within the same lot. Sampling variation and human errors can contribute to the differences in test results between two samples. However, as long as this difference between the results of two samples is within the acceptable tolerance limit, which has been established based on statistical procedure, then this difference is considered nonsignificant and acceptable. The tolerance is the largest nonsignificant difference between two values such as the results of two purity tests or a test value and a labeled value (Elias et al., 2000).

Table 4 in the AOSA Rules far Testing Seeds and Tables 3.1, 3.2, and 3.3 in the ISTA Rules for Seed Testing (2001) show purity analysis tolerances applicable to pure seed, other crop seed, weed seed, and inert matter on nonchaffy (e.g.. soybean) and chaffy seeds (e.g., bentgrass). Special tolerances are used for mixed seed lots where pure seed components are different in size and weight (for example, a seed lot consists of a mixture of orchardgrass 15 percent and tall fescue 85 percent). Tables 6 and 7 in the AOSA Rules for Testing Seeds (2002) show the special tolerance for any component of purity analysis.

The general procedure for using purity analysis tolerances is to average the two values (e.g., labeled value and a result of a test to check the labeled value). This average is located in column A or B in the AOSA Table 4, and the corresponding tolerance is in column C or D. The average of the two values is located in column 1 or 2 in the ISTA Tables 3.1, 3.2, 3.3, and the corresponding tolerances are located in columns three through six. If the difference between the first and second purity analysis values is equal to or less than the percentage of tolerance found in column C or D, then the difference between the two values is not significant and acceptable or "within tolerance" (Table 1.4.3. and 1.4.4).

Noxious weed seed tolerances are different from purity analysis tolerances. It follows Poisson distribution and is listed in Table 8 of the AOSA Rules for Testing Seeds, 2002 (Table 1.4.5.).

SEED VIABILITY TESTING

Viability refers to the capability of seeds to germinate and produce a new plant. It is an important element of seed quality. For a seed to germinate, grow, develop a new plant, multiply, and complete the life cycle, it has to be viable or "alive." It is difficult or impossible, however, to determine whether a seed is alive or dead from external physical appearance. Two seeds may appear alike and have the same size, shape, and color, yet one of them may be alive and the other dead. Hence, seed viability testing is used to determine the percentage of viable seed in a given lot. Seed viability is one of the indications of potential performance in the field. In general, at similar environmental conditions a lot with high viability percentage can be expected to produce better seedling emergence than another lot with low viability. Seed viability indicates that a seed contains structures and substances that give it the capacity to germinate under favorable conditions in the absence of dormancy (Copeland and McDonald, 2001). The germination test and the tetrazolium test are the two most important viability tests used by seed laboratories.

Why Does Seed Lose Viability?

Many factors contribute to loss of seed viability (germinability), including the following:

1. Weather conditions during seed maturation: Stress conditions of drought, excess water, extreme temperatures, etc., contribute to reduce of loss of seed germinability.
2. Nutrient deficiency and pesticide injury during seed development and maturation.
3. Environmental conditions after physiological maturity, during harvest, drying cleaning, storage, and handling: Seed are stored on the plants in the field after physiological maturity (i.e., maximum dry weight of seeds) until the proper time of harvest (i.e., suitable seed moisture content). Any adverse conditions during this period such as heavy rainfall or frost damage can affect seed viability. Overdrying the seed makes them more susceptible to mechanical damage and may reduce their capabilities to germinate, or may even kill the embryos. Improper storage conditions lead to seed deterioration and eventual loss in viability.

Germination Testing

One of the most important components of seed quality is germination. The germination test is designed to evaluate viability and germinability of seeds under favorable germination conditions. If seed is viable and not dormant, it will germinate; hence, the percentage of germination is a good indication for the viability level of a seed lot. A standard germination test measures the ability of seed to germinate and produce normal seedlings under standardized conditions and is the most widely accepted test to determine seed viability. According to the AOSA Rules for Testing Seeds (2002), germination is defined as the emergence and development of the essential structures (i.e., shoot and root axis) from the embryo, which are indicative of the ability of a seed to produce a normal plant under favorable conditions. Germination percentage is used for labeling purposes to provide information for marketing seeds. It also protects consumers from buying poor-quality seeds and to compare between the qualities of two seed lots.

The objective of the germination test is to determine the maximum germination potential of a seed lot under favorable standardized laboratory conditions of temperature, moisture, light, and air. Detailed germination testing procedures for each seed species are defined in the Rules for Testing Seeds of AOSA (2002) and in the International Rules for Seed Testing (ISTA, 2001). These procedures are determined based on research and refereed studies to provide assurance that the prescribed germination method is repeatable within and among seed testing laboratories with an acceptable level of reliability.

If test conditions are standardized, then difference among test results should be due to random sampling variation and must fall within acceptable tolerance limits, which is determined according to certain statistical procedures (Miles, 1963). Otherwise, the discrepancy between test results is considered significant due to factors other than random sampling variation.

Because the standard germination test is performed under favorable conditions, which are seldom encountered in the field, it is not always correlated with field emergence, where seeds may germinate under adverse weather conditions (Tekrony and Egli, 1977). As field conditions become near optimum, the correlation between standard germination test and field emergence increases (Suryatmana et al., 1980). In less than optimal field conditions, seed vigor tests may be more indicative of field performance than the standard germination test (AOSA, *Seed Vigor Handbook,* 1983; Anfirund and Schneiter, 1984; Miles and Copeland, 1980).

Germination Testing Requirements

The following factors are considered when conducting a standard germination test:

1. *Substrata:* Planting media used for germination tests must provide adequate moisture for the germinating seeds. Substratum must be

nontoxic for seedlings, free from fungi or any microorganism that may affect the growth and development ofthe seedlings. Nontoxic substratum test (i.e., phytotoxic test as described in the AOSA Rules for Testing Seeds, 2002) can be performed to ensure that substratum is free from toxic chemicals. If phytotoxic symptoms are evident (root abnormalities), a comparative retest in sand or soil may be necessary.

According to the Seed Testing Rules of both AOSA and ISTA, filter paper, blotters, paper towels, creped cellulose paper (0.3-inch thick Kimpak or equivalent), sponge rok sand, vermiculite, terralite, or a mixture of 50 percent sand and perlite or vermiculite, or soil are acceptable substrata for germinating various seeds.

2. *Moisture:* Germination tests should be moistened frequently to ensure that an adequate moisture supply is available all the time. In general, excessive moisture should be avoided, as it restricts aeration of the seed. Some seed species, however, require high moisture levels for germination.
3. *Temperature:* Single constant temperatures (e.g., 25°C) or alternation of temperatures (e.g., 20 to 30°C; the first for approximately 16 hours in dark and the second for 8 hours in light) are used in germinating various kinds of seeds. Table 3 in the AOSA Rules for Testing Seeds and Table 5A of the International Seed Testing Rules prescribe the optimum temperatures for various seed crops. Alternating temperatures mimics the natural day and night cycle, stimulates germination, and helps in breaking dormancy.
4. *Light:* Some seed species require light for germination or breaking dormancy as prescribed in Table 3 of the AOSA Rules, 2002, and Table 5A of the ISTA Rules, 2001. On the other hand, light can have an inhibitory effect on the germination of other species. Seeds should be illuminated for at least 8 hours in every 24 at approximately 75 to 125 foot candles.
5. *Chemicals:* The following chemicals are known to have a positive effect on breaking dormancy and stimulating germination of some seed species:

   —Potassium nitrate (KNO_3) (0.2 percent)
   —Gibberellic acid (GA_3)
   —Ethephon (2-chloroethyl phosphoric acid) (0.0029 percent)
   —Ethylene gas (5 mL of ethylene per cubic foot of germinator space)

   Check Tables 3 of AOSA and 5A of ISTA for more details on the kind of seeds that respond to each of these chemical treatments.
6. *Prechilling:* Also called stratification, it is the practice of exposing imbibed seeds to cool temperatures (usually between 5 to 10°C) for a period of time (a few days, weeks, or months depending on the

species) prior to germination at warmer temperature in order to break dormancy (Copeland and McDonald, 2001). Table 3 of the AOSA and Table 5A of ISTA specify the low temperature and the period of prechilling time for various seed species.

7. *Duration of the germination test:* Number of days for first count and final count for germination tests of various kinds of seeds are given in Table 3 of the AOSA and Table 5A of ISTA. The specified test period does not include the prechilling duration. A deviation of one to three days from the specified time is allowed in the first count and two days is allowed in the final count for positive, clear evaluation of seedling structures. For swollen or "just-starting-to-germinate" seeds of some families (i.e., Convolvulaceae, Geraniaceae, Malvaceae, and Fabaceae) which often possess hard seeds, an additional five days beyond the prescribed final count period is allowed. Seed analysts are allowed to terminate the test when a maximum germination of a sample has been attained, even before the end of the final count listed in the seed testing rules.

If the viability of a seed lot is low, the reason for the low germination should be reported. It may be caused by dormancy or the lot may contain poor quality seeds.

Seed of some species have hard seed coats, which acts as a physical barrier that prevents water uptake and, consequently, germination. These seeds are considered viable; therefore, the percentage of hard seed should be reported in addition to the percentage of germination. The sum of germination and hard seed provides the viable seed percentage.

In some tree and shrub seeds, especially those that have not been prechilled, seed may not germinate. To determine whether seed is alive, they must be cut open to observe the internal structures. If they are fully developed and the seed has firm tissues with proper coloring, such seed are considered viable, while shriveled, decayed, and discolored tissues or seed lacking an embryo are considered nonviable.

When to Retest?

According to the AOSA Rules for Testing Seeds (2002), a retest is required if one or more of the following conditions are encountered:

1. Improper test conditions
2. Error in seedling evaluation
3. The presence of fungi or bacteria
4. Inaccuracies in counting and reporting results

5. When seedlings show abnormalities as a result of chemical treatments, exposure to chemicals, or toxicity from any source (The retest may be made in soil.)
6. When the range of 100-seed replicates of a given test exceed the maximum tolerance in Table 2 of the AOSA Rules
7. When two tests are not within tolerance

Evaluation of Seedlings

To determine the percentage of germination of a sample, it is necessary to evaluate the essential structures of seedlings. Understanding the internal and external structure of plants is important for accurate evaluation of seedlings. In order to standardize the evaluation, AOSA has developed a *Seedling Evaluation Handbook*, 1992, revised 2002, which includes seedling evaluation criteria for several selected families. Examples of normal seedlings of dicotyledons and monocotyledons are shown in Figures 20.6 and 20.7.

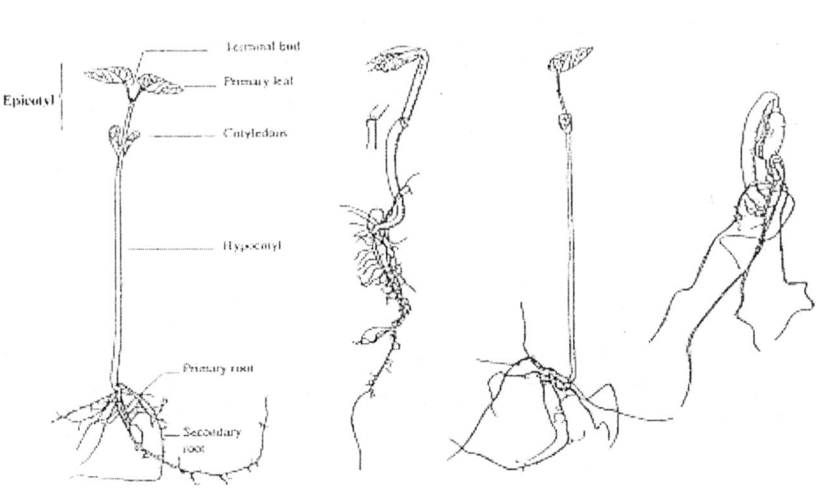

Hypocotyl break due to towel test

One leaf missing, remaining leaf and terminal bud undamaged

Hypocotyl thickened due to towel test

FIGURE 20.6. Seedling evaluation: Normal seedlings. Normal seedlings of garden bean *(Phaseolus vulgaris)*. Type of germination: Epigeal (cotyledons grow above soil surface). (*Source:* From AOSA Seedling Evaluation Handbook, 2002.)

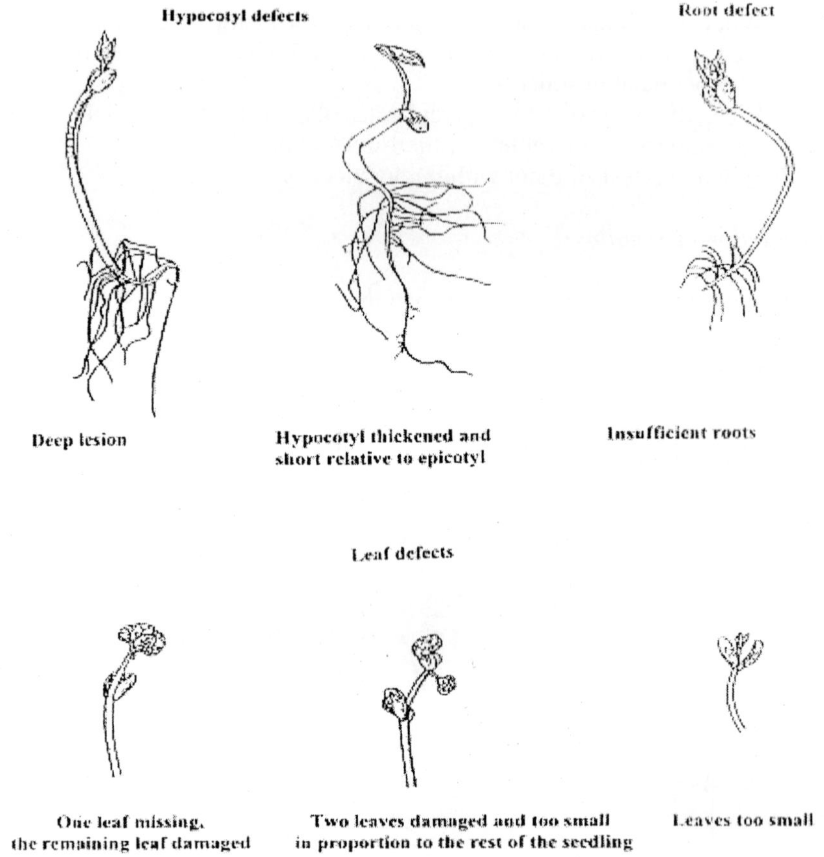

FIGURE 20.7. Seedling evaluation: Abnormal seedlings. Abnormal seedlings of garden bean *(Phaseolus vulgaris).* Type of germination: Epigeal (cotyledons grow above soil surface). (*Source:* From AOSA. Seedling Evaluation Handbook, 2002.)

Seedling Structures

Three main structures have to be identified and examined for proper evaluation of seedlings:

1. *The root system:* Roots play an essential role in plants. They anchor the plant in the ground and absorb water and nutrients from the soil to

the growing plant. Therefore, it is important for each plant to have healthy root system. The root system is developed from the embryonic root (radicle), which is located in the basal end of the embryo and is usually the first structure to emerge from the seed. After emergence, it is referred to as the primary root. In some species of the grass family (i.e., Poaceae), the primary root is indistinguishable from other roots which develop from the scutellar node region and are referred to as seminal roots (AOSA, *Seedling Evaluation Handbook,* 1992). Secondary roots develop from the primary root. If roots emerge from structures other than the primary root (i.e., stem), they are referred to as adventitious roots.

2. *The shoot system:* It includes the hypocotyl, epicotyl, mesocotyl, and primary leaf/leaves. The hypocotyl is the part of the seedling axis between the root and the cotyledons. It serves as a transition structure for transporting water and nutrients from the root to the epicotyl (i.e., the portion of the seedling above the cotyledons). In epigeal "above-the-earth" type of germination (as in beans), the cotyledons are raised above the soil surface by elongation of the hypocotyl. In hypogeal "below-the-earth" type of germination (as in pea and corn), the cotyledons, seutellum, or endosperm tissues remain below the soil surface; hence, there is little or no hypocotyl elongation. In this case, the first embryonic (primary) leaf/leaves become the first photosynthetic structures rather than the cotyledons (AOSA, *Seedling Evaluation Handbook,* 1992). In monocotyledons, the mesocotyle is the part of seedling axis between the base of the plant (i.e., scutellum and root system) and the coleoptile (sheathlike structure), which encloses the first embryonic leaf to protect it as it pushes up through the soil. Elongation of the mesocotyle is suppressed by light.
3. *The storage tissues:* Cotyledons are the food storage tissues of the embryo in dicotyledons. In the grass fiamily, Poaceae, the cotyledon is called the scutellum. Other nutritive storage tissues are the perisperm, which originates from the nucellus tissues (as in *Beta vulgaris*) and the female gametophyte (as in gymnosperms). Some species have more than one storage tissue. The function of cotyledons and the other nutritive tissues is to supply the growing seedling with nutrients until it becomes capable of producing its own food (i.e., autotrophic). In most species, the storage tissues shrivel and drop off as their reserve nutrients are depleted.

Seedlings Classification

When evaluating a germination test, seedlings are classified into the following categories:

1. *Normal seedlings:* Seedlings are classified as normal if they have the shoot structures (with no or only slight defects) that will allow the development of a normal plant under favorable conditions.
2. *Abnormal seedlings:* If seedlings have defects which will prevent them from developing into mature plants under favorable conditions, they are classified as abnormal. Among abnormalities are weak, stubby, or missing roots, missing epicotyl, deep, open cracks extending into the conducting tissues, and less than half of the cotyledon tissue remaining attached.
3. *Hard seed:* When seed remain hard at the end of the test period because they have not absorbed water due to an impermeable seed coat, they are classified as hard seed. Such hard seeds are considered viable but should be reported separately in addition to the germination percentage.
4. *Swollen seed:* When seed have imbibed water but have not germinated, they are called swollen and may need additional time for germination to occur.
5. *Dormant seed:* They are viable seeds that fail to germinate even in the presence of favorable conditions. Methods of breaking dormancy for various seed species such as prechilling treatments and scarification are listed in the AOSA and ISTA rules for testing seeds.
6. *Dead seed:* At the end of test period, if seed are decayed, soft, discolored, or covered with fungi and bacteria, they are classified as dead. Special care should be given to avoid misclassifying dead seed as hard, swollen, or dormant seed.

Tetrazolium (TZ) Test for Measuring Seed Viability

In addition to the germination test, the TZ test is a common test that is widely used to measure seed viability. It is a biochemical test that was developed by German scientist George Lakon in the early 1940s to distinguish between viable and dead tissues of seed embryos on the basis of dehydrogenase enzyme activity (respiration enzymes). Upon seed hydration, the activity of dehydrogenase enzymes increases, resulting in the release of hydrogen ions, which reduce the colorless tetrazolium salt solution (2,3,5-triphenyl tetrazolium chloride) into a red compound called formazan.

Formazan stains living cells red, while dead cells remain colorless. The viability of seeds is interpreted according to the staining pattern of seeds. Examples for TZ staining patterns of viable and nonviable seeds of monocots (e.g., corn) and dicots (e.g., bean) are given in Figures 20.8 and 20.9).

The four main steps of this test are as follows:

1. Seed hydration, by soaking the seeds in water for a specific period of time depending on the species (This is to activate the hydrolytic enzymes [dehydrogenases] and stimulate respiration.)
2. Cutting or puncturing seeds to allow for the penetration of the TZ solution to the internal tissues
3. Staining the seeds by soaking them in 0.0 to 1.0 percent tetrazolium solution for a specific period of time to allow for staining the viable tissues in the seed
4. Evaluating the seed according to the pattern of staining

The AOSA Tetrazolium Testing Handbook, 2000, provides detailed test procedures and interpretation for the staining patterns of seeds of different species.

This test provides rapid evaluation of seed viability even for dormant seeds. It must be performed by experienced seed analysts in order to reduce the level of subjectivity in interpreting the staining pattern of seeds and increase the precision of test results.

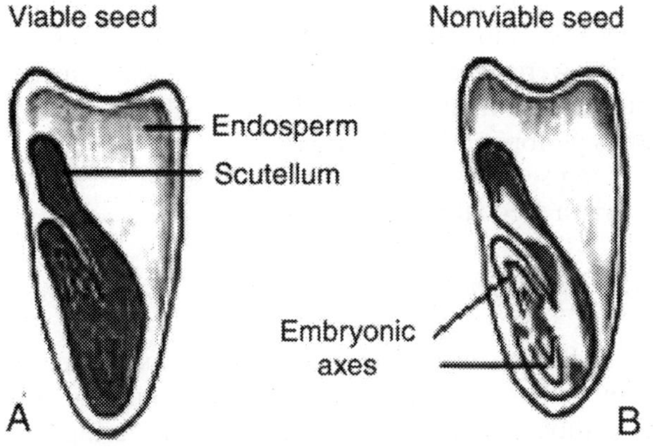

FIGURE 20.8. TZ staining pattern of (A) viable seed of corn *(Zea mays)*, embryo entirely stained; and (B) nonviable seed, extended areas of the embryonic axes unstained.

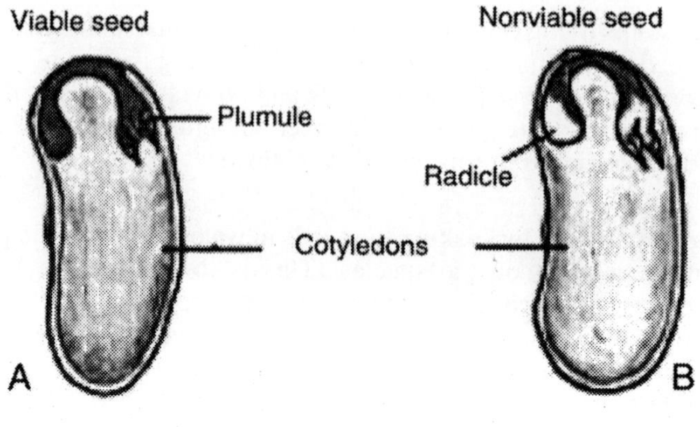

FIGURE 20.9. TZ staining pattern of (A) viable seed of bean *(Phaseolus vulgaris)*, embryo entirely stained; and (B) nonviable seed, extended areas of the embryonic axis unstained.

TESTING FOR SPECIAL SEED ATTRIBUTES

Ahhough purity and germination are the most widely accepted tests for labeling and seed certification purposes, additional special tests can provide useful information about the quality of seeds for producers and buyers. Tests such as chromosome count (ploidy), fluorescence, and grow-out tests have been developed to identify cultivars that possess certain characteristics. Other tests were developed to detect mechanical damage, insect damage, and internal abnormalities in seeds, all of which can affect the quality of seed and its performance in the field.

The following sections discuss samples of some tests that supply information on certain seed attributes.

Mechanical Damage and Seed Coat Integrity

Mechanical damage and seed coat integrity influence seed susceptibility to invasion by pathogens, leakage of nutrients from seed, and, ultimately, seed germinability. Mechanical damage can occur during harvesting, conditioning, or handling seeds. As the size of seed increases (e.g., large-seeded legumes) and the seed moisture content decreases, seeds become more susceptible to mechanical damage. The following tests are used to assess the physical damage in seeds.

Fast Green Test

If there is a crack in the seed coat, the fast green chemical will penetrate the seed coat and stain only the endosperm but not the seed coat of lighter-colored seeds such as corn and wheat. Seeds are soaked in a 0.1 percent fast green solution for 30 seconds and then completely washed. If the seed coat has any fractures, they will become apparent.

Indoxyl Acetate Test

Lesions in the seed coat of light-colored legumes turn purplish-green when soaked in 0.1 percent solution of indoxyl acetate. This test is conducted by soaking the seeds in a 0.1 percent solution of indoxyl acetate prepared in 95 percent ammonia for 10 seconds and then allowing the seed to air dry (French et al., 1962). Lesions in the seed coat which are difficult to detect by the naked eye turn to purplish-green and become visible against the light seed coat background.

Ferric chloride tests and sodium hypochlorite test are also used to assess the mechanical damage of the seed coat in legumes.

Examining the Internal Tissues of Seed

X-Ray Test

The X-ray test is used to detect the following:

1. Internal abnormalities in seeds
2. Empty seeds and absence of embryos
3. Mechanical and internal damages
4. Insect infestation

All of the factors cited might affect the physical purity and impair germination capacity of seeds. The X-ray is particularly useful in revealing the inner seed structure within the hard seed coat (e.g., tree seeds and native species) and showing any developmental deficiencies. Although the X-ray test is not a viability test, it provides useful, quick, and accurate information on the quality of seeds. An X-ray machine is used to examine the internal tissues of seeds (Photo 20.7).

X-ray results should include the following:

1. Percentage of filled seeds
2. Percentage of empty seeds

3. Percentage of insect-damaged seeds
4. Percentage of physically (mechanically, internally) damaged seeds (Radiographic Analysis of Agricultural and Forest Tree Seed, AOSA Handbook No. 31)

Determination of Ploidy Level

Ploidy by Cytometry

The flow cytometer (Photo 20.8) has been adapted to distinguish among species which are different in ploidy levels (i.e., diploid, tetraploid, etc.) by measuring the nuclear DNA content in living cells of plant tissues.

The ploidy by cytometry test has certain advantages, which are listed here:

- It is accurate: It differentiates between the nuclear DNA of plant cells with a high level of precision.
- It is objective: There is no subjectivity in interpreting the test results. Therefore, results are repeatable within and among labs.
- It is flexible: The ploidy level can be determined in any stage of plant growth, e.g., seedlings, mature leaves, etc.
- It is quick: This is particularly helpful when a large number of samples need to be evaluated.
- It is easy to perform.

The main disadvantage is that the flow cytometer equipment is relatively expensive.

Chromosome Count

This test differentiates among species based on the difference in the number of chromosome in the nucleus of plant cells. It is performed by counting the chromosomes in the stained root tip squash of young seedlings. It involves germinating seed, maceration and staining of root tip squash of young seedlings, and microscopic evaluation (*Cultivar Purity Testing Handbook*, AOSA, 1991).

Testing for Seed Chemical Composition

Seeds are major sources of protein, fat, carbohydrates, and vitamins fur humans and animals. Other seed chemical constituents are used to manufacture

PHOTO 20.7. The X-ray test is used to examine the internal tissues of seed, particularly useful for native species, tree, and shrub seeds.

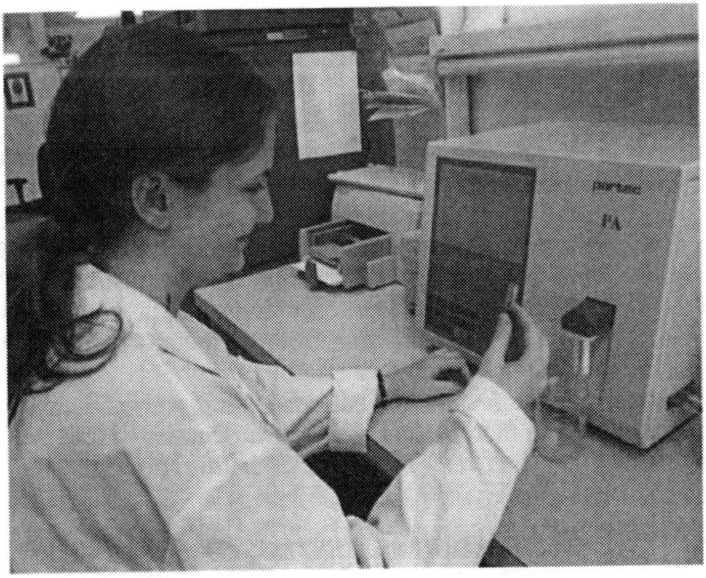

PHOTO 20.8. The adaptation of the flow cytometry technique for ploidy determination was implemented in 1999. (*Source:* Courtesy Sabry Elias, Oregon State University Seed Laboratory.)

medicine and industrial products such as clothes, soaps, detergents, cosmetic products, lubricants, etc.

Testing for the chemical composition of seeds is important to determine the value of seed for various usages. Following are some methods for the determination of protein, fat, and carbohydrates in seed:

1. *Protein determination:* The major methods for protein determination are (1) Kjeldahl, (2) Dumas, and (3) near-infrared reflectance spectroscopy (NIR) (Bullock and Moore, 1992).
2. *Fat determination:* The most common determination techniques are (1) solvent extraction, (2) nuclear magnetic resonance spectroscopy (NMR), and (3) NIR (Bullock and Moore, 1992).
3. *Carbohydrate determination:* Starch is the major form of carbohydrates in seeds. Amylose and amylopectin are the two main polysaccharides in cereal starches. Several methods are used to determine starch content in seeds, among which are (1) quantitative conversion of starch to glucose, using enzymes; (2) colorimetric determination of amylase; and (3) gel permeation chromatography (GPC) of native or debranched starches (Morrison, 1992).

SEED VIGOR TESTING

Vigor testing not only measures the percentage of viable seed in a sample, but also reflects the ability of those seeds to produce normal seedlings under less than optimum or stressed growing conditions similar to those which may occur in the field. Seeds may be classified as viable in a germination test which provides optimum conditions to the growing seedlings; however, they may not be capable of continued growth and the ability to complete their life cycle under field conditions. Seeds may start to lose vigor before they lose their ability to germinate; therefore, vigor testing is an important practice in seed production programs.

For more details on factors that determine seed vigor, relationship between seed vigor and field performance, measurements of seed vigor, and common seed vigor tests, refer to Chapter 21 in this book.

PATHOLOGICAL TESTING OF SEED

Pathological testing is part of seed quality evaluation. Planting infected seed can affect stand establishment, seed composition, yield, and the potential storability of seed. Planting infected seed is a path for disease transmission into uninfected fields; therefore, seed health testing is required for the international seed trade to limit the introduction of new diseases into new areas.

Recently, several seed health organizations have been established to develop and standardize seed health tests and develop an accreditation system for seed health testing laboratories. Among these organizations are the National Seed Health System (NSHS) in the United States and the International Seed Health Initiative (ISHI), which are working jointly with the International Seed Testing Association (ISTA).

Seed Pathogens

Seed pathogens can be found inside the seed, as is the case with mosaic virus inside the soybean embryo and endophyte fungi in tall fescue. Pathogens can also be on the seed surfaces or in a free form mixed with seed as in *Sclerotina* and ergot (*Claviceps* spp.) in soybean and wheat. Other pathogens exit in the soil and infect seeds (e.g., nematodes). Pathogens may transmit diseases to the growing plant as is the case with downy mildew and anthracnose fungi; others do not transmit the disease into the growing seedlings, e.g., *Aspergillus, Penicillium,* and *Phytophthora megasperma.* Parasitic weed seeds of dodder (*Cuscuta* spp.) and broomrape (*Orobanche* spp.) can cause economic loss in alfalfa and red clover, respectively.

In general, four main microorganism groups are associated with seeds:

1. *Fungi:* Undifferentiated plants lacking chlorophyll, produce thread-like structures called mycelia and reproductive tissues called spores. Examples of fungi are *Alternaria, Fusarium,* and *Cladosporium.* Fungi cause more plant diseases than do bacteria or viruses.
2. *Bacteria:* Unicellular microscopic plants that lack chlorophyll and multiply by fission. Examples include *Agrobacterium* (crown gall), *Bacillus* (seed rot), and *Pseudomonas* (blight).
3. *Viruses:* Submicroscopic organisms consisting of nucleic acid surrounded by proteins. Examples are mosaic virus diseases.
4. *Nematodes:* Microscopic wormlike organisms live in water, soil, or as parasites in seed and plant cells.

Detecting Pathogens in Seeds

The first and most important step in seed health testing is to accurately identify pathogens associated with seeds. There are several methods of detecting microorganisms in seeds:

1. *Visual examination:* Diseased seeds may be discolored or covered by mycelia. Pathogens structures can be removed from seeds and identified using a compound or stereomicroscope. For example, *Phomopsis* often cover soybean seeds with white mycelia, while *Fusarium* is characterized by pink mycelia on corn seeds. Ergot is detected by the presence of large dark sclerotina. Identification of fungal pathogens requires considerable knowledge and experience in mycology.
2. *Wash test:* It is useful when spores are carried loose on seed surface and cannot be identified visually. Seeds are soaked in sterile water, spores suspended in solution by centrifugation, stained, and examined by a microscope.
3. *Grow-out test:* It is performed by planting seeds to be tested in an appropriate sterile medium under favorable germination environment. Seedlings are then closely observed for disease symptom development. This test is simple, inexpensive, and suitable for testing individual seeds.
4. *Blotter and agar tests:* When it is difficult to visually identify pathogenic structures, seeds are germinated to promote fungal or bacterial growth, to make pathogen identification possible.
   —*Blotter test:* It is similar to the standard germination test using a moistened blotter with sterile water and germinating seeds for a specified period of time, after which pathogen structures can be identified.
   —*Agar test:* This test involves disinfecting seeds to be tested by growing them on agar media. Pathogens can be identified after development of fungal characteristics and the physical characteristics of the bacterial colony such as color, size, edge shape, and texture.
5. *Plant injection test:* This test is performed by soaking seeds to be tested in sterile water for a few hours, after which the leachate is injected into a young healthy seedling. The plant is then observed for development of disease symptoms.
6. *Bacteriophage technique:* This test is performed by adding phage viruses to seeds grown on agar media. If a bacterial organism is present, a clear plaque area appears on the agar due to attack on the bacteria by the virus.
7. *ELISA:* The principle of the enzyme linked immunosorbent assay (ELISA) test is that each microorganism (e.g., fungus, virus, or bacterium) has a unique protein (antigen) structure. Certain antibodies recognize and become bound to specific antigens. An enzyme is used to determine the recognition process between specific antigens and respective antibodies resulting in color reaction which can be red or interpreted by eye or color sorter (i.e., spectrophotometer). A fluorescence

marker can be attached to a pathogen-specific antibody for easier detection of a specific pathogen.

8. *Molecular techniques:* DNA-based seed health tests have been developed in the past few years. Polymerase chain reaction (PCR) is a technique used to produce multiple copies of a specific sequence of a pathogen DNA. The specific DNA sequence is then identified through electrophoresis technique.
9. *Other serological tests:* Carroll (1979) described the following serological techniques for detecting seed pathogens: (1) double diffusion test, (2) radial diffusion test, and (3) latex flocculation.

Control of Seedborne Diseases

Copeland and McDonald (2001) suggested the following practices to prevent and/or control seedborne disease:

Preharvest Control

The following factors should be considered:

1. The use of disease-resistant cultivars
2. Planting disease-free seed
3. Isolation of seed fields from sources of potential infection
4. Crop rotation
5. Treatment of seeds with fungicide and antibiotics
6. Hand rouging of diseased plants
7. Spraying seed fields with bacteriocides and fungicides to control disease buildup
8. Chemical or biological control of insect vectors
9. Avoiding overhead irrigation which favors the spread of diseases

Postharvest Control

Preharvest control is the better prevention practice seedborne disease. However, the following practices can upgrade the phytosanitary quality of seed after harvest:

1. Separation of diseased seed and foreign material
2. Surface disinfectant by chemical seed treatment
3. Hot-water treatment
4. Treatment of seeds with antibiotics and fungicides

TESTING GENETICALLY MODIFIED SEEDS

"FlavrSavr" tomato was the first genetically modified seed (GMS) to be developed in 1992 by Calgene, USA, to delay tomato ripening and maintain its freshness for a longer time. Since then, many genetically modified cultivars of various crops, including corn, soybean, rape seed, cotton, wheat, potato, beet, and others, have been developed. Currently, genetically modified crops (also called transgenic crops) are utilized on a large scale in North America and several other areas. Transgenic seeds are developed using molecular biology techniques (e.g., gene insertion into the genome of the plant to be modified) to add desired traits to an existing cultivar. Examples of the traits introduced to existing crops through genetic modification (Anonymous, 1998) follow:

- Tolerance to herbicides, e.g., glyphosate, tolerant soybean, corn, cotton, canola, and creeping bentgrass
- Resistance to bacteria and viruses, e.g., ring rot; late blight resistant potato, PLRV/PVY/TRV resistant potato; BCTV resistant tomato; and Nepovirus resistant grape
- Lepidopteran resistance in rice, corn, soybean, and cotton; phytophthora and coleopteran resistant potato
- Drought, salt, and aluminum tolerance in creeping bentgrass
- Modification of carbohydrate metabolism, protein quality, and fatty acid profile in some crops, e.g., corn, soybean, and canola

Genetically modified seeds can be identified by detecting either the inserted genetic material at the DNA level, the mRNA transcribed from the newly introduced gene, or the resulting protein, metabolites, or phynotype. The following methods are used in testing genetically modified seeds (Lubeck, 2002; Kjellsson and Strandberg, 2001; Freeman et al., 2000; Lipp et al., 1999; MacCormick et al., 1998; Zagon et al., 1998):

1. Polymerase chain reaction to detect the inserted DNA. The PCR-based methods are the most sensitive and precise GMS identification assays. They also provide quantitative analyses.
2. ELISA and the "Dipstick" method to detect the resulting protein. The ELISA method is faster and less expensive than the PCR-based methods. The Dipstick method does not provide quantitative analysis. The immunoassay methods can be considered as screening methods since the same target protein can be found in different GMS or different varieties of the same species. The immunoassay methods, however, will

not work on denatured protein. Also, foreign proteins may vary in expression levels in different plant tissues, making quantitative analysis difficult (Lipp et al., 1999; Stave et al., 2000).

3. Bioassay to detect the resultant phynotype (e.g., herbicide bioassay). In general, herbicide bioassays are inexpensive and reliable in identifying GMS in a sample. Normally, 400 seeds are tested per sample. Only viable seeds should be tested.

Other methods that can be used in GMS analysis are mass spectroscopy, chromatography, near-infrared spectroscopy, and DNA chip technology (microarrays).

Criteria for Testing Genetically Modified Seeds

The analytical methods which are used for GMS tests must

1. detect all genetically modified seeds in the sample,
2. avoid false positives and false negatives,
3. provide quantitative information, and
4. be reliable and reproducible.

The issues of concern for GMS testing are (1) sampling, including field and laboratory sample size, and (2) confidence limit of detecting the GM materials (Peccoud and Jacob, 1998). Validation and standardization of GMS testing methods are yet to be achieved. Further research and proficiency testing are needed (Kay and Van den Eede, 2001).

QUALITY ASSURANCE IN SEED TESTING

The main objective of quality assurance systems in seed testing laboratories is to provide dependable, repeatable seed quality information to customers based on uniform standards. The standard operation procedures and work instructions for seed tests (e.g., germination, purity) are based on specified, uniform standards, usually set by the ISTA or AOSA rules for testing seeds. Test procedures other than those in the ISTA and AOSA rules may be used as long as they are specified in the quality assurance manual of each laboratory. The main concept is that each laboratory should write down in the quality assurance manual procedures for what they do and then follow what they wrote.

When different laboratories follow quality assurance programs that are based on the same principles (e.g., ISO 9001), they are expected to have similar results if the same test (e.g., germination test) has been performed on the same sample. Differences in test results between seed analysts or laboratories following the same quality assurance program should not exceed the allowable tolerance stated in the tolerance table for such test. This should result in customer satisfaction and reduce the need to retest seed samples as a result of using unstandardized test protocols, uncalibrated equipment, or untrained seed analysts. All of these can be avoided by using a quality assurance program.

Components of the Quality Assurance Program

The quality assurance program for seed laboratories should include the following documents:

- Flow chart for laboratory organization and testing procedures
- Standard operational procedures and work instructions for all tests performed in the laboratory (References for the source of procedures for each test should be mentioned and kept in records.)
- All seed laboratory staff including seed analysts, supervisors, clerks, laboratory manager, etc., and the responsibility of each employee
- Records of all operations, calibration of equipments, test results, purchases of materials and equipment, reports of refereed studies, and other laboratory activities
- References for all test procedures, such as ISTA or AOSA rules for testing seeds, seed training manuals, etc.

Heinz Schmidt (2000) listed the following as good reasons to introduce quality assurance systems:

- To achieve better understanding and consistency of all laboratory activities
- To improve the already existing quality system
- To reduce errors
- To increase productivity
- To reduce cost and improve profit margin
- To minimize customer complaints
- To maximize customer satisfaction
- To improve documentation
- To reduce liability

- To reduce stress, fear, and anger
- To improve quality awareness (i.e., prevention not correction of errors)
- To increase self-responsibility and staff empowerment
- To improve communication
- To facilitate national and international trade
- To meet customer expectations of having quality assurance system
- To provde assurance to customers about repeatable, reliable, and quality seed test results
- To provide quality tests in a cost-effective way

SEED QUALITY TESTING AND SEED CERTIFICATION

The objective of any seed certification program is to ensure high-quality, genetically pure seed or propagating material of a certain variety in the market system This helps farmers reduce the risk of buying poor-quality seed. In seed certification programs, the seed is increased according to established standards of previous field-crop rotation history, isolation from potential cross-pollination crops, field inspections, seed purity and germination minimum standards (seed testing), and record keeping of the pedigree in stocks.

Certified seed is required to be labeled with the following information: variety and kind; lot number; pure seed, other crop seed, inert matter, and weed seed percent; noxious weeds (name and number/pound); germination percent, hard seed (if applicable), and total viable seed (germination percent + hard seed percent); date tested; and net weight.

The original seed of a newly developed variety (breeder seed) is increased and the quality is maintained through three generations (also called classes): foundation, registered, and certified seed. Breeder seed is the original source of all certified seed classes. Foundation seed is planted with breeder seed in limited quantities. Registered seed is planted with foundation seed, providing larger quantities of seed. The certified seed is planted with registered or foundation seed and is produced in sufficient amounts for public trade.

The blue tag which identifies certified seed gives confidence to farmers that this meets minimum quality standards. Certification programs increase the opportunity of marketing and exporting seed and provide trust between seed buyers and sellers as the information regarding the quality of seed is determined by a third party and is printed on the label of each bag of seed.

SEED TESTING ORGANIZATIONS

The main objective of the seed industry is to ensure enough supply of quality seed of improved varieties at a reasonable price at any time. With today's global seed market, seed quality testing should be standardized for faster national and international trade. In the past century, many organizations were developed to assist in standardization of seed testing rules, marketing, and development of uniform laws that facilitate the movement of seeds nationally and internationally.

Seed organizations are developed to regulate rules and laws concerning standardization of seed testing, regulate relationships between seed seller and buyer, and movement of seeds from the place of production to the place of consumption at the state, national, and international levels.

Among the organizations that are concerned with the development of rules for testing seeds and standardization of seed testing protocols on a national and global scale are the following:

- *International Seed Testing Association (ISTA):* For developing uniform seed testing protocols and rules for various countries in the world. They are located in Zurich, Switzerland. <www.seedtest.org>
- *Association of Official Seed Analysts (AOSA):* For developing standardized seed testing rules for the United States. A harmonization committee within the association has been developed to harmonize rules with ISTA. <www. aosaseed.com>
- *Society of Commercial Seed Technologists (SCST):* An organization to prepare qualified seed analysts and test seeds according to the seed testing rules of AOSA, ISTA, or other seed testing rules (e.g., Canadian rules). <www.seedtechnology.net>

Seed Trade Organizations

The following group of organizations focus on developing uniform laws for marketing seeds on national, regional, and international levels:

- *European Economic Community (EC):* Regulate seed marketing laws for member countries in the European Community.
- *Organization for Economic Cooperation and Development (OECD):* An international seed certification organization for setting minimum seed quality standards for various seed species. <www.oecd.org>
- *American Seed Trade Association (ASTA):* It is concerned with developing uniform seed marketing laws for export and import of seeds from and to the United States. <www.amseed.com>

REFERENCES

Abdul-Baki, A. A. and J. D. Anderson. 1972. Physiological and biochemical deterioration of seed. In T. Kozlowski (ed.), *Seed Biology*, Volume 2 (pp. 283-315). New York: Academic Press.

Anfirund, M. N. and A. A. Schneiter. 1984. Relationship of sunflower germination and vigor tests to field performance. *Crop Science* 24: 341-344.

Anonymous. 1998. Seed products of the future. *Seed and Crops Digest* (June/July): 38-39.

Association of Official Seed Analysts. 1983. *Seed Vigor Handbook*. Las Cruces, NM: AOSA.

Association of Official Seed Analysts. 1991. *Cultivar Purity Testing Handbook*. Las Cruces, NM: AOSA.

Association of Official Seed Analysts. 1992. *Seedling Evaluation Handbook*. Las Cruces, NM: AOSA.

Association of Official Seed Analysts. 2000. *Tetrazo Testing Handbook*. Las Cruces, NM: AOSA.

Association of Official Seed Analysts. 2001. *Uniform Classification of Weed and Crop Seeds*. Handbook 25. Las Cruces, NM: AOSA.

Association of Official Seed Analysts. 2002. *Rules for Testing Seeds*. Las Cruces, NM: AOSA.

Bullock, D. and K. Moore. 1992. Protein and fat determination in corn. In H. F. Linskens and J. F. Jackson (eds.), *Seed Analysis—Modern Methods of Plant Analysis*, New Series, Volume 14 (pp. 182-197). New York: Springer-Verlag.

Carroll, T. W. 1979. Methods of detecting seedborne plant viruses. *J. Seed Technol.* 4(2): 82-95.

Ching, T. M. 1973. Biochemical aspects of seed vigor. *Seed Sci. Technol.* 1: 73-88.

Copeland, L. O. 1977. High quality seed. Extension Bulletin E-1161. East Lansing, MI: Cooperative Extension Service, Michigan State University.

Copeland, L. O. and M. B. McDonald. 2001. *Principles of Seed Science and Technology*, Fourth Edition. Dordrecht, the Netherlands: Kluwer Academic Publisher.

Crop Protection Compendium. 1999. CD-ROM. New York: CAB International.

Delouche, J. C. 1963. Seed deterioration. *Seed World* 92(4): 14-15.

Delouche, J. C. 1971. Determinants of seed quality. SC Proc. Seed Technology Lab. Mississippi State University.

Delouche, J. C. 1980. Environmental effects on seed development and seed quality. *Hort Sci.* 15(6): 13-18.

Dickson, M. H. 1980. Genetic aspects of seed quality. *Hort. Sci.* 15: 771-774.

Edje, O. T. and J. S. Burris. 1970. Physiological and biochemical changes in deteriorating soybean seeds. *Proc. Asso. Off. Seed Anal.* 60: 158-166.

Elias, S. G. and L. O. Copeland. 1991. Effect of preharvest sprouting on germination, storability and field performance of red and white wheat seed. *J. Seed Technol.* 15: 67-78.

Elias, S. G. and L. O. Copeland. 1994. The effect of storage conditions on canola (*Brassica napes.* L) seed quality. *J. Seed Technol.* 18: 21-24.

Elias, S. G. and L. O. Copeland. 2001. Physiological and harvest maturity of canola in relation to seed quality. *Agron. J.* 93: 1054-1058.

Elias, S. G., H. Lui, O. Schabenberger, and I. O. Copeland. 2000. Reevaluation of tolerances for noxious weed seeds. *Seed Technol.* 22: 5-14.

Freeman, W. M., Robertson, D. J., and Vrana, K. E. (2000}. Fundamentals of DNA hybridization arrays for gene expression analysis. *BioTechniques* 29: 1042-1055.

French, R. C., J. A. Thompson, and C. H. Kingsolver. 1962. Indoxylacetate as an indicator of cracked seed coats of white beans and other light colored legume seeds. *Journal of American Society for Horticultural Science* 80: 377-386.

Gill, N. S. and C. J. Delouche. 1973. Deterioration of corn seed during storage. *Proc. Assoc. Off Seed Anal.* 63: 33-50.

Gomez, K. A. and A. A. Gomez. 1984. *Statistical Procedures for Agriculture Research.* New York: John Wiley & Sons, Inc.

Harrington, J. F. 1977. The effect of seed deterioration on the growth of barley. *Ann. Appl. Biol.* 87: 485-494.

International Seed Testing Association. 2001. International Rules for Seed Testing. Seed Sci. & Technol. 29. Zurich, Switzerland: Author.

Kay, S. and G. Van den Eede. 2001. The limits of GMO detection. *Nature Biotechnology* 19: 405.

Kjellsson, G. and Strandberg, M. 2001. Monitoring and surveillance of genetically modified higher plants: Guidelines for procedures and analysis of environmental effects. Basel, Switzerland: Birkhauser Verlag.

Kneebone, W. R. 1976. Some genetic aspects of seed vigor. *J. Seed Technol.* 1(2): 86-97.

Likhatchev, B. S., G. V. Zelensky, Y. G. Kiashko, and Z. N. Shevchenko. 1984. Modeling of seed aging. *Seed Sci. Technol.* 12: 385-393.

Lipp, M., P. Brodmann, K. Pietsch, J. Pauwels, and E. Anklam. 1999. IUPAC collaborative trial study of a method to detect the presence of genetically modified organisms in soy beans and maize. *J AOAC Int.* 82: 923-928.

Lubeck, M. 2002. Detection of genetically modified plants: Methods to sample and analyze GMO content in plants and plant products. Available online at <www.sns.dk/erhvogadm/biotek/detection.htm>.

MacCormick, C. A., H. G. Griffin, H. M. Underwood, and M. J. Mason. 1998. Common DNA sequences with potential for detection of genetically manipulated organisms in food. *J. Appl. Microbiol.* 84: 969-980.

Maguire, J. D. 1977. Seed quality and germination. In A. A. Khan (ed.), *The physiology and biochemistry of seed dormancy and germination.* Amsterdam: North Holland Publishing Company.

McDaniel, R. G. 1973. Genetic factors influencing seed vigor: Biochemistry of heterosis. *Seed Sci. Technol.* 1: 25-50.

McDonald, M. B., R. Danielson, and T. Gutormson. (eds.). 1993. *Seed Analyst Training Manual.* Las Cruces, NM: Association of Official Seed Analysts.

McGee, D. C. 1988. *Maize Diseases: A Reference Source for Seed Technologists.* St. Paul, MN: APS Press.
Meyer, D. L. 1998. Seed quality problems commonly encountered in the laboratory for vegetable and flower seeds. *Seed Technol.* 20: 136-161.
Miles, D. F. and L. O. Copeland. 1980. The relationship of seed vigor and field performance in soybean. *Agron. Abstr.*, p. 111.
Miles, S. R. 1963. Handbook of tolerances and of measures of precision for seed testing. *Proc. Int. Seed Test. Ass.* 28(3): 535-686.
Morrison, W. R. 1992. Analysis of cereal starches. In H. F. Linskens and J. F. Jackson (eds.), *Seed Analysis—Modern Methods of Plant Analysis,* New Series, Volume 14 (pp. 199-215). New York: Springer-Verlag.
Peccoud, J. and C. Jacob. 1998. Statistical estimations of PCR amplification rates. In F. Ferre (ed.), *Gene Quantification* (pp. 111-128). New York: Birkhauser.
Roberts, E. H. 1973. Loss of seed viability: Chromosomal and genetical aspects. *Seed Sci. Technol.* 1: 515-527.
Roos, E. E. 1980. Physiological, biochemical, and genetic changes in seed quality during storage. *Hort. Sci.* 15(6): 781-784.
Schmidt, H. 2000. *Quality assurance workshop notes.* Salem, OR.
Shah, F. S., C. E. Watson, N. D. Meredith, P. A. Bohn, and B. Martin. 2001. Effect of bean ladder usage on mechanical damage during soybean seed conditioning. *Seed Technology* 23: 92-97.
Stave, J. W., K. Magin, H. Schimmel, T.S. Lawruk, P. Wehling, and A. Bridges. 2000. AACC collaborative study of a protein method for detection of genetically modified corn. *Cereal Foods World* 45: 497-501.
Suryatmana, G., L. O. Copeland, and D. F. Miles. 1980. Comparison of laboratory indices of seed vigor with field performance of navy bean. *Agran. Abstr,* 113.
Tekrony, D. M. and D. B. Egli. 1977. Relationship between laboratory indices of soybean seed vigor and field emergence. *Crop Sci.* 17: 573-577.
USDA. 2001. State noxious-weed seed requirements recognized in the administration of the Federal Seed Act, USDA. Distribution No. 02071.
Zagon, M., Schauzu, H., Broll, K., Bogl, W., and Winkler, D. 1998. Methods for the detection of genetic modifications in transgenic organisms. bGVV Report 06/1998. ISBN 3-93167530-0.

Chapter 21

Seed Vigor and Its Assessment

Alison A. Powell

CONCEPT OF SEED VIGOR

The concept of seed vigor was first recognized in 1876 by Friedrich Nobbe who introduced the term *Triebcraft*, meaning "driving force" (Perry, 1973), and there have been many reviews of the subsequent development of the concept in the twentieth century (for example, AOSA, 1983; McDonald, 1994; Dornbos, 1995). However, an appreciation of vigor in the modern context probably began at the 1950 International Seed Testing Association (ISTA) Congress, where there were many discussions of the concepts behind the germination test in the United States and in Europe. The U.S. researchers believed that for a true expression of the emergence potential of a seed lot, the germination test should be completed in soil. On the other hand, European scientists conducted the germination test under optimum conditions for a species, in order to maximize the production of normal seedlings that had the potential for emergence. As a result of their discussions, the germination test is now completed under standard, optimum conditions for each species. However, they also decided that any tests performed in, or related to, germination in soil should be called "seedling vigor tests" (Franck, 1950). At the same time the ISTA Biochemical and Seedling Vigor Committee was established to define seedling vigor and standardize methods for its determination. This later became the ISTA Vigor Test Committee.

The debate regarding the use of soil in germination tests had highlighted a limitation to the germination test. Although the results of the standard germination test give a good correlation between germination and field emergence in favorable conditions, germination can fail to indicate the ability of a seed lot to establish a crop in poor field conditions. Instances have been described in a wide range of species where seed lots having equally high laboratory germinations show large differences in field emergence. This has been shown to be a particular problem in grain legumes (Powell, Matthews, and Oliveira, 1984). Early observations in 114 seed lots of peas having over 80 percent germination (Clark and Little, 1955) revealed a range in emergence from 31 to 85 percent. Similar observations have been made in soybean (Oliveira et al., 1984), where emergence ranged from 17 to 90 percent, in field bean (62 to 85 percent emergence; Hegarty, 1977), and in seed lots of *Phaseolus vulgaris,* which gave emergence percentages from 52 to 97 percent and 34 to 95 percent on two sowing dates (Powell et al., 1986b). In small-seeded vegetable species, Matthews (1980) showed a range of field emergence in two to six sowings of commercial seed lots from nine species. Similarly, a range of emergence has been reported in calabrese (Hegarty, 1974, 1976), carrot (Hegarty, 1971, 1976), and onion (Bedford and Mackay, 1973; Hegarty, 1976), and in a range of vegetables and cucurbit species (Perry, 1973). Differences in the emergence of germinable seed have also been demonstrated in sugar beet (Perry, 1973; Akeson and Widner, 1980; Matthews, 1980; Lovato and Balboni, 1996), and maize (Nijenstein, 1986; Bekendam et al., 1987), among a number of other species.

This failure of the germination test to predict differences in field emergence, particularly in poor field conditions, suggested that a further physiological aspect to seed quality exists. This has come to be referred to as seed vigor. Seed lots having high germination but poor emergence are referred to as low vigor seeds, whereas those giving good emergence are termed high vigor seeds.

It soon became apparent, however, that differences in emergence in poor field conditions are not the only characteristic of seeds differing in vigor. Delouche and Caldwell (1960) noted that vigor is also reflected in the rate of germination and seedling growth, and that this applies in both favorable and unfavorable conditions for germination and emergence. This has been emphasized on many occasions (Pollock and Roos, 1972; AOSA, 1983; Dornbos, 1995; ISTA, 2001). Perry (1987a) noted that high vigor seeds of cabbage and Brussels sprouts produced larger and more uniform seedlings than low vigor seeds, indicating both more rapid and synchronous germination. Similarly, both Finch-Savage and McKee (1990) and Powell et al. (1991) noted that as seed vigor of cabbage and cauliflower decreased, the mean emergence time, spread of emergence times, and coefficient of variation of seedling transplant weight increased.

Differences in vigor are also reflected in seed longevity. A high vigor seed lot has good storage potential and retains high germination potential during storage, whereas low vigor seed lots show poor storage potential and may show a rapid decline in germination (Delouche and Baskin, 1973; Hampton, 1992). This is particularly the case when seeds are stored in less than ideal conditions, such as at raised relative humidity or temperature. However, even in controlled storage conditions of low temperature and moisture content, seed lot response to storage may depend on the initial vigor of the lot (Powell and Matthews, 1984a,b; Wang and Hampton, 1991). In both dehumidified storage and under two uncontrolled storage conditions, there was a significant correlation between the initial seed vigor of onions (Powell and Matthews, 1984a) and Brussels sprouts (Powell and Matthews, 1984a) and their germination after up to 33 and 48 months, respectively.

There have also been considerations of the impact of vigor on crop yield. Clearly, when low vigor leads to reduced crop establishment, this can have a direct effect on yield. In cereals, the effect is limited if emergence is only slightly reduced, due to the ability of the plant to tiller and hence produce compensatory growth. However, if emergence is markedly reduced, so is yield. In vegetable crops, the impact of reduced establishment is often on marketable yield, since particular plant densities are required to produce specifically sized crops for the market (Salter, 1985). In this case, the rate and uniformity of emergence are also critical to the production of a product that is uniform in size and stage of maturity (Salter, 1985).

If the plant populations produced by high and low vigor seeds are the same, the impact of vigor on yield of annual crops depends largely on the nature of the harvested material (TeKrony and Egli, 1991). In crops harvested in the vegetative state (e.g., lettuce, cabbage, radish, carrot), evidence from both stored and commercial seed revealed a frequent yield response to use of high vigor seed. Similarly, crops such as peas and tomatoes, which are harvested during early reproductive growth, also revealed a yield response to vigor. TeKrony and Egli (1991) suggested that this relationship between vegetative growth and yield indicates that vigor affects the ability of the plant community to accumulate dry matter. In contrast, in crops harvested at full reproductive maturity, i.e., the grain crops in which only the seeds are harvested for yield, there was no relationship between vigor and yield. In this case, they proposed that once the vegetative size of a community has increased there is no effect of vegetative growth on yield. Therefore, if vegetative growth is adequate, there is no effect of vigor on yield, as seen in the grain crops.

The varied expressions of vigor in the laboratory and field have created difficulties in agreeing on a suitable definition of vigor that incorporates all its characteristics. As a result, the definition has been modified many times, as described in a number of reviews (Heydecker, 1972; Perry, 1973; McDonald, 1975, 1980; AOSA, 1983). Clear definitions have now been agreed upon by the two organizations that are responsible for the development and standardization of seed testing. The Association of Seed Analysts (AOSA) definition was adopted in 1980 (McDonald, 1980) and states, "Seed vigor comprises those seed properties that determine the potential for rapid, uniform emergence, and development of normal seedlings under a wide range of field conditions" (p. 784). The ISTA definition that was accepted into the ISTA Rules in June 2001 (ISTA, 2001) is based on that proposed by Perry (1978):

> Seed vigor is a sum of those properties that determine the activity and level of performance of seed lots of acceptable germination in a wide range of environments.
>
> Seed vigor is not a single measurable property, but is a concept describing several characteristics associated with the following aspects of seed lot performance.
>
> i. Rate and uniformity of seed germination and seedling growth
> ii. Emergence ability of seeds under unfavorable environmental conditions
> iii. Performance after storage, particularly the retention of the ability to germinate
>
> A vigorous seed lot is one that is potentially able to perform well even under environmental conditions that are not optimal for the species.

CAUSES OF DIFFERENCES IN SEED VIGOR

The vigor of a seed lot is determined initially by its genotype and subsequently by the influence of a range of factors on this genotype. These factors include seed aging, sometimes referred to as deterioration, imbibition damage, the conditions during seed development, and seed size. The most significant of these factors is seed aging, which will therefore receive the most emphasis in this chapter.

Aging

As long ago as 1933, Goss (1933) related the process of deterioration to the impairment of seed vigor. Seed deterioration, or aging, can be described as the accumulation of deleterious changes within the seed until the ability to germinate is lost (Powell, Matthews, and Oliveira, 1984). The aging of a seed lot, in terms of loss of germination, is described in the seed survival curve (see Figure 21.1). This illustrates that over a period of time the germination of a population of seeds initially shows a very slow decline (Figure 21.1), followed by a rapid fall in germination, and a period when only a few seeds within the population retain the ability to germinate. During the slow decline in germination, it is difficult to differentiate between the germination of samples of seeds at different points on the survival curve. All

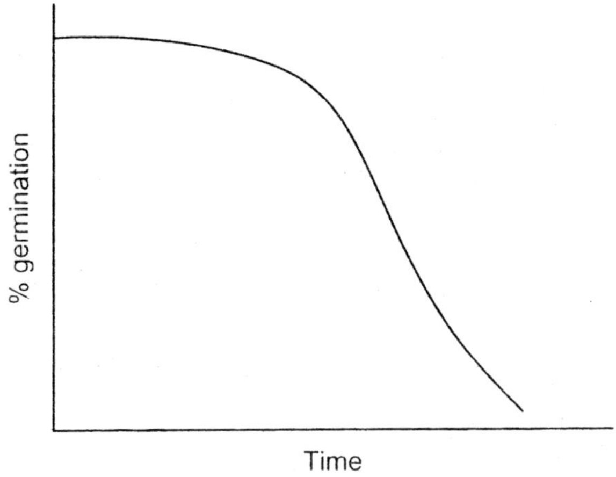

FIGURE 21.1. Seed survival curve.

samples have high germination, with any small differences in germination not discernable on the basis of a sample of 400 seeds used in a germination test. However, as the population moves along the survival curve, the seeds are aging, i.e., deleterious changes are accumulating, until the proportion of seeds within the population that are incapable of germinating increases and germination falls. Thus, at the beginning of the initial slow decline in germination, seeds can be described as physiologically young seeds; these are high vigor seeds, whereas at the end of this phase they are physiologically old or low vigor seeds.

A large body of evidence supports the idea of seed aging as a major cause of vigor differences, from comparisons of commercial and both naturally and artificially aged seeds. In addition, germination after a precise period of deterioration relates to seed vigor (Matthews, 1980). Thus, following a period of controlled deterioration or aging at the same high seed moisture content and high temperature, a sample from a physiologically young seed lot (e.g., see lot A, Figure 21.2) would retain a high germination. However, physiologically older seed lots (e.g., lots B and C) would show reduced germination. The germination after controlled deterioration reflects the initial extent of deterioration in the seed, which is not evident on the basis of the germination test. Furthermore, the germination after the period of controlled deterioration reflects seed vigor both in terms of field emergence (Matthews, 1980) and storage potential (Powell and Matthews, 1984a,b), supporting the hypothesis that vigor differences can be attributed to the extent of aging in the seed.

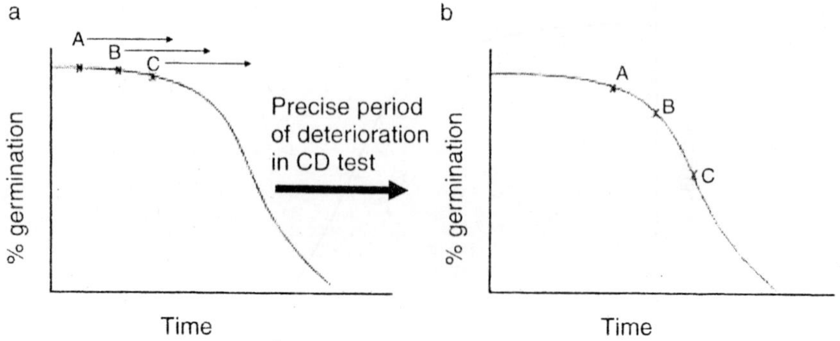

FIGURE 21.2. The effect on two seed lots of a period of controlled deterioration at the same seed moisture content and temperature.

The deleterious changes that occur during the process of aging have been described in detail on many occasions (including Abdul-Baki, 1980; Powell Don, et al., 1984; Priestley, 1986; Dornbos, 1995; McDonald, 1999). The first change that is believed to occur is the deterioration of cell membranes, probably due to the oxidation of fatty acid chains within the phospholipids of the membranes (McDonald, 1999). This leads to a reduction in the integrity of cell membranes and the first clear expression of deterioration, an increase in solute leakage from the seeds (Powell and Matthews, 1977). The deterioration of membranes can be linked to many of the subsequent deleterious changes that are observed (Powell, Don, et al., 1984), given the key role that membranes play in cell function. Electron microscope studies have revealed the disruption of subcellular organization, reduced enzyme activity, a decrease in the rate and efficiency of respiration, and an overall reduction in synthesis of macromolecules. Finally, before the germination of a population begins to decline, an increase in the incidence of chromosome abnormalities, observed in the root tips of germinating aged seeds, suggests damage to DNA. The accumulation of all these deleterious changes contributes to a gradual decline in seed vigor with age and subsequently results in the loss of the ability to germinate.

There is some debate regarding when seeds reach maximum seed quality on the plant. For many years maximum seed quality, in terms of both germination and vigor, was said to be achieved at, or close to, physiological maturity (Harrington, 1972), when seeds have reached their maximum dry weight. However, work on wheat and barley (Ellis and Piéto Filho, 1992), pepper (Demir and Ellis, 1992a,b), soybean (Zanakis et al., 1994), and *Phaseolus vulgaris* (Sanhewe and Ellis, 1996) has suggested that seeds do not reach maximum quality until after so-called physiological maturity. They therefore proposed that the time of maximum dry weight should be referred to as mass maturity. In the above work probit analysis of seed survival curves was used to obtain an estimate of initial seed quality at the different stages of seed development, which is referred to as K_i. Maximum K_i occurred after maximum dry weight was reached. Similarly, Ferguson (1993) observed that seed vigor of combining peas continued to increase after seed reached maximum dry weight, as revealed by a reduction in seed leachate conductivity (see Figure 21.3). In contrast, in work on a range of nine agricultural and horticultural crops, TeKrony and Egli (1997) demonstrated that maximum germination occurred at or before physiological

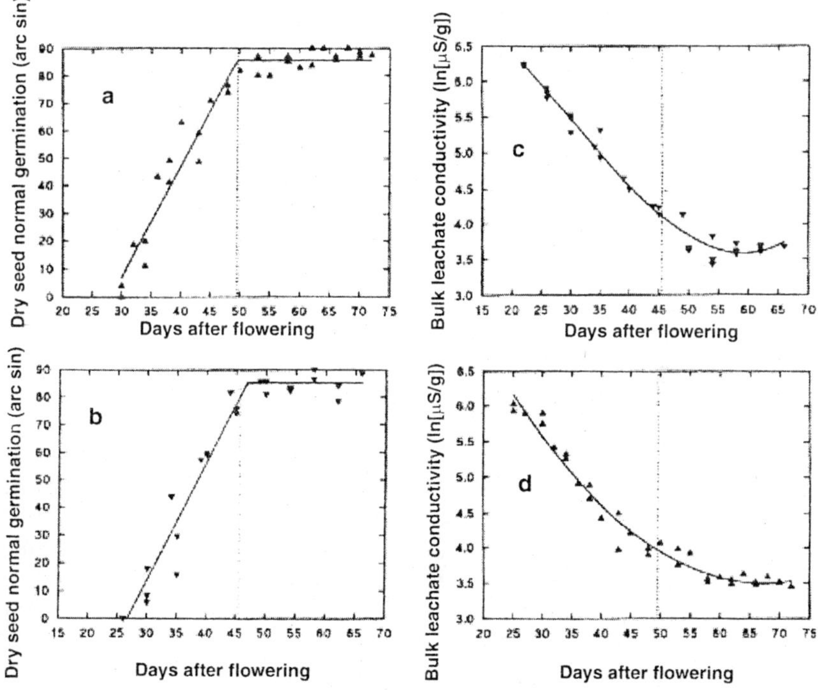

FIGURE 21.3. Changes in normal germination of dried seed (a,b) and in bulk leachate conductivity (c,d) during seed development and maturation in combining pea cv. Boharyr, after an early (March 26, 1991; a,c) and a late sowing (April 23, 1991; b,d): a. $y_1 = 4.06x - 114.7$, $y_2 = 85.9$, $r^2 = 0.887$, df = 27; b. $y_1 = 4.34x - 116.9$, $y_2 = 84.6$, $r^2 = 0.920$, df = 24; c. $y = 10.21 - 0.199x + 0.00147x^2$, $r^2 = 0.974$, df = 30; d. $y = 9.97 - 0.198x + 0.00153x^2$, $r^2 = 0.974$, df = 27. Broken vertical lines represent end of seed fill (physiological maturity).

maturity in all crops except tomato. Maximum seed vigor also occurred at or slightly before physiological maturity in crops harvested as dry seeds, but after physiological maturity for crops harvested as fleshy fruits (tomato, pepper).

Whatever the exact timing of the achievement of maximum seed quality, it is clear that this occurs close to the time at which the seeds no longer receive dry matter from the mother plant, due to the severance of vascular connections between the plant and the seed. At this time the seed has become an independent unit, and seed aging can take place from this point onward. Aging may therefore initially occur while the seed is still on the plant (postmaturation and preharvest), both during and after postmaturation drying. The degree of preharvest deterioration is dependent on the climatic factors temperature and moisture (humidity/rainfall), their importance being reflected in the use of the term "weathering" to describe this phase of seed aging.

Weathering is a serious seed production problem in areas where frequent rainfall, or high humidity, combining with high temperature results in rapid losses of vigor and germination in the field (Delouche, 1980). The influence of the weathering on seed quality can occur both during desiccation and in the phase after the seeds have dried down to harvest maturity.

During the desiccation phase, Green et al. (1965) reported that early-sown soybean, which matured during hot dry weather conditions, produced seed having lower laboratory germination and field emergence than late-sown cultivars which reached maturity after the hot dry conditions. TeKrony et al. (1980) also attributed a decrease in seed vigor in one cultivar to high temperature during the desiccation period. High temperatures, alone, or in combination with heavy watering, were also detrimental to seed quality in *Phaseolus vulgaris* (Siddique and Goodwin, 1980). Humid conditions during the maturation phase have been seen to reduce seed quality in lima beans and in cotton *(Gossypium hirsutum)* (Christiansen and Justus, 1963; Woodruff et al., 1967). Furthermore, cotton seeds from earlier opening bolls which were exposed to the field environment for a longer period before harvest had a lower vigor than seeds from later-opening bolls (Caldwell, 1972; Woodruff et al., 1967).

Most work on preharvest deterioration has concentrated on the period after the seeds have dried down to harvest maturity and has examined the effects of delayed harvest on vigor and germination (Fields and King, 1963; Green et al., 1966; Nangju, 1977, 1979). Climatic conditions led to a marked decrease in the vigor of soybean seeds within four days of harvest maturity (TeKrony et al., 1980), and the rate of loss was related to the mean air temperature, relative humidity, and precipitation per day. Green et al. (1966) also found that deterioration after harvest maturity was particularly marked after a period of rain. In peas, periods of repeated rain, followed by drying, had been found to increase the proportion of "bleached" low vigor seeds, and a decline in quality was exacerbated by delayed drying (Fields and King, 1963).

The greatest incidence of aging arises, however, during seed storage. Storage commonly refers to commercial storage, but it should be recognized that aging can also take place during brief, temporary periods of storage, for example, after harvesting and before and during processing. The major factors that influence seed aging and hence a decline in vigor and subsequently germination are seed moisture content and temperature. Harrington (1972) described their effects in his "Rules of Thumb," which state that a 1 percent increase in seed moisture content or a 5°C increase in storage temperature will double the rate of aging, i.e., halve the storage life. The relationship between seed moisture content, storage temperature, and storage life is described more accurately in the viability equation (Roberts, 1986). This includes four species constants, since they appear to be applicable to different seed lots and cultivars within a species. Three constants indicate the effects of moisture content (one constant) and temperature (two constants) on longevity, relative to a fourth constant, which indicates the intrinsic longevity of the species and cultivars within a species. Finally the viability equation includes a measure of initial seed quality (K_i).

The importance of seed moisture content in determining the rate of seed aging means that the efficient drying of the seed at or after harvest is an important first step in reducing the rate of aging and preventing a decline in vigor. However, the low water potential of the dry seed, and, hence, high matric potential, means that it will readily absorb moisture from the air (Roberts, 1972). Thus, if the relative humidity of the seed storage is high, seeds will absorb moisture and the rate of aging will increase.

The impact of periods of storage at high seed moisture content/relative humidity (rh) and raised temperature on the rate of aging and decrease in germination and vigor has been described on many occasions (Roberts, 1972; Copeland and McDonald, 2001). The response of maize to increased seed moisture during storage for 90 days at 26 to 27°C was demonstrated by Moreno et al. (1988) who noted that in seeds stored at a mc of 16.5 percent, germination fell to 15 percent, compared with 97 percent germination when stored at 14.2 percent mc. Similarly, Joao Abba and Lovato (1999) reported that more rapid deterioration of maize occurred during storage at 90 to 95 percent rh and 30°C where the seed mc increased to 17 percent, than at 45 to 50 percent rh and 25°C when the seed mc remained below 11.5 percent. In both cases the decrease in germination was preceded by a fall in seed vigor, as revealed by the cold and accelerated aging tests and plumule length (Joao Abba and Lovato, 1999). In three temperate forage grasses *(Lolium perenne, L. multiflorum, Lolium × boucheanum)* and one forage legume *(Trifolium repens),* Lewis et al. (1998) found that deterioration was most rapid during storage when there was no control of either temperature or relative humidity (granary storage). All four species showed a decline in vigor, revealed by the rate of hypocotyl growth before germination declined. These reductions in vigor and germination took place more slowly when seeds were stored in a refrigerator, i.e., with temperature control but no humidity control, while the best storage conditions were in a seed store with both temperature (2°C) and humidity (10 to 20 percent) control. In soybean, six months of storage at a range of temperatures (0 to 55°C) and relative humidities (45 to 84 percent) revealed an increase in the rate of deterioration at both increased temperature and relative humidity (Nkang and Umoh, 1996). More rapid deterioration of pepper occurred as seed moisture content increased from 5.8 to 24.8 percent (Sundstrom, 1990), and greater deterioration of sunflower was observed during storage at 95 percent rh, 28°C, than at 85 percent rh, with least deterioration at 50 percent rh (Halder and Gupta, 1980).

Imbibition Damage

A second factor that has been shown to have a major effect on seed vigor is imbibition damage. Imbibition damage is a phenomenon that has been clearly shown to influence vigor in a range of temperate and tropical grain legumes (Powell, Matthews, and Oliveira, 1984), although it may also occur in small-seeded dicotyledonous crops (Thornton and Powell, 1992; Draper and Keefe, 1990). Imbibition damage arises when water rapidly enters the cotyledons during imbibition, leading to cell death and high solute leakage from the seeds (Powell and Matthews, 1978). The instantaneous occurrence of damage within two minutes of the beginning of imbibition suggests that it results from physical damage, possibly through the disruption and disorganization of cell membranes. Certainly, the extensive loss of cellular material and enzymes from the seeds (Powell and Matthews 1981a; Duke and Kakefuda, 1981) indicates extensive membrane disruption. Imbibition damage is greater at low temperatures (Powell and Matthews, 1978), possibly because the membrane components are held more rigidly and are therefore more sensitive to physical damage.

The observation of imbibition damage has highlighted the role of the testa in protecting the cotyledons from the damaging effect of water uptake (Powell and Matthews, 1977). In addition, it has questioned the evidence used by Simon and Raja Harun (1972) to support their hypothesis of changes in membrane conformation during imbibition. They proposed that the initial high leakage of solutes that they observed from pea embryos (seed minus testa) resulted from membranes of the dry seed being in the porous hexagonal state. The subsequent decline in leakage was explained by the formation of the normal bilamellar membrane following hydration. However, following the observation of imbibition damage, Powell and Matthews (1977) suggested that the initial high leakage during imbibition resulted from the death of the outer cells of the cotyledons due to imbibition damage and that the decline in leakage represented loss of solutes from within the seed. Indeed, they went on to demonstrate that the pattern of leakage observed by Simon and Raja Harun (1972) occurs from completely dead embryos and from any spherical structure containing ions (Powell and Matthews, 1981a).

In some species, such as peas and soybean, the presence of an intact testa limits the incidence of imbibition damage (Powell and Matthews, 1979; Oliveira et al., 1984). In these species, the incidence of damage to the testa influences the extent of imbibition damage and the vigor of the seeds. Thus, seed lots showing extensive testa damage imbibe rapidly and exhibit a high incidence of imbibition damage (Powell and Matthews, 1979, 1981b). These lots emerge poorly in the field (Powell and Matthews, 1980; Oliveira et al., 1984) and therefore have low vigor. In contrast, seed lots with little testa damage imbibe slowly and show little imbibition damage. The high emergence of these slowly imbibing lots indicates that they have high vigor. This emphasizes the importance of the integrity of the testa in determining the vigor of some grain legume species and hence the need to minimize damage to the testa during harvest and processing.

In other grain legume species, however, there is a genotypic component to the susceptibility of seeds to imbibition damage. Cultivars of *Phaseolus vulgaris* (Powell, Don, et al., 1984; Powell, Matthews, and Oliviera, 1984), chickpea (*Cicer ciceris;* Legesse, 1991), long bean (*Vigna sesqupedalis;* Abdullah et al., 1991), and cowpea (*Vigna unguicularis;* Legesse and Powell, 1992), in which the testa is partially or completely unpigmented, imbibe more rapidly and show greater levels of imbibition damage compared with cultivars having pigmented testae. As a result, the vigor of unpigmented cultivars is reduced, leading to poorer field emergence than is found in pigmented cultivars (Powell, Matthews, and Oliveira, 1984). The close association between pigmentation and slower rates of imbibition has been demonstrated in studies on seed development (Legesse and Powell, 1996) and using isogenic lines of peas, differing only in the A gene for testa pigmentation (Powell, 1989). In some species, for example *Phaseolus vulgaris* (Powell et al., 1986a) and chickpea (Legesse and Powell, 1996), the close adherence of the pigmented testa to the cotyledons appears to limit the rate of water movement within the seed. In cowpea, however, testa permeability also has a role (Legesse and Powell, 1996). Identification of the factor associated with pigmentation that leads to reduced rates of water uptake could lead to the introduction of this characteristic to unpigmented cultivars through breeding programs, thereby increasing the seed vigor of these cultivars.

Interaction of Imbibition Damage and Aging

The two major causes of reduced vigor in grain legumes—aging and imbibition damage—also interact, with aged seeds being more susceptible to imbibition damage (Powell, 1985; Asiedu and Powell, 1998). Thus, pea seeds aged for up to 7.5 h at high moisture content and temperature showed complete vital staining after imbibition with the testa intact (see Table 21.1). However, when the testa of the aged seeds was scarified before imbibition, leading to more rapid water uptake, the extent of living tissue declined markedly, indicating the incidence of imbibition damage. This decrease in living tissue associated with rapid imbibition was greater in aged seeds.

Similarly, five cultivars of cowpea stored at high relative humidity and temperature for months or days showed a greater reduction in vital staining following rapid imbibition after storage (Asiedu and Powell, 1998). Cultivars having white and brown testae had almost complete vital staining after storage (see Table 21.2) when imbibed between moist cloths. Only at the later storage times was there a decrease in tissue, which reflected seed aging during storage (Table 21.2). When seeds imbibed rapidly in water before storage, there was little evidence of imbibition damage, since most

TABLE 21.1. Comparison of the percentage of pea cotyledons having complete vital staining with tetrazolium chloride following seed aging at 20 percent seed moisture content and 45°C and subsequent imbibition either intact or after scarification of the testa. Imbibition took place either in water or in a 30 percent solution (w/v) polyethylene glycol 6,000 (PEG).

Aging time (h)	Imbibition in water		Imbibition in PEG (mean of intact and scarified)
	Intact seeds	Scarified seeds	
0	90	48	85
2	70	18	97
4	88	10	95
7.5	90	14	100

Source: Adapted from Powell, 1985.

TABLE 21.2. Comparison of the percentage of cowpea cotyledons having complete tetrazolium staining following seed storage at either (a) 30°C 75 percent rh or (b) 40°C 100 percent rh. Seeds from unpigmented and pigmented cultivars were imbibed before staining either between moist cloths (slow imbibition) or in water (rapid imbibition).

Storage conditions/	Unpigmented seeds (mean of three cultivars)		Pigmented seeds (mean of two cultivars)	
time	Slow	Rapid	Slow	Rapid
(a) 30°C 75 percent rh (months)				
0	95	88	100	97
1	83	61	100	95
2	75	40	100	87
3	43	7	95	65
4	8	0	90	48
5	3	0	85	30
6	0	0	80	30
(b) 40°C 100 percent rh (days)				
0	90	85	100	100
1	78	73	100	100
2	75	68	100	100
3	73	13	97	90
4	68	0	97	83
5	47	0	95	67
6	30	0	85	47

Source: Adapted from Asiedu and Powell, 1998.

cotyledons stained completely. However, as seeds aged during storage, the extent of living tissue following rapid imbibition decreased, even in seeds that stained well after slow imbibition (Table 21.2). This reflected an increase in imbibition damage in the aged seeds. It has been proposed that the increased susceptibility of aged seeds to imbibition damage arises because membranes that have been weakened by physiological deterioration are more sensitive to physical damage during imbibition (Powell, 1986; Asiedu and Powell, 1998).

These observations of aging and imbibition damage in stored cowpea seeds emphasize a further point. Both in cowpea (see Table 21.2; Asiedu and Powell, 1998) and long bean *(Vigna sesquipedalis)* (Abdullah et al., 1992), cultivars having pigmented testae have been shown to have greater longevity during storage at high humidity than cultivars having unpigmented testae (Abdullah et al., 1992; Asiedu and Powell, 1998). This was attributable to more rapid water uptake from the humid atmosphere by the unpigmented eultivars, leading to a higher seed moisture content and hence more rapid aging during storage. Differences in water uptake, both from water and from humid atmospheres, in cultivars differing in testa pigmentation can therefore contribute to a decline in seed quality.

The uptake of moisture by seeds is therefore critical in determining seed quality and can have its influence at a number of stages of seed production (see Figure 21.4). Moisture uptake may occur on the plant after harvest maturity due to delayed harvest, or subsequently during storage in poor storage conditions. In such cases, the increase in moisture content leads to seed aging and reduced vigor (Figure 21.4). In addition, in grain legumes, the incidence of testa damage during harvest results in more rapid water uptake, imbibition damage, and hence a reduction in vigor. Finally, the interaction of a period of aging with rapid water uptake results in a greater incidence of imbibition damage and even lower vigor.

Factors During Seed Development

Seed production conditions can influence seed quality during seed development, although in general their effect tends to be on seed yield rather than seed quality. Where low fertility occurs, the number and weight of seeds produced is reduced, but the seeds produced are as vigorous and germinable as those produced under fertile conditions. A number of studies show an association between nutrition and seed quality, but these are the exception rather than the rule. Thus, foliar nutrition of wheat and sunflower improved seed vigor, with nitrogen being the most effective element (Szirtes and Szirtes, 1980). A similar observation was made in wheat by Bulisani and Warner (1980), while Austin and Longden (1965) found no influence of nitrogen level on seed vigor. Hadavizadeh and George (1989) noted that pea seed vigor increased only when high nitrogen was accompanied by medium inputs of phosphorus. In contrast, reduced vigor was found in bleached pea seeds produced at high nitrogen levels.

Climate can also influence seed vigor during seed development. Austin (1972) demonstrated that temperantre could affect subsequent performance in various ways. High temperature during seed fill led to the production of small poor quality seeds in *Phaseolus vulgaris* (Siddique and Goodwin, 1980), and in peas, high temperature at the wrinkled pod stage (70 to 80 percent moisture content) produced seeds having the highest incidence of hollow heart, and hence low vigor (Myers, 1948). On the other hand, freezing

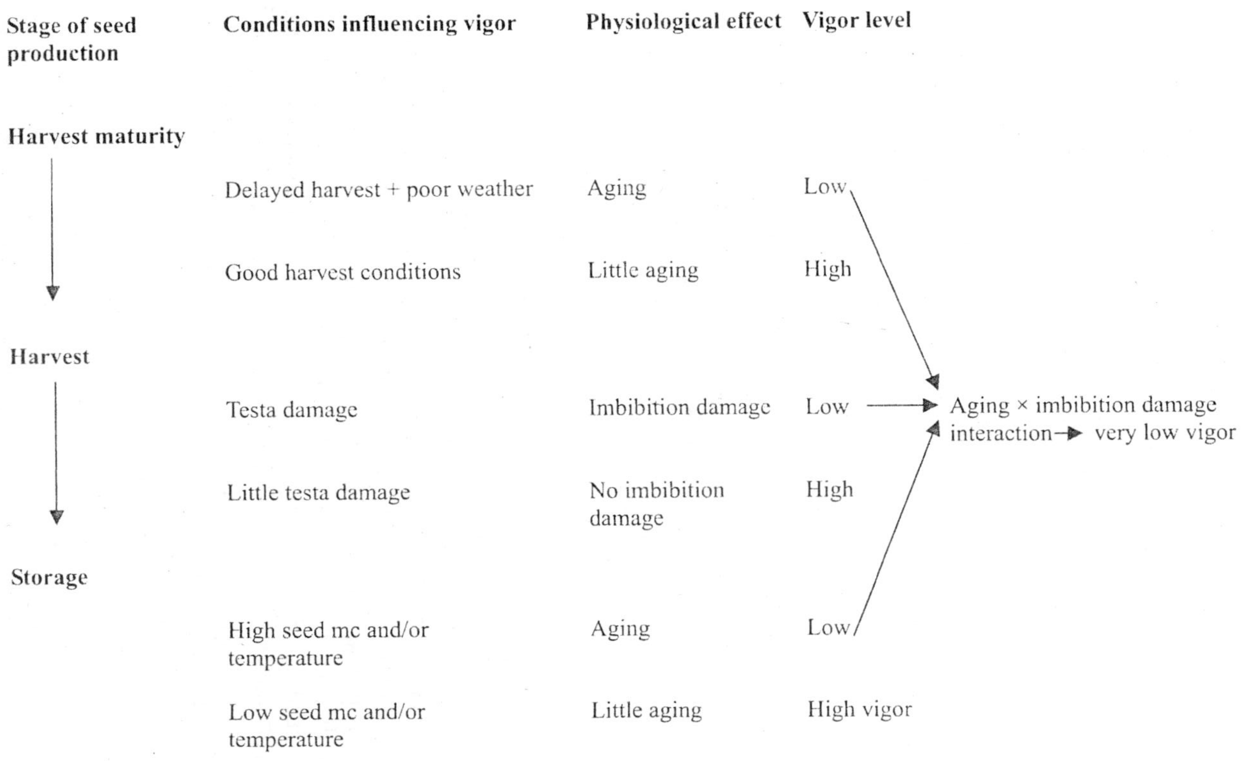

FIGURE 21.4. Major factors influencing vigor at stages of seed production.

temperatures during development may also impair performance (Judd et al., 1982). Reduced water availability may interrupt seed development, and Delouche (1980) reported that temporary but severe drought leads to the production of light, shrivelled, and misshapen soybean seed. However, the effect of drought on seed quality is less than that on seed yield (Dornbos et al., 1989), and Viera et al. (1992) failed to detect any effect of drought stress during development on the germination or vigor of soybean. Conditions during the period of seed fill therefore have comparatively little influence on the vigor and germination of seeds.

Seed Size

Seed size is often considered to be a factor that can influence vigor. However, no consistent association has been shown between small or large seeds and either high or low vigor (Powell, 1988). Furthermore, most reports have focused on the relationship between seed size and germination and/ or seedling growth, rather than a specific vigor assessment. In some reports, large-seeded soybean had higher germination, vigor, and yield potential than did small seeds (Burris, 1973; Burris et al., 1971; Fontes and Ohlrogge, 1972), whereas others found no effect of seed size in soybean (Singh et al., 1972; Johnson and Leudders, 1974). Hunter and Kannenberg (1972) and Shieh and McDonald (1982) also found little effect of seed size on the germination of either hybrid maize or two maize inbreds, respectively, while Hong et al. (1982) reported more rapid germination of larger maize seeds. On the other hand, more rapid germination has been observed in large seeds of two *Chenopodium* species, tomatoes, and barley (Hall and MacNeill, 1982) and Gabrielson et al. (1982), reported an increase in plant density when large seeds of crimson clover, KY-31 fescue, and soybean were used. In cowpea, large seeds produced larger seedlings, although improved emergence occurred only at deep sowings (Lush and Wien, 1980).

Thus, seed size may relate to the rate of germination or emergence, total germination/emergence, and seedling size in specific species and circumstances. However, a consistent association between seed size and seed vigor has not been demonstrated.

ASSESSMENT OF SEED VIGOR

Requirements of a Vigor Test

Many authors have clearly identified the requirements for a vigor test (Heydecker, 1972; McDonald, 1980; Matthews, 1981; Perry, 1987a). Matthews (1980) emphasized that a test should first of all have a good theoretical background. This would ensure that the outcome of any test could be explained or, alternatively, unexpected results would be more easily identified and examined further. More commonly, however, the emphasis has been on the technical requirements for a test. These are that it should be simple to complete, and hence applicable in a wide range of seed testing environments, and also rapid, to enable prompt reporting of results. The results of the test should also have been shown to correlate with a practical expression of seed vigor, such as emergence in the field or glasshouse, or seed longevity in store. The test should also, preferably, provide a quantitative method of assessment to avoid subjective assessments and hence make standardization easier. More important, if the test is to be used in many laboratories, it should be repeatable both within and between laboratories; thus, results from different laboratories would be comparable. Finally, and a critical aspect in the running of any seed testing laboratory, the test should be economically practical.

Nature of Vigor Tests

As described earlier, seed aging is a major general cause of differences in seed vigor, with many clearly identifiable outcomes. As a result, the theoretical background of most vigor tests lies in seed aging and involves the application of one or more of the consequences of aging. In this discussion of seed vigor assessment, the link between the tests and the aging process will be emphasized and the tests discussed in relation to the other criteria that have been identified as important in the development of vigor tests.

Vigor tests can be divided into three broad categories that reflect different aspects of the aging process, namely, physiological tests, biochemical tests, and tests that apply the whole aging process. Within these groupings, the tests can also be described as being either direct or indirect. Direct tests measure differences in vigor in terms of an aspect of growth, often under stress conditions. An advantage of direct tests is that they evaluate all factors that influence vigor at once, e.g., mechanical injuries, morphological factors, aging. They also often apply conditions similar to those in the field and hence have a psychological advantage in terms of their interpretation. Indirect tests usually apply more controlled test conditions and are more reproducible. They are also less time-consuming and may use less equipment. However, they do not evaluate all vigor factors at once.

Physiological Tests

Physiological tests are largely direct tests and can be divided into two groups. First are those tests that measure the impact of aging on an aspect of germination and early seedling growth. These reveal the reduced rate and uniformity of germination that is found in aged seeds. Furthermore, the slower germination of aged seeds will lead to the production of smaller seedlings, when seedling size is assessed at a particular point in time. Second, physiological tests apply stresses to the seed that reflect the type of stress that may be encountered in the field. In this case, low vigor seeds are revealed by the reduced tolerance of aged seeds to stress.

Tests Based on Germination

These are essentially a modification of the germination test. Most simply, vigor can be assessed as the rate of germination based on the first count from the standard germination test, an assessment of seed vigor that is often used. However, there are also more precisely defined assessments of vigor based on germination. These include the so-called seedling growth rate test and the seedling growth test (AOSA, 1983; ISTA, 1995).

Seedling growth rate. This test is applied to soybean and maize (AOSA, 1983; ISTA, 1995). It is based on the standard germination test, but in this case the moisture content of the towels is more clearly defined (approximately 30 g water per towel) in order to give more standardized growth conditions. At the end of the test, normal, abnormal, and dead seeds are counted and the normal seedlings are cut free from the seed or cotyledons and dried at 80°C for 24 h. The average dry weight per seedling is then calculated. Seedling dry weight, measured at a specific time after the seed were set to germinate, essentially reflects the time taken to germinate, since aging does not influence the actual rate of seedling growth (Finch-Savage, 1987; Wheeler and Ellis, 1991). Thus, a high seedling dry weight reflects rapid germination and high vigor, a low dry weight, slower germination and a lower vigor level. The disadvantages of this test are that it is time-consuming to complete the dry weight measurements and growth may be different between cultivars. Furthermore, the rate of germination is influenced by even small variations in temperature. Although many laboratories may have sophisticated systems for accurate temperature control of incubators/germination rooms, this is not universally the case. Hence, repeatability of these tests is not easily achieved.

Seedling growth test. This test has been applied to cereals (barley, wheat, oats, and maize) and to grasses (ryegrass and fescue). Again, this is a modification of the rolled towel germination test (ISTA, 1995). However, in this case, a series of parallel lines are drawn along the length of the paper towel, 1 cm apart. After completion of the germination test the numbers of normal and abnormal seedlings, and dead seeds are counted. In addition, the number of shoots from normal seedlings between each pair of parallel lines are counted and a vigor rating calculated as

$$\text{Vigor rating} = \frac{(n \times 1) + (n \times 2) \ldots + (n \times 15)}{N},$$

where n = number of shoot tips within the pair of parallel lines; 1, 2 . . . 15 = distance (cm) of the top line of the pair from the baseline; and N = total number of seeds originally set to germinate.

Similar to the growth rate test, the results reflect the time that individual seeds take to germinate, with a higher vigor rating being produced from rapidly emerging seeds that have a longer period for growth within the test period.

Perry (1987a) suggested that problems of repeatability in this test, due to variations in temperature, could be overcome by using a high vigor reference lot on each test occasion. Thus, the growth of the reference lot would be adjusted to ten and that of other lots adjusted on a proportional basis, i.e.,

$$L = \frac{Y \times 10}{C},$$

where L = mean plumule length, Y = value of test sample, and C = value of the control.

Stress Tests

Stress tests measure directly the physiological effects of one or more of the stresses that seeds might encounter in the field environment. Differences in seed vigor are identified as a result of the reduced tolerance of aged, i.e., low vigor, seeds to stress.

Hiltner test. The Hiltner test was initially developed to detect seed-borne infection by *Fusarium* spp. (Hiltner and Ihssen, 1911). However, it was subsequently developed as a vigor test when it was shown to reflect defects that prevented normal seedling growth, other than those resulting from seed-borne infection (Hiltner and Ihssen, 1911; Eggebrecht, 1949). The principle of the test is that seeds having low vigor cannot withstand physical stress during germination. Hence, seeds are sown within layers of sterile brick grit (crushed black stone) or coarse sand with a particle size of 2 to 3 mm. This provides a mechanical barrier through which seeds must penetrate in order to emerge. The emergence of normal seedlings is considered to indicate seed vigor status.

The Hiltner test has been applied within Europe for vigor testing of cereals (Hampton, 1992) but is little used elsewhere (AOSA, 1983; Hampton,

1992). Although it is reasonably reproducible, the relationship with field emergence has been variable (ISTA, 1995). Furthermore, the test is expensive, time-consuming, and requires a great deal of space. There may also be difficulties in obtaining supplies of brick grit and in the washing and drying of the grit (Fritz, 1965).

Cold test. The cold test is probably the best known of the stress tests and is the most widely used test for maize in North America and Europe (TeKrony, 1993; Ferguson, 1993; Hampton, 1992). The development of the cold test has been suggested (AOSA, 1983) as critical to the evaluation of vigor, as it established the idea of emergence as a means of vigor evaluation and of using environmental stresses to assess it.

The test sets seeds to germinate at low temperature, most commonly 10°C, for 7 days, followed by a further period of 4 to 7 days at 25°C (AOSA, 1983; ISTA, 1995). Both the AOSA (1983) and ISTA (1995) suggest that the test should be conducted with the seed in contact with field soil, although not all workers support this view (Burris and Navratil, 1979; Bruggink et al., 1991; Loeffler et al., 1985). The test is often completed in paper towels or shallow trays, although the AOSA (1983) suggest that deep boxes may be used as an alternative. The conditions of the test therefore simulate the adverse conditions of early spring sowings, and the ability of seeds to produce normal seedlings reflects the vigor of the seed lot. A close association between the cold test results and field emergence has been shown in maize, and this relationship has also been evaluated for a number of other species (ISTA, 1995).

The results of the cold test are commonly believed to express the interaction among genotype, soilborne pathogens, seed quality, and seed treatment. Thus, the different genotypes will show differing sensitivity to chilling injury, which will influence their susceptibility to pathogens. Reduced seed quality, for example, due to aging, will further increase this susceptibility, while seed treatments may reduce it. Thus, the interaction of these factors would reflect emergence potential in the field. There is, however, an alternative explanation for the damage that occurs at low temperature. Powell and Matthews (1978) suggested that the so-called imbibitional chilling injury may result from the greater sensitivity of seeds to imbibition damage at low temperatures. In support of this, they provided evidence that a number of grain legumes failed to show damage during imbibition at low temperatures if water uptake occurred slowly (Powell and Matthews, 1978). In addition, the presence of excess water in many cold tests will result in rapid imbibition. The cold test may, therefore, reflect the susceptibility of the large maize embryo to imbibition damage at low temperatures and, when the test includes soil, the subsequent impact on fungal infection.

Many problems have been highlighted in association with the standardization of the cold test (Nijenstein, 1995; Nijenstein and Kruse, 2000). Aspects of the test identified as influencing the results have included the presence or absence of soil (Nijenstein, 1995; Burris and Navratil, 1979; Bruggink et al., 1991; Loeffler et al., 1985), the length of the cold period (Nijenstein, 1995; Nijenstein and Kruse, 2000), the temperature of the cold period (Nijenstein, 1995; McDonald, 1999; Bruggink et al., 1991), and the availability of oxygen (Nijenstein, 1995; AOSA, 1983). However, Byrum and Copeland (1995) noted, in a comparison of ten maize seed lots, that the cold test was as repeatable as the germination test. Subsequently, Nijenstein and Kruse (2000) reported that when specific directions were given to 20 seed laboratories from four climatic zones, the variation in the results of the cold test was reduced to half that found when laboratories used their own version of the test. Gutormson (1995) also reported good repeatability when the cold test was carried out by laboratories in the U.S. midwest which carry out the test on many seed lots on a routine basis.

Nijenstein and Kruse (2000) concluded that the standardization of the cold test would be possible, particularly if further attention was paid to particular aspects of the test. Two aspects that they highlighted were (1) the need to standardize oxygen supply, since the rolled towel and tray methods gave rise to less variation than the box method, and (2) a requirement for further standardization of seedling classification, since in their study, the percentage of abnormal seedlings differed among laboratories in each test run. They also pointed out that warmer countries preferred a test having a seven-day cold period and accepted sand as a substrate. On the other hand, colder countries required more stress to be placed on the seed during the test in order to identify even small differences in vigor that could be important in determining emergence in their more stressful sowing conditions. Thus, these countries used a ten-day cold period and insisted on the use of soil as the test substrate. Despite identification of the potential for standardization of the cold test, the implementation of a standard procedure may be difficult, since comments to Nijenstein and Kruse (personal communication) indicated that laboratories would not adopt a standard procedure. The preference of individual laboratories was to continue to implement their own in-house method.

Cool temperature germination. This test was developed specifically for *Gossypium hirsutum* and involves the assessment of normal germination at 18°C in rolled towels in which the moisture content is controlled (ISTA, 1995). Cotton can germinate at temperatures from 12 to 40°C, but 18°C is below the optimum for germination, which is 25 to 34°C (Cole and Wheeler, 1974). As a result, the seeds germinate more slowly than would be the case at the optimum temperature for germination. Thus, the test in effect evaluates the rate of germination in different seed lots.

Complex stress vigor test. This test was developed initially for wheat and subsequently for maize, and it simulates the anaerobic and cold conditions of sowing encountered in temperate zones of Europe, North America, and Asia. It is therefore possibly of little relevance elsewhere. The test was developed in Hungary and has consistently identified high and low vigor lots so that more desirable plant populations have been achieved (Barla-Szabo and Dolinka, 1988; Barla-Szabo et al., 1990).

The test imposes temperature stress and oxygen deficiency by soaking seeds first for 48 h at 20°C (wheat) or 25°C (maize), followed by 48 h at 2°C (wheat) or 5°C (maize) (ISTA, 1995). The seeds are then set to germinate between paper towels, and after 96 h, normal and abnormal seedlings and dead seeds are assessed. The length of normal seedlings is measured; the average length of the five longest seedlings is calculated and multiplied by 0.25. On the basis of this derived value, the seedlings of each lot are divided into two groups. High vigor seedlings are longer than the derived value; low vigor seedlings are shorter than this value. Rapid germination, leading to the production of longer seedlings following stress, is indicative of high vigor.

Biochemical Tests

The second category of vigor tests includes those that can be classified as biochemical tests. These are indirect tests, which largely measure one of the consequences of aging.

Conductivity Test

The conductivity test has been well developed for garden peas but has also been shown to apply to a wide range of grain legumes (e.g., soybean, French bean, *Phaseolus vulgaris;* mung bean, *Phaseolus mungo;* field bean, *Vicia faba*) as well as some other species (ISTA, 1995). The test measures the leakage of solutes, particularly potassium ions, into water. Hence, an increase in the electrical conductivity of seed soak water indicates high solute leakage. Loss of solutes from seeds reflects the integrity of cell membranes and/or the presence of dead tissues within the seed. Membrane integrity and the incidence of dead tissue are both influenced by the two major factors determining seed vigor in grain legumes, namely, seed aging and the incidence of imbibition damage, and also by the greater sensitivity of aged seed to imbibition damage. Thus, the measurement of conductivity assesses the incidence of the major causes of reduced vigor in grain legumes.

The conductivity test for garden peas is one of the two ISTA-validated vigor tests (ISTA, 2001). The method is therefore standardized (ISTA, 2001) and has been shown to be reproducible and repeatable. It is simple and involves soaking replicates of 50 seeds in 250 ml deionized or distilled water for 24 h at 20°C, after which the electrical conductivity of the seed soak water is measured. Seed lots that show high levels of solute leakage, i.e., high conductivity, are low vigor lots; high vigor seed lots exhibit low leakage and low conductivity (Bradnock and Matthews, 1970; Matthews and Bradnock, 1967, 1968). The correlation between conductivity test results and the emergence of peas has been proven many times (ISTA, 1995).

A number of factors may influence the repeatability of the conductivity test, namely, the cleanliness of equipment, water quality, soaking temperature, initial seed moisture content, and seed size (Hampton, 1995a,b; McDonald, 1999). However, the variability that these factors may create has been overcome by the close prescription for the method of completing the test (ISTA, 2001) and the expression of the results in terms of seed weight.

The conductivity test is commonly used in the assessment of the quality of seed peas. For example, in the United Kingdom, all peas sown for the production of the so-called vining peas for freezing or canning would be vigor tested using the conductivity test (Biddle, personal communication). Conductivity tests have also been applied to detect vigor differences in many other grain legumes and indeed some other species (ISTA, 1995). However, the methods used for other species have not been standardized and the relationships with field emergence not consistently proven.

The bulk conductivity test provides an average evaluation of the conductivity of several replicates of bulk samples of seeds. Wide variations between replicates could occur, however, as a result of high leakage from individual seeds that show extensive physical or physiological deterioration. In this case, an evaluation of the leakage from individual seeds may be desirable. Commercial instruments are available to measure the conductivity of the leachate from individual seeds (Steere et al., 1981) by placing seeds in a tray containing 100 individual cells. After soaking the seeds in the cells, a specially designed electrode head fits over the tray. Instruments initially designed for this purpose (e.g., ASA, 610, Automatic Seed Analyser, Agro Sciences Inc, USA; ASAC 1000, Neogen Inc., USA) measured the current passing through the leachate, expressed in μamps, while conductivity measurements are expressed in μS. Research with soybean seeds has shown a highly significant correlation between current (μamps) and conductivity (μS) readings for individual seeds (McDonald and Wilson, 1979, 1980; Loeffler, 1981). More recent models of this instrument (G2000, Wavefront

Inc., Ann Arbor, Michigan) measure the conductivity of the leachate in $\mu S \cdot cm^{-1}$.

Studies with pea, soybean, cotton, field bean, maize, and small-seeded crops have demonstrated that analysis of the single seed leachate conductivity, using the instruments described previously, evaluated the seed vigor of many crops. The same basic procedure and precautions should be followed with this instrument as outlined for the bulk conductivity test (ISTA, 2001). However, the instrument does not adjust readings to take account of seed weight. It is therefore recommended that each of the 100 seeds be weighed prior to testing so that the average reading can be recorded per gram of individual seed weight (Hepburn et al., 1984), i.e., as $\mu amps \cdot cm^{-1}$ per gram or $\mu S \cdot cm^{-1}$ per gram, depending on the instrument used.

The measurement of the leachate conductivity of single seeds therefore offers the opportunity to determine the extent of variability in solute leakage within a seed lot. It can identify whether high leakage results from the presence of a few highly damaged seeds or whether there is an overall high level of leakage within the seed population. In particular, it could be used to determine the incidence of seed coat damage that is not discernible visibly but which influences the rate of water uptake and hence imbibition damage, and the incidence of hard or slowly imbibing seeds that show low levels of leakage.

Measurement of Enzyme Activity

Enzyme activity is known to decline as seeds age. However, only the measurement of dehydrogenase activity and glutamic acid decarboxylase activity have been practically applied in vigor tests. Other assessments of enzyme activity are largely confined to research on seed vigor.

Tetrazolium test of dehydrogenase actvity. The tetrazolium test is commonly used to assess the viability of seed (Lakon, 1950a,b; Steiner and Werth, 1974; Moore, 1973, 1976; AOSA, 1983; ISTA, 1999) but can also be used to detect vigor differences. The test evaluates the presence and location of living tissue within the seed through the reaction between a solution of 2,3,5 triphenyl tetrazolium chloride (TTC) with active dehydrogenase enzymes. The TTC is reduced by the enzymes to produce the red, stable substance triphenyl formazan (Cottrell, 1948; Roberts, 1951). Hence, the presence of red staining indicates living tissue; dead tissue remains unstained. Vigor is evaluated through the appraisal of sound, weak, and dead tissues as they relate to subsequent seedling development (AOSA, 1983; Moore, 1985). The tetrazolium test to detect vigor differences has been applied to cereals in general (Lakon, 1950a,b) and wheat (Perry, 1987b), and in the United States to a range of crops including soybean, cotton, pea and clover (AOSA, 1983). The procedure for completing the test is the same as that described for the viability test (AOSA, 1983; ISTA, 1999). The pattern and intensity of staining is then used to classify viable seeds as high or low vigor (Lakon, 1950a,b; Moore, 1985) or as high, medium, or low vigor (AOSA, 1983; ISTA, 1995). The limitations of the tetrazolium test are two-fold. First, it requires experienced analysts for the preparation of the material and for the evaluation of the results. Second, the evaluation of the staining is subjective, and although extensive training might help to eliminate variations in evaluation, Hampton and Coolbear (1990) questioned the routine application of this test in vigor assessment.

Glutamic acid decarboxylase activity (GADA). Many of the observations of reduced enzyme activity associated with seed aging have been accompanied by a decline in germination, rather than a decrease in the vigor of germinable seeds (Abdul Baki and Anderson, 1972). In contrast, the activity of glutamic acid decarboxylase has been clearly shown to decrease in maize before germination begins to fall, that is, as vigor declines (Grabe, 1964). However, the link between GADA and vigor was not clear in soybean (Burris et al., 1969), beans (James, 1968), or peanut (Baskin and Delouche, 1971). Furthermore, although GADA was correlated with the storage potential of maize (Grabe, 1965), no relationship has been established between GADA and field emergence. There has been little or no evaluation of the repeatability of GADA as a vigor test. In addition, the more complex nature of enzyme extraction and assessment of activity suggests that it is not particularly suited to application as a routine test but is more likely to be used as a means of evaluating vigor within research programs.

Other Biochemical Assessments of Vigor

Other biochemical assessments of seed vigor include the measurement of the activities of groups of enzymes, i.e., processes or the concentration of particular compounds in the seed. Examples include analyses of adenosine triphosphate (ATP) production or content (Ching 1973a,b; Yaklich et al., 1979; Lunn and Madsen, 1981), malate (Perl and Kretschmer, 1988) content, and of the products of lipid peroxidation (volatile aldehydes) (Fielding and Goldsworthy, 1982; Harman et al., 1982; Wilson and McDonald, 1986). Measurements of metabolic activities or processes within the seed include those made on respiration (Wahab and Burris, 1971; Abdul-Baki and Anderson, 1973; Carver and Matthews, 1975), glucose utilization (Abdul Baki, 1980), and DNA synthesis (Clay et al., 1981). In some cases these measurements have been linked with an expression of vigor such as field emergence. For example, Carver and Matthews (1975) linked the low vigor and poor field emergence of peas with reduced oxygen uptake compared with high vigor seed lots. However, this is the exception rather than the rule. Furthermore, the more complex technical nature of these assessments makes them more suited to research than to routine application.

Tests Applying the Whole Aging Process

The third group of vigor tests evaluates the effect of a period of aging on germination. Each of the three tests involved is based on the manipulation of the rate of seed aging by holding seeds at a raised temperature and moisture content or relative humidity. The seeds therefore age more rapidly and changes in germination, normally seen over a period of months or years, occur within days or even hours.

Accelerated Aging Test

This test was initially developed by Delouche and Baskin (1973) to assess the storage potential of a number of species. However, the results have subsequently been shown to reflect vigor in more general terms and hence reveal both the field emergence and storage potential of a seed lot (AOSA, 1983; ISTA, 1995).

In the accelerated aging test, seeds are held at high relative humidity (close to 100 percent) at raised temperature (41 to 45°C) for 48 to 72 h (AOSA, 1983; ISTA, 1995, 2001). The seeds absorb moisture from the humid atmosphere, reaching moisture contents ranging from 28 to 45 percent depending on the species (ISTA, 1995). At this high moisture content and the raised temperature, the seeds age rapidly. After the period of aging, the seeds are used in a standard germination test. The germination after accelerated aging is compared with the standard germination of the seed lot to provide an assessment of vigor. Where the AA germination is similar to the standard germination, the seed lot is high vigor. If the AA germination is less than the standard germination, the lot is medium or low vigor. This vigor assessment enables seed lots to be ranked in terms of their emergence and storage potential (ISTA, 1995).

A number of variables have been described that can influence the results of an AA test, in particular the sample size, temperature, seed moisture content, and timing of the test (TeKrony, 1993). However, there has been considerable effort to evaluate the effects of these variables and to refine techniques to eliminate their effects (McDonald and Phaneendranath, 1978; Tomes, 1985; Tomes et al., 1988). As a result, the AA test method for soybean has been standardized. The AA test for soybean is therefore recommended by the AOSA (1983) and appears in the ISTA rules as one of the two validated vigor tests (ISTA, 2001). Both the AOSA (1983) and ISTA (1995) also publish details of suggested AA test conditions for a range of other species. However, there has not been extensive comparative testing to provide data on the repeatability of the test as applied to these species or to confirm the relationship between the test results and an expression of vigor.

A modification of the AA test has been proposed (Jianhua and McDonald, 1996) for application to seeds of small-seeded crops. They (Jianhua and McDonald, 1996) demonstrated that when the flower seeds, *Impatiens wallerana,* were exposed to a humidity close to 100 percent, the moisture content rose rapidly to a sufficiently high level that seed aging occurred so rapidly that all seeds had a very low germination after the aging period. Thus, it was difficult to detect any differences in vigor of the seed lots. However, when seeds were held over saturated solutions of KCl or NaCl, giving relative humidities of 87 percent and 76 percent, respectively, the seed moisture contents rose to only 9 to 11 percent, compared with 77 percent when seeds were held over water (close to 100 percent relative humidity). This led to slower rates of aging during the test and clearer differentiation between the seed lots.

Controlled Deterioration Test

The controlled deterioration (CD) test was developed to detect differences in the vigor of seed lots of small-seeded vegetable species (Matthews, 1980). The test was developed following the observation that during accelerated aging of such species, different seed lots took up moisture from the high relative humidity at different rates, as exemplified by Powell (1995). Thus, after one day of accelerated aging at 100 percent relative humidity and 45°C, the moisture content of 21 seed lots of onion ranged from 11.76 to 23.98 percent (Powell, 1995). This meant that the lots were held at different raised moisture contents at the high temperature, and hence experienced differing degrees of deterioration. As a result, the germination after AA ranged from 1 to 90 percent and there was a significant correlation between seed moisture content and subsequent germination. The aim of the CD test is to expose each seed lot to exactly the same amount of deterioration or aging. Thus, in this test, the seed moisture content of all seed lots to be tested is accurately raised to the same moisture content at the start of the test (Powell, 1995; ISTA, 1995). This is achieved through imbibition of seeds of known moisture content on moist germination papers and frequent weighing to determine when the required moisture content has been reached.

Seeds are then sealed in a foil packet and the moisture allowed to equilibrate throughout the seed overnight at 5 to 10°C before being placed in a waterbath at 45°C for 24 h. After the period of deterioration, seed germination is tested.

The precise period of deterioration during CD effectively moves each seed lot along the seed survival curve by the same amount (see Figure 21.2). Thus, a physiologically young seed (high vigor) will retain a high germination after CD, while a physiologically older seed (low vigor) will show reduced germination. The differences in germination after CD allow seed lots to be ranked in terms of their vigor. CD test results have been shown to relate to emergence both in the field (Matthews, 1980) and glasshouse (Powell et al., 1991), and to seed storage potential under commercial conditions (Powell and Matthews, 1984a,b).

The limitations of the CD test lie in the precision required in the raising of the seed moisture content and in the frequent weighing to determine whether the desired weight and moisture content has been reached. However, the frequency of weighing can be reduced with experience, and comparative tests both in the United Kingdom (Powell, Don, et al., 1984) and among a number of ISTA laboratories (Hampton, 1998; TeKrony, 2001b) have shown good repeatability both within and between laboratories.

Determination of K_i (Initial Theoretical Viability)

The determination of K_i has been proposed as an accurate measure of initial seed quality by Ellis and Roberts (1980) and has been used in many research programs as an assessment of quality beyond standard germination. It has not, however, been proposed as a potential vigor test by either ISTA or AOSA.

The basis of this test is the normal distribution of seed deaths with time that gives rise to the seed survival curve. This means that the survival curve can be re-expressed in terms of probit germination. This gives a straight line that can be extrapolated to the y-axis to give what is referred to as the initial theoretical viability or K_i. The K_i of a seed lot is therefore determined by aging seeds at a specific moisture content and temperature. Seed samples should be taken on at least five occasions during aging, and it is essential that all three phases of the survival curve are included within the samples (Wilson et al., 1989; Tang et al., 1999). The germination of each sample is tested and converted to probits, and then calculation of the regression of probit germination against storage time allows calculation of K_i.

K_i is believed to be more accurate than the standard germination test in assessing seed quality because it is based on at least five germination tests of 400 seeds per lot, rather than the 400 seed sample of the germination test. Hence, it may more truly reflect initial seed quality, possibly reflecting both germination and vigor. However, the features of K_i that lead to this accuracy are also limitations in the routine application of the test. For each sample to be tested, determination of K_i requires manipulation of seed moisture content, an aging period, and a germination test for at least five samples. This requires not only precision, as in the CD test, but also considerable time in the aging and germination of each sample. There may also be problems in achieving the required amount of aging to produce a survival curve within a sensible time period for a test. The time period to produce a survival curve can, of course, be shortened by increasing seed moisture content. However, once seed moisture content has been raised above a certain value, specific to each species, metabolic repair processes may be activated that counteract the aging process (Walters, 1998; McDonald, 1999). Thus, a survival curve produced at high moisture contents over a short period is unlikely to give an accurate K_i.

STANDARDIZATION OF VIGOR TEST PROCEDURES

Standardization of vigor test procedures is important if results are to be compared within and between laboratories. It is also important that if disputes should arise over the quality of seed, previous tests of seed quality should have been completed in a recognized, standard manner that are accepted by all parties involved in seed quality evaluation.

The requirements for precision and standardization of test methods, to ensure repeatability and reproducibility of the results, have in the past been seen as a problem for seed vigor testing (AOSA, 1983; McDonald, 1995; Hampton, 1995b). Assessment of the precision of a vigor test has been perceived as difficult (AOSA, 1983), since there is no standard against which a result can be compared, such as a standard concentration, and problems may be associated with the effects of genotype, seed size, pesticides, etc. Problems associated with standardization have also been identified, including sampling, the subjective nature of many assessments, seed moisture content, and variations in environmental variables such as temperature and substrate moisture content. For example, subjectivity may influence seedling vigor classification and tetrazolium staining assessment, and even the cold test and accelerated aging test involve the evaluation of normal and abnormal seedlings. In terms of the test conditions, many do not influence the level of germination itself, but substrate moisture and even a 1°C increase or decrease in temperature will influence the rate of germination.

However, the recognition that such factors can influence test results has also been a catalyst to research on specific vigor tests to identify the variables likely to have an impact on test results. This has assisted in the development of closely prescribed test procedures. These can then be evaluated in comparative testing programs, involving a number of laboratories, which assess the within and between laboratory variability. Comparative tests can also be designed to evaluate the ruggedness of a test procedure, i.e., to determine the importance of differences in procedure at possible critical stages of the test to the test results.

Two vigor tests for which the influence of test procedures and variables have been investigated in detail are the conductivity test for peas (Hampton, 1995a) and the accelerated aging test for soybean (TeKrony, 1993). As a result, standardization of these tests has been achieved, leading to their acceptance by ISTA as validated vigor tests (see next section).

CURRENT STATUS OF VIGOR TESTING

Vigor testing took a major step forward within ISTA in June 2001, when the membership voted to include a seed vigor chapter within the Rules for Seed Testing. This was, accordingly, published in the Rules Amendments 2001 (ISTA, 2001). The Rules Amendments include an annex to the seed vigor chapter that gives details of the precise methods to be used to carry out the two ISTA-validated vigor tests, namely the conductivity test for *Pisum sativum* and the accelerated aging test for *Glycine max*. The description of these tests as being "validated" indicates that they have undergone extensive international comparative testing to establish the reproducibility and repeatability of the described methods. The data from the comparative tests also form the basis of tolerance tables for the results of the tests, which appear in the Rules Amendments 2001 (ISTA, 2001). As a further requirement for validation, the relationship(s) between the test results and an expression of vigor has also been clearly established. Thus, the results of the conductivity test have been related to the field emergence of peas, and those of the accelerated aging test to both the field emergence and storage potential of soybean. As a further step toward the application of these vigor tests, the ISTA Vigor Committee, with the Proficiency and Statistics Committees, has established proficiency tests for laboratories to demonstrate their competence in carrying out these two tests and hence to include these within their ISTA accreditation.

Other vigor tests appear in the ISTA Handbook of Vigor Test Methods (ISTA, 1995) as suggested vigor tests. These differ from validated tests in that they have revealed potential for identifying differences in seed vigor but have not undergone extensive comparative testing in a validation procedure. The Association of Official Seed Analysts also produces a Seed Vigor Testing Handbook (AOSA, 1983), in which the procedures to carry out seven suggested vigor tests are described. In North America, however, vigor tests are not found in the Rules for Seed Testing.

The development of vigor tests from the stage of being a suggested test to that of validated status will, within ISTA, follow the Method Validation Program that has been developed for all proposed new ISTA test methods. Two possible validation procedures may be followed. The first of these, multilaboratory validation, involves six to eight partcipating laboratories and is appropriate for vigor tests, while the second, performance-based validation aims to evaluate the performance of proprietary test methods, such as those used for seed health testing. Acceptance of a test by the Vigor Committee after validation leads to its description as an ISTA approved method, and following acceptance by an ISTA membership vote, as an ISTA official method.

Following the development of the method validation programs, any test developer can submit a proposal for test validation to the Vigor Committee through the ISTA Secretariat. The current Vigor Committee aims to apply the programs to extend the species base for both the conductivity and accelerated aging tests. In addition, the comparative testing program that has been applied to the controlled deterioration test between the last two ISTA Congresses (1995-1998, 1998-2001) will be developed within the validation program to advance the controlled deterioration test toward validation.

PRESENTATION AND INTERPRETATION OF VIGOR TESTS

A clearly perceived problem in the presentation of vigor test results is that the nature of the tests for different species differs and hence the way in which results are expressed are different (AOSA, 1983). A conductivity test therefore describes the electrical conductivity of a seed leachate in terms of $\mu S \cdot cm^{-1}$ per gram, while the results for accelerated aging are given as a germination percentage. Thus, the consumer must understand the nature of the vigor test that is completed for a particular species. The AOSA (1983) also suggested that there was a further problem in that the results from a sample express vigor only as it is at a particular point in time. They suggested that the vigor test results would not be valid if the seeds were subsequently stored under poor storage conditions, and hence the vigor of the whole lot subsequently fell rapidly. However, the vigor test results would remain valid in good storage conditions. This argument is not one that is confined to seed vigor. The same would be true of a germination test result, particularly if the germination was high, but the seed lot lay well toward the end of the initial slow decline of the survival curve. In this case, a short period of rapid deterioration would result in a standard germination percentage that is completely unrelated to the germination test result. A germination test will often be repeated within three to six months. The ISTA Vigor Committee have recommended that, since vigor falls before germination, repeat vigor tests should be completed more frequently than this.

Education of the consumers into the nature of vigor tests is therefore essential if they are to understand the presentation of vigor test results. Education is also essential for consumers to be able to interpret the results of vigor tests. Neither the AOSA nor ISTA have a responsibility to interpret test results; their role lies in the standardization of the test methods. McDonald (1995) emphasized that it is the role of the consumer, not of the seed analysts, to interpret vigor test results. The consumer must appreciate that a vigor test does not predict or forecast emergence, as this will vary depending on field conditions (McDonald, 1995; Ferguson Spears, 1995). Thus, the consumer should consider the conditions under which the seed will be sown. High vigor seed will increase the potential for good emergence in poor conditions but will not guarantee it. The AOSA (1983) advises the consumer to compare the vigor and standard germination results for more than one seed lot, if possible, before buying. This would allow the consumer to select a lot having higher vigor.

The consumer therefore has to exercise judgment regarding the vigor of the seed and the conditions under which he or she will sow it. This is currently the approach taken by producers of garden (vining) peas for processing in the United Kingdom. All such seeds that are sown in the United Kingdom will have undergone a conductivity test (Biddle, 2001, personal communication). The results of these tests may be divided into four categories in relation to their suitability for sowing (Bedford, 1974). These categories are not used by growers as strict advice on sowing time, but as guidelines to the suitability of a seed lot for sowing, which are judged alongside such factors as the date of sowing and soil conditions. Thus, the growers use the vigor of the seed to judge the element of risk at sowing for emergence. These guidelines for conductivity results and field emergence of peas can be considered to be applicable only under United Kingdom conditions. Similar guidelines that would assist in interpretation of vigor test results could be produced elsewhere for the conductivity and other vigor tests. Organizations with an interest in the emergence and storage potential of seeds such as growers groups, processing companies, and agricultural extension services, may become involved in guideline development.

APPLICATION OF VIGOR TESTS

A number of authors have considered how the results of vigor tests may be applied (AOSA, 1983; Hampton and Coolbear, 1990; Dornbos, 1995; Hampton, 1995a,b; Ferguson Spears, 1995). The two areas in which vigor testing can probably have the most immediate impact are seed production and quality control, and to help meet market demands.

Application of vigor tests at different stages of seed production can assist seed producers to identify where any reduction in seed quality has occurred during the production process, whether it be while the seed is still on the plant (weathering), during harvest and processing, or during storage. Thus, guidelines can be developed to minimize any reduction in vigor in subsequent years. In addition, companies may use vigor results in their in-house quality control. This could involve preventing the sale of particularly low vigor seed, or providing guidance regarding the storage potential of seed lots. Companies could avoid storing seeds that will rapidly lose their vigor, and subsequently the ability to germinate during storage, and can select seed for storage that have good storage potential.

Identification of seed having low vigor will also assist in meeting the demands of the market for the production of uniform produce at similar levels of maturity, which will enable once-over harvesting. Indeed, in the United Kingdom, the demands of supermarkets for specifically sized produce has led transplant producers and vegetable growers to be more demanding of the seed companies for information on their seed quality. High vigor seed that germinates rapidly and uniformly will produce a crop that meets such standards, whether this is established directly or by transplants, whereas the slow and asynchronous germination of low vigor seeds will not.

Other possible uses of vigor tests are in marketing, research, and plant breeding. Application in marketing and promotion of seeds has largely been avoided by seed companies, and the American Seed Trade Association has advised against their use in this way until tests are standardized and the interpretation and use of vigor tests has been resolved. However, Hampton and Coolbear (1990) commented that although this caution was understandable, growers may become more demanding as the market becomes more competitive, leading to a requirement for vigor test information.

There is a clear application for vigor tests in research on seed quality, both to identify where there are critical limitations to the production of high quality seeds and in the improvement of seed quality through seed treatments such as priming. For example, seed companies may wish to determine the initial quality of their seed before deciding whether priming treatment of a seed lot is likely to be worthwhile. Treatment of low vigor or aged seed has, generally, had a greater impact on subsequent seed performance (Brocklehurst and Dearman, 1983; Basu, 1994; Goldsworthy et al., 1982). This has additional implications if there is a subsequent period of storage, since Powell et al. (2000) reported that although both an aerated hydration and PEG priming treatment enhanced the longevity of low vigor seeds, the longevity of high vigor seeds was markedly reduced after treatment.

Finally, there are some possible current applications within plant breeding. Genotypes may be selected in relation to their timing of maturity and harvest in relation to the environmental conditions for good seed production; there may be selection for morphological traits associated with high vigor, such as hypocotyl growth in soybean; and selections for seed size may take place in species in which this characteristic has been associated with high vigor.

CONCLUSIONS

Seed vigor has come a long way since its initial recognition in 1876. However, progress in understanding vigor has not followed what might be considered to be a logical route. The initial interest in vigor was in the development of tests in which it could be evaluated, which was the obvious consequence of the objectives of the AOSA and ISTA in evaluating seed quality. As a result, most of the vigor tests that were initially developed were developed empirically, with little or no theoretical background to the test. Many were modifications of the germination test or simulated the field conditions to which seed would be exposed.

Only later did researchers begin to question the physiological basis of vigor. As a result of their work, seed aging is now considered to be the major factor determining vigor in most species, although it is accepted that the fundamental cause of aging itself can be debated. In grain legumes, and possibly other dicotyledons, imbibition damage is a second major cause of vigor differences, which can have its effect alone or by interacting with seed aging. With this knowledge of the physiological basis of vigor differences, we can now take a retrospective view of the vigor tests developed over the years (see Table 21.3). In all cases, it is evident that there is a role for seed aging and/or imbibition damage as the theoretical basis of the tests. This may be through the measurement of the direct effect of aging, for example, the reduced tolerance of aged seeds to stress, as seen in the cold test and Hiltner test, or the reduced germination rate of aged seeds, as in the seedling growth rate and seedling growth tests. Alternatively, the tests may measure the consequence of aging and/or imbibition damage indirectly, as in the conductivity test and tetrazolium staining. The two fundamental determinants

TABLE 21.3. Summary of the physiological basis of seed vigor tests.

Vigor test status	Vigor test	Physiological basis of the test	Physiological expression of vigor differences in vigor test
ISTA validated tests	Accelerated aging (AA)	Seed aging	Reduced germination of low vigor seeds after AA
	Conductivity	Seed aging; imbibition damage; interaction of aging and imbibition damage	Higher solute leakage in low vigor seeds
ISTA suggested tests	Seedling growth rate	Seed aging	Slower germination of low vigor seed leads to production of smaller seedlings
	Seedling growth		
	Hiltner	Seed aging	Reduced tolerance of low vigor seeds to stress during germination
	Cold test	Seed aging; imbibitional chilling injury/imbibition damage?	Reduced tolerance of aged (low vigor) seeds to low temperature and/or greater susceptibility to imbibitional chilling injury/imbibition damage leads to reduced germination
			Germination also reduced by increased susceptibility to fungal infection when field soil is used in the test
	Cool temperature germination	Seed aging	Greater reduction in germination rate of aged seeds at low temperature, results in lower final germination germination test
	Complex stress	Seed aging	Reduced tolerance of aged (low vigor) seeds to stress results in reduced germination rate and hence reduced seedling growth at time of assessment
	Tetrazolium staining	Seed aging; physiological disorders; damage due to pests and diseases; mechanical damage	Reduced staining of injured tissues
	Controlled deterioration (CD)	Seed aging	Reduced germination of aged (low vigor) seeds after period of CD
Other tests	GADA	Aging	Reduced enzyme activity in low vigor seeds
	K_i	Aging	Reduced initial seed quality (K_i) in low vigor seeds

of seed vigor differences form the theoretical background that is important to the development of vigor tests, both in the early, empirically developed tests and in the more recently developed tests, such as accelerated aging, controlled deterioration and K_i.

Vigor tests are now widely applied. In a recent survey of seed testing laboratories (TeKrony, 2001b), 65 percent of the responding AOSA laboratories and 63 percent of the ISTA laboratories conducted vigor tests, with a total of almost 700,000 vigor tests conducted annually by the 104 laboratories. With increased demands from the agricultural and horticultural industries for high quality seeds, the requirements for assessments of seed vigor are likely to increase. There is therefore an increasing need for both national and international agreement on the standardization of vigor testing methods, their application, and interpretation.

REFERENCES

Abdul-Baki, A.A. (1980). Biochemical aspects of seed vigour. *HortScience* 15: 765-771.

Abdul-Baki, A.A. and Anderson, J.D. (1972). Physiological and biochemical deterioration of seeds. In T.T. Kozlowski (ed.), *Seed biology*, Volume II (pp. 283-315). London: Academic Press.

Abdul-Baki, A.A. and Anderson, J.D. (1973). Relationship between decarboxylation of glutamic acid and vigor in soybean seed. *Crop Science* 13: 227-232.

Abdullah, W.D., Powell, A.A., and Matthews, S. (1991). Association of differences in seed vigour in long bean (*Vigna sesquipedalis* L. Fruhw) with testa colour and imbibition damage. *Journal of Agricultural Science, Cambridge* 116: 259-264.

Abdullah, W.D., Powell, A.A., and Matthews, S. (1992). Prediction of the storage potential of long bean (*Vigna sesquipedalis* Fruhw) seed in the tropics. *Seed Science and Technology* 20: 142-147.

Akeson, W.R. and Widner, J.N. (1980). Laboratory packed sand test for measuring vigor of sugarbeet seed. *Crop Science* 20: 641-644.

AOSA (1983). *Seed vigour testing handbook.* Contribution No. 32 to the Handbook on Seed Testing. Lincoln, NE: Association of Official Seed Analysts.

Asiedu, E.A. and Powell, A.A. (1998). Comparisons of the storage potential of cultivars of cowpea (*Vigna unguiculata*) differing in seed coat pigmentation. *Seed Science and Technology* 26: 299-308.

Austin, R.B. (1972). Effects of environment before harvesting. In E.H. Roberts (ed.), *Viability of seeds* (pp. 209-252). Syracuse, NY: Syracuse University Press.

Austin, R.B. and Longden, P.C. (1965). Effects of nutritional treatments of seed-bearing plants on the performance of their progeny. *Nature* 205: 819-820.

Barla-Szabo, G. and Dolinka, B. (1988). Complex stressing vigour test: A new method for wheat and maize seeds. *Seed Science and Technology* 16: 63-73.

Barla-Szabo, G., Bosci, J., Dolinka, B., and Odiemah, M. (1990). Diallel analysis of seed vigour in maize. *Seed Science and Technology* 18: 721-729.

Baskin, C.C. and Delouche, J.C. (1971). Differences in metabolic activity in peanut seed of different size classes. *Proceedings of the Association of Official Seed Analysts* 61: 73-77.

Basu, R.N. (1994). An appraisal of research on wet and dry physiological seed treatments and their applicability with special reference to tropical and sub-tropical countries. *Seed Science and Technology* 22: 107-126.

Bedford, L.V. (1974). Conductivity tests in commercial and hand harvested seed of pea cultivars and their relation to field establishment. *Seed Science and Technology* 2: 323-335.

Bedford, L.V. and Mackay, D.B. (1973). The value of laboratory germination and weight measurements in forecasting emergence of onion and carrot seed in the field. *Journal of the National Institute of Agricultural Botany* 13: 50-62.

Bekendam, J., Kraak, H.L., and Vos, J. (1987). Studies on field emergence and vigour of onion, sugar beet, flax and maize seed. *Acta Horticulturae* 215: 83-94.

Bradnock, W.T. and Matthews, S. (1970). Assessing field emergence potential of wrinkle-seeded peas. *Horticultural Research* 10: 50-58.

Brocklehurst, P.A. and Dearman, J. (1993). Interactions between seed priming treatments and nine seed lots of carrot, celery, and onion: I. Laboratory germination. *Annals of Applied Biology* 102: 577-584.

Bruggink, H., Kraak, H.L., and Bekendam, J. (1991). Some factors affecting maize (*Zea mays* L.) cold test results. *Seed Science and Technology* 19:15-23.

Bulisani, E.A. and Warner R.L. (1980). Seed protein and nitrogen effects upon seedling vigor in wheat. *Agronomy Journal* 72: 657-662.

Burris, J.S. (1973). Effect of seed size on seedling performance of soybean: II. Seedling growth and photosynthesis, and field performance. *Crop Science* 13: 207-210.

Burris, J.S, Edje, O.T., and Wahab, A.H. (1969). Evaluation of various indices of seed and seedling vigor in soybeans, *Glycine max* (L.) Merr. *Proceedings of the Association of Official Seed Analysts* 59: 73-81.

Burris, J.S. and Navratil, T. (1979). Relationship between laboratory cold test methods and field emergence in maize inbreds. *Agronomy Journal* 71: 985-988.

Burris, J.S., Wahab, A.H., and Edje, O.T. (1971). Effects of seed size on seedling performance in soybeans: I. Seedling growth and respiration in the dark. *Crop Science* 11: 429-436.

Byrum, J.R. and Copeland, L.O. (1995). Variability in vigor testing of maize (*Zea mays* L.) seed. *Seed Science and Technology* 23: 543-549.

Caldwell, W.P. (1972). Relationship of pre-harvest environmental factors to seed deterioration in cotton. Doctoral dissertation. Mississippi State University.

Carver, M.F.F. and Matthews, S. (1975). Respiratory measurements as indicators of field emergence ability in peas. *Seed Science and Technology* 3: 871-879.

Ching, T.M. (1973a). Adenosine triphosphate content and seed vigor. *Plant Physiology* 51: 400-402.

Ching, T.M. (1973b). Biochemical aspects of seed vigor. *Seed Science and Technology* 1: 73-88.

Christiansen, M.N. and Justus, N, (1963). Prevention of field deterioration of cotton seed by an impermeable seed coat. *Crop Science* 3: 439-440.

Clark, B.E. and Little, H.B. (1955). The quality of seeds on sale in New York as revealed by tests completed in 1954. New York State Agricultural *Experimental* Station Bulletin No. 770.

Clay, W.F., Buxton, D.R., and Katterman, R.H. (1981). Cottonseed germination related to DNA synthesis following chilling stress. *Crop Science* 17: 342-344.

Cole, D.F. and Wheeler, J.E. (1974). Effect of pre-germination treatments on germination and growth of cottonseed at sub-optimum temperatures. *Crop Science* 14: 431-435.

Copeland, L.O. and McDonald, M.B. (2001). *Principles of seed science and technology* (Fourth edition). Boston: Kluwer Academic Publishers.

Cottrell, W.F. (1948). Tetrazolium salt as a seed germination indicator. *Annals of Applied Biology* 35: 123-131.

Delouche, J.C. (1980). Environmental effects on seed development and seed quality. *Horticultural Science* 15: 775-780.

Delouche, J.C. and Baskin, C.C. (1973). Accelerated ageing techniques for predicting the storability of seed lots. *Seed Science and Technology* I : 427-452.

Delouche, J.C. and Caldwell, W.P. (1960). Seed vigor and vigor tests. *Proceedings of the Official Seed Analysts* 50: 124-129.

Demir, I. and Ellis, R.H. (1992a). Changes in seed quality during seed development and maturation in tomato. *Seed Science Research* 2: 81-87.

Demir, I. and Ellis, R.H. (1992b). Development of pepper *(Capsicum annuum)* seed quality. *Annals of Applied Biology* 121: 385-399.

Dornbos, D.L. (1995). Seed vigor. In A.S. Basra (ed.), *Seed quality, basic mechanisms and agricultural implications* (pp. 45-80). Binghamton, NY: Food Products Press.

Dornbos, D.L., Mullen, R.E., and Shibles, R.M. (1989). Drought stress effects during seed fill on soybean seed germination and vigor. *Crop Science* 29: 476-480.

Draper, S. and Keefe, P.D. (1990). Seed vigour in species of field vegetables: Prediction by means of machine vision. *Project News Supplement, Spring 1990:* 2-3. Horticultural Development Council.

Duke, S.H. and Kakefuda, G. (1981). Role of the testa in preventing cellular rupture during imbibition of legume seeds. *Plant Physiology* 67: 449-556.

Eggebrecht, H. (1949). *Die Untersuchung von Saatgut.* Handbuch der Landwirtsehaftlichen Versuchs-und Untersuchungsmethodik. Band V. Hamburg, Germany: Neumann-Neudamm.

Ellis, R.H. and Pieto Filho, C. (1992). The development of seed quality in spring and winter cultivars of barley and wheat. *Seed Science Research* 2: 9-15.

Ellis, R.H. and Roberts, E.H. (1980). Toward a rational basis for testing seed quality. In P.D. Hebblethwaite (ed.), *Seed production* (pp. 605-635). London: Butterworths.

Ferguson, A.J. (1993). The agronomic significance of seed quality in combining peas (*Pisum sativum* L.). Unpublished doctoral thesis, University of Aberdeen, United Kingdom.

Ferguson Spears, J. (1995). Introduction to seed vigour testing. In H.A. van de Venter (ed.), *Seed vigour testing seminar* (pp. 1-9). Zurich: International Seed Testing Association.

Fielding, J.L. and Goldsworthy, A. (1982). The evolution of volatiles in relation to ageing in dry wheat seed. *Seed Science and Technology* 10: 277-282.

Fields, R.W. and King, T.H. (1963). Influence of storage fungi on deterioration of stored pea seed. *Phytopathology* 52: 336-339.

Finch-Savage, W.E. (1987). A comparison of seedling emergence and early seedling growth from dry sown natural and fluid-drilled pre-germinated onion (*Allium cepa* L.) seeds in the field. *Journal of Horticultural Science* 62: 39-47.

Finch-Savage, W.E. and McKee, J.M.T. (1990). The influence of seed quality and pre-germination treatment on cauliflower and cabbage transplant production and field growth. *Annals of Applied Biology* 116: 365-369.

Fontes, J.A.N. and Ohlrogge, A.J. (1972). Influence of seed size and population on yield and other characteristics of soybeans (*Glycine max* L.). *Agronomy Journal* 64: 833-836.

Franck, W.J. (1950). Introductory remarks concerning a modified working of the international rules for seed testing on the basis of experience gained after the world war. *Proceedings of the International Seed Testing Association* 16: 405-430.

Fritz, T. (1965). Germination and vigour tests of cereal seed. *Proceedings of the International Seed Testing Association* 20: 923-927.

Gabrielson, F.C., Cross, E.A., Bradshaw, D.K., and Carter, O.C. (1982). Seed size influences on germination and plant growth of *Kobe lespedeza* and other species used for surface mine revegetation. *Reclamation and Revegetation Research* 1: 271-281.

Goldsworthy, A., Fielding, J.L., and Dover, M.B.J. (1982). "Flash imbibition": A method for the re-invigoration of aged wheat seed. *Seed Science and Technology* 10: 55-65.

Goss, W.L. (1933). Twenty-five years of germination testing. *Proceedings of the Association of Official Seed Analysts* 1933: 275-278.

Grabe, D.F. (1964). Glutamic acid decarboxylase activity as a measure of seedling vigor. *Proceedings of the Association of Official Seed Analysts* 54: 100-109.

Grabe, D.F. (1965). Prediction of the relative storability of corn seed lots. *Proceedings of the Association of Official Seed Analysts* 55: 92-96.

Green, D.E., Pinnell, E.L., Cavannah, L.E., and Williams L.F. (1965). Effect of planting and maturity date on soybean seed quality. *Agronomy Journal* 57: 165-168.

Green, D.E., Cavannah, L.E., and Pinnell, E.L. (1966). Effect of seed moisture content, field weathering and combine cylinder speed on soybean seed quality. *Crop Science* 6: 7-10.

Gutormson, T.J. (1995). Soil cold test—USA perspective. In H.A. van de Venter (ed.), *Seed vigour testing seminar* (pp. 29-33). Zurich: International Seed Testing Association.

Hadavizadeh, A. and George, R.A.T. (1989). The effect of mother plant nutrition on the seed yield and seed vigour in pea (*Pisum sativum* L.) cultivar Sprite. *Acta Horticulturae* 253: 55-61.

Halder, S. and Gupta, K. (1980). Effects of storage of sunflower seeds in high and low relative humidity on solute leaching and internal biochemical changes. *Seed Science and Technology* 8: 317-321.

Hall, B. and MacNeill, M.M. (1982). A note on the influence of seed size on seedling development. *The Plantsman* 4: 58-60.

Hampton, J.G. (1992). Prolonging seed quality. *Proceedings of the 4th Australian Seeds Research Conference* pp. 181-194.

Hampton, J.G. (1995a). Conductivity test. In H.A. van de Venter (ed.), *Seed vigour testing seminar* (pp. 10-28). Zurich: International Seed Testing Association.

Hampton, J.G. (1995b). Methods of viability and vigor testing: A critical appraisal. In A.S. Basra (ed.), *Seed quality, basic mechanisms and agricultural implications* (pp. 81-118). Binghamton, NY: Food Products Press.

Hampton, J.G. (1998). Report of the Vigour Test Committee 1995-1998. *Seed Science and Technology* 26(Supplement 1): 265-269.

Hampton, J.G. and Coolbear, P. (1990). Potential versus actual seed performance—Can vigour testing provide an answer? *Seed Science and Technology* 18: 215-228.

Harman, G.E., Nedrow, B.L., Clark, B.E., and Mattick, L.R. (1982). Association of volatile aldehyde production during germination with poor soybean and pea seed quality. *Crop Science* 22: 712-716.

Harrington, J.F. (1972). Seed storage and longevity. In T. Kozlowski (ed.), *Seed biology*, Volume III (pp. 145-245). London: Academic Press.

Hegarty, T.W. (1971). A relation between field emergence and laboratory germination in carrots. *Journal of Horticultural Science* 46: 299-305.

Hegarty. T.W. (1974). Seed duality and field emergence in calabrese and leeks. *Journal of Horticultural Science* 49: 189-196.

Hegarty, T.W. (1976). Field establishment of some vegetable crops: Response to a range of soil conditions. *Journal of Horticultural Science* 51: 133-146.

Hegarty, T.W. (1977). Seed vigour in field beans (*Vicia faba* L.) and its influence on plant stand. *Journal of Agricultural Science, Cambridge* 88: 169-173.

Hepburn, H.A., Powell, A.A., and Matthews, S. (1984). Problems associated with the routine application of electrical conductivity measurements of individual seeds in the germination testing of peas and soyabeans. *Seed Science and Technology* 12: 403-413.

Heydecker, W. (1972). Vigour. In E.H. Roberts (ed.), *Viability of seeds* (pp. 209-252). London: Chapman and Hall.

Hiltner, L. and Ihssen, G. (1911). Uber das schlechte Auflaufen and die Auswinterung des Getreides infolge Befalls des Saatgutes durch Fusarium. *Landirtschaftliches. Jahrbuch fur Bayern:* 20-60: 231-278.

Hong, C.K., Han, S.K., Ree, D.W., and Kim, K.S. (1982). Effect of seed size on growth and yield of hybrid maize. *Research Report of the Office of Rural Development Suweon* 24: 188-192.

Hunter, R.B. and Kannenberg, L.W. (1972). Effects of seed size on emergence, grain yield and plant height in corn. *Canadian Journal of Plant Science* 53: 252-256.

ISTA (1995). J.G. Hampton and D.M. TeKrony (eds.), *Handbook of vigour test methods* (Third edition). Zurich: International Seed Testing Association.

ISTA (1999). International rules for seed testing. *Seed Science and Technology* 27: Supplement, Rules.

ISTA (2001). Rules amendments 2001. *Seed Science and Technology* 29, Supplement 2.

James, E. (1968). Limitations of glutamic acid decarboxylase activity for estimating viability in beans (*Phaseolus vulgaris* L.). *Crop Science* 8: 403-404.

Jianhua, Z. and McDonald, M.B. (1996). The saturated salt accelerated aging test for small-seeded crops. *Seed Science and Technology* 23: 123-131.

Joao Abba, E. and Lovato, A. (1999). Effect of seed storage temperature and relative humidity on maize (*Zea mays* L.) seed viability and vigour. *Seed Science and Technology* 27: 101-114.

Johnson, D.R. and Leudders, V.D. (1974). Effects of planted seed size on emergence and yield of soybeans (*Glycine max* L. Merr). *Agronomy Journal* 66: 117-118.

Judd, R., TeKrony, D.M., Egli, D.B., and White, G.M. (1982). Effect of freezing temperatures during soybean maturation on seed quality. *Agronomy Journal* 74: 645-650.

Lakon, G. (1950a). Die 'Triebkraft' der Samen and ihre Feststellung nach dem Topographischen Tetrazolium-Verfahren. *Snatgnt-Wirtsch* 2: 37-41.

Lakon, G. (1950b). Weitere Forschungen uber das Topographische tetrazolium-Verfahren ubd die Ermittlung der Triebkraft. *Proceedings of the International Seed Testing Association* 16: 254-261.

Legesse, N. (1991). Genotypic comparisons of imbibition in chickpea *(Cicer arietinnm)* and cowpea *(Vigna anguicalata)*. Unpublished doctoral thesis, University of Aberdeen, United Kingdom.

Legesse, N. and Powell, A.A. (1992). Comparisons of water uptake and imbibition damage in eleven cowpea culitvars. *Seed Science and Technology* 20: 173-180.

Legesse, N. and Powell, A.A. (1996). The association between the development of seed coat pigmentation during maturation of grain legumes and reduced rates of imbibition. *Seed Science and Technology* 24: 23-32.

Lewis, D.N., Marshall, A.H., and Hides, D.H. (1998). Influence of storage conditions on seed germination and vigour of temperate forage species. *Seed Science and Technology* 26: 643-655.

Loeffler, T.M. (1981). The bulk conductivity test as an indicator of soybean seed quality. Unpublished master's thesis, University of Kentucky, Lexington, Kentucky.

Loeffler, T.M., Meier, J.L., and Burris, J.S. (1985). Comparison of two cold test procedures for use in maize drying studies. *Seed Science and Technology* 13: 653-658.

Lovato, A. and Balboni, N. (1996). Seed vigour in maize (*Zea mays* L.): Two-year laboratory and field test compared. *Italian Journal of Agronomy* 1: 1-6.

Lunn, G. and Madsen, E. (1981). ATP levels of germinating seeds in relation to vigour. *Physiologia Plantarum* 53: 164-169.

Lush, W.M. and Wien, H.C. (1980). The importance of seed size in early growth of wild and domesticated cowpea. *Journal of Agricultural Science, Cambridge* 94: 177-182.

Matthews, S. (1980). Controlled deterioration: a new vigour test for crop seeds. In P.D. Hebblethwaite (ed.), *Seed production* (pp. 647-660). London: Butterworths.

Matthews, S. (1981). Evaluation of techniques for germination and vigour studies. *Seed Science and Technology* 9: 543-551.

Matthews, S. and Bradnock, W.T. (1967). The detection of seed samples of wrinkle-seeded peas (*Pisum sativum* L.) of low planting value. *Proceedings of the International Seed Testing Association* 32: 553-563.

Matthews, S. and Bradnock, W.T. (1968). The relationship between seed exudation and field emergence in peas and French beans. *Horticultural Research* 8: 89-93.

McDonald, M.B. (1975). A review and evaluation of seed vigor tests. *Proceedings of the Association of Official Seed Analysts* 65: 109-139.

McDonald, M.B. (1980). Assessment of seed quality. *HortScience* 15: 784-788.

McDonald, M.B. (1994). The history of seed vigor testing. *Journal of Seed Technology* 17: 93-100.

McDonald, M.B. (1995). Standardization of seed vigour tests. In H.A. van de Venter (ed.), *Seed vigour testing seminar* (pp. 34-52). Zurich: International Seed Testing Association.

McDonald, M.B. (1999). Seed deterioration: Physiology, repair and assessment. *Seed Science and Technology* 27: 177-237.

McDonald, M.B. and Phaneendranath, B.R. (1978). A modified accelerated aging seed vigor test for soybeans. *Journal of Seed Technology* 3: 27-37.

McDonald, M.B. and Wilson, D.O. (1979). An assessment of the standardization and ability of the ASA-610 to rapidly predict soybean germination. *Journal of Seed Technology* 54: 1-11.

McDonald, M.B. and Wilson, D.O. (1980). ASA-610 ability to detect changes in soybean seed quality. *Journal of Seed Technology* 5: 56-66.

Moore, R.P. (1973). Tetrazolium staining for assessing seed quality. In W. Heydecker (ed.), *Seed ecology* (pp. 347-366). London: Butterworths.

Moore, R.P. (1976). Tetrazolium seed testing developments in North America. *Journal of Seed Technology* 1: 17-30.

Moore, R.P. (1985). *Handbook on tetrazolium testing*. Zurich: International Seed Testing Association.

Moreno, E., Benavideo, C., and Ramirez, J. (1988). The influence of hermetic storage on the behaviour of maize seed germination. *Seed Science and Technology* 16: 427-434.

Myers, A. (1948). Hollow heart: An abnormal condition of the cotyledons of *Pisum sativum* L. *Proceedings of the International Seed Testing Association* 14: 35-37.

Nangju, D. (1977). Effect of date of harvest on seed quality and viability of soyabeans. *Journal of Agricultural Science, Cambridge* 89: 107-112.

Nangju, D. (1979). Seed characters and germination in soybean. *Experimental Agriculture* 15: 385-392.

Nijenstein, J.H. (1986). Effects of some factors influencing cold test germination of maize. *Seed Science and Technology* 14: 313-326.

Nijenstein, J.H. (1995). Soil cold test—European perspective. In H.A. van de Venter (ed.), *Seed vigour testing seminar* (pp. 34-52). Zurich: International Seed Testing Association.

Nijenstein, J.H. and Kruse, M. (2000). The potential for standardisation in cold testing of maize (*Zea mays* L.). *Seed Science and Technology* 28: 837-851.

Nkang, A. and Umoh, E.O. (1996). Six month storability of five soybean ctiltivars, as influenced by stage of harvest, storage temperature and relative humidity. *Seed Science and Technology* 25: 93-99.

Oliveira, M. de A., Matthews, S., and Powell, A.A. (1984). The role of split seed coats in determining seed vigour in commercial seed lots of soyabean, as measured by the electrical conductivity test. *Seed Science and Technology* 12: 659-668.

Perl, M. and Kretschmer, M. (1988). Biochemical activities and compounds in seeds: Possible tools for seed quality evaluation. *Annals of Botany* 62: 61-68.

Perry, D.A. (1973). Interacting effects of seed vigour and the environment. In W. Heydecker (ed.), *Seed ecology* (pp. 311-323). London: Butterworths.

Perry, D.A. (1978). Report of the vigour test committee, 1974-1977. *Seed Science and Technology* 6: 159-181.

Perry, D.A. (1987a). Seedling growth and evaluation tests. In *Handbook of vigour test methods* (Second edition) (pp. 10-20). Zurich: International Seed Testing Association.

Perry, D.A. (1987b). Topographical tetrazolium test. In *Handbook of vigour test methods* (Second edition) (pp. 57-60). Zurich: International Seed Testing Association.

Pollock, B.M. and Roos, E.E. (1972). Seed and seedling vigor. In T.T. Kozlowski (ed.), *Seed biology*, Volume 1 (pp. 314-388). NewYork: Academic Press.

Powell, A.A. (1985). Impaired membrane integrity—A fundamental cause of seed quality differences in peas. In P.D. Hebblethwaite, M.C. Heath, and T.C.K. Dawkins (eds.), *The pea crop* (pp. 383-395). London: Butterworths.

Powell, A.A. (1986). Cell membranes and seed leachate conductivity in relation to the quality of scads for sowing. *Journal of Seed Technology* 10: 81-100.

Powell, A.A. (1988). Seed vigour and field establishment. *Advances in the Research and Technology of Seeds* 11: 29-61.

Powell, A.A. (1989). The importance of genetically determined seed coat characteristics to seed quality in grain legumes. *Annals of Botany* 63: 169-175.

Powell, A.A. (1995). The controlled deterioration test. In H.A. van de Venter (ed.), *Seed vigour testing seminar* (pp. 73-88). Zurich: International Seed Testing Association.

Powell, A.A., Don, R., Haigh, O., Phillips, G., Toulon, J.H.B., and Wheaton, O.E. (1984). Assessment of the repeatability of the controlled deterioration vigour test both within and between laboratories. *Seed Science and Technology* 12: 641-647.

Powell, A.A. and Matthews, S. (1977). Deteriorative changes in pea seeds (*Pisum sativum* L.) stored in humid or dry conditions. *Journal of Experimental Botany* 28: 225-234.

Powell, A.A. and Matthews, S. (1978). The damaging effect of water on dry pea embryos during imbibition. *Journal of Experimental Botany* 29: 1215-1229.

Powell, A.A. and Matthews, S. (1979). The influence of testa condition on the imbibition and vigour of pea seeds. *Journal of Experimental Botany* 30: 193-197.

Powell, A.A. and Matthews, S. (1980). The significance of damage during imbibition to the field emergence of pea (*Pisum sativum* L.) seeds. *Journal of Agricultural Science, Cambridge* 95: 35-38.

Powell, A.A. and Matthews, S. (1981a). A physical explanation for solute leakage during imbibition of dry pea embryos. *Journal of Experimental Botany* 32: 1045-1050.

Powell, A.A. and Matthews, S. (1981b). The significance of seed coat damage in the production of high quality legume seeds. *Acta Horticulturae* 111: 227-233.

Powell, A.A. and Matthews, S. (1984a). Application of the controlled deterioration test to detect seed lots of Brussels sprouts with low potential for storage under commercial conditions. *Seed Science and Technology* 12: 649-657.

Powell, A.A. and Matthews. S. (1984b). Prediction of the storage potential of onion seed under commercial storage conditions. *Seed Science and Technology* 12: 641-647.

Powell, A.A., Matthews, S., and Oliveira, M. de A. (1984). Seed quality in grain legumes. *Advances in Applied Biology* 10: 217-285.

Powell, A.A., Oliveira, M. de A., and Matthews, S. (1986a). The role of imbibition damage in determining the vigour of white and coloured seed lots of dwarf French bean *(Phaseolus vulgaris). Journal of Experimental Botany* 57: 716-722.

Powell, A.A., Oliveira, M. de A., and Matthews, S. (1986b). Seed vigour in cultivars of dwarf French bean *(Phaseolus vulgaris)* in relation to the colour of the testa. *Journal of Agricultural Science, Cambridge* 109: 419-425.

Powell, A.A., Thornton, J.M., and Mitchell, J.A. (1991). Vigour differences on Brassica seed and their significance to emergence and seedling variability. *Journal of Agricultural Science, Cambridge* 116: 369-373.

Powell, A.A., Yule, L.J., Jing, H.-C., Groot, S.P.C., Bino, R.J., and Pritchard, H.W. (2000). The influence of aerated hydration seed treatment on seed longevity, as assessed by the viability equations. *Journal of Experimental Botany* 51: 2031-2043.

Priestley, D.A. (1986). *Seed ageing: Implications for seed storage and persistence in the soil.* Ithaca, NY: Comstock Publishing Associates.

Roberts, E.H. (1972). Storage environment and the control of viability. In E.H. Roberts (ed.), *Viability of seeds* (pp. 14-58). London: Chapman and Hall.

Roberts, E.H. (1986). Quantifying seed deterioration. In M.B. McDonald (ed.), *Physiology of seed deterioration* (pp. 101-123). Madison, WI: Crop Science Society of America.

Roberts, L.W. (1951). Survey of factors responsible for reduction of 2, 3, 5-triphenyl tetrazolium chloride in plant meristems. *Science* 113: 692-693.

Salter, P.J. (1985). Crop establishment: Recent research and trends in commercial practice. *Scientific Horticulture* 36: 32-47.

Sanhewe, A.J. and Ellis, R.H. (1996). Seed development and maturation in *Phaseolus vulgaris:* II. Post-harvest longevity in air-dry storage. *Journal of Experimental Botany* 47: 959-965.

Shieh, W.J. and McDonald, M.B. (1982). The influence of seed size, shape and treatment on inbred corn seed quality. *Seed Science and Technology* 10: 307-313.

Siddique, M.A. and Goodwin, P.B. (1980). Seed vigour in bean (*Phaseolus vulgaris* L. cv. Apollo) as influenced by temperature and water regime during development and maturation. *Journal of Experimental Botany* 31: 313-323.

Simon. E.W. and Raja Harun, R.M. (1972). Leakage during seed imbibition. *Journal of Experimental Botany* 23: 1076-1085.

Singh, J.N., Tripathi, S.K., and Negi, P.S. (1972). Note on the effect of seed size on germination, growth and yield of soybeans. *Indian Journal of Agricultural Science* 42: 83-86.

Steere, W.C., Levengood, W.C., and Bondie, J.M. (1981). An electrical analyser for evaluating seed germination and vigour. *Seed Science and Technology* 9: 567-576.

Steiner, A.M. and Werth, H. (1974). Ein Vergleich des ISTA—und AOSA—Verfahrens des Topographischen Tetrazolium Tests auf Lebensfahigkeit sowie die Tetrazoliu-Triebkraftbestimmung bei Getreide. *SAFA-Saatgutwirtschaft* 26: 536-538.

Sundstrom, F.J. (1990). Seed moisture influence on tabasco pepper seed viability, vigour and dormancy during storage. *Seed Science and Technology* 8: 179-185.

Szirtes, A.G. and Szirtes, J. (1980). Experiments on inducing physiological seed vigour. *Novenytermeles Crop Production* 29: 151-162.

Tang, S., TeKrony, D.M. Egli, D.B., Cornelius, P.L., and Rucker, M. (1999). Survival characteristics of corn seed during storage: I. Normal distribution of seed survival. *Crop Science* 39: 1400-1406.

TeKrony, D.M. (1993). Accelerated aging test. *Journal of Seed Technology* 17: 110-120.

TeKrony, D.M. (2001a). ISTA seed vigour survey—2000. *ISTA News Bulletin* 122: 14-15.

TeKrony, D.M. (2001b). Report of the Vigour Test Committee 1998-2001. *Seed Science and Technology* 29 (Supplement 1): 259-263.

TeKrony, D.M. and Egli, D.B. (1991). Relationship of seed vigor to crop yield: A review. *Crop Science* 31: 816-822.

TeKrony, D.M. and Egli, D.B. (1997). Accumulation of seed vigour during development and maturation. In R.H. Ellis, M. Black, A.J. Murdoch, and T.D. Hong (eds.), *Applied aspects of seed biology,* Proceedings of the Fifth International Workshop on Seeds (pp. 369-384). Dordrecht, the Netherlands: Kluwer Academic Publishers.

TeKrony, D.M., Egli, D.B., and Phillips, A.D. (1980). Effect of field weathering on the viability and vigor of soybean seed. *Agronomy Journal* 72: 749-753.

Thornton, J.M. and Powell, A.A. (1992). Short term aerated hydration for the improvement of seed quality in *Brassica olerecea* L. *Seed Science Research* 2: 41-49.

Tomes, L.J. (1985). The use of accelerated aging as a vigor test in soybeans. Unpublished master's thesis, University of Kentucky. Lexington, Kentucky.

Tomes, L.J., TeKrony, D.M., and Egli, D.B. (1988). Factors influencing the tray accelerated aging test for soybean seed. *Journal of Seed Technology* 12: 37-53.

Vieira, R.D., TeKrony, D.M., and Egli, D.B. (1992). Effect of drought and defoliation stress in the field on soybean seed germination and vigor. *Crop Science* 32: 471-475.

Wahab, A.H. and Burris, J.S. (1971). Physiological and chemical differences in low and high quality soybean seeds. *Proceedings of the Association of Official Seed Analysts* 61: 57-58.

Walters, C. (1998). Understanding the mechanisms and kinetics of seed ageing. *Seed Science Research* 8: 223-244.

Wang, Y.R. and Hampton, J.G. (1991). Seed vigour and storage in 'Grasslands Pawera' red clover. *Plant Varieties and Seeds* 4: 61-66.

Wheeler, T.R. and Ellis, R.H. (1991). Seed quality, cotyledon elongation at suboptimal temperatures, and the yield of onion. *Seed Science Research* 1: 57-67.

Wilson, D.O. and McDonald, M.B. (1986). A convenient volatile aldehyde assay for measuring soybean seed vigor. *Seed Science and Technology* 14: 259-268.

Wilson, D.O., McDonald, M.B., and St. Martin, S.K. (1989). A probit planes method for analyzing seed deterioration data. *Crop Science* 29: 471-476.

Woodruff, J.M., McCain, F.S., and Hoveland, C.S. (1967). Effect of relative humidity, temperature and light intensity during boll opening on cotton seed quality. *Agronomy Journal* 59: 441-444.

Yaklich, R.W., Kulik, M.M., and Anderson, J.D. (1979). Evaluation of vigor tests in soybean seeds: Relationship of ATP, conductivity, and radioactive tracer multiple criteria tests to field performance. *Crop Science* 19: 806-810.

Zanakis, G.N., Ellis R.H., and Summerfield, R.J. (1994). Seed quality in relation to seed development and maturation in three genotypes of soyabean (*Glycine max*). *Experimental Agriculture* 30: 139-156.

Chapter 22

Diagnosis of Seedborne Pathogens

Emily Taylor
Jayne Bates
David Jaccoud

INTRODUCTION

The importance of seed as a potential source of inoculum of plant pathogens was reported over 200 years ago in wheat seed (Neergaard, 1977; Noble, 1979). Diagnosis of seedborne diseases is essential for a number of reasons. First, seed certification schemes require a measure of good quality seed, as indicated by the absence of pathogens; also, if disease is present, information about inoculum level can help to make accurate decisions regarding the appropriate use of seed treatments. In addition, diagnosis of seed pathogens for quarantine purposes helps to avoid the spread of disease to new regions (Maddox, 1998). The international seed trade has increased the growing problem of seed as a source for the dissemination of important plant pathogens. To prevent the spread of diseases that may have a devastating impact on agricultural production, scientists have identified important seedborne pathogens for each country, and the appropriate phytosanitary measures have been introduced (Richardson, 1996). The most effective ways of controlling seedborne disease are through plant breeding and seed health testing. Seed health tests are also a valuable tool for investigating the issues related to seedborne disease development.

At present there is widespread prophylactic use of seed treatments in the United Kingdom, Germany, and France, despite the fact that this confers

This research was supported by DEFRA and the HGCA. The authors would like to thank Vincent Mulholland (SASA, Scotland) for the *M. nivale* primers, which he designed and tested. The Scorpion primers were designed, synthesized, and tested by Oswel DNA service (Southampton, United Kingdom). The technical assistance of the staff in the Seeds Diagnostic Unit at NIAB for maintenance and preparation of single spore fungal isolates is also appreciated. Assistance with the figures was done with excellent help from Cathy Carlin at NIAB.

little or no benefit when applied to uninfected seed. There is, therefore, scope to reduce fungicide usage for environmental and financial reasons. Scandinavian countries have implemented agricultural policies to minimize the use of seed treatments and apply them only where absolutely needed (Brodal, 1999). To make informed management decisions about the appropriate use of seed treatment, it is important to understand the development of seedborne disease, including the amount of seedborne inoculum, extent of inoculum transmission from seed to seedling, rate of disease increase in the crop. and the amount of reestablishment of seedborne inoculum (Paveley et al., 1997).

Most seedborne pathogens cannot be detected by naked-eye examination and contaminate the seed by invisible spores or presence of mycelial infection. Many traditional and modern methods are available for detection of pathogens in different crops, and new ones continue to be developed. The method selected for a seed health test is not necessarily the most modern or fashionable. The main criteria are that the test must be specific, sensitive, accurate, reproducible, rapid, easy to perform, and cost-effective. Results should be repeatable within and between samples (within limits of sampling error) and the test sensitive enough to detect tolerated thresholds. Economics dictate that the cost of the test must be appropriate for the importance of the crop and the ability of the producers to pay. This means that the cost benefit of routine seed health testing compared with the cost of seed treatment or potential loss of yield is balanced.

In addition to economic factors, the seed health method selected for the detection of a seedborne pathogen depends on the type of pathogen and the association between the pathogen and the seed (Maude, 1996; Neergaard, 1977). Current seed health testing is based on standardized methods described by International Seed Testing Association (ISTA) working sheets. The size of the sample to be analyzed in a seed health test is determined by the epidemiological pattern of a specific pathogen. Important factors include the disease-causing threshold, the level of its transmission by the seeds, and the sensitivity of the method. In general, the majority of the standard ISTA methods are performed with a working sample of between 200 and 1,000 seeds selected randomly from a seed sample.

Conventional diagnostic methods include agar plate tests, seedling bioassay, and microscopic observation. More recent nucleic acid–based assays have an advantage in that they are highly specific, sensitive, and rapid with potential to being automated, leading to high throughput (Reeves, 1995). However, only a limited number of these have been developed into commercial seed health tests (e.g., Bates et al., 2001). This may be partly due to technical difficulties, cost of instrumentation, PCR (polymerase chain reaction) license and royalties, as well as the time and organization needed for test standardization.

A knowledge-based strategy to reduce seed treatment use based on seed health testing will increase the demand for seed tests. For some crops e.g., winter cereals, there is a very short period between harvest and sowing of the following crop. New molecular diagnostic technologies offer the potential for automated high throughput screening with sensitive and specific detection of pathogens, often with detection of more than one pathogen in a single test by multiplex PCR and/or the use of fluorescent probes or primers. Sensitivity can be such that infection can be detected long before visual symptoms appear. This chapter describes the wide range of conventional and molecular techniques used for seed diagnostics with examples from the literature and our own laboratories.

CONVENTIONAL SEED HEALTH DIAGNOSTIC METHODS

Direct Visual Examination

This is the most basic examination method among those available. The examination is performed with no equipment, relying on the macroscopic structures of the pathogen or typical symptoms exhibited. In some cases, the seedborne pathogen structures can be found mixed with the seed and not fixed on the surface. Classical examples of this method are for the detection of sclerotia from *Claviceps purpurea,* the causal agent of wheat ergot, and *Sclerotinia sclerotiorum,* which is associated with soybean, black bean, or sunflower seeds.

Fruiting bodies of some seedborne pathogens can be associated with crop debris from the host plant as well as with the seeds, for example, the presence of perithecia from *Gibberella zea* (teleomorphic stage of *Fusarium graminearum* Schwabe), the agent of scab of wheat (see Photo 22.1).

Other symptoms—including the presence of spots (e.g., black point in wheat caused by *Bipolaris sorokiniana;* black powdery mass (e.g., infected wheat seeds by teliospores of *Tilletia indica* Mitra); shrivelling and discoloration (e.g., light pink regions in wheat seeds infected by *Fusarium graminearum* [see Photo 22.1]); black or brown spots caused by *Colletotrichum lindemuthianum* in beans (see Photo 22.2); a purple spot *(Cercospora kikuchii)* on the soybean seed coat (Photo 22.3); and white lines on the maize seed coat (Photo 22.4)—are examples of how direct examination can help the seed health quality assessment (Machado et al., 2002; Neergaard, 1977).

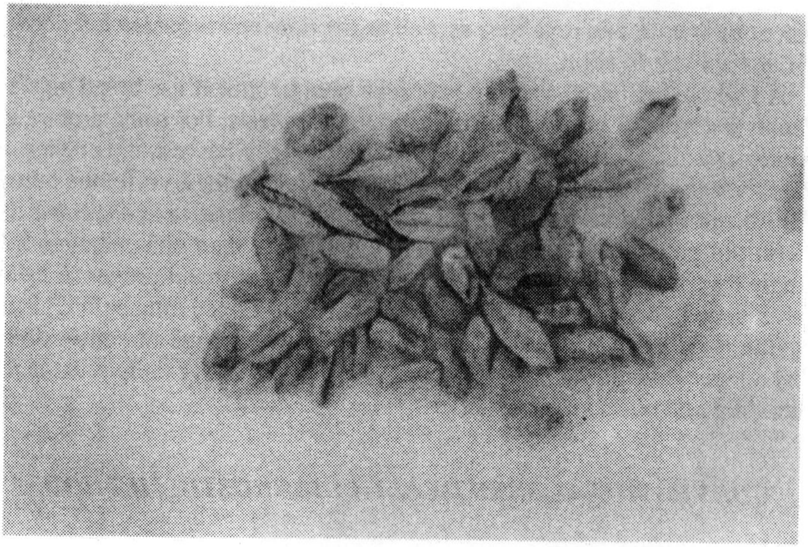

PHOTO 22.1. Wheat seeds infected by *Gibberella zeae (Fusarium graminearum)*, showing a light purple color and also black bodies indicative of the presence of perithecia.

PHOTO 22.2. Bean seeds infected by *Colletotrichum lindemuthianum*. Brown spots can be seen over the seed coat.

PHOTO 22.3. Soybean seeds with purple spots caused by the infection of the seedborne fungus *Cercospora kikuchii.*

PHOTO 22.4. Maize seeds infected by *Fusarium moniliforme,* showing characteristic white lines on the seed coat from mycelial growth of this pathogen.

Sediment Suspension Examination

This technique is carried out with pathogens that produce propagules or other structures that normally are superficially located on the seed coat. The inspection of a seed wash suspension using a microscope will allow the detection of these superficial pathogen structures. The sediment suspension examination can detect spores from different fungal sources, such as oospores of the genus *Peronospora manshurica* in infected soybean (Machado, 2002a), spores of *Sphacelotheca reiliana* in maize (McGee, 2002), *Tilletia caries* (Rennie, 2002) and *Tilletia indica* on wheat, and the nematode *Aphelenchoides bessey* in rice.

Staining Methods

Staining methods are normally used to detect fungi that cannot be grown in artificial media, i.e., they are obligate parasites. Spores of these seedborne fungi, such as the agents of "loose smut" *(Ustilago tritici)* on barley or "downy mildew" *(Peronosclerospora sorghi)* on maize (Maude, 1996), can be detected by this method.

Growing-Out Test

This method relies on planting seeds in field plots or in boxes normally filled with sterile sand. The pathogen of interest is allowed to develop during seedling growth. After a determined period, the seedlings are examined for symptoms produced by seedborne fungi such as *Colletotrichum lindemuthianum* and *Rizoctonia solani* in beans. This test is also used to the detection of seedborne bacteria and fungi (Maude, 1996).

Blotter Test

This method is one of the most commonly used in seed pathology laboratories worldwide. It is reliable, cheap, and easy to perform but demands a high level of technical expertise from the seed analyst. The production of characteristic mycelia, fruiting bodies, or symptoms of the disease of interest on the seedlings indicates that the seed sample is infected.

The seeds are normally incubated on two or more damp blotting papers placed inside a plastic container called a "gerbox" (Photo 22.5) or in petri dishes. The period of incubation, temperature, light, and moisture conditions are determined by the type of fungi (Neergaard, 1977; Maude, 1996; Machado et al., 2002; Machado and Langerak, 2002). Where suppression

PHOTO 22.5. Soybean seeds incubated in a gerbox under appropriate light and temperature conditions for the identification of seedborne pathogens.

of seed germination is required, so that better observation of the symptoms may be observed, this can be achieved by freezing the sample for a short time or using 2,4-D (2,4-dichlorophenoxyacetic acid). More recently, a water restriction technique has been used successfully for the same purpose (Machado, 2002a).

Roll Paper Method

This method is primarily used to detect the fungus *Colletotrichum lindemuthianum* in bean seeds (Machado, 2002b). After surface sterilization using sodium hypochlorite (1 percent), the seeds are evenly distributed over two moistened blotter paper sheets (Photo 22.6) and covered by another sheet of paper. The papers, with the seeds, are rolled to form a tube and incubated for approximately 8 days inside a black plastic bag in a germination chamber at 20°C. After this incubation period, fungi can be found on infected bean seeds as a characteristic dark and depressed lesion on the cotyledons (Photo 22.7). The presence of *Rhizoctonia solani* also can be observed as a reddish brown lesion produced over the hypocotyledon (Photo 22.7).

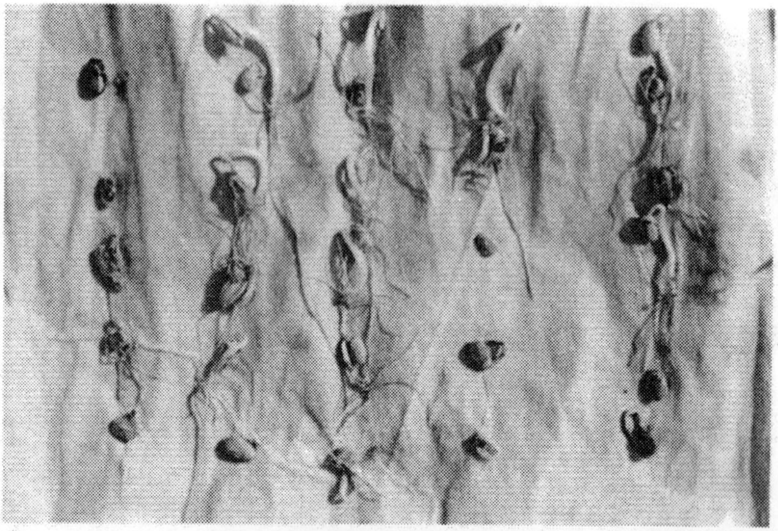

PHOTO 22.6. A general view of bean seeds incubated by using the "roll paper" method for the identification of seedborne pathogens.

PHOTO 22.7. Bean seeds after an incubation period using a "roll paper" method showing a black and depression lesion from *Colletotrichum lindemuthianum* on the seed coat. There is a brown lesion caused by *Rhizoctonia solani* which can be observed on the hypocotyl.

Agar Method

The agar plate method is a highly recommended and widely used method for the detection and identification of a range of seedborne pathogens of many crops. This method gives an indication about the viability of the pathogen inoculum in a seed sample. In general, the seeds are surface sterilized and distributed under aseptic conditions in petri dishes usually containing PDA (potato dextrose agar) or other specific media. The plates are then incubated for a specific photoperiod regime, which is, in general, alternating dark with 12 hours under NUV (near-ultraviolet light) or fluorescent light each day at 20°C for 7 to 8 days in order to allow the development of the characteristic growth morphology of the pathogen (Langerak, 2002; Neergaard, 1977; Maude, 1996; Machado et al., 2002).

Selective or Semiselective Media

This is used for the detection of bacteria. The first step requires their extraction from the seed sample. The bacteria are then cultured using selective media prior to the use of specific tests for their identification (Neergaard, 1977; Maude, 1996; Schaad, 1997). Alvarez and Kaneshiro (1999) reported a modification of the agar plating test using semi-selective media for the detection of *Clavibacter michiganensis* subsp. *michiganensis* in tomato seeds. Other comparative tests to detect bacteria in infected seeds were reported by the ISTA/PDC Bacteriology Working Group (Schaad, 1997), such as the agar plating test for *Pseudomonas syringae* pv. *phaseolicola* in *Phaseolus vulgaris; Xanthomonas campestris* pv. *translucens* in Poaceace, and *Xanthomonas campestris* pv. *campestris* in Cruciferae (Koenraadt, 1997). A selective medium method to detect the seedborne bacteria *Clavibacter michiganensis* ssp. *nebraskensis* causal agent of the "Goo's bacterial wilt and blight" in corn in the United States was reported by Shepherd et al. (1999).

Flow Cytometry

Flow cytometry is a technique in which characteristic parts of biological particles or cells are identified when passed individually through a sensor in a liquid stream. The particles are identified by fluorescent dyes and light scattering (Chitarra et al., 2002). They detected the plant pathogenic bacteria *Xanthomonas campestris* pv. *campestris,* the agent of "black rot" in seed extracts of *Brassica* species, using fluorescent antibodies and flow cytometry. They concluded that flow cytometry could be used for the detection and quantification of *X. campestris* pv. *campestris* bacterial cells by labelling with a mixture of specific FITC-labelled monoclonal antibodies in crude seed extracts.

Immunological Seed Health Techniques

Immunological techniques for the detection of seedborne pathogens have been widely used for several decades (Maude, 1996). These techniques are based on a result of a binding reaction between an antigen, which is found in the pathogen of interest, and a specific antibody.

The most widely used immunological technique in seed diagnostics is double antibody sandwich enzyme-linked immunosorbent assay (DAS-ELISA). A primary or "capture" antibody is bound to the wells of a microtitre plate or a membrane in a lateral flow device or dipstick kit. A few drops of infected seed soak/wash extract are added to the microtitre plate or lateral flow device, and any target antigens present are bound to the capture antibody. Unbound seed soak/wash extract is removed by a washing step. A second antibody, which is covalently linked to an enzyme, is then added to the sample. The second antibody also binds to the target antigen, forming an antibody-antigen-antibody sandwich. A second washing step is used to remove any further unbound material. Finally, a substrate is added that reacts with the enzyme-linked antibody to produce a color change reaction which can be quantified by spectrophotometry.

DAS-ELISA is most commonly used for identifying bacterial and viral infections that are not possible to detect by microscopic analysis. Immunological techniques such as DAS-ELISA may not be as sensitive or specific as molecular techniques, but they are relatively easy to perform and do not require much expensive equipment except for a spectrophotometer. Lateral flow devices and dipstick diagnostic kits provide a rapid on-site detection system as an alternative to a laboratory test (Figure 22.1).

Immunocapture-PCR combines molecular and immunological techniques to provide a specific and sensitive test method that has been successfully used for detection of seedborne viruses (Nolasco et al., 1993). A capture antibody is bound to magnetic beads and used to concentrate the pathogen in a seed soak/wash. This is accomplished by using a magnet to separate the antibody bound to the target organism from the seed liquor. This step also is efficient for the removal of PCR inhibitors that are contained within the seed liquor. A PCR can then be performed on the concentrated sample.

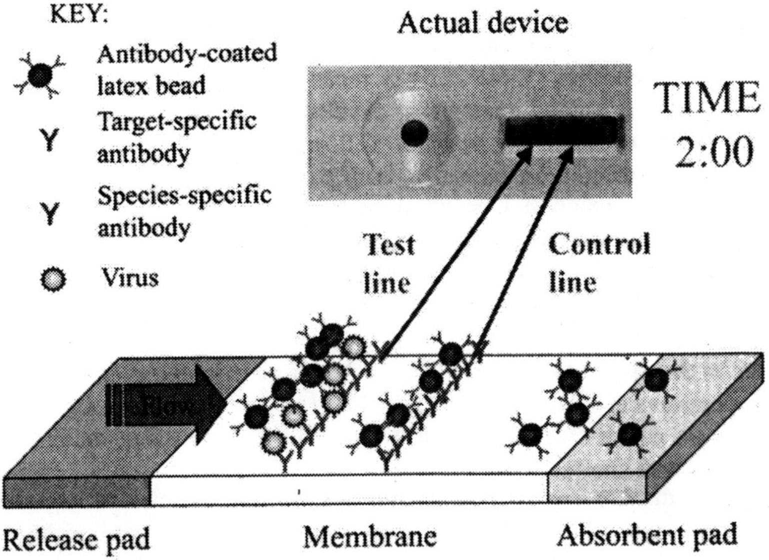

FIGURE 22.1. Schematic diagram of the one-step lateral flow assay (Pocket diagnostic) developed at CSL, York, United Kingdom. (*Source:* Image kindly provided by Dr. Chris Danks, CSL, York. United Kingdom, Crown Copyright.)

MOLECULAR METHODS FOR SEED HEALTH TESTING

Seed Health Testing by Conventional PCR

For detection of seedborne pathogens using PCR-based methods, some DNA sequence information of the pathogen of interest is required. If no prior sequence information is available in the nucleic acid databases (e.g., GenBank and EMBL), there are numerous ways of obtaining sequence data that can be used for primer design. These include screening for genetic variation using random amplified polymorphic DNA (RAPD) PCR products (e.g., Nicholson et al., 1996; Taylor et al., 2001) and sequencing internal transcribed spacer (ITS) regions of the ribosomal genes using universal primers (e.g., Konstantinova et al., 2002; Willits and Sherwood, 1999). Sequences of the ITS1 and ITS2 regions can often show variation even between very closely related species (Stevens et al., 1998). Primers can be designed to be extremely specific and amplify a single pathogen from other very closely related species (Bates and Taylor, 2001), or more general to amplify a range of organisms (Willits and Sherwood, 1999).

Numerous molecular markers have been developed for conventional PCR seed diagnostics (Table 22.1). A recent paper by Konstantinova et al. (2002) provides a good example of developing a highly specific conventional PCR seed health test based on ITS sequence differences. The genus *Alternaria* is ubiquitous, including both pathogenic and saprophytic species, several of which may occur on the same host. A highly specific test is required to differentiate these closely related species, as they may differ in pathogenicity and/or require different seed treatments. Conventional methods rely on lengthy incubations followed by morphological identification of spores, which can lead to misidentification. Using sequence differences in the rDNA ITS regions of *A. alternata, A. radicina,* and *A. dauci,* Konstantinova et al. (2002) were able to differentiate the three species isolated from carrot seed with results comparable to existing methods but within a much shorter time.

Although PCR can provide a rapid test for the presence of any single plant pathogen, for some seed diagnostic tests it is necessary to quantify the level of seed inoculum to assess the need for seed treatment. Using conventional PCR methods, this can be achieved using competitive PCR. In this type of assay, a constant amount of "competitor" DNA is added to each PCR, and quantification of the pathogen DNA is achieved by comparing the ratio between the amounts of competitor and pathogen products amplified. The products must be of different size so as to be separated by agarose gel electrophoresis; however, amplification kinetics of the competitor must be as similar as possible to the DNA of interest. Nicholson et al. (1996) have used this method to quantify *Microdochium nivale* infection in wheat seedlings. They designed the competitor template such that it contained 5' and 3' termini identical to the full-length PCR product but on amplification produced a truncated version. Some optimization of this method is required to determine the optimal amount of competitor to include in the PCR.

The main advantage of using conventional PCR assays over traditional seed testing methods is the reduction in time taken, often with improved sensitivity, specificity, and reliability. Disadvantages include the need to run agarose gels with the potential for cross-contamination between samples and experiments. There are also limitations in quantification, as conventional competitive PCR is based on end-point analysis which reduces the dynamic range of sample concentrations that can be calculated. The emergence of real-time PCR in plant pathogen diagnostics is now beginning to remove many of these disadvantages.

TABLE 22.1. Molecular markers for conventional and real-time PCR.

Seed/tuber-borne pathogen	PCR protocol	Primer sequence origin	Reference
Alternaria alternata	Conventional PCR	ITS	Konstantinova et al., 2002
Alternaria brassicae	Conventional PCR	ITS	Iacomi-Vasilescu et al., 2002
Alternaria brassicola	Conventional PCR	ITS	Iacomi-Vasilescu et al., 2002
Alternaria dauci	Conventional PCR	ITS	Konstantinova et al., 2002
Alternaria japonica	Conventional PCR	ITS	Iacomi-Vasilescu et al., 2002
Alternaria radicina	Conventional PCR	ITS	Konstantinova et al., 2002
Ascochyta rabiei	Conventional PCR/RFLP	ITS	Phan et al., 2002
Colletotrichum gloeosporiodes	Conventional PCR	ITS	Mills et al., 1992
Diaporthe phaseolorum	Real-time QPCR/TaqMan	ITS	Zhang et al., 1999
Fusarium avenaceum	Conventional PCR	ITS 2	Schilling et al., 1996
Fusarium culmorum	Competitive QPCR	RAPDs	Nicholson et al., 1998
Fusarium graminearum	Real-time PCR	tri5 gene	Schnerr et al., 2001
Fusarium moniliforme	Conventional PCR	Genomic clones	Murillo et al., 1998
Fusarium poae	Conventional PCR	RAPDs	Parry and Nicholson, 1996
Fusarium spp.	Conventional PCR	RAPDs	Nicholson et al., 1996
Leptosphaeria maculans	Competitive QPCR	ITS 1 and 2	Mahuku et al., 1995
Leptosphaeria maculans	Conventional PCR	RAPDs	Voigt et al., 1998
Microdochium nivale	Competitive QPCR	RAPDs	Nicholson et al., 1996
Microdochium nivale	Real-time QPCR/ SYBR Green I	28 S rRNA	Authors, unpublished
Phomopsis longicolla	Real-time QPCR/TaqMan	ITS	Zhang et al., 1999
Phytophthora infestans	Conventional/Nested PCR	Repetitive sequences	Niepold and Schober-Butin, 1997
Phytophthora infestans	Conventional PCR	ITS	Trout et al., 1997
Phytophthora infestans	Real-time QPCR/TaqMan	Nuclear satellite DNA	Bohm et al., 1999
Phytophthora infestans	Conventional PCR	ITS 2/RAPDs	Kim and Lee, 2001
Phytophthora spp.	Conventional PCR	ITS	Ristaino et al., 1998
Pyrenopeziza brassicae	Conventional PCR	MAT-1/MAT-2	Foster et al., 1999
Pyrenophora graminea	Real-time PCR/ SYBR Green I	RAPDs	Taylor et al., 2001
Pyrenophora teres	Real-time QPCR/ SYBR Green I	ITS	Bates et al., 2001
Pyrenophora teres	Real-time QPCR/Scorpion primers	ITS	Bates and Taylor, 2001
Stagonospora nodorum	Conventional PCR	ITS	Beck and Ligon, 1995
Tilletia barclayana	Conventional PCR	ITS/mtDNA	McDonald et al., 1999
Tilletia controversa	Conventional PCR	ITS/mtDNA	McDonald et al., 1999
Tilletia indica	Conventional PCR	mtDNA	Ferreira et al., 1996
Tilletia laevis	Conventional PCR	ITS/mtDNA	McDonald et al., 1999
Tilletia tritici	Conventional PCR	ITS/mtDNA	McDonald et al., 1999

TABLE 22.1 *(continued)*

Seed/tuber-borne pathogen	PCR protocol	Primer sequence origin	Reference
Ustilago hordei	Conventional PCR	ITS	Willits and Sherwood, 1999
Ustilago nuda	Real-time QPCR/ Hyb probes	ITS	Authors, unpublished
Clavibacter michiganensis subsp *sepedonicus*	Real-time-TaqMan-BIO-PCR	Genomic DNA	Schaad et al., 1999
Clavibacter michiganensis subsp *sepedonicus*	Conventional mPCR	ITS	Pastrik, 2000
Pseudomonas syringae pv. *phaseolicola*	BIO-PCR/Nested PCR	tox gene	Schaad et al., 1995
Ralstonia solanacearum	Conventional PCR	16 S rDNA/ITS	Seal et al., 1999
X. axonopodis pv. *manihotis*	Conventional/Nested PCR	Pathogenicity gene	Ojeda and Verdier, 2000
X. campestris pv. *campestris*	Conventional PCR	GumD gene	Authors, unpublished
X. campestris pv. *phaseoli* var. *fuscans*	Conventional PCR	RAPDs	Toth et al., 1998
Potato leafroll virus	Real-time RT-PCR/ Mol. beacons	Coat protein	Klerks et al., 2001
Potato leafroll virus	RT-PCR	Capsid gene	Singh et al., 1995
Potato mop top virus	RT-PCR/TaqMan	Coat protein	Mumford et al., 2000
Potato virus A	RT-PCR	NIb gene	Cerovska et al., 1998
Potato virus Y	RT-PCR-ELOSA	Coat protein	Chandelier et al., 2001
Potato virus Y	Real-time RT-PCR /Mol. beacons	Coat protein	Klerks et al., 2001
Potato virus Y	CF RT-PCR	Coat protein	Walsh et al., 2001
Tobacco rattle virus	RT-PCR/TaqMan	RNA 1	Mumford et al., 2000

Abbreviations: QPCR = quantitative PCR; RT-PCR = reverse transcriptase PCR; CF RT-PCR = competitive fluorescent RT-PCR; ELOSA = enzyme-linked oligosorbent assay.

Seed Health Testing by Real-Time PCR

Real-time PCR is the direct detection of PCR amplicons using either a double stranded DNA-binding fluorescent dye (e.g., SYBR Green I) or a specific fluorescent probe (e.g., LightCycler hybridization probes, TaqMan probes, Molecular beacons, etc.). The methods of quantifying nucleic acids by competitive PCR (e.g., Nicholson et al., 1996) are based on end-point analysis where PCR amplification has (or has almost) reached a plateau phase (Figure 22.2). Real-time PCR instruments are able to continually monitor the amount of product throughout the log-linear phase of amplification, which is the most informative part of the PCR. Quantification of target DNA in unknown samples can be achieved by direct comparison to

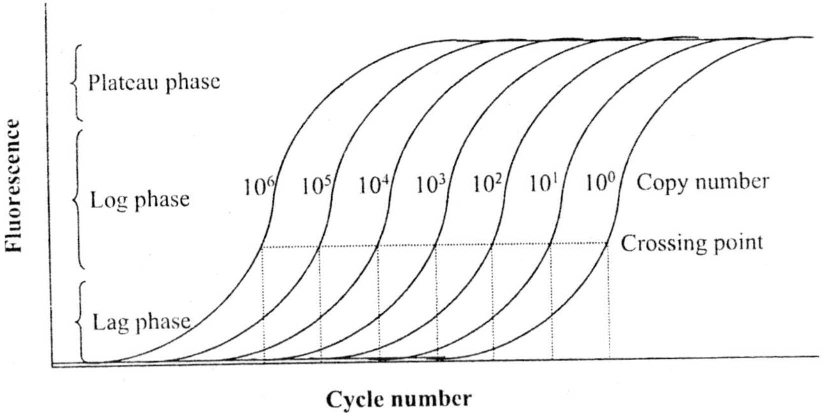

FIGURE 22.2. Schematic profile of quantitative real-time PCR. Amplification is detected by fluorescence of a specific probe or reporter dye. The first few cycles show a lag phase, where product can just be detected above background. This is followed followed by a log-linear phase of amplification. It is at this stage where there is greatest discrimination between samples of different starting target concentration (copy number). Toward the final few cycles, a plateau phase is reached as the PCR reagents become limiting. This is the point measured by conventional PCR using agarose gel electrophoresis and there is little discrimination between samples with low or high starting target concentration.

standards amplified in parallel reactions (Ririe et al., 1997). A fluorescent reporter dye such as SYBR Green I, which binds double-stranded DNA, is included in the PCR mix. Fluorescence of the dye increases with the amount of product amplified and is recorded once per cycle. A standard curve can be generated by plotting the known concentrations of the standard samples against the cycle threshold (the cycle at which fluorescence of the sample rises significantly above background and indicates the start of amplicon detection). In some rapid cycle instruments, such as the LightCycler (Roche Diagnostics Limited, Lewes, United Kingdom), the PCR and analysis can be performed in less than 30 minutes due to (1) rapid heat exchange in the sample capillaries and (2) no need for post-PCR analysis such as electrophoresis, giving distinct advantages over conventional methods. This closed-tube system also reduces the risk of cross-contamination.

In our laboratory we have applied this technique to the quantification of a number of cereal pathogens in an attempt to shorten the time required for advisory testing (Taylor et al., 2001; Bates et al., 2001, 2002). Conventional agar plate test methods for cereal fungal pathogens take about one week to complete, whereas a PCR test can be performed within 24 hours. Figure 22.3

shows the quantification of *Microdochium nivale* DNA extracted from infected wheat seed. The DNA used as the standard was extracted from fungal cultures and the concentration measured by a fluorometric assay before dilution to appropriate concentrations. Pathogen DNA from infected seed was extracted using a CTAB method (Dr. S. Edwards, Harper Adams Agricultural College, United Kingdom, personal communication). PCR was carried out using primers specific for *M. nivale* designed by Dr. V. Mulholland of the Scottish Agricultural Science Agency, Edinburgh, United Kingdom. Figure 22.3, part A shows the fluorescence curves of the four standards (ranging from 2 pg to 2 ng DNA) and three seed samples of 0.5 percent, 6.6 percent, and 8.6 percent infection with *M. nivale;* Figure 22.3, part B, shows the resulting standard curve from which the template concentration of the unknowns is calculated. We are currently carrying out a number of experiments to correlate the amount of DNA extracted from the infected seeds and the level of infection, as measured by the agar plate test.

An added feature of real-time detection is the ability to measure the melting profile of the PCR products (Ririe et al., 1997) (Figure 22.3, part C). At the end of the cycling reactions, the products are gradually melted by increasing the temperature while continually measuring the fluorescence. The temperature at which the products melt is determined by their sequence composition and length and is characteristic of the product. It provides a check that specific amplification has occurred. Several real-time PCR instruments are also able to detect dyes of different emission spectra simultaneously, so there is the possibility of developing multiplex PCRs to measure more than one pathogen in a single reaction (Woo et al., 1997). This would further increase sample throughput and become more cost-effective.

Real-Time PCR Detection Using Specific Fluorescent Probes

SYBR Green I is a nonspecific dye for detection of product amplification, hence nonspecific products such as primer-dimers will also be detected. As indicated previously, a number of different types of fluorescent probes are available that are designed to hybridize with the amplified sequence and therefore directly detect the product of the specific primers used for amplification.

TaqMan probes are dual-labelled fluorogenic probes with a reporter dye at the 5' end (e.g., 6-carboxy-fluorescein, FAM) and a quencher (e.g., 6-carboxy-tetramethylrhodamine, TAMRA) at the 3' end (Figure 22.4, part A). The 3' end is also phosphorylated to prevent extension by *Taq* DNA polymerase. The proximity of the quencher to the reporter dye blocks fluorescence in the intact probe. The probe will hybridize to specific product as

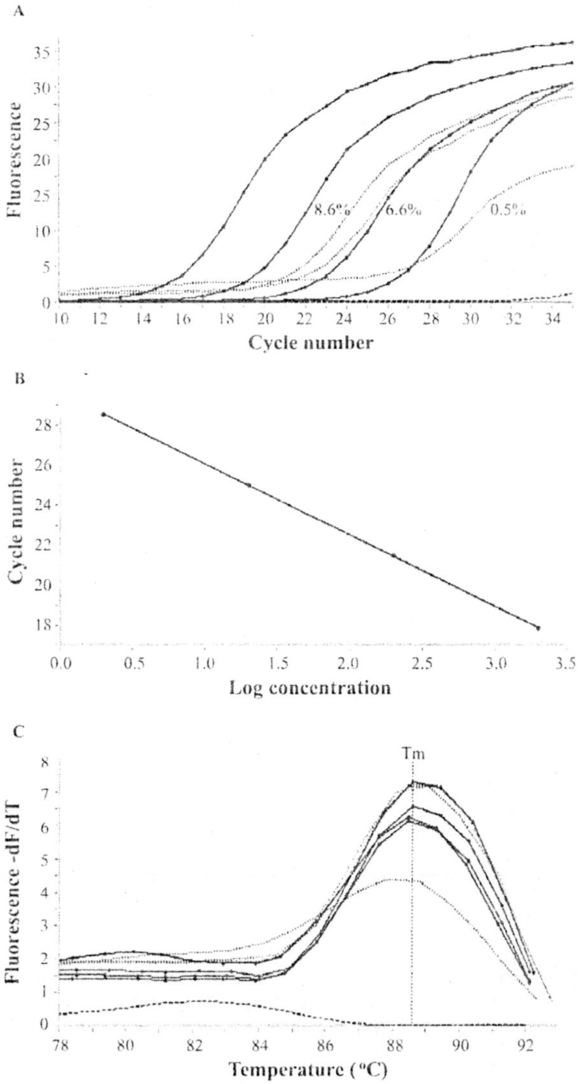

FIGURE 22.3. Quantitative PCR of *Microdochium nivale* extracted from wheat. A shows amplification of four standards (lines with points) and three samples extracted from infected wheat seed. B shows the standard curve derived from plotting the cycle number at the chosen crossing point (see Figure 22.2) against the concentration of target in the standard samples. C shows the melting curves of the products.

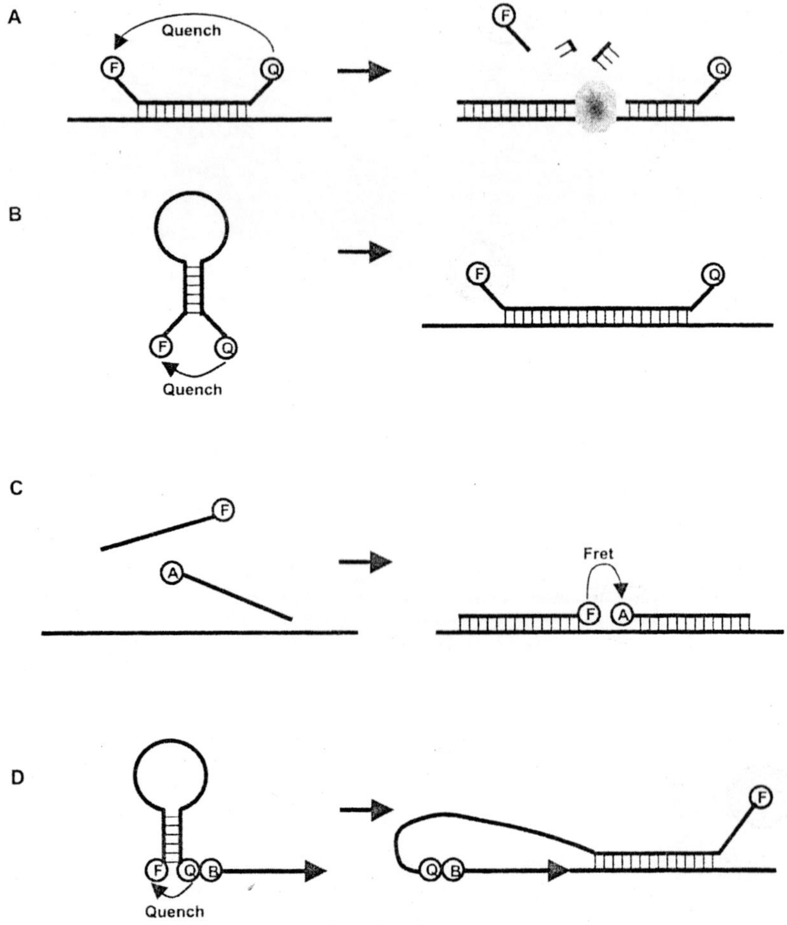

FIGURE 22.4. Fluorescent probes used in real-time PCR. A: TaqMan probes; B: Molecular beacons; C: LightCycler Hybridization probes; D: Scorpion primers. Further details are given in the text. F = fluorophore; Q = quencher; A = acceptor; B = PCR blocker.

it is amplified; a signal is produced during the following amplification cycle as the 5'-exonuclease activity of *Taq* digests the probe, releasing the quencher from the reporter dye. This type of probe has been used to detect *Phytophthora infestans* in potato tubers (Böhm et al., 1999) as well as *Diaporthe phaseolorum* and *Phomopsis longicolla* from soybean seeds (Zhang et al., 1999), among others.

Molecular beacons are similar to TaqMan probes in that they are dual labelled in a similar manner; however, they also contain self-complementary arm sequences at the 5' and 3' ends unrelated to the target sequence (Figure 22.4, part B). These arms hold the molecular beacon in a stem-loop conformation in the absence of target, thus bringing the reporter dye and quencher in close proximity. In the presence of target, the hairpin unfolds and the probe hybridizes to the target sequence in a more thermodynamically favorable reaction, thus distancing the quencher from the reporter and allowing fluorescence. Klerks et al. (2001) describe the use of this method for detection of Potato leafroll virus and Potato virus Y in tubers.

In our laboratory we have used LightCycler hybridization probes for real-time detection of *Ustilago nuda* in barley seed. Hybridization probes are a pair of probes that hybridize adjacent to each other on the amplified target (Figure 22.4, part C). On doing so, light absorbed by the fluorophore is transferred to an acceptor dye by fluorescent resonance energy transfer (FRET) and emitted at a different wavelength which is then detected.

Ustilago nuda can lead to significant loss of yield in infected barley crops. Current methods of identification involve microscopic examination of extracted embryos, a procedure that requires considerable skill and is extremely labor-intensive and costly. For seed certification in the United Kingdom, it is necessary for *U. nuda* to be detected to a level of 0.2 percent infection. This, combined with the fact that infection resides within the seed embryo and not on the surface, means that a rigorous extraction technique and extremely sensitive detection system are required for a molecular assay.

The major barrier in preventing direct transfer of the conventional PCR reaction to the LightCycler in the case of *U. nuda* detection is the large amount of nontarget DNA extracted from the seed. This causes background fluorescence due to the nonspecific binding of SYBR Green I dye and reduces the sensitivity for detecting the specific product. An alternative method for detection is to use a specific probe for the amplified target that can be measured directly and independently of the background DNA. Primers specific for *Ustilago* spp. (Willits and Sherwood, 1999) together with a pair of LightCycler hybridization probes have allowed us to specifically measure *U. nuda* DNA extracted from whole seeds of 0.2 percent infection level, which is the limit for seed certification.

Another interesting probe system known as Scorpion primers combine a fluorescent label with a specificity sequence and primer in a single molecule to form a detection system that has unimolecular kinetics (Whitcombe et al., 1999) (Figure 22.4, part D). Scorpions can be combined with amplification refractory mutation system (ARMS) assays, so that both allelic variants can be tested in the same tube. In our laboratory we have used this system to discriminate between *P. teres* and *P. graminea,* which are closely related seedborne pathogens of barley seed (Bates and Taylor, 2001). There is a need to distinguish these two pathogens at seed level, as there are different disease management strategies for each. *Pyrenophora graminea* is strictly seed transmitted, so if it is found present in U.K. seed above the disease threshold of 2 percent it is treated with fungicide. *Pyrenophora teres*, however, is mainly spread through crop debris, so foliar fungicide sprays are used to control this disease unless a high level of infection is present in the seed.

Sequence information from the ITS region of *Pyrenophora* spp. (Stevens et al., 1998) indicated that there was little consistent sequence differences between species. There was, however, one base difference in the ITS 1 region between *P. graminea* and *P. teres*. This single base difference was used to design and develop ARMS Scorpion primers that would be able to distinguish the two species in a single tube. The single base mismatch was placed at 3' end of the Scorpion primer and fluorescence increase over time was real-time detected on the LightCycler for the matched product. Only a marginal increase in fluorescence was observed for the singly mismatched product (Figure 22.5). This result demonstrates the high specificity of the assay in being able to distinguish a single base change.

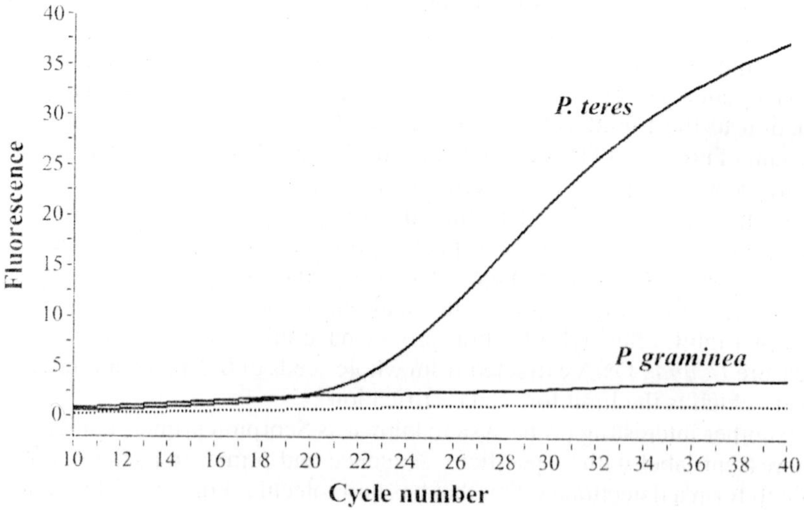

FIGURE 22.5. Scorpion primer detection of *Pyrenophora teres*. The primer is able to distinguish a single base pair mismatch between *P. teres* and *P. graminea*. The dotted line is a no template control.

In addition to detection and quantification, probes can be used discrimination of a single base pair mismatch during melting curve analysis and can also be used for genotyping single nucleotide polymorphisms (SNPs) and mutation analysis. They also present the option of carrying out multiplex reactions if different acceptor fluorophores are used for each probe pair.

BIO-PCR

BIO-PCR is a method particularly used for improving sensitivity of detection from bacterial pathogens. It is a technique whereby infected material, such as seed, is washed and the liquor from this wash plated onto selective medium. After a short incubation period (two to three days) to allow for growth of the target organism, the colonies are washed from the plates and an aliquot is removed for PCR analysis. An advantage of this protocol is that only living infectious material may be detected, due to the incubation step in the BIO-PCR procedure. This may be an important factor for deciding the appropriate disease control strategy to be applied and avoids any false positives from dead cells.

A good example of this procedure is the detection of *Clavibacter michiganensis* ssp. *sepedonicus* in potato tubers (Schaad et al., 1999). *C. michiganensis* ssp. *sepedonicus* is often difficult to detect, as it is slow growing and present as low levels of inoculum. A real-time BIO-PCR method was developed using primers designed from genomic DNA fragments isolated by subtracted hybridization (Mills et al., 1997). This technique greatly increased the sensitivity by 100- to 1000-fold compared with conventional PCR. This was partly due to growth of the pathogen and also due to the removal of PCR inhibitors produced by the tubers. This rapid method is suitable for high-throughput routine detection in a seed-testing laboratory due to the use of sophisticated equipment such as an ABI Prism 7700 Sequence Detection System (Applied Biosystems).

DNA Chips

Due to the advances in research for developing diagnostic tests in seed borne pathogens and the increasing need for high-throughput methods, interest in the use of DNA chips for seed diagnostics is beginning. DNA biochips can be divided into two groups: those that consist of high-density probes on plate-based substrates, and integrated biochip devices that include DNA arrays and integrated circuit (IC) microchip sensors for detection of a signal (Vo-Dinh, 2001). Samples of DNA, in the form of spots, are "printed" on the slide using an "arrayer," which consists of a high-speed robotic arm fitted with pins which can precisely position the DNA. This procedure is similar to that used to print computer chips. Each DNA spot may represent a molecular diagnostic marker for a specific plant disease. In this way multiple diseases can be screened for in a single experiment. This would be especially useful for quarantine purposes where a seed lot needs to be screened for many diseases before being certified and allowed entry into a country different from its origin.

CONCLUSIONS

An increase in new molecular seed health tests that are able to give more rapid but cost-effective results compared with conventional tests may help to change the present use of prophylactic seed treatment in Europe. With new tests being capable of a higher throughput, a more targeted use of seed treatment is possible, i.e., the use of seed treatment only on seed lots that are infected at or above the disease threshold. This has cost benefits to the farmer and the environment. Molecular tools for identifying seedborne fungal pathogens can also be used for obtaining information on disease thresholds and other epidemiological data, leading to a greater understanding of seedborne disease.

For the future of DNA diagnostics to reach mass markets, tests will need to become much simpler, cheaper, and more reliable (Hodgson, 1998), and this may be realized by DNA chip technology. Detection of hybridization on a DNA chip can be achieved through mass spectrometry, which avoids the expense and time required for fluorescence labelling. Costs of tests may also be reduced substantially through the avoidance of the use of PCR, which incurs a licensing and royalties fee. New systems are already in the process of development as alternatives to PCR: Molecular Tools (Baltimore, United States) plans the use of a linear amplification system based on the expression of RNA (Hodgson, 1998).

REFERENCES

Alvarez, A.M. and Kaneshiro, W.S. (1999). Detection and identification of *Clavibacter michiganensis* subsp. *michiganensis* in tomato seed. In J.W. Sheppard (ed.), *Proceedings of the Third International Seed Testing Association Plant Disease Committee* (pp. 93-97). Seed Health Symposium, Zurich.

Bates, J.A. and Taylor, E.J.A. (2001). Scorpion ARMS primers for SNP real-time PCR detection and quantification of *Pyrenophora teres*. *Molecular Plant Pathology* 2:275-280.

Bates, J.A., Taylor, E.J.A., Gans, P.T., and Thomas , J. (2002). Determination of relative proportions of *Globodera* species in mixed populations of potato cyst nematodes using PCR melting product peak analysis. *Molecular Plant Pathology* 3: 153-161.

Bates, J.A., Taylor, E.J.A., Kenyon, D.M., and Thomas, J.E. (2001). The application of real-time PCR to the identification, detection and quantification of *Pyrenophora* species in barley seed. *Molecular Plant Pathology* 2:49-57.

Beck, J.J. and Ligon, J.M. (1995). Polymerase chain reaction assays for the detection of *Stagonospora nodorum* and *Septoria tritici* in wheat. *Phytopathology* 85: 319-324.

Böhm, J., Hahn, A., Schubert, R., Bahnweg, G., Adler, N., Nechwatal, J., Oehlmann, R., and Oßwald, W. (1999). Real-time quantitative PCR: DNA determination in isolated spores of the mycorrhizal fungus *Glomus mosseae* and monitoring of *Phytophthora infestans* and *Phytophthora citricola* in their respective host plants. *J. Phytopathology* 147:409-416.

Brodal, G. (1999). Seed health in the Nordic and Baltic countries. In J.W. Sheppard (ed.), *Proceedings of the Third International Seed Testing Association Plant Disease Committee* (pp. 154-160). Seed Health Symposium, Zurich.

Cerovska, N., Petrzik, K., Moravec, T., and Mraz, I. (1998). Potato virus A detection by reverse transcription-polymerase chain reaction. *Acta Virologica* 42:83-85.

Chandelier, A., Dubois, N., Baelen, F., De Leener, F., Warnon, S., Remacle, J., and Lepoivre, P. (2001). RT-PCR-ELOSA tests on pooled sample units for the detection of virus Y in potato tubers. *Journal of Virological Methods* 91:99-108.

Chitarra, L.G., Langerak, C.J., Bergevoet, J.H., and van den Bulk, R.W. (2002). Detection of the plant pathogenic bacterium *Xanthomonas campestris* pv. *campestris* in seed extracts of *Brassica* sp.: Applying fluorescent antibodies and flow cytometry. *Cytometry* 47:118-126.

Ferreira, M.A.S.V., Tooley, P.W., Hatziloukas, E., Castro, C., and Schaad, N.W. (1996). Isolation of a species-specific mitochondrial DNA sequence for identification of *Tilletia indica,* the Karnal bunt of wheat fungus. *Applied and Environmental Microbiology* 62:87-93.

Foster, S.J, Singh, G., Fitt, B.D.L., and Ashby, A.M. (1999). Development of PCR based diagnostic techniques for the two mating types of *Pyrenopezia brassicae* (light leaf spot) on winter oilseed rape (*Brassica napus* ssp. *oleifera*). *Physiological and Molecular Plant Pathology* 55:111-119.

Hodgson J. (1998). Shrinking DNA diagnostics to fill the markets of the future. *Nature Biotechnology* 16:725-727.

Iacomi-Vasilescu, B., Blancard, D., Guenard, M., Molinero-Demilly, V., Laurent, E., and Simoneau, P. (2002). development of a PCR-based diagnostic asssay for detecting pathogenic *Alternaria* species in cruciferous seeds. *Seed Science and Technology* 30:87-95.

Kim, K. and Lee, Y.S. (2001). Selection of RAPD markers for *Phytophthora infestans* and PCR detection of *Phytophthora infestans* from potatoes. *Journal of Microbiology* 39:126-132.

Klerks, M.M., Leone, G.O.M., Verbeek, M., van den Heuvel, J.F.J.M., and Schoen, C.D. (2001). Development of a multiplex AmpliDet RNA for the simultaneous detection of *Potato leafroll virus* and *potato virus Y* in potato tubers. *Journal of Virological Methods* 93:115-125.

Koenraadt, H. (1997). Comparative Test for the Detection of *Xanthomonas campestris* pv. *campestris* in Crucifer Seeds. In J. D. Hutchins and J. R. Reeves (eds.), *Seed Health Testing: Progress Toward the 21st Century* (pp. 205-209). Wallingford, UK: CAB International.

Konstantinova, P., Bonants, P.J.M., Gent-Pelzer, M.P.E., van de Zouwen, P., and van den Bulk, R. (2002). Development of specific primers for detection and identification of *Alternaria* spp. in carrot material by PCR and comparison with blotter and plating assays. *Mycological Research* 106:23-33.

Langerak, C.J. (2002). Modified blotter to detect *Phoma lingam* in brassica seeds. In J.C. Machado, C.J. Langerak, and D.S. Jaccoud-Filho (eds.), *Seed-Borne Fungi: A Contribution to Routine Seed Health Analysis* (pp. 63-64). Lavras, Brazil: ISTA.

Machado, J.C. (2002a). Modified blotter test to detect *Botrytis cinerea* on sunflower seeds. In J.C. Machado, C.J. Langerak, and D.S. Jaccoud-Filho (eds.), *Seed-Borne Fungi: A Contribution to Routine Seed Health Analysis* (pp. 73-74). Lavras, Brazil: ISTA.

Machado, J.C. (2002b). "Roll paper method" for the detection of *Colletotrichum lindemuthianum* in common bean seeds. In J.C. Machado, C.J. Langerak, and D.S. Jaccoud-Filho (eds.), *Seed-Borne Fungi: A Contribution to Routine Seed Health Analysis* (pp. 64-65). Lavras, Brazil: ISTA.

Machado, J. C. (2002c). Salt agar medium to detect storage fungi in seeds. In J.C. Machado, C.J. Langerak, and D.S. Jaccoud-Filho (eds.), *Seed-Borne Fungi: A Contribution to Routine Seed Health Analysis* (pp. 77-79). Lavras, Brazil: ISTA.

Machado, J. C. (2002d). Washing test to detect *Peronospora manshurica* in soybean seeds. In J.C. Machado, C.J. Langerak, and D.S. Jaccoud-Filho (eds.), *Seed-Borne Fungi: A Contribution to Routine Seed Health Analysis* (pp. 71-72). Lavras, Brazil: ISTA.

Machado, J.C. and Langerak, C.J. (2002). General Incubation Methods for Routine Seed Health Analysis. In J.C. Machado, C.J. Langerak, and D.S. Jaccoud-Filho (eds.), *Seed-Borne Fungi: A Contribution to Routine Seed Health Analysis* (pp. 49-59). Lavras, Brazil: ISTA.

Machado, J.C., C.J. Langerak, and Jaccoud-Filho, D.S. (2002) *Seed-Borne Fungi: A Contribution to Routine Seed Health Analysis* (pp. 74-75). Lavras, Brazil: ISTA.

Maddox, D.A. (1998) Implications of new technologies for seed health testing and the worldwide movement of seed. *Seed Science Research* 8: 277-284.

Mahuku, G.S., Goodwin, P.H., and Hall, R. (1995). A competitive polymerase chain reaction to quantify DNA of *Leptosphaeria maculans* during Blackleg development in oilseed rape. *Molecular Microbe Interactions* 8:761-767.

Maude, R.B. (1996). *Seedborne Diseases and Their Control, Principles & Practice.* Wallingford, UK: CAB International.

McDonald, J.G., Wong, E., Kristjansson, G.T., and White, G.P. (1999). Direct amplification of PCR of DNA from ungerminated teliospores of *Tilletia* species. *Canadian Journal of Plant Pathology* 21:78-80.

McGee, D. (2002) Washing seeds for the detection of *Ustiago maydis* and *Sphacelotheca reiliana* in maize. In J.C. Machado, C.J. Langerak, and D.S. Jaccoud-Filho (eds.), *Seed-Borne Fungi: A Contribution to Routine Seed Health Analysis* (pp. 70). Lavras, Brazil: ISTA.

Mills, D., Russell B.W., and Hanus, J.W. (1997). Specific detection of *Clavibacter michiganensis* subsp. *sepedonicus* by amplification of three unique DNA sequences isolated by subtractive hybridisation. *Phytopathology* 87:853-861.

Mills, P.R., Sreenivasprasad, S., and Brown, A.E. (1992). Detection and differentiation of *Colletotrichum. gloeosporioides* isolates using PCR. *FEMS Microbiology Letters* 98:137-143.

Mumford, R.A., Walsh, K., Barker, I., and Boonham, N. (2000). Detection of *Potato mop top virus* and *Tobacco rattle virus* using a multiplex real-time fluorescent reverse-transcription chain reaction assay. *Phytopathology* 90:448-453.

Murillo, I., Cavallarin, L., and Segundo, B.S. (1998). The development of a rapid PCR assay for detection of *Fusarium moniliforme. European Journal of Plant Pathology* 104:301-311.

Neergaard, P. (1997). *Seed Pathology,* Volume 1. London: McMillan Press.

Nicholson, P., Lees, A.K., Maurin, N., Parry, D.W., and Reazanoor, H.N. (1996). Development of a PCR assay to identify and quantify *Microdochium nivale* var. *nivale* and *Microdochium nivale* var. *majus* in wheat. *Physiological and Molecular Plant Pathology* 48: 257-71.

Nicholson, P., Simpson, D.R., Weston, G., Rezanoor, H.N., Lees, A.K., Parry, D.W., and Joyce, D. (1998). Detection and quantification of *Fusarium culmorum* and *Fusarium graminearum* in cereals using PCR assays. *Physiological and Molecular Plant Pathology* 53:17-37.

Niepold, F. and Schöber-Butin, B. (1997). Application of the one-tube PCR technique in combination with a fast DNA extraction procedure for detecting *Phytophthora infestans. Microbial Research* 152:345-351.

Noble, M. (1979). Outilne of the history of seed pathology. In *Proceedings of The Latin American Workshop on Seed Pathology,* 1 (pp. 3-17). Londrina, Brazil.

Nolasco, G., de Blas, C., Torres, V., and Ponz, F. (1993). A method combining immunocapture and PCR amplification in a microtitre plate for the detection of plant viruses and subviral pathogens. *Journal of Virological Methods* 45:201-218.

Ojeda, S. and Verdier, V. (2000). Detecting *Xanthomonas axonopodis* pv. *manihotis* in cassava true seeds by nested polymerase chain reaction assay. *Canadian Journal of Plant Pathology* 22:241-247.

Parry, D.W. and Nicholson, P. (1996). Development of a PCR assay to detect *Fusarium poae* in wheat. *Plant Pathology* 45:383-391.

Pastrik, K.H. (2000). Detection of *Clavibacter michiganensis* subsp. *sepedonicus* in potato tubers by multiplex PCR with coamplification of host DNA. *European Journal of Plant Pathology* 106:155-165.

Paveley, N.D., Rennie, W.J., Reeves, J.C., Wray, M.W., Slawson, D.D., Clark, W.S., Cockerell, V., and Mitchell, A.G. (1997). Cereal seed health strategies in the UK. In J.D. Hutchins and J.C. Reeves (eds.), *Seed Health Testing: Progress Toward the 21st Century* (pp. 95-106). Oxon, UK: CAB International.

Phan, H.T.T., Ford, R., Bretag, T., and Taylor, P.W.J. 2002) A rapid and sensitive polymerase chain reaction (PCR) assay for detection of *Ascochyta rabiei*. *Australaian Plant Pathology* 31:31-39.

Reeves J.C. (1995). Nucleic acid techniques in testing seedborne diseases. In J.H. Skerritt and R. Appels (eds.), *New Diagnostics in Crop Sciences* (pp. 127-151). Oxon, UK: CAB International.

Rennie, W.J. (2002). Washing seeds of wheat to detect *Tilletia* caries. In J.C. Machado, C.J. Langerak, and D.S. Jaccoud-Filho (eds.), *Seed-Borne Fungi: A Contribution to Routine Seed Health Analysis* (pp. 75-76). Lavras, Brazil: ISTA.

Richardson, M, J. (1996). *An Annotated List of Sèed-borne Diseases*, Fourth Edition. Zurich, Switzerland: International Seed Testing Association.

Ririe, K.M., Rasmussen, R.P., and C.P. Wittwer (1997) Product differentiation by analysis of DNA melting curves during the polymerase chain reaction. *Analytical Biochemistry* 245:154-160.

Ristaino, J.B., Madritch, M., Trout, C.L., and Parra, G. (1998). PCR amplification of ribosomal DNA for species identification in the plant pathogen genus *Phytophthora*. *Applied and Environmental Microbiology* 64:948-954.

Schaad, N. (1997). ISTA/PDC Bacteriology Working Group: Comparative Tests and Working Sheets. In J. D. Hutchins and J. R. Reeves (eds.), *Seed Health Testing: Progress Toward the 21st Century* (pp. 201-203). Wallingford, UK: CAB International.

Schaad, N.W., Berthier-Schaad, Y., Sechler, A., and Knorr, D. (1999). Detection of *Clavibacter michiganensis* subsp. *sepedonicus* in potato tubers by BIO-PCR and an automated real-time fluorescence detection system. *Plant Disease* 83:1095-1100.

Schaad, N.W., Cheong, S.S., Tamaki, S., Hatziloukas, E., and Panopoulos, N.J. (1995). A combined biological and enzymatic amplification (Bio-PCR) technique to detect *Pseudomonas-syringae* pv. *phaseolicola* in bean seed extracts. *Phytopathology* 85:243-248.

Schilling, A.G., Moller, E.M., and Geiger, H.H. (1996). Polymerase chain reaction-based assays for species-specific detection of *Fusarium culmorum, F. graminearum*, and *F. avenaceum*. *Phytopathology* 86:515-522.

Schnerr, H., Niessen, L., and Vogel, R.F. (2001). Real time detection of the *tri5* gene in *Fusarium* species by lightcycler (TM)-PCR using SYBR ((R)) green I for continuous fluorescence monitoring. *International Journal of Food Microbiology* 71:53-61.

Seal, S.E., Taghavi, M., Fegan, N., Hayward, A.C., and M. Fegan (1999). Determination of *Ralstonia (Pseudomonas) solanacearum* rDNA subgroups by PCR tests. *Plant Pathology* 48:115-120.

Shepherd, L.M., McGee, D., and Block, C.C. (1999). A selective medium method to detect seedborne *Clavibacter michiganensis* subsp. *nebraskensis* in corn. In J.W. Sheppard (ed.), *Proceedings of the Third International Seed Testing Association Plant Disease Committee* (pp. 115-118). Zurich: Seed Health Symposium.

Singh, R.P., Kurz, J., Boiteau, G., and Bernard, G. (1995). Detection of potato Leafroll virus in single aphids by the reverse transcritpion-polymerase chain-reaction and its potential epidemiological application. *Journal of Virological Methods* 55:133-143.

Stevens, E.A., Blakemore, E.J.A., and Reeves, J.C. (1998). Relationships amongst isolates of *Pyrenophora* spp. based on the sequences of the ribosomal DNA spacer regions. *Molecular Plant Pathology on-line.* Available online at <http://www.bspp.org.uk/mppol/1998/1111stevens>.

Taylor, E.J.A., Stevens, E.A., Bates, J.A., Morreale, G., Lee, D., Kenyon, D.M., and Thomas, J.E. (2001). Rapid-cycle PCR detection of *Pyrenophora graminea* from barley seed. *Plant Pathology* 50:347-355.

Toth, I.K., Hyman, L.J., Taylor, R., and Birch, P.R.J. (1998). PCR-based detection of *Xanthomonas campestris* pv. *phaseoli* var. *fuscans* in plant material and its differentiation from *X.c.* pv. *phaseoli*. *Journal of Applied Microbiology* 85:327-336.

Trout, C.L., Ristaino, J.B., Madritch, M., and Wangsomboondee, T. (1997). Rapid detection of *Phytophthora infestans* in late blight-infected potato and tomato using PCR. *Plant Disease* 81:1042-1048.

Vo-Dinh, T. (2001). DNA chips: Technology and applications. *Clinical Laboratory International*, October: 12-14.

Voigt, K., Schleier, S., and Wostemeyer, J. (1998). RAPD-based molecular probes for the blackleg fungus *Leptosphaeria maculans (Phoma lingam)*: Evidence for pathogenicity group-specific sequences in the fungal genomes. *Journal of Phytopathology* 146:567-576.

Walsh, K., North, J., Barker, I., and Boonham, N. (2001). Detection of different strains of potato virus Y and their mixed infections using competitive fluorescent RT-PCR. *Journal of Virological Methods* 9:167-173.

Whitcombe, D., Theaker, J., Guy, S.P., Brown, T., and Little, S. (1999). Detection of PCR products using self-probing amplicons and fluorescence. *Nature Biotechnology* 17:804-807.

Willits, D.A. and Sherwood, J.E. (1999). Polymerase chain reaction detection of *Ustilago hordei* in leaves of susceptible and resistant barley varieties. *Phytopathology* 89:212-217.

Woo, T.H.S., Patel, B.K.C., Smythe, L.D., Symonds, M.L., Norris, M.A., and Dohnt, M.F. (1997). Identification of pathogenic *Leptospira* genospecies by continuous monitoring of fluorogenic hybridisation probes during rapid-cycle PCR. *Journal of Clinical Microbiology* 35: 3140-3146.

Zhang, A.W., Hartman, G.L., Curio-Penny, B., Pedersen, W.L., and Becker, K.B. (1999). Molecular detection of *Diaporthe phaseolorum* and *Phomopsis longicolla* from soybean seeds. *Phytopathology* 89: 796-804.

Chapter 23

Seed Quality in Vegetable Crops

S. D. Doijode

INTRODUCTION

Seeds are tiny plant parts containing valuable genetic material and are capable of giving new life in the form of young seedlings on germination under favorable conditions. This characteristic helps in large-scale use of seeds in both propagation and genetic conservation. Seeds are widely used in vegetable production, and those seeds that give uniform seedlings and maintain adequate populations in the field are preferred for cultivation.

Fresh vegetables form an important part of the human diet, and of late there is an increased demand for them from all over the world. They are popular in temperate, tropic, and subtropic zones. A specific climatic condition favors the optimum growth and successful production of vegetable crops, and hence particular kinds of vegetable are popular in a given region. Vegetable crops produce different plant parts, including tender and succulent foliage, roots, tubers, flowers, and fruits, which are used fresh as well as cooked. They are valued mainly as an important source of vitamins and minerals. Some vegetable crops have medicinal properties as well. Many of them are cultivated as annual crops, while some are biennial and a few are perennial crops. These are predominantly grown through seed. The vegetable seed industry is rapidly expanding, and there is a great demand for good-quality seeds all over the world.

Seed quality normally consists of the genetic, physiological, and physical characteristics of seeds, and it determines crop performance and yield in association with potential of the cultivar. It plays a major role in cultivation of vegetable crops and is an effective means of achieving sustainable production. Availability, quality, and cost of seeds influence the total production. High-quality seeds are genetically and physically pure, vigorous, and free from insect pests and disease-causing organisms. They also meet the demands of diverse agro-climatic conditions for cultivation. It is believed that quality seeds with high viability and vigor contribute nearly 30 percent to total production. They are greatly valued, especially in hybrids and protected cultivation of vegetable crops. Such seeds germinate faster, ensure an adequate plant population, withstand certain biotic and abiotic stresses, and produce high-quality yields. Plant uniformity is an expression of high seed quality achieved by high vigor of seeds. Normally, low-quality seeds are not used for sowing (Roberts, 1986).

Farmers select good-quality seeds worldwide for raising healthy crops and receiving higher returns. The importance of good-quality seeds was well-known even during ancient farming history. It is clear that the concept of quality is not new to the farming community, and even ancient farmers were conscious about purity, hygiene, and marketing of quality seeds. Demand for good seeds was increasingly felt under changing agricultural scenarios in different parts of world (Nene, 1999).

The genetic potential of a genotype determines the plant characteristics, quality, and final yield. Its genetic purity declined with genetic contamination and drift in fields. Further, it was governed by various extraneous factors such as unfavorable climatic conditions, inadequate nutrition, irrigation, and isolation, improper rogueing of plants in the field, and invasion by insect pests and pathogens. The physical purity of seeds is the indication of presence or absence of impurities such as weed and other crop seeds and inert matter. Quality seeds have high vigor and normally withstand unfavorable growing conditions. High-vigor seeds germinate early and establish better even under water stress or in hard soil. High viability and vigor ensure good seedling establishment and optimum plant population. Seed health determines the continuous performance of seedlings in the field. Seed quality is influenced by several factors during seed formation, development, maturation, growth, harvesting, extraction, drying, cleaning, grading, packing, and storage. It is essential to produce good-quality seeds and maintain this quality until seeds are sown for successful production of vegetable crops. In cabbage, low-quality seed is associated with low rate of germination and emergence, reduced seedling fresh weight, and an increase in number of abnormal seedlings (Liou et al., 1989). Further, germinating power and yield of the crop raised from stored seeds at ambient conditions steadily declines. The crop yield is lower in plants raised from older seeds as compared to fresher seeds (Frohlich and Henkel, 1964).

FACTORS AFFECTING SEED QUALITY

During Seed Production

Production of quality seeds is a boon to the vegetable seed industry. The main emphasis in seed production is to produce quality seeds that are true to type, uniform, and of high vitality as essential parameters. The sowing of poor seeds results in defective seedling stands and more unhealthy plants in the field. Quality seed can be harvested only when proper attention is given to all the factors that affect such quality at every stage of seed production. This spans the entire process, from selection of planting material to harvesting of crops.

Climate

The climate of the seed production area affects seed quality. Factors such as temperature, relative humidity, rainfall, sunshine hours, and wind velocity determine the growth, development, and health of the plant. Extreme climatic factors such as excess rain or drought during flowering affect the seed set and bring low seed yields. Seed quality is also affected by environmental stress during seed filling stages. The cropping seasons differ markedly with respect to rainfall distribution, bright sunshine hours, and temperature. Fluctuations in environmental conditions affect the physiological process and thereby the seed quality. Seed production of cole crops, namely, cabbage, cauliflower, kohlrabi, and Brussels sprout, are confined to temperate conditions. These crops need vernalization for bolting. Temperatures less than 10°C for six weeks are usually ideal for induction of flowering. Similarly, onions need cool weather during the initial stages of crop growth but require warm and dry conditions at the time of seed maturation and harvesting. Plants grown at 15 to 16°C yielded more high-quality seeds than those grown at 22 to 23°C (Gray and Steckel, 1984). Tomato plants grow better at temperatures of 18 to 27°C, and fruit and seed setting is affected by high temperature and low humidity. Temperatures above 38°C affect the fruit set. Okra crops prefer warm conditions for optimum growth. The prevalence of rains during fruit development affects the seed quality. There are two main types of radishes, small and large. Small radishes grow better under a cooler climate, while large ones withstand fluctuations of temperature. Carrot seeds are normally produced at higher altitudes and in cooler regions, which helps in profuse growth of flowers. French bean and peas also produce quality seeds under cooler conditions. Cucumber, melon, and gourds prefer hot and dry or humid climate for growth. They grow better in warm seasons and sunny locations. Muskmelon seeds produced in humid conditions are of inferior quality. Thus, quality seeds are produced during prevalence of favorable climatic conditions for growth, flowering, and fruiting of plants.

Cultural Practices

Plant growth is completed in one or two seasons in different vegetable crops. They are typically raised through seeds. Seed yield increases with good planting material and following optimum cultural practices. Seeds are either sown directly in the field or seedlings are raised in a nursery and transplanted after four to six weeks at appropriate spacing in field. Well-drained humus and fertile soils are most suited for vegetable cultivation. Initially support is provided to the young plant in the form of stakes, arches, etc., as in tomato, pod vegetables, and cucurbits. Further, timely weeding, irrigation, and adequate fertilizers must be provided. In garden peas, seed quality improves with application of nitrogen and phosphorus (Hadavizadeh and George, 1989). Nitrogen and phosphorus application significantly improves the seed vigor, as they form important constituents of proteins and phospholipids, which affect cell membrane integrity and thereby seed vigor (Padrit et al., 1996). The deficiency of calcium affects the seed quality in muskmelon (Frost and Kretchman, 1989). More research is needed to correlate seed quality parameters with nutrients. It is believed that good health of plants determines the status of the offspring. Irrigation is ensured especially in dry season and during critical stages such as seed filling. Plants are attacked by several insects and pathogens, and they are protected by using suitable pesticides. Weeds are removed to avoid competition and contamination with the main seed crop. Intercultural operation not only eliminates the weeds but also promotes root growth in the soil.

Isolation Distance and Rogueing

Genetic purity is the major parameter of seed quality. It is affected by genetic contaminations and genetic drift in the field. Pollen contamination is prevented by removing weeds or related plants and also by providing isolation distance. The distance is about 50 m for self-pollinated crops such as tomato, garden peas, and French bean, and 1,000 to 1,600 m for cross-pollinated vegetable crops such as cole crops, onion, beet root, and cucurbits. The contamination is more relevant during the flowering season; therefore, frequent inspection and removal of off-type plants can help in maintaining genetic purity.

Seed Maturity

The time of harvesting of crops largely influences the quality of seeds. Most often matured seeds are heavier, lustrous, and give high germination and high seedling vigor. Seed maturity coincides with that of maturity of fruit, and at this stage seeds attain the maximum dry weight. A series of morphological, physiological, and functional changes occur from fertilization until seeds mature. The level of fruit maturity affects both germination and vigor of seeds and determines seed longevity (Eguchi and Yamada, 1958). Seed viability is higher at the mature stage, and it reduces with early or late harvest of crop. Therefore, it should be harvested at the right stage for getting high-quality seeds. In onion, seed maturity is associated with rapid reduction in chlorophyll and water content, increase in dry weight, as well as germination. Seeds of vegetable crops normally attain maximum quality on completion of the seed filling phase; thereafter the viability and vigor declines. However, Ellis et al. (1993) recommended leaving some more time after mass maturity. Pepper seeds are allowed to mature for a short period within red fruits after harvest for maximum germination (Sanchez et al., 1993). The onset of both germinability and desiccation tolerance occur either just before or at mass maturity (Demir and Ellis, 1992). Seeds attain maximum vigor and viability when fruit cracks in okra and fruit became fully red in pepper and tomato (Doijode, 1983). Early-harvested seeds show low viability and poor seedling vigor, while matured seeds retain high quality for longer periods. The mature seeds of eggplant show higher seed storability than immature seeds (Eguchi and Yamada, 1958). In cucurbits, especially watermelon, fruits are kept on the vine for a comparable period of time since the seed coat completes growth before the embryo, which might inhibit the germination process in immature seeds. Germination rate is increased in immature seeds by removal of the seed coat (Nerson, 1991). Seed quality improves in fresh seeds following ripening, as observed higher germination in seeds of unripe fruits after ripening (Demir and Ellis, 1993). Seed maturity is also estimated by measuring chlorophyll fluorescence, as the amount of chlorophyll is directly related to the degreening process and maturity.

Fruit and Seed Position on Mother Plant

Fruit setting and seed arrangements vary among different vegetable crops. In muskmelon, fruit position on the mother plant influences the seed quality. There is reduction in fruit weight, number of seeds per fruit, seed weight, and viability with increasing fruit numbers on the plant

(Incalcaterra and Caruso, 1994). Likewise, seeds from early pickings of tomatoes showed higher seed quality. Pepper seeds from fruit formed on the main stem and on the first and secondary branches had high seed weight and germination percentage. Further, plants from these seeds show early, are more productive, and have better disease resistance than plants from seeds of fruit borne on third or fourth order branches appearing later. In okra, seed quality is affected by the fruit's nodal position; the best quality seeds obtained from fruits borne at lower nodes. Seeds separated from fruits borne up to the seventh node showed higher germination percentage and seedling vigor. Further, seed storability reduced with increase in fruits' nodal position. Seeds from lower nodal-borne fruits possess greater germination capacity and vigor. Petrov and Dojkov (1970) obtained the best quality seeds from lower- borne fruits in eggplant. In carrot, seed germination and 1000-seed weight were greater for seeds of primary umbels than from seeds of secondary umbels (Szafirowska, 1994).

Seed position in fruit also affects the seed quality. In pepper, seeds from the basal portion showed higher viability and vigor than those from the middle and tip portion of the fruit (Doijode, 1991). Similarly, seeds from the basal portion of eggplant fruit showed good seed quality. These seeds must be utilizing adequate photosynthetates from the mother plant, thereby showing higher seed germination and vigor. Fruits and/or seeds formed earlier are normally bold and relatively large due to a source and sink relationship within the plant, which was further modified by presence of adequate nutrients. Subsequently formed fruits and seeds are smaller in size and of poor quality.

Insect Pests and Diseases

Plants often are infested with several insects and pathogens, some of them carried by seeds. Seed carries pathogens internally or externally, and they multiply on sowing, affecting seed quality and crop productivity. Infested seeds carry insects and/or pathogens, which cause shriveling, discoloration, decay, low germination, and low vigor, and affect the productivity. The invasion of pathogens depends on injury to seed, availability of food material, seed moisture, storage temperature, and aeration that occur either in field or in storage. Insects are predominantly found in seeds of pod vegetables such as cowpea, French bean, and pea. Maggots may feed on endosperm and also damage the embryo, resulting in death of the seed.

Disease expression is the result of interaction among a susceptible host, pathogen, and environment. Seeds extracted from infected, rotten fruits are of inferior quality. The microorganisms are parasitic or saprophytic, affect germination, produce heat and toxins during storage, and induce various biochemical changes. According to Christensen and Kaufmann (1969), fungal pathogens are mainly responsible for seed deterioration and reduction of germination potential in stored seeds. Fungi such as *Alternaria, Aspergillus, Penicellium, Chaetomium, Cladosporium,* and *Rhizophus* spp. are commonly associated with seeds. Seed treatment with suitable pesticides improves the seed quality. Also, proper drying of seeds and storage at low temperature lowers insects and microbial activity. Collect seeds from healthy plants, avoid injury during handling and harvesting, and treat with insecticides and fungicides to maintain good seed health.

During Seed Storage

Seeds are preserved for a few days to years depending on needs for different vegetable crops. Farmers like to maintain high seed quality until next growing season, while breeders aim at maintaining genetically viable and stable seed material for a very long period, avoiding repeated growing in field, which is a cumbersome and labor-intensive process. The main objective of seed storage is to retain high seed quality. Seeds start deteriorating immediately after separation from the mother plant. In an ideal storage method, the process of deterioration is lowered so that seeds remain viable for relatively longer periods.

Seeds are broadly classified into orthodox and recalcitrant based on drying and storage behavior. Those seeds retain that viability on drying and are able to store at low temperature are called orthodox seeds, while seeds whose viability is affected on drying and are unable to preserve at subzero temperatures are called recalcitrant. Seeds of many vegetable crops belong to the orthodox group except seeds of *Sechium edule*, which exhibits recalcitrant storage behavior. Orthodox seeds are relatively easier to handle and capable of preserving high seed quality in terms of viability and vigor for longer periods. However, quality of seeds is affected by changes in factors such as seed moisture, storage temperature, relative humidity, and oxygen content during storage.

Seed Moisture

Seed moisture is an important component in determination of storage life of seeds. It plays a major role in seed deterioration. Seed moisture is high immediately after fertilization and decreases during seed development, maturation, and ripening. It attains equilibrium with atmospheric relative humidity on harvesting. Further, seed moisture fluctuates depending on chemical composition of seeds, relative humidity, and temperature of the atmosphere. The water evaporates from seeds into the atmosphere, and a moisture gradient is created that causes moisture to move toward the surface. The embryo may get damaged at higher rate of evaporation and resulting in loss of viability. Therefore, seeds are to be well protected during drying, preferably at low temperature (15°C) and low relative humidity (15 percent RH). Sometimes during drying a little pulp is retained along with seeds, as in the case of pepper seeds, which helps in translocation of metabolites at higher temperature. In garden peas rapid and excessive removal of seed moisture affected the seed vigor and also yield raised from such seeds (Pavelkova and Curiova, 1984). In okra high seed moisture tends to show little or no dormancy, while reduction to 4 to 6 percent develops hardseededness. It causes very slow uptake of water and is overcome by seed scarification or germinating seeds at alternate temperature (20 to 30°C for 16 to 8 h, respectively) (Medina et al., 1972). Normally, 6 to 7 percent of moisture is maintained in several vegetable seeds, while 5 percent moisture is retained in seeds rich in oils. Harrington (1972) reported that seed storability reduced by one-half for every rise in 1 percent moisture, which implies a 5 to 14 percent moisture range. Vegetable seeds are kept at low moisture level (<10 percent) in sealed containers with small amount of insecticides and in clean, dry, and cool places for longer storage life (Laurico, 1984). Horky (1991) reported that seed moisture content of 5.6 to 7.5 percent in cabbage, carrot, and lettuce has helped in maintaining high seed quality for seven years at 0°C. High-moisture seeds are largely affected by high temperature and sealed conditions. They favor increased metabolic and fungal activity; therefore, low-moisture seeds are preserved in moisture-proof containers and at low temperature for maintaining good seed quality for longer period.

Storage Temperature

Temperature influences the seed quality during storage. Seed deterioration is enhanced at higher temperature and is further accelerated in presence of high seed moisture. High temperature is congenial for greater enzymatic, insect, and microbial activity in seed, thus affecting the seed quality in terms of germination, seedling vigor, and storage life. According to Harrington (1972) storage life of seeds reduces by one-half for every rise in 5°C temperature, and it covers from 0 to 50°C. Orthodox seeds tolerate freezing temperature and thus maintain high seed quality especially when packed in moisture-proof containers. Seed storage at –20°C rather than 20°C is beneficial to seed survival, and at the latter temperature there was significant loss of viability in carrot, lettuce, and onion (Ellis et al., 1996). Among low-temperature seed, deterioration is greater at 5°C than –18°C and least at –196°C (Stanwood and Sowa, 1995). High seed quality is maintained in cluster bean, French bean, onion, pepper, and radish for 20 years by preserving seeds at –20°C in laminated aluminum foil pouches (Doijode, unpublished data).

Relative Humidity

Seed quality is affected by presence of high relative humidity as seeds absorb moisture from the atmosphere. The rate of moisture absorption depends on chemical composition, vapor pressure gradient, temperature, and duration. Dry seeds on exposure to humid conditions not only gain moisture but also stimulate sprouting in seeds and promote growth of fungi. Tomato seeds exposed to high relative humidity gave low germination, low vigor, and affected membrane integrity (Berjak and Villiers, 1973), while eggplant, onion, pepper, and tomato seeds retained high viability at low relative humidity (<40 percent) even after four years of storage under ambient conditions (Doijode, unpublished data). Very low humidity also damages structure of seed, such as rupturing of seed coat. It is suggested not to store seeds at very dry conditions, especially relative humidity less than 10 percent (Waiters, 1998). High humidity under temperature fluctuations of tropical climates is deleterious to seeds, affecting the seed quality by rapid seed deterioration that results in loss of viability. At higher relative humidity, seeds of many species release volatile substances during storage such as aldehydes, alcohols, hydrocarbons, monoterpenoids, methanol, ethanol, acetal, dehyde and acetone. These gases reduce seed vigor. The evolution and effect of gases has been observed from –3.5 to 23°C. Therefore, it is recommended to maintain low relative humidity during storage (Zhang et al., 1993). Seed can be protected from high humidity by using desiccants or moisture-vapor-proof containers such as aluminum cans or laminated aluminum foil pouches in different vegetable crops.

Oxygen

Seeds require oxygen for different metabolic processes, and the rate of metabolism increases at higher level of oxygen content—resulting in greater chemical reactions, breakdown of food material, and insect and fungal activity, thereby promoting seed deterioration. The deleterious effects of oxygen are prominently expressed in the presence of a high level of seed moisture (Justice and Bass, 1978). The removal of air from storage containers helped in retaining high seed quality during storage. Onion seeds stored in partial vacuum exhibited twice the percentage of germination than in air after 18 months of storage in ambient conditions (Doijode, 1988b). Likewise, in peas high oxygen content in the environment affected seed quality and reduced storability of seeds (Roberts and Abdalla, 1968). Barton (1960) reported that seeds sealed in an oxygen-rich environment lose more seed quality than seeds stored in partial vacuum. Cabbage seeds also maintain high seed quality under partial vacuum (Isely and Bass, 1960), while lettuce seeds did not perform well under vacuum conditions (Bass, Clark, and James, 1962), which indicates that response varies in different crops. Pepper seeds maintain high seed quality in presence of carbon dioxide, while muskmelon, okra, and onion seeds perform better in a nitrogen-rich atmosphere (Doijode, unpublished data).

Storage Insect Pests and Diseases

Several insects and fungal pathogens affect seed quality in storage by damaging seed embryo and causing discoloration and decaying of seed material. Such seeds are unfit for sowing. The damage is more severe under greater insect or microbial populations, higher seed moisture, and higher storage temperatures. Insects feed on different kinds of seeds and damage the growing point. It is common in seeds of pod vegetables that are affected by pulse beetle (*Callosobruchus chinensis* L.) during storage. Its activity is rapid at high seed moisture and temperature, and it is controlled by use of fumigants such as ethylene dibromide or malathion (Hunje et al., 1990). Microorganisms are present on the surface or inside seed tissue. In onion, reduction of seed viability and vigor during storage is attributed to production of fungal metabolites (Gupta et al., 1989). Seed treatment with fungicides effectively controls the infestation and helps in retaining high seed quality.

IMPROVEMENT OF SEED QUALITY

Seed quality parameters such as viability and vigor differ from seed to seed, but for practical purposes a seed lot is considered to be a single unit represented by sample seeds for determining seed quality. Seeds of vegetable crops in general are a highly valued commodity and are needed in a small quantity for sowing. Initially they are sown in a nursery and seedlings are then transplanted into the field. These seeds demand high physical purity, grading, and impressive packing for protection and market promotion.

Seed Cleaning

Seed lots are comprised of pure seeds, broken seeds, other crop seeds, weed seeds, and inert matter after harvest. The seed quality improves with proper cleaning, either manually or by using machines. Seeds are cleaned by simple hand methods of rubbing, sieving, and winnowing. This removes contaminants such as chaffs, stems, stones, broken seeds, and weed seeds from seeds; separation of similar types of seeds of other varieties is difficult, especially those having similar appearance and physical characteristics. In seed processing units seeds are cleaned with the help of high-speed machines using physical parameters of seeds such as size, length, shape, density, surface texture, and electrical properties.

Seed Size

Heavy, bold, lustrous, and fairly uniform seed lots are selected for sowing as they reduce disparity among seeds during sowing and germination. Seed size variation is observed in several vegetable crops, and more often in seeds of pod vegetables. This has attributed to both genetic and environmental means. Late-formed and undernourished seeds are normally smaller in size. Removal of lighter seeds from a seed lot improves the seed quality. In carrot and onion, heavy seeds performed better and gave higher yield by 40 percent (Usik, 1980), which was also seen in radish (Lee and Nichols, 1978). The heavier seeds emerge faster and better even in greater soil depths (Townsend, 1979). Seedlings that emerge from large French bean seeds are more vigorous than those of small seeds, and such large seeds remain viable for longer periods (Doijode, 1984). Large seeds of okra gave higher germination than small seeds after 24 months of ambient storage (Nakagawa et al., 1991). Wester (1964) also reported that large lima bean seeds produced tall, vigorous plants and gave higher yield than small seeds from the same lot. In asparagus, large and medium-sized seeds showed higher germination and vigor (Belletti, 1985). Large seeds are preferred in eggplant for better seedling vigor in field, which is attributed to the availability and metabolism of reserve food (Pollock and Roos, 1972). Seed size influences germination, seedling growth, and yield; therefore, bold and large seeds are selected for vegetable cultivation.

Seed Color

Seed lots are comprised of various colored seeds. Seed sorting based on color improves the seed quality. Color of seed is genetically controlled;

however, it is regulated by different physiological processes. Seed color is an important varietal attribute, and it changes with age and deterioration. Initially seeds are lustrous, with prominent color, which fades with increased storage duration or due to storage under improper conditions. West and Harris (1963) reported that seeds exhibit certain changes in seed coat color during storage and showed lower seed viability and vigor in several vegetable crops. Seed color is commonly used as an indicator of seed quality in radish. In okra, seed coat color varies from green to gray and black. The black seeds are considered to be of poor quality. Further, seed germination is rapid and higher seedling growth occurs from green than black seeds (Baruah and Paul, 1996). Cabbage seeds are yellowish red, light brown, and dark brown, and the latter gave high germination (Mihajlovski and Kostov, 1998). Electronic devices are widely used for rapid separation of different colored seeds in seed lots.

Seed Coating

In general, vegetable seeds are small and irregular in shape, making them difficult to handle during sowing. It is a common practice to coat seeds with natural or synthetic resins to improve their size and appearance and also for identification of seed product during marketing and any dispute or litigation. The resins are applied alone or mixed with nutrients, growth substances for stimulating seedling growth, fungicides for protection against pathogens, and unique dye for identification and labeling of particular seed lots. Cabbage seed coated with resins prevented premature seed germination and extended storage life of seeds. Resins do not affect germination. Seed coating with alginate or alginate with abscisic acid enhanced seed longevity by three or four times. Alginate coating reduces rate of dehydration of primordial and thereby extends storage life of coated seeds. Vinamul 8330 is an effective resin for preventing premature germination, and it dries immediately, facilitating easy handling and usage during storage (Elaine and Shiel, 1980). Seed coating is widely used in developed nations and is becoming popular in developing countries.

Seed Packaging

Vegetable seeds are normally packed in small containers, which are effective for maintaining seed quality for various periods. These are convenient for handling and transport of seed material, and for promotion of marketing of seed material using attractive packages. Different kinds of packages are available for protecting seeds from various extraneous factors such as high humidity, insects, microorganism, etc. The choice of packaging material depends on kind, quantity, and cost of seeds and duration of storage. Paper bags are widely used for packing, but seed quality is affected on storage in humid environments and also on longer duration, due to excessive intake moisture. Cucurbit seeds stored in paper bags showed poor viability and vigor (Villareal et al., 1972). Polyethylene and laminated aluminum foil pouches are widely used for packing small quantities of seeds and effective for maintaining high seed quality. Laminated pouches are relatively impermeable to moisture content and retain high viability for longer periods. The packages are labeled suitably with ink for identification of seed material. It is believed that volatile substances such as formaldehyde are released from ink or paper and glue used while packaging, thereby affecting the seed quality, especially in lettuce seeds which are more sensitive to such volatile substances. Therefore, packaging material used should protect seeds from penetration of volatile gases and maintain high seed quality.

Seed Sorting

Seeds are also sorted based on X-ray photography as well as chlorophyll fluorescence methods where very high precision is required.

X-Ray Photography

Seeds are exposed to X-rays to visualize conditions of the internal structure without cutting and removal of the seed coat. It indicates the condition of the embryo and the presence of any insects or foreign bodies damaging the seed and thereby affecting the seed quality. Seedling performance is predicted in accordance with embryo size and amount of endosperm and free space available which enables selection of high quality seeds (van der Burg et al., 1994).

Chlorophyll Fluorescence (CF) Method

Presence or absence of chlorophyll in seeds indicates the nature of seed quality and also level of maturity. The amount of chlorophyll in seeds is directly related to the degreening process and thus maturity. Seeds with the lowest amount of chlorophyll fluorescence had the highest number of normal seedling and high percentage of germination. It is possible to increase percentage of germination of normal seedlings in a seed lot by removing seeds with very high chlorophyll fluorescence signals (Jalink et al., 1998).

Advantages of the CF method are high sensitivity, nondestructive nature, and rapid detection of quality seeds.

MAINTENANCE OF SEED QUALITY

Seed quality, particularly in terms of viability and vigor, decreases with increase in storage period and is further affected by improper seed handling and storage. Seed quality can be maintained for a short period in fruit as fruit storage and for a longer period in seeds as seed storage.

Fruit Storage

Mature or ripe fruits provide congenial conditions for short-term preservation of seeds, which is unfavorable for germination and induces dormancy in seeds. Tokumasu (1975) reported that seeds of *Brassica carnua* remain dormant in the pod, and dormancy disappeared rapidly on separation from the pod. Fruit storage is the most common practice in cucurbits, where seeds are stored as such in fruit until next growing season, which preserves the seed quality. Farmers of hilly regions also follow the use of fruit storage for pepper. Fruits are stored in natural conditions or under controlled conditions. Pepper fruits stored in cloth bags maintain their seed quality for three years under ambient conditions (Doijode, 1996). Fruits stored under low temperature retain quality for longer periods. High seed vigor can be preserved for six weeks in ripe tomatoes when fruits are kept at 11.5°C and 57 percent relative humidity (Baldo et al., 1988). Seed quality is affected only when fruits are preserved at very low temperatures. Petrov and Dojkov (1970) reported that seeds in underripe fruits of eggplant were particularly susceptible to freezing temperatures and became dormant; however, it is overcome by warming seeds for a short period. The fruit maturity determines the length of storage. Matured fruits of watermelon stored for 48 months retain high seed viability, but germination was adversely affected in seeds of immature fruits (28 dpa) (Nerson and Paris, 1988). Fruit storage is a low-cost tool for seed preservation and provides natural packaging for maintenance of seed quality. Such fruits are easily handled and transported to distant places without affecting the seed quality.

Seed Storage

Seed storage is primarily aimed at maintaining the seed quality for a fairly good period with good percentage of viability and vigor. Seed storage is the only viable available method for safe, convenient, and inexpensive preservation of high-quality planting material and genetic material in different vegetable crops. Also, it does not alter genetic material on very long storage, so it is widely used as means of conservation of germplasm in gene bank. Various storage conditions such as temperature, relative humidity, packaging, and presence or absence of insects and pathogens affect seed storability. Seed quality is preserved for short to long periods based on requirements and prevalence of required storage conditions. Seed storage requirements and different methods have been addressed at length in different vegetable crops (Doijode, 2001b). The various factors responsible for seed deterioration are monitored in storage for preserving high seed quality in vegetable crops.

Storage with Desiccant

High seed moisture is major factor responsible for seed deterioration, and its removal by a desiccant such as silica gel or quick lime is an economical means of monitoring seed quality. Grubben (1978) recommends a method for small-scale storage of vegetable seeds in the tropics using silica gel at 1:10 proportion of the seed quantity in an airtight jar. These desiccants are not very useful for storage under freezing temperature, but they can preserve seed quality effectively in fluctuating temperatures (Horky, 1991). Seed storage with desiccants is inexpensive and useful in maintaining seed quality for a short period, particularly until the next growing season as demanded by the farming community. Silica gel is a widely used desiccant in storage. It is a strong moisture absorbent, has a self-indicator property, is economical, and is nontoxic to seeds. It is packed in cloth bags and stored along with seeds in storage containers. The color changes from blue to white on moisture absorption, and it is reusable after drying. Seed storage with desiccants is a practical and effective means of preserving high seed quality in different vegetable crops.

Sealed Storage

Seeds are normally produced in large quantity in various vegetable crops; however, the total is relatively less compared to that of field crops such as cereal, pulses, and oilseeds. They are stored in cloth, jute, paper and polyethylene bags, glass and aluminum containers, and laminated pouches involving aluminum foil, polyethylene, and paper in open or sealed conditions. Seed quality is preserved better under sealed storage than open storage. Well-dried seeds are normally packed in sealed containers. Cabbage seeds having low moisture retain high seed quality as compared to seeds with high moisture in sealed containers (Mackay and Flood, 1970). Onion seeds exhibited high longevity when stored in sealed containers at 20°C as compared to those stored in unsealed containers (Tronickova, 1965). Seed storage in sealed containers maintains the seed quality by providing protection to seeds against high atmospheric humidity, insects, and pathogens. Open storage is generally used for bulk handling of seed material using dehumidifiers. Sealed storage of seeds preserves the quality even under ambient conditions for a short period; however, for longer periods it must be combined with low temperature.

Cold Storage

The use of cold storage is the most common and successful means of maintaining high seed quality in several crops including the vegetable crops. Normally low temperature up to –20°C is used for seed storage. In rare instances, ultra-low temperature is employed using liquid nitrogen since its limited availability and high cost prohibits the usage. The temperature in cold storage is selected based on purpose and duration of storage material. Low temperatures (10 to 0°C) are used for short- to medium-term storage while subzero temperatures (–20°C) for long-term storage as prevalent in seed banks. Well-dried seeds are placed in moisture-proof containers such as laminated aluminum foil pouches for longer storage. Cauliflower seeds stored at –20°C retained their original germination for five years and in the same period seeds stored at 15 to 20°C declined in viability markedly (Tronickova, 1965). Onion seed quality was preserved for 15 years both at 5 and –20°C (Doijode, 1998), as were other vegetable crops such as amaranth, bottle gourd, eggplant, longmelon, pepper, pumpkin, and tomato for 15 years and bell pepper, cluster bean, French bean, and radish for 20 years (Doijode, unpublished data). Kretschmer and Waldhor (1997) reported that carrot seeds packed in plastic bags stored at –20°C preserved the seed quality for 11 years. Further, cabbage seeds under similar temperature in airtight jars maintained viability for 23 years (Reitan, 1977). Normally, seeds stored in cold storage do not require any plant protection measures, as prevalent conditions are unfavorable for growth and development of insects and fungi.

Modified-Atmosphere Storage

In modified-atmosphere storage, seeds are packed in an airtight container after creating a partial vacuum or replacing air with inert gases such as nitrogen and carbon dioxide. Tao (1992) reported that vacuum storage of seeds is beneficial to a small number of crops, and the seed quality is affected in the presence of high seed moisture and storage under warm conditions. Even survival of low-moisture seeds is better in a nitrogen environment than in the air, while the reverse is true in high-moisture seeds (Rao and Roberts, 1990). According to Martin et al. (1960) high-moisture seeds stored at high level of oxygen and low level of carbon dioxide result in low germination, while high carbon dioxide and low levels of oxygen tend to maintain seed quality. Lettuce seeds stored with carbon dioxide reduced seed deterioration and preserved seed quality (Harrison and McLeish, 1954). Seed storage in a partial vacuum maintained the seed quality for three years in pepper (Doijode, 1993); in a nitrogen atmosphere for five years in onion (Doijode. 2000b), for seven years in muskmelon (Doijode, 2000a), amaranth (Doijode, 2001a), and watermelon (Doijode, unpublished data); and in carbon dioxide for four years in eggplant (Doijode, unpublished data). Seed storage in a modified atmosphere preserves the seed quality for a shorter period in low-moisture seeds and is beneficial to use especially in absence of a cold-storage facility.

Moist Storage

Seed storage under moist or imbibed conditions is useful for storage of recalcitrant seeds like seeds of *Sechium edule* and *Telfaria occidentalis*, wherein loss of seed moisture affects the seed quality. The fully imbibed lettuce seeds maintain the seed quality for ten months under dark conditions at 25°C (Powell et al., 1983). In moist storage seeds are normally treated with fungicides, packed in perforated thin polyethylene bags, and stored at low temperature (15°C) for maintaining seed quality.

Storage of Pregerminated Seeds

Seeds of vegetable crops, especially hybrids and transgenic species, are expensive and required in only small quantities for sowing. It is common practice in dicot vegetable crops to protect and use costly seedlings for gap filling. Normally germinated seeds are used under unavoidable postponement of sowing; hence, they are preserved for a very short period. The sprouted seeds remain viable for few days and are used immediately for sowing. The sprouted seeds of cabbage, lettuce, and tomato remain viable for five days (Sunil, 1991) and cucumber for ten days (Sunil and Mabesa, 1991). These are packed in glass bottles or laminated aluminum foil pouches and stored at 10°C. Similarly, germinated pepper seeds preserve the seed quality for 63 days when placed in thin polyethylene film with vacuum or nitrogen and stored at 7°C (Ghate and Chinnan, 1987).

REVIVAL OF SEED QUALITY

Seed quality in terms of viability and vigor decreased gradually during storage. The quality can be revived to a certain extent by using a physical stimulus such as electromagnetic force or a chemical stimuli such as seed priming, treatment with water, chemicals, vitamins, and growth substances in different vegetable crops. The quality revival is more effective in low- and medium-vigor seeds, and such seeds performed well under field conditions and gave higher yields.

Magnetic Stimulus

A physical force such as magnetic stimulus improves the seed quality in certain vegetable crops, but their action is not fully known. Seed quality is improved in stored onion seeds by exposing seeds to an electromagnetic field (108 oersteds) for 30 minutes (Alexander and Doijode, 1995). A magnetic stimulus promotes the physiological process and reduces the seed deterioration. More research is needed for wider applications.

Seed Priming

Seed priming is the treatment of seeds before sowing with an osmotic solution. The commonly used and most effective chemical is polyethylene glycol (PEG). Seeds are soaked in osmotic solution that allows seeds to imbibe water and prevents actual germination. These primed seeds are dried to their original level of moisture and stored. This gives early and uniform emergence of seedlings as compared to unprimed seeds and repairs the cellular system during imbibed conditions (Burgass and Powell, 1984). Primed seeds performed better even under unfavorable soil and climatic conditions. The osmotic regulation with PEG slowed seed deterioration and improved the seed quality. Brocklehurst et al. (1987) reported that seed priming with PEG resulted in rapid and uniform emergence of seedling. These are vigorous and speeded up the germination process even at lowest temperature of sowing (Giulianini et al., 1992).

Water

Soaking seeds in water followed by drying improved the seed quality in different vegetable crops. Seeds are soaked in water for 2 to 5 hours (h) or exposed to humid atmosphere for 24 to 48 h followed by drying to original weight, which improves the seed germination and seedling vigor and reduces physiological deterioration during subsequent storage. Such seeds performed well under field conditions and gave higher yield in carrot (Kundu and Basu, 1981) and lettuce (Dowdles, 1960). The water soaking activates the enzyme network and eliminates the deposition of toxic substances from seeds. Tomato seeds soaked in water for 24 h followed by storage at 12°C for 10 days resulted in higher percentage of germination and fruit yield (Petrikova, 1989). The revival of seed quality is attributed to enhancement of metabolic activity (Berrie and Drennan, 1971) and repair of cellular systems during the hydration phase (Villiers and Edgcumbe, 1975). Simple soaking of seeds in water improves their quality as well as protects and promotes the use of expensive, low-vigor seed material.

Chemicals

Seed treatment with salt solution improved the seed quality in certain vegetable crops. Soaking of pepper and tomato seeds in salt solutions followed by drying improved the germination (Woodstock, 1969). Carrot seeds soaked in sodium thiosulphate (0.00001 M) or disodium phosphate (0.0001 M) reduced deterioration in subsequent storage (Kundu and Basu, 1981). Likewise, a year-old lettuce seed improved in quality with applications of potassium iodide (O.000IM), p-hydroxy benzoic acid (0.0001 M), or tannic acid (0.00001 M) treatment (Basu et al., 1979). The chemicals lower the deterioration process by counteracting the free radical damage. In onion, seeds soaked in EDTA (ethylenediaminetetraacetic acid 0.1 M) improved the seed quality in stored seeds (Doijode, 1988a). According to Demopolous (1973), EDTA prevents the formation of the radical center on unsaturated fatty acids and checks the deterioration. These chemicals are helpful in extending the storage life and enhancing the vigor by repair of cellular system (Villiers and Edgcumbe, 1975).

Vitamins

Seed quality is affected by production and accumulation of free radicals in seeds during oxidation of lipids. Low-vigor seeds showed higher level of free radicals than high-vigor seeds. Seed treatment with vitamins such as vitamins C and E prevents oxidation of lipids and thereby reduces seed deterioration to a certain extent (Sathiyamoorthy and Nakamura, 1995).

Growth Substances

Seed treatment with growth substances enhances the seed quality even in stored seeds. Growth promoters such as auxins, indole acetic acid (100 ppm) in okra (Kumar et al., 1996), naphthalene acetic acid (100 ppm) in eggplant (Suryanarayana and Rao, 1984), gibberellic acid (50 ppm) in bell pepper (Mostafa et al., 1982), in eggplant (Demir et al., 1994), and in okra (Kumar et al., 1996), and kinetin improved seed germination and vigor, while inhibitor such as abscisic acid delayed and reduced seed germination (Styer and Cantliffe, 1977).

CONCLUSION

Seed quality is one of the major aspects in crop production. High seed quality comprising high genetic and physical purity, viability, vigor, and seed health is prerequisite for successful cultivation of vegetable crops. The importance of seed quality was known earlier to farmers but its need has been felt more intensely under present conditions owing to greater dependence on agriculture for high food production and security and for greater employment opportunities. Further, vegetable crops provide nutritional security as they are rich sources of vitamins and minerals, which are an important part of human diet. Good-quality seeds are required for raising a healthy crop, and the same is collected on harvest and used for growing the next crop; this cycle then repeats. Several factors are associated with seed quality; some of them are known while others are yet to be explored. Good quality seed begins with selection of viable, bold, lustrous, healthy seed which is able to grow under biotic and abiotic stresses in field, that emerges into a healthy plant, finally terminating with harvest of quality seeds. To achieve the higher seed quality much more research is needed in designing and developing simple, practical, effective, and low cost tools for production, maintenance, and revival of seed quality in different vegetable crops.

REFERENCES

Alexander, M.P. and Doijode, S.D. 1995. Electromagnetic field a novel tool to increase germination and seedling vigor of conserved onion *(Allium cepa)* and rice *(Oryza sativa)* seeds with low viability. *Plant Genetic Resources Newsletter* 104:1-5.

Baldo, N.B., Fonollera, V.C., Vallador, D.M., and Panilan, D.E. 1988. Influence of length and storage temperature on the seed viability and seedling growth of tomato. *CMUJ. Sci* (The Philippines) 1:48-58.

Barton, L.V. 1960. Storage of seeds of *Lobelia cardinalis* L. *Boyce. Thompson Inst. Contrib.* 20:395-401.

Baruah, G.K.S. and Paul, S.R. 1996. Effect of seed coat color on germination and seedling vigor of okra *(Abelmoschus esculentus)* in Assam. *Seed Tech. News.* 26:3-4.

Bass, L.N., Clark, D.C., and James, E. 1962. Vacuum and inert gas storage of lettuce seeds. *Proc. Assoc. Off. Seed Anal.* 52:116-122.

Basu, R.N., Pan, D., and Punjabi, B. 1979. Control of lettuce seed deterioration. *Indian J. Pl. Physiol.* 22:247-253.

Belletti, P. 1985. Correlation of weight and external surface of seed to the percentage and rate of germination in *Asparagus plumisus* var. *nanus. Asparagus Res. Newsletter* 3:15.

Berjak, P. and Villiers, T.A. 1973. Ageing in plant embryo: II. Age induced damage and its repair during early germination. *New Phytol.* 71:135-144.

Berrie, A.M.M. and Drennan, D.S.H. 1971. The effect of hydration-dehydration on seed germination. *New Phytol.* 70:135-142.

Brocklehurst, P.A., Dearman, J., and Drew, R.L.K. 1987. Recent developments in osmotic treatment of vegetable seeds. *Acta Hort.* 215:193-200.

Burgass, R.W. and Powell, A.A. 1984. Evidence for repair processes in the invigoration of seeds by hydration. *Ann. Bot.* 53:753-757.

Christensen, C.M. and Kaufmann, H.H. 1969. *Grain storage, the role of fungi in quality loss.* Minneapolis: University of Minnesota Press.

Demir, I., Ellialtioglu, S., and Tipirdamaz, R. 1994. The effect of different priming treatments on repairability of aged eggplant seeds. *Acta Hort.* 362:205-221.

Demir, I. and Ellis, R.H. 1992. Changes in seed quality during seed development and maturation in tomato. *Seed Sci. Res.* 2:81-87.

Demir, I. and Ellis, R.H. 1993. Changes in potential seed longevity and seedling growth during seed development and maturation in marrow. *Seed Sci. Res.* 3:247-257.

Demopolous, H.B. 1973. Control of free radicals in biologic system. *Fed. Proc.* 32:1903-1908.

Doijode, S.D. 1983. Studies on vigor and viability of seeds at different stages of fruit development in tomato. *Singapore J. Pri. Indus.*11:106-109.

Doijode, S.D. 1984. Effect of seed size on the longevity of seeds in French bean. *Singapore J. Pri. Indus.* 12:62-69.

Doijode, S.D. 1988a. Effect of pretreatment on the germination of onion *(Allium cepa)* seeds. *Die Gartenbauwissenschaft* 53:101-102.

Doijode, S.D. 1988b. Studies on partial vacuum storage of onion *(Allium cepa)* and bell pepper (*Capsicum annuum* L.) seeds. *Veg. Sci.* 15:126-129.

Doijode, S.D. 1991. Influence of seed position in fruit on seed viability and vigor during ambient storage of chilli fruits. *Capsicum Newsletter* 10:62-63.

Doijode, S.D. 1993. Influence of partial vacuum on the storability of chilli *(Capsicum annuum)* seeds under ambient conditions. *Seed Res. Spl* 1:322-326.

Doijode, S.D. 1996. Effect of packaging and fruit storage on seed viability, vigor, and longevity in chilli (*Capsicum annuum* L). *Yeg. Sci.* 23:36-41.

Doijode, S.D. 1998. Conservation of *Allium cepa* germplasm in India. *Natl. Symp. Veg. Varanasi Abstr* 3:2.

Doijode, S.D. 2000a. Modified atmosphere storage for conservation of genetic diversity in muskmelon (*Cucumis melo* L). *Natl. Sem. Hitech. Hort. Bangalore* p. 14.

Doijode, S.D. 2000b. Modified atmosphere storage of onion (*Allium cepa* L) seeds. *Proc. Natl. Symp. Onion Garlic. NHRDF Nashik, India,* pp. 162-166.

Doijode, S.D. 2001a. Seed germplasm conservation with modified atmosphere storage in amaranth (*Amaranthus* spp). *Indian J. Plant Genetic Resources.* 14:288-289.

Doijode, S.D. 2001b. *Seed storage of horticultural crops.* Binghamton, NY: Food Products Press.

Dowdles, D. 1960. Germinating lettuce in summer. *Qd. Agric. J.* 86:774.

Eguchi, T. and Yamada, H. 1958. Studies on the effect of maturity on longevity in vegetable seeds. *Natl. Inst. Agri. Sci. Bull. Ser. E Hort.* 7:145-165.

Elaine, M.S. and Shiel, R.S. 1980. Coating seeds with polyvinyl resins. *J. Hort. Sci.* 51:371-373.

Ellis, R.H., Demir, L., and Pieta, F.C. 1993. Changes in seed quality during seed development in contrasting crops. *Proc. 4th Inter. Workshop on Seeds, Angers, France.* 3:897-904.

Ellis, R.H., Hong, T.D., Astley, D., Pinnegar, A.E., and Kraak, H.L. 1996. Survival of dry and ultra dry seeds of carrot, groundnut, lettuce, oilseed rape and onion during five years of hermetic storage at two temperatures. *Seed Sci. Technol.* 24:347-358.

Frohlich, H. and Henkel, A. 1964. The problems of the duration of viability and the quality of outdoor cucumber seeds. *Die Gartenbauwissenschaft* 11:130-132.

Frost, D.J. and Kretchman, D.W. 1989. Calcium deficiency reduces cucumber fruit and seed quality. *J. Am. Soc. Hort. Sci.* 114:552-556.

Ghate, S.R. and Chinnan, M.S. 1987. Storage of germinated tomato and pepper seeds. *J. Am. Soc. Hort. Sci.* 112:645-651.

Giulianini, D., Nuvoli, S., Pardossi, A., and Tognoni, F. 1992. Pregermination treatment of tomato and pepper seeds. *Colture Protette* 21:73-79.

Gray, D. and Steckel, J.R.A. 1984. Vialbility of onion *(Album cepa)* seeds as a infuenced by temperatures during seed growth. *Ann. Appl. Biol.* 104:375-382.

Grubben, G.J.H. 1978. Vegetable seeds for the tropics. *Royal Tropical Institute Amsterdam Bull.* 301:40.

Gupta, R.P., Mehra, U., and Pandey, U.B. 1989. Effect of various chemicals on viability of onion seeds in storage. *Seed Res.* 17:99-101.

Hadavizadeh, A. and George, R.A.T. 1989. The effect of mother plant nutrition on seed yield and seed vigor in pea *(Pisum sativum)* cultivars IV. International Symposium on Seed Research in Horticulture. *Acta Horic.* 253:55-61.

Harrington, J.F. 1972. Seed storage and longevity. In T.T. Kozlowski (ed.), *Seed biology,* Volume 3 (pp. 145-245). New York: Academic Press.

Harrison, B.J. and McLeish, J. 1954. Abnormalities of stored seeds. *Nature* 173:593-594.

Horky, J. 1991. The effect of temperature on the long term storage of dry seeds of some selected vegetables. *Zahradonietvi* 18:29-33.

Hunje, R.V., Kulkarni, G.N., Shashidhar, S.D., and Vyakaranahal, B.S. 1990. Effect of insecticides and fungicides treatment on cowpea seed quality. *Seed Res.* 18:90-92.

Incalcaterra, G. and Caruso, P. 1994. Seed quality of winter melon *(Cucumis melo* var. *indorus)* as influenced by the position of fruits on mother plant. *Acta Hort.* 362:113-116.

Isely, D. and Bass, L.N. 1960. Seeds and packaging material. *Proc. 14th Hybrid Corn Indust. Res. Conf.*, pp. 101-110.

Jalink, H., Franadas, A., Schoor, R., and Bino, J.B. 1998. Chlorophyll fluorescence of the testa of *Brassica oleracea* seeds as an indicator of seed maturity and seed quality. *Proc. Seed Biol. Symp. Piracicaba, Sau Paulo, Brazil*, pp. 24-27.

Justice, O.L. and Bass, L.N. 1978. *Principles and practices of seed storage*. Washington, DC: USDA.

Kretschmer, M. and Waldhor, O. 1997. Seed storage at –20°C of lettuce, dwarf bean and carrot. *Gemuse Munchen* 33:504-555.

Kumar, S., Singh, P., Katiyar, R.P., Vaish, C.P., and Khan, A.A. 1996. Beneficial effect of some plant growth regulators on aged seeds of okra under field conditions. *Seed Res.* 24:11-14.

Kundu, C. and Basu, R.N. 1981. Hydration-dehydration treatment of stored carrot seeds for maintenance of vigor, viability and productivity. *Scientia Hort.* 15:117-125.

Laurico, N.C. 1984. Study on the indigenous techniques used by farmers in storing vegetable seeds for planting. *Central Luzon State Univ. Sci. J.* 5:167-168.

Lee, S.K. and Nichols, M.A. 1978. Some aspects of seed size and plant spacing on the maturity characteristics of radish. *Acta Hort.* 72:191-199.

Liou, T.D., Waaenvoort, W.A., Kraak, H.L., and Karssen, C.M. 1989. Aspect of low vigor of cabbage seeds. *J. Agric. Res. China.* 38:429-437.

Mackay, D.B. and Flood, R.J. 1970. Investigations in crop seed longevity: I. The viability of *Brassica* seeds stored in permeable and impermeable containers. *J. Natl. Inst. Agril. Bot.* 12:84-99.

Martin, J.A., Seen, T.L., and Crawford, J.H. 1960. Response of okra seed to moisture content and storage temperature. *Proc. Am. Soc. Hort. Sci.* 75:490-494.

Medina, P.V.L., Medina, R.M.T., and Shimoya, C. 1972. Okra seed coat anatomy and the use of chemicals to hasten germination. *Revista Ceres* 19:385-394.

Mihajlovski, M. and Kostov, T. 1998. Some quality properties of the cabbage seed related to the color of the seed coat. *Macedonian Agril. Rev.* 45:1-2.

Mostafa, H.A.M., Mohamedian, S.A., and Nassar, S.M. 1982. A study on improving germination of old seeds of sweet pepper *(Capsicum annuum). Res. Bull. Fac. Agr. Ain. Shams University* 1808:17.

Nakagawa, J., Zanin, A.C.W., and Pizzigatti, R. 1991. Effect of seed size and storage on seed quality in okra cv Amarelinho. *Horticultura-Brasileira* 9:84-86.

Nene, Y.L. 1999. Seed health in ancient and medieval history and its relevance to present day agriculture. *Asian Agril. History* 3:157-184.

Nerson, H. 1991. Fruit age and seed extraction procedures affect germinability of cucurbit seeds. *Seed Sci. Technol.* 19:185-195.

Nerson, H. and Paris, H.S. 1988. Effects of fruit age, fermentation and storage on germination of cucurbit seeds. *Scientia Hort.* 35:15-26.

Padrit, J., Hampton, J.G., Hill, M.J., and Watkin, B.R. 1996. The effects of nitrogen and phosphorus supply to the mother plant on seed vigor in garden pea (*Pisum sativum* L) cv Pania. *J. Appl. Seed Production* 14:41-45.

Pavelkova, A. and Curiova, S. 1984. Effect of severe dehydration on the biological quality of pea seeds. *Genetika a Slechteni* 20:251-256.

Petrikova, K. 1989. Results of the effect of four temperatures on tomato seed swelling. *Acta University Agric. Fac. Hort.* 4:47-51.

Petrov, H. and Dojkov, M. 1970. Influence of some factors on seed quality and germinability in eggplants. *Gradinarstvo.* 128:29-31.

Pollock, B.M. and Roos, E.E. 1972. Seed and seedling vigor. In T.T. Kozolowski (ed.), *Seed biology,* Volume I (pp. 313-387). New York: Academic Press.

Powell, A.D., Leung, D.W.M., and Bewley, J.D. 1983. Long term storage of dormant Grand Rapids lettuce seeds in the imbibed state: Physiological and metabolic changes. *Planta* 159:182-188.

Rao, N.K. and Roberts, E.H. 1990. The effect of oxygen on seed survival and accumulation of chromosome damage in lettuce *(Lactuca sativa). Seed Sci. Technol.* 18:229-238.

Reitan, A. 1977. Storage of cabbage seeds. *Forskning og Forsok i Landburket* 28:487-495.

Roberts E.H. 1986. Quantifying seed deterioration. In McDonald, M.B. and Nelson, C.J. (eds.), *Physiology of seed deterioration* (pp. 101-123). Madison, WI: Crop Science Society of America.

Roberts, E.H. and Abdalla, F.H. 1968. The influence of temperature, moisture and oxygen on period of seed viability in barley, broad bean and peas. *Ann. Bot.* 32:97-117.

Sanchez, V.M., Sundstrom, F.J., McClure, G.N., and Lang, N.S. 1993. Fruit maturity, storage and post harvest maturation treatment affects bell pepper (*Capsicum annuum* L) seed quality. *Scientia Hort.* 54:191-201.

Sathiyamoorthy, P. and Nakamura, S. 1995. Free radical induced lipid peroxidation in seeds. *Israel J. Plant Sci.* 43:295-302.

Stanwood, P.C. and Sowa, S. 1995. Evaluation of onion *(Album cepa)* seeds after 10 years of storage at 5, –18°C and –196°C. *Crop Sci.* 35:852-856.

Styer, R.C. and Cantliffe, D.J. 1977. Effect of growth regulators on storage life of onion seeds. *Proc. Fl. St. Hort. Soc.* 90:415-418.

Sunil, G.D.J.L. 1991. Storage of pregerminated vegetable seeds. *Laguna Coll. Tech. Bull,* p. 123.

Sunil, G.D.J.L. and Mabesa, R. 1991. Storage techniques for extending the viability of pregerminated cucumber seeds. *Trop. Agril.* 147:19-31.

Suryanarayana, V. and Rao, V.K. 1984. Effect of growth regulators on seed germination in okra, tomato and brinjal. *Andhra Agric. J.* 31:220 224.

Szafirowska, A.I. 1994. The correlation between mother plant architecture, seed quality and field emergence of carrot. *Acta Hort.* 354:93-97.

Tao, K.L. 1992. Should vacuum packing be used for seed storage in gene bank? *Pl. Genetic Resources Newsletter* 88/89:27-30.

Tokumasu, S. 1975. Prolonged dormancy in seed preserved in harvested fruits of *Brassica* vegetables. *Scientia Horticulturae* 3:267-273.

Townsend, C.E. 1979. Association among seed weight, seedling emergence and planting depth in *Cicer milkevetch. Agron. J.* 71:410-414.

Tronickova, E. 1965. The effect of sub-zero temperatures on the biology of some vegetable seeds. *Ved. Pr. ustred. vyzk. Ust rostl Vyroby v praze Ruzyni,* 9:27-39.

Usik, G.E. 1980. Effect of seed size on emergence and yield in onion and carrot. *Intensifik Ovosch Kishinev Mold,* pp. 30-33.

van der Burg, W.J., Aartse, J.W., van Zwol, R.A., Jalink, H., and Bino, R.J. 1994. Predicting tomato seedling morphology by x-ray analysis of seeds. *J. Am. Soc. Hort. Sci.* 119:258-263.

Villareal, R.L., Balagedan, J.B., and Castro, A.D. 1972. The effects of packing materials and storage conditions on the vigor and viability of squash *(Cucurbita maxima),* patola *(Luffa acutangula),* and upo *(Lagenaria siceraria)* seeds. *Philippines Agriculturist* 56:59-76.

Villiers, T.A. and Edgcumbe, D.J. 1975. On the cause of seed deterioration in dry storage. *Seed Sci. Technol.* 3:761-764.

Waiters, C. 1998. Ultra dry technology: Perspective from the National Seed Storage Laboratory USA. *Seed Sci. Res.* 8:11-14.

West, S.H. and Harris, H.C. 1963. Seed coat colors associated with physiological changes in alfalfa crimson and white clover. *Crop Sci.* 3:190-193.

Wester, R.E. 1964. Effect of size of seed on plant growth and yield of Fordhook 242 bush lima bean. *Proc. Am. Soc. Hort. Sci.* 84:327-331.

Woodstock, L. 1969. Biochemical tests for seed vigor. *Proc. Inter. Seed Test Assoc.* 34:253-261.

Zhang, M., Liu, Y., Toril, L, Sasaki, H., and Esashi, Y. 1993. Evolution of volatile compounds by seeds during storage periods. *Seed Sci. Technol.* 21:359-373.

Chapter 24

Vegetable Hybrid Seed Production in the World

David Tay

INTRODUCTION

Vegetables consist of many species and cultivars, and can be classified in many ways such as by botanical family and species, cultivar group, the edible part, climatic region, user and ethnic origin, life cycle, and nutrition. In seed production, vegetables can be categorized into open-pollinated, F1 hybrid, and clonally propagated cultivars. The trend of F1 hybrid seed usage in vegetable crops is increasing global in term of species, cultivars, and volume of seed used. Most of the seed of our main vegetables (e.g., tomato, sweet pepper, eggplant, cucumber, squash, pumpkin, melon, watermelon, brassicas such as cabbage, cauliflower, broccoli, Chinese cabbage, and radish, and onion) are of F1 hybrid cultivars. The popularity of F1 hybrid cultivars is due to their vigor, uniformity, disease resistance, stress tolerance, and good horticultural traits, including earliness and long shelf-life, expressed and therefore giving consistently stable high yields. From the breeder's point of view, hybrid seed production is a fast and convenient way to combine desirable characters (i.e., fruit size and color, plant type, and disease resistance) of a vegetable, and as a means to control intellectual property rights through control and protection of the parental lines by the breeders. The latter was the main reason Japanese seed companies applied to protect their cultivars in the 1940s and 1950s (Shinohara, 1984).

In F1 hybrid vegetable seed production, vegetables can be divided into two groups: the hand-pollinated and the gene-control-pollinated species. The genetic control system can be due to the self-incompatible system, in which pollen of the same plant or flower cannot pollinate itself, or to the male-sterile genetic system, in which a female plant has no male organ, a deformed organ or no functional pollen to pollinate itself. When no such genetic control system is found or when it is not introduced into inbred parental lines, tedious hand emasculation and pollination have to be used to produce F1 seed. In both the gene-control system and hand-pollinated species, sufficient field or female flower isolation has to be maintained to obtain high genetic purity in seed. Table 24.1 gives examples of these two groups of species and the trend of F1 hybrid cultivar adoption in the world.

TABLE 24.1. Vegetables with both F1 hybrid and open-pollinated cultivars showing their adoption trends in the world and their F1 seed production method.

Vegetables	**F1**	**OP**
Asparagus *(Asparagus officinalis)*	Mainly (di)	Old cultivars
Beet and chard *(Beta vulgaris)*	Increasingly (di)	Constant
Bitter gourd *(Momordica charantia)*	Increasingly (h)	Local cultivars
Broccoli *(Brassica oleracea)*	Mainly (si)	Local cultivars
Cabbage *(Brassica oleracea)*	Mainly (si)	Local cultivars
Carrot *(Daucus carota)*	Increasingly (cms)	Local cultivars
Cauliflower *(Brassica oleracea)*	Mainly (si)	Local cultivars
Celery *(Apium graveolens)*	New (gms)	Mainly
Chinese cabbage *(Brassica rapa)*	Mainly (si)	Local cultivars
Chinese mustard *(Brassica juncea)*	Increasingly (si)	Local cultivars
Cucumber *(Cucumis sativa)*	Mainly (h)	Local cultivars
Eggplant *(Solanum melongena)*	Mainly (h)	Local cultivars
Gourd *(Benincasa hispida)*	Increasingly (h)	Local cultivars
Leek *(Allium porrum)*	New (cms)	Mainly
Luffa (*Luffa angulata* and *L. cylindrica*)	Increasingly (h)	Local cultivars
Melons *(Cucumis melo)*	Mainly (h)	Local cultivars
Okra *(Abelmoschus esculantus)*	Increasingly (h)	Local cultivars
Onion *(Allium cepa)*	Mainly (cms)	Old cultivars
Pakchoi and petsai *(Brassica rapa)*	Increasingly (si)	Old cultivars
Peppers *(Capsicum annuum)*	Mainly (h)	Local cultivars
Pumpkin *(Cucurbita moschata)*	Increasingly (h)	Old cultivars
Radish *(Raphanus sativus)*	Mainly (si)	Old cultivars
Spinach *(Spinacia oleracea)*	Mainly (di)	Local cultivars
Sweet corn *(Zea mays)*	Mainly (h & cms)	Local cultivars
Tomato *(Lycopersicum esculentum)*	Mainly (h)	Old cultivars
Turnip *(Brassica rapa)*	Mainly (si)	Old cultivars
Watermelon *(Citrullus lanatus)*	Mainly (h)	Old cultivars
Zucchini *(Cucurbita pepo)*	Mainly (h)	Old cultivars

F1: F1 cultivars; OP: open-pollinated cultivars; (di): dioecious; (h): hand-pollinated hybrids; (cms): cytoplasmic male-sterile system hybrids; (gms): genetic male-sterile system hybrids; and (si): self-incompatibility system hybrids.

THE GENE-CONTROL POLLINATION F1 VEGETABLE SEED PRODUCTION SYSTEM

The vegetables with a highly developed self-incompatibility system are those in the family Crucifereae. They include *Brassica oleracea* (brussels sprouts, cabbage, cauliflower, broccoli, kohlrabi, and kale), *Brassica rapa* (Chinese cabbage, turnip, and a range of Asian leafy brassicas), and *Raphanus sativus* (see Table 24.1). The genetics of the self-incompatibility system in cruciferous crops are so well developed that they consist of a series of genes (loci) and alleles. Vegetable breeders have been very successful in using them for decades in F1 hybrid seed breeding. Hybrid seed production of sweet corn, carrot, and onion are based on male sterility gene systems, and the genetic control can be either clear-cut male sterility genes or the interaction of a male sterility gene with a cytoplasmic factor. Brassica breeders have been trying to use a cytoplasmic male sterility system instead of the standard incompatibility system. Some of the difficulties encountered were poor vigor, chlorosis, floral abnormalities, and the reduction in nectary gland size and decreasing function of these glands, which inhibits bee activity resulting in poor pollination and reduced seed set (Opena et al., 1988). The progenies of the male sterility progeny in *Brassica juncea* also gave rise to young leaf yellowing symptom.

THE HAND-POLLINATED F1 VEGETABLE SEED PRODUCTION SYSTEM

Most of the seed of F1 hybrid vegetables are produced by hand pollination as indicated in Table 24.1. The method in principle is simple, as it involves the manual emasculation of the pollen-producing organ (the anthers), followed by hand pollination with pollen of the male parent, then preventing other pollen from contaminating the pollinated flowers. However, it is labor intensive and requires a team of skillful growers and many dedicated pollinators with good eyesight, gentle hands, a lot of patience and commitment, and the ability to follow instructions accurately. The main task of a seed producer is the management of the production system and business. To be cost-effective, this system works only in species for which a single pollination of a female flower will produce many seeds. This is the case for all the solanaceous crops and cucurbits. On the contrary, in legumes the small number of seeds per flower/pod prevents hand pollination to be efficient, and thus no hybrid beans have been produced to date. In this case the use of gene-control pollination has to be exploited. Similarly, if a good gene-control pollination system is available in tomato and pepper, their seed production could be transformed into less intensive large-field production systems as in the brassicas and sweet corn.

DISTRIBUTION OF F1 VEGETABLE SEED PRODUCTION IN THE WORLD

The two systems of hybrid vegetable seed production have different production requirements. The gene-control pollination system requires suitable climatic conditions, good growers with mechanized farms, and high standards of seed quality and seed health control. Many of these locations are in developed agricultural countries such as the United States, Canada, Europe, Australia, and New Zealand. On the other hand, the intensive hand-pollination system demands efficient, low-cost pollination teams, small intensive fields with constant supervision, techniques to harvest and extract pollen from the male parent, efficient techniques and machinery to extract, clean, and dry the valuable hybrid seed produced, and quarantine certification, in addition to the basic requirements. These conditions are usually found in less-developed countries where horticulture is progressive and a contract agreement is respected and kept. The contracting seed company through stringent contract specifications and regular supervision of the seed production field and the harvested seed is able to control and maintain high seed quality.

The distribution of hybrid vegetable seed production in the world is therefore limited to some specific regions (Figure 24.1) where the climate, weather, and availability of good growers are the deciding factors. Vegetables can be classified into three categories according to their temperature requirements as follows:

1. Low-temperature species (e.g., brassicas, radish, carrot, and spinach) which require a low temperature of 8° to 15°C of vernalization to bolt, flower, and set seed;
2. Moderate-temperature species (e.g., tomato, sweet pepper, and zucchini) which require a temperature of around 25°C in the day and 15° to 20°C at night for optimum seed production. The diurnal temperature difference is desired to obtain the best result. Too low a temperature causes low seed set and pollen production, and too high a temperature causes flower abscission, low pollen production and viability, and pest and disease problems; and
3. High-temperature species (e.g., okra, cucurbits, sweet corn, and tropical vegetables) which require a temperature of 20°C and above.

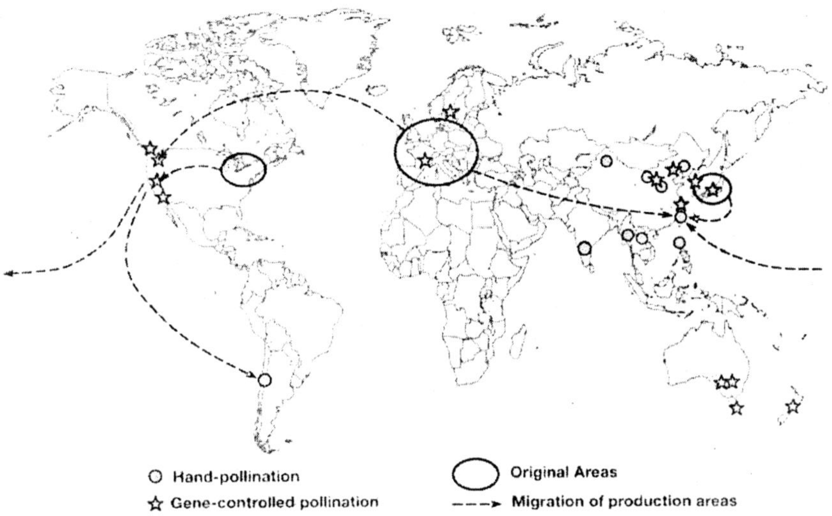

FIGURE 24.1.The main past and present F1 hybrid vegetable seed production areas in the world.

This temperature requirement is attained at different latitudes by a combination of climatic season and elevation above sea level. For instance, in tomato, a moderate-temperature-requirement vegetable, seed production can be carried out equally well in the winter season of a subtropical regions (e.g., Taiwan and Southeast Asia) or in the summer season in temperate regions such as northeast and northwest China, and in Chile. In the tropics, seed production is sometimes achieved in highlands of 500 to 2,000 m altitude where the cool temperature is suitable for the moderate-temperature-loving vegetables to produce seed. Photoperiodic reaction is not a concern, as most of the modern cultivars of the moderate temperature requirement species are day-neutral plants and thus insensitive to photoperiod. The low-temperature-requirement vegetables often require a specific period of cold vernalization in their growth phase to induce bolting and flowering. Seed of these species are often produced in the higher latitudes or coastal regions e.g., the northwest United States. However, some of the heat-tolerant cultivars of these cold-loving vegetables (e.g., some cruciferous crops and carrot) require less vernalization time and also possess a shorter day photoperiod requirement, and thus their seed production has been successfully carried out in subtropical areas.

Hand-pollinated F1 hybrid seed occurs mainly in East Asia (China and Taiwan) and Southeast Asian highlands (northern Thailand and northern

Philippines), India (Karnataka and Andhra Pradesh), Mexico, and Chile. Gene-controlled pollinated seed production is found largely in the United States (California—brassicas, onion, and radish; Idaho—brassicas, onion, and radish; Oregon—brassicas, onion, radish, and turnip; and Washington State—brassicas, carrot, onion, radish, and turnip), Canada (British Columbia—crucifers), Denmark (crucifers), Australia (southeast Australia including Tasmania—crucifers and onion), New Zealand (crucifers), and in East Asia, China (northeast, north, and northwest China—crucifers, central and southeast China), Korea (crucifers), Japan (crucifers), and Taiwan (crucifers).

DEVELOPMENT OF HAND-POLLINATED HYBRID VEGETABLE SEED PRODUCTION IN THE WORLD

Post–World War II Period

Following the success of F1 hybrid sweet corn breeding in the United States in the 1940s and 1950s, other F1 hybrid vegetables were developed. The end of World War II could be said to mark the beginning of the global hybrid vegetable seed industry when the seed production technologies were spread from the United States and Japan to other countries. Japan was already exporting hybrid tomato and eggplant in the 1950s. The main recipient of this technology was Taiwan, which was to develop into one of the most successful countries in producing hand-pollinated F1 hybrid vegetable seeds. The combination of United States and Japanese know-how in Taiwan has resulted in an efficient production system. During the peak of the industry in the 1970s, Taiwan produced most of the world's F1 hybrid tomato seed, as well as watermelon, melon, sweet pepper, eggplant, and cucumber (unpublished personal survey with major international seed company personnel and Taiwanese seed growers in 1983). Hot pepper and other lesser cucurbits such as bitter gourds, luffa, pickling melon, and sponge gourds were also introduced in the early 1980s (based on seed catalog of Known-You Seed, 1982). Almost all of the hybrid seed produced were for U.S., Japanese, and European seed companies.

Some of the international companies decided to establish offices and research farms, or joint partnerships with local Taiwanese seed companies. For example, Petoseed (presently, Seminis) acquired Wann Shiang to form Peto Wann Shiang, which later served as Petoseed's technical base for its entry into other Asian countries including China, Thailand, and India. On the other hand, Sluis & Groot (presently, Syngenta) decided to enter into partnership with Ching Choung to form Fu Lan, which then operated similarly to other local seed companies. Sakata Seed of Japan decided to establish its own research farm there. The prominent local seed companies were Known-You Seed and Evergrow Seed, and they are still the main vegetable seed companies there. Known-You has been the winner of several All-America Selections awards in past years including most recently in 2001 for its melon hybrids.

Seed production areas were concentrated in southern Taiwan in Kaohsiung, Pintung, Chaiyi, and Tainan Prefecture with approximately 40, 40, 14, and 6 percent of the total production, respectively (see Figure 24.2) (unpublished personal survey with major international seed company personnel and Taiwanese seed growers in 1983). Several factors contributed to this success and development, and the four main factors are as follows:

1. The cool dry climate with plenty of sunshine and good irrigation water during the winter season is ideal for growing moderate-temperature-requirement crops such as tomato, melon, and pepper and thus their seed production. In general, the day and night temperature difference at this period is optimal for good fruit-set. In addition, the dry climate allows the soil moisture to be accurately controlled by furrow irrigation for optimal growth and seed development.
2. The seed crop fits well into the existing paddy rice-based cropping system which ensures better farmland utilization and pest and disease control, especially soil-borne pathogens such as bacterial and fungal wilts, root-knot nematode, viruses, and others. This is important because some of the parental inbred lines, due to their inbreeding or polyploidy origin as in the case of seedless watermelon seed production, are often not so vigorous, so specific cultural techniques have to be used to grow them successfully. In addition, every new field has to be tried out to identify the species and best hybrid combinations that it can produce.
3. The general crop management skill of the seed growers was high, including abilities in raising seedlings, irrigating, fertilizing, pruning, staking, and general field cleanliness and hygiene. The growers had the economic capability to invest in labor and farm inputs in this intensive undertaking. Skillful and patient pollinators with delicate and stable hands were available. Hence, a successful hand-pollinated hybrid seed production enterprise depends both on the availability of technical knowledge at the actual operational level of seed production and the ability and willingness of the growers to participate in this high-risk, high-capital cost, and slow-turnover investment.

4. An efficient contract seed production system was developed and established between the international seed companies and specialized local seed contract production companies. Those international companies even after setting up their own offices in Taiwan continued to use some of the services of the local production companies and their area seed agents to implement their seed production targets. The contract system is summarized in Figure 24.3.

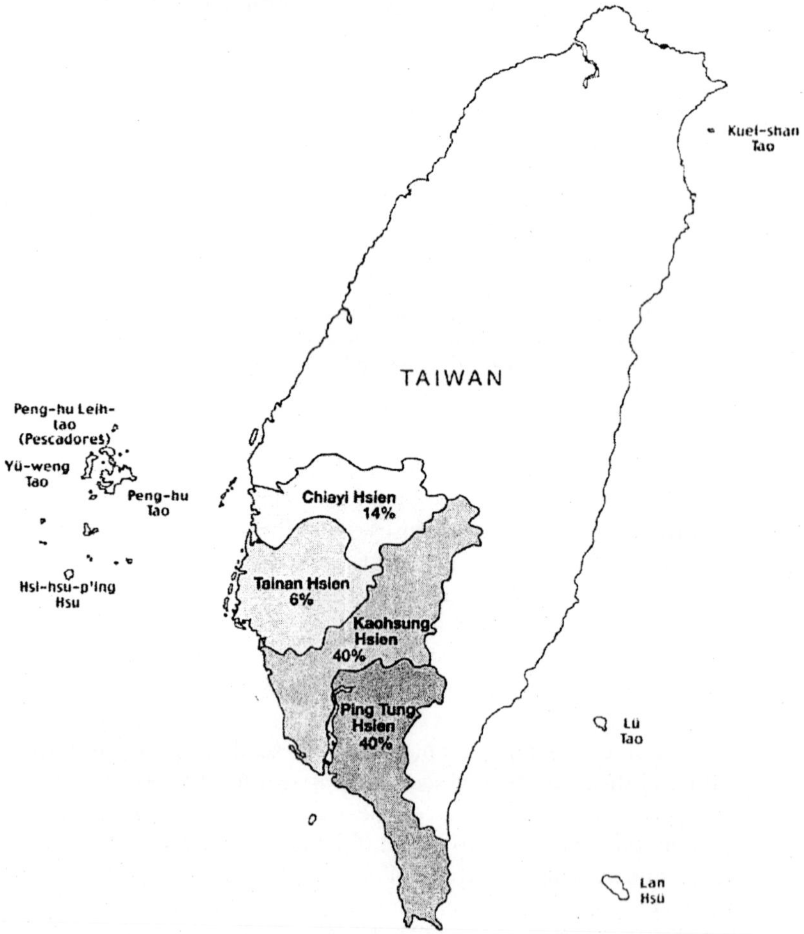

FIGURE 24.2.The main F1 hybrid tomato seed production areas in Taiwan in the 1980s.

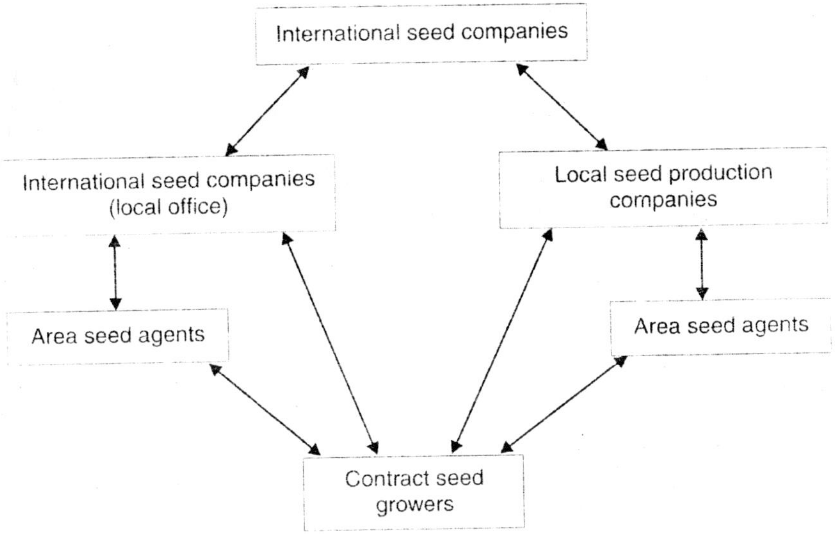

FIGURE 24.3. Summary of contract system of F1 hybrid vegetable seed production in Taiwan.

Both local and overseas companies in Taiwan were equipped with appropriate modern machinery and had highly trained technical staff who worked directly with the growers or through their area seed agents. The responsibilities of the area agents and company staff were to recruit potential seed growers, to negotiate the production contracts, and to train and retain good growers. The contract covered agreements on the amount of seed to be produced, seed price, time of delivery, terms of payment, and seed standards including germination rate, genetic purity, and moisture content. Sometimes, certain production inputs (e.g., mechanized wet seed extractors) had to be loaned to growers depending on the contract agreement. Once a contract was signed, company personnel issued seeds of the parental lines and provided expertise on seed production ranging from crop management practices to pollination and seed processing techniques to the growers. The most important duties of the seed company staff and agents were to ensure the contract quality and quantity of the seed produced under their supervision. The seed standard of tomato was a genetic purity of at least 98 percent, germination rate of above 85 percent, and moisture content of less than 8 percent. Normally, a grower would be paid around 45 days after seed submission to the seed company. Due to the close supervision of seed company's staff and agents, only occasionally would a seed lot be rejected. Through the contract production system the seed company technical staff could also plan and locate seed production fields to ensure sufficient isolation distances between species, and each company usually would develop its own production areas and respect those of another company. This contract production system has led to the development of the "seed village" concept where the whole village will be trained to produce hybrid vegetable seed.

From Early 1980s to Present

The era from the 1960s to 1980s marked the beginning of large-scale hand-pollinated hybrid vegetable seed production in Taiwan and in the world. When the cost of production started to increase in the mid-1980s in Taiwan, due mainly to its manufacturing industry competing for labor, both local and international companies started to seek new production sites in other countries, including China, Thailand, the Philippines, and India. For example, Peto Wann Shiang was to dispatch and station its Taiwanese technicians in China and India for the entire seed production season.

In China, several very successful production locations have been established in northeast, north, and northwest China, with some locations in the area of south and central China. China has positioned itself as one of the main hand-pollinated F1 hybrid vegetable seed production countries in the world. The expansion of Chinese domination is to a great extent the result of the Chinese government policy to commercialize its seed industry from a centrally planned operational system where seed production targets were handed down from the top into a more open business and profit-oriented industry. The state-owned seed companies are facilitated to operate in a reasonably independent manner and they seek outside funds such as World Bank loans to modernize and to train a new class of market-oriented managers. They seek and forge corporation among themselves and with international companies to build competitive advantages in the market (Tay, 1997).

In Thailand, vegetable hybrid seed production is concentrated in the northeast, where cool winter weather is drier and thus has less disease and pest problems as compared to northern Thailand. However, the cooler weather in the highlands of northern Thailand continues to offer suitable locations both for seed production and plant breeding stations (personal observations in the 1980s to 1990s). This industry has shown signs of stabilization and consolidation due to the increasing wages paid to hand pollinators. In the Philippines, seed production is concentrated in Northern Luzon and is not expanding, and has stabilized as in Thailand (personal observations in the 1980s to 1990s). In India, vegetable production is expanding at a rapid rate, with a threefold increase during the past 50 years. A large area is now planted with F1 hybrid cultivars and thus hybrid vegetable seed production is growing at a rapid rate. The production is mainly concentrated in Karnataka and Andhra Pradesh (personal communication with international vegetable seed companies). All of these new production locations in East, Southeast and South Asia are based on Taiwanese technology and the rice paddy-based cropping system.

In the western hemisphere, Mexico—due to its proximity to the United States and the availability of hand labor—is also established in Baja California. In Chile, the reverse season in the southern hemisphere and suitable climates have provided specific advantages that are not found elsewhere, and a substantial hybrid vegetable seed production industry has been established.

F1 HYBRID VEGETABLE SEED PRODUCTION: A CASE STUDY ON TOMATO IN TAIWAN

The goal of a seed grower is to produce good quality seed and to make a profit by increasing seed yield and reducing labor and other farm input costs. To achieve this, both the growers and the seed companies have developed their own specific techniques to be more competitive than others. Many growers, because of their experience in managing their own fields, have modified the standard management recommendations to suit their fields. Some of the modified techniques are unsuitable for other fields, and others have no obvious advantages over the standard methods.

The following are the standard techniques in use for commercial F1 tomato hybrid seed production in Taiwan.

Crop Management Practices

In tomato, higher fruit yield generally gives higher seed yield, but this applies only within each of the genetic types. For example, the Roma type generally gives fewer seed per fruit as compared to medium-size globe fruit type. Most reciprocal crosses in tomato express almost the same hybrid vigor, and therefore the better seed yielding line is always used as a female. Optimal crop management practices should therefore be adopted to grow a healthy and vigorous crop with a more concentrated flowering period to reduce pollination days and increase fruit set. The application of more phosphorus and potassium fertilizers is reported to increase seed filling and thus seed yield.

1. *Planting season.* The optimal planting season is from September to October at the onset of the cool season. This will allow the fruit to ripen in the cool, dry month of February, thus reducing disease occurrence and facilitating seed drying.
2. *Soil and location.* Well-drained sandy loams are selected, but fields with deep sandy soil along the banks of rivers have been successfully used when fertilization and irrigation are well managed. Fields immediately after rice paddy culture are selected because they are relatively free of pests and diseases, and they should be away from commercial tomato production areas.
3. *Female to male parent ratio.* The ratio of female to male parents is normally four to six females to one male depending on the flowering ability, pollen productivity, and fertility of the male.
4. *Synchronizing sowing and planting.* Depending on the flowering date difference between the male and female parents, sowing dates are adjusted. However, the male is normally sown one to two weeks earlier in order to produce enough pollen for pollination when the females are ready. This also provides more time for the off-types to express themselves, thus more accurate roguing and better genetic purity.
5. *Stagger planting.* Stagger planting is commonly practiced to spread out the concentration of workers at any one time. This eases the problem of shortage of experienced pollinators.
6. *Transplanting.* Seedlings are raised in small plastic pots or flats. The male and female parents are planted in separate fields or in different sections of the same field. They are usually planted in double rows per bed with within row spacing of about 0.40 to 0.45 m and between row of 0.80 m for the female. The bed height is normally about 0.25 to 0.30 m. The wide spacing of the female rows allows easy movement for pollinators during emasculation and pollination.
7. *Staking and pruning.* Depending on the availability of labor and capital, staking of the female plants helps to facilitate hand emasculation and pollination, and if the weather is wet it also reduces diseases. Seed yield will therefore increase. A comparison of the two staking systems—tee-pee over a raised bed and tee-pee over a furrow—and no staking is given in Table 24.2.
8. *Purity of the parental lines.* The parental lines are supplied by the seed companies, and seed genetic purity is usually high. Optimum crop management is used to allow all the plants to express their full potential so that all off-types can be recognized easily during field inspection and roguing from the beginning of the production season before they contaminate the production. Extra precaution and effort are emphasized in the male parent because when contamination occurs it

could be extensive, as it is not limited to an individual as in the case of a female plant. Several rounds of field inspection are done at different growth stages including seedling, transplanting, growing, flowering, etc. to rogue all suspected off-types and volunteer plants. Every plant is inspected because of the high demand in cultivar purity of better than 98 percent. In addition, nonemasculated flowers and selfed fruits on the female plants are removed during emasculation and pollination rounds, and at harvesting.

Emasculation and Pollination

The tomato style and stigma are normally enclosed by the anther cone, and the stigma is often at the same level or below the tip of anther cone. As the result, tomato is predominantly self-pollinated with only about 2 percent of natural out-crossing. It is therefore safe to leave the female flowers uncovered after emasculation and pollination when isolation distance of at least 50 m between two lines is provided in hybrid seed production. However, usually greater isolation distance is planned for in the field.

TABLE 24.2. Comparison of the three staking systems in tomato F1 hybrid seed production.

Staking method	Advantage	Disadvantage
Tee-pee over a raised bed	• Easier hand emasculation and pollination • Fewer diseases • No direct soil compaction on the plants	• Extra stake cost and labor for staking • Extra weeding of furrows • Wet working path in wet weather and after irrigation
Tee-pee over a furrow	• Easier hand emasculation and pollination • Fewer diseases • Dry working path in wet weather and after irrigation • No weeding of furrows once covered by the tomato	• Extra stake cost and labor for staking • Direct soil compaction on the plants
No staking	• Low capital cost and less risk • No direct soil compaction on the plants	• Emasculation and pollination more difficult • More diseases • Extra weeding of furrows • Wet working path in wet weather and after irrigation

1. *Emasculation.* The female flowers are emasculated, usually starting from the second cluster up with a forcep at two days before anthesis. Too-early emasculation can damage the bud, and too-late stage increases the chances of selfing. The whole anther cone can be taken out or each anther is removed individually. Usually, emasculation is done in the afternoon after the emasculated flowers from the previous day were pollinated.
2. *Pollen collection.* Male parent flower buds that will open the following day are picked in the afternoon, the anthers are separated that night to dry, and the pollen is extracted the following morning for pollination. Once the anthers split, pollen can then be shaken out in a closed container; the pollen is separated from other flower parts by sieving through a 300-mesh screen. Alternatively, buds that will open that day are collected, allowed to open, and the pollen extracted for pollination. Tomato pollen can be stored in a cool dry place for several weeks in moisture-proof containers with calcium chloride.
3. *Pollination.* Pollination is done first thing in the morning. The emasculated flowers are pollinated on the day of flowering by using either the little finger dipped with viable pollen or with a special small vial that has a ring to be worn on the pollinator's finger. It is important to introduce enough pollen onto the stigma to ensure high seed set because normally pollination is done only once in a flower and one pollen can give only one seed. Two to three sepals are cut with a small pair of scissors to indicate that the flower has been pollinated and to identify the hybridized fruit during harvesting. Generally, setting of four to five fruits per cluster and five to eight clusters are sufficient to give good yield. However, individual plants with up to 80 pollinated fruits are not uncommon in a good season. Pollination one to two days before blooming gives low fruit set and seed yield, whereas one day after blooming has no detrimental effect. The pollination period normally last about a month to a month and half with about 40 to 60 workers per hectare per day.
4. *Postpollination cultural management.* Immediately after the final round of pollination the female plants are pruned to remove new growth, first to prevent formation of new flowers and thus selfed fruits, and second to reduce competition for nutrients with the pollinated fruits. The field is given another top-dressing which is followed by furrow irrigation if needed. Pest and disease control are rigorously implemented until harvesting. All selfed fruits found are removed throughout the season to avoid seed contamination.

Harvesting, Seed Extraction, and Processing

1. *Harvesting.* Fully matured fruits with full color are harvested and any selfed fruits found should be discarded. The fruits can be immediate seed extraction or kept in a cool place for three to four days for post-harvest maturation before extraction. However, some crosses with no seed dormancy may have to be harvested a little earlier than full fruit maturity and should not allow for postharvest incubation because seed could start to germinate in the fruit. A simple field test to determine seed readiness for harvesting is to cut a fruit with a sharp knife and if the seed are not being cut they are ready for harvesting.
2. *Seed extraction and processing.* Most growers use a mechanical "wet" seed extractor which can process about 2 to 3 t of fruits per hour. A wet seed extractor consists of a feeder, a cutting component, fruit pulp and seed separating cylinder, seed strainer cylinder, and a water-jet flashing system to assist the seed and pulp separation. The seed with the mucilagous coating is separated from the pulp. This mucilagous coating can be removed by acid treatment or fermentation. The acid method consists of mixing thoroughly in proportion of 7 ml of commercial grade concentrated hydrochloric acid (19 to 21 percent) in every kg of wet extracted seed for 20 minutes. Washing is done immediately when treatment is completed using the decantation washing process. Alternatively, the fruits are squashed in the bags manually by stepping on them until all the fruits are broken and the seeds have come out of the fruits. The slurry of the squashed tomato in the bags is then allowed to ferment in a nonstaining container away from direct sun and rain. No water should be added during squashing and fermenting. The fermentation process may take two to three days depending on the ambient temperature. Warm days allow faster fermentation. The fermentation is indicated by the bubbling CO_2 gas releases, and the slurry swells and produces heat. Fermentation completion is recognized by the reduced bubbling activity, cooling of the slurry, the swollen slurry subsiding to its initial volume, and the clearing up of the supernatant of the slurry. The seed may then be acid treated to disinfect the seed and to improve the seed color.
3. *Seed drying.* The washed seed are bagged into nylon netting bags and spin-dried with a standard laundry centrifuge before being spreading out into a thin layer of less than 0.5 cm thickness on fine-mesh netting trays for drying. In solar drying, partial shade is provided during hot midday hours with a layer of clear fine-mesh netting. The seed are turned regularly to allow uniform drying and to break down the seed

clumps into individual seed. Dried seed of about 7 percent seed moisture wet basis are packed in multiple layers of plastic bag for submission to seed companies.

4. *Quality testing.* The individual seed companies control their own seed quality. Seed germination is carried out following ISTA seed testing method and rules. In the field genetic purity is done in the greenhouse or field by observing for specific morpho-physiological markers, disease resistance markers, and DNA, isozyme, and protein markers.

A well-managed seed production field coupled with optimal climatic conditions will give a seed yield of 140 to 200 kg·ha^{-1}. The success of this seed industry depends on a set of complementary technical, environmental, and social inputs and considerations. The lack of a single attribute will negate the others, resulting in poor performance.

FUTURE OF HAND-POLLINATED F1 VEGETABLE HYBRID SEED PRODUCTION

The present hand-pollinated F1 vegetable hybrid seed production industry is entering a stabilizing phase following the rush to look for new locations after Taiwan became to costly to be competitive in the early 1980s. This stabilizing phase will become even more settled in the coming years as both China and India, the current main production countries, still possess the flexibility and cushion in terms of available new locations and human resources to provide for new expansion and cost-effective pollinators, respectively. As the adoption of F1 hybrid cultivar vegetables is growing globally, the vegetable seed industry will thus continue to explore and move to new, better, and more cost-effective areas and countries.

REFERENCES

Opena, R.T., G.G. Kuo, and J.Y. Yoon (1988). Breeding and seed production of Chinese cabbage in tropics and subtropics. Technical Bulletin No. 17. AVRDC, Shanhua, Tainan.

Shinohara, S. (1984). *Vegetable seed production technology of Japan—Elucidated with respective variety development histories, particulars,* Volume 1. Tokyo, Japan: Shinohara's Authorized Agricultural Consulting Engineer Office.

Tay, D. (1997). China: A seed industry in transition. *Asian Seed* 4(1):16-17.

Chapter 25

Practical Hydration of Seeds of Tropical Crops: "On-Farm" Seed Priming

D. Harris
A. Mottram

Failure to establish an adequate stand precludes the fulfillment of maximum yield potential in any crop (Monteith and Elston, 1983; Harris, 1992). Resowing a poorly established crop is possible if time and resources are available, although yield and cost penalties are often associated with late sowing in many systems (e.g., Harrington et al., 1993; Ortiz-Monasterio et al., 1994; Chiduza et al., 1995). In marginal areas, many resource-poor farmers cannot afford replacement seed without recourse to loans, and failure to establish a crop can be the prelude to a spiral of indebtedness (Harris et al., 1998). Unfortunately, achieving good crop establishment is a difficult task for such farmers, particularly in semiarid areas where seedbeds dry out and heat up rapidly.

Many authors have demonstrated that fast germination and emergence are associated with vigorous seedling growth and, often, higher yields (Okonwo and Vanderlip, 1985; Austin, 1989; Carter et al., 1992; Harris, 1992, 1996). In temperate, more developed agriculture and horticulture, seed technologists have developed the concept of seed priming (seed hardening, seed conditioning) to promote fast, vigorous, synchronous seedling emergence (see Parera and Cantliffe, 1994 for a review). Such controlled hydration, followed by rapid re-desiccation, of seeds is expensive, however, and is usually available to farmers only as a value-added service provided by seed suppliers. The technique achieves its effect by stimulating enzyme sys-

---

We wish to thank all colleagues for providing seed to test, in particular the DR&SS, Zimbabwe (maize), KRIBHCO, India (wheat, chickpea), WARDA, West Africa (rice), and ICRISAT (sorghum and pearl millet). This chapter is an output from a project (Plant Sciences Research Programme R6395) funded by the UK Department for International Development and administered by the Centre for Arid Zone Studies for the benefit of developing countries. The views expressed are not necessarily those of DFID.

tems and other cellular components of germination, and is generally targeted toward systems where seed is sown into cool, moist seedbeds, i.e., low temperature is the main constraint.

In contrast, seeds of many tropical crops are often sown into hot, drying seedbeds where lack of moisture, not low temperature, is the main constraint. The objective of presowing seed treatments in these circumstances must be to hydrate seed, to minimize time spent imbibing moisture under suboptimal conditions in the soil. In addition, most farmers in developing countries use seed saved from their previous crops (e.g., Turner, 1994) and operate in environments where agriculture support services are minimal. Any seed treatments must be simple, low cost, and low risk. Harris (1996) suggested "on-farm" seed priming, soaking seeds, generally overnight, in water before surface drying them and sowing as usual. Recent farmer-participatory research has demonstrated a wide range of benefits for farmers using the technique. Initial concerns that seeds would be damaged by rapid hydration in free water (e.g., Ellis et al., 1982; Abdullah et al., 1991) have proved unfounded. In vitro germination tests provided "safe limits"—the maximum length of time for which farmers should prime seeds and which, if exceeded, could lead to seed or seedling damage—for a number of crops (Harris et al., 1999; Massawe et al., 1999). In this chapter, a more detailed investigation of the relation between the germination response to seed priming and ranges of varieties of tropical crops is pursued.

Much of the work described here was designed to provide background information and simple, robust recommendations to farmers and development workers involved in testing seed priming in a number of countries and farming systems. The germination and response to priming seeds of varieties of a number of tropical crops were determined in an incubator at constant temperature (see Table 25.1). Crops normally sown under summer conditions were tested at 30°C (and also at 40°C for pearl millet), whereas winter-sown crops were tested at 20°C. After preliminary experiments to determine optimum priming durations for each crop, 50 seeds of each variety were either immersed in distilled water or kept dry at the same temperature. Soaked seeds were removed from the water, surface dried, and set to germinate at the same temperature on moist filter paper in petri dishes or lidded plastic containers, depending on seed size. For each species, experimental design was a randomized block design with a factorial combination of variety and seed treatment in each of four blocks. Seed lots were observed every four or six hours when germinated seeds were counted and removed. Time to 50 percent germination ($t_{50\%}$) was calculated for each seed lot by linear interpolation between counts and final emergence (G) was noted when germination had ceased.

TABLE 25.1. Linear regression parameters for 13 crop × environment combinations of the relation between time saved by priming (y) and time for 50 percent germination of nonprimed seeds (x).

No. in Figure 25.1	Crop (priming time)	b	a	R^2	n	Sig.	Comments
1	Wheat at 20°C (12 h)	0.3833	3.78	0.6133	17	0.01	Indian varieties (Includes slow germination in saline conditions)
2	Chickpea at 20°C (8 h)	1.0734	–37.6	0.7166	4	ns	Indian varieties
3	Cowpea at 30°C (8 h)	0.5317	1.9	0.7239	6	0.05	Nigerian varieties
4	Maize at 30°C (12 h)	0.6760	–11.6	0.7321	17	0.01	Zimbabwe hybrid varieties (One variety did not respond and was not included.)
5	Rice at 30°C (12 h)	–0.0094	7.33	0.0013	10	ns	WARDA
6	Rice at 30°C (24 h)	0.0883	9.86	0.0345	11	ns	WARDA
7	Mung bean at 20°C (8 h)	0.8332	–21.4	0.9207	4	ns	Indian varieties
8	Mung bean at 30°C (8 h)	1.7081	–25.6	0.7237	5	0.10	Indian varieties
9	Mustard at 20°C (8 h)				2		No regression—only two data points
10	Sorghum at 30°C (8 h)	0.0932	5.4	0.0179	12	ns	Indian varieties (ICRISAT)
11	Pearl millet at 30°C (8 h)	0.7936	–6.8	0.7796	11	0.01	Indian varieties (ICRISAT)
12	Pearl millet at 40°C (8 h)	0.2339	–2.13	0.2016	11	ns	Indian varieties (ICRISAT)
13	Finger millet at 30°C (8 h)	0.6605	–6.49	0.7203	7	0.02	Indian varieties
	Overall (plotted line in Figure 25.1)	0.4890	–3.73	0.7900	13	0.02	

The germination rate of varieties at constant temperature varied widely within each crop (see Table 25.1). However, priming for eight to twelve hours ("overnight") resulted in significant reductions in $t_{50\%}$ in most varieties of all crops, with no significant differences in G between primed and nonprimed seed lots of any given crop or variety. The average time saved varied from only four hours in pearl millet at 40°C to more than 20 hours in wheat, chickpea, cowpea, and maize (see Figure 25.1). Time saved was more than just the soaking period in most cases except the very fast germinating seeds of pearl millet, finger millet, and sorghum, and the special case of rice in which priming for 12 hours was suboptimal (compare points 5 and 6 on Figure 25.1). Harris and Jones (1997) reported positive responses to 24 h priming for a range of rice cultivars, although farmers in India still preferred to prime for 12 h as it was more practical to do so (Harris et al., 1999).

On average, slower-germinating crops responded better to seed priming. Mean time saved by priming was linearly related to $t_{50\%}$ over all crop combinations (Figure 25.1). A similar inverse relation between intrinsic germination rate and response to priming also held over a range of cultivars for some crops, e.g., wheat, maize, pearl millet, and finger millet (Table 25.1). In a sample of 17 maize cultivars, $t_{50\%}$ ranged from 35 to 65 h, and priming for 12 h reduced that range to 23 to 32 h (Figure 25.2). In wheat, priming for

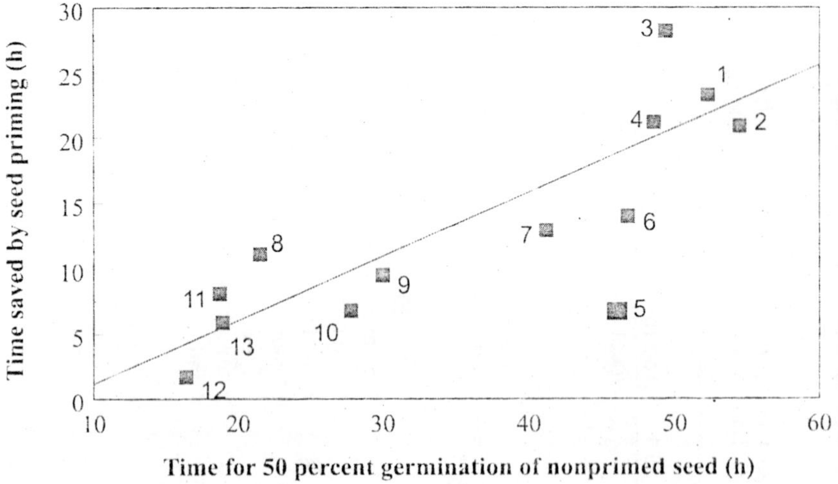

FIGURE 25.1. Relation between time saved by seed priming and the time for 50 percent germination of nonprimed seed for 13 species × treatment combinations, listed in Table 25.1. Values are means of a number of cultivars for each combination. Regression line is in Table 25.1.

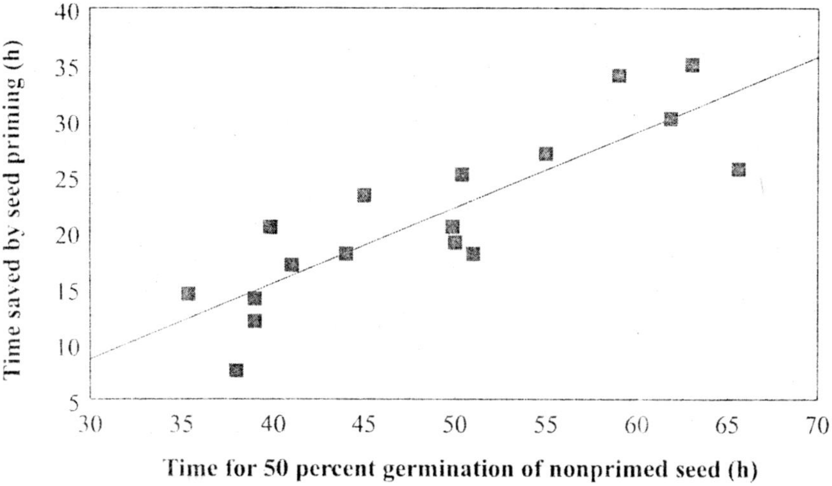

FIGURE 25.2. Relation between time saved by seed priming and the time for 50 percent germination of nonprimed seed for 17 cultivars of maize. Regression line is number 4 in Table 25.1.

12 h reduced $t_{50\%}$ from 45 to 90 h down to 23 to 53 h. Slow germination of seeds in 150 mM NaCl solution was accelerated the most by priming with distilled water (Figure 25.3). For pearl millet at 30°C, priming reduced the range of $t_{50\%}$ from 16 to 21 h down to 10 to 12 h (Figure 25.4), whereas there was no significant relationship between time saved and $t_{50\%}$ at 40°C (Table 25.1). This small-seeded crop is intrinsically fast to germinate, and it seems that, for the range of cultivars chosen, almost all differences in $t_{50\%}$ are due to differences in the time required for imbibition. Figure 25.5 shows a significant relation between time saved by priming and seed weight for pearl millet at 30°C (but not at 40°C—data not shown). Larger seeds have a lower ratio of surface area to volume than that of small seeds, so rates of influx of water are slower.

The data presented here are not intended to provide a comprehensive summary of the response to seed priming of all varieties of all crops. Varieties and crops to be tested were chosen for a number of reasons, but primarily because of their relevance to, and familiarity for, farmers and extension workers in various parts of the developing world. In vitro tests were designed to identify which crops and varieties would, or would not, respond to priming and to determine safe limits and practical recommendations for field use. In this respect, the results are striking.

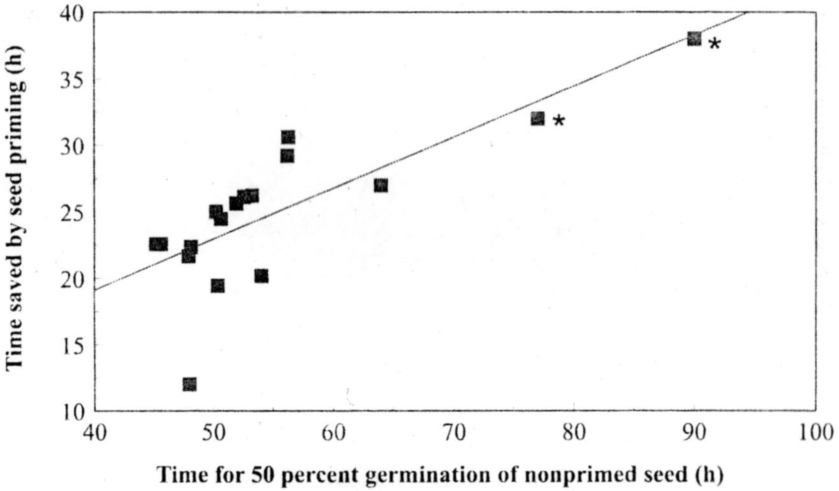

FIGURE 25.3. Relation between time saved by seed priming and the time for 50 percent germination of nonprimed seed for 15 cultivars of wheat. Symbols marked * represent cv. KRL-14 germinating in 150 mM NaCl solution. Regression line is number 1 in Table 25.1.

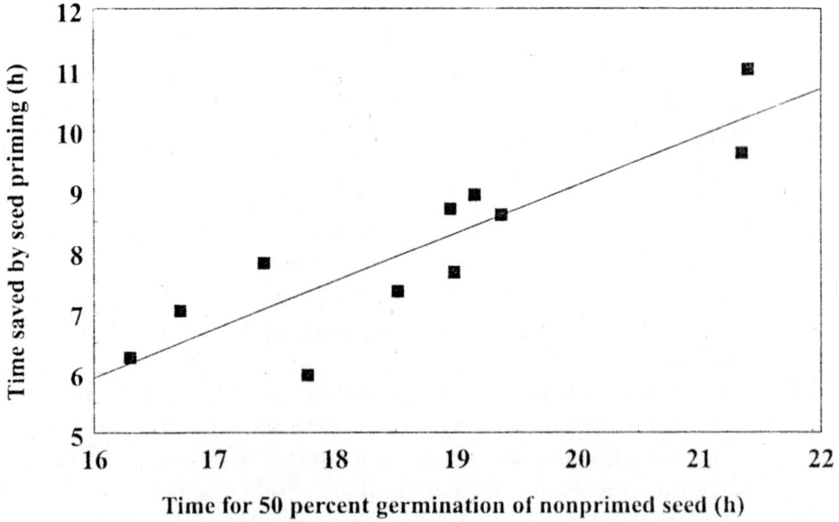

FIGURE 25.4. Relation between time saved by seed priming and the time for 50 percent germination of nonprimed seed for 11 cultivars of pearl millet at 30°C. Regression line is number 11 in Table 25.1.

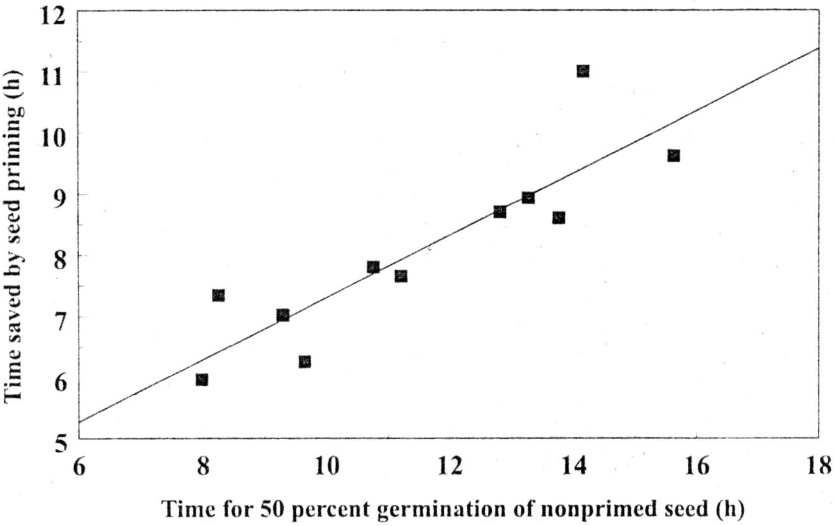

FIGURE 25.5. Relation between time saved by seed priming at 30°C and 1,000-grain weight of 11 cultivars of pearl millet. Regression equation is $y = 0.51x + 2.22$; $R^2 = 0.77$.

Apart from one exception, germination of all lines of all crops tested was accelerated by priming seed in water, with no negative effects on G. The germination of one variety of maize, SC 501, was not accelerated by seed priming, and this is being investigated further, since it has been observed to respond well in the field (Harris, Rashid, et al., 2002). Farmers in India, in the course of on-farm participatory trials with maize, rainfed rice, chickpea, and wheat, have demonstrated that overnight soaking for these crops sped up emergence by one to three days, hastened flowering and crop maturity and increased yields (Harris et al., 1998, 1999). Similar positive effects with wheat, maize (Rashid et al., 2002), and barley (Rashid, unpublished data) have been demonstrated under moderately saline conditions when fresh water was used to prime seeds. It is noteworthy that these crops all appear on the right-hand side of Figure 25.1 (points 1, 2, 4, 5, and 6), i.e., they are intrinsically slow to germinate, or are germinating under adverse conditions, but respond markedly to priming.

The response to priming of three other, smaller-seeded, crops in Table 25.1 has been evaluated in the field. Mungbean (points 1 and 8 in Figure 25.1) has consistently shown large, positive yield responses to priming for 6 to 8 hours, both on station and in farmers' fields in Pakistan (Harris and Rashid, 2002). Harris (1996) in Botswana and Harris, Pathan, et al. (2001)

in Zimbabwe reported that overnight priming led to faster, better establishment and early growth of sorghum (point 10, Figure 25.1) despite the intrinsically fast germination of this species. Even the very small seeds of finger millet (point 13, Figure 25.1) were responsive to priming and gave average yield increases of 14 percent in on-station trials over two years in eastern India (Kumar et al., 2002).

Other investigations, not reported here, of small numbers of varieties of other crops, e.g., pigeonpea, cowpea, sunflower, and cotton, suggest that varietal differences in response to rapid hydration may be more common in other crops and that the safe limit for soaking may not always be as farmer-friendly as "overnight." Damage following rapid hydration of seed has been reported in several crops, particularly hard-seeded legumes (e.g., Ellis et al., 1982; Abdullah et al., 1991) although it is uncertain in these crops if such damage is worse overall in the field than the consequences of poor crop establishment in the absence of priming. Massawe et al. (1999) have reported positive responses to priming in Bambara groundnut in controlled environments.

More rapid germination is clearly possible if the crops tested here are soaked in water overnight; farmers with limited resources are easily capable of doing this successfully, and early establishment has led to a wide range of yield and productivity benefits (Harris et al., 1998, 1999; Harris, Pathan, et al., 2001; Harris, Raghuwanshi, et al., 2001).

The wide range of intra-specific germination rates was reduced a great deal by seed priming, and time saved was greater in varieties that were inherently slow to germinate (Figure 25.1). This suggests that much of the difference between varieties in germination performance is related to seed characteristics associated with imbibition. Because the ratio of seed surface area to volume is smaller in larger seeds, replotting time saved by priming versus seed weight (Figure 25.5) gives a reasonable relation for pearl millet and confirms the importance of imbibition. In addition, since all varieties of each crop were tested at the same temperature, the priming-related reduction in within-species variability (Figures 25.2 through 25.5) raises important questions for crop models based on the concept of thermal time. Any predicted differences in the performance of varieties based on differences in germination characteristics under "normal" conditions are likely to be spurious if seeds are primed. Indeed, farmers' reports that primed crops can flower earlier and mature up to ten days earlier than identical nonprimed stands (Harris et al., 1998, 1999; Harris, Tripathi, and Joshi, 2002; Musa et al., 1999, 2001) suggest that thermal time characteristics may be significantly modified throughout the life of the crop by the rapid emergence and establishment promoted by seed priming, although the mechanism responsible for this effect is not yet known.

These in vitro data and the results of on-farm trials in India (Harris et al., 1998, 1999; Harris, Raghuwanshi, et al., 2001; Harris, Tripathi, and Joshi, 2002), Bangladesh (Musa et al., 1999, 2001), Nepal and Pakistan (Harris and Rashid, 2002; Harris, Raghuwanshi, et al., 2001; Harris, Rashid, et al., 2002; Rashid et al., 2002), and Zimbabwe (Harris, Pathan, et al., 2001) should reassure farmers, researchers, and policymakers that, at least for the major crops investigated here, on-farm seed priming is a safe, effective way to improve crop performance in the tropics.

REFERENCES

Abdullah, W.D., Powell, A.A., and Matthews, S. (1991). Association of differences in seed vigour in long bean *(Vigna sesquipedalis)* with testa colour and imbibition damage. *Journal of Agricultural Science, Cambridge* 116: 259-264.

Austin, R.B. (1989). Maximising crop production in water-limited environments. In F.W.G. Baker (ed.), *Drought resistance in cereals* (pp. 13-26). Wallingford, UK: CAB International.

Carter, D.C., Harris, D., Youngquist, J.B., and Persaud, N. (1992) Soil properties, crop water use and cereal yields in Botswana after additions of mulch and manure. *Field Crops Research* 30: 97-109.

Chiduza, C., Waddington, S.R., and Rukuni, M. (1995). Evaluation of sorghum technologies for smallholders in a semi-arid region of Zimbabwe (Part I): Production practices and development of an experimental agenda. *Journal of Applied Science in Southern Africa* 1: 1-10.

Ellis, R.W., Osei-Bonsu, K., and Roberts, E.H. (1982). Desiccation and germination of seed of cowpea *(Vigna unguiculata). Seed Science and Technology* 10: 509-515.

Harrington, L.W., Fujisaka, M.L., Morris, P.R., Hobbs, H.C., Sharma, R.P., Singh, R.B., Chaudhary, M.K., and Dhiman, S.D. (1993). *Wheat and Rice in Karnal and Kurukshetra Districts, Haryana, India: Farmers' Practices, Problems and an Agenda for Action* (pp. 1-54). Mexico, D.F.: HAU, ICAR, CIMMYT and IRRI.

Harris, D. (1992). Seedbeds and crop establishment. In M. Kronen (ed.), *Proceedings of the Second Annual Scientific Conference of the SADCC/ODA Land & Water Management Programme* (pp. 165-172). P. Bag 00108, Gaborone, Botswana.

Harris, D. (1996). The effects of manure, genotype, seed priming, depth and date of sowing on the emergence and early growth of *Sorghum bicolor* (L.) Moench in semi-arid Botswana. *Soil & Tillage Research* 40 (1/2): 73-88.

Harris, D. and Jones, M. (1997). On-farm seed priming to accelerate germination in rainfed, dry-seeded rice. *International Rice Research Notes* 22 (2): 30.

Harris, D., Joshi, A., Khan, P.A., Gothkar, P., and Sodhi, P.S. (1999). On-farm seed priming in semi-arid agriculture: Development and evaluation in maize, rice and chickpea in India using participatory methods. *Experimental Agriculture* 35: 15-29.

Harris, D., Khan, P.A., Gothkar, P., Joshi, A., Raguwanshi, B.S., Parey, A., and Sodhi, P.S. (1998). Using participatory methods to develop, test and promote on-farm seed priming in India. In *International Conference on Food Security and Crop Science* (pp. 298-299). November 3-6, 1998, Hisar, India. Abstracts.

Harris, D., Pathan, A.K., Gothkar, P., Joshi, A., Chivasa, W., and Nyamudeza, P. (2001). On-farm seed priming: Using participatory methods to revive and refine a key technology. *Agricultural Systems* 69 (1-2): 151-164.

Harris, D., Raghuwanshi, B.S., Gangwar, J.S., Singh, S.C., Joshi, K.D., Rashid, A., and Hollington, P.A. (2001). Participatory evaluation by farmers of "on-farm" seed priming in wheat in India, Nepal and Pakistan. *Experimental Agriculture* 37 (3): 403-415.

Harris, D. and Rashid, A. (2002). The potential of on-farm seed priming for increasing mungbean production. Presentation made at the 2nd International Mungbean Conference, 23-25 May 2002, ICRISAT, Hyderabad, India.

Harris, D., Rashid, A., Hollington, P.A., Jasi, L., and Riches, C. (2002). Prospects of improving maize yields with "on-farm" seed priming. In N.P. Rajbhandari, J.K. Ransom, K. Adikhari, and A.F.E. Palmer (eds.), *Sustainable Maize Production Systems for Nepal* (pp. 180-185). Proceedings of a maize symposium held December 3-5, 2001. Kathmandu, Nepal. Kathmandu: NARC and CIMMYT.

Harris, D., Tripathi, R.S., and Joshi, A. (2002). "On-farm" seed priming to improve crop establishment and yield in dry direct-seeded rice. In S. Pandey, M. Mortimer, L. Wade, T.P. Tuong, K. Lopez, and B. Hardy (eds.), *Proceedings of the International Workshop on Direct Seeding in Asian Rice Systems: Strategic Research Issues and Opportunities* (pp. 231-240). January 25-28, 2000, Bangkok, Thailand. Los Banos, Philippines: International Rice Research Institute.

Kumar, A., Gangwar, J.S., Prasad, S.C., and Harris, D. (2002). "On-farm" seed priming increases yield of direct-sown finger millet *(Eleusine coracana)* in India. *International Sorghum and Millets Newsletter* 43: 90-92.

Massawe, F.J., Collinson, S.T., Roberts, J.A., and Azam-Ali, S.N. (1999). Effect of pre-sowing hydration on germination, emergence and early seedling growth of bambara groundnut (*Vigna subterranea* L. Verdc). *Seed Science and Technology* 27: 893-905.

Monteith, J.L and Elston, J. (1983). Performance and productivity of foliage in the field. In J.E. Dale and F.L Milthorpe (eds.), *The Growth and Functioning of Leaves* (pp. 409-518). Cambridge, UK: Cambridge University Press.

Musa, A.M., Harris, D., Johansen, C., and Kumar J. (2001). Short duration chickpea to replace fallow after *aman* rice: The role of on-farm seed priming in the High Barind Tract of Bangladesh. *Experimental Agriculture* 37 (4): 509-521.

Musa, A.M., Johansen, C., Kumar, J., and Harris, D. (1999). Response of chickpea to seed priming in the High Barind Tract of Bangladesh. *International Chickpea and Pigeonpea Newsletter* 6: 20-22.

Okonwo, J.C. and Vanderlip, R.L. (1985). Effect of cultural treatment on quality and subsequent performance of pearl millet seed. *Field Crops Research* 11: 161-170.

Ortiz-Monasterio. I., Dhillon, S.S., and Fischer, R.A. (1994). Date of sowing effects on grain yield and yield components of irrigated spring wheat cultivars and relationships with radiation and temperature in Ludhiana, India. *Field Crops Research* 37: 169-184.

Parera, C.A. and Cantliffe, D.J. (1994). Presowing seed priming. *Horticultural Reviews* 16: 109-141.

Rashid, A., Harris, D., Hollington, P.A., and Khattak, R.A. (2002). On-farm seed priming: A key technology for improving the livelihoods of resource-poor farmers on saline lands. In R. Ahmad and K.A. Malik (eds.), *Prospects for Saline Agriculture* (pp 423-431). The Netherlands: Kluwer Academic Publishers.

Turner, M.R. (1994). Trends in India's seed sector. Paper presented at Asian Seed 94, Chiang Mai, Thailand, September 27-29.

Chapter 26

Seed Technology in Plant Germplasm Conservation

David Tay

INTRODUCTION

Plant germplasm conservation is the safekeeping of seed or other plant propagules for present and future use to improve our crops and for biological research. Plant genetic materials are more easily maintained as seed by taking advantage of the quiescent phase in the plant life cycle. At optimal storage conditions of low temperature and dry environment, seed can stay alive for decades or even thousands of years, as in the case of the Canadian arctic lupin (Porsild et al., 1967). However, this applies only to species with "orthodox" seed where seed moisture content can be dried down to as low as 5 percent (wet basis) for storage; it does not apply to "recalcitrant" seeds, which die upon dehydration. Most of our important crops have orthodox seed. Sophisticated facilities known as gene banks have been built to provide these favorable storage conditions. There are some 1,300 gene banks in national programs and the International Research Centers of the Consultative Group on International Agricultural Research (CGIAR). The United Nations Food and Agriculture Organization (FAO) in 1998 estimated that some 6.1 million accessions of our food, fiber, and industrial crops are under conservation in gene banks around the world (FAO, 1998). The U.S. National Plant Germplasm System (NPGS), the U.S. national gene-bank system, currently conserves more than 450,000 accessions of 4,474 species in its collection (USDA, 2005).

Plant genetic resource conservation has developed during the past fifty years into a discipline in which specialized courses are being taught in universities and technicians are trained. For example, the MSc Course on Plant Genetic Resources Conservation and Utilization at the University of Birmingham, England, has been taught since early 1970s and was supported by FAO and the CGIAR International Board for Plant Genetic Resources (presently, the International Plant Genetic Resources Institute [IPGRI]). An

International Commission on Genetic Resources for Food and Agriculture—Global System on Plant Genetic Resources (FAO, 2005a), based at FAO Headquarters in Rome, Italy, administers and coordinates international policies in the exchange and conservation of plant germplasm, including the recently established International Treaty on Plant Genetic Resources for Food and Agriculture (FAO, 2005b). The FAO Seed and Plant Genetic Resources Service (FAO, 2005c) provides technical support to national programs and the International Commission in terms of gene-bank accreditation, technical advisory assistance, and training. The IPGRI (IPGRI, 2005), also based in Rome, provides research leadership and coordination in plant genetic resources conservation and development, including training for national programs in agricultural crops but excluding ornamental plant species.

A seed gene bank typically consists of seed storage freezers operating at −18°C for long-term storage of permanent collections, often termed "base" collections, and seed storage coolers maintained at 2°C and low relative humidity (r.h.) of below 40 percent for "active" collections, the seed used for distribution to researchers. Many gene banks also have a short-term seed storage facility maintained at around 15°C and 40 percent r.h., as well as an accompanying seed processing facility including seed dryers, seed processing equipment, and seed packaging equipment. The short-term seed storage room is usually integrated with the seed drying system where seed that have been dried but not yet cleaned for entering into storage are temporarily kept. This provides a relatively safe storage environment free from insect pest infestation. It therefore functions as a buffer working space between seed harvesting and storage. These operational conditions have been the standard since the 1980s (FAO/IPGRI, 1994).

Knowledge and skills in seed science and technology therefore play an important role in running an efficient gene bank. Management decisions are based on the knowledge and ability of gene-bank staff to produce the best quality seed for storage by preventing seed from deterioration during seed production, harvesting, processing, drying, and storage. A seed is a living unit and starts to deteriorate immediately after it reaches physiological maturity—usually the stage when a seed has achieved its maximum dried weight and, theoretically, has the best seed quality. In spite of this, usually a stage before the fruit is practically ready and/or ripened for harvesting is often referred to as seed physical maturity (or crop maturity). The rate of seed deterioration depends on two environmental factors: storage temperature and seed moisture content. Every decision in a gene bank from seed production to putting seed into storage is to minimize the period that a seed lot is exposed to unsuitable high temperature and seed moisture content. With a given operational budget and human resources, all decisions made have to ensure that the greatest number of accessions with the highest quality seed are being put into storage. This paper describes the gene-banking steps and the seed science and technology knowledge and skills required to achieve this.

SEED SCIENCE AND TECHNOLOGY IN GENE BANKS

Knowledge and skills in seed science and technology are the foundation to run an efficient seed gene bank. From germplasm acquisition to seed multiplication, and seed storage to germplasm distribution, seed science and technology knowledge and skills have to be applied when making decisions. The quality of a seed lot accession has to meet the minimum standards of the gene bank, otherwise seed regeneration has to be repeated (FAO/IPGRI, 1994). These seed standards include

1. genetic quality—to ensure that the original genetic composition of an accession is maintained;
2. physiological quality—to ensure that the physiological state of a seed lot is optimum for maximum storage period;
3. physical quality—to ensure that a seed lot has no mechanical injuries thus affecting its storage life and also free from other unwanted extraneous matters; and
4. seed health quality—to ensure that a seed lot is not carrying infectious diseases, pests, and weeds.

Plant genetic resources conservation is therefore multidisciplinary, and gene-bank personnel have to possess adequate knowledge and experience not only in seed science and technology but also in agronomy or horticulture, plant genetics, plant physiology, plant protection, agricultural engineering, management, etc., to function efficiently.

Genetic Quality

Seed genetic quality in germplasm conservation means minimal changes in the genetic composition of an accession during seed regeneration and conservation. The changes could be due to loss in variability during seed regeneration as the result of either genetic drift and conscious and unconscious selection, or to contamination by foreign pollen in the seed production field. In recent years, the latter is becoming a focal issue in relation to contamination by genetically modified genes. At NPGS repositories, utmost care is taken during seed regeneration to prevent this. Mutation is another concern, but it should be considered as part of crop evolutionary dynamics.

The strategy adopted to reduce the amount of genetic change of an accession is to use as large a plant population as possible from randomly selected seed and seedlings during seed regeneration. Seed production environments and cultural practices should aim at getting all the randomly selected seedlings to grow into mature plants and to produce seed. In cross-pollinating species, random pollination between plants of an accession should be provided in isolation and any off-type plants should be removed before flowering to prevent contamination. At harvesting, the same amount of seed from each plant of an accession is strived for to form a representative composite sample for storage. From seed production to packaging, mechanical contamination should be prevented and accurate seed labeling ensured.

Physiological Quality

The physiological quality of a seed lot determines its storage life and vigor. The two most important factors influencing this parameter are the state of seed physiological maturity during seed development and the degree of seed weathering incurred after seed physiological maturity. Seed weathering is the effect of injuries on seed due to extreme weather conditions such as too high moisture and high or cold temperature extremes causing the affected seed to deteriorate at a quick rate. The resulting injuries also encourage disease infection and thus augment further seed deterioration. Seed of high physiological quality have better seed vigor and are able to be stored for a longer period. Thus, every seed multiplication step has to promote seed filling and complete maturation. In some species, such as cucurbits, postharvest fruit incubation must be provided to allow seed maturation. In warm, high precipitation conditions, seed weathering is frequently the result of pregermination to a stage that drying the seed back to low moisture content for storage will either kill the seed or adversely shorten their storage life. The extreme wet and warm environment, also, promotes pathogens growth and thus seed injuries. In germplasm conservation, seed of some accessions have no dormancy and are so precocious that they germinate in fruits that are not fully ripened or ready for harvesting. Usually this information is not available until it has happened. A monitoring procedure by sampling seed heads during seed harvesting rounds may be adopted to determine the correct harvesting stage. Freezing weather causes frost injury to seed with high seed moisture content.

Another intrinsic genetic problem is the different stages of seed maturation within an inflorescence or a plant. When harvesting is done in one pass, the harvested seed is a mixture of seed that have been weathered, seed that are just right for harvesting, and seed that are still immature. The ways to reduce these differences in seed maturity and weathering are to shorten the pollination period, for example, to only two to four weeks depending on the species, as in the brassicas, and/or to conduct several rounds of selective harvesting. The former can be achieved by topping all the inflorescences or flowers on a cut-off date.

Seed should be dried to safe seed moisture content immediately after harvesting to avoid seed deterioration, and this is one of the most common causes of seed quality problems in a gene bank that does not have a good seed drying facility. As in commercial seed production, seed regeneration for germplasm conservation also requires a cool, dry location during pollination, seed filling, and maturation to obtain good quality seed.

The foundation of seed storage is to understand the seed life curves, i.e., the seed viability and vigor curves. These are the typical sigmoid life curve, as shown in Figure 26.1, which has a flat section at the top left, followed by

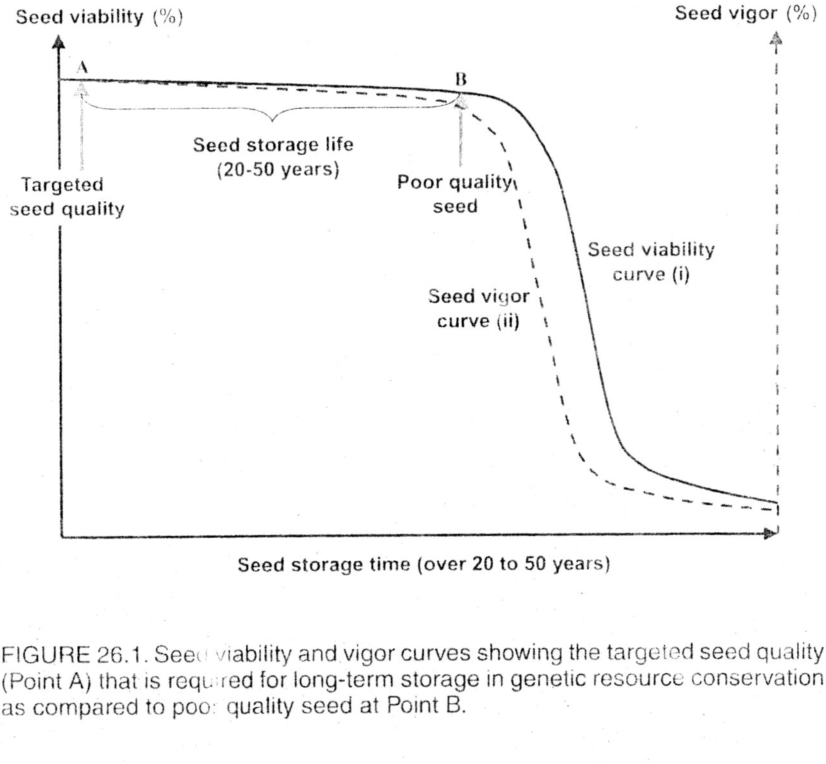

FIGURE 26.1. Seed viability and vigor curves showing the targeted seed quality (Point A) that is required for long-term storage in genetic resource conservation as compared to poor quality seed at Point B.

a sharp drop, and then flattened out at the bottom right. The seed viability curve (Curve [i] in Figure 26.1) is usually obtained from standard germination tests (ISTA, 1999) over storage time, and the vigor curve (Curve [ii] in Figure 26.1) is usually an estimated and projected curve based on one of the many vigor tests (Hampton et. al, 1995). The aim is to regenerate seed in the best seed production conditions so that the physiological state of the seed lot is as close to the original physiological seed quality as possible as indicated by point A in Figure 26.1, the targeted seed quality required for successful long-term seed storage as compared to poor quality seed at point B. Point A may be estimated by comparing the difference between the result of standard germination test and a vigor test. If the difference is small and the viability is high, the seed lot quality should be close to point A, and if the difference is large even though the standard germination is high, it should be at point B. It is therefore useful to perform these analyses before a seed lot is put into long-term storage in order to estimate its storage life. Based on these data the interval between germination tests in a seed viability-monitoring program during storage can be established for individual seed lots. Storing seed with seed quality at point B is a waste of gene-bank resources because in a short period of time the viability is going to fall rapidly to an unsafe level. If the seed viability-monitoring interval is not frequent enough to detect the sharp drop in viability of an accession, the accession will lose a significant portion of its gene pool by the next round of monitoring.

Physical Quality

Physical seed quality in genetic resource conservation is mainly related to mechanical damage during seed harvesting, threshing, drying, and processing. Threshing is one of the main causes of seed injuries, and it happens when seed are threshed at either too high or too low seed moisture content, and with a drum thresher the threshing drum is too fast and/or the concave clearance is not enough. The problem could be avoided in a gene bank by manual threshing because the seed lots are usually small. Damaged seed show broken cotyledons in seedlings during standard germination tests, and seed coat damage can be detected by staining tests, e.g., fast green stain. In wet seed extraction the problems are usually due to over-postharvest maturation, excessive fermentation, and too fast or too slow drying following seed extraction.

The other requirements are that the seed must be free of other unwanted materials including soil, dust, and stone, noxious weeds and other seeds, damaged and broken seed, and inert matter. Seed containers are labeled both inside and outside and checked by at least two persons to avoid labeling mistakes.

Health Quality

A plant germplasm collection is an assembly of a crop genetic variability from different parts of the world. If the quarantine procedure is not meticulous during germplasm introduction, a collection of plant pathogens and diseases, especially systemic microbes, may also be assembled. A seed regeneration field provides an ideal place for the different strains, races, and species of pathogens to intercross, select, and establish. As the main purpose of a germplasm collection is for distribution to researchers all over the world, it therefore provides an opportunity for the assembled diseases to be distributed to new, clean areas. Stringent plant quarantine regulations and protocols are in place throughout the world by national plant quarantine authorities and the International Plant Protection Convention of FAO (FAO, 1997). Special precautions have to be considered and put in place during seed regeneration. For example, clean fields isolated from any potential source of disease inoculum have to be used, and all diseased plants and seed in the field are discarded to prevent contamination. A disease prevention program is often more practical and effective than trying to control an infestation. For diseases that show clear symptoms on the seed, e.g., soybean mosaic virus and purple seed stain in soybean, the infected seed can be identified and discarded manually. However, seed of many seed-borne diseases cannot be visually inspected and specific laboratory pathological tests have to be applied. In the case of some viruses, heat and chemical treatment can be applied; for example, tomato seed are treated at 78°C dry heat before seed distribution or acid and trisodium phosphate solution treatment following seed extraction to eliminate tomato mosaic virus.

GENE-BANK MANAGEMENT SYSTEM

Gene banks are mandated to conserve and distribute the widest representation of their targeted crop gene pools. These genetic variations can be local landraces, traditional varieties, wild forms, related wild species, genetic stocks, inbred lines, and even modern cultivars. The functions of a gene bank therefore include germplasm acquisition, regeneration, conservation, characterization, evaluation, documentation, and distribution. All these sequential procedures demand sound knowledge and skills in seed science and technology among gene-bank personnel. This is described in the following section to show the close connection between plant germplasm conservation activities and the seed science discipline. Table 26.1 provides a summary of the conservation tasks in relation to seed knowledge and skills needed to accomplish them.

Germplasm Collection

Seed sampling techniques (Bould, 1986) have to be applied when collecting germplasm from farmers' seed bags to ascertain that the collected sample is representative of the farmers' variety. When collecting seed in a standing field, techniques used in field inspection for seed certification are applied in order to capture genetic representations from the whole field, instead of only from a corner of the field. In addition, only matured and good quality seed are collected. When collecting germplasm in another country knowledge in international seed conventions including the Convention on Biological Diversity (CBD) (CBD, 1992) and Convention on International Trade in Endangered Species of Wild Fauna and Flora (CITES) (CITES, 1979), the country germplasm collection and transfer policies and regulations, agreement and permits, seed import and export permits, and phytosanitory certification are essential. Seed permits and phytosanitory certificates are part of seed health and plant quarantine requirements.

Germplasm Regeneration and Storage

Germplasm regeneration is the multiplication of new lots of good quality seed for depositing into long-term storage. This has to be done when an accession is first acquired and the number of seed is not enough for putting into storage and/or the history of the seed lot is unknown for direct storage. Also, when the seed viability of an accession has fallen below the gene bank's threshold standard during storage a new lot of seed has to be produced to replace the deteriorated lot. During germplasm regeneration the goal is to maintain an accession's original genetic integrity and to produce high quality seed. The wide variability of germplasm materials among and within accessions requires that each accession be treated a bit differently as needed within a reasonable uniform practical regeneration program. Seed regeneration guidelines covering growing season, seed germination, crop rotation, seedling raising, cultural techniques, pollination, harvesting, threshing, cleaning, drying, packaging, and storage have to be prepared as part of a gene-bank operation manual. Some germplasm characterization and

TABLE 26.1. Plant genetic resource conservation and distribution procedures in relation to seed science and technology.

Gene-bank procedure	Seed science and technology knowledge
Acquisition	
1. Collecting in seed market	1. Sampling strategy, sample size, and seed health
2. Collecting in farmer's field	2. As in (1)
3. Collecting in another country	3. As in (1) + plant collection and transfer agreements, import and export permits, and phytosanitory certification
4. Exchange and donation	4. As in (1) or (3) with another country
Regeneration	
5. Seed germination and seedling raising	5. Germination and vigor testing, dormancy breaking, stratification and vernalization treatment, etc., and random selection technique
6. Direct sowing	6. As in (5) and field rotation systems
7. Seedling transplanting	7. Random selection and field rotation
8. Synchronizing flowering	8. Site selection, cultural, vernalization, and photoperiod treatments
9. Pollination	9. Isolation, pollination (insect and hand) and pest control
10. Harvesting	10. Number of times, seed extraction techniques
11. First-stage drying	11. Drying methods, driers and m.c. testing
12. Seed threshing and cleaning	12. Hand versus machinery
13. Final drying	13. Slow drying and equipment and m.c. testing
14. Packaging	14. Hermetic versus nonhermetic, container types, equipment, and packaging environment
Conservation	
15. Base and active collection	15. Seed storage
16. Viability monitoring	16. Sampling, number of seed, monitoring interval
17. Seed for regeneration	17. Threshold viability, seed sampling

TABLE 26.1 *(continued)*

Gene-bank procedure	Seed science and technology knowledge
Characterization and evaluation	
18. Seed germination	18. As in (5)
19. Sowing	19. As in (6)
20. Seedling transplanting	20. As in (7)
Distribution	
21. Withdrawing sample	21. Prepacked versus recloseable container
22. Packaging for shipment	22. Protecting against temperature and moisture
23. Documentations	23. Barcode versus manual

evaluation activities may be incorporated with the regeneration program to make it more cost-effective. This is achieved as follows.

Growing and Preharvest Handling

The aim is to achieve the targeted number of plants per accession as resources permit in order to capture the maximum proportion of the targeted genepool. At the Asian Vegetable Research and Development Center, 30 randomly selected plants (7.5m × 2 rows) were used in tomato and pepper, 160 plants (4 m × 2 rows) in soybean and mungbean, and 150 to 200 plants (15m × 6 rows) in Chinese cabbages (Tay, 1988). Seed dormancy breaking treatments, seed stratification, and hard seed scarification are applied when needed to ensure uniform germination, thus preventing genetic drift. Seedlings are also randomly selected for transplanting. The most suitable climate, growing season, and planting date, and field plot, crop rotation system, and land preparation are used to ensure that all the randomly selected plants grow and produce seed to contribute to the next generation. Optimal cultural practices are used to ensure high seedling and field survival rate, as well as to prevent genetic drift. Pollination cages are used to provide isolation for cross-pollination species, and in this case pollinators such as honeybees, bumblebees, and flies are used. Flowering is synchronized by vernalization and photoperiod treatment and cultural manipulation to promote more concentrated flowering and opportunity for cross-pollination between plants of an accession and thus also more concentrated seed maturation. Self-pollinated species with high rate of out-crossing, such as peppers, also require isolation but do not need pollinators. All off-types are removed, and to do this properly one has to decide whether the variation is part of the genepool or is a mix. This is straightforward for known pure-line cultivars but becomes not so clear-cut in open-pollinated landraces and wild populations. Inbred lines with self-imcompatibility systems have to be maintained using hand pollination, as in maize, and bud pollination or carbon dioxide treatment in brassicas.

Harvesting

Seed regeneration should be planned so that harvesting occurs in a cool, dry season to minimize field weathering. The aims are to harvest from every plant in order to maintain the targeted multiplication plant population and to sample the same amount of seed from each plant to form a representative composite sample for storage. Multiple harvestings have to be done for species with different stages of seed maturation, and seed head should be harvested immediately after physiological maturity to obtain high quality seed. Harvesting a mixture of mature and immature seed together will be more difficult to clean because of unfilled seed and more trash. Accurate labeling and crosschecking should be followed to prevent mechanical contamination. Pest-infested and diseased seed should not be harvested. Any free water due to morning dew or rain on the harvested seeds and pods has to be dried immediately. Some species, such as brassicas and yardlong bean (*Vigna unguiculata* subsp. *sesquipedalis*), can be harvested with stems and individual pods, respectively, when the seed are matured green and then allow for postharvest ripening under shade to avoid bad weather.

Threshing

Threshing is an important cause of mechanical injuries and thus affects seed quality and storage life. The common injuries are broken seed or visible cracks and micro-cracks on seed; in germination test seedlings show broken organs. Such damage is inflicted when threshing is done at too low seed moisture content (less than 12 percent in beans), the speed of thresher is too fast, and/or the concave clearance is too narrow, thus causing seed crack. Then again, too high seed moisture content (more than 16 percent) causes seed bruising. For germplasm purposes, a single-head thresher with rubber-lined spikes and spring-loaded belt thresher that is easy to clean between seed lots is most practical and useful because of small seed lots.

Manual threshing has often proved to be more gentle and efficient. The recommendation is to dry the harvested pods and flower heads in sun or with a high-flow drier until crispy before threshing.

Wet Seed Extraction

Seed of most solanaceous and cucurbit vegetables have to be extracted from fleshy fruits. In some species, postharvest maturation before seed extraction is required to allow seed maturation. In precocious varieties with no seed dormancy fruits may have to be harvested before fully mature. Fermentation method is effective for the extraction of small seed lots as in germplasm conservation (Tay, 1990). Hot peppers can be extracted wet, instead of dry threshing with a modified meat mincer (Tay, 1991a). The clean wet seed is spin-dried with a domestic cloth spin-drier and dried immediately to prevent germination.

Processing

Seed processing is more efficiently done by hand winnowing because of the many small seed lots being handled in a gene bank and often the seed lots are reasonably clean. Any extraneous materials and unwanted seed can also be removed by hand. A sieve set and a seed blower are used for difficult seed lots with light, fine contaminates.

Seed Drying

Unlike drying seed for commercial use, seed drying for long-term storage in germplasm conservation has to be gentle and usually involves a two-phase drying process in which phase I covers the period from harvesting to threshing and processing, and phase II is the final drying before packaging for storage. The efficiency of a drier depends on its temperature, drying air relative humidity (r.h.), airflow and seed moisture content, and its distribution in the seed. The drying capability of driers with only temperature-controlled feature depends on outside air temperature and r.h.

In most crops, phase I drying is probably the most critical stage, especially when harvesting has be done in wet weather. The moistened seed with all the harvested plant parts must be dried as fast as possible to remove free surface water to prevent the seed from commencing germination in the moisture condition. This is efficiently accomplished with a high-volume airflow batch drier at a temperature below 35°C. In humid places a dehumidified drier with good air circulation may have to be used to avoid raising the drying temperature to an unsafe level. In places with good sunshine solar drying is efficient before threshing. The main impediment is to prevent accidental mixing of accessions when seed containers are moved around.

Phase II drying is usually done with dehumidified low-temperature drier, and the IPGRI recommendation is 15°C and 15 percent r.h. (FAO/IPGRI, 1994). If the temporary storage room is maintained at 15°C and below 40 percent r.h. it could be used as a buffered intermediate drier between phase I and II drying. At this temperature and low humidity most insect pests stop reproduction.

Seed Packaging and Storage

Seed for long-term storage at −18°C must be hermetically sealed at low seed moisture content of 4 to 6 percent (wet basis) for most crops. Seed with high seed moisture content in subzero temperatures suffer from freezing injury. At freezing temperatures, it is difficult to control the r.h. of the air. However, long-term storage rooms should be maintained frost-free. The containers used could be metal cans or aluminum-polyethylene (Al-PE) pouches depending on the species. The quality of an Al-PE pouch depends on the thickness of its component layers and the type of plastic used, and the seal quality depends on the materials, sealing temperature, pressure, and time. Seed with sharp appendices may puncture or cause pinholes in Al-PE pouches. Nonhermetic packaging is recommended for the storage of active collection at 2 to 5°C and below 40 percent r.h. The types of containers commonly used are screw-top glass and plastic jars, and metal cans. These containers allow easy withdrawal of seed for distribution. During seed withdrawal the seed container is allowed to equilibrate in the packaging room at 20°C and 40 percent r.h. to prevent moisture condensation on seed when the container is opened (Tay, 1988).

Seed Viability Monitoring

A good seed storage facility will not improve a seed lot quality, as it only helps to slow down its deterioration rate. Seed under storage therefore have to be monitored at set interval so that seed lots deteriorated below the viability threshold limit could be withdrawn in time for regeneration to prevent genetic erosion in storage. Standard germination procedure (ISTA, 1999) is adopted but with fewer seed because of the small seed lots. Monitoring interval at every three to five years in the first fifteen years for lots with high initial seed quality in long-term storage followed by more frequent interval after that. The main shortfalls in the current system are that seed are being consumed in germination tests and that it is time-consuming and costly to do. Ellis et al. (1980) proposed a sequential seed sampling method to cut down seed used, but it has not been adopted because of the increasing labor involved in implementing the method. Presently, research efforts concentrate on finding nondestructive ways to evaluate seed viability in storage.

BASIC GENE-BANK DESIGN

The basic gene bank design consists of a seed storage facility, a tissue culture laboratory for in vitro collections, a seed drying system, a seed laboratory for seed processing and testing, a germplasm research laboratory, a packing-cum-distribution laboratory, and a data documentation facility. The principle in the design of a gene-bank storage system is the stepwise removal of outside air moisture and temperature. This is accomplished by the packaging room operating at 20 ± 2°C and 40 percent r.h. as an anteroom for the slow dryer (15 ± 2°C and 15 percent r.h.) and temporary seed cooler (15 ± 2°C and 40 percent r.h.). The temporary storage cooler in turn is the anteroom for the medium-term seed cooler, operating at 2 ± 2°C and 40 percent r.h., and the medium-term seed cooler the anteroom for the long-term seed freezer at –18 ± 2°C. The stepwise removal of air moisture will allow the frost-free requirement of the long-term seed freezer to be achieved more easily. The use of prefabricated insulated panels has proved to be more efficient and cost-effective (Tay, 1991b). The use of a series of smaller units is more economical and flexible to operate than to have a large single room, because each small unit could be turned on only when seed are in it.

Construction Considerations

The whole shell and doors of a seed storage facility must be fully insulated and moisture proofed. An air-pressure equalizer has to be installed in a seed freezer to prevent a suction effect on its door due to differences in air pressure as the inside air is cooled down. Prefabricated steel- or aluminum-polyurethane sandwiched-panel unit fulfils these requirements and the space between the freezer and building roof provides a ventilated insulation air layer. The seed storage facility is placed away from direct sun, i.e., facing the north side in the northern hemisphere. In hot places a roof with large over-hanged eaves and light-color outside walls are recommended to provide shade and reflect the sun, respectively. Machinery room should be well ventilated to prevent heat accumulation. The structural design should withstand earthquake and cyclonic wind in areas subjected to these natural disasters.

Machinery Considerations

Some of the basic design considerations include whether to use standard cooling and dehumidifying machinery available on the market as compared to the more expensive custom-built machinery, to combine refrigerating and dehumidifying machinery versus independent separate units, to use refrigerated-reheating dehumidifiers or chemical dehumidifiers, etc. All the seed storage units should have a simple independent power cut-off system, to avoid heat building up when the main electronic control fails, and a continuous temperature and r.h. monitoring and alarm system. Emergency electricity generators have to be provided to run the long-term and medium-term storage units when the main grid fails.

Workplace Safety Considerations

A fire alarm and an emergency light system both inside and outside the seed storage units and laboratories, and fire-fighting equipment and emergency exits have to be strategically positioned. Fire-fighting protocol and fire escape drills need to be rehearsed regularly. The seed storage unit door lock should have an over-right safety device from inside so that when a person is being accidentally locked inside the lock can be released. Only authorized personnel are allowed to enter the seed vaults and to sign in/out on a logbook with job description. It is recommended for personnel to work in pairs with appropriate clothing in a long-term seed freezer.

GENE-BANK RESEARCH PROGRAM

Genetic erosion in a gene bank, i.e., loss of genetic material while under storage, is one of the most concerning issues during conservation in recent years (FAO, 1998). The cause is attributed mainly to the lack of operational resources. Research on conservation methodology and gene-bank management is therefore critical to streamline the operation of a gene bank and to make it more cost-effective and attractive in funding. Little information is available on the loss and shift in genetic variability of an accession during each cycle of seed regeneration. The exact plant population required during seed regeneration in order to retain most of the original diversity is not known for most species. Controlled pollination and isolation in the seed production of cross-pollinated and out-crossing species are tedious and costly. The efficient use of insect pollinators in small pollination cages with different species of bees and flies in different crops has not been systematically evaluated. For example, preliminary observations at the Ornamental Plant Germplasm Center in its 2002 seed production program showed the effectiveness and robustness of bumblebees in small pollination cages and greenhouses as compared to honeybees. The goal to get all viable seed sown to germinate, and all randomly selected seedlings to grow and set an equal amount of seed is difficult to achieve because of uneven seed dormancy in a seed lot and the need to synchronize flowering to promote concentrated flowering and inter-crossing. Efficient and timely seed harvesting, drying, threshing, and cleaning of many small seed lots of many species are often a challenge in the management and operation of a gene bank. Many techniques used especially in seed companies are developed in-house and unpublished. Seed production technology is therefore an important area of research. The effect of seed production methods used on seed vigor is another area that requires further research. Presently, most procedures in use are somewhat experiential.

A quick, low-cost, accurate seed vigor test for predicting seed quality for use in the planning and programming of seed viability monitoring interval for each accession during storage is lacking. Similarly, a quick, low-cost, nondestructive viability test is necessary during viability monitoring. The use of ultra-dried seed storage and ultra-low-temperature cryopreservation technique is under experimentation, and more genotypes and species should be evaluated to study their usefulness and efficiency. Research on seed biochemistry and physiology, cellular and molecular biology, and seed anatomy and microstructure will contribute new knowledge and understanding on the many empirical practices. Seed research will become more a science than technology. Finally, this chapter deals only with gene banking of species with orthodox seed. As plant genetic resource conservation diversifies into other species, especially trees and tropical plants, where many of them have recalcitrant seed, the problems and techniques applied are different from those discussed here. For those species, in vitro slow-growth and cryopreservation techniques (Engelmann and Takagi, 2000) are the most promising conservation methods and have been successfully applied in many species. Further research on physiology, biochemistry, biophysics, and anatomy at the cellular level will provide new knowledge to extend these techniques for use in more species and even in orthodox seed. The croyopreservation of orthodox seed at the USDA National Center for Genetic Resources Preservation (NCGRP, 2005) has proved promising (personal observation), and it may be the way to conserve all species in the future.

REFERENCES

Bould, A. (1986). *Handbook of Seed Sampling,* First Edition. International Seed Testing Association, Zurich, Switzerland.

CBD (1992). Convention on Biological Diversity. Secretariat of the Convention on Biological Diversity, Montreal, Canada.

CITES (1979). Convention on International Trade in Endangered Species of Wild Fauna and Flora. CITES Secretariat, Chemin des Anémones, CH-1219 Châtelaine, Geneva, Switzerland.

Ellis, R.H., Roberts, E.H., and Whitehead, J. (1980). A new, more economic and accurate approach to monitoring the viability of accessions during storage in seed banks. *Plant Genetic Resources Newsletter* 41: 3-18.

Engelmann, F. and Takagi, H. (2000). Cryopreservation of tropical plant germplasm: Current research progress and application. Proceedings of the Japan International Research Center for Agricultural Sciences/International Plant Genetic Resources Institute Joint International Workshop, held in Tsukuba, Japan, October 20-23, 1998.

FAO (1997). The New Revised Text of the International Plant Protection Convention 1997. Food and Agriculture Organization of the United Nations, Rome, Italy.

FAO (1998). The State of the World's Plant Genetic Resources for Food and Agriculture. Food and Agriculture Organization of the United Nations, Rome, Italy.

FAO (2005a). Commission on Genetic Resources on Food and Agriculture. http://www.fao.org/ag/cgrfa/PGR.htm. Food and Agriculture Organization of the United Nations, Rome, Italy.

FAO (2005b). International Treaty on Plant Genetic Resources for Food and Agriculture. http://www.fao.org/ag/cgrfa/itpgr.htm. Food and Agriculture Organization of the United Nations, Rome, Italy.

FAO (2005c). Seed and Plant Genetic Resources Service—AGPS. http://www.fao.org/WAICENT/FAOINFO/AGRICULT/AGP/AGPS/ default.htm. Food and Agriculture Organization of the United Nations, Rome, Italy.

FAO and IPGRI (1994). Genebank Standards. Food and Agriculture Organization of the United Nations, Rome, Italy. and International Plant Genetic Resources Institute, Rome, Italy.

Hampton J.G., TeKrony D.M., and the Vigour Test Committee (1995). *Handbook of Vigour Test Methods,* Third Edition. International Seed Testing Association, Zurich, Switzerland.

IPGRI (2005). International Plant Genetic Resources Institute. http://www.ipgri.org. IPGRI, Rome, Italy.

International Seed Testing Association (1999). International Rules for Seed Testing 1999. Seed Science and Technology. Vol.27, Supplement, 1999. ISTA, Zurich, Switzerland.

NCGRP (2005). NCGRP Home Page. http://www.ars-grin.gov/ars/NoPlains/FtCollins/nsslmain.html. National Center for Genetic Resources Preservation. USDA, Fort Collins, CO.

Porsild, A.E., Harrington, C.R., and Mulligan, G.A. (1967). *Lupinus arcticus* Wats. grown from seeds of Pleistocene age. *Science,* New Series, 158(3797): 113-114.

Tay, C.S. (1988). Present status, management and utilization of tropical vegetable genetic resources at AVRDC. In S. Suzuki (ed.). *Crop Genetic Resources of East Asia* (pp. 41-51). International Board for Plant Genetic Resources, Rome, Italy.

Tay, D.C.S. (1990). Vegetable seed production technology for small vegetable farmers. In *Vegetable Production Training Manual* (pp. 116-131). AVRDC, Taiwan.

Tay, D.C.S. (1991a). Extraction of seeds of hot peppers using a modified meat mincer. *Hortscience* 26: 1334.

Tay, D.C.S. (1991b). Prefabricated cold store—Advantages and design. *Plant Genetic Resources Newsletter* 85: 19-23.

USDA (2005). National Plant Germplasm System. http://www.ars-grin.gov/npgs. United States Department of Agriculture, Beltsville, MD.

Index

Page numbers followed by the letter "f" indicate figures; those followed by the letter "t" indicate tables.

Index

Page numbers followed by the letter "f" indicate figures; those followed by the letter "t" indicate tables.